CARBON-BASED MAGNETISM

An Overview of the Magnetism of Metal Free Carbon-Based Compounds and Materials

CARBON-BASED MAGNETISM

An Overview of the Magnetism of Metal Free Carbon-Based Compounds and Materials

edited by

Tatiana Makarova
Institutionen för fysik
Umeå universitet, Sweden
Ioffe Physico-Technical Institute
St. Petersburg, Russia

Fernando Palacio
Instituto de Ciencia
de Materiales de Aragon
CSIC - Universidad de Zaragoza
Zaragoza, Spain

ELSEVIER

Amsterdam – Boston – Heidelberg – London – New York – Oxford – Paris – San Diego
San Francisco – Singapore – Sydney – Tokyo

ELSEVIER B.V.
Radarweg 29
P.O. Box 211, 1000 AE
Amsterdam, The Netherlands

ELSEVIER Inc.
525 B Street
Suite 1900, San Diego
CA 92101-4495, USA

ELSEVIER Ltd.
The Boulevard
Langford Lane, Kidlington,
Oxford OX5 1GB, UK

ELSEVIER Ltd.
84 Theobalds Road
London WC1X 8RR
UK

First edition 2006

Library of Congress Cataloging in Publication Data
A catalog record is available from the Library of Congress.

British Library Cataloguing in Publication Data
A catalogue record is available from the British Library.

ISBN-13: 978-0-444-51947-4
ISBN-10: 0-444-51947-5

Transferred to digital print 2007
Printed and bound by CPI Antony Rowe, Eastbourne

Contents

Editor's Preface vii

Foreword ix

1. **T. Sugawara and M.M. Matsushita**
Localized Spins Exchange-coupled with Itinerant Electrons in Organic π-electronic System 1

2. **P.M. Lahti**
Magneto-structural Correlations in π-Conjugated Nitroxide-based Radicals: Hydrogen-bonds and Related Interactions in Molecular Organic Solids 23

3. **H. Murata and H. Nishide**
Persistent High-spin Organic Polyradicals 53

4. **K. Mukai**
Verdazyl-based Magnetic Systems 75

5. **Y. Hosokoshi and K. Inoue**
Quantum Spin Magnetism in Nitroxide-based Compounds 107

6. **K. Takeda and M. Mito**
The Magnetism of π-Orbitals under Pressure 131

7. **J. Campo, J. Luzón, F. Palacio, and J. Rawson**
Spin Density Distribution and Interaction Mechanisms in Thiazyl-based Magnets 159

8. **K. Awaga, W. Fujita, T. Tanaka, H. Matsuzaki, and H. Okamoto**
Magnetic Bistability in Molecule-based Magnetic Materials 189

9. **K. Yamaguchi, S. Yamanaka, and Y. Kitagawa**
The Nature of Effective Exchange Interaction 201

10. **A. Omerzu and M. Tokumoto**
Magnetism of Fullerene Charge-transfer Complexes 229

11. **A.V. Nikolaev and B. Verberck**
Diamagnetism of Diamond and Graphite 245

12. **K. Wakabayashi**
Electronic and Magnetic Properties of Nanographites 279

13. **K. Kusakabe**
Flat-band Ferromagnetism in Organic Crystals 305

14. **A. Oshiyama and S. Okada**
Magnetism in Nanometer-scale Materials that Contain No Magnetic Elements 329

15. **F. Guinea, M.P. López-Sancho, and M.A.H. Vozmediano**
Interactions and Disorder in 2D Graphite Sheets 353

16. **P.O. Lehtinen, A.V. Krasheninnikov, A.S. Foster, and R.M. Nieminen**
The Magnetic Nature of Intrinsic and Irradiation-induced Defects in Carbon Systems 371

17. **T. Enoki and K. Takai**
Unconventional Magnetic Properties of Nanographite 397

18. **Y. Kopelevich, S. Moehlecke, and R.R. da Silva**
Ferromagnetic and Superconducting Instabilities in Graphite 417

19. **P. Esquinazi, R. Höhne, K.-H. Han, D. Spemann, A. Setzer, M. Diaconu, H. Schmidt, and T. Butz**
Induced Magnetic Order by Ion Irradiation of Carbon-based Structures 437

20. **A.V. Rode, A.G. Christy, E.G. Gamaly, S.T. Hyde, and B. Luther-Davies**
Magnetic Properties of Novel Carbon Allotropes 463

21. **A.N. Andriotis, M. Menon, R. M. Sheetz, and E. Richter**
McConnel Model for the Magnetism of C_{60}-based Polymers 483

22. **J. Ribas-Ariño and J.J. Novoa**
The Mechanism of the Magnetic Interaction in Polymeric-C_{60} Fullerenes 501

23. **J.A. Chan, B. Montanari, and N.M. Harrison**
Ferromagnetism in Defective Polymerised C_{60} 523

24. **T.L. Makarova**
Ferromagnetic Carbonaceous Compounds 541

Subject Index 563

Editors' Preface

One of the most intriguing phenomena observed in nature, magnetism is relevant to physics and geology, biology and chemistry. Magnets play a crucial role in our life as we conceive it, for they are ubiquitous basic components of almost every instrument we depend upon. From electricity generators to cars, recording tapes and computers, from sensors, detectors and actuators to domestic appliances and countless everyday gadgets magnets are everywhere. Traditional magnets, made of iron- or nickel based materials, are heavy and unfriendly for environment. In the past fifteen years metal-free magnets made from carbon-based compounds have been discovered and, more recently, there have been reports on the observation of magnetism in carbon, a very light and biocompatible element. Metal-free carbon structures exhibiting magnetic ordering represent a new class of materials and open a novel field of research that may lead to many new technologies.

No physical law prohibits magnetic ordering in the structures containing only light elements like carbon. However, there are strong reasons why high-temperature ferromagnetism in carbon is hard to expect and it is a major challenge to produce it. A major requisite for magnetism in an all-carbon structure is the presence and stability of carbon radicals. The occurrence of radicals, which can introduce an unpaired spin, is cut down by the strong ability of pairing all valence electrons in covalent bonds. Moreover, magnetic ordering in free-radical based compounds can be hampered by the often loose and unfavorable packing of the molecules in the solid. Bulk crystalline graphite is a strong diamagnet, with the magnetic susceptibility second only to superconductors. However, graphite containing certain defects can exhibit spontaneous magnetization.

The large and growing number of publications and patents in magnetism of metal-free materials show that the last decade has witnessed a number of impressive findings important to both basic science and technology, with application ranging from nanoscale electronics to cancer treatment. In recent years researchers have found a handful of metal-free compounds that can become magnetic at temperatures close up to the liquid nitrogen. On the other hand, magnetism at room temperature has been reported in specially treated pure carbon structures - graphite and fullerenes.

This book provides a comprehensive review of relevant experimental data and theoretical concepts related to the magnetism of metal-free carbon systems. These systems include carbon-based compounds, namely organic radical magnetic systems, and magnetic materials based on carbon structures. By gathering together chemists and physicists, theorists and experimentalists in a most valuable ensemble of leading experts, essentially every major experimental achievement and modern hypothesis on the physics nature of magnetic ordering in carbon-based systems is discussed. It also highlights some problems that have yet to be solved and points out new avenues for research. Moreover, this volume also aims to contribute to the understanding of the fundamental properties of carbon, the main element of living creatures.

The first 10 chapters discuss magnetic ordering mechanisms, compounds and properties in π-electron systems and analyze molecular magnets with spins residing only in π-orbitals. In this part of the book, different approaches to construct organic and polymeric magnets and the chemical groups of free radicals, including fullerene-based compounds, that have proved to be the most efficient to provide examples of organic magnets are discussed in combination with the physics of these magnetic compounds. Molecular magnets exhibit unique possibilities as quasi-ideal systems to investigate quantum spin magnetism; they are also quite soft materials which properties can dramatically change under mild pressures, and more recently magneto-structural bistability has also been discovered in some of these compounds opening new research and applied possibilities. Finally, a theoretical analysis on the nature of magnetic exchange interactions in these compounds is also included.

The second part of the book starts with the description of magnetic properties of basic carbon allotropies: graphite and diamond. Next, the influence of the defect structures is discussed and mechanisms for localized magnetism coming from the graphite defect states are described. A variety of nanometer-scale structures consisting solely of non-magnetic elements is shown to exhibit magnetic ordering. Experimental results include the observation of ferromagnetic and superconducting instabilities in graphite, unconventional magnetic properties of nanographite, examples of room temperature magnetic ordering in carbon allotropies, including fascinating novel carbon material: carbon nanofoam. Controlled transformation of diamagnetic graphite into ferromagnetic phase is achieved by means of proton bombardment.

As a whole, the 24 review articles in this book, while not exhaustive, summarize the most significant recent and ongoing exciting scientific developments in the subject and provide explanations on what might cause magnetic behaviour in a structure made solely of carbon. This book would not have been possible without the enthusiastic collaboration of the contributors. Their time, work and dedication deserve our deepest acknowledgement. And we offer special thanks to Larisa Zaitseva for technical editing.

T. L. Makarova and F. Palacio

Foreword

Magnetic Properties of otherwise Non-Magnetic Materials

If only 30 years ago a research team had said that carbon or graphitic compounds showed ferromagnetism, the most probable answer of the scientific community would have been: *the experiments that these results are based have not been carefully performed and some small particles of magnetic materials may be contained in the sample. This was not a careful experiment.* The story would have been finished and the paper rejected. In fact this happens with many papers related to graphitic compounds and other materials. Even today reviewers of scientific journals play the same "*cantinela*", and this happens even if competent scientists, in their best will and with hard work, explain in their papers that careful chemical analysis has been done and that the magnetic moment observed in the experiment cannot be explained by the detected amount of magnetic impurities. Notice that even if there are magnetic impurities, they will never show spontaneous magnetic moment if they are sufficiently diluted. To have spontaneous magnetic moments one needs not only impurities but aggregation of them with a reasonable size above the limit of superparamagnetism.

However, and fortunately, more and more competent scientists are taking the lead and insisting, after many years of hard work, that the magnetism they are observing is not due to magnetic impurities, or aggregations of them, formed by atoms of angular momentum larger than two. They are dealing with systems containing only *s* or *p*-electrons and observing long range order. This growing experimental evidence is thanks to the resolution and precision of the experimental techniques and the refinement of theories. This book uses these new techniques to further penetrate the essence of the magnetism in *s* and *p*-electrons systems. In 1928 Heisenberg wrote: ***"Long range [magnetic] order in an infinite lattice would be found only when the lattice contained heavy [i.e. metallic] elements."*** Well, in recent years long range magnetic order has been unambiguously characterized in organic compounds, which only contain *s*- and *p*-electrons. Heisenberg statement is true in a perfect infinite lattice. Real lattices with aggregated clusters or patches of defects can be far more complex. The world is bigger and more diverse that an imagined proposal.

It is especially the reduction in size of the systems studied that helps to establish the existence of magnetic properties in materials that otherwise are non magnetic. Let us put it this way, to justify magnetism in perfect graphite becomes a cumbersome work. However, what is the problem with understanding magnetism in the case of presence of clusters or patches of defects that do not contain the perfect sp^2 hybridization but instead have combinations of electrons at the defects that are not have perfectly paired because they possess different hybridizations?. If this situation exists, then the unpaired electrons will have a magnetic moment that can couple defective regions through exchange and exhibit long range order. Notice that exchange coupling is not exclusive of electrons with angular momentum equal or larger than 2. We mean long range in the sense of ordering in the region of the defects that may extend for several tens of nanometres, sizes larger than the superparamagnetic limit of the system at hand. If the number of patches of defects is large enough, to be detected by the more and more precise developed techniques, a net spontaneous magnetization will be observed. This can be quite small, of the order of 10^{-3}-10^{-4} Bohr magneton per atom, since it corresponds to the average over the whole sample but is considerably larger at the defect patches. In addition the critical temperature may be very large because this is related to the critical temperature of the patches that depends of the exchange coupling. One can have large magnetic moment and small critical temperature and vice-versa or combinations of both. Magnetism requires basically unpaired electrons with atoms carrying a net magnetic moment that can be coupled via exchange with other electrons. This coupling can be between itinerant or localized electrons. One may argue that it is needed, not only coupling but also large density of states at the Fermi level. However this is not a problem on edges or defects that can have flat bands with large density of states at the Fermi level. No difference exists in this case between *s* and *p* electrons and other of larger angular momentum. The above picture of the patches in graphite is just an example of how defects may dramatically affect magnetism even when only *s*- and *p*-electrons are involved. Also the existence of flat bands in the breaking symmetry points of edges and defects seems to be explained theoretically, as it is done in this book. These are some proposals but there are others developed from full calculations and more sophisticated theories that may explain such a phenomena and related observations described in this book.

For about two decades chemists have started a thorough project of using molecular chemistry techniques to develop new kinds of magnets based on molecules rather than on metals and its oxides. The idea behind this is the challenge of creating new classes of materials from which new exciting properties may be expected. This, combined with the fast growing nanotechnology techniques, permits us to explore the world of molecular magnets based on *s*- and *p*-electrons in detail. The experiments reported here open expectation on this point and it is not necessary to mention the impact in technology that will have the existence of magnetic properties at high temperatures in a class of materials based on molecular magnets conformed by light atoms. ***The magnetic plastics!***

It seems that with the new nanotechnology techniques one may be able to discover more and more materials exhibiting unpredicted magnetic properties, given the small dimensions of the systems studied. In my opinion there will be a tremendously large quantity of materials with all kinds of electrons and molecular binding forces. In this sense interconnection of small systems that show new magnetic properties will open new routes to all types of devices and technologies where physicists, chemists and computing theorists will have a say if they work in close collaboration.

Precisely in this book the readers, the newly initiated in the field as well as those that are already working on it, can enjoy a group of experienced scientists (experimentalists and theorists) that try to express their findings and opinions of developments in a field that I would like to call "*Magnetism in otherwise non-magnetic materials*", although its real title is **Carbon-Based Magnetism: an overview of the magnetism of metal free carbon-based compounds and materials.** That is to say materials that 20 years ago the mere mention of their magnetic properties and long range order would have been a leitmotif for excommunication. However, nowadays this is not the case and this field represents an exciting one which everyday attracts more and more addicts. In this book you can find experimental work from supramolecular approaches to *s*- and *p*-electrons magnetism to all possible imaginations on magnetic properties of carbon in all its allotropic forms, including the fullerenes and related material passing by verdazyl and thiazyl -based magnets.

From the point of view of theory the reader can enjoy: Exchange-coupled localized spins with itinerant electrons in organic *p*-electronic system; Ab-initio calculations in organic magnets, for there is always a need to prove the improvement of computers and how much we can understand with numbers; Electronic properties of quasi 2D graphite sheets; Edge states and edge-localized magnetic states in the nanographite, Flat-bands ferromagnetism in organic crystals, etc.

The editors have tried to complete a comprehensive book that in a scholarly and systematic way contributes furthermore to confirm the magnetism in these otherwise nonmagnetic materials. I am most convinced, in a reasonable scientific way that this phenomenon exists. However I have no doubt that there will be other scientists still sceptical. To them, this book is a special invitation for contrasting their vision with the careful analysis made in the chapters ahead. Maybe they will still end with some hesitations but they will certainly gain some light which will hopefully make them to adopt another perspective for looking at the problem. To the newcomers to the field it is a good encyclopaedic and consulting book which can provide them with many new ideas.

Finally I am tempted to say that one should think that stranger things have been seen. And remember that Jorge Luis Borges in his admirable little book "El Aleph" saw inside a little brilliant sphere, one centimetre radius, whose centre is in all parts and the surface nowhere, the sphere that has been revolving thinkers and philosophers during generations, where the whole universe and in particular, for example, a pack of Spanish card in a shop window of Samarqand, and all the grains of sand of all the deserts of the world, together with a white horse running in the Caspian sea. All that can be seen in a little sphere: El Aleph.

In conclusion this book shows that magnetism in otherwise non magnetic materials, ***it is.*** Now remains to think deeply in the technological applications that no doubt the interesting observations described throughout this book should have, and applied them.

N. García

Laboratorio de Física de Sistemas Pequeños y Nanotecnología
Consejo Superior de Investigaciones Científicas, Madrid, Spain

Carbon-Based Magnetism
T. Makarova and F. Palacio (Editors)

1

Localized Spins Exchange-coupled with Itinerant Electrons in Organic π-electronic System

T. Sugawara and M.M. Matsushita

Department of Basic Science, The University of Tokyo, Japan

1 Preface

Various functions of organic materials, such as conductivity, magnetism, non-linear optics, *etc.*, are derived from the characteristic π-electronic structure. Whereas a one-dimensional π-conjugated system (e.g. polyacetylene) exhibits an electrical conductivity upon doping, a two-dinesional π-conjugated system (e.g. graphite) exhibits a semi-metallic property (Fig. 1). On the other hand, arrangement of sp^2 carbons in a cross-conjugated manner affords a branched π-system that exhibits magnetism due to the intrinsic appearance of unpaired electrons. Thus, it can be said that the nature of a π-conjugated system depends largely on its topology. Taking account of such a topological nature of the π-system, several high spin molecules have been created and significance of these high spin molecules as one-dimensional ferromagnets has been documented [1–3]. Nowadays extremely high spin molecules spin quantum number of which exceeds 40 have been reported [4].

Itinerant electrons are known to play an important role in aligning electron spins in inorganic ferromagnets, such as 3d metals or rare-earth metals (Refer Section 3). In order to construct molecule-based conducting magnets, combination of a conducting unit and a spin-carrying unit within a molecule produces a most efficient building block. Recently, a novel high spin species, which is derived from so-called "spin-polarized donor", has been proposed [5–9]. A spin-polarized donor consists of a donor unit and a radical one, and these units are connected in a cross-conjugating manner.

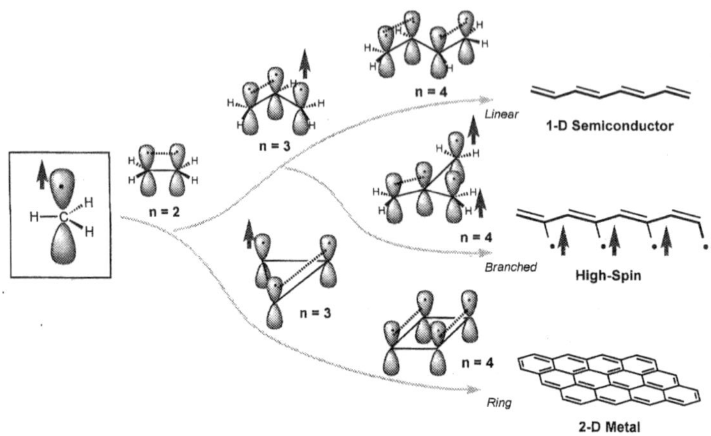

Fig. 1-1. π-Topolgy versus electronic properties.

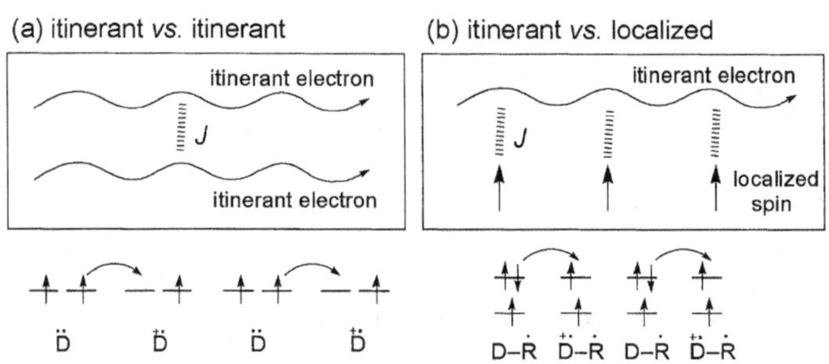

Fig. 1-2. Electronic structure of spin-polarized donor.

2 High Spin Hydrocarbons as a Model for One-dimensional Organic Ferromagnet

2.1 Spin-alignment in Poly(meta-phenylenemethylene)

Before discussing the magnetic interaction between localized spins and itinerant electrons in an organic π-system, the authors revisit spin alignment in organic π-conjugated molecule. It is worthwhile to look at the magnetic property of polycarbenes as a model for one-dimensional organic ferromagnets [10]. Mataga discussed possible "ferromagnetic states" of some hypothetical hydrocarbons (Fig. 2-1) [11]. Among them, a series of meta-substituted polycarbenes, *m*-phenylenebis(phenylmethylene), have been explored experimentally as high spin molecules [1–3] Huckel MO of the *meta*-substituted polycarbene has the following characteristics. According to the topological symmetry of alternant hydrocarbons, it has degenerate non-bonding π MOs as well as

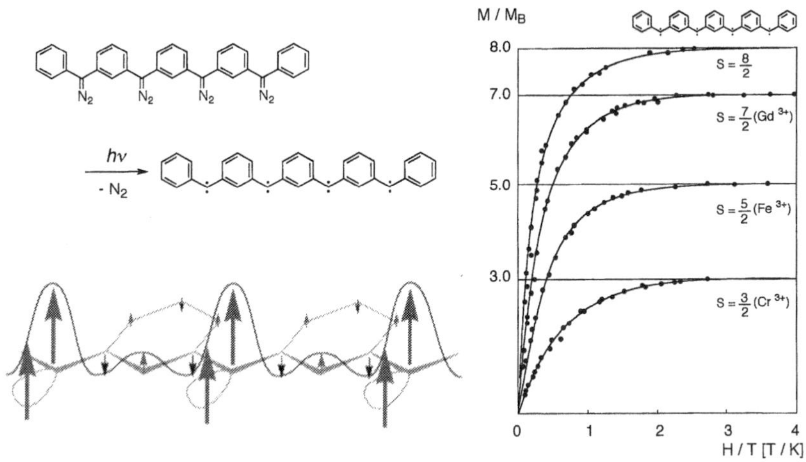

Fig. 2-1. Spin-alignment of poly(*meta*-phenylenemethylene).

non-bonding σ-type MOs, the orbitals being orthogonal to each other. The energy difference between these two types of orbitals is smaller compared with one-center exchange integral $J(n|\pi) = <\pi n|r_{12}|n\,\pi>$. For example, the energy difference between triplet and singlet states of diphenylmethylene is estimated to be –5 kcal/mol, which corresponds to –2J (J = 2.5 kcal/mol or 900 cm^{-1}). Therefore all unpaired electrons are expected to be parallel according to π-topology, resulting in the high spin multiplicity in the ground state. The electronic structure of the high spin state was rationalized by means of Hubbard Hamiltonian calculations [12].

3 Spin Alignment Based on Itinerant Electron

3.1 Spin Alignment by RKKY Mechanism

Ferromagnetic spin alignment is frequently encountered in inorganic materials, for example, 3*d* metals, rare-earth metals, ferromagnetic diluted alloys, *etc.* [13]. In these materials, itinerant electrons are known to play an important role in aligning electron spins. In 4*f* metals, for example, Gd, Tb, and Dy, 4*f* electron spins are parallel on the basis of Hund's rule. When an itinerant 5*s* electron approaches to the lattice point where the transition metal ion is located, it exchange-couples with 4*f* spins ($\boldsymbol{S}_i$) localized on the metal ions. Thus, probability of finding an α or a β spin of the itinerant electron is modulated by the localized 4*f* spins. As a result, the spin-polarization of 5*s* conduction electrons is expressed by a spin wave, and the modulated spin wave is transmitted as a damping oscillation to the next lattice point where 4*f* spins reside with the separation of *r*. If the phase of the spin wave matches at each lattice point, all the 4*f* spins align to the same direction, achieving a ferromagnetic spin-ordering (Fig. 3-1). This mechanism of the spin-ordering is known as RKKY theory (Ruderman, Kittel, Kasuya, Yoshida) [14], and the theory is applied successfully to rationalize the ferromagnetic spin ordering in 4*f*

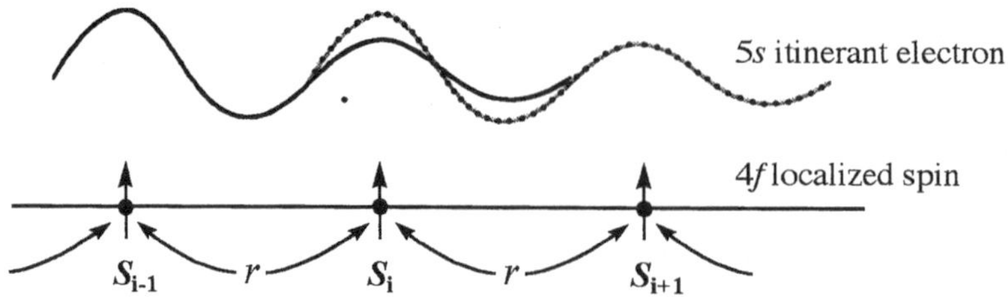

Fig. 3-1. Mechanism of spin alignment based on RKKY theory in $4f$ metal.

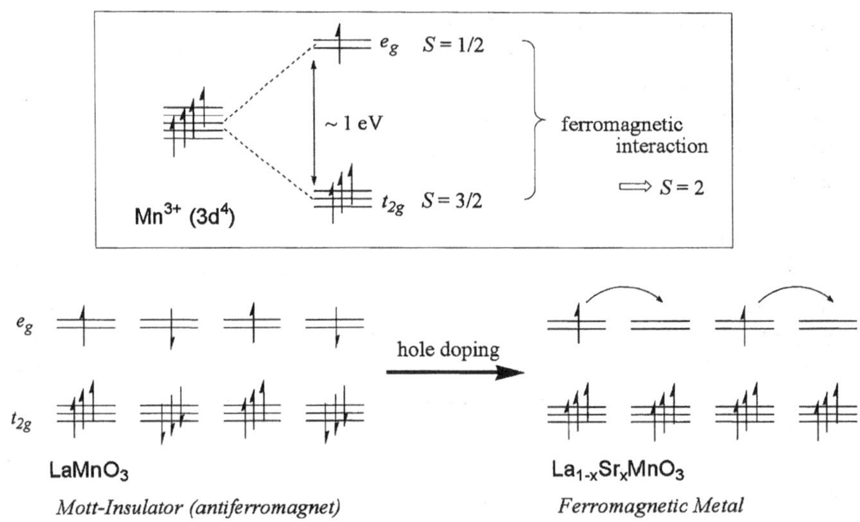

Fig. 3-2. Double exchange mechanism of spin alignment in $La_{1-x}Sr_xMnO_3$.

metals or ferromagnetic diluted alloys. In $3d$ metals, however, RKKY theory can not be applied in a straightforward manner because of the itineracy of $3d$ electrons. Nevertheless this mechanism is important to discuss the basis of the magnetic interaction between localized spins and the itinerant electrons.

3.2 Spin Alignment by Double-exchange Mechanism

In order to consider the spin-alignment in organic conducting materials, a "double exchange" mechanism [15], operating in metal oxides $La_xSr_{1-x}MO_3$ (M=Mn, Co), should be referred to here [16]. Although $LaMO_3$ (M=Mn, Co) is a Mott insulator and antiferromagnet, it becomes conducting when electrons in e_g orbitals are partially doped with Strontium, and the local spins (electrons in t_{2g} orbital) on Mn or Co line up ferromagnetically through itinerant electrons (electrons in e_g orbital) based on the double exchange mechanism. Nameny, the spin-polarized electron in the e_g orbital can migrate along the metal array, aligning the localized spins to α at each metal site (Fig. 3-2). It would be extremely interesting to construct a similar electronic structure using organic π-donor radicals as discussed in Section 5.

4 Spin Alignment via Charge Transfer Interaction

4.1 Spin Alignment by McConnell's type II Mechanism and its Modification

In this section, the authors introduce ideas of spin-alignment using a spin-dependent charge-transfer interaction. McConnell proposed a possibility of spin-alignment in a charge transfer complex between a donor with degenerated SOMOs and an acceptor [17]. The spin-alignment in an alternate stack of donors and acceptors is achieved by contribution of the reverse CT state, in which two electrons in the degenerate SOMOs are in parallel based on Hund's rule of the molecular level (Fig. 4-1(a)).

Thereafter, experimentalists modified this conceptual model by providing their model compounds. Breslow proposed a cation radical of a donor with degenerate HOMOs (Fig. 4-1(b)). He prepared CT complexes of a donor of high symmetry ($\geq$ C3) and an acceptor, their redox potentials being adjusted finely for the charge transfer of the sufficient degree [18] Although cation radicals of these symmetrical donors afforded a ground state triplet species, the complex did not exhibit any ferromagnetic interactions. Miller *et al.* examined the crystal structure of such a type of complexes precisely and revealed that the symmetry of the donors are lowered due to Jahn-Teller distortion [19]. Since the degeneracy of SOMOs are removed, two unpaired electrons couple antiferromagnetically.

Wudl proposed a concept of an organic ferromagnetic metal consisting of donors with degenerated SOMOs (Fig. 4-1(c)) [20]. In a columnar stacking of partially oxidized donors, the unpaired electrons residing in two degenerated SOMOs are

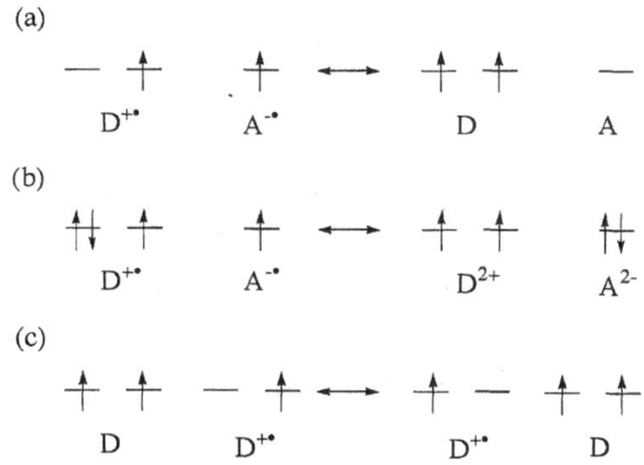

Fig. 4-1. Models of intermolecular spin-alignment contributed by the virtual intramolecular high-spin state; (a) McConnell model (b) Bleslow's modification (c) Wudl's modification.

ferromagnetically coupled according to Hund's rule, and the electron of α spin, for example, can migrate along the donor stack. Although his proposal is attractive, the model compounds which precisely fit with this mechanism have not been reported.

4.2 Spin-alignment in Cross-cyclophane Twin Donor

As described in the previous section, (Section 4-1), Wudl et al. proposed a unique electronic structure, which was constructed by the stacking of a ground state triplet diradical donor (D^0) with degenerate HOMOs in a mixed-valence state ($\cdots D^0D^+D^0D^+D^0D^+\cdots$) as shown in Fig. 4-1(c) [20] A similar electronic structure may be realized if two singly oxidized dimeric donors (D^+-D^0) are in resonance with a doubly oxidized state (D^+-D^+) of a ground state triplet and a neutral one (D^0-D^0) as a charge-tansferred state (Fig. 4-2) [21].

From the above viewpoint, it is interesting to prepare a dimeric donor that exhibits a ferromagnetic intramolecular coupling in the singly oxidized state by virtue of the charge transfer interacton. As a model compound which satisfies the above requirment, a novel TTF-based dimeric donor (**CPTD**), in which two equivalent donor units are constrained in a cross-cyclophane type geometry with four trimethylenedithio chains, has been prepared [22] Generally speaking, unpaired electrons in the dication-diradical of a dimeric donor are likely to be coupled antiferromagneitcally. In the dication-diradical of **CPTD**, however, the through-space interaction between singly oxidized donor units should be negligibly small, because two donor units are fixed in an orthogonal orientation by trimethylene chains. Therefore, there is a chance for the dimeric donor to manifest a ferromagnetic coupling if the through-bond interaction (vide infra) operates efficiently [23].

CPTD afforded ion radical salts, CPTD·Br·TCE_2 (TCE = 1,1,2-trichloroethane), by electro-crystallization in the presence of tetra-*n*-butyl ammonium bromide [24]. An X-ray crystallographic analysis of the salt (space group: $P4_1$) reveals that cation radicals

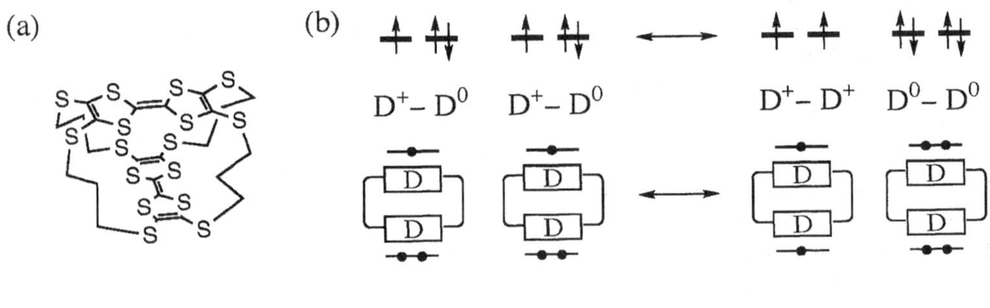

Fig. 4-2. (a) Molecular structure of a cyclophane-type dimeric donor **CPTD** (b) Spin alignment in a cation radical salt of a dimeric donor with degenerate HOMOs.

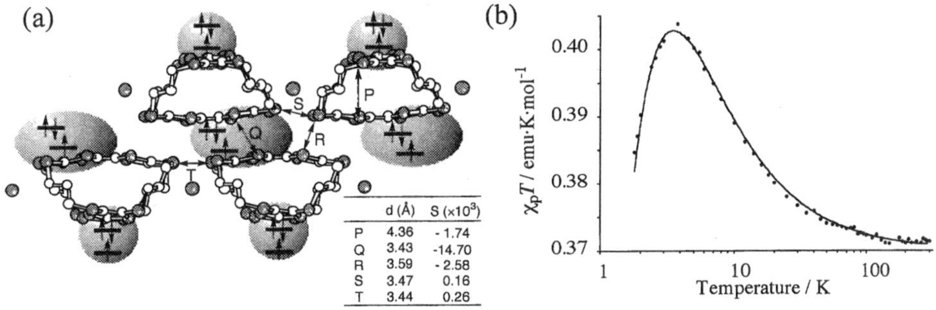

Fig. 4-3. (a) Crystal structure of **CPTD**·Br·TCE_2. [Crystal data: *Tetragonal*, Space group: $P4_1$, a = 12.3395(5), c = 29.6494(8) Å, V = 4514.5098 Å^3, Z = 4, D_{calc}= 1.724 g·cm^{-1}, R = 0.050]. (b) Temperature dependence of the product of paramagnetic susceptibility and temperature ($\chi_P T$) of **CPTD**·Br·TCE_2.

of CPTD are stacked nearly parallel to the c axis and they are arranged in arrays along the a- and b-axes, respectively; these axes are equivalent to each other. The overlap integrals at sites from P to T are also listed (Fig. 4-3(a)). The temperature dependence of the χ_PT value was analyzed in terms of a one-dimensional ferromagnetic Heisenberg chain (J/k_B = +1.6K) with the antiferromagnetic coupling (J/k_B = –0.1K) with adjacent chains (z = 4) (Fig. 4-3(b)) [25]. Judging from the calculated overlap integrals, the unpaired electrons are mainly distributed in a region where HOMO and SOMO interact intermolecularly (at site Q). Since these regions are located above and below CPTD, the ferromagnetic interaction detected here may be ascibed to the intramolecular origin.

In order to obtain information on the spin-correlation in the doubly oxidized state of CPTD, the PM3 calculation was carried out on the dication-diradical species (D^+-D^+). There are nearly degenerated SOMOs and most of the coefficients of SOMO are localized on each donor unit, respectively. This means that the intramolecular electronic interaction between SOMOs should not be significantly large. However, it is to be noted that there are small coefficients at the middle carbon atom of the trimethylene chains (Fig. 4-4(a)). Therefore, the ferromagnetic coupling between unpaired electrons derived from the CT state (D^+-D^+) because SOMOs of D^+-D^+ have a space-sharing nature (refer Section 5-1) [26]. The spin polarization mechanism along the trimethylenedithio chains is also considered to be operating (Fig. 4-4(b)) [27]. Although the salt could be regarded as a Mott insulator at low temperatures where the magnetic interaction became predominant, the conductivity of the single crystal of the ion radical salt exhibited semiconducting behavior at elevated temperatures ($\sigma_{//ab}$ = 5.7×10^{-3} S·cm^{-1}, $\Delta E_{//ab}$ = 0.24 eV; $\sigma_{//c}$ = 1.3×10^{-4} S·cm^{-1}, $\Delta E_{//c}$ = 0.28 eV) [24].

The result discussed above may be the first experimental evidence for the appearance of the ferromagnetic coupling assisted by the intermolecular CT interaction in the conductive salt [28]. Although a suitable chemical modification is needed, this cyclophane donor can be classified as a new class of compounds in the research for conducting magnetic materials.

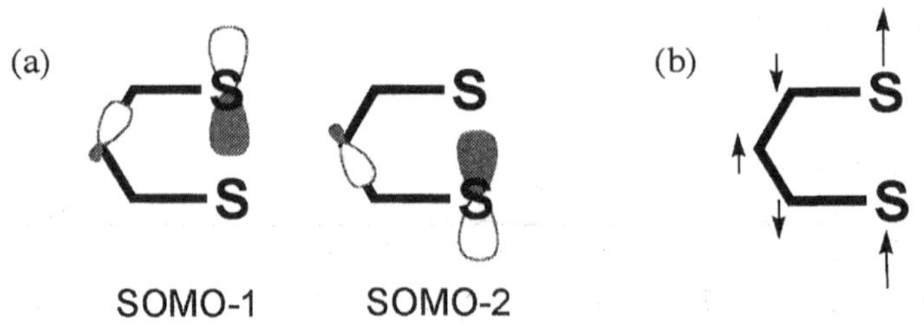

Fig. 4-4. Ferromagnetic coupling mechanisms of $\mathbf{1}^{2+}$ based on the through-bond interaction; (a) two degenerate SOMOs sharing coefficients on the middle carbon of the chain; (b) a long-range ferromagnetic coupling through π–σ and σ–σ spin polarization.

4.3 Yamaguchi's Model of Organic Ferromagnetic Model

Yamaguchi proposed organic magnetic materials consisting of charge transfer complexes composed of a donor carrying a radical unit and an acceptor [29]. If the charge transfer complex crystallizes in alternate stacked columns, a ferrimagnetic spin system is expected due to the antiferromagnetic interaction between the donor and the acceptor (Fig. 4-5(a)), while a ferromagnetic metal may be constructed, if the complex forms a segregated columnar stacking (Fig. 4-5b) due to the ferromagnetic coupling between the itinerant electron and the localized spin on the radical unit.

According to Yamaguchi's proposal, several radical donors composed of a donor unit and a radical unit have been reported [30]. In most of radical donors, however, a donor and a radical unit are not connected through a conjugating system. As a result, an effective exchange interaction is negligibly small, if any, between the local spin and the π spin which is generated on the donor site in their singly oxidized state.

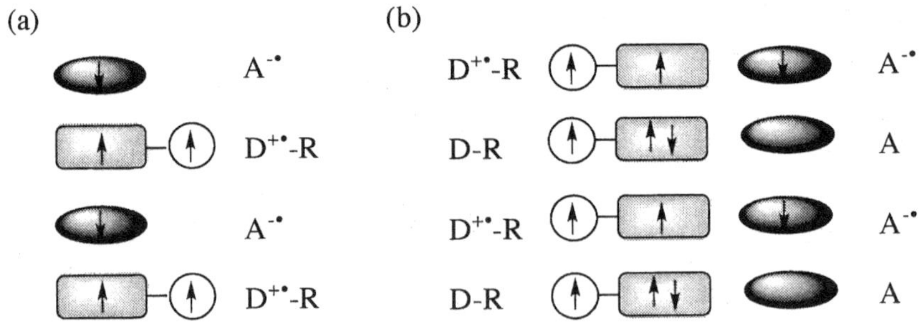

Fig. 4-5. Arrangements of radical donors and acceptors; (a) an alternate stacked column, (b) a segregate stacked column.

5 Design of Spin-polarized Donors

5.1 Spin-polarization in Trimethylenemethane

Trimethylenemethane (**TMM**) is a prototypical multi-centered diradical [31]. A close examination shows that the two degenerated NBMOs share coefficients with the peripheral carbon atoms (Fig. 5-1(a)). Incidentally, a molecular orbital, of which energy level is same as that of the π-atomic orbital of an isolated carbon atom, is termed NBMO (non-bonding molecular orbital). As shown in Fig. 5-1, the distribution of coefficients of the degenerated NBMOs of **TMM** can be considered as a space-sharing type. The electrons in NBMOs of **TMM** can, therefore, occupy the same space simultaneously, resulting in a large exchange repulsion, especially when the spins are anti-parallel [32]. This situation strongly suggests that **TMM** has the triplet ground state. Indeed, the triplet ground state spin multiplicity of **TMM** was determined experimentally, and the energy difference between triplet and singlet states was estimated to be 16 kcal·mol^{-1} (0.69 eV, 8000K) [33].

In the triplet configuration in **TMM**, the two α-spins, residing in SOMOs, are distributed in the periphery of the molecule. The distribution of α-spin affects the spin distribution of NHOMO (next highest occupied molecular orbital) in the following way. While the α-spin in NHOMO(α) resides at each carbon atom almost equally, the β-spin in NHOMO(β) avoids existing at the peripheral carbon atoms and tends to reside at the central carbon: Here, NHOMO(α), NHOMO(β) represent unrestricted Hartree-Fock (UHF) type orbitals (Fig. 5-1(b)). This inhomogeneous spin distribution caused by the interaction between relevant MOs is called *'spin polarization'*. When the spin distribution of each orbital of **TMM** is summed up, the spin density is depicted as shown in Fig. 5-1(b). It should be noted that the central carbon has a *'negative spin density'*.

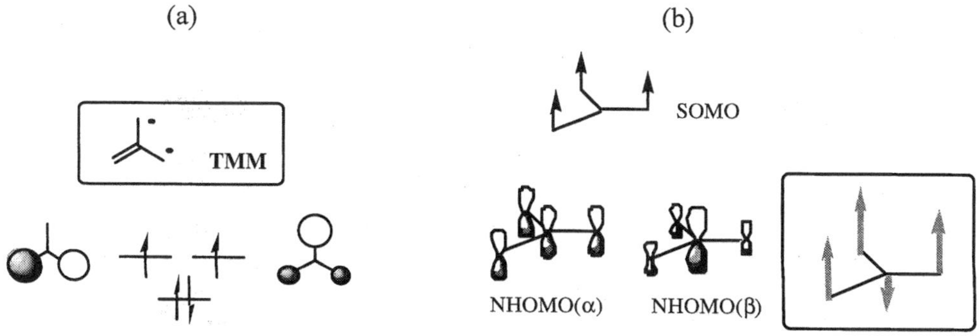

Fig. 5-1. (a) Degenerate NBMOs of **TMM**, (b) Spin polarization in **TMM**.

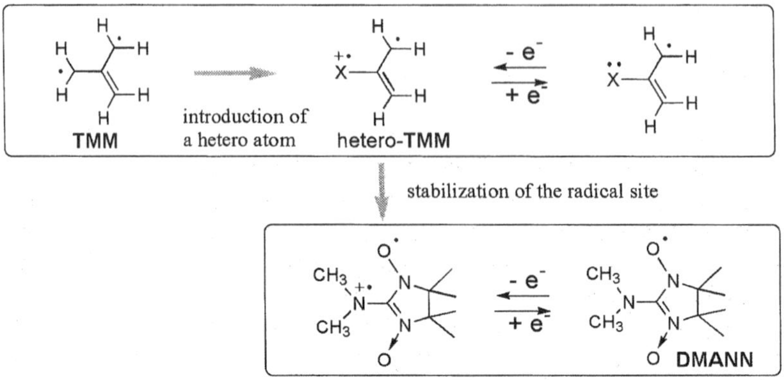

Fig. 5-2. Design of a radical donor affording a triplet cation diradical upon one-electron oxidation.

5.2 Molecular Design and Electronic Structure of Spin-polarized Donor

If a hetero atom with a lone pair of electrons is introduced to the nodal carbon of allyl radical, it gives rise to a hetero-analogue of **TMM**, especially when it is singly oxidized to a cation diradical. Since the electronic structure of the cation diradical resembles that of **TMM**, its ground state is considered to be triplet. In order to enhance the kinetic stability of the allyl radical part, it is replaced with a nitronyl nitroxide (NN), which can be regarded as a stable hetero-analogue of allyl radical. According to the above chemical modifications, dimethylamino nitronyl nitroxide, **DMANN**, was prepared (Fig. 5-2) [5, 7]. These radical donors turns out to afford ground state triplet cation diradicals upon one-electron oxidation.

5.3 Electronic Structure of Triplet Cation Diradical Derived from Spin-polarized Donor

The electronic features of a spin-polarized donor can be interpreted by a perturbational molecular orbital method, provided that the radical donor is composed of a donor site and a radical site and that they interact mutually. The electronic structure of a radical donor, in general, may be represented by either (a) or (b) as shown in Fig. 5-3.

In case (a), one-electron oxidation will remove the electron from a singly occupied molecular orbital (SOMO) of the normal radical donor to afford a closed-shell cationic species [30]. This electronic structure is, of course, unsuitable for generating a triplet cation diradical. In case (b), the highest occupied molecular orbital (*homo*) of the donor site interacts with the next highest molecular orbital (*nhomo*) of the radical site, not interacting with the singly occupied molecular orbital (*somo*) of the radical site, because of the symmetry mismatch of the relevant partial molecular orbitals. This electronic interaction raises the energy level of HOMO of the radical donor. As a result, HOMO is placed above SOMO. Such an exotic electronic structure can be maintained, if the on-site Coulombic repulsion of SOMO is larger than the orbital energy difference (ΔE)

between HOMO and SOMO. The donor radical of such an electronic structure is classified as a spin-polarized donor.

Recently Okada *et al* succeeded to isolate cation diradical salt of the spin-polarized donor, the donor part of which is N, N-diphenylphenazine (Fig. 5-4). The cation diradical was stable even at room temperature and it was found that the ferromagnetic coupling was larger than 700K on the basis of the temperature dependence of the magnetic susceptibility of the salt [34]. This result is strong evidence for the validity of the spin-polarized donor and proves the strong exchange coupling between the localized spin on the radical unit and the generated π-spin, delocalizing over the entire molecule.

Using thianthrene as a donor core, 2-thianthrenyl nitoronyl nitoroxide, 2,7- and 2,8-thianthrene bis(nitronyl nitroxide) were prepared. Their ground state triplet and quartet spin-multiplicities of the singly oxidized species of these donor radicals, respectively, were proved by ESR and spin-nutation measurements [35].

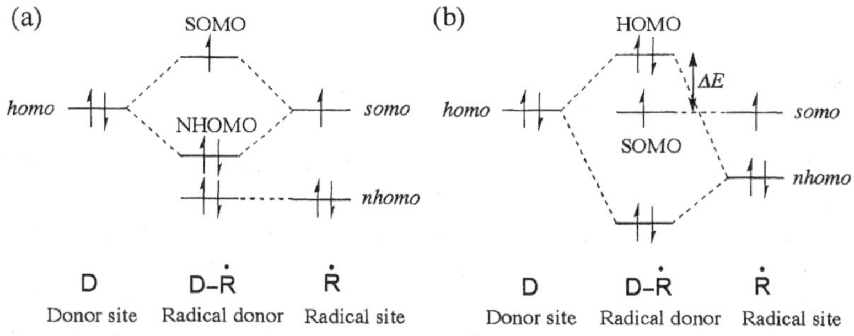

Fig. 5-3. Schematic drawing for the electronic configuration of a radical donor : (a) *homo* of the donor site interacting with *somo* of the radical site; (b) *homo* of the donor site interacting with *nhomo* of the radical site.

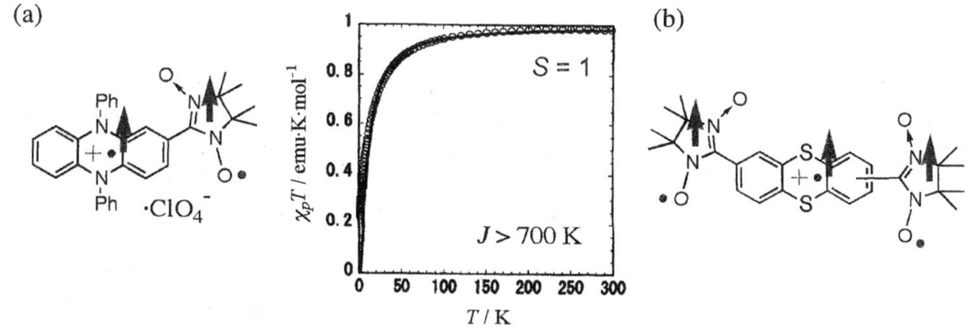

Fig. 5-4. (a) Temperature dependence of the magnetic susceptibility of the ion-diradical salt of a pyradine-based spin polarized donor. (b) Ground state cation triradical of thianthrene bis(nitonyl nitorxide).

6 Molecular and Crystal Design for Organic Ferromagnetic Metal

6.1 Design of TTF-based Spin-polarized Donor

During this decade, molecule-based materials exhibiting both conductivity and magnetism have been investigated extensively [36, 37]. Several π-d systems consisting of organic donors and transition metal counter ions have expanded a wing of organic conducting materials, such as paramagnetic superconductors, (anti)ferromagnetic metals, *etc.* [37]. If these two types of electrons are coupled magnetically, all spins in the bulk material have a chance to align on the basis of the double exchange mechanism [15]. As a building block for such a ferromagnetic conductor, several TTF-based spin-polarized donor have been reported [38–41].

The electronic structure of a spin-polarized donor is depicted by UHF-desicription in Fig. 6-1(a). It is to be noticed that the energy level of HOMO(β) is higher than SOMO(α). Since the on-site Coulomb (U) of HOMO of a TTF skeleton is small, the electron in HOMO is expected to show itineracy when the mixed valent state is realized in the donor radical column (Fig. 6-1(b)). Thus the spin-alignment of the unpaired electron in SOMO(α) is achieved, coupling with the spin-polarized (β) itinerant electrons.

As a prototypal donor-radical, a TTF derivative carrying a nitronyl nitroxide group (**TTF-NN**) was prepared (Fig. 6-2) [38]. Although a triplet ESR spectrum was observed in an iodine-doped sample of **TTF-NN**, the triplet signal was found to be a thermally populated one ($J = -100$K). The result was attributed to the twisting around a C-C bond between of the NN group from the TTF plane. In order to remove such a steric repulsion, a *p*-phenylene group was inserted between the donor site and the radical site. The modified donor radical, **TTF-PN**, afforded a ground state triplet cation diradical (**TTF-PN$^{+\bullet}$**) upon one-electron oxidation [39].

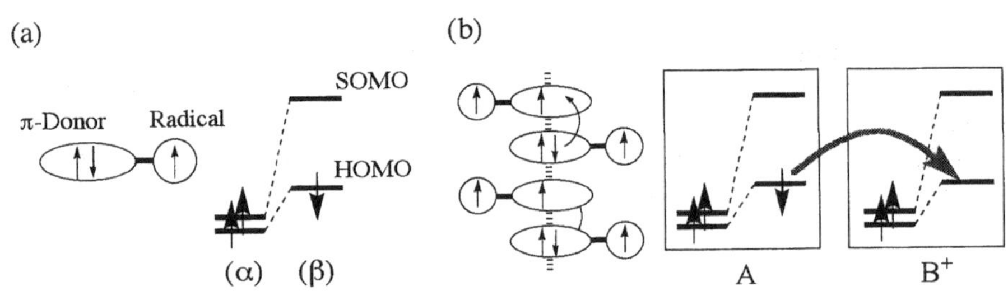

Fig. 6-1. (a) Electronic structure of a spin-polarized donor by UHF description. (b) Stack of spin-polarized donors in a mixed valence state ($A{\cdot}A^{+}{\cdot}A{\cdot}A^{+}$...) and spin-dependent intermolecular electron transfer.

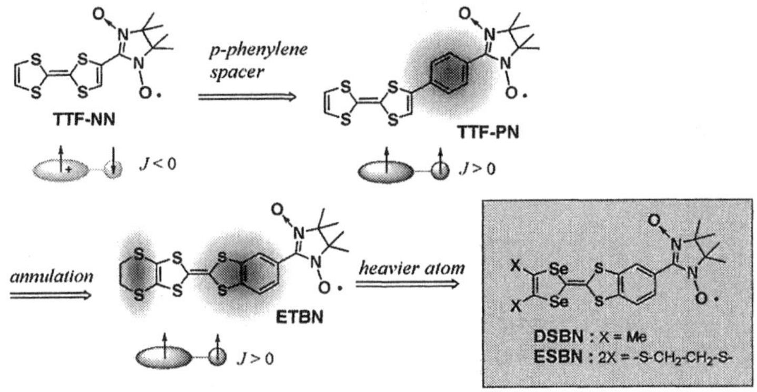

Fig. 6-2. Exploration of TTF-based spin-polarized donors.

6.2 Conductivity and Magnetism of Ion Radical Salts of TTF-based Spin Polarized Donors

In order to increase the kinetic stability of the singly oxidized species, chemical modification was carried out on a TTF-based spin-polarized donor. A benzo-annulated derivative, **ETBN**, was found to show good crystallinity, and it gave birth to single crystals of a neutral donor (Fig. 6-2) [40]. In the crystal, the arrangement of donor molecules are related with rotatory inversion axes (space group: *R*-3), and the crystal exhibits a ferromagnetic intermolecular interaction ($\theta\sim$ +0.5K). This novel donor was found to give crystalline ion-radical salts of the 2:1 donor to counter-ion ratio through a galvanostatic electro-crystallization [41]. The salt was a semiconductor (E_a = 0.16 eV) with a room temperature conductivity of 10^{-2} S·cm^{-1}. At the same time, it showed a paramagnetic behavior with a Curie constant of 0.73 emu·K·mol^{-1} for an $\mathbf{ETBN_2}\cdot ClO_4$ unit at room temperature, indicating that all the radical sites remained intact during the electro-crystallization. Although the π-spins on the donor sites are conductive, the density of the charge carrier is still low in this salt. Therefore, the spin polarization of the π-spin caused by the local spin on the radical site can not be transmitted to the neighboring local spins (Fig. 6-3).

The above result, however, promises that magnetic ordering of the local spins may be observed, if the carrier concentration in the ion-radical salt becomes reasonably high. Recently diselena-analogue of **ETBN, DSBN** and **ESBN**, have been synthesized by replacing inner sulfur atoms by selenium ones (Fig. 6-2). It was found that the singly oxidized species of both **DSBN** and **ESBN** exhibited outstanding kinetic stability compared with previous TTF-based donor radicals. The conductivity and magnetism of charge transfer complexes and an ion radical salt of **DSBN** and **ESBN** will be investigated extensively in the near future.

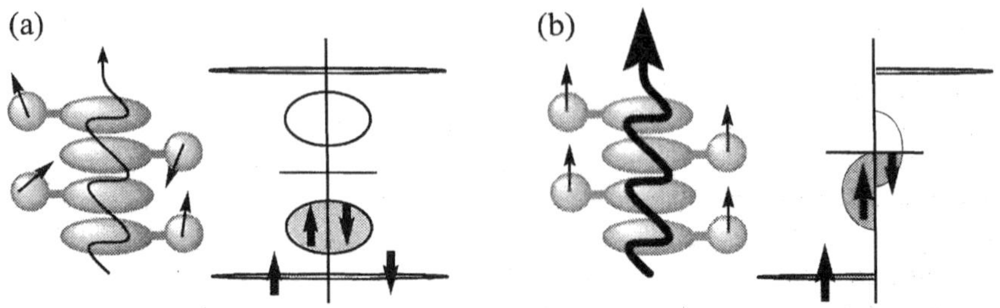

Fig. 6-3. Possible two-types of electronic structures of partially oxidized columner stack of spin-polarized donors. (a) Paramagnetic semiconductor (b) Ferromagnetic metal.

7 Conducting and Magnetic Properties of Networked Gold Nanoparticles Connected by Spin-polarized Molecular Wires

7.1 Design of Pyrrole-based Spin-polarized Donor

As described in the previous section (Sections 5, 6), a spin-polarized donor tends to donate a β-spin electron, provided that the spin of an unpaired electron is α. If a molecular wire, such as an oligothiophene, is connected to the spin-polarized donor core, only a β-spin electron can pass through the molecular wire. A schematic drawing of this idea is depicted in Fig. 7-1. According to the above idea, 1-pyrrolylphenylNN and its derivative bearing thienyl group at 2- and 5-positions have been designed, and their ground sate triplet spin multiplicity was confirmed [42].

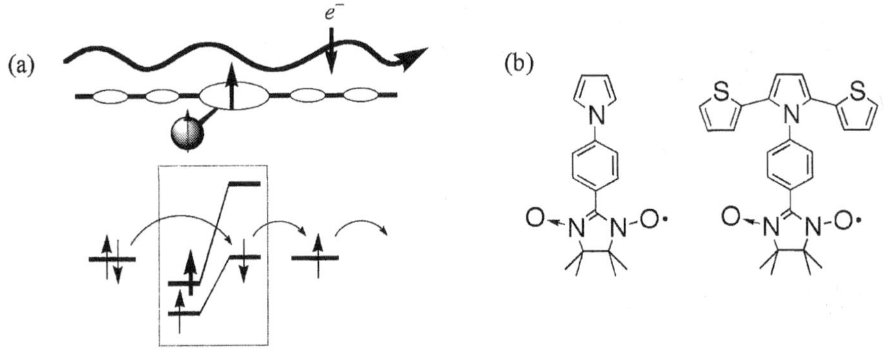

Fig. 7-1. (a) Schematic drawing of the concept of spin-rectification. (b) Pyrrole-based spin-polarized donors.

7.2 Electronic Property of Nano-sized Gold Nano-particles

As a challenge towards the ultimate goal to realize coexistence of conductivity of magnetism which is, in particular, originated from organic spins, a network structure of gold nanoparticles with a diameter of 4 nm has been constructed by connecting them with spin-polarized molecular wires.

Interestingly, the electronic structure of gold nanoparticles is size-dependent and a gold nano-particle with the average size of 4 nm shows a plasmon resonance peak at about 520 nm, indicating a metallic electronic structure [43–46]. These small nano-particles exhibit a Coulomb staircase in the I-V plot due to the presence of the Coulomb blockade [47, 48]. This is because the capacitance of the nano-particle is so small that the charging energy $Ec = e^2/C$, that is a repulsive energy caused by the addition of an extra electron, amounts to 60 meV. This is larger than the thermal energy k_BT, e.g. 20 meV at 300K. Such a small nanaoparticle must operate as a single electron transistor, if a tunnel barrier of the nanoparticle is larger than ca. h/e^2 =26 kΩ. Therefore it would be extremely intriguing to connect these nanoparticles in a networked manner through the conducting molecular wires.

Recently a network structure of gold nanoparticles with the average diameter of 4 nm has been constructed, using molecular wires of oligothiophene 3mer, 9mer, or 15mer, respectively. It is found that the network as a whole operates as a semiconductor with small activation energies of around 20 meV. Moreover, the contribution of electron tunneling becomes predominant at lower temperatures [49]. These results stimulated us to prepare a network structure composed of gold nanoparticles and spin-polarized molecular wire. For this purpose, a pyrrole-based spin-polarized donor with two disulfide groups at α,α'-positions of the terminal thienyl ring was prepared [50]. This molecular wire (**TPNT-wire**) is expected to interconnect the gold nanoparticle and also chemisorb on gold electrodes through thiol groups introduced at both terminals of the wire [42, 51, 52]. A unique point in this network is that a local spin resides on a π-conjugating molecular wire, whereas conduction electrons tunnel among gold nanoparticles in the network.

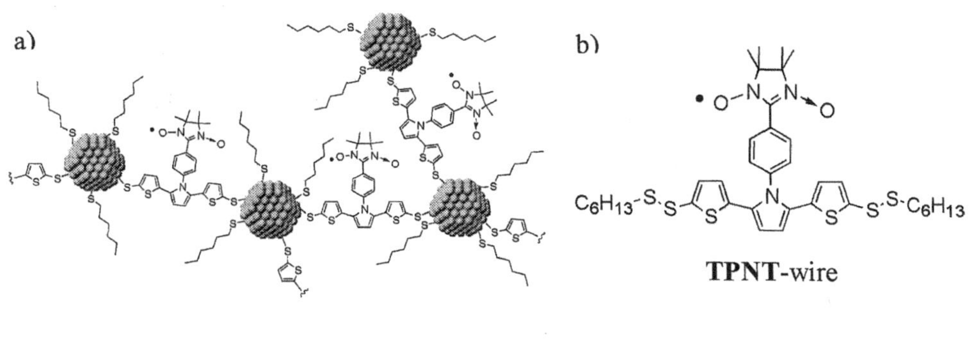

Fig. 7-2. (a) Schematic scheme of networked gold nanoparticle with spin-polarized molecular wire. (b) Spin-polarized molecular wire with protected thiol groups at terminals.

7.3 Preparation and Characterization of the Network Structure of Gold Nanoparticles Connected with Spin–polarized Molecular Wire

Preparation of the network structure of gold nanoparticles connected with spin–polarized molecular wire is practically the same as the oligothiophene network [53–55]. Namely, dip of the interdigitated electrodes into a solution of the molecular wire (0.31 mM), followed by the addition of a solution of surfactant-stabilized gold nanoparticles (11.25 mM), forms a networked nanoparticles connected with spin-polarized wires on the electrodes. If electrodes with a gap of 100 μm is used, it requires at least 1.8×10^4 nanoparticles to link the electrodes.

In order to confirm the structure of the networked gold nanoparticles, a sample was measured by transmission electron microscope (TEM). It turned out that the network consists of granules of gold nanoparitcles (the average diameter of 100 nm) interconnected by the **TPNT**-wires (Figure 7-3a) [56, 57]. The expanded TEM image showed that the granule is composed of gold nanoparticles with the average diameter of 4 nm with the separation of about 1.6 nm, corresponding with the length of the **TPNT**-wire (Fig. 7-3(b)).

In order to confirm that gold nanoparticle were connected with the **TPNT**-wire, temperature dependence of the magnetic susceptibility of the network was measured by means of a SQUID apparatus. Judging from the magnitude of Curie constant (5.55 emu·K·mol^{-1}), about 15 spins are on a nanoparticle in average. The result is consistent with the value estimated by the elemental analysis (C 4.38%, H 0.64%, N 0.28%, ash 91.3%, indicates that @Au1750:wire:*n*-hexanethiol:TOAB = 1:17:17:22). Since one wire is shared with two particles, at least 30 wires are coated on one nanoparticle.

(a) (b)

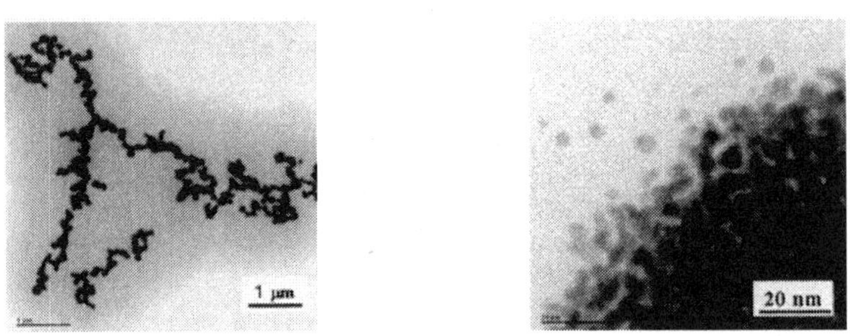

Fig. 7-3. (a) TEM micrograph for a sample of a networked gold nanoparticles cast on carbon-coated Cu-grid The average diameter of each granule is 100 nm. (b) A magnification of a granule composed of 4 nm gold nanoparticles.

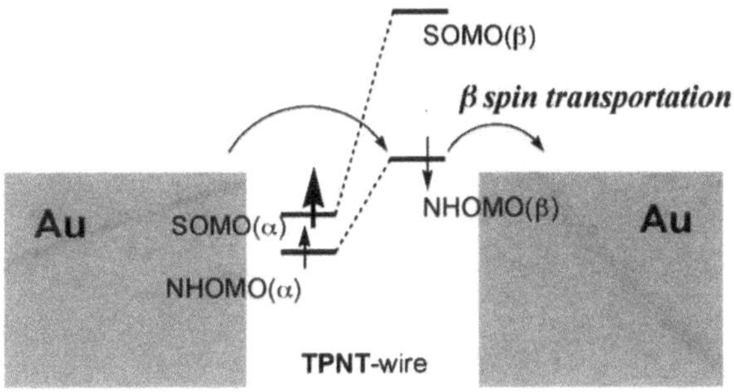

Fig. 7-4. Schematic drawing of spin-dependent electron transportation through a spin-polarized molecular wire.

7.4 Conducting Property of Network of Gold Nanoparticles

Temperature dependence of the conductance of the networked gold nanoparticles with the **TPNT**-wire was measured, using interdigitated electrodes with a gap of 100 μm. The non-linear trace of the network in the Arrhenius plot strongly suggests that the contribution of the electron tunneling through a molecular wire becomes significant at lower temperatures than 20 [53–55].

Figure 7-4 is a schematic figure depicting the energy levels of frontier orbitals of the **TPNT**-wire near the Fermi level of a gold particle. Since the electronic structure of the radical wire is spin-polarized, the highest occupied orbital should be SOMO (β) in the UHF description, provided that the singly occupying electron is α. Under such a circumstance, the tunneling current is expected to be spin-polarized to β. Therefore, it is extremely interesting to measure the tunneling current, paying an attention to the effect of application of an external magnetic field.

7.5 Negative Magneto-resistance of on the Spin-polarized Network

In order to examine the external magnetic field on the networked gold nanoparticles connected with the TPNT-wire, the conductance of the network was measured using electrodes with a gap of 2 μm; the gap is connected with at least 360 nanoparticles.

The effect of an external magnetic field on the resistance of the network was examined by applying the magnetic field to the parallel direction in reference to the wires so that the Lorentz force did not operate. It was fond that the network made of spinless wires exhibited a small positive magneto-resistance at 4K under the external magnetic field in the range of 0–6 T (Fig. 7-5(a)). To the contrast, the magnetic field effect on the spin-polarized wires exhibited a negative magneto-resistance (Fig. 7-5(b)). It means that the resistance of the wire decreases under the external magnetic field presumably because the fluctuation of the spins on the molecular wires is suppressed by the external magnetic field.

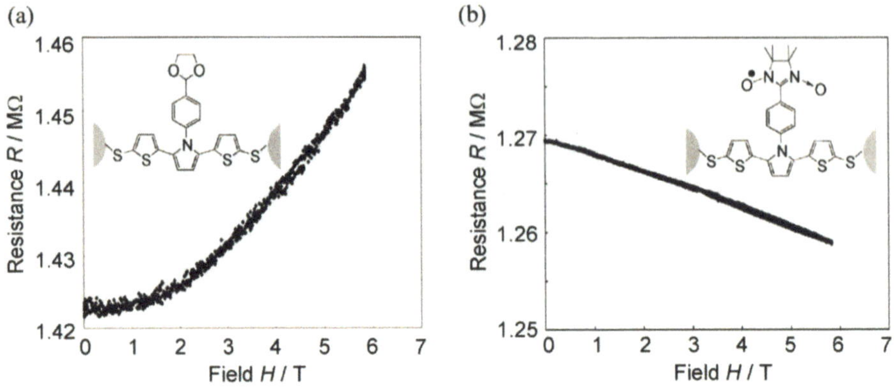

Fig. 7-5. Magneto-resistance of networked gold nanoparticles with (a) a spinless molecular wire and (b) a spin-polarized molecular wire at 4.2K.

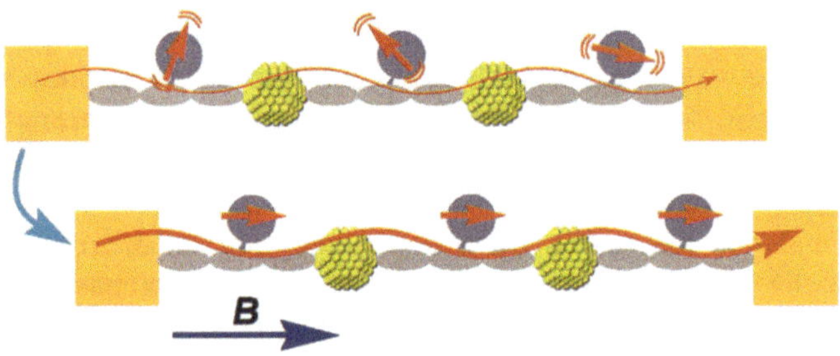

Fig. 7-6. External magnetic field effect on the tunneling current of spin-polarized network.

Although the detail of the magneto-resistance effect on the network has not been fully understood yet, this is the first example that the negative magneto-resistance was observed in an organic-inorganic nano-composite in which a local spin is located on the organic molecular wire (Fig. 7-6). The network composed of such small gold nanoparticles connected with spin-polarized molecular wires in a 3-dimensional manner should exhibit exotic non-linear conducting behaviors, including the Coulomb blockade, a memory effect, and should give rise to important data for developing molecule-based spin-electronic devices.

8 Summary

The former part of this chapter described the spin-alignment of localized spins of organic radicals through the-spin polarized π-conjugating electrons. Cautious examination of this spin-ordering mechanism leads to the creation of a new class of donor radicals, designated as spin-polarized donors. Pyrrole-type spin-polarized donor, in particular, produced a spin-polarized molecular wire with protected-thiols groups at both terminals. Combination of the spin-polarized wire and metallic gold nanoparticles gave birth to a coexistence system of itinerant electrons and organic localized spins. It was found that this networked gold nanoparticle connected by the spin-polarized molecular wire exhibited the negative magnetoresistance with the external magnetic field was applied. Such an organic and inorganic hybrid system must give insight into the spin-ordering problem in graphites and other carbon-based materials.

References

1. (a) K. Itoh, Chem. Phys. Lett. 1 (1967) 235; (b) E. Wasserman, R.W. Murray, W.A. Yanger, A.M. Trozzolo, and G. Smolinsky, J. Am. Chem. Soc. 89 (1967) 5076.
2. (a) C.A.Jr. Hutchison and B.E. Kohler, J. Chem. Phys. 51 (1969) 3327; (b) R.J. Anderson, and, B.E. Kohler, J. Chem. Phys. 65 (1978) 2451.
3. (a) T. Sugawara, S. Bandow, K. Kimura, H. Iwamura, and K. Itoh, J. Am. Chem. Soc. 106 (1984) 6449; (b) T. Sugawara, S. Bandow, K. Kimura, H. Iwamura, and K. Itoh, J. Am. Chem. Soc. 108 (1986) 368; (c) Y. Teki, T. Takui, K. Itoh, H. Iwamura, and K. Kobayashi, J. Am. Chem. Soc. 108 (1986) 2147; (d) I. Fujita, Y. Teki, T. Takui, T. Kinoshita, and K. Itoh, J. Am. Chem. Soc. 112 (1990) 4074; (e) K. Matsuda, N. Nakamura, K. Takahashi, K. Inoue, N. Koga, and H. Iwamura, J. Am. Chem. Soc. 117 (1995) 5550.
4. A. Rajca, J. Wongsriratanakul, and S. Rajca, Science 294 (2001) 1503.
5. H. Sakurai, R. Kumai, A. Izuoka, and T. Sugawara, Chem. Lett. (1996) 879.
6. T. Sugawara, Mol. Cryst. Liq. Cryst. 334 (1999) 257.
7. H. Sakurai, A. Izuoka, and T. Sugawara, J. Am. Chem. Soc. 122 (2000) 9723.
8. T. Sugawara, A. Izuoka, and R. Kumai, "Spin Alignment via Conduction Electron" in 'Molecular Magnetism - New Magnetic Materials', edited by K. Itoh and M. Kinoshita, Tokyo: Kodansha and Gordon&Breach, 2000, pp 239-251.
9. T. Sugawara, H. Sakurai, and A. Izuoka, "Electronically Controllable High Spin System Realized by Spin-Polarized Donors" in 'Hyper-Structured Molecules II : Chemistry, Physics, and Applications', editted by H. Sasabe, Amsterdam: Gordon&Breach, 2000, pp. 35-57.
10. (a) J. Higuchi, J. Chem. Phys. 38 (1963) 1237; (b) J. Higuchi, J. Chem. Phys. 39 (1963) 1847; (c) J. Higuchi, Bull. Chem. Soc. Jpn. 43 (1970) 3773.
11. (a), N. Mataga, Theor. Chim. Acta 10 (1968) 372; (b) H.C. Longuet-Higgins and A.J. Pople, Proc. Phys. Soc. (London) A68 (1955) 591; (c), A.A. Ovchinnikov, Theoret, Chim. Acta, 47 (1980) 297.
12. (a) K. Nasu, Phys. Rev. B: Solid State, B33 (1986) 330; (b) Y. Teki, I. Fujita, T. Takui, T. Kinoshita, and K. Itoh, J. Am. Chem. Soc. 116 (1994) 11499.
13. (a) R. L. Carlin, 'Magnetochemistry' Berlin: Springer-Verlag, 1986; (b) S. Chikazumi, 'Physics of Magnetism', New York: John Wiley and Sons, 1964; (c) 'Theory of Magnetism in Transition Metals', Edited by W. Marshall, New York: Academic Press, 1967; (d) R.J. Elliott, 'Magnetism IIA', New York: Academic Press, 1966; (e) T. Kasuya, 'Magnetism II B', New York: Academic Press, 1966; (f) K. Yoshida, 'Progress in Low Temperature Physics' vol. 4, Amsterdam: North Holland 1964.
14. (a) M. A. Ruderman and C. Kittel, Phys. Rev. 96 (1954) 99; (b) T. Kasuya, Prog. Theor. Phys. 16 (1956) 45; (c) T. Kasuya, Prog. Theor. Phys. 16 (1956) 58; (d) K. Yoshida, Phys. Rev. 106 (1957) 893.

15. (a) G.H. Jonker and J.H. van Santen, Physica 16 (1950) 337; (b) C. Zener, Phys. Rev. 82 (1951) 403; (c) P.W. Anderson, and H. Hasegawa, Phys. Rev. 100 (1955) 675; (d) P.G. de Gennes, Phys. Rev. 118 (1960) 141.
16. (a) T. Kimura, Y. Tomioka, H. Kuwahara, A. Asamitsu, M. Tamura, and Y. Tokura, Science 274 (1996) 1698; (b) V. Kiryukhin, D. Casa, J.P. Hill, B. Keimer, A. Vigliante, Y. Tomioka, and Y. Tokura, Nature 386 (1997) 813; (c) M. Imada, A. Fujimori, and Y. Tokura, Rev. Mod. Phys. 70 (1998) 1039.
17. McConnell, H.M. Proc. Robert A. Welch Found. Conf. Chem. Res. 11 (1967) 144.
18. (a) R. Breslow, Pure Appl. Chem. 54 (1982) 927, (b) T.J. Lepage, and R. Breslow, J. Am. Chem. Soc. 109 (1987) 6412.
19. J.S. Miller, D.A. Dixon, J.C. Calabrese, C. Vazquez, P.J. Krusic, M.D. Ward, E. Wasserman, and R.L. Harlow, J. Am. Chem. Soc. 112 (1990) 381.
20. E. Dormann, M. Nowak, K.A. Williams, R.O.Jr. Angus, and F. Wudl J. Am. Chem. Soc. 109 (1987) 2594.
21. A. Izuoka, R. Kumai, T. Tachikawa, and T. Sugawara Mol. Cryst. Liq. Cryst. 218 (1992) 213.
22. (a) J. Tanabe, T. Kudo, M. Okamoto, Y. Kawada, G. Ono, A. Izuoka, and T. Sugawara, Chem. Lett. 579 (1995); (b) K. Takimiya, Y. Shibata, K. Imamura, A. Kashihara, Y. Aso, T. Otsubo, and F. Ogura, Tetrahedron Lett. 36 (1995) 5045; (c) K. Takimiya, K. Imamura, Y. Shibata, Y. Aso, and T. Otsubo, J. Org. Chem. 62 (1997) 5567.
23. (a) R. Hoffmann, Acc. Chem. Res. 4 (1971) 1; (b) R. Gleiter, Angew. Chem. Int. Ed. Eng. 13 (1974) 696.
24. (a) A. Izuoka, J. Tanabe, T. Sugawara, Y. Kawada, R. Kumai, A. Asamitsu, and Y. Tokura, Mol. Cryst. Liq. Cryst. 306 (1997) 265; (b) J. Tanabe, G. Ono, A. Izuoka, T. Sugawara, T. Kudo, T. Saito, M. Okamoto, and Y. Kawada, Mol. Cryst. Liq. Cryst. 296 (1997) 61.
25. D.D. Swank, C.P. Landee, and R.D. Willett, Phys. Rev. B20 (1979) 2154.
26. W.T. Borden, Mol. Cryst. Liq. Cryst. 232 (1993) 195.
27. H.M. McConnell, C. Heller, T. Cole, and R.W. Fessenden, J. Am. Chem. Soc. 82 (1960) 766.
28. A. Izuoka, J. Tanabe, T. Sugawara, T. Kudo, T. Saito, and Y. Kawada, Chem. Lett. 2002, 910.
29. (a) K. Yamaguchi, H. Nishimoto, T. Fueno, T. Nogami, and Y. Shirota, Chem. Phys. Lett. 166 (1990) 408; (b) K. Yamaguchi, M. Okumura, T. Fueno, and K. Nakasuji, Synth. Metals 41 (1991) 3631.
30. (a) T. Sugano, T. Fukasawa, and M. Kinoshita, Synth. Metals 41 (1991) 3281; (b) T. Sugimoto, S. Yamaga, M. Nakai, M. Tsujii, H. Nakatsuji, and N. Hosoito, Chem. Lett. 1817 (1993) (c) T. Ishida, K. Tomioka, T. Nogami, K. Yamaguchi, W. Mori, and Y. Shirota, Mol. Cryst. Liq. Cryst. 232 (1993) 99; (d) S. Nakatsuji, A. Hirai, J. Yamada, K. Suzuki, T. Enoki, and H. Anzai, Mol. Cryst. Liq. Cryst. 306 (1997) 409, (e) S. Nakatsuji, S. Satoki, K. Suzuki, T. Enoki, N. Kinoshita, and H. Anzai, Synth. Metals 71 (1995) 1819; (f), S. Nakatsuji, N. Akashi, K. Suzuki, T. Enoki, N. Kinoshita, and H. Anzai, Mol. Cryst. Liq. Cryst. 268 (1995) 153.
31. (a) R.J. Baseman, D.W. Pratt, M. Chow, and P. Dowd, J. Am. Chem. Soc. 98 (1976) 5726; (b) W.T. Borden, H. Iwamura, and J.A. Berson, Acc. Chem. Res. 27 (1994) 109; (c) W.T. Borden, "Diradicals [as reactive intermediates], Reactive Intermediates 2", New York: Wiley, 1981, pp. 175-209.
32. D. Feller, W.T. Borden, and E.R. Davidson, J. Chem. Phys. 74 (1981) 2256.

33. P.G. Wenthold, J. Hu, R.R. Squires, and W.C. Lineberger, J. Am. Chem. Soc. 118 (1996) 475.
34. S. Hiraoka, T. Okamoto, M. Kozaki, D. Shiomi, K. Sato, T. Takui, and K. Okada, J. Am. Chem. Soc. 126 (2004) 58.
35. A. Izuoka, M. Hiraishi, T. Abe, T. Sugawara, K. Sato, and T. Takui, J. Am. Chem. Soc. 122 (2000) 3234.
36. (a) H. Fujiwara, H-J. Lee, H.-B. Cui, H. Kobayashi, E. Fujiwara, and A. Kobayashi, Adv. Mater. 16 (2004) 1765; (b) H. Akutsu, J. Yamada, and S. Nakatsuji, Chem. Lett. 32 (2003) 1118.
37. (a) E. Coronado, J. R. Galan-Mascaros, and C. J. Gomez-Garcia, Nature 408 (2000) 447; (b) H. Fujiwara, H. Kobayashi, E. Fujiwara, and A. Kobayashi, J. Am. Chem. Soc. 124 (2002) 6816; (c) T. Mallah, C. Hollis, S. Bott, M. Kurmoo, and P. Day, J. Chem. Soc. Dalton Trans. 859 (1990); (d) J. Yamaura, K. Suzuki, Y. Kaizu, T. Enoki, K. Murata, and G. Saito, J. Phys. Soc. Jpn. 65, 2645 (1996) (e), H. Kobayashi, H. Tomita, T. Naito, A. Kobayashi, F. Sakai, T. Watanabe, and P. Cassoux, J. Am. Chem. Soc. 118 (1996) 368.
38. R. Kumai, M.M. Matsushita, A. Izuoka, and T. Sugawara, J. Am. Chem. Soc. 116 (1994) 4523.
39. J. Nakazaki, M.M. Matsushita, A. Izuoka, and T. Sugawara, Tetrahedron Lett. 40 (1999) 5027.
40. Y. Ishikawa, T. Miyamoto, A. Yoshida, Y. Kawada, J. Nakazaki, A. Izuoka, and T. Sugawara, Tetrahedron Lett. 40 (1999) 8819.
41. J. Nakazaki, Y. Ishikawa, A. Izuoka, T. Sugawara, and Y. Kawada, Chem. Phys. Lett. 319 (2000) 385.
42. (a) J. Nakazaki, I. Chung, M. M. Matsushita, T. Sugawara, R. Watanabe, A. Izuoka, and Y. Kawada, J. Mater. Chem. 13 (2003) 1011; (b) J. Nakazaki, I. Chung, R. Watanabe, T. Ishitsuka, Y. Kawada, M. M. Matsushita, and T. Sugawara, Internet Electron. J. Mol. Des. 2 (2003) 112; (c) J. Nakazaki, M.M. Matsushita, A. Izuoka and T. Sugawara, Mol. Crust. Liq. Cryst. 306 (1997) 81.
43. J. Fink, C. J. Kiely, D. Bethell, and D. J. Schiffrin, Chem. Mater. 10 (1998) 922.
44. M. Brust, M. Walker, D. Bethell, D. J. Schiffrin, and R. Whyman, J. Chem. Soc. Chem. Commun. 801 (1994).
45. K.H. Su, Q.H Wei, X. Zhang, J. J. Mock, D. R. Smith, and S. Schultz, Nano Lett. 3 (2003) 1087.
46. P. Nordlander, C. Oubre, E. Prodan, K. Li, and M.I. Stockman, Nano Letters 4 (2004) 899.
47. J. Park, A.N. Pasupathy, J.I. Goldsmith, C. Chang, Y. Yaish, J.R. Petta, M. Rinkoski, J.P. Sethna, H.D. Abruña, P.L. McEuen, and D.C. Ralph, Nature 417 (2002) 725.
48. B.M. Quinn, and K. Kontturi J. Am. Chem. Soc. 126 (2004) 7168.
49. S. Taniguchi, M. Minamoto, M.M. Matsushita, and T. Sugawara, *to be published.*
50. M. Minamoto, M.M. Matsushita, and T. Sugawara, Polyhedron, *in press.*
51. G.A. Prinz, Science 282 (1998) 1660.
52. (a) M.M. Matsushita, N. Ozaki, T. Sugawara, F. Nakamura, and M. Hara, Chem. Lett. (2002) 596; (b) G. Harada, H. Sakurai, M.M. Matsushita, A. Izuoka, and T. Sugawara, Chem. Lett. (2002) 1030.
53. F.P. Zamborini, L.E. Smart, M.C. Leopold and R.W. Murray, Analytica Chimica Acta 496 (2003) 3.
54. W.P. Wuelfing, S.J. Green, J.J. Pietron, D.E. Cliffel, and R.W. Murray, J. Am. Chem. Soc. 122 (2000) 11465.

55. (a) M. Brust, D. Bethell, D.J. Schiffrin, and C.J. Kiely, Adv. Mater. 7 (1995) 795; (b) T. Ogawa, K. Kobayashi, G. Masuda, T. Takase, and S. Maeda, Thin Solid Films 393 (2001) 374.
56. M.M. Maye, J. Luo, I.-I.S. Lim, L. Han, N.N. Kariuki, D. Rabinovich, T. Liu, and C.J. Zhong, J. Am. Chem. Soc. 125 (2003) 9906.
57. M.M. Maye, S.C. Chun, L. Han, D. Rabinovich, and C.J. Zhong, J. Am. Chem. Soc. 124 (2002) 4958.

Carbon-Based Magnetism
T. Makarova and F. Palacio (Editors)

2

Magneto-structural Correlations in π-Conjugated Nitroxide-based Radicals: Hydrogen-bonds and Related Interactions in Molecular Organic Solids

P.M. Lahti

Department of Chemistry, University of Massachusetts, Amherst, MA 01003 USA

1 Introduction

This chapter discusses various aspects of the role of hydrogen-bonding in molecular magnetic materials. The hydrogen-bond is one of the strongest and most directional non-covalent interactions in chemistry, yet has subtle and variable influence upon exchange interactions in molecular materials incorporating unpaired electrons. Some systems have exchange interactions that appear to have direct electronic propagation through hydrogen-bonds, while others have hydrogen-bonding that provides excellent crystal lattice assembly but no clear electronic exchange effect. Considerable effort continues to be placed upon discerning and defining the role of hydrogen-bonding in various molecular magnetic materials.

This chapter is intended to provide the reader with a useful introduction to the present level of understanding of this aspect of exchange and magnetism in organic solids. The major focus is on magnetic and exchange effects in π-conjugation organic nitroxide-based radicals, where hydrogen-bonding is important either to organize the molecules in the solid state, or as direct electronic exchange linkage, or both. The chapter does not aim to have comprehensive scope as a review of all aspects and examples of hydrogen-bonding in organic molecular magnetic materials. Rather, it sets forth a series of examples that illustrates the fascinating array of exchange behavior observed in organic systems with hydrogen-bonding, as well as some of the difficulties in interpreting structure-property relationships linking hydrogen-bonding and exchange. The reader is strongly encouraged to consult the various works cited throughout this chapter for more details.

2 Exchange and Crystal Assembly in Phenolic Nitronynitroxides

The attachment of phenolic groups to radical fragments provided some of the first examples of hydrogen-bond guided crystal assembly in organic magnetic materials. The phenolic OH is not as strongly directional as some hydrogen-bonding groups, but has proven an effective moiety for assembling radicals in a manner that promotes solid-state exchange interaction networks.

Veciana and coworkers have extensively studied hydrogen-bonding effects on exchange in phenolic nitronylnitroxides, including number of mono- and dihydroxyphenyl nitronylnitroxides: **4HOPhNN** [1], **3HOPhNN**,[1d-e, 2] **2HOPhNN**,[1c, 2] **34diHOPhNN** [2b, 2d, 3], and **RSNN** [2, 4]. Table I summarizes the reported behaviors of these systems. Among these, this chapter will focus on one of the first and most-studied, **4HOPhNN**. Despite its structural simplicity, its crystallography exhibits a network incorporating both O-H···O-N hydrogen-bonds and N-O···H-C (nitronylnitroxide methyl) close contacts that form a 2-D network among the molecules. Solution ESR hyperfine splitting studies show significant delocalization of nitronylnitroxide spin density onto the phenol ring, with an upper limit for the hydroxyl proton hyperfine of 0.002 mT (sign not determined). The χT vs. T data fit quite well to both 1-D and 2-D planar Heisenberg models composed of S = 1/2 spin units. Although the crystallographic contacts network is suggestive of 2-D interactions, dc magnetic measurements alone could not confirm this, since reasonable fits were obtained for both 1-D (J/k(chain) = +1.0K, J/k(interchain) = −0.27K) and 2-D behavior (J/k(2D plane) = +0.63K, J/k(interplane) = −0.56K for a 2-neighbor model).[1a] In an array of single-crystal ESR studies, variations in peak width and g-factor with orientation and temperature are consistent with 2-D FM sheet-like behavior in the crystallographic *ab* plane (the close contact network). Solid-state NMR and computational studies were carried out to probe the spin density distributions in the solid state.[1d-e] Ac-susceptibility likewise was in accord with the 2-D model of exchange. μMR experiments even indicate[1c] a possible cross-over from 2-D to 3-D ordering behavior.

The balance of information indicates that **4HOPhNN** acts as a quasi-2-D or 3-D organic ferromagnet, with both O-H···O-N hydrogen-bonds and N-O···H-C contacts acting as important electronic contributors to the overall magnetic behavior. The multidimensional behavior of **4HOPhNN** depends on the lack of a single, dominating exchange mechanism, so the hydrogen-bonds can become electronically "crucial" as part of a network, as the authors point out. The Veciana group work on **4HOPhNN** is an excellent example of how a variety of experimental studies is often necessary to define the magnetic nature and exchange mechanisms in an organic magnetic material.

Sugawara and coworkers have described detailed studies of **HQNN** [5–6], (both α- and a β- crystallographic forms) and magneto-structural correlations for **RSNN**[6]. **α-HQNN** crystallographically exhibits a complex network of intramolecular and intermolecular bifurcated O-H···O-N hydrogen-bonds and N-O···H-C (nitronylnitroxide methyl group) contacts, with temperature dependence of its magnetic susceptibility that fits well to a ST (dimeric) model having J/k = +0.93K, θ = +0.46K. Low temperature heat capacity measurements show a sharp anomaly at 0.42K consistent with 3-D ordering of a true organic ferromagnet composed of S = 1/2 units. The various data suggest that pairs of **HQNN** molecules ferromagnetically couple, with longer-range coupling between dimers at low temperatures to give ferromagnetic (FM) ordering. **β-HQNN** combines double hydrogen-bonded ribbons with π-stacking: its magnetic behavior fits the ST model with J/k = +5.0K, θ = −0.32K. The negative Weiss constant shows the presence of modest antiferromagnetic (AFM) exchange interactions at low temperatures, although there is still a ferromagnetic exchange contribution to the overall magnetic behavior of this phase. **RSNN** exhibits crystallography that is very similar to that of **β-HQNN** [2a, 2d, 6], but its magnetic behavior analyzed with the same ST model gives [6] J/k = +10.0 K, θ = −4.0K. The larger negative θ shows strengthened AFM exchanges by comparison to **HQNN**. Figure 1 summarizes the spin polarization mechanisms that appear to give the observed behavior. Deuterium substitution on the OH groups of **HQNN** decreases the FM interaction, supporting an electronic role for the OH groups in the overall magnetic behavior [5b, 6]. NMR measurements were used to deduce a negative spin density on the OH group that is *ortho* to the nitronylnitroxide group in **HQNN**. This experimental data – supported by semiempirical PM3/UHF computations [7] – were interpreted to support

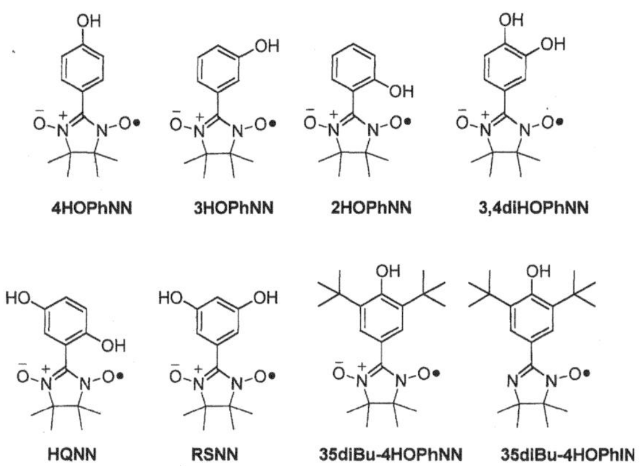

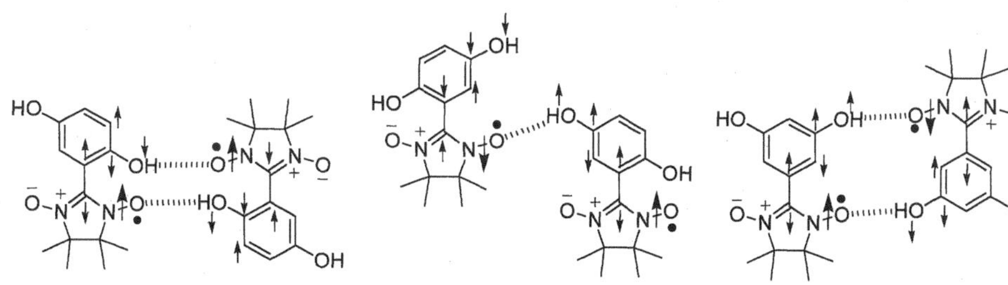

α-HQNN spin-polarized exchange pathway

β-HQNN spin-polarized exchange pathway

RSSN spin-polarized exchange pathway

Fig. 1. Exchange pathways in phenolic hydrogen bonded radicals.

Table 1.
Magnetic behavior of phenolic nitronylnitroxides.

Radical	Crystallography	Magnetic Behavior	Exchange Strength
2HOPhNN	3D network CH···ON contacts NO···methyl contacts	ferromagnet (T_c ~ 0.41K)	Curie-Weiss θ = +0.62K [1c]
3HOPhNN	hydroxyl···ON cyclic dimmers aryl CH···ON contacts π-stacking	AFM	Curie-Weiss θ < 0K [2a-b]
4HOPhNN	2D network hydroxyl···ON contacts NO···methyl contacts	quasi-2D ferromagnet	J_{1D}/k = 1.01K [1a] $J_{interchain}/k$ = –0.27K; J_{2D}/k = 0.63K [1a] $J_{interplane}/k$ = –0.56K
34diHOPhNN	double hydroxyl···ON contacts	alternating 1-D chains	J_{1D}/k = 9.28, (–)3.78K [3] with α = 2.48 $J_{interchain}/k$ = 0.34K
α-HQNN	hydroxyl···ON contacts NO···methyl contacts	ferromagnet (T_c = 0.45K)	J_{ST}/k = 0.93K [6] θ = +0.46K
β-HQNN	hydroxyl···ON contacts π-stacking	AFM	J_{ST}/k = 5.0K [6] θ = –0.32K
RSNN	double hydroxyl···ON contacts π-stacking	AFM	J_{ST}/k = 10.0K [6] θ = –4.0K
35diBu-4HOPhNN	2D network hydroxyl···ON contacts NO···methyl contacts	1-D AFM chains	J_{1D}/k = –3.0K[8] θ = –0.38K
35diBu-4HOPhIN	defective 2-D network hydroxyl···ON contacts NO···methyl contacts	multiple AFM spin pairing?	(see text) [8]

extended, close-contacts, spin-polarization exchange mechanisms for HQNN and RSNN, such that the OH group is electronically important as well as providing of crystallographic scaffolding through hydrogen-bonding. In **α-HQNN**, the hydrogen-bond *ortho* to the transannular bond appears crucial to the dimeric FM interaction. By comparison, the *meta* hydrogen-bonds in **β-HQNN** appear to provide AFM exchange at low temperatures, while a McConnell II π-stack between the nitronylnitroxide units (not shown) provides the FM dimer exchange mechanism. The magneto-structural behaviors of the two polymorphs thus are quite different. In **RSNN**, the same motifs are present as in **β-HQNN**, with presumed FM coupled nitronylnitroxide π-stacking and AFM exchange through ribbons of hydrogen-bonded O-H···O-N contacts. The stronger AFM exchange in **RSNN** is attributed to double hydrogen-bonding in the ribbons (similar to that in **α-HQNN**, but through a *meta* OH), by comparison to the single hydrogen-bonded chain in **β-HQNN**. The complexity of the close atomic contacts in the crystallography of these systems makes magnetostructural analysis quite difficult, but it is clear that the molecular connectivity and geometry of the hydrogen-bonds that are directly in contact with major spin density sites (the NO groups) are quite important to the overall magnetic behavior.

Taylor and Lahti [8] found that sterically hindered *tert*-butyl groups in **35diBu-4HOPhNN** altered the hydrogen-bond geometry relative to that of **4HOPhNN**, but did not result in magnetic or crystallographic isolation of the radical spin sites. As in **4HOPhNN**, the OH group of **35diBu-4HOPhNN** participates in a network of hydrogen-bond type O-H···O-N and N-O···H-C (nitronylnitroxide methyl) contacts (Fig. 2), giving a sheet-like motif much like that of **4HOPhNN**. Although the N-O···H-C contacts for the two are quite similar, the O-H···O-N contacts of **35diBu-4HOPhNN** differ from those found in **4HOPhNN**. In **4HOPhNN**, the O-H···O-N is nearly collinear and shorter, with r(O)H···O(N) = 1.6–1.7 Å, consistent with a relatively typical hydrogen-bond. In **35diBu-4HOPhNN**, the O-H···O(N) contact is bent at a 134–136° angle, and is longer, with r(O)H···O(N) = 2.2–2.3 Å. In addition, the N-O unit lies above the phenyl-OH plane. In this geometry, it seems that a N-O singly occupied molecular orbital (SOMO) rather than a lone pair is involved in the close contact, hence the elongated O-H···O(N) distance. Still, infrared spectroscopy of **35diBu-4HOPhNN** shows broadening of the OH stretching frequency, like a typical hydrogen-bond.

Dc magnetic measurements of **35diBu-**4HOPhNN down to 1.8K are consistent with AFM 1-D Heisenberg chain behavior with $J/k = -3$K, unlike the quasi-2D ferromagnetic behavior of **4HOPhNN**. The different geometries of the O-H···ON contacts seem crucial to the different exchange behaviors between **4HOPhNN** and **35diBu-4HOPhNN**, since the

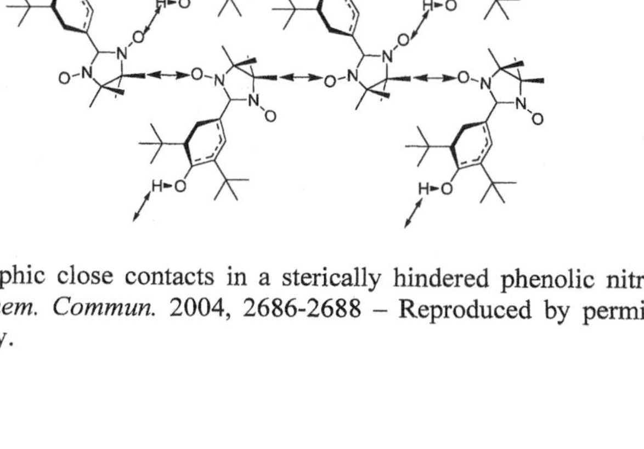

Fig. 2. Crystallographic close contacts in a sterically hindered phenolic nitronylnitroxide. From Taylor & Lahti, *Chem. Commun.* 2004, 2686-2688 – Reproduced by permission of The Royal Society of Chemistry.

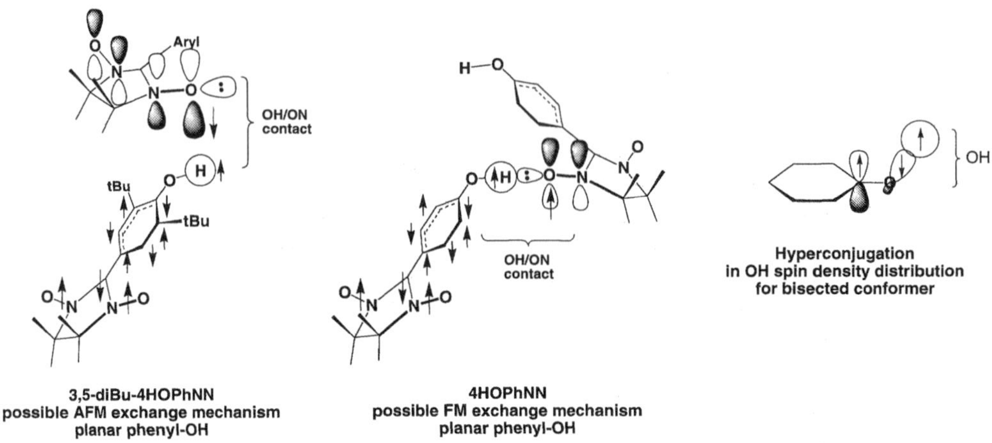

Fig. 3. A possible OH-mediated exchange mechanism in phenolic nitronylnitroxides as a function of relative intermolecular geometry and phenolic OH conformation.

N-O···H-C nitroxide methyl contacts and other crystal motifs are very similar between the two. UB3LYP/6-31G* computations show that the near-linear O-H···ON contact in **4HOPhNN** favors an FM intermolecular interaction by $2J/k = 1K$, while the bent O- H···ON contact in **35diBu-4HOPhNN** favors an AFM interaction by $2J/k = (-)8K$. This is in qualitative agreement with the experimental exchange behavior.

The AFM interaction in **35diBu-4HOPhNN** may be assisted by directional overlap of the NO SOMO with the OH orbitals. Figure 3 shows a qualitative representation, using the spin polarizations of isolated molecules (see the following paragraph) to represent a possible polarization exchange pathway. The polarization on the OH atoms represented in this figure is different from that suggested by detailed work [1d-e] from the Veciana group, but use of their spin polarization pattern seems to lead to expectation of FM exchange in **35diBu-4HOPhNN**. Incorporation of individual molecules into the solid lattice may alter spin density expectations by comparison to isolated molecules, so further investigation is required to identify why **4HOPhNN** and **35diBu-4HOPhNN** have similar crystallography but rather different magnetic behaviors.

Although the spin density of the OH proton in the phenolic nitronylnitroxides is computed to be quite small (ρ <0.1%), the closeness of the OH to ON contact makes it very important to the overall exchange behavior. Sugawara's semiempirical PM3-UHF computations found the OH proton spin density to increase and match the spin density sign of the ipso π-carbon (hyperconjugation mechanism, F. 3; see also footnote 21 in ref. 6) as the OH rotates to bisect the phenyl ring in a phenolic nitronylnitroxide. We have found that UB3LYP/6-31G* density functional theory (DFT) computations [7] on an isolated molecule of **4HOPhNN** using a crystallographic model geometry yield a similar trend (Fig. 4), although the computed spin density on the OH proton is small and slightly

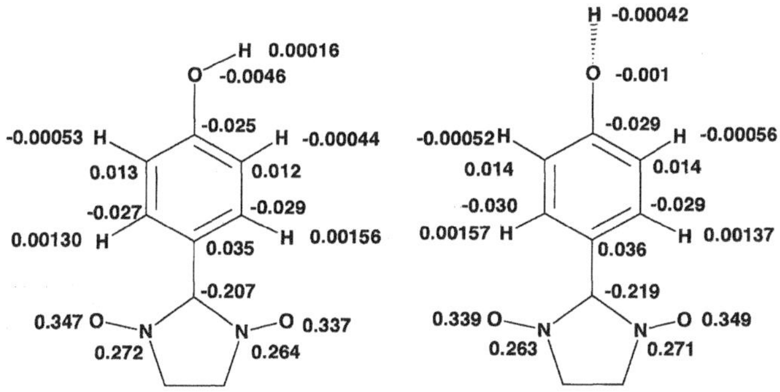

Fig. 4. UB3LYP/6-31G* Mulliken spin densities for a model geometry taken from the crystallographic structure of **35diBu-4HOPhNN**, with replacement of methyl groups by hydrogen atoms. The Ph-OH bond was rotated without further geometric change.

positive (+0.02%, same as the nitroxide NO polarization) when the OH is coplanar with the phenyl ring at the crystallography-derived geometry. Computational studies of model chains and further experiments will likely be important to help clarify possible changes in spin density distribution going from isolated molecules to the solid-state in **35diBu-4HOPhNN**, and to augmenting the studies of **4HOPhNN** by Veciana and coworkers [1d-e].

The iminoylnitroxide **35diBu-4HOPhIN** was also reported by Taylor and Lahti [8]. This system exhibits substantial crystallographic disorder of its iminoylnitroxide ring (interannular torsion), even though the molecular packing was very similar to that of **35diBu-4HOPhNN**. The iminoylnitroxide crystal lattice forms a network of OH···ON and N-O···H-C (iminoylnitroxide methyl groups) that is very similar to that of the nitronylnitroxide, but defective, due to randomized placement of the NO unit and the imidazole N. Magnetically, **35diBu-4HOPhIN** shows fairly strong AFM behavior consistent with multiple exchange mechanisms. This is not surprising, since multiple types of contacts occur between sites of significant spin density, due to the iminoylnitroxide disorder. A double Bleaney-Bowers fit to the data – two spin-pairing mechanisms below 50K – yields $2J_1/k = (-)0.9$ (P = 65% of pairs), $J_2/k = (-)20K$ (35% of pairs). The fractional amounts of the two exchange contributions are consistent with the disorder of the iminoylnitroxide unit in the lattice.

In the case of the phenolic nitronylnitroxides, the preponderance of evidence supports direct electronic participation by hydrogen-bonding OH groups in exchange and magnetic behavior. A variety of behaviors is possible, because of the sensitivity of the exchange mechanisms to the geometry and number of contacts between atoms with small spin density with the NO groups, when no dominating NO···ON contacts are found. Overall, the use of phenolic OH···ON contacts provides a good strategy for assembling radicals into organic magnetic materials [2c].

3 Bio-inspired Molecular Recognition for Crystal Assembly

Biological chemistry abounds with examples where multipoint hydrogen-bonding favors fascinatingly complex and specific folding in proteins and nucleic acids. Chemists have recently started to take advantage of this fact to craft nonbiological examples of molecular assembly through hydrogen-bond based recognition [9].

A basic idea of molecular recognition is that incorporation of complementary hydrogen-bonding donor and acceptor groups into different molecules can lead to a preferred "fit" of multi-point hydrogen-bonding between the molecules. The resulting donor-acceptor ($D \rightarrow A$) dyads therefore can self-assemble in solution to bring together whatever other functionalities are attached to the dyad subunits. In biology, perhaps the best-known examples are the **C-G** and **T-A** complementary base pairs in DNA. While **C** can and does hydrogen-bond with other molecules, it has a preferred, multi-point complementary fit with **G**. A sufficient number of complementary fits along an extended chain allows such complex assembly as the classic double-helix assembly of DNA in solution, although other geometries are also possible with different complementary hydrogen-bonding patterns. In addition to "natural" $D \rightarrow A$ dyad pairings from biological systems, a number of artificially designed recognition dyads have been designed and utilized in recent years.

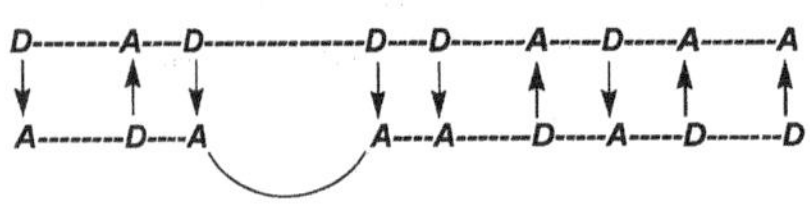

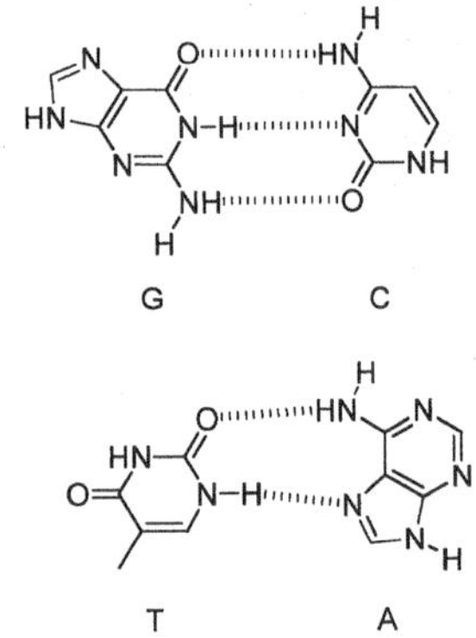

Molecular recognition in solution is crucial to life as we know it, but can be even stronger in the solid state, since the solvent sphere is not present. Hydrogen-bonding motifs that are too weak to produce large association constants in solution may be quite sufficient to produce donor-acceptor heteromolecular assemblies in the solid state, so long as the interaction is strong enough to overcome a tendency for donor and acceptor to crystallize separately. This strategy offers much promise for designing heterospin assemblages in organic molecule-based magnetic materials.

3.1 Uradinyl-functionalized Radicals for Molecular Recognition

Among the bio-recognition moieties attached to radicals in the small number of cases designed for molecular magnetism, thymine (**T**) has been much studied. It is omplementary to cytosine (**C**) in DNA molecular recognition as shown earlier above, and has been utilized in three-point hydrogen-bonded recognition dyads based upon 2,6-diaminopyridines (X = CR) and diamino-*s*-triazines (X = N) [9].

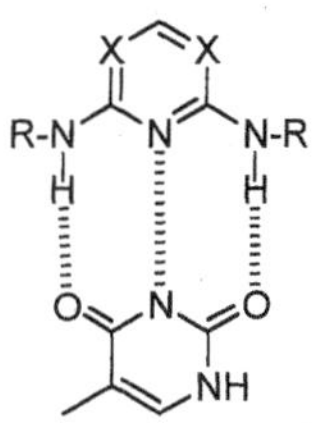

thymine

In 1999, Feher et al.[10] described the synthesis and partial characterization of a uradinyl unit (the generic moiety corresponding to thymine) functionalized in radical **Ur-5-NN** with a nitronylnitroxide attached at the 5-position. ESR studies show a small amount of delocalization of the radical spin onto the uradinyl unit. UV-vis studies suggest homomolecular aggregation of the radicals in solution. Crystallography of **Ur-5-NN** exhibits 1-D hydrogen-bonded ribbons, a common motif in uradinyl-based molecules. The torsion between the uradinyl unit and the spin-bearing nitronylnitroxide ring is 65°, which would severely limit spin delocalization in the solid state. The torsion is presumably due to dipolar repulsion between the nitronylnitroxide oxygen atoms and the lone pair electrons on the uradinyl 4-carbonyl group. The dc magnetic susceptibility behavior as a function of temperature is not straightforward to interpret, since the χT vs. T plot shows a sudden decrease at about 5K that did not fit simple magnetic models attempted by the investigators. A phase change inducing antiferromagnetic exchange seems likely for this system.

During the same time that Feher et al. were working on **Ur-5-NN**, we were working on the same molecule and its isomer, **Ur-6-NN**. After the other group's publication of their work, we focused on **Ur-6-NN**, which we anticipated to have a higher degree of planarity than **Ur-5-NN** and possibly more spin delocalization, due to favorable internal hydrogen-bonding and dipolar interactions.

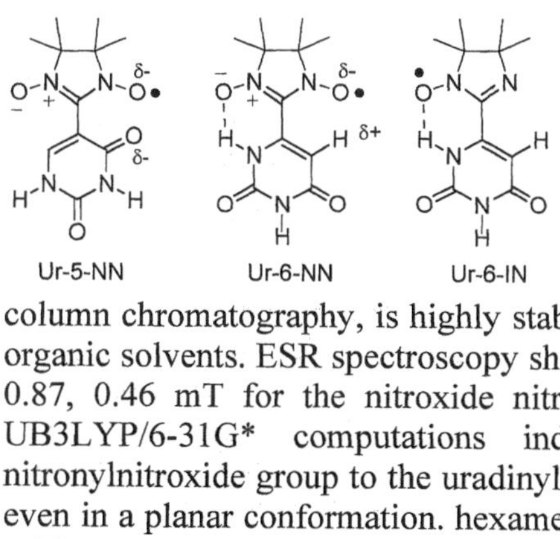

Ur-5-NN Ur-6-NN Ur-6-IN

Taylor *et al.* [11] did not isolate **Ur-6-NN** under any of the conditions attempted. However, preferential formation of **Ur-6-IN** occurred upon *in situ* aqueous $NaIO_4$ oxidation of the adduct of 2,3-bis(hydroxylamino)-2,3-dimethylbutane hydrogensulfate with uradinyl-6-carboxaldehyde. **Ur-6-IN** can be purified by column chromatography, is highly stable and highly soluble in water as well as various organic solvents. ESR spectroscopy shows fairly typical nitrogen hyperfine constants of 0.87, 0.46 mT for the nitroxide nitrogen and the imidazole nitrogen, respectively. UB3LYP/6-31G* computations indicate very small delocalization from the nitronylnitroxide group to the uradinyl ring (maximum 2% onto the uradinyl 5-carbon), even in a planar conformation. hexamers in methanol, with no such aggregates in acetic acid. Although the structures of these aggregates is not certain, it seems likely that ribbons analogous to Neither fluid nor frozen solution ESR experiments showed behavior consistent with major aggregation. However, ESI-MS studies shows aggregates of **Ur-6-IN** up tothose formed in crystalline **Ur-5-NN** are also formed by **Ur-6-IN** during the ESI-MS volatilization process (Fig. 5), and that these ribbons are not favored in the highly hydrogen-bond participating acetic acid.

Upon crystallization, **Ur-6-IN** forms dimeric pairs involving two-point hydrogen-bonds, not ribbons. The favorable intramolecular interaction of nitroxide oxygen with

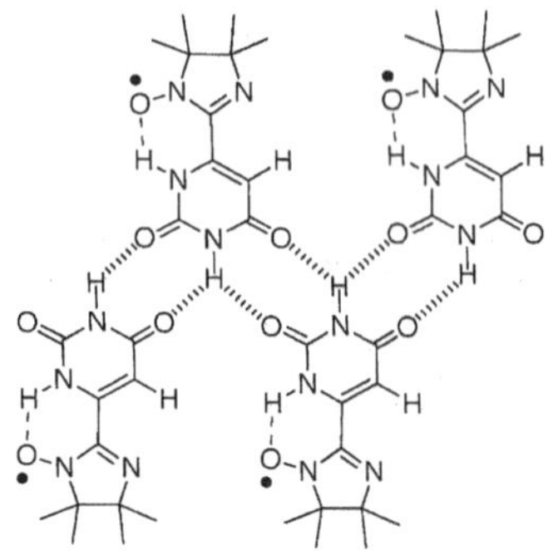

Fig. 5. Hypothetical hydrogen-bond assisted ribbon formed by **Ur-6-IN** during ESI-MS.

the NH group at the 1-position does planarize the molecule as hoped; the interannular torsion is only 14.5°. Due in part to the planarity, **Ur-6-IN** forms stacks of dimmers (Fig. 6). The dc magnetic susceptibility vs. temperature data show a maximum at about10K, attributed to an AFM exchange interaction between the radical imidazole nitrogens in the crystal lattice that are only 3.3 Å apart, and bear about 15% of the radical spin density based on the hyperfine splitting. These are the only close contacts between sites of major spin density in the crystal structure (see space filled portions of the diagram). The structure suggests that the exchange should be dimeric, and indeed the susceptibility versus temperature behavior for **Ur-6-IN** gives an excellent fit to a Bleaney-Bowers pairing model for S = 1/2 spin carriers with $2J/k = -14K$.

In summary, both **Ur-5-NN** and **Ur-6-IN** exemplify hydrogen-bonding that provides good crystal lattice scaffolding, but not significant direct electronic exchange. However, further extrapolation is possible for assembly and crystal scaffolding to include heteromolecular assembly, as described below.

In related work, Shiomi *et al.* [12] recently reported the synthesis, crystallography, and magnetic behavior of radical **Cy-NN**, which attaches a nitronylnitroxide to a cytosine moiety using an sp^3-hybridized CH_2 group. Due to this design feature, no significant spin density is expected or found in the cytosine group, making cytosine a molecular assembly and crystal scaffolding unit, not an electronic exchange unit. Like **Ur-5-NN**, this system forms doubly hydrogen-bonded ribbons in the solid state (Figure 7). Its magnetic behavior is dominated by AFM exchange coupling. There are two conformers of the molecule in the unit cell, of which only one exhibits a significant exchange interaction. The magnetic behavior as a function of temperature can be fit to a 50% paramagnetic plus 50% Bleaney-Bowers S = 1/2 spin pairing model ($2J/k = -38K$) with a mean-field term of $\theta = -1.5K$. The spin pairing is ascribed to an inversion-type close contact between NO units in adjacent hydrogen-bonded chains (half of the molecules in the adjacent chains have r(NO···ON) = 3.28 Å). The synthesis of this molecule should allow future exploration of **C-G** heteromolecular, heterospin exchange pairings.

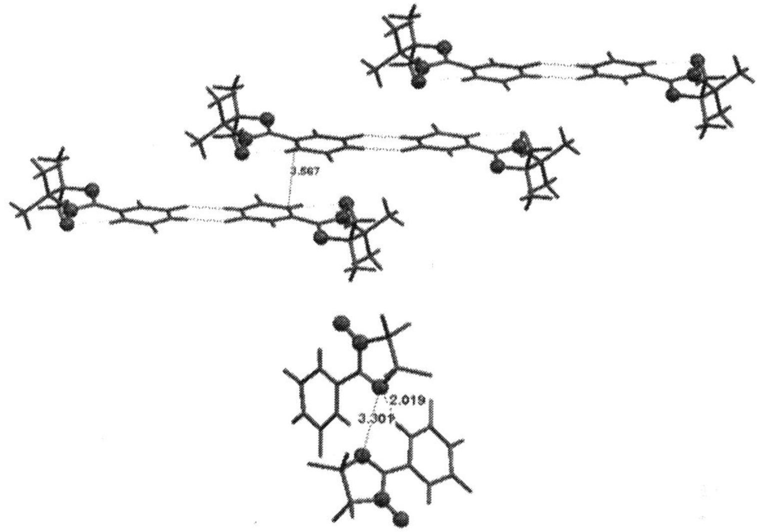

Fig. 6. Close crystallographic contacts in **Ur-6-IN**. Distances in angstroms.

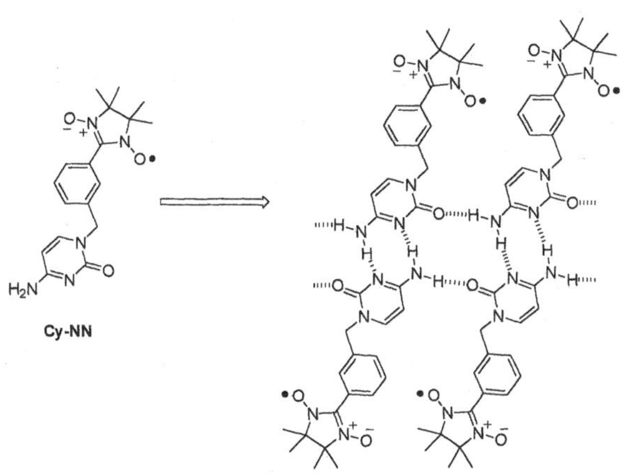

Cy-NN

Fig. 7. Hydrogen-bond assisted ribbon formation in **Cy-NN** (see ref 12).

3.2 Bioinspired, Heteromolecular Recognition Pairing for Molecular Magnetism

As mentioned above, thymine and uradinyl-based systems form dyads with 2,6-diaminopyridines and related systems in solution. Taylor et al. recently showed [13] that mixtures of **Ur-6-IN** with 2,6-dipropylaminopyridine (**DAP**) exhibit such recognition-based heteromolecular dyad formation in solution. ^{1}HNMR measurements show strong paramagnetic downfield shifts of the propylamido NH groups of **DAP** in titrations with **Ur-6-IN**; these protons are closest to the IN group for a three-point hydrogen-bonded dyad geometry. The titration data yield an association constant of 220 M^{-1} at 33 °C. ESI-MS measurements on the titration mixture in methanol showed not only the **Ur-6-IN/DAP** dyad, but also a **Ur-6-IN/(DAP)$_2$** triad. The structure of the latter is not clear (a possible structure is shown), but there are two sites for multipoint hydrogen-bonding of the uradinyl moiety with **DAP**.

Work is ongoing in our laboratories to investigate the scope of heterospin exchange interactions in solution and in the solid-state using the uradinyl/**DAP** recognition dyad. Others have investigated related, recognition-inspired radicals. In addition to the work on **CyNN** [12] mentioned above, nucleoside-based systems **UrNIT** [14] and **PurNIT** [15] bearing *tert*-butylnitroxide groups have been studied in solution. The **PurNIT** systems can be isolated as solids that exhibit pH-dependent ESR spectroscopy with variable spin delocalization. Hopefully, bio-recognition inspired heteromolecular assemblies derived from these and similar systems will provide a rich new area for solid-state organic magnetic materials, as well as providing potential site-specific spin-bearing molecular probes.

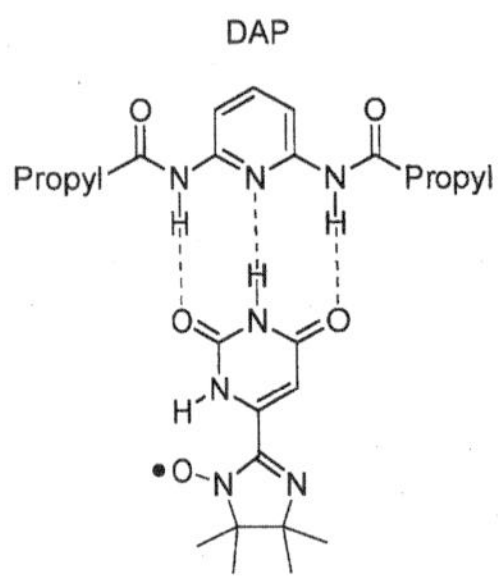

Ura-6-IN

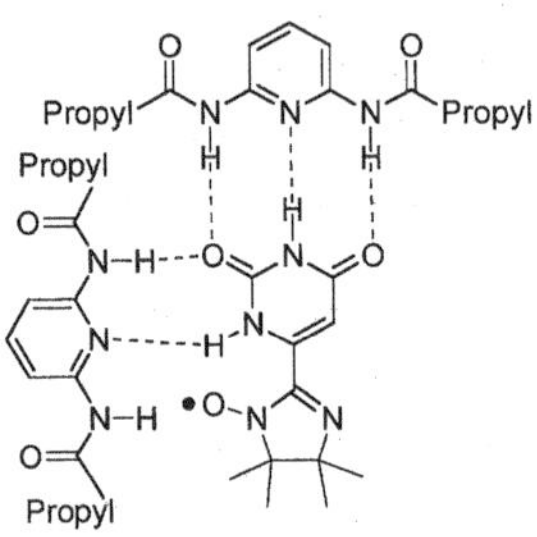

Ura-6-IN / (DAP)$_2$

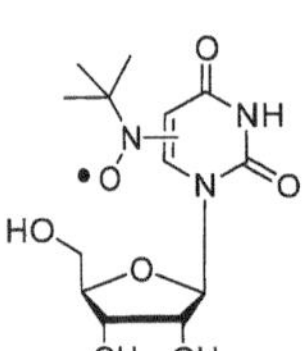

UrNIT

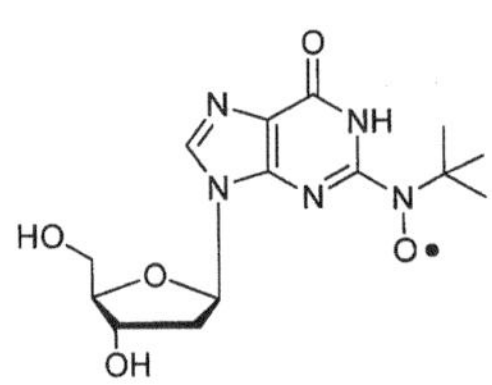

PurNIT

4 Exchange and Crystal Assembly in Nitronylnitroxides with Hydrogen-bonding Azaarenes

Hydrogen-bonding in the aza-heterocycles that incorporate NH groups can be very useful for forming chains and aggregates. Kahn's group has studied triazole-linked nitronylnitroxides [16], including **TzNN** [16a], which crystallizes to form hydrogen-bond type chains of NH···ON-C-NO···HC contacts, where ON-C-NO is the nitronylnitroxide moiety of one **TzNN** molecule in close contact with the NH of another plus a nitroxide methyl group of a third **TzNN**. The overall effect is to form an -A-B-A-B- chain of alternating NH···ON and NO···HC contacts (dashed lines in the structural scheme), which interestingly shows a fairly strong 1-D FM Heisenberg chain magnetic behavior with a weak AFM interchain interaction; the data fit well for $J_{intrachain}/k = +21K$, $zJ'_{interchain}/k = -2.8K$. This is a fairly strong intrachain interaction, which the authors attributed to exchange mechanisms analogous to those invoked [1a-b] by Veciana and coworkers for **4HOPhNN**. Neutron diffraction studies suggest [16b-c] an important role for close atomic contacts between CH and ON groups in an isomer of **TzNN** that does not allow strong NH hydrogen-bonding (N-CH_3 instead of C-CH_3 on the triazole ring).

However, in **TzNN** itself, a possible alternative exchange mechanism may involve the ability of its NH···ON hydrogen-bonds to organize the radicals to form chains of nitroxide NO···C contacts (double-headed arrows in Fig. 8), where the C-atom is the central, negatively spin polarized central atom of the next nitronynitroxide in the chain; r([N]O···C) = 3.127 Å. These contacts would provide an excellent FM spin polarized chain, and are in fact quite analogous to contacts found and identified in Yoshioka's work on **ImNN** described later, below. Thus, is it not completely clear whether these hydrogen-bonds are primarily crystallographic scaffolding, or electronic exchange linkers, or both. Given that the hydrogen-bonding chain is an alternating type (both NH···ON and NO···HC contacts), and that the NO···C contacts are all the same, the excellent fit to a simple 1-D Heisenberg chain model with modest interchain mean field term seems consistent with the NO···C contact chain model, unless NH···ON and NO···HC exchange strengths are coincidentally almost identical.

Catala *et al.* described [17] magnetostructural studies of **PzNN** and **PzIN**, which incorporate the pyrazole group (Fig. 8). Solution ESR spectroscopy shows only small or virtually no spin delocalization from the radical groups onto the pyrrazole ring – less than 1%, based on an 0.1 G hyperfine splitting in **PzNN**. Higher concentration solutions

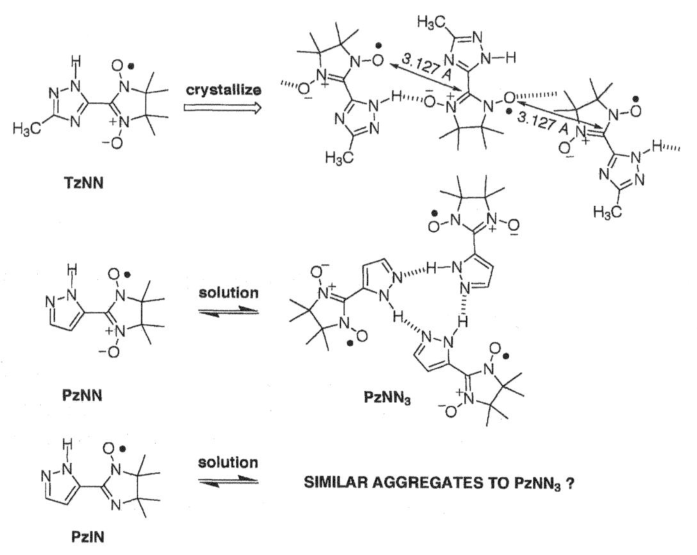

Fig. 8. Azole-based radicals that show evidence of hydrogen-bonding interactions.

of **PzNN** show extra lines consistent with aggregate formation, and simulations are consistent with formation of a significant amount of trimer, the structure of which was suggested as cyclic **PzNN$_3$**. Frozen solution studies show a half-field band consistent with an S > 1/2 species, again consistent with aggregation. No resolvable additional lines are found at room temperature for **PzIN**, but frozen solution studies again show a half-field band consistent with an S > 1/2 aggregate. Using an equilateral triangle trimer model to fit the solution ESR half-field band intensity versus temperature behaviors, Catala et al. estimated the aggregates to have FM exchange coupled quartet ground states by $J/k = 2K$ for both systems.

Both **PzNN** and **PzIN** crystallize as nearly planar molecules, which nominally can help the small amount of delocalization that occurs from the radical units to the pyrazole rings. **PzNN** forms cyclic tetramers linked by NH···N and NH···ON contacts; the tetramers also π-stack. **PzIN** forms chains of NH···N hydrogen-bond contacts, with interchain iminoylnitroxide to methyl NO···HC contacts. In dc magnetic susceptibility studies, both show AFM downturns in their χT vs. T plots. Curie-Weiss analysis yields Weiss constants of $\theta = -2.1K$ and $-3.1K$ for **PzNN** and **PzIN**, respectively. Further magneto-crystal analysis was not given in this study, understandably since correlating the solution phase FM interactions with the overall solid-state AFM behavior is not straightforward in these systems.

Benzimidazole and imidazole are also effective groups for organizing organic nitronylnitroxides in a crystal lattice with hydrogen-bonding. Yoshioka and coworkers have reported relatively detailed magnetostructural studies for a number of systems based on these and related units. Their initial work described **ImNN** and **BimNN** [18–21]. ESR hyperfine analysis of both **ImNN** and **BImNN** shows that spin density delocalization from the nitronylnitroxide group onto the imidazole is very small. **ImNN** forms hydrogen-bonded chains from imidazole NH donor groups to imidazole N: acceptors, a common motif in imidazoles and benzimidazoles. These crystallographic chains are linked through McConnell II type stacks between the pendant nitronylnitroxide groups, such that pairs of head-to-tail NO···ON contacts are formed. These contacts should provide an AFM exchange mechanism, and indeed the temperature dependence of the magnetic behavior of **ImNN** fits a Bleaney-Bowers AFM spin pairing mechanism with $2J/k = -177K$. Therefore, the hydrogen-bonds in **ImNN** seem to act as crystallographic scaffolding, not electronic exchange linkers.

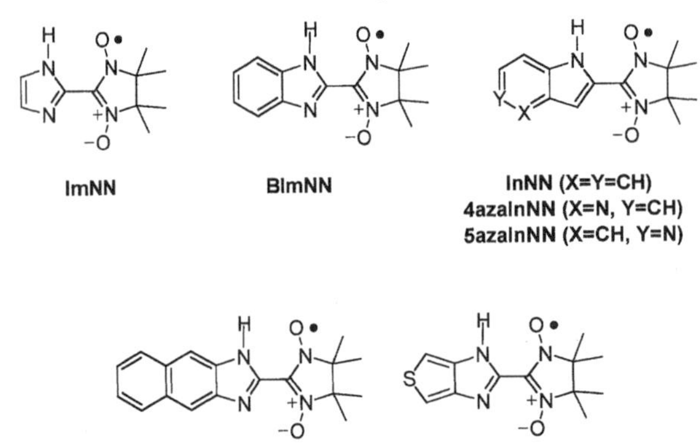

ImNN **BImNN** **InNN (X=Y=CH)** **4azaInNN (X=N, Y=CH)** **5azaInNN (X=CH, Y=N)**

NaphImNN **ThienImNN**

Table 2.
Aza-aryl nitronylnitroxide molecular magnetic behavior.

Radical	Crystallography	Magnetic Behavior	Exchange Strength
TzNN	H-bonded chains NH···ON contacts NO···methyl contacts NO···C(nitroxide) contacts	1-D FM chains	$J_{1D}/k = +21$ K [16a]
PzNN	π-stacked H-bonded tetramers, NO···methyl contacts	paramagnetic with weak AFM exchange	$\theta = -2.1$K [17]
PzIN	H-bonded chains, NO···methyl contacts	paramagnetic with weak AFM exchange	$\theta = -3.1$K [17]
ImNN	H-bonded chains, NO···ON stacking	dimer spin pairs	$2J_{ST}/k = -177$K [19-20]
MeImNN	(not determined; See ref 20)	1-D FM chains	$J_{1D}/k = +18$K [19-20]
BImNN	H-bonded chains NH···ON contacts NO···methyl contacts NO···C(nitroxide) contacts	1-D FM chains	$J_{1D}/k = +17$K [19-20]
InNN	NH···ON contacts NO···C(nitroxide) contacts	1-D FM chains	$J_{1D}/k = +12$K [21]
NaphImNN	H-bonded chains NH···ON contacts NO···ON stacking	1-D AFM chains	$J_{1D}/k = -14$K [22]
4azaInNN	H-bonded chains NH···ON contacts NO···methyl contactsπ-stacks	half-dimer spin pairs, half paramagnetic	$2J_{ST}/k = -92$K [23]
5azaInNN	NH···ON contacts NO···methyl contacts π-stacks, "brick-wall" overall network	metamagnetic(?) below 3K	$\theta = +0.6$ K[24] for T > 3K
ThienImNN	(not determined, See ref 25)	1-D AFM chains (?)	$J_{1D}/k = +13$ K[25]

By contrast, **BImNN** forms chains of benzimidazole NH to nitronylnitroxide NO hydrogen-bond contacts that are not collinear (∠[NH···O(N)] ~ 111°), but are fairly tight with r(NO···HN) ~ 2 Å (Fig. 9). There are no N···HN hydrogen bonds, in contrast to **ImNN**. The temperature dependence of the magnetic susceptibility for **BImNN** is consistent with a 1-D Heisenberg FM coupled chain having $J/k = +17$K; a small downturn of the χT vs. T plot at low temperatures indicates a weak AFM exchange between the FM coupled chains. The FM chain behavior was attributed to good McConnell II overlap between the nitronylnitroxide NO units with the central carbon atom in the ON-C-NO unit of the next molecule in a chain, where r(NO···C) = 3.17 Å, similar to crystallographic chains found in **TzNN** (see Fig. 8). The exchange strengths

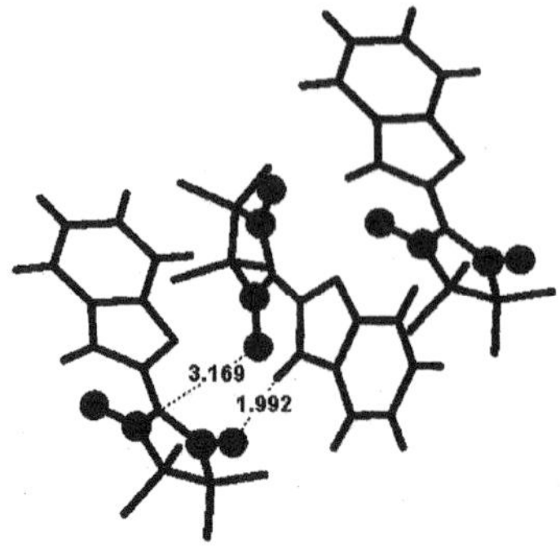

Fig. 9. Crystallographic close contacts in **BImNN**. Distances in angstroms. See ref 19a.

of the 1-D chains in **BImNN** and **TzNN** are even similar (see Table 2). The authors note that the NH group for **BImNN** may also play a direct FM exchange linker role in the exchange through NH···ON contacts, making the situation somewhat unclear as in the case of **TzNN**.

As an extension of their initial studies, Yoshioka's group also studied indole based **InNN**, which cannot form hydrogen-bonded chains analogous to those in **ImNN**. Similar NH···ON contacts to those in **BImNN** are found, as well as a chain of NO···C contacts contacts with a somewhat longer distance of r(NO···C) = 3.21 Å. **InNN** also shows magnetic behavior consistent with Heisenberg 1-D FM exchange coupled chains where $J/k = +12$K. The longer NO···C contact distance is consistent with the weaker FM exchange constant by comparison to **BImNN**.

The Yoshioka group has further expanded on the hydrogen-bonding strategy described above to make radicals **NaphImNN**,[22] **4azaInNN**,[23] **5azaInNN**,[24] and **ThienImNN**.[25] Table 2 summarizes the magnetic behaviors of these systems by comparison to those of **ImNN**, **BImNN**, and **InNN**. The results show that even a set of similar molecules with a reasonably reproducible crystallographic motif and apparently similar exchange interactions can give a variety of overall magnetic behaviors, due to the interplay of small exchange interactions between sites of small spin density. The reader is referred to the original papers for full details of these studies.

5 Exchange and Crystal Assembly in *Tert*-butylnitroxides with Hydrogen-bonding Azaarenes

Our group has pursued the magnetostructural analysis of benzimidazole substituted *tert*-butylnitroxides in hopes of achieving some of the interesting exchange and assembly characteristics observed for the benzimidazole and imidazole nitronylnitroxides, with the greater level of spin delocalization expected for a π-conjugated nitroxide.

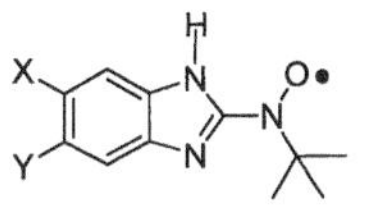

5(6)-Cl-BIm-NIT X,Y = Cl,H
5,6-di-Me-BIm-NIT X,Y = Me
BABI X,Y = H

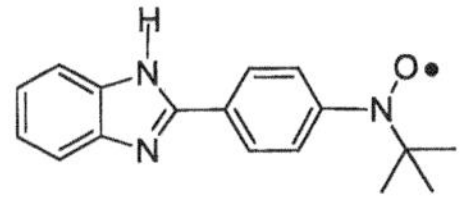

BIm-Ph-NIT

Stable radicals **5(6)-Cl-BIm-NIT** and **5,6-di-Me-BIm-NIT** crystallize into hydrogen-bonded dimers that form canted π-stacks without significant additional close contacts between the nitroxide units [26–27]. ESR solution hyperfine splitting and density functional computations show significant delocalization (over 12%) of the nitroxide spin onto the imidazole ring, and some onto the benzenoid ring (about 1–5%). Two possible dominating exchange interactions in the dimer pairs seem plausible (Fig. 9): a FM interaction that follows the pathway of closest atomic contacts through the cycle of hydrogen-bonded N-H···O-N contacts, and an AFM interaction that is directly mediated by the close (3.44 Å) through-space N-O···O-N approach. Both **5(6)-Cl-BIm-NIT** and **5,6-di-Me-BIm-NIT** show strong downturns in magnetic susceptibility at about 40 K as temperature decreases. They fit Bleaney-Bowers S = 1/2 spin-pairing models with $2J/k = -44$ and -48K, respectively.

The antiferromagnetic exchange behavior of **5(6)-Cl-BIm-NIT** and **5,6-di-Me-BIm-NIT** indicates that direct through-space exchange dominates over the possible close-contacts, FM mechanism. One can explain this by using computed spin densities [26], and a dipolar expression for interaction between unpaired spins, viz.,

$$\Delta E \sim \rho_1\rho_2 / R_{12}^3$$

where ρ_1 and ρ_2 are interacting isotropic spin densities and R_{12} is the distance between them. Using this expression with a simplified structural model derived from the crystallographic geometry in the dimers (r[(N)O···O(N)] = 3.43 Å, r[(N)H···O(N)] = 2.03 Å; see Fig. 10), the interaction directly between the nitroxide units is estimated to be about tenfold larger than the interaction between the pair of NH···ON interacting units.

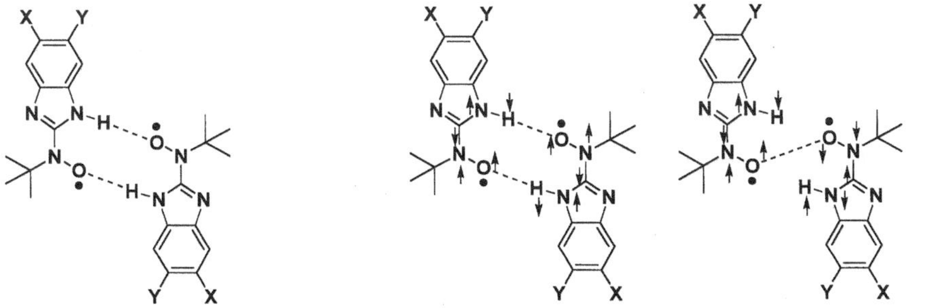

5(6)-BIm-NIT; X = Cl, Y = H
5,6-di-Me-BIm-NIT; X = Y = CH$_3$

close-contacts mechanism
ferromagnetic exchange

through-space mechanism
antiferromagnetic exchange

Fig. 9. Possible exchange mechanisms in dimeric hydrogen-bonded radicals, based on close interatomic contacts. See ref 26.

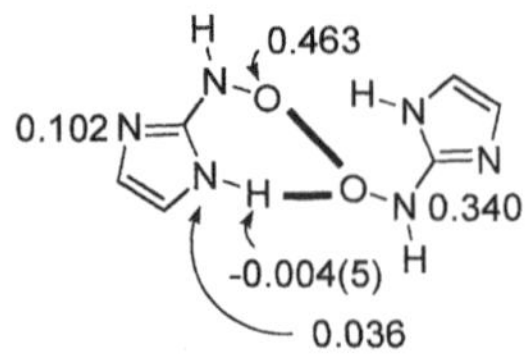

Fig. 11. UBLYP/cc-VDZ computed[7] Mulliken spin densities on a portion of a model dimer for **5(6)-Cl-BIm-NIT** or **5,6-di-Me-BIm-NIT**. r[(N)O···O(N)] = 3.43 Å, r[(N)H···O(N)] = 2.03 Å.

Clearly, direct through-space interactions can dominate over close atomic contacts in a magnetic lattice, if the direct interactions are between large enough or diffuse enough spin density sites. Here, a possible FM exchange effect of N-H···O-N contacts in **5(6)-Cl-BIm-NIT** and **5,6-di-Me-BIm-NIT** is overwhelmed by the direct electronic N-O···O-N exchange path. Nagashima et al. interpreted the magnetic behavior of **NaphImNN** (see the previous section) in similar terms,[22] since a moderately strong AFM exchange is observed and attributed to direct N-O···O-N contact, even though N-H···O-N close contacts are also present that seem to favor a FM exchange pathway. In the case of **NaphImNN**, a deuterated sample showed virtually no change in magnetic behavior [22], further evidence for a lack of electronic exchange contribution of the N-H(D)···O-N contacts in that case.

One must note that for such cases, the absolute sign of the spin density on the NH unit is very important to help understand solid-state exchange mechanisms involving an imidazole-type NH. Solution ENDOR studies on **BABI** have recently confirmed [28] that the sign of the spin density on the proton of the NH group is indeed negative, as expected by spin polarization, and as postulated [26b, 27, 29] in magnetostructural analyses of the benzimidazole *tert*-butylnitroxides.

By comparison to **5(6)-Cl-BIm-NIT** and **5,6-di-Me-BIm-NIT**, the less substituted benzimidazole-2-*tert*-butylnitroxide (**BABI**) radical has a much more complex crystal structure (Fig. 12), although its molecular spin density distribution is quite similar based upon DFT computations and ESR hyperfine splitting analysis. Instead of cyclic NH···ON hydrogen-bonds, **BABI** forms [26, 28, 29] NH···N hydrogen-bonded chains from NH donor groups of one molecule to acceptor imidazole nitrogen atoms of neighbors. The molecules in the chain are alternately canted in a sawtooth manner, with the chains forming canted stacks in the perpendicular direction. The nitroxide groups in each chain are aligned parallel, due to a favourable dipolar interaction between the NH and NO groups. This NH/NO interaction also gives a high degree of nitroxide to benzimidazole coplanarity (torsion = 8–9°), which should promote spin delocalization.

A herringbone interaction occurs from the 5/6-CH bonds in each benzene ring to the π-cloud of a benzene ring in a neighboring chain, such that each chain is paired with a parallel chain to place the nitroxide units on the exterior and the benzene rings in the interior. When the double-chain motif is combined with the perpendicular canted stacking motif, the result is a bilayer that is rich in spin density at the nitroxides. Each bilayer is symmetry-related to another through inversion. The closest approach between

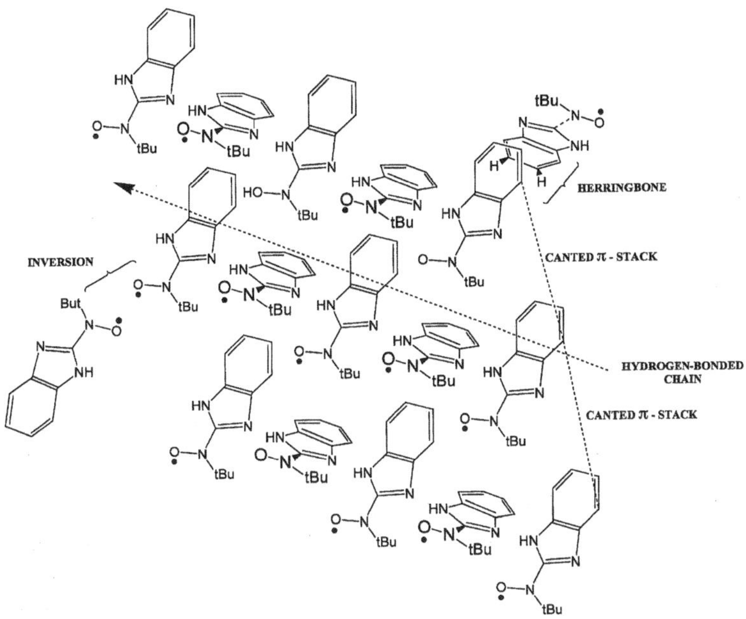

Fig. 12. Crystallographic packing and contacts in **BABI**.

the major nitroxide spin density sites is 4.86 Å across the inversion center, not especially close, but close enough for some level of expected interaction. The hydrogen-bonded chains provide close-contacts between sites with some spin density, due to nitroxide delocalization onto the imidazole ring. This crystal arrangement thus provides a number of contacts between spin density sites, some of which are quite close together but with small spin density (*e.g.*, the NH···N hydrogen-bonds), and some not so close but with large spin density (the NO···ON inversion center contacts).

The dc magnetic susceptibility versus temperature for **BABI** reaches a maximum at about 3K, then decreases somewhat. Efforts to fit the observed behavior to an S = 1/2 spin pairing model were not satisfactory, nor were fits to 1-D AFM chain models [28]. The best fits were to 2-D Heisenberg AFM planar models, particularly [30] a bilayer model with $(J/k)_{intra}$ = (–)1.2.1.4K, $(J/k)_{inter}$ = (–)1.9-1.3K. This was very satisfying, since cryogenic heat capacity studies of **BABI** shows a broadened anomaly below 5K whose shape is consistent with 2-D AFM spin ordering.[30] Analysis of the lattice-subtracted magnetic C_p vs. T data for **BABI** also fits 2-D Heisenberg AFM planar behavior, with a Néel ordering temperature of T_N = 1.7K. Further fitting of the 0.2–0.5K data by a spin-wave (magnon) model ($C_p \propto T^{d/n}$) gives a dimensionality parameter $d = 2.74$ ($n = 2$ for an antiferromagnet), indicating a 2-D to 3-D cross-over of the ordering at very low temperatures.

2-D or higher dimensional magnetic ordering phenomena are not common in organic systems. The low molecular symmetry and typical small number of interacting neighbors in an organic crystal lattice make 1-D phenomena most common for longer

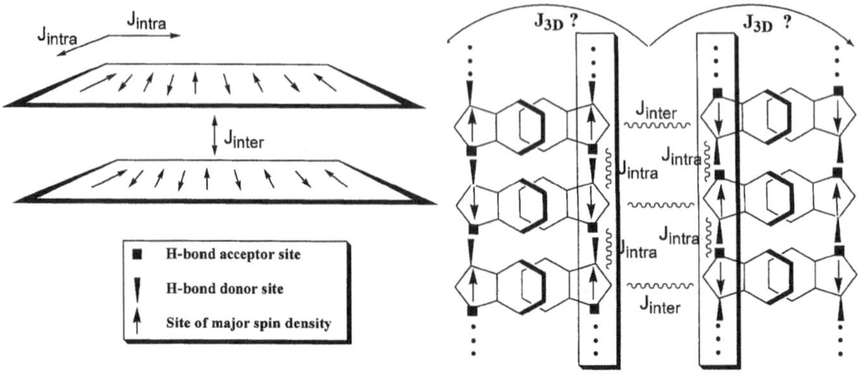

Fig. 13. 2-D spin bilayer antiferromagnetic exchange in **BABI**, with a possible reason for the onset of 3-D antiferromagnetism at below about 0.5K.

range interactions. **BABI** is not a highly symmetrical molecule, so its behavior as a true Heisenberg 2-D antiferromagnet shows the potentials for utilizing extended networks of modest exchange interactions to make purely organic, magnetically ordered solids. The relatively high (for a pure organic) Néel temperature and high dimensionality of exchange behavior for **BABI** appear to be a due to a combination of its hydrogen-bonding chains, π-stacking, and herringbone interactions. No single exchange interaction dominates in the magnetic lattice. In particular, the NO···ON contact distance between the largest spin sites is fairly long at 4.86 Å. As a result, exchange in multiple dimensions is competitive, despite the structural differences of the exchange mechanisms in the lattice. The magnetic bilayer motif is consistent with a double-"slab" of large spin density in the nitroxides that are aligned in plane. The slab is held together by the canted π-stacking plus hydrogen-bonded chains ($(J/k)_{inter}$), and connected by $(J/k)_{inter}$ through inversion centers to a neighboring plane of high spin density. The bilayers are then separated by the region of the herringbone interactions, which have low spin density, but possibly enough to account for the onset of 3-D ordering at very low temperatures. These interactions are summarized in Fig. 13.

The tendency of the benzimidazole unit to induce packing motifs like those found in **BABI** is supported by the case of **BIm-Ph-NIT**, whose crystal structure looks very similar [27]. Both **BABI** and **BIm-Ph-NIT** have bilayer motifs where planes formed by canted, π-stacked, hydrogen-bonded benzimidazole rings are held together by a herringbone interaction region (hatched lines in Fig. 14). The bilayers are about 11.66 Å wide in **BABI**, 15.40 Å wide in **BIm-Ph-NIT**. The NO···ON interaction in **BIm-Ph-NIT** contact is shorter than that in **BABI** (4.27 vs. 4.86 Å), and has a somewhat different relative geometry, but is otherwise similar. Notably, the nitroxide groups all orient in the same direction within a hydrogen-bonded chain, even though the nitroxides are not close to the aligning influence of NH units.

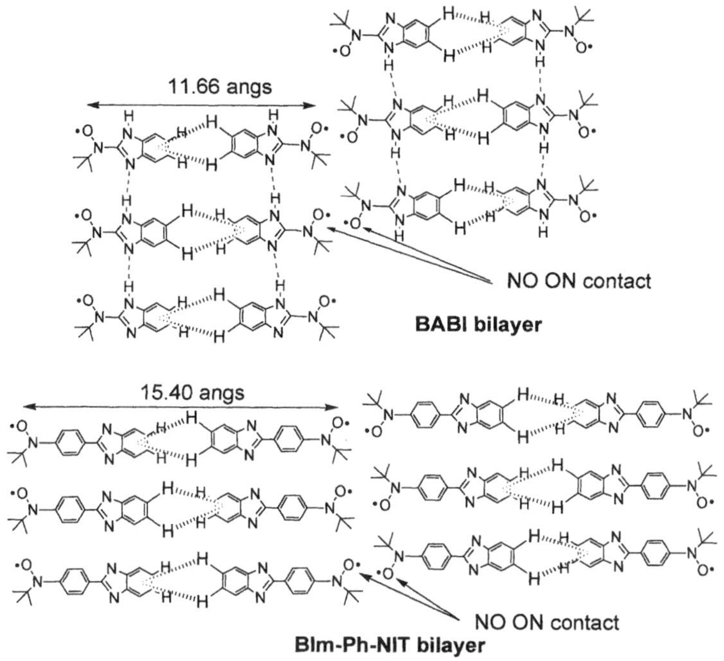

Fig. 14. Crystallographic herringbone interactions related to bilayer widths in **BABI** and **BIm-Ph-NIT**.

The similarity of the crystal packing of **BIm-Ph-NIT** to that of **BABI** suggests the possibility of similar magnetic behavior. In fact, **BIm-Ph-NIT** does exhibit antiferromagnetic exchange in its susceptibility vs. temperature behavior that is most consistent with a 2-D planar Heisenberg model having $J/k = (-)15$K, like **BABI**. This molecule is still under investigation to have a clearer understanding of its exchange behavior, but shows that benzimidazole can be an effective crystal lattice scaffolding for organizing radicals in the solid state, in a manner that allows some control over crystal structure and magnetic behavior.

We continue to study hydrogen-bonding azaarenes functionalized with *tert*-butylnitroxides. For example, Ruiz [31] has made **7azaBImNIT** and **ImNIT** by synthetic routes analogous to those used to make the benzimidazole *tert*-butylnitroxides, and characterized them by solution ESR spectroscopy. These systems, however, were more reactive and harder to isolate as pure solids than the other systems described in this section, so they have not been subjected to magnetostructural analysis.

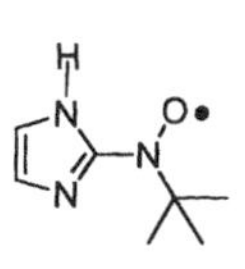

ImNIT

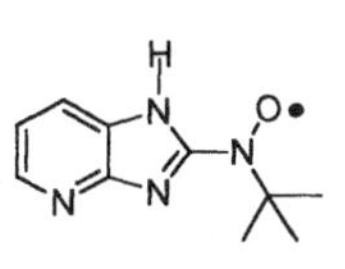

7azaBImNIT

6 Exchange and Crystal Assembly in Cyclic Aza-Radicals Bearing Carboxylic Acid Groups

One of the best-known hydrogen-bonding moieties is the carboxylic acid group, which forms complementary dyads in gaseous, solution, and solid states. This interaction is strong (60-70 kJ/mol [32]) and directional, ideal for forming dyads and chains. It has also seen use in assembling components of molecular magnets.

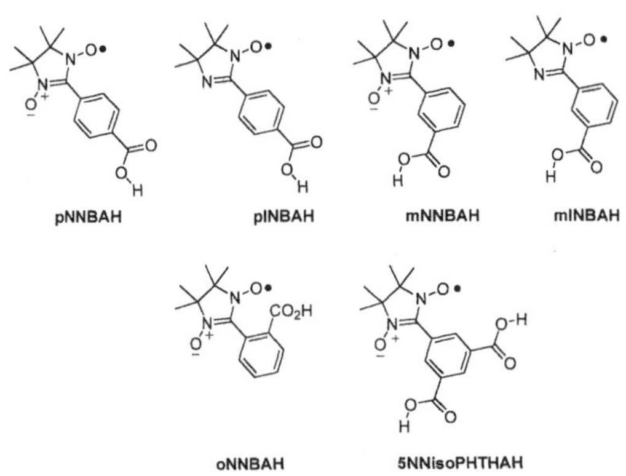

Several benzoic acid variants of nitronylnitroxides and iminoylnitroxides have been studied. pNNBAH forms classic dimers due to CO_2H two-point hydrogen-bonds. Its variable temperature magnetic susceptibility is consistent with AFM spin pairing where $2J_{ST}/k = -6.0K$. Interestingly, the dimeric salt where CO_2H units are replaced with CO_2Li units (with methanol in the crystal lattice) exhibits FM spin pairing with $2J_{ST}/k = +31.8K$. The corresponding Na^+ and K^+ salts are essentially paramagnetic, with small Weiss constants of $\theta = -0.4$ and $-1.0K$, respectively [33]. pINBAH forms hydrogen-bonded dimeric rings involving two radicals and two water molecules (Fig. 15); the rings are further linked into chains by further hydrogen-bonding through the water molecules (see below) [34]. Magnetically, it acts as a paramagnet above 20K, with a Weiss constant of $\theta = +1.38K$; there is a slight upturn in its χT versus T plot that could not be fit to any low-dimensional model considered.**mNNBAH** has been incorporated into coordination complexes with metal cations [35], but its own magnetic

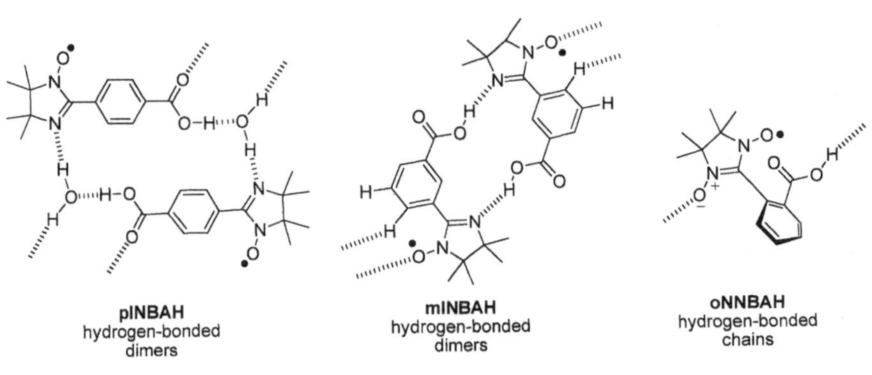

Fig. 15. Crystallographic hydrogen-bonding patterns in some benzoic acid functionalized radicals. See the text for source references.

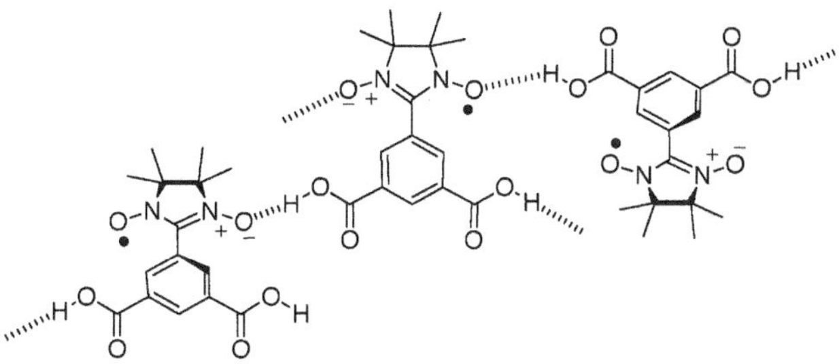

Fig. 16. Crystallographic hydrogen-bonding in **5NNisoPHTHAH**. See ref 36.

behavior has apparently not been reported. **mINBAH** forms cyclic, head-to-tail dimers involving the $-CO_2H$ and the iminoylnitroxide imidazole nitrogen (Fig. 15); the dimers form chains based on a double NO···HC contact involving the nitroxide group and aryl C-H bonds *para* to the $-CO_2H$ [34]. Magnetically, **mINBAH** shows a modest rise in its χT versus T plot, with a strong downturn below 10K. Its reciprocal magnetic susceptibility versus temperature behavior above 20K fits a Curie-Weiss expression with a fairly large $\theta = +5.24K$; however, the χT downturn of clearly shows the presence of an AFM exchange mechanism that is not identified. **oNNBAH** forms hydrogen-bonded chains involving the $-CO_2H$ and one NO group per molecule, where the $-CO_2H$ and the nitronylnitroxide plans canted strongly relative to one another (Fig. 15) [34]. The **oNNBAH** reciprocal magnetic susceptibility versus temperature behavior above 20K fits a Curie-Weiss paramagnet expression with $\theta = -0.81K$. The authors suggest that the modest variations in magnetic behavior for **pINBAH, mINBAH**, and **oNNBAH** are attributable to very small modifications of spin density distribution in the radical moieties due to the hydrogen-bonding. There is no resolvable hyperfine in the solution ESR spectra of these radicals, other than the nitrogen atoms, so it is difficult to probe this suggestion.

5NNisoPHTHAH forms ribbons of NO···HOOC hydrogen-bonds, which are further interconnected into a 3-D network (Fig. 16) [36]. However, its magnetic behavior as a function of temperature was essentially paramagnetic, with a small FM Weiss constant $\theta < +0.5K$. Despite the presence of an extended network of crystallographic contacts, this system has little interspin exchange.

Finally, although this chapter is specifically aimed at nitroxide-incorporating radicals and close relatives, it is worth noting that verdazyl radicals with attached carboxylic acid units have been incorporated into coordination complexes, but their magnetic properties as pure organic systems have not yet been described [37].

7 Exchange and Crystal Assembly in Aryl *Tert*-butylnitroxides Bearing Carboxylic Acid Groups

Maspoch et al. recently reported the synthesis, solution phase spectroscopy, and molecular magnetism of the *tert*-butylnitroxylbenzoic acid.**NOA** [38]. Frozen solution

ESR spectroscopy shows triplet ESR peaks (including a half-field transition for S > 1/2) with temperature dependent behavior that was consistent with a ferromagnetically coupled pair of radicals having $J/k \sim 2K$. If one assumes the radical pair forms a typical hydrogen-bonded carboxylic acid dyad geometry in solution, triplet ground state behavior is expected by radical π-spin polarization, assuming that the carboxylic acid moiety is an effective mediator of exchange in a through-bond manner. However, the solid-state magnetic behavior shows overall AFM exchange, with a small downturn of the χT vs. T plot and a Weiss constant of $\theta = -4.1K$. The authors attribute the solid-state AFM behavior to π-stacking of the hydrogen-bonded dyads of **NOA** in the crystal lattice that overwhelms any intra-dyad FM exchange mechanism.

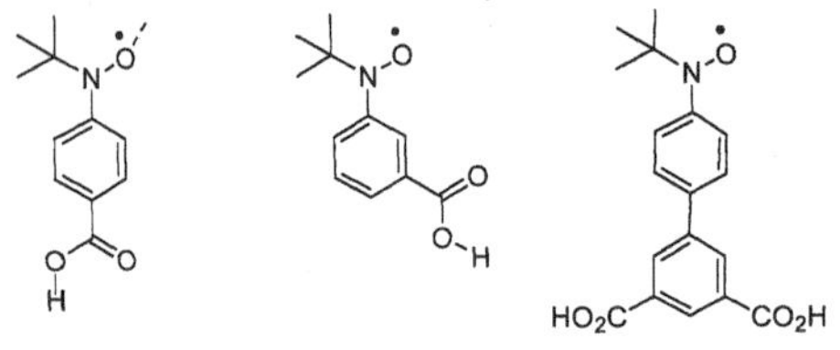

NOA 3HOOCPhNIT BDA

Baskett and Lahti carried out a similar study of **3HOOCPhNIT** [39]. Unlike the 4-isomer, this system shows no evidence of radical pairing in solution ESR, even though through-bond spin polarization considerations seem to favor triplet state radical pair formation (Fig. 17). However, unlike the case of **NOA**, the hydrogen-bonded dyads of **3HOOCPhNIT** could have very different nitroxide configurations, so radical pair ESR spectral features may be too broadened to see in conventional X-band spectroscopy.

Crystallographically, 3HOOCPhNIT is similar to NOA. Both molecules exhibit carboxylic acid hydrogen-bonded dyads that form 1-D chains through nitroxide to aryl CH contacts. Unlike NOA, 3HOOCPhNIT does not form π-stacks. Still, like NOA, **3HOOCPhNIT** exhibits AFM exchange in variable temperature magnetic susceptibility measurements. There are no close contacts between sites of major spin density (nitroxide units) in either molecule, so some sort of spin polarized exchange mechanism

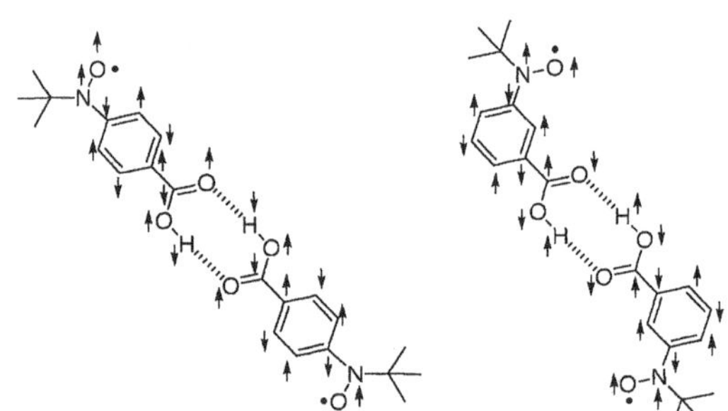

NOA 3HOOCPhNIT

Fig. 17. Spin polarization mechanisms that could yield FM exchange in isomeric *tert*-butylnitroxylbenzoic acids.

is required for both. Exchange through the carboxylic acid units or through the nitroxide to aryl CH contacts (CH···ON) are likely candidates in **3HOOCPhNIT**. A Curie-Weis $1/\chi$ vs. T plot for **3HOOCPhNIT** reproducibly shows a change in slope at about 10K, which complicates the magnetostructural analysis. The higher temperature portion of the plot yields a Weiss constant of $\theta = -3.8K$, the lower temperature portion $\theta = -1.5K$. The crystal structure of **3HOOCPhNIT** has two different chains of CH···ON contacts (A and B in the Fig. 18) with different degrees of nitroxide to benzene ring torsion (0.8° vs. 22.0°). Either low temperature structural changes (conformational or phase change) or differences in exchange through or within the chains could cause the observed change in slope at low temperature. None of the simple models applied to **3HOOCPhNIT** gave a satisfactory fit to the magnetic data, including spin-pairing, 1-D chain, spin ladder, and four-spin fits. UB3LYP/6-31G* computational modelling [7] suggested that AFM exchange interactions could occur through the chain contacts, qualitatively consistent with the experimental behavior. By contrast, hydrogen-bonded dyad models predicted that FM exchange would occur through the dyad connection.

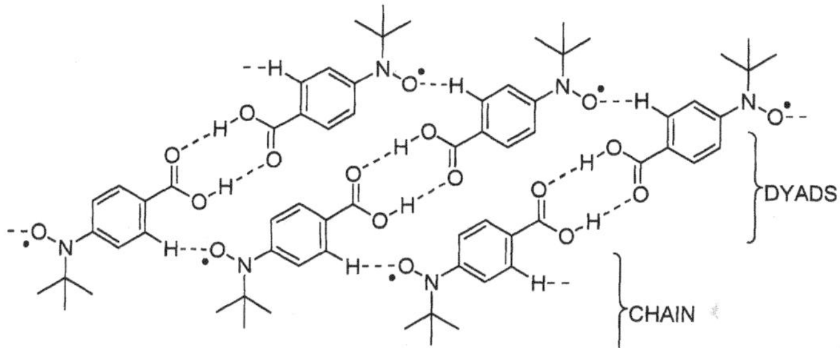

NOA

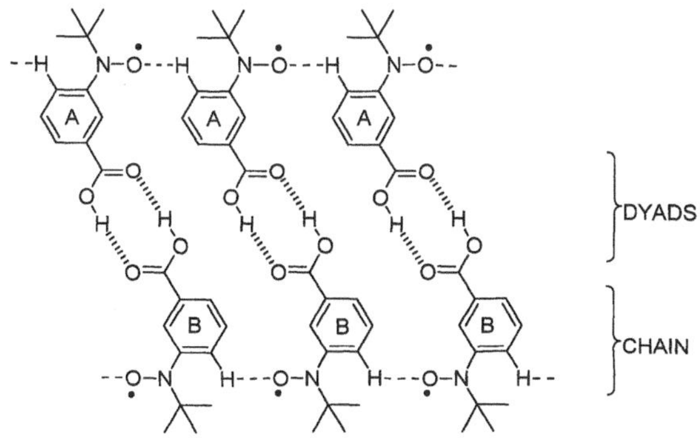

3HOOCPhNIT

Fig. 18. Chain and dimer hydrogen-bonding patterns in isomeric *tert*-butylnitroxylbenzoic acids.

The solid state magnetic behaviors of both **3HOOCPhNIT** and **NOA** are complicated by multiple close contacts between sites of small spin density, and a lack of dominating close contacts between sites of large spin density. Assignment of exchange mechanisms to specific crystallographic contact features is not completely clear. The strongest evidence for direct electronic exchange through the carboxylic acid moiety in these systems is the frozen solution ESR spectral study of **NOA**. The possibility that the **NOA** solution ESR radical pair triplet spectra arise from through-space exchange instead of a through-bond spin polarization mechanism is worth consideration.

Carboxylic acid groups can provide good crystallographic scaffolding, even if not being electronic components of the overall spin-exchange lattice. The β-form of 4-(*N*-*tert*-butyl-*N*-aminoxyl)-1,1'-biphenyl-3',5'-dicarboxylic acid (**β-BDA-NIT**) exhibits 1-D ribbon chains of hydrogen-bonds along the isophthalic acid portion of the molecule [40]. Variable temperature magnetic susceptibility studies show the molecule to exhibit 1-D Heisenberg antiferromagnetic chain behavior with J/k = (–)3.9K, but solution phase ESR studies and UB3LYP/6-31G* computations show virtually no spin density on the isophthalic acid ring. Instead, the hydrogen-bonded ribbons assemble **β-BDA-NIT** into an array where nitroxide to aryl CH···ON contacts propagate in chains perpendicular to the ribbons (Fig. 18). The aryl CH units have about 3–4% spin density, and so can generate appreciable exchange through close interatomic contact with the

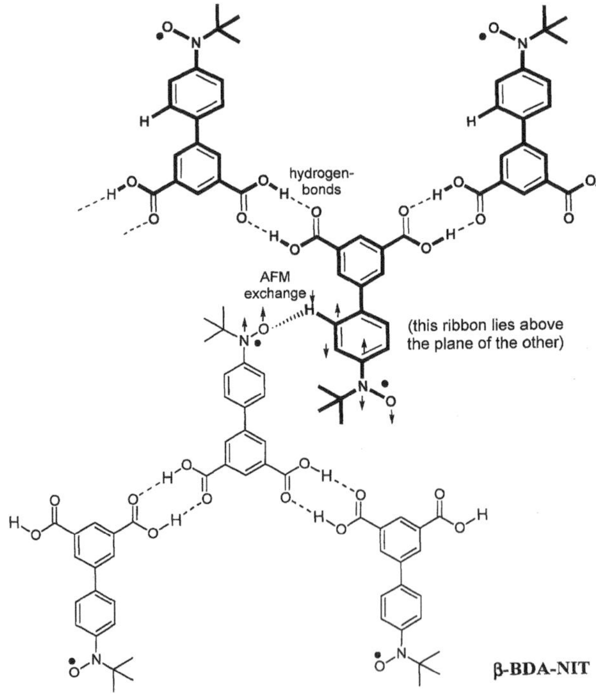

Fig. 18. Hydrogen-bonded ribbon chains in **β-BDA-NIT**, which propagate perpendicular to a set of nitroxide to aryl HC contacts. Reproduced with permission from *Chem. Mater.* 15 (2003), 2861-2863. Copyright 2003 Am. Chem. Soc.

large nitroxide spin density site. Assuming a typical spin polarization of the π-system, one expects antiferromagnetic exchange along the chain, consistent with the observed behavior. Although **β-BDA-NIT** has multiple sets of crystallographic contacts due to hydrogen-bonds and similar interactions, only the contacts that assemble significant spin density sites appear to be electronically important. In **β-BDA-NIT**, the role of the –CO_2H hydrogen-bonds is solely to arrange the radical spins in the lattice.

The ability of carboxylic acid hydrogen-bonds to act as dependable crystal scaffolding for radicals gives possibilities for complex magnetic spin arrays from simple organic molecules. Two fascinating examples are **PTMDC** [41] and **PTMTC** [42]. **PTMDC** self-assembles upon crystallization into large, solvent-incorporating nanotubes in the solid state. Solvent incorporated into the lattice cavities can be driven out by heating, and the remaining framework remains crystalline and stable up to 275°C according to thermal analysis. Magnetic analysis shows that the spins are only weakly AFM coupled in the lattice, but the crystalline structure is very interesting, with implications for molecular design. Recent work with **PTMTC** includes organizing it into hexagonal nanochannels (Metal-Organic Radical Open-Frameworks, by the authors' nomenclature) using 4,4'-bipyridine and Cu(II) or Co(II) ions.[43–44] Further details are beyond the scope of this chapter, but shows that prospects are bright for incorporation of carboxylic acid derivatives into molecular magnetic materials.

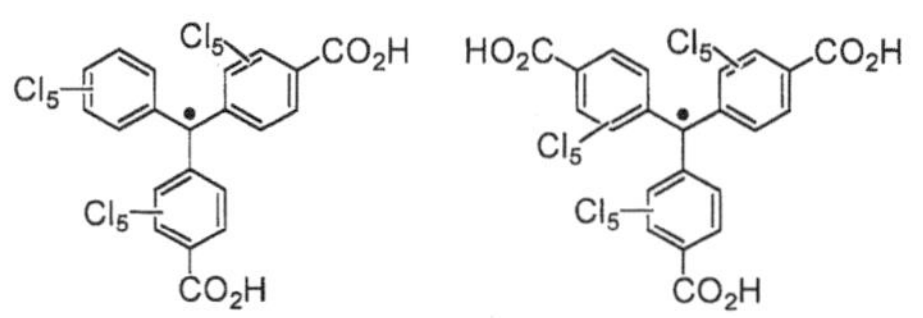

PTMDC **PTMTC**

8 Conclusion

This chapter hopefully serves as a useful overview of some of the aspects of strong hydrogen-bonds in both crystal assembly and molecular magnetism of pure organic radicals. Theoretical and computational studies of this important area abound [1d-e, 45], and experimental work to date – as exemplified above – shows that further work is needed to clarify situations where hydrogen-bonds can act as electronic exchange linker units, as well as arranging molecules into electronically interesting lattice structures without being electronic exchange linkers themselves. Further advancement of molecular magnetic design will, in all likelihood, continue to make use of hydrogen-bonding motifs in the solid state to obtain reproducible and dependable crystallographic "synthons".

Acknowledgements

I am extremely grateful to my graduate student coworkers, whose hard work allowed me the opportunity to discuss the work described in this chapter: Ahmet Inceli (initiated work on nitroxides and nitronylnitroxides in my group), Burak Esat (first work in my group on hydrogen-bonding nitronylnitroxides), Jacqueline Ruiz (Ferrer) (first work in my group on hydrogen-bonding *tert*-butylnitroxides), Lora (Michalak) Field (carboxylic acid functionalized nitronylnitroxides), Paul Serwinski (molecular recognition in iminoylnitroxides), Patrick Taylor (molecular recognition and phenolic hydrogen-bonding in iminoylnitroxides and nitronylnitroxides), Martha Baskett (carboxylic acid functionalized *tert*-butyl nitroxides), and Safo Aboaku (amido functionalized nitronylnitroxides). I am grateful for many helpful and stimulating interactions with major collaborators for work described herein: Prof. Fernando Palacio (University of Zaragoza, Spain), Prof. Michio Sorai (now emeritus, Osaka University, Japan), Profs. Takeji Takui and Kazunobu Sato (Osaka City University, Japan), and Prof. Vincent Rotello (University of Massachusetts Amherst). The National Science Foundation has provided the major support for this work for over 12 years. I also am grateful for support and assistance from the Fullbright España Comision, and the Japan Society for the Promotion of Science.

References

1. (a) J. Cirujeda, E. Hernandez-Gasio, C. Rovira, J.-L. Stanger, P. Turek, and J. Veciana, J. Mater. Chem. 5 (1995) 243; (b) E. Hernandez, M. Mas, E. Molins, C. Rovira, and J. Veciana, Angew. Chem. Int. Ed. Engl. 32 (1993) 882; (c) J.L. Garcia-Munoz, J. Cirujeda, J. Veciana, and S.F. Cox, J. Chem. Phys. Lett. 293 (1998) 160; (d) H. Heise, F.H. Koehler, F. Mota, J.J. Novoa, and J. Veciana, J. Am. Chem. Soc. 121 (1999) 9659; (e) J. Cirujeda, J. Vidal-Gancedo, O. Juergens, F. Mota, J.J. Novoa, C. Rovira, and J. Veciana J. Am. Chem. Soc. 122 (2000) 11393.
2. (a) J. Cirujeda, M. Mas, E. Molins, F. Lanfranc de Panthou, J. Laugier, J.G. Park, C. Paulsen, P. Rey, C. Rovira, and J. Veciana, Chem. Commun. (1995) 709; (b) J. Veciana, J. Cirujeda, C. Rovira, and J. Vidal-Gancedo, Adv. Mater. 7 (1995) 221; (c) J. Cirujeda, C. Rovira, and J. Veciana, Synth. Met. 71 (1995) 1799; (d) J. Cirujeda, E. Hernandez-Gasio, F. Lanfranc de Panthou, J. Laugier, M. Mas, E. Molins, C. Rovira, J.J. Novoa, P. Rey, and J. Veciana, Mol. Cryst. Liq. Cryst. Section A. 271 (1995) 1.
3. J. Cirujeda, L.E. Ochando, J.M. Amigo, C. Rovira, J. Rius, J. Veciana, Angew. Chem. Int. Ed. Engl. 34 (1995) 55.
4. Spectroscopic studies of RSNN were also reported by, L.Y. Chiang, R.B. Upasani, and J.W. Swirczewski, MRS Sympos Proc. 247 (1992) 435.
5. (a) T. Sugawara, M.M. Matsushita, A. Izuoka, N. Wada, N. Takeda, and M. Ishikawa, Chem. Commun. (1994) 1723; (b) M.M. Matsushita, A. Izuoka, and T. Sugawara, Mol. Cryst. Liq. Cryst., Section A 279 (1996) 139.
6. M.M. Matsushita, A. Izuoka, T. Sugawara, T. Kobayashi, N. Wada, N. Takeda, and M. Ishikawa, J. Am. Chem. Soc. 119 (1997) 4369.
7. Computed using Gaussian 03: M.J. Frisch, G.W. Trucks, H.B. Schlegel, G.E. Scuseria, M.A. Robb, J.R. Cheeseman, J.A.J. Montgomery, T. Vreven, K.N. Kudin, J.C. Burant, J.M. Millam, S.S. Iyengar, J. Tomasi, V. Barone, B. Mennucci, M. Cossi, G. Scalmani, N. Rega, G.A. Petersson, H. Nakatsuji, M. Hada, M. Ehara, K. Toyota, R. Fukuda, J. Hasegawa, M. Ishida, T. Nakajima, Y. Honda, O. Kitao, H. Nakai, M. Klene, X. Li, J.E. Knox, H.P. Hratchian, J.B. Cross, C. Adamo, J. Jaramillo, R. Gomperts, R.E. Stratmann, O. Yazyev, A.J. Austin, R. Cammi, C. Pomelli, J.W. Ochterski,

P.Y. Ayala, K. Morokuma, G.A. Voth, P. Salvador, J.J. Dannenberg, V.G. Zakrzewski, S. Dapprich, A.D.Daniels, M.C. Strain, O. Farkas, D.K. Malick, A.D. Rabuck, K.Raghavachari, J.B. Foresman, J.V. Ortiz, Q. Cui, A.G. Baboul, S. Clifford, J. Cioslowski, B.B. Stefanov, G. Liu, A. Liashenko, P. Piskorz, I. Komaromi, R.L. Martin, D.J. Fox, T. Keith, M.A. Al-Laham, C.Y. Peng, A. Nanayakkara, M. Challacombe, P.M.W. Gill, B. Johnson, W. Chen, M.W. Wong, C. Gonzalez, J.A. Pople, Gaussian 03, Rev B03 , Gaussian, Inc.: Pittsburgh, PA, (2003).

8. P. Taylor and P.M. Lahti Chem. Comm. (2004) 2686.
9. Cf. for example G. Cooke, V. M. Rotello, Chem. Soc. Rev. 31 (2002) 275, and references therein.
10. R. Feher, D.B. Amabilino, K. Wurst, J. Veciana Mol. Cryst. Liq. Cryst. Sci. Technol., Section A 334 (1999) 333.
11. P. Taylor, P.R. Serwinski, and P.M., Lahti, Chem. Commun. (2003) 1400.
12. D. Shiomi, M. Nozaki, T. Ise, K. Sato, and T. Takui, J. Phys. Chem. B 108 (2004) 16606.
13. P. Taylor, P.M. Lahti, J.B. Carroll, and V.M. Rotello, Chem. Commun. (2005) 895.
14. (a) M. Aso, T. Ikeno, K. Norihisa, M. Tanaka, N. Koga, and H. Suemune, J. Chem. Soc. Perkin Trans. 2 (2000) 1637; (b) M. Aso, T. Ikeno, K. Norihisa, M. Tanaka, N. Koga, and H. Suemune, J. Org. Chem. 66 (2001) 3513.
15. T. Kaneko, M. Aso, N. Koga, and H. Suemune, Org. Lett. 7 (2005) 303.
16. (a) A. Lang, Y. Pei, L. Ouahab, and O. Kahn, Adv. Mater. 8 (1996) 60; (b) B. Gillon, M.A. Aebersold, O. Kahn, and L. Pardi, Chem. Phys. 250 (1999) 23.
17. L. Catala, R. Feher, D.B. Amabilino, K. Wurst, and J. Veciana, Polyhedron 20 (2001) 1563.
18. N. Yoshioka, M. Irisawa, N. Aizawa, T. Aoki, H. Inoue, and S. Ohba, Mol. Cryst. Liq. Cryst. Section A 286 (1996) 487.
19. (a) N. Yoshioka, M. Irasawa, Y. Mochizuki, T. Kato, H. Inoue, and S. Ohba, Chem. Lett. (1997) 251; (b) H. Nagashima, M. Irisawa, N. Yoshioka, and H. Inoue, Mol. Cryst. Liq. Cryst. Sci. Technol., Sect. A 376 (2002) 371.
20. N. Yoshioka, N. Matsuoka, M. Irisawa, S. Ohba, and H. Inoue, Mol. Cryst. Liq. Cryst. Sci. Technol., Section A 334 (1999) 239.
21. N. Yoshioka, H. Inoue, "Crystal Control in Organic Radical Solids," in Lahti, P. M. (ed.) Magnetic Properties of Organic Materials, (Marcel Dekker, New York, NY, (1999)) p. 566.
22. H. Nagashima, H. Inoue, and N. Yoshioka, J. Phys. Chem. B 108 (2004) 6144.
23. H. Nagashima, N. Hashimoto, H. Inoue, and N. Yoshioka, New. J. Chem. 27 (2003) 805.
24. H. Nagashima, S. Fujita, H. Inoue, and N. Yoshioka, Cryst. Growth Des. 4 (2004) 19.
25. H. Nagashima, H. Inoue, and N. Yoshioka, Polyhedron 22 (2003) 1823.
26. J.R. Ferrer, P.M. Lahti, C. George, P. Oliete, M. Julier, and F. Palacio, Chem. Mater. 13 (2001) 2447.
27. P.M. Lahti, J.R. Ferrer, C. George, P. Oliete, M. Julier, and F. Palacio, Polyhedron 20 (2001) 1465.
28. D. Baba, R.D. Rahimi, T. Koto, K. Sato, D. Shiomi, T. Takui, P.M. Lahti, and J. Ruiz, unpublished results (2005).
29. J.R. Ferrer, P.M. Lahti, C. George, G. Antorrena, and F. Palacio, Chem. Mater. 11 (1999) 2205.
30. Y. Miyazaki, T. Sakakibara, J.R. Ferrer, P.M. Lahti, G. Antorrena, F. Palacio, and M. Sorai, J. Phys. Chem. B 106 (2002) 8615.
31. J. Ruiz "Synthesis, Isolation, and Magnetic Properties of Hydrogen-bonded Tert-Butyl Nitroxides", Ph. D. Dissertation, University of Massachusetts, Amherst, MA ((2001)).
32. (a) T. Miyazawa and K.S . Pitzer, J. Am. Chem. Soc. 81 (1959) 74; (b) A.W. Pross, F. van Zeggeren, Spectrochim. Acta 16 (1960) 563; (c) D.M. Mathews, R.W. Sheets, J. Chem. Soc. A (1969) 2203; (d) R. Gaufres, J. Maillols, and V. Tabacik, J Raman

Spectrosc. 11 (1981) 442; (e) D.J. Frurip, L.A. Curtiss, and M.J. Blander, J. Am. Chem. Soc. 102 (1980) 2610; (f) L. Turi and J.J. Dannenberg, J. Phys. Chem. 97, (1993) 12197.
33. K. Inoue and H. Iwamura, Chem. Phys. Lett. (207) (1993) 551.
34. C. Stroh, F.M. Romero, N. Kyritsakas, L. Catala, P. Turek, and R. Ziessel, J. Mater. Chem. 9 (1999) 875.
35. U. Schatzschneider, T. Weyhermuller, and E. Rentschler, Inorg. Chim. Acta 337 (2002) 122.
36. O. Félix, M.W. Hosseini, A. de Cian, J. Fischer, L. Catala, and P. Turek, Tetrahedron Lett., (1999) 2943.
37. (a) O.M. Polumbrik and E.I. Zaika, Organic Reactivity (Tartu) 14 (1977) 375; (b) T.M. Barclay, R.G. Hicks, M.T. Lemaire, L.K. Thompson, and Z. Xu, Chem. Commun. (2002) 1688.
38. D. Maspoch, L. Catala, P. Gerbier, D. Ruiz-Molina, J. Vidal-Gancedo, K. Wurst, C. Rovira, and J. Veciana, Chem. Eur. J. 8 (2002) 3635.
39. M. Baskett and P. M. Lahti, Polyhedron, submitted for publication.
40. L.M. Field and P.M. Lahti, Chem. Mater. 15 (2003) 2861.
41. D. Maspoch, N. Domingo, D. Ruiz-Molina, K. Wurst, J. Tejada, C. Rovira, and J. Veciana, J. Am. Chem. Soc. 126 (2004) 730.
42. D. Maspoch, D. Ruiz-Molina, N. Domingo, K. Wurst, G. Vaughan, J. Tejada, C. Rovira, and J. Veciana, Angew. Chem. Int. Ed. Engl. 43 (2004) 1828.
43. D. Maspoch, D. Ruiz-Molina, K. Wurst, N. Domingo, M. Cavallini, F. Biscarini, J. Tejada, C. Rovira, and J. Veciana, Nature Materials 2 (2003) 190.
44. D. Maspoch, D. Ruiz-Molina, K. Wurst, C. Rovira, and J. Veciana, Chem. Commun. (2004) 1164.
45. For some examples, see (a) J. Zhang, R. Wang, M. Baumgarten, Mol. Cryst. Liq. Cryst. 306 (1997) 705; (b) J. Zhang, M. Baumgarten, Chem. Phys. 222 (1997) 1; (c) A. Oda, T. Kawakami, G. Maruta, S. Takeda, W. Mori, K. Yamaguchi, M.M. Matsushita, A. Izuoka, and T. Sugawara, Mol. Cryst. Liq. Cryst. 334 (1999) 1345; (d) A. Oda, T. Kawakami, S. Takeda, W. Mori, M.M. Matsushita, A. Izuoka, T. Sugawara, and K. Yamaguchi, Mol. Cryst. Liq. Cryst., 306 (1997) 151.

Carbon-Based Magnetism
T. Makarova and F. Palacio (Editors)

3

Persistent High-spin Organic Polyradicals

H. Murata and H. Nishide

Department of Applied Chemistry, Waseda University, Japan

1 Introduction

Purely organic magnetic molecules are expected to display molecular-based magnetism ascribed to the interaction of the p-electrons of C, O, N, and S [1], and there has been interest in preparing magnetically-responsive, purely organic polymers that possess both the inherent properties of organic polymers, such as film forming, and a usable magnetic property, such as superparamagnetic and ferromagnetic properties. These magnetic polymers could also be examined for use in new or advanced magnetic devices based on a molecular system with an electrical and/or optical property.

Theoretical and experimental studies on π-conjugated and alternant, but non-Kekulé-type organic molecules bearing multiple radical centers have successfully revealed a correlation between the molecular connectivity or substitution positions of the radical centers on the conjugated skeleton and the spin multiplicity or spin quantum number (S) in a ground state (GS) of the molecules [2–4]. Based on their conclusions, some such non-Kekulé molecules are expected to exhibit a stabilized high-spin state even at room temperature based on the intramolecular and strong through-bond spin-exchange interaction between the multiple unpaired electrons on radical centers. These types of high-spin organic radical molecules could be extended to the corresponding non-Kekulé-type oligo- and poly-radical molecules. A very high-spin state or S that is proportional to the degree of polymerization is, at least theoretically, predicted in the polyradical molecules, which may correspond to a classical magnetic domain.

One of the advantages of organic polyradicals as a single (macro)molecule is, besides their physical property and moldability, that the size, shape, and radical concentration of the molecules are designable and can be constructed by conventional macromolecular chemistry to produce the polymers with a two- or three-dimensional topology such as a globular, disk-like, or star-shaped structure with a nanometer-size. When such single-molecular polyradicals themselves become magnetically responsible, they are potentially a new class of magnetic materials (even they are not a ferromagnet). For example, such high-spin polyradicals are quite promising as a nm-sized and magnetically responsible dot.

The highest spin alignment (S) value of macromolecules has been improved year by year and has now reached about 5000 for the highly cross-linked, π-conjugated and non-Kekulé polyradical involving a triarylmethine radical as the spin source, which was recognized for the first organic macromolecule behaving as a superparamagnet below 10K [5]. However, the triarylmethine radical can survive only at low temperature. Chemical stability and feasibility of the high-spin polyradicals are indispensable for evaluating the high-spin alignment in organic polyradical molecules at room temperature and crucial for the development of the polyradicals as practical materials. In order to realize a persistent high-spin polyradical, the proper selection of both the backbone structure and radical species is essential. The radical species is required not only to provide a strong spin-exchange interaction between the unpaired electrons, but also a sufficient chemical stability or life-time at room temperature.

In this chapter, we describe the molecular designing of persistent and high-spin organic polyradicals, review the aminium polyradicals already synthesized, and demonstrate the examples of high-spin organic polyradicals at room temperature.

2 Spin Alignment of Non-Kekulé and Non-Disjoint Molecules

π-Electrons play a vital role in determining the high-spin character of organic polyradicals, and the π-conjugated backbone has been thoroughly studied in order to connect the radical's spins via the through-bond intramolecular ferromagnetic spin-exchange interaction. The design of a π-conjugated, but non-Kekulé and non-disjoint type molecule is a requisite for the ferromagnetic spin-exchange interaction or spin alignment in a π-conjugated polyradical [6]. By using the non-Kekulé type stilbene diradical 1 as an example, the designing of a non-disjoint type molecule is described (Fig. 1). An effective overlap of the two nonbonding molecular orbitals (NBMOs) on the *p*, *m'*-isomer of the 1 diradical favors the parallel spin alignment of the two unpaired electrons to yield a triplet ground state, which is called "non-disjoint" connectivity. On the other hand, the complete separation of the two NBMOs on the *m*, *m'*-isomer of 1 results in a very small exchange interaction between the two unpaired electrons, consequently, the almost complete degeneration of the triplet and singlet states, which is called disjoint connectivity of the NBMOs of the diradical.

An unpaired electron's or a spin source, which is to be introduced into the π-conjugated backbone, should be carefully selected from the list of radical species to provide both a sufficient spin-exchange interaction and chemical stability of the unpaired electrons. Fig. 2 shows the trade-off relation between the exchange interaction or triplet-singlet energy gap ($\Delta E_{t\text{-}s}$) and the chemical stability or half-life (τ) of the non-Kekulé and non-disjoint *p*, *m'*-substituted stilbene diradicals. For example, the spin

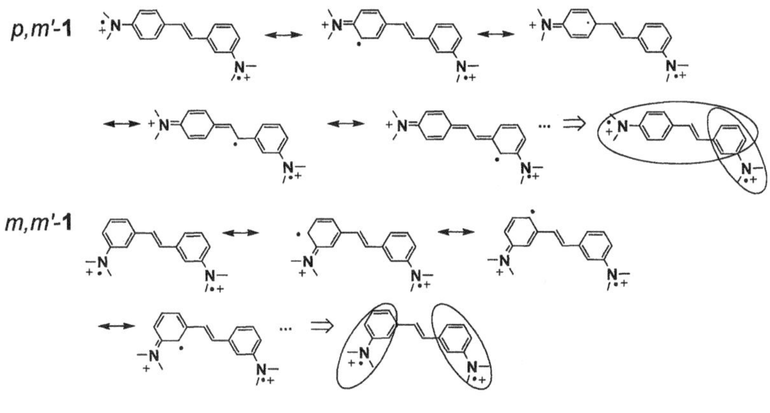

Fig. 1. Non-disjoint and disjoint type connectivities of the diradical **1**.

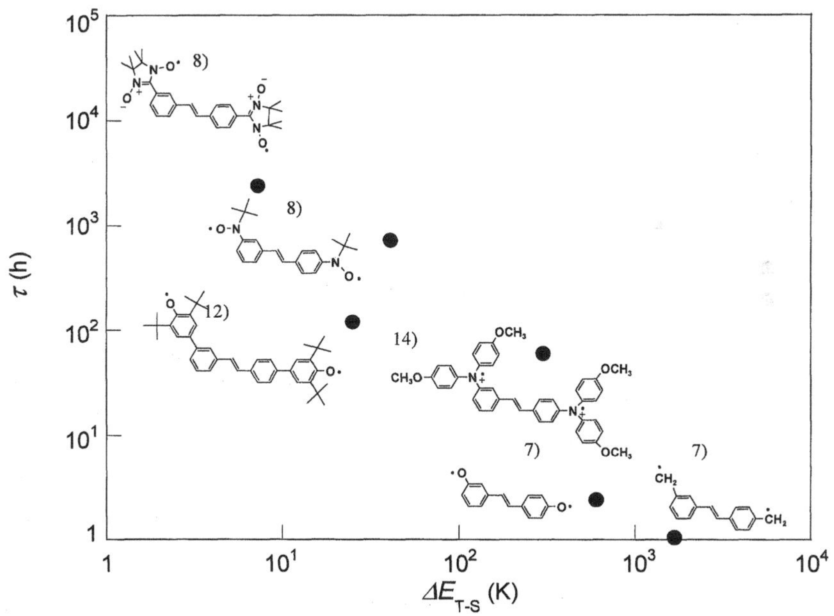

Fig. 2. Relationship between the triplet-singlet energy gap $\Delta E_{\text{T-S}}$ and the half life τ for the *p*, *m*'-substitued stilbene diradicals.

density of the diarylmethine radical's unpaired electron is delocalized into the π-conjugated stilbene skeleton, which enhances the exchange interaction between the two unpaired electrons' spins on the biradical molecule [7]. However, this spin density delocalization increases the radical reactivity of the sterically unprotected skeleton, which results in the chemical degradation of the molecule. On the other hand, the spin density is almost localized on the oxygen atom of the nitronyl nitroxide and galvinoxyl radical group [8]. These radicals have a significant chemical stability, which is an advantage of the synthetic pathway of the polyradicals and during handling the radical as a material. However, the localized spin reduces the exchange interaction.

Among the organic radical species, it is known that the spin density of the triarylaminium radicals delocalizes into the three attached aryl groups, which would efficiently work in the spin exchange interactions [9]. In addition, the triarylaminium cationic radicals derived from the *para*-substituted triphenylamines are chemically quite stable, and had been often studied even as an oxidizing reagent and a catalyst in redox reactions [10]. The triarylaminium cationic radical is a favorable candidate for the spin source to be utilized in a chemically durable and high-spin polyradical at room temperature.

Besides the selection of the radical species, the choice of the backbone skeleton is also crucial to produce a high-spin alignment. In parallel to the cross-conjugated polyradicals, such as Rajca's highly cross-linked, non-Kekulé and non-disjoint type triarylmethine, there is another synthetic procedure, which focuses on a π-conjugated linear polymer bearing multiple pendant radical groups in a non-Kekulé and non-disjoint fashion. The pendant radical groups are attached to one π-conjugated backbone that satisfies the non-Kekulé and non-disjoint connectivity among the NBMOs of the radicals' unpaired electrons. Such linear polyradicals bearing pendant spin sources possess the following advantages [11]. First, a strong spin-exchange interaction occurs through the π-conjugated backbone, and potentially even works over a long range. That is, the exchange interaction is not sensitive to the defects that are unavoidable for macromolecular polyradicals. In addition, the spins are expected to interact not only with their neighboring spins but also with more remote spins. Second, the precursors of these types of polyradicals can be synthesized via a one-pot polymerization, and they and the polyradicals themselves are soluble in common solvents; these are very favorable for obtaining well-defined samples to study. Third, various radical species including the chemically stable triarylaminium radicals can be introduced into a polyradical as the pendant group. Such feasibility would be indispensable for the future application of organic polyradicals as a material.

Poly(1,2-phenylenevinylene) is characterized as being a relatively coplanar backbone with an extended π-conjugation even after the introduction of the pendant radical groups. The radical-substituted poly(phenylenevinylene)s are soluble in common solvents. An intramacromolecular high-spin alignment among the pendant unpaired electrons has been realized by synthesizing poly(1,2-phenylenevinylene)s that were 4-substituted with the 3,5-di-*tert*-butyl-4-oxyphenyl 2 [12] and *N*-*tert*-butyloxyamino 3 [13].

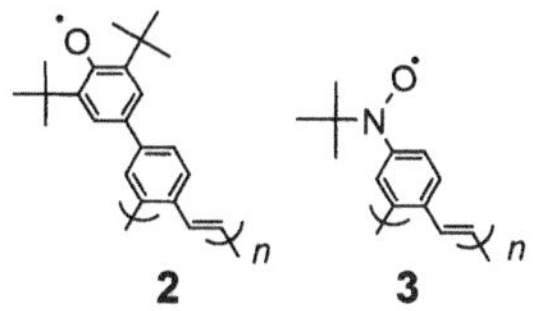

Scheme 1

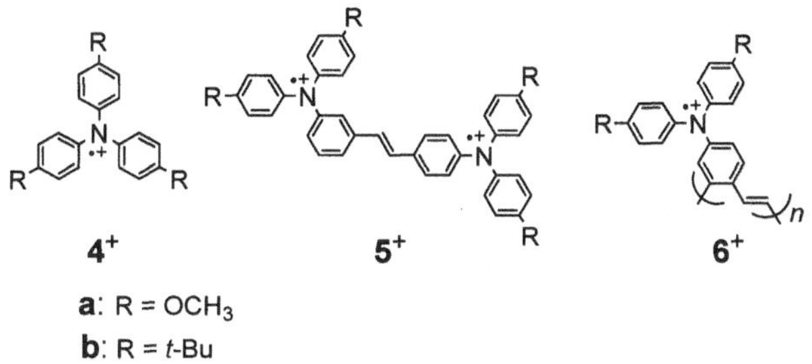

Scheme 2

Along with this molecular strategy, a pendant-type poly(triarylaminium cationic radical), poly(4-diphenylaminium-1,2-phenylenevinylene) 6^+, and its dimer models, the stilbene diradicals 5^+, were designed as the target molecules of persistent high-spin organic polyradicals.

The stilbene-based triarylaminium diradicals 5^+ were synthesized by coupling 3,4'-dibromostilbene and diphenylamine using a Palladium-catalyst and oxidizing with chemical oxidants [14]. 5^+ was isolated as a dark blue powder: The half-life of the radical was 7 days for the powder $5a^+$, which was much longer than those reported for *m*-phenylene-coupled oligo(triarylamine)s (e.g., half-life 0.8 h). The cyclic voltammogram of 5 in the methylene chloride solution was reversibly recorded in the repeated potential sweeps at room temperature. This result means that the cationic radical is stable even in solution and that the radical generation is not accomplished by a subsequent side reaction, such as dimerization, to form a benzidine derivative. The overall redox potentials (vs Ag/AgCl) were 0.75 and 0.93 V for 5a and 5b, respectively, which were appropriately anodically shifted in comparison with the potentials of the corresponding mono triarylamines (0.65 and 0.89 V for 4a and 4b, respectively). The differential pulse voltammetry on 5 gave two oxidation responses ascribed to the two amine sites. It suggested an interaction between two aminium radical sites through the stilbene backbone.

By considering both the oxidation potential of 5 and the counter anion species, $NOBF_4$, of which the oxidizing potential has been reported to be 1.58 V, was chosen as the chemical oxidant of 5. $NOBF_4$ was solubilized with 18-crown-6-ether in methylene chloride, and when added to the 5 solution, it turned to deep blue (λ_{max} = 769 and

689 nm for $5a^+$ and $5b^+$, respectively) which is characteristic of a triphenylaminium cationic radical [15, 16]. The spin concentration increased beyond 0.9 spin/amine unit for $5a^+$.

The frozen glass of 5^+ gave a strong $\Delta M_s = \pm 2$ forbidden transition at $g = 4$ ascribed to a triplet species. The doubly integrated signal intensities of the transition for $5a^+$ and $5b^+$ obeyed Curie's law in the temperature range of 5–100 K. This result suggests that the triplet species is a stable ground state with a large triplet-singlet gap for 5^+ (however, it does not rule out the possibility of a degenerate singlet-triplet state). The normalized plots of magnetization (M/M_S) for $5a^+$ were close to the Brillouin curve for $S = 2/2$ supporting the triplet ground state for the diradical $5a^+$. The ratio of the effective magnetic moment (μ_{eff}) and Bohr magneton (μ_B) lay close to 2.83 (theoretical value for $S = 2/2$) even in the high temperature range up to 300K. The stability of the triplet ground state ($2J$) was estimated to be significantly large ($> k_BT$) for the diradical 5^+. The very strong spin-exchange interaction for 5^+ is also reasonable by taking into account both the strong spin-exchange coupling effect of the 3,4'-stilbene linker, which has already been reported using other diradical species [12], and the spin density delocalization in the triarlyaminium cationic radical indicated by ESR. The NBMOs of the 3,4'-disubstuted stilbene 5^+ effectively overlapped each other on the stilbene framework, which could realize both the efficient spin-exchange interaction between the aminium cationic radicals and the chemical stability of the diradical even at room temperature. Poly(cationic radical)s derived from the 3,4'-bis(diarylamino)stilbenoid units are expected to show a strong intramolecular spin-exchange coupling.

3 π-conjugated Polymers Bearing Aminium Cationic Radicals for High-spin Alignment at Room Temperature

As an extension of the diradical 5^+, the pendant-type poly(aminium cationic radical)s based on poly(1,2-phenylenevinylene), 6^+, were synthesized with well-defined chemical structures [17, 18]. The pendant triarylaminium cationic radical groups are attached to one π-conjugated poly(1,2-phenylenevinylene) backbone at its 4 position of the repeating unit that satisfies the non-Kekulé structure and non-disjoint connectivity among the non-bonding molecular orbitals of the radicals' unpaired electrons.

A head-to-tail linkage structure of the radical group-bearing monomer unit is essential for the non-Kekulé and non-disjoint requisite or the following magnetic study; this was established through the polycondensation via the Heck reaction of the 4-radical precursor-substituted 2-bromostyrene. The Heck reaction is the arylation of an olefin with an aryl bromide using a palladium catalyst [19–21], and more favorable than the Wittig reaction to yield poly(phenylenevinylene)s with a relatively high molecular weight and an all *trans*-vinylene structure. The bromostyrene derivatives as the monomer, 7a and 7b, were polycondensed using the catalyst of palladium acetate and tri(*o*-tolyl)phosphine in the presence of triethylamine and lithium chloride as the base and the source of chloride ion, respectively, in DMF solution at 70°C for 72 h [18].

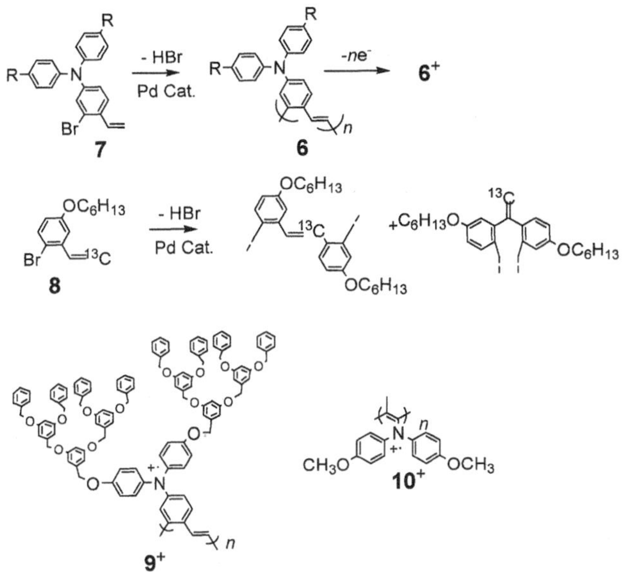

Scheme 3

These polymerization conditions were selected for the selective *β*-arylation of the vinyl group with the aryl bromide of 7, based on the ^{13}C-labeled control reaction using 2-bromo-5-hexyloxy(*2-*^{13}*C*)styrene 8 as the monomer. The ^{13}C-NMR spectrum of the polymer indicated the complete poly(1,2-phenylenevinylene) structure without any *α*-arylation defects, supporting the head-to-tail linked primary structure of 6, at least, based on the ^{13}C-NMR analysis.

The degree of polymerization (DP) measured by GPC using polystyrene standards and by the terminal bromine contents coincided with each other (e.g., 12.5 and 12.8, respectively, for 6a), which supported the well-defined chemical structure of 6. The UV/vis absorption and strong photoluminescence of 6 suggested that π-conjugation developed on the backbone. The polymers 6 were soluble in the common solvents such

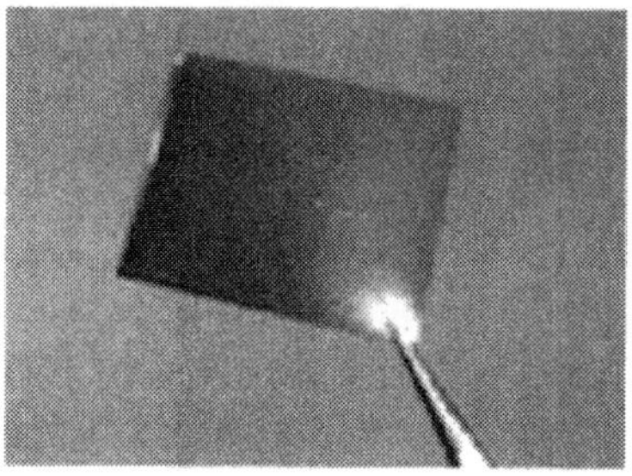

Fig. 3. Film of poly(4-dianisylamino-1,2-phenylenevinylene) **6a**.

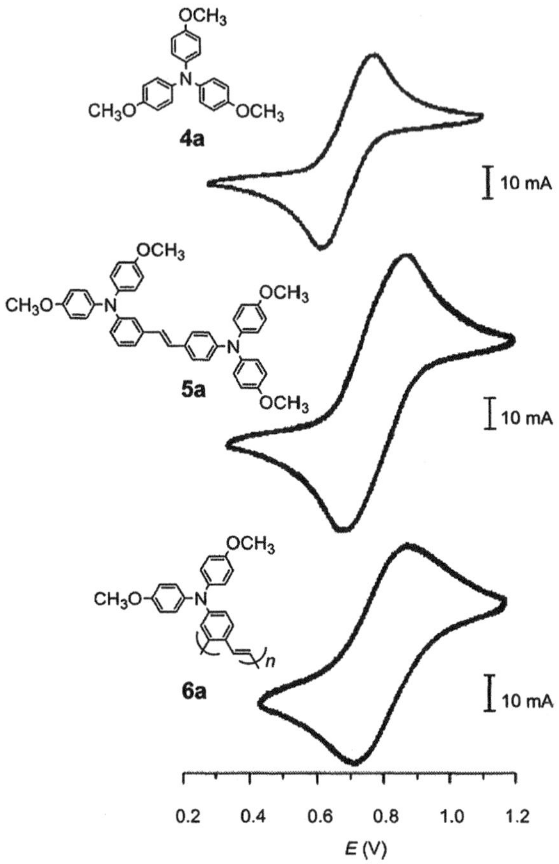

Fig. 4. Cyclic voltammograms for the monomer **4a**, the dimer **5a**, and the polymer **6a** in 1mM CH_2Cl_2 solution with 0.1M $(C_4H_9)_4NBF_4$, 100 mV/s.

as methylene chloride, chloroform, THF, benzene, toluene, triethylamine, and DMF. The polymer has a film-forming ability, of which the example of 6 is shown in Fig. 3.

The cyclic voltammetry of 6 in the methylene chloride solution with tetrabutylammonium tetrafluoroborate as the electrolyte gave unimodal waves which were repeatedly recorded from the potential sweeps of, e.g., 100 times at room temperature (Fig. 4). The overall redox potentials were 0.81 and 0.92 V for 6a and 6b (a resonance shift attributed to the electron donating methoxy and *t*-butyl group), respectively. The voltammogram result means that the triarylamine groups are electrochemically oxidized to the aminium cationic radicals and reduced to the amines without any subsequent chemical side reactions. This also suggested that the electrostatic repulsion was not strong among the adjacent aminium cationic radical groups along the π-conjugated backbone to produce a separation in the redox waves. The redox potential between the amine and the aminium cationic radical increased in the order of the monomer 4a (0.68 V) < the dimer 5a (0.78 V) < the polymer 6a

(0.81 V). The potential separations between the oxidation and reduction peaks were larger than the theoretical value (58 mV) for an electrochemically reversible reaction and increased as the sweep rate increased during the cyclic voltammetry. The cathodic shift in the redox potential and the broadening or constancy of the redox wave for the dimer, and especially the polymer, could be explained by the delocalization effect of the formed aminium cationic radical in the π-conjugation of the dimer and polymer.

A nitrosonium ion salt, e.g., $NOBF_4$ or $NOPF_6$, with the oxidizing potential of 1.58 V [22], was chosen by considering both the oxidation potential of 6 and the counter anion species. The radical or spin concentrations for $6a^+$ and $6b^+$ were estimated to be 0.89 and 0.56 spin/monomer amine unit, respectively. The aminium cationic radical salt of $6a^+$ with a larger sized anion, such as hexafluorophosphate (PF_6^-), did not show any deterioration even after one week at room temperature in air (the half-life estimated by the ESR signal intensity >one month). This half-life was much longer than that for the aminium cationic radical salt of $6a^+$ with BF_4^- (half-life ca. 2 weeks).

The ESR study at cryogenic temperatures gave a $\Delta M_s = \pm 2$ (half-field and forbidden) transition at 158 mT for 6^+. The signal intensity of the $\Delta M_s = \pm 2$ transition for $6a^+$ was proportional to the reciprocal of temperature in the range between 4.5 and 100K, and followed Curie's law. This result means that the triplet state ascribed to the $\Delta M_s = \pm 2$ peak is a stable ground state with a large triplet-singlet gap for $6a^+$ (However, it does not rule out the possibility of a degenerate singlet-triplet state).

The PF_6^- salt of 6^+ was obtained as a blue powder by pouring the methylene chloride solution of 6^+ oxidized with a small excess of $NOPF_6$/18-crown-6-ether into dimethyl ether. The salt of 6^+ was isolated in the presence of poly(styrenesulfonate tetrabutylammonium) for the magnetic measurement to suppress any antiferromagnetic interaction through space. Poly(styrenesulfonate tetrabutylammonium) formed the polycationic/polyanionic complex with 6^+ through an electrostatic interaction or a polymer salt formation (Fig. 5). The isolated sample was subjected to static magnetic susceptibility (χ) and magnetization (M) measurements using a SQUID magnetometer. χ was normalized to the χ_{mol} values using the radical concentration in the sample determined by the saturation magnetization (M_s) of the M vs magnetic field (H) plots.

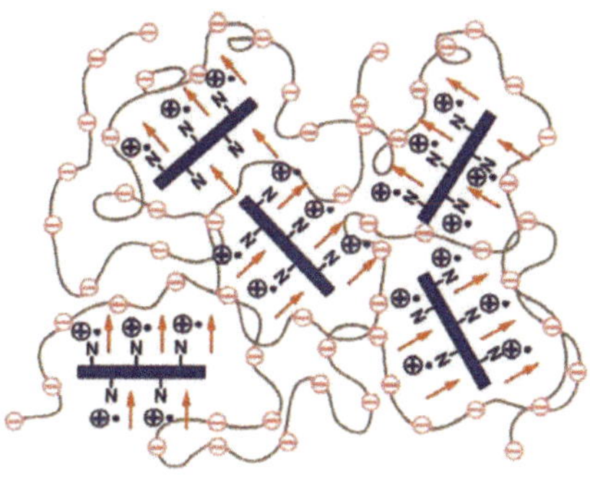

Fig. 5. High-spin poly(4-dianisylaminium-1,2-phenylenevinylene) **6a$^+$** complexed with poly(styrenesulfonate tetrabutylammonium).

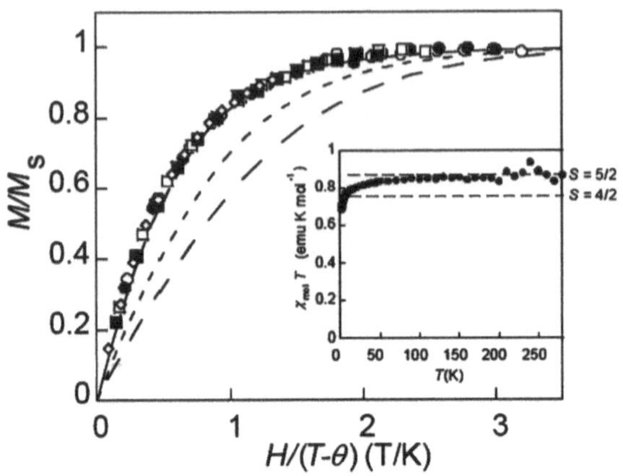

Fig. 6. Normalized plots of magnetization (M/M_s) vs the ratio of the magnetic field and temperature ($H/(T - \theta)$) for the isolated poly(4-dianisylaminium-1,2-phenylenevinylene) **6a**$^+$ (DP = 12.8) with spin concentration = 0.63 spin/unit at 1.8 (○), 2 (●), 2.5 (□), 3 (■), and 5 K (◊). Theoretical Brillouin curves for S = 1/2 (dashed line), 2/2 (broken line), and 4/2 (solid line). Inset; $\chi_{mol}T$ vs T Plots for the isolated **6a**$^+$.

The normalized plots of magnetization (M/M_{sat}) for the isolated 6a$^+$ are close to the Brillouin curves for 4/2 and 5/2, indicating a high-spin ground state for 6a$^+$ (Fig. 6). These results concluded that a strong spin-exchange and intramolecular spin-alignment are realized between the unpaired electrons of the poly(arylaminium cationic radical)s, by taking into account both the molecular weight (DP = 13) and the spin concentration (0.63 spin/unit) of 6a$^+$. The $\chi_{mol}T$ plots leveled off in the temperature range from 50 to 280K (Inset of Fig. 6). The flat $\chi_{mol}T$ plots denote the high-spin ground state for poly(arylaminium cationic radical) 6a$^+$ (S = (4.5)/2) and significantly large triplet-singlet energy gap (~ k_BT).

The magnetic susceptibility of 6$^+$ measured at room temperature by the NMR shift method was also reported [18]. The χ_{mol} values were calculated based on the Evans equation [23], measuring the resonance frequency separation of the cyclohexane standard-peak for a concentration series of the 6$^+$ solution. The calculated $\chi_{mol}T$ values for 6a$^+$ and 6b$^+$ in the solution were 0.86 and 0.60 emu·K·mol^{-1} at 300K [18], respectively; they corresponded to S = (4.9)/2 and (2.8)/2. These results conclude that the idea of a linear polyradical pendantly substituted with arylaminium cationic radicals in the non-Kekulé and non-disjoint fashion realized an observed high S value for a purely organic material even at room temperature.

For the high-spin poly(arylaminium cationic radical) 6$^+$, however, a through-space and antiferromagnetic interaction worked between the high-spin molecules, which sacrificed the high-spin behavior especially at low temperature. The through-space interaction had been partially reduced by diluting the high-spin molecule with diamagnetic solvents or the specific polyanionic polymer as described above, but the dilution produced a significant decrease in the content of the spin active molecule in the sample.

Dendrons, the subpart of dendrimers, possess a regular branch-like three-dimensional structure [24]. The dendron moiety has been studied to cover and sterically isolate a core active site with its bulky structure [25]. For example, dendron-combined porphyrins and ferrocenes have been reported to examine the isolation of the redox active core. A dendron-combined poly(4-diphenylaminium-1,2-phenylenevinylene) 9^{+} with a well-defined chemical structure was synthesized to suppress the through-space magnetic interaction of the high-spin molecule [26], and Fig. 7 illustrates the isolation of the poly(phenylenevinylene) backbone by the bulky dendron (Fig. 7).

The 3,5-bis(3,5-bis(benzyloxy)benzyloxy)benzyloxy group was selected as the dendron moiety and prepared using a convergent approach via the Willianson reaction [27]. The dendron-coupled bromostyrene derivative was synthesized and polymerized using the catalyst of palladium and tri(*o*-tolyl)phosphine. The brownish yellow polymer 9 was quite soluble in the common solvents such as methylene chloride, benzene, and THF, due to the introduction of the bulky dendron group. The molecular weight of 9 measured by GPC and the terminal bromine content were 7400 and 7000, respectively, which supported the primary structure of 9. The chloroform solution of 9 exhibited a yellowish green photoluminescence: The λ_{em} for 9 was 507 nm with DP = 4.3 (the λ_{em} = 506 nm for 6a with DP = 4.1). The strong photoluminescence suggested that the phenylenevinylene backbone maintains its π-conjugation. The cyclic voltammograms of 9 were reversibly recorded more than 1000 times in the repeated potential sweeps at room temperature. The cyclic voltammogram for 9 was broader than that of 6. These results show that the amine or the aminium radical moiety is efficiently covered by the bulky dendron to form a chemically very stable cationic polyradical.

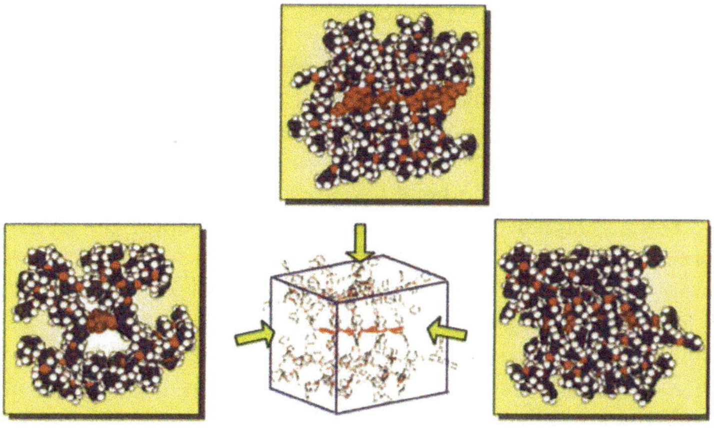

Fig. 7. Optimized structure of the dendron-combined poly(4-diphenylamino-1,2-phenylenevinylene) **9**.

The triarylaminium cationic radical salt, 9^+, was also prepared by oxidizing the corresponding triarylamine 9 with $NOPF_6$. The spin concentration of 9^+ was estimated to be 0.78 spin/monomer amine unit. The color of the 9 solution turned from pale yellow to deep blue, which agreed with the UV/vis absorption change for the aminium cationic radical generation of 6. The aminium cationic radical 9^+ in the solid state was stable even after several weeks at room temperature. The PF_6^- salt of 9^+ was isolated as a blue powder: χ and M for the neat 9^+ sample were reported. The $\chi_{mol}T$ plots for the dendron combined aminium polyradical 9^+ were flat even at low temperature, although the plots for the neat sample of the non-dendron type $6a^+$ decreased below 50K. This result indicated that the antiferromagnetic and through-space interaction was suppressed by introducing the dendron moieties at the *para* position of the triphenylaminium radical. The M/M_s plots vs the effective temperature (T-θ) gave a θ value which is the mean field parameter for intermolecular magnetic interactions [28]. θ was estimated to be -0.08K and -1.5K for the dendron-combined polyradical 9^+ and the non-dendron type $6a^+$, respectively, which supported the reduced intermolecular antiferromagnetic interaction of the dendron-combined 9^+. The bulky dendron moiety produced an isolated high-spin polyradical without any intermolecular interactions. The normalized plots of magnetization (M/M_s for 9^+) were close to the Brillouin curve for S = 2/2 indicating a triplet ground state. Taking into account both the molecular weight or DP and the spin concentration of 9^+, all the radical spins in each polymer are considered to be aligned inside the dendron-covering. S would be expected to proportionally increase with the DP of the poly(aminium cationic radical) 9^+, after optimizing the generation of the dendron or replacing with other more flexible dendron species.

Besides the poly(phenylenevinylene) backbone, polyacetylene, poly(diacetylene), poly(arylene), poly(aryleneethynylene), and polythiophene are available as a π-conjugated backbone of the pendant polyradicals (Table 1). Poly(aryleneethynylene) is characterized by a sterically compact and a hydrogen-free structure with an ethynylene bridge [29]. Poly(aryleneethynylene) looks like the favorite choice as a π-conjugated backbone of the polyradicals, when they are sterically crowded and suffer considerable twisting upon conjugation. The hydrogen-free bridging structure is also a benefit for the chemically unstable radical species. Poly(2-radical precursor-substituted phenyleneethynylene)s have been synthesized from *p*-bromophenylacetylene using a palladium catalyst with a head-to-tail linked structure, however, they often involved defect structures such as a head-to-head linked or a diethynylene defect. Polythiophene is known to possess a highly developed and rigid conjugation and to take a relatively planar conformation even after the introduction of the substitution [30]. However, polythiophene is easily doped because of its small HOMO-LUMO energy gap, which leads to side reactions after the radical generation. Polyacetylene is also one of the favorable candidates as a π-conjugated backbone because of the head-to-tail monomer linkage, the high molecular weight, and the small repeating unit; the number of carbons between the substituted radicals is three, which could enhance the interaction between the pendant unpaired electrons. High molecular weight and substituted polyacetylenes have been reported to be obtained via the polymerization or the repeated head-to-tail bonding of the corresponding substituted acetylenic monomer using a catalyst, such as

Table 1.
Characteristic features of π -conjugated backbone polymers.

	H-T Polymn	Mol. Wt.	Planarity	Repeating Unit	Conjugation
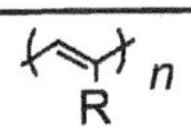	Yes	Very High	Distorted	Very Small	Short
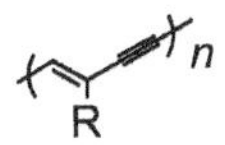	None	Very High	Planar	Large	Long
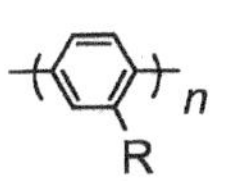	None	Low	Distorted	Large	Short
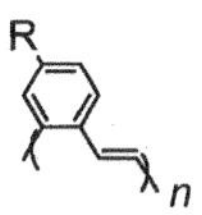	Yes	Low	Planar	Large	Med
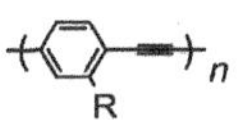	Yes	Low	Distorted	Large	Short
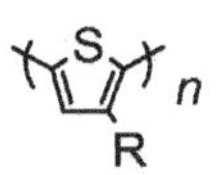	Yes	High	Planar	Small	Long

niobium (V) chloride, and a rhodium complex [31]. The substituted polyacetylenes are soluble in common solvents, and their chemical structure and purity could be rigorously defined. It has also been reported that polyphenylacetylenes bearing an amino group form a one-handed helix structure upon the addition of a chiral acid. Such helical poly(amino-substituted acetylene)s would bring about the possibility of a chiral and high-spin polyradical molecule.

Poly(dianisylaminiumacetylene) 10^+ was designed as the π-conjugated linear polyacetylene bearing pendant diphenylaminium radicals [32]. In 10^+, the pendant aminium radicals are π-conjugated with the polyacetylene backbone, in order to satisfy the non-Kekulé and non-disjoint connectivity. The acetylenic monomer, 1-bis(4-methoxypheny)amino-2-methylacetylene, was polymerized with niobium(V) chloride to give the precursor polymer 10 with the molecular weight of 2100. The polyradical 10^+ was prepared from the precursor polymer through the oxidation with $SbCl_5$ (the spin concentration was 0.56 spin/amine unit). 10^+ displayed $S = 4/2$ at low temperature, which revealed the strong spin-exchange interaction between radicals. S could be expected to increase with the DP of the poly(aminium cationic radical) 10^+ by utilizing the recently reported effective catalyst and/or co-catalysts for the acetylenic monomers with bulky substitutions [33].

4 *m*-phenylene-connected and Cross–conjugated Arylaminium Cationic Radicals

As has been already described, the aminium cationic radical is a favorable candidate as the spin source to be utilized in chemically durable and high-spin pendant-type polyradicals. From the same viewpoint, a series of oligomeric, cyclic, and polymeric aminium radicals has also been derived from the *m*-phenylene-connected and cross–conjugated poly-arylamines.

Bis- and tris(cationic radical)s, 11 and 12, of the *m*-bis(diphenylamino)- and 1,3,5-tris(diphenylamino)benzenes have been reported to be chemically persistent radicals and could be isolated under ambient conditions [34]. 11 and 12 have been studied to reveal triplet and quartet ground states, resvectively. The more recent research of Tanaka and Takui [35] on the bi- and tri-aminium radicals using a field-swept 2D electron spin transient nutation method clearly described their triplet and quartet states and that the ground state high spins originate from the topological pseudodegeneracy of the π-HOMOs in the polyheteroatomic system. Blackstock *et al.* indicated the triplet ground state of the 2,7-diaminonaphthalene diradical 13 with the half-field ESR signal and Evans NMR shift methods and also suggested a spin alignment at room temperature [36].

The *p*-phenylenediamine-based cationic radical 14 was also reported as one of the stable aminium cationic radicals [37]. A series of linear and branched derivatives, such as 15 and 16, have been synthesized, and their multi redox processes and triplet and quartet ground states were fully investigated [38]. Macrocyclic arlyaminium radicals 17–19 have been prepared by Tanaka *et al.* [39], Hartwig *et al.* [40], and Blackstock *et al.* [41]. using the palladium-catalyzed condensation of arylhalides and arylaimes. X-ray crystallography of these derivatives showed an alternate *meta-para*-linked macrocyclic structure. The electronic and magnetic properties of these radicals were investigated by ESR spectroscopy and splution phase magnetic susceptibility, which concluded a triplet state derived from two *m*-phenylene-connected aminium cationic radicals. The use of these types of radicals as a spin-source unit is pertinent for the exploitation of a high-spin ladder-like polymer. Janssen *et al.* prepared oligo(1,4-phenyleneethynylene)s 20 and oligo(1,4-phenylenevinylene)s 21 substituted with stable *p*-phenylenediamine radical cations [42]. The UV/vis/near-IR experiments of 20 and 21 showed that the radical cations are delocalized in the pendant *p*-phenylenediamine units of the conjugated oligomers. They suggested a triplet ground state diradical with a large triplet-singlet energy gap, because the ESR intensity of the $\Delta M_s = \pm 1$ and $\Delta M_s = \pm 2$ signals of 20 and 21 obeyed Curie's law.

Hartwig *et al.* applied the palladium-catalyzed amination reaction to the syntheses of linear polymers containing the triarylaminium cationic radicals [43]. The *meta*-linked and linear arylamine 22 was synthesized along with the exponential growth strategy employing the palladium-catalyzed aryl perfluorosulfonate amination. 23 and 24 had a

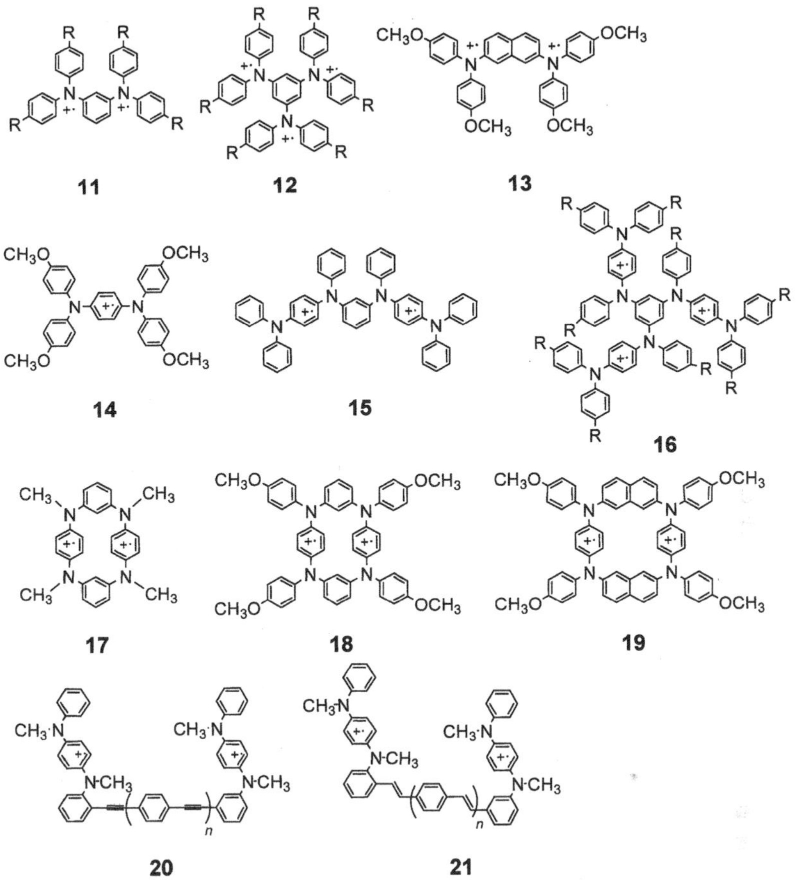

Scheme 4

regiodefined primary structure with a molecular weight of 10^4. The oxidized forms of these polymers were shown to be paramagnetic by the Evans method and displayed a chemical stability at room temperature in solution. Bushby *et al.* then extended these bi- and triradicals into a network backbone [44]. The two-dimensionally extended triarylamine-based polymer 25 with the molecular weight of 3.5×10^4 displayed $S = 8/2$ with a spin concentration of 0.6. Hartwig *et al.* synthesized a triphenylamine-based dendrimer 26 with a high glass-transition temperature and low redox potential. 26 formed delocalized cationic radicals displaying a half-life of 1 h [45]. Blackstock *et al.* reported the poly(arylamine) dendrimer 27 which displayed a unique redox behavior with a gradient [46]. The three interior *p*-phenylenediamine moieties of 27 were classified as a quartet structure, and 27 itself corresponded to a precursor of the high-spin poly(aminium cationic radical).

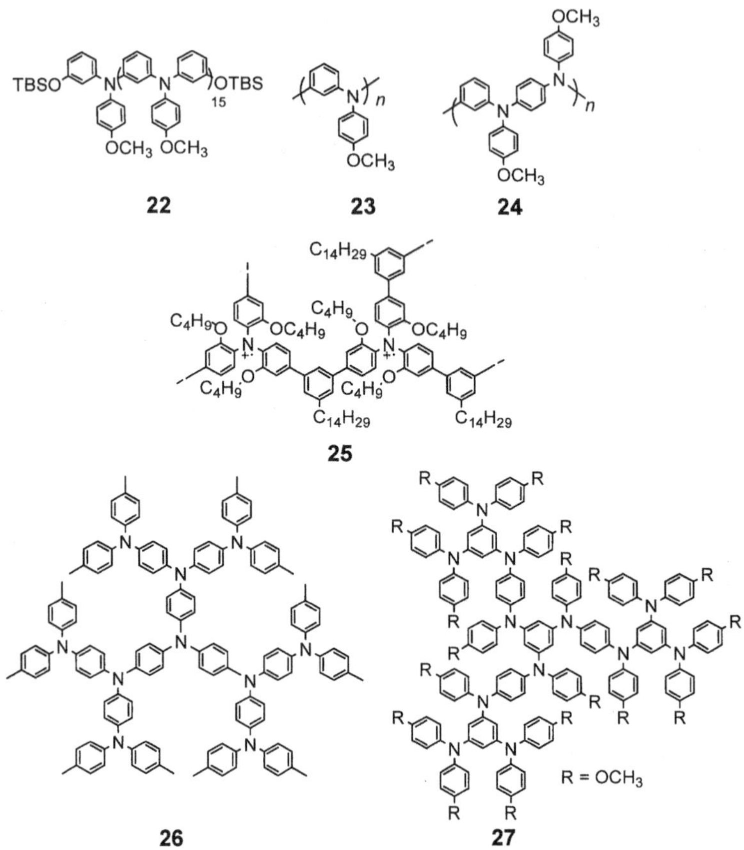

Scheme 5

It was concluded that the triarylaminium cationic radicals, which have both an efficient spin distribution of the spin density and a chemical stability, are the spin sources for synthesizing high-spin organic molecules even at room temperature. However, there still remains some difficulty with the quantitative radical generation. This is derived from the electrostatic repulsion between the cationic radicals. A low radical generation, especially for the *m*-phenylene-connected and cross-conjugated molecules, leads to a shutdown of the π-conjugation at the radical defect. In order to reduce the static repulsion, cationic radicals are required to be built-in the π-conjugated coupler with a moderate distance. For the stilbene diradicals 5^{+} and poly(4-diphenylaminium-1,2-phenylenevinylene)s 6^{+}, a sufficient distance between the arylaminium cationic radicals could reduce the static repulsion and bring about a high radical generation.

5 Two-Dimensionally Extended Poly(aminium cationic radical)s

In Sections 3 and 4, the stability of the arylaminium cationic radicals was described even after the extension to the poly(arylaminium cationic radical)s. Especially, both the sufficient life-time and the high-spin alignment related to the π-conjugated structure for the pendant-type poly(arylaminium cationic radical)s have been demonstrated in Section 3. The pendant-type polyradicals have been recently extended to their two-dimensional homologues, i.e., branched and dendric polyradicals [47–49]. The term two-dimensional means that the poly(phenylenevinylene) backbone is connected in three directions or that the π-conjugated backbone is cross-linked to form a quasi lattice. Such (pseudo-)two-dimensional extensions successfully brought about an increase in *S*; the extensions diminished the damage in a spin defect by guaranteeing multiple pathways of the ferromagnetic interaction in the two-dimensional frameworks. To realize a high-spin alignment at room temperature for organic polyradicals using such a two-dimensional strategy, the pendant-type poly(4-diphenylaminium-1,2-phenylenevinylene) 9^+ was also extended to a hyperbranched poly(arylaminium cationic radical), poly[1,2,(4)-phenylenevinyleneanisylaminium] 28^+ [50]. The hyperbranched polymer 28^+ has a π-conjugated, but non-Kekulé and non-disjoint hyperbranched poly(phenylenevinylene) skeleton, in which the spins are two-dimensionally connected, so that a more cooperative spin alignment is expected.

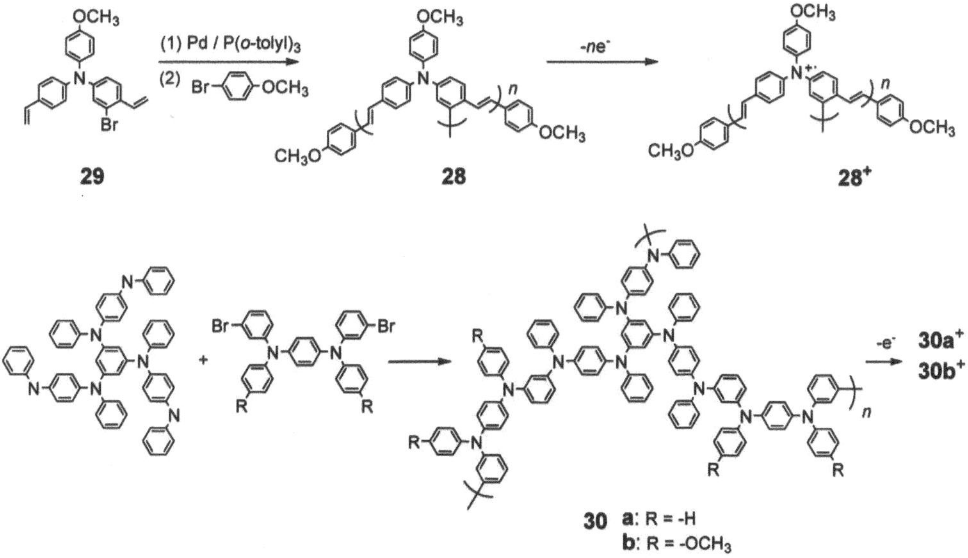

Scheme 6

The hyperbranched and head-to-tail linked poly(phenylenevinylene), poly[1,2,(4)-phenylenevinyleneanisylamine] 28, was prepared via a one-pot palladium-catalyzed polymerization of a ABB'-type monomer, *N*-(3-bromo-4-vinylphenyl)-*N*-(4-vinylphenyl)-4-methoxyaniline 29 (A = bromo group, B = vinyl group, and B' = another vinyl group). The hyperbranched polymer was characterized by ^{1}H-NMR, IR, UV/vis, and fluorescence spectroscopies. For example, the integration rate of two vinyl groups (B and B') was almost unity indicating a highly branched polymer structure. The IR, UV/vis, and fluorescence spectra indicated a transvinylene structure and the developed π-conjugation of 28 in spite of the crowded, hyperbranched structure.

The polymer was end-capped with the anisyl group to exclude any side reactions of the vinyl groups in the following radical generations. The molecular weight was 5200–7000 (DP = 13–17). The dilute solution of the hyperbranched polymer 28 was transferred to a mica surface and subjected to atomic force microscopy, AFM. The AFM produced a ca. 15 nm-sized and globular image of the polyradical 28. The high solubility of 28 in common solvents also supported the globular structure of the polymer.

The hyperbranched polymer 28 was converted to the corresponding poly[1,2,(4)-phenylenevinyleneanisyl- aminium] 28^+ with an equivalent of thianthrene radical tetrafluoroborate in acidic dichloromethane. The spin yield of 28^+ was determined by the tetrabutylammoniumiodide titration, which is based on the reduction or quenching of the aminium cationic radical by I^- [51]. The spin yield determined from the titration was 0.65. The half-life of 28^+ estimated by the ESR signal intensity was 1 week at room temperature. χ_{mol} of 28^+ was measured by the NMR shift method; the calculated $\chi_{mol}T$ value was 1.15 (1.125 = theoretical value of S = 7/2) at room temperature. A high-spin ground state of 28^+ at room temperature was concluded. The high $\chi_{mol}T$ value in the temperature range of 303–343K (30–70°C) was also reported; a very strong spin-exchange interaction (J > 340K) was maintained for the hyperbranched poly(arylaminium cationic radical) 28^+.

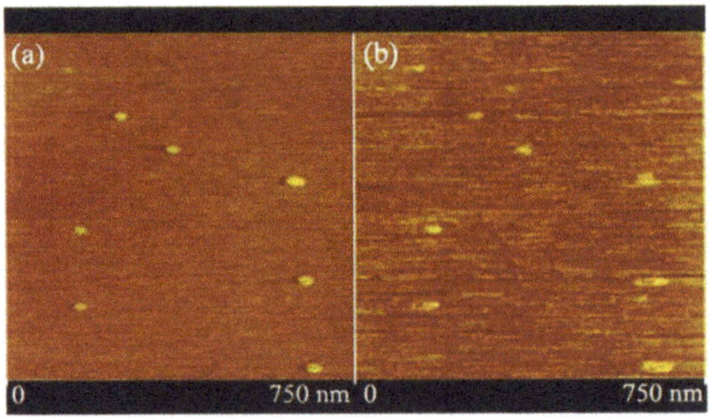

Fig. 8. AFM tapping mode (a) and MFM non-contact-mode (b) images of the hyperbranched poly[1,2,(4)-phenylenevinyleneanisylaminium] **28**$^+$ with the molecular weight of 5200.

Magnetic force microscopy (MFM) was applied at room temperature in air to study both the molecular image and the magnetic response of the high temperature and persistent high-spin polyradical 28^+. MFM is almost the same method as AFM, except it uses a magnetic tip [52]. The AFM of the polyradical 28^+ monomolecularly dispersed on a mica substrate gave a 15 nm-sized and globular image of 28^+. The first scan is in a tapping mode to recognize the surface (AFM with a magnetized tip), and the second scan in the same area uses a noncontact mode (with ca. 20 nm above the surface in this experiment) to monitor the local magnetization of the sample. The MFM (Fig. 8 (b)) clearly indicated a magnetic response exactly on the molecular position of 28^+. It was concluded that MFM is an effective tool to detect a very weak magnetic signal of such high-spin organic polymers at room temperature in air and that the poly(arylaminium cationic radical) 28^+ behaves as a single molecular-based magnetic dot with a nanometer-size. The details of the MFM measurements and the proof of the intrinsic nature of the signal is given in Ref. 49b.

In Section 4, it was described that the use of the *p*-phenylenediamine-based cationic radical is an efficient spin source for a high-spin organic molecule. An integration of the *p*-phenylenediamine-based cationic radical into a two-dimensional framework has been attempted. A *p*-phenylenediamine-based two-dimensional cationic polyradical was designed, in which an effective interaction would occur through a 1,3-phenylene or 1,3,5-benzenetriyl coupler [53]. The networked aromatic polymer was prepared by the palladium-catalyzed amination of two multi-functionalized monomers. The reaction of the trifunctionalized *sec*-amine monomer and the aryl dibromide yielded solvent-soluble but high molecular weigh (ca. 10^4) polymers, 30a and 30b. The chemical oxidation of 30 with $NOPF_6$ gave the corresponding polyradicals. The spin quantum numbers of $30a^+$ and $30b^+$ with spin concentrations of 0.66 and 0.75 were $S = 9/2$ and $8/2$, respectively. The high-spin result was considered to reflect the effective extension of a π-conjugated system in the polyradical framework. AFM measurement of 30a on mica revealed that the average vertical distance of the single polymer decreased from 1.43 to 0.83 nm after the radical oxidation and indicated higher π-coplanarity for the polyradical. A high-spin and networked poly(arylaminium cationic radical) was realized with a substantial stability and an intramolecular interaction of eight or nine spins.

The linear and π-conjugated polyradicals pendantly substituted with arylaminium cationic radicals in a non-Kekulé and non-disjoint fashion provided high *S* values for a purely organic polyradical molecule at room temperature. The design and synthesis of the two-dimensionally extended poly(arylaminium cationic radical)s represented a promising route toward the development of very high-spin polymers, although the quantitative radical generation without any chemical defects needs to be determined in advance. The magnetic response of the high-spin polyradical detected by MFM indicated as a nanometer-size dot at room temperature. An additional modification, such as the introduction of hydrogen bonding groups into a polyradical molecule, would be an effective way to realize an intermolecular spin alignment. On the other hand, the purely organic high-spin polyradicals described in this Chapter will be able to be applied as new magnetic materials while they do not display ferromagnetism. The persistent and high-spin organic polyradical molecule has the advantages of a nanometer-size, designable molecular shape, and a tunable spin concentration with a radical generation step. This multi-valued expression of spin information upon the organic polyradical molecule is in contrast to the magnetic performance of conventional inorganic materials. Some of the high-spin organic polyradicals are reversibly oxidized and reduced electrochemically, which has been recently successfully employed as the electrode active material of rechargeable batteries [54]. The high-spin organic polyradical molecules with chemical stability, flexibility, plasticity, and moldability will open a new field of carbon-based magnetic materials.

References

1. (a) Magnetic Properties of Organic Materials, ed. by P.M. Lahti, New York: Marcel Dekker, 1999; (b) Molecular Magnetism: New Magnetic Materials, ed. by K. Itoh and M. Kinoshita, Tokyo and Amsterdam: Kohdansha and Gordon & Breach, 2000; (c) π-Electron Magnetism From Molecules to Magnetic Materials, ed. by J. Veciana Berlin: Springer, 2001.
2. K. Itoh, Pure Appl. Chem. 50 (1978) 1251.
3. H. Iwamura and, N. Koga, Acc. Chem. Res. 26 (1993) 346.
4. A. Rajca, Chem. Rev. 94 (1994) 871.
5. (a) A. Rajca and J. Wongsriratanakul, Science 294 (2001) 1503; (b), A. Rajca, S. Rajca, and J. Wongsriratanakul, J. Am. Chem. Soc. 121 (1999) 6308.
6. (a) W.T. Borden and, E.R. Davidson, J. Am. Chem. Soc. 99 (1977) 4587; (b) W.T. Borden, in Magnetic Properties of Organic Materials, ed. by P.M. Lahti, New York: Marcel Dekker, 1999, pp. 61-102.
7. N. Yoshioka, P. M. Lahti, T. Kaneko, Y. Kuzumaki, E. Tsuchida, and H. Nishide, J. Org. Chem. 59 (1994) 4272ю
8. H. Nishide, Y. Hozumi, T. Nii, and E. Tsuchida, Macromolecules 30 (1997) 3986.
9. (a) R.I. Walter, J. Am. Chem. Soc. 77 (1955) 5999; (b) E.T. Seo, R.F. Nelson, J.M. Fitsch, L.S. Marcoux, D.W. Leedy, and R.N. Adams, J. Am. Chem. Soc. 88 (1966) 3498, (c) L. Hagopian, K. Günyer, and R.I. Walter, J. Phys. Chem. 71 (1967) 2290, (d) S. Sasaki and M. Iyoda, Chem. Lett. (1995) 1011.
10. Connelly N. G. and Geiger W. E. , Chem. Rev. 96 (1996) 877ю
11. H. Nishide, Adv. Mater. 7 (1995) 937.
12. (a) H. Nishide, T. Kaneko, T. Nii, K. Katoh, E. Tsuchida, and K. Yamaguchi, J. Am. Chem. Soc. 117 (1995) 548, (b) H. Nishide, T. Kaneko, T. Nii, K. Katoh, E. Tsuchida, and P. M. Lahti, J. Am. Chem. Soc. 118 (1996) 9695.
13. (a) T. Kaneko, S. Toriu, Y. Kuzumaki, H. Nishide, and E. Tsuchida, Chem. Lett. (1994) 2135; (b) H. Nishide, T. Kaneko, S. Toriu, Y. Kuzumaki, E. Tsuchida, Bull. Chem. Soc. Jpn. 69 (1996) 499.
14. (a) T. Michinobu, M. Takahashi, and E. Tsuchida, H. Nishide, Chem. Mater. 11 (1999) 1969; (b) T. Michinobu, E. Tsuchida, and H. Nishide, Bull. Chem. Soc. Jpn. 73 (2000) 1021.
15. (a) W. Schmidt and E. Steckhan, Chem. Ber. 113 (1980) 577; (b) E. Steckhan Top. Curr. Chem. 142 (1987) 1.
16. (a) F.A. Neugebauer, S. Bamberger, and W.R. Groh, Chem. Ber. 108 (1975) 2406.
17. M. Takahashi, T. Nakazawa, E. Tsuchida, and H. Nishide Macromolecules 32 (1999) 6383.
18. H.Murata, M. Takahashi, K. Namba, N. Takahashi, and H. Nishide, J. Org. Chem. 69 (2004) 631.

19. R.F. Heck, Org. React. 27 (1982) 345; (b) A. Greiner and W. Heitz, Makromol. Chem. Rapid Commun. 9 (1988) 581.
20. (a) C.A. Merlic and M.F. Semmelhack, J. Organomet. Chem. 391 (1990) C23; (b) A.S. Carlstroem and T. Frejd, Acta Chem. Scand. 46 (1992) 163.
21. M. Ludwig, S. Strömberg, M. Svensson, and B. Åkermark, Organometallics 18 (1999) 970.
22. J. K. Kochi, Acc. Chem. Res. 25 39 (1992).
23. (a) D.F. Evans, J. Chem. Soc. (1959) 2003; (b) D.H. Live and S.I. Chan, Anal. Chem. 42 (1970) 791.
24. (a) S. M. Grayson and J. M. J. Fréchet, Chem. Rev. 101 (2001) 3819; (b) K. Yamamoto, M. Higuchi, S. Shiki, M. Tsuruta, and H. Chiba, Nature 415 (2002) 509; (c) M. Jikei and M. Kakimoto, J. Polym. Sci. Part A: Polym. Chem. 42 (2004) 1293.
25. (a) R. Sadamoto, N. Tomioka, and T. Aida, J. Am. Chem. Soc. 118 (1996) 3978; (b) K. W. Pollak, J. W. Leon, J. M. Fréchet, M. Maskus, snd, H. D. Abruña, Chem. Mater. 10 (1998) 30; (c) F. Vögtle, M. Plevoets, M. Nieger, G.C. Azzellini, A. Credi, L. De Cola, V. De Marchis, M. Venturi, and V. Balzani, J. Am. Chem. Soc. 121 (1999) 6290; (d) D.L. Stone, D.K. Smith, and P.T. McGrail, J. Am. Chem. Soc. 124 (2002) 856.
26. H. Murata, Y. Yonekuta, and H. Nishide, Org. Lett. 6 (2004) 4889.
27. C. J. Hawker and J. M. J. Fréchet, J. Am. Chem. Soc. 112 (1990) 7638.
28. (a) J. Veciana, C. Rovira, N. Ventosa, M. I. Crespo, and F. Palacio, J. Am. Chem. Soc. 115 (1993) 57; (b) N. Fukita, M. Ohba, H. Okawa, K. Matsuda, and H. Iwamura, Inorg. Chem. 37 (1998) 842.
29. H. Nishide, T. Kawasaki, R. Morikawa, and E. Tsuchida, Polym. Preprints, Jpn. 45 (1996) 451.
30. (a) M. Miyasaka, T. Yamazaki, E. Tsuchida, and H. Nishide, Macromolecules 33 (2000) 8211; (b) M. Miyasaka, T. Yamazaki, E. Tsuchida, and H. Nishide, Polyhedron 20 1157 (2001).
31. (a) H. Nishide, N. Yoshioka, M. Igarashi, and E. Tsuchida, Macromolecules 21 (1988) 3319; (b) H. Nishide, N. Yoshioka, K. Inagaki, T. Kaku, and E. Tsuchida, Macromolecules 25 (1992) 569; (c) H. Nishide, T. Kaneko, N. Yoshioka, H. Akiyama, M. Igarashi, and E. Tsuchida, Macromolecules 26 (1993) 4567; (d) N. Yoshioka, H. Nishide, T. Kaneko, H. Yoshiki, and E. Tsuchida, Macromolecules 25 (1992) 3838; (e) Y. Miyra, M. Matsumoto, Y. Ushitani, Y. Teki, T. Takui, and K. Itoh, Macromolecules 26 (1993) 6673.
32. H. Murata, R. Takada, D. Miyajima, and H. Nishide, to be submitted.
33. (a) R. Nomura, S.M. Abdul Karim, H. Kajii, R. Hidayat, K. Yoshino, and T. Masuda, Macromolecules 33 (2000) 4313; (b) J. Chen, J.W.Y. Lam, C.C.W. Law, and B.Z. Tang, Macromolecules 36 (2003) 1108.
34. (a) K. Yoshizawa, A. Chano, A. Ito, K. Tanaka, T. Yamabe, H. Fujita, and J. Yamauchi, Chem. Lett. (1992) 369; (b) K. Yoshizawa, A. Chano, A. Ito, K. Tanaka, T. Yamabe, H. Fujita, J. Yamauchi, and M. Shiro, J. Am. Chem. Soc. 114 (1992) 5994; (c) K.R. Stickley and S.C. Blackstock, J. Am. Chem. Soc. 116 (1994) 11576; (d) K.R. Stickley and S.C. Blackstock, Tetrahedron Lett. 36 (1995) 1585; (e) M. Yano, M. Furuichi, K. Sato, D. Shiomi, A. Ichimura, K. Abe, T. Takui, and K. Itoh, Synth. Met. 85 (1997) 1665.
35. (a) K. Sato, M. Yano, M. Furuichi, D. Shiomi, T. Takui, K. Abe, K. Itoh, A. Highchi, K. Katsuma, and Y. Shirota, J. Am. Chem. Soc. 119 (1997) 6607; (b) A. Ito, H. Ino, K. Tanaka, K. Kanemoto, and T. Kato, J. Org. Chem. 67 (2002) 491.
36. T.D. Selby and S.C. Blackstock, J. Am. Chem. Soc. 121 (1999) 7152.

37. T.D. Selby and S.C. Blackstock, Chem. Mater. 14 (2002) 1685.
38. (a) M.M. Wienk and R.A.J. Janssen, Chem. Commun. (1996) 267; (b) M.M. Wienk and R.A.J. Janssen, Synthetic Metals 85 (1997) 1725; (c) M.M. Wienk and R.A.J. Janssen, J. Am. Chem. Soc. 119 (1997) 4492 (d) K.R. Stickey, T.D. Selby, and S.C. Blackstock, J. Org. Chem. 62 (1997) 448.
39. (a) A. Ito, Y. Ono, and K. Tanaka, New J. Chem. (1998) 779; (b) A. Ito, Y. Ono, and K. Tanaka, J. Org. Chem. 64 (1999) 8236; (c) A. Ito, Y. Ono, and K. Tanaka, Angew. Chem. Int. Ed. 39 (2000) 1072.
40. (a) S.I. Hauck, K.V. Lakshmi, and J.F. Hartwig, Org. Lett. 1 (1999) 2057.
41. T.D. Selby and S.C. Blackstock, Org. Lett. 1 (1999) 2053.
42. P.J. van Merus and R.A. J. Janssen, J. Org. Chem. 65 (2000) 5712.
43. (a) J.F. Hartwig, Angew. Chem. Int. Ed. 37 (1998) 2046; (b) F.E. Goodson and J.F. Hartwig, Macromolecules 31 (1998) 1700; (c) J. Louie and J.F. Hartwig, Macromolecules 31 (1998) 6737; (d) F.E. Goodson, S.I. Hauck, and J.F. Harywig, J. Am. Chem. Soc. 121 (1999) 7527.
44. (a) R.J. Bushby, D.R. McGill, K.M. Ng, and N. Taylor, J. Mater. Chem. 7 (1997) 2343; (b) R.J. Bushby and D.J. Gooding, Chem. Soc., Perkin Trans. 2 (1998) 1069; (c) R.J. Bushby, D. Gooding, and M.E. Vale, Phil. Trans. R. Soc. Lond. A 357 (1999) 2939.
45. J. Louie and J.F. Hartwig, J. Am. Chem. Soc. 119 (1997) 11695.
46. T.D. Selby and S.C. Blackstock, J. Am. Chem. Soc. 120 (1998) 12155.
47. H. Nishide, M. Miyasaka, and E. Tsuchida, Angew. Chem. Int. Ed. 37 (1998) 2400.
48. H. Nishide, T. Ozawa, M. Miyasaka, and E. Tsuchida, J. Am. Chem. Soc. 123 (2001) 5942.
49. (a) H. Nishide, M. Miyasaka, and E. Tsuchida, J. Org. Chem. 63 (1998) 7399; (b) M. Miyasaka, Y. Saito, and H. Nishide, Adv. Funct. Mater. 13 (2003) 113.
50. E. Fukuzaki and H. Nishide, J. Am. Chem. Soc. , to be submitted.
51. L. Eberson and B. Larsson, Acta Chem. Scand. B41 (1987) 367.
52. N. J. Dinard, Nanoscale Characterization of surface and Interfaces, Weinheim: VCH, 1994; (b) D. Sarid, Scanning Force Microscopy, New York: Oxford University Press, 1994.
53. T. Michinobu, J. Inui, and H. Nishide, Org. Lett. 5 (2003) 2165.
54. (a) H. Nishide, S. Iwasa, Y.-J. Pu, T. Suga, K. Nakahara, and M. Satoh, Electrochem. Acta 50 (2004) 827.

Carbon-Based Magnetism
T. Makarova and F. Palacio (Editors)

4

Verdazyl-based Magnetic Systems

K. Mukai

Department of Chemistry, Faculty of Science, Ehime University, Matsuyama 790-8555, Japan

1 Introduction

The study of molecular magnetism of organic radical crystal has attracted much attention, since the discovery of bulk ferromagnetism in the β-phase of *p*-nitrophenyl nitronyl nitroxide (*p*-NPNN) (T_C = 0.60K) in 1991 [1–3]. Magnetic properties of organic radical crystals have been studied extensively after that. However, the examples are almost limited to nitroxide and nitronyl nitroxide radicals [4, 5], because of the high stability and the easiness of the synthesis of these radicals.

The verdazyl radicals are well known as one of the representative stable free radicals which can be isolated as solvent-free pure radicals in crystalline state. 1,3,5-Triphenylverdazyl (TPV, Fig. 1) and its derivatives have been prepared by Kuhn et al. in 1964 [6]. The magnetic properties of these verdazyl radical crystals were studied many years before (1970–1982) [7–9]. However, notable magnetic properties such as ferromagnetism, antiferromagnetism etc. have not been found for these verdazyl radicals at that time.

Neugebauer et al. have reported the preparation of two kinds of new types of verdazyl radicals, that is, (i) 3-phenyl-1,5-dimethyl-6-oxo- and –thioxo-verdazyl (PDOV and PDTV) radicals in 1988 [10] and (ii) 1,3,5-triphenyl-6-oxo- and –thioxo-verdazyl (TOV and TTV) radicals in 1993 [11] and their derivatives, in which methylene group (>CH_2) of TPV is replaced by 6-oxo and –thioxo groups (>C=O and >S=O) (see Fig. 1). The stabilities of the former (PDOV and PDTV) radicals are lower than that of TPV, and the preparation of single crystals is not easy in usual case.

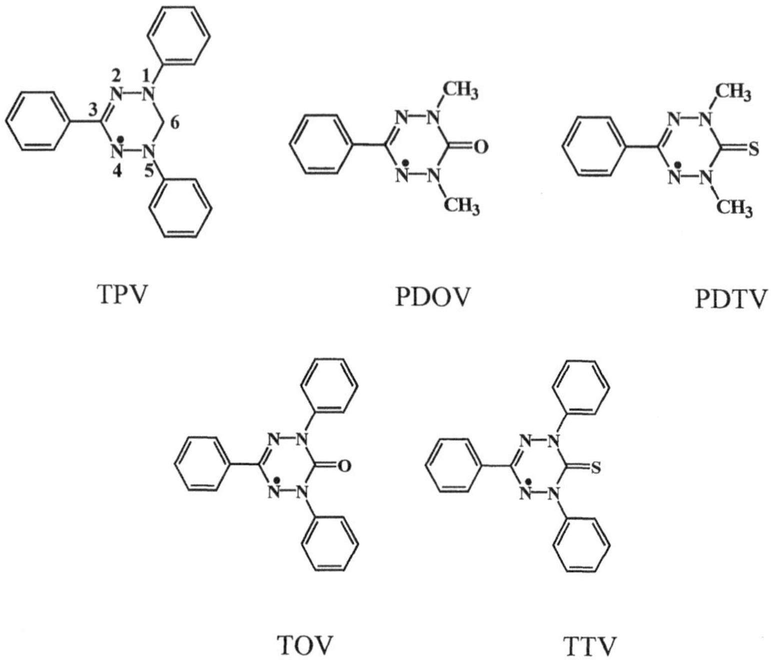

Fig. 1. Molecular structures of five different types of verdazyl radicals.

However, we can isolate those as pure radicals in crystalline state. The latter radicals (TOV and TTV) are very stable, and we can purify it by recrystallization under the air. We can expect strong intermolecular exchange interaction ($2J/k_B$) for these verdazyl radical crystals, because of their molecular planarity, relative lack of bulky substituents, and delocalization of unpaired π-electron. Further, we can prepare many derivatives because of the easiness of introduction of substituents.

The magnetic property of these verdazyl radical crystals has been studied recently, and several interesting magnetic properties such as ferromagnetism [12–14], antiferromagnetism [15], weak ferromagnetism [16–19], spin-Peierls transition [20, 21], and spin frustration [22] have been reported. The doping effects of magnetic and nonmagnetic impurities on the above magnetic properties have also been studied. Further, the molecular magnetic conductors and photo-excited high spin systems based on verdazyl radicals, and the transition metal complexes of verdazyl radicals have been studied. In this chapter, the details of the above magnetic properties obtained for verdazyl radicals will be introduced. However, the ferrimagnetism, spin ladder, and Halden gap based on verdazyl radical crystals have not been reported, as far as we know.

2 General Magnetic Properties of Verdazyl Radical Crystals

Generally, the magnetic susceptibility (χ_M) of these verdazyl radicals shows a broad maximum and the temperature dependence of χ_M was explained by one-dimensional (1D) antiferromagnetic (AFM) Heisenberg nonalternating or alternating chain model [14, 15, 20, 23–26]. The corresponding spin Hamiltonian is given by

$$H = -2J_1 \sum_{i=1}^{N/2} S_{2i} S_{2i+1} - 2J_2 \sum_{i=1}^{N/2} S_{2i-1} S_{2i} \qquad (1)$$

where J_1 and J_2 are the nearest-neighbor exchange integrals. Negative values of J_1 and J_2 corresponding to AFM coupling are appropriate to the free radicals considered here. α (= J_2/J_1) is a parameter which conveniently indicates the degree of alternation and α = 1 corresponds to the uniform limit, that is, a nonalternating AFM chain. α = 0 corresponds to an isolated dimer (spin-pair) system and in this case the susceptibility should be interpreted by the singlet-triplet equilibrium (S-T) model (Eq. (2)).

$$\chi_{S\text{-}T}(T) = (N_0 g^2 \mu_B^2 / k_B T)\, [1 / (3 + e^{-2J/k_B T})] \qquad (2)$$

For instance, the χ_M's of *p*-MDpOV, *p*-CyDpOV, *p*-MDpTV and *p*-MeDpTV radical crystals (see Fig. 2) show a broad maximum (T_{max}) [24], and can be interpreted in terms of a 1D AFM Heisenberg nonalternating or alternating chain model [25, 26]. The values of $2J/k_B$ obtained are summarized in Table 1.

On the other hand, the χ_M's of *p*-MeDpOV, *p*-NDpOV and *p*-CyDpTV follow the Curie-Weiss law with positive Weiss constant in the temperature region between 10 and 300K. A plot of $1/\chi_M$ against T is no longer linear at low temperature. The low temperature behavior of the χ_M of these radical crystals was analyzed on the basis of the

Table 1.
Magnetic properties of 3-(4-R-phenyl)-1,5-diphenyl-6-oxo- and -thioxo-verdazyl radical crystals.

Radical	Magnetism	Values of T_{max}, θ, and $2J/k_B$
p-MDpOV	1D AFM Heisenberg Linear Chain	T_{max} = 17K, $2J/k_B$ = –27.3K
p-MeDpOV	Curie-Weiss	θ = +2.5K, $2J/k_B$ = +7.0K
p-CyDpOV	1D AFM Heisenberg Linear Chain	T_{max} = 17K, $2J/k_B$ = –27.3K
p-NDpOV	Curie-Weiss	θ = +1.5K, $2J/k_B$ = +5.0K
p-MDpTV	1D AFM Heisenberg Linear Chain	T_{max} = 6.7K, $2J/k_B$ = –10.9K
p-MeDpTV	1D AFM Heisenberg Alternating Chain	T_{max} = 56K, $2J_1/k_B$ = –92.0K, $2J_2/k_B$ = –64.4K
p-CyDpTV	Curie-Weiss	θ = +2.9K, $2J/k_B$ = +7.0K
p-NDpTV	Curie-Weiss	θ = –0.9K

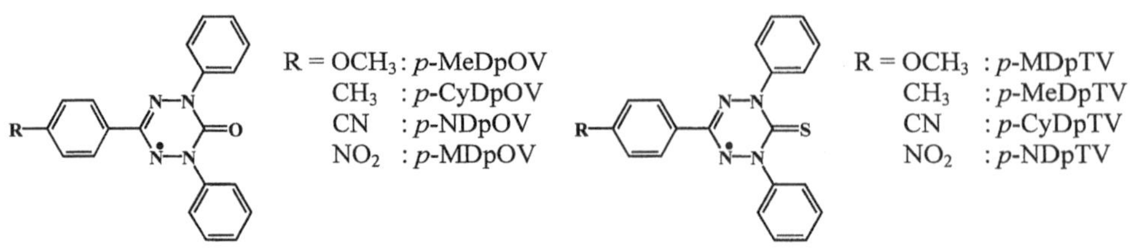

Fig. 2. Molecular structures of 3-(4-R-substituted-phenyl)-1,5-diphenyl-6-oxo- and -thioxoverdazyl radicals.

quasi-1D ferromagnetic (FM) Heisenberg model. The susceptibility (χ_{1D}) is given by

$$\chi_{1D} = \frac{Ng^2\mu_B^2}{4kT}\left\{1+\left(\frac{J}{kT}\right)^{\alpha}\right\} \tag{3}$$

with $\alpha = 1$ for $k_BT/J > 1$ [25]. For lower temperatures ($k_BT/J < 1$), however, α depends on T ($\alpha \to 4/5$ for $T \to 0$). In fact, above 4.2K the χ_M's of these verdazyls are well reproduced by Eq. (3) with positive values of $2J/k_B$ (see Table 1).

A 1D magnet with localized spins is one of the simplest systems of many-body problems from which exact or correct physical quantities have been derived. For 1D Heisenberg antiferromagnets with $S = 1/2$, the physical properties theoretically expected of these antiferromagnets have been experimentally checked with abundant real magnetic compounds. The magnetic properties of *p*-MDpOV, *p*-CyDpOV and *p*-MDpTV belong to this group, and can be explained by the 1D AFM Heisenberg linear chain model. However, for 1D ferromagnetic systems, the situation rather differs from the case of antiferromagnets, because we can rarely get the 1D quantum ferromagnetic substances. Consequently, only a few examples of a 1D Heisenberg ferromagnet have been reported. *p*-MeDpOV, *p*-NDpOV and *p*-CyDpTV radicals belong to this group. Further, surprisingly, three of eight verdazyls studied exhibit an intermolecular FM interaction, and behave as a quasi-1D ferromagnet. Ab initio MO calculations have been carried out for 5 kinds of 6-oxo-verdazyl radicals [24]. The energy level of β-NHOMO is higher than that of α-SOMO in *p*-MDpOV and *p*-MeDpOV radicals, and the energy level of β-NHOMO is close to that of α-SOMO in TOV and *p*-NDpOV (see Table 2 in ref. 24). The result indicates that in these verdazyl radicals the exchange interaction within the molecule is great enough to stabilize the α-SOMO. In other words, the spin correlation causes a large spin polarization effect in 6-oxo-verdazyl radicals. This feature closely resembles that in the FM organic radical, galvinoxyl [27, 28] and β-*p*-NPNN radicals [2, 3]. Thus, the FM interaction of the verdazyl radicals is considered to originate mainly in the β-SOMO-β-NHOMO charge transfer interaction.

Table 2.
Organic weak ferromagnets.

Radical	T_N/K	M_S/emu mol^{-1}	Magnetic properties above T_N
TPV	1.78	3	Curie-Weiss (θ = –8K), $2J/k_B$ = –10.8K
TOV	5.6	60.9	Curie-Weiss (θ = –14K)
p-CyDpTV	0.41	31.1	Curie-Weiss (θ = +2.9K), $2J/k_B$ = 7.0K, zJ'/k_B = –0.19K
$Li^+[TCNQF_4]^-$	12	50	Curie-Weiss (θ = –30K)
p-NCC_6F_4CNSSN	35.5	8.4	$2J/k_B$ = –168K
MATMP	0.15	1.6	$2J/k_B$ = 2.80K, $2J'/k_B$ = –0.18K
AOTMP	0.66	5.6	$2J/k_B$ = –5.0K

3 Ferromagnetism in Verdazyl Radical Crystals

The ac susceptibility of *p*-CDTV (Fig. 3(a)) was measured in the temperature region 1.7–300K, using ac susceptometer (see Fig. 3 in ref. 14) [12, 14]. Above 20K, the ac susceptibility of *p*-CDTV follows the Curie-Weiss law with a positive Weiss constant of +3.0K. The low-temperature heat capacity, C_P, of *p*-CDTV was measured in the external fields of 0–30 kOe [14]. The magnetic heat capacity, C_m, of *p*-CDTV exhibits a sharp λ-like peak at 0.67K in zero external field, corresponding to a three-dimensional magnetic phase transition, as shown in Fig. 3(b). In zero field and in the paramagnetic state, C_m of *p*-CDTV keeps the values comparable to C_m(max) = 0.134R = 1.12 J/(mol·K) above 2K, which is a characteristic feature of the isotropic 1D Heisenberg ferromagnet [25]. This is ascertained by a theoretical calculation of the magnetic heat capacity of a 1D FM Heisenberg chain with J/k_B = 6.0K. The application of external field gives a sensitive effect on the heat capacity; even in the small field H = 0.25 kOe, the sharp peak is crushed down to leave a vague trace of it nearly at T_C, and another broad hump of the heat capacity appears above T_C. This hump grows up toward higher temperatures as the field is increased. Such a field dependence in the heat capacity of *p*-CDTV is well explained as a characteristic feature of the 1D Heisenberg ferromagnet [25]. These results indicate that the *p*-CDTV behaves as a quasi-1D ferromagnet above the transition temperature T_C.

The low temperature ac susceptibility (χac) of the p-CDTV was also measured down to about 0.1K at ac field H (ν) < 10 Oe (ν = 164 Hz), using the Hartshorn bridge method. The χac of p-CDTV increases rapidly below 1K and reaches a maximum at T = 0.68K under zero external field (see Fig. 4 in ref. 14). The appearance of the peak of χ_{ac} may be due to the time-dependent effects of the ac method. At low temperatures, the susceptibility seems to reach a limiting value which corresponds to the inverse of demagnetization field factor for a ferromagnet. The fluctuation of the magnetic moments is suppressed by the external fields and the peak becomes smaller. Below 0.5K, the susceptibility shows a plateau which seems to be a character in the

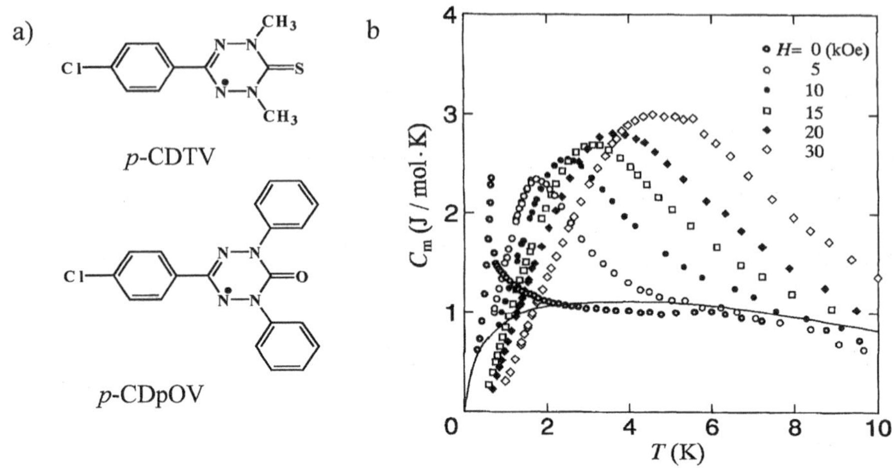

Fig. 3. (a) Molecular structures of *p*-CDTV and *p*-CDpOV. (b) Field dependence of magnetic heat capacity (C_m) of *p*-CDTV. The solid line represents the theoretical results for the isotropic Heisenberg ferromagnet with $J/k_B = 6.0$K.

ferromagnetically ordered state, as demonstratively seen in the results for the field 500 Oe. The small hump around 1 K is due to the 1D short range ordering effects in the paramagnetic region. The susceptibility for the higher fields than 100 Oe is almost suppressed as observed in the case of the FM *p*-NPNN (β-phase) [2]. The result suggests that the *p*-CDTV crystal undergoes a bulk FM transition at $T_C = 0.68$K. The values of an intra- and inter-chain exchange interactions (J and zJ') were obtained by the detailed analysis of the temperature dependence of χ_{ac}, giving $2J/k_B = +12.0$K and $zJ'/k_B = 0.21$K or $|zJ'|/J = 3.8\times10^{-2}$, where z is the number of neighboring chains. Similar measurements were performed for *p*-CDpOV (see Fig. 3(a)). *p*-CDpOV also behaves as a quasi-1D Heisenberg ferromagnet with $T_C = 0.21$K and $2J/k_B = +11.0$K and $zJ'/k_B = 0.03$K above T_c.[13] *p*-CDpOV has higher one-dimensionality than *p*-CDTV.

4 Antiferromagnetism in Verdazyl Radical Crystal

The magnetic properties of organic radical crystals have been studied extensively, generally indicating the AFM intermolecular interaction in these radical crystals. However, as far as we know, the examples of real antiferromagnets that show the AFM long-range order are very limited. Furthermore, they are nitroxides [29] and nitronyl nitroxide [2, 3], except for the case of the 1,3-bisdiphenylene-2-(4-chlorophenyl)allyl (*p*-Cl-BDPA) radical, which is known as an example of the first organic antiferromagnet [30]. For this carbon-centered radical the highest Neel temperature $T_N = 3.25$K has been observed.

The ac susceptibility (χ_{ac}) of NTV (Fig. 4(a)) follows the Curie-Weiss law with a positive Weiss constant (θ) of +3.1K, as reported by Allemand et al [31]. The low-temperature χ_{ac} shows a sharp peak at T_{max} = 1.16K and decreases toward 0K with its magnitude of about 2/3 of the peak value of χ_{ac} at 1.16K, as in ordinary powder crystal susceptibilities of antiferromagnets (see Fig. 7 in ref. 15). The result suggests that NTV undergoes an AFM transition at T_N = 1.16K [15].

The low-temperature heat capacity (C_P) of NTV radical crystals was measured in the applied fields H = 0, 480, and 1530 Oe. A distinct peak that corresponds to a three-dimensional magnetic phase transition is detected at T_N = 1.10K, and a flat plateau follows in the higher temperatures above T_N. The intrinsic magnetic heat capacity (C_m) of NTV is shown in Fig. 4(b). It is noted that the plateau of the heat capacity is reproduced by the theory for the 1D Heisenberg ferromagnet [25] with $2J/k_B$ = 8.0K as in the case of the magnetic susceptibility ($2J/k_B$ = 7.0K). The sharp heat capacity peak is stable even in external fields up to 1.53 kOe, as seen in the inset of the Fig. 4(b). This field dependence of the peak makes a good contrast to the sensitive field dependence in the *p*-NPNN (β-phase) [2, 3], *p*-CDTV [12, 14] and *p*-CdpOV [13] where the peak becomes a round maximum even in a field of 1.0 kOe. Rather, the behavior of NTV in the fields resembles the case of the AFM γ phase of *p*-NPNN [2, 3]. In zero field and in the paramagnetic state, C_m of NTV keeps the values comparable to C_m(max) = 0.134R = 1.12 J/(mol·K) above 3K, which is a characteristic feature of the isotropic 1D Heisenberg ferromagnet [25]. Further evidence for the 1D interaction in

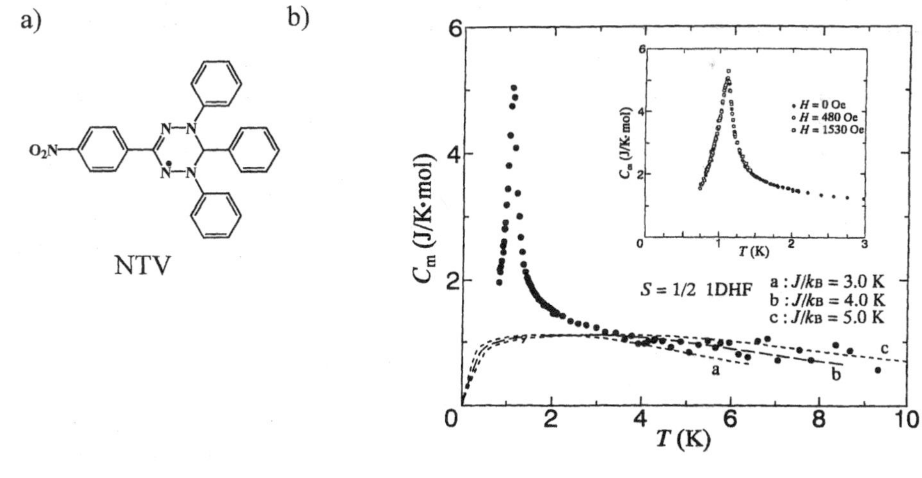

Fig. 4. (a) Molecular structure of NTV. (b) Temperature dependence of the magnetic heat capacity (Cm) of NTV (closed circles). Theoretical results for the isotropic 1D Heisenberg ferromagnet (1DHF) are drawn for 2J/kB = +6.0, 8.0, and 10.0K. The inset shows the field dependence of the magnetic heat capacity (Cm).

NTV is provided by its crystal structure determination. The result of crystal structure analysis of NTV indicates that the NTV radical molecules are stacked along the *a* axis, building up a stairlike chain structure. As a conclusion, NTV is an antiferromagnet with a Neel temperature of $T_N = 1.16$K and behaves as a quasi-1D Heisenberg ferromagnet with the intrachain exchange interaction of $2J/k_B = +7.0$K above the transition temperature T_N [15].

5 Weak Ferromagnetism in Verdazyl Radical Crystals

Weak ferromagnetism due to spin canting has been detected in several organic radical crystals, such as nitroxide (AOTMP and MATMP) [32], dithiadiazolyl radical, p-NCC_6F_4CNSSN [33, 34], and $Li^+[TCNQF_4]^-$ radical salt [35], as listed in Table 2. At the high temperature region, the susceptibility of these organic weak ferromagnets, except for MATMP, follows a Curie-Weiss law with a negative Weiss constant. Above T_N, these radicals behave as a quasi-1D antiferromagnet. On the other hand, MATMP shows a positive Weiss constant, and behaves as a 1D Heisenberg ferromagnet above $T_N = 0.15$K. However, the result of the detailed study of magnetic property of MATMP performed by using a single crystal indicated that the weak ferromagnetism of MATMP cannot be explained by the Dzyaloshinsky-Moriya (D-M) type interaction [36]. Palacio et al. [34] reported that the β-phase of the dithiadiazolyl radical, p-NCC_6F_4CNSSN, orders as a weak ferromagnet at 35.5K, a temperature that is at least one order of magnitude higher than that of any other purely organic free-radical compound. However, the origin of the anisotropy in these organic weak ferromagnets has not been well understood.

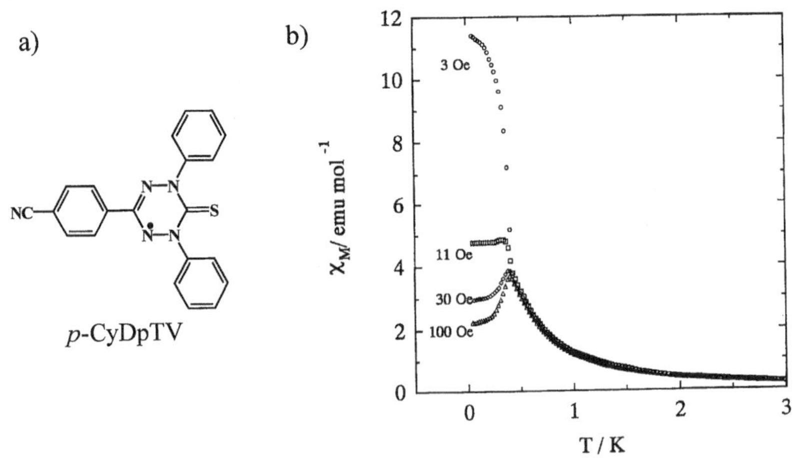

Fig. 5. a) Molecular structure of *p*-CyDpTV; b) Temperature dependence of the magnetic susceptibility (χ_M) of *p*-CyDpTV measured at the applied magnetic fields of 3, 11, 30 and 100 Oe.

Weak ferromagnetism has also been reported for three kinds of verdazyl radicals (TPV, TOV, and *p*-CyDpTV) (see Figs. 1 and 5) [16–19]. TPV is a first example of weak ferromagnetism observed for organic radical crystal [16]. High-quality sizable single crystals of TPV have been prepared and their magnetic properties investigated by neutron-diffraction and magnetization measurements. Neutron-diffraction peaks corresponding to long-range antiferromagnetic ordering have been observed below the Neel temperature of 1.78K. Kremer et al. [17] found that TOV is a weak ferromagnet below T_N = 4.9K with spin-canting angle (φ) of 0.26°. The anomalous temperature dependence of the weak ferromagnetism at low temperature is qualitatively explained by a four-sublattice model with two kinds of Dzyaloshinsky-Moriya (D-M) vectors of opposite sign on the ac-plane [18]. Further, detailed study of magnetic property of TOV has been performed by using single crystal of TOV [37].

Figure 5(b) shows the temperature dependence of χ_M of *p*-CyDpTV at the external fields 3, 11, 30 and 100 Oe [19]. By decreasing the temperature from 300K, the χ_M of *p*-CyDpTV increases gradually until about 0.5K being independent of external fields. The χ_M of *p*-CyDpTV shows a maximum at T = 0.40K and 0.38K under the external field of 100 and 30 Oe, respectively. A sharp increase of χ_M was observed near 0.41K in the case of the external field of 3 Oe, corresponding to the onset of weak ferromagnetism associated with three-dimensional AFM ordering, as described below.

Figure 6 shows the magnetization curves at 0.05, 0.3, 0.4 and 1.5K. At 1.5K the remanence was not observed and the magnetization was almost proportional to the magnetic field up to 100 Oe. At 0.4, 0.3 and 0.05K the extrapolation of the linear part of the magnetization to zero field shows a finite magnetization. The low-field magnetization curve is approximately expressed by the sum of the FM moment (M_S) and the linear contribution ($\chi_M H$), i.e., $M(H) = M_S + \chi_M H$. From this, the magnitude of

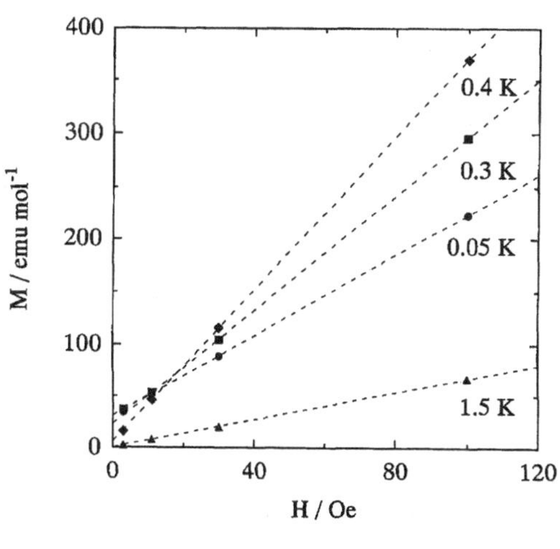

Fig. 6. Magnetization (M) as a function of magnetic field (H) in the range of 0~100 Oe at 0.05, 0.3, 0.4 and 1.5K in *p*-CyDpTV.

the spontaneous magnetization (M_S) was estimated to be 31.1, 22.9, and 6.4 emu/mol at 0.05, 0.3, and 0.4K, respectively. These values correspond to the order of about 10^{-3} μ_B/molecule, which is a reasonable value for the usual organic [16, 17, 32–35] and inorganic [36] weak ferromagnets. These results clearly indicate that the *p*-CyDpTV is a weak ferromagnet.

Similarly, the value of spontaneous magnetization (M_S) was calculated for each temperature, and plotted against temperature (see Fig. 4 in ref. 19). This is a first observation of temperature dependence of M_S for weak ferromagnet, as far as we know. As shown in Fig. 4 in ref. 19, spontaneous magnetization (M_S) increases rapidly at 0.41K, followed by gradual increase down to 0.05K. The result suggests that the magnetic ordering occurs in the present system below $T_N = 0.41$K. This increase of M_S is due to the appearance of canted moments. From the value of spontaneous magnetization at 0.05K, we can estimate the canting angle $\varphi = 0.32°$ between the neighboring sublattice moments. An antisymmetric spin coupling of D-M type may contribute to the spin canting in *p*-CyDpTV, because the fine structure energy (or the single ion anisotropy energy in inorganic weak ferromagnets) is absent for $S = 1/2$ spin systems like organic mono radicals. The temperature dependence of $M_S(T)$ is very similar to that observed for usual inorganic ferromagnets, such as Ni. In fact, at temperatures near T_N, $M_S(T) \propto [1 - (T/T_N)]^{\beta}$ with the $T_N = 0.40$K and critical index $\beta = 0.36$, in agreement with a value of 0.36 expected for an isotropic 3D Heisenberg spin system [38].

6 Spin Peierls Transition in Verdazyl Radical Crystals

6.1 Spin Peierls Transition in Verdazyl Radical Crystal (p-CyDOV)

The spin-Peierls (SP) transition is well known as one of the most interesting magnetic behaviors observed for organic radical salts, such as TTF-CuBDT and MEM-$(TCNQ)_2$ [39, 40]. The SP transition occurs when a system of uniform AFM Heisenberg linear chains undergoes a transformation to a system of dimerized or alternating AFM linear chains, due to spin-phonon coupling between the 1D spin and the 3D phonon systems. Below the transition temperature (T_{SP}), the ground state is spin singlet (nonmagnetic), and a finite energy gap opens in the excitation spectrum. The most striking feature in SP transition is that the magnetic susceptibilities exponentially drop to zero below T_{SP}. The SP transition was also observed for the inorganic compound $CuGeO_3$ ($T_{SP} = 14.2$K) [41].

The doping effects of magnetic (Mn^{2+} ($S = 5/2$), Ni^{2+} ($S = 1$) and Co^{2+} ($S = 3/2$)), and nonmagnetic (Zn^{2+} ($S = 0$) and Mg^{2+} ($S = 0$)) impurities have been studied for the SP transition of $CuGeO_3$ (Cu^{2+} ($S = 1/2$)) [42–44]. Similar studies have been performed for the Si-doped $CuGeO_3$ ($CuGe_{1-x}Si_xO_3$) [42, 45]. The SP transition temperature ($T_{SP} = 14.2$K) decreased rapidly by increasing the concentration (x) of impurities and disappeared at around $x = 0.01 \sim 0.04$ depending on the kinds of impurities. Further, it has been found that the AFM phase transition appears in the doped spin-Peierls systems below 5K, and the AFM long-range order and SP state coexist in the doped SP systems below the critical concentration (x_C).

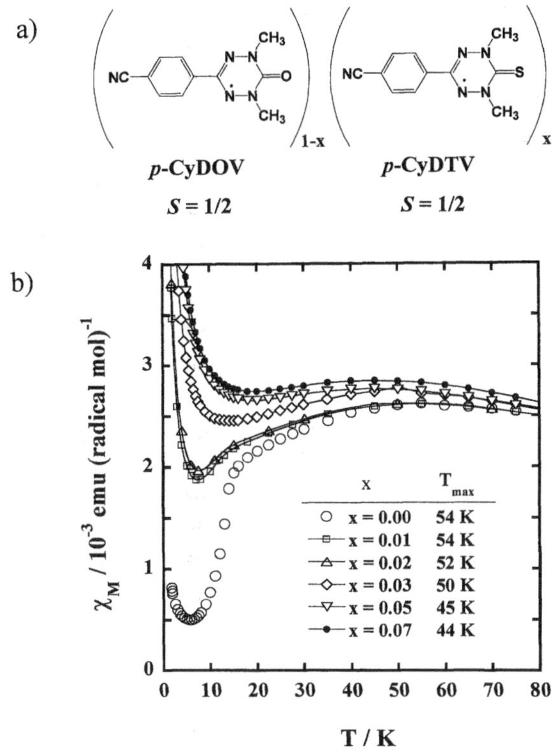

Fig. 7. (a) Molecular structures of *p*-CyDOV (S = 1/2) and *p*-CyDTV (S = 1/2) radicals. (b) Magnetic susceptibility ($\chi_M(x, T)$) of the doped *p*-CyDOV radical system, $(p\text{-CyDOV})_{1-x}(p\text{-CyDTV})_x$ (x = 0 - 0.07), in the temperature range of 1.8–80K.

The χ_M of organic neutral verdazyl radical crystal (*p*-CyDOV) (Fig. 7(a)) with S = 1/2 exhibits the characteristic properties of the SP transition [20, 21]. The χ_M shows a broad maximum at 54K, and decreases abruptly at SP transition temperature T_{SP} = 15.0K (Fig. 7(b)). The χ_M above 15.0K can be well explained by the 1D Heisenberg linear-chain model [25] with AFM exchange interaction of $2J/k_B = -84$K between neighboring spins. The magnetic energy gap $\Delta(0)$ at T = 0K and the spin-lattice coupling constant λ calculated from T_{SP} and $2J/k_B$ are 26K and 0.33, respectively [20]. High-field magnetization data (H = 0 ~ 35 T) that provide clear evidence for a SP transition were reported, and the magnetic phase diagram was determined for the *p*-CyDOV crystal [21].

6.2 Doping Effect of Paramagnetic Impurity (p-CyDTV) in Spin-Peierls System (p-CyDOV): Finding of Antiferromagnetic Order at Low Temperature

The doping effects of magnetic (*p*-CyDTV) impurities have been studied for the SP transition of *p*-CyDOV, where the >C=O group in *p*-CyDOV is replaced by the >C=S group in *p*-CyDTV [46, 47]. The Debye-Scherrer X-ray diffraction pattern of the *p*-CyDOV radical crystal ($x = 0$) is quite different from that of the *p*-CyDTV crystal ($x = 1$) (see Fig. 3 in ref. 46). However, the X-ray diffraction patterns of doped-radical crystals with $x = 0.01 \sim 0.07$ are quite similar to that of the pure *p*-CyDOV crystal ($x = 0$), indicating that the crystal structure of doped-radical crystals ($x = 0.01 \sim 0.07$) is almost the same as that of the pure *p*-CyDOV ($x = 0$) at room temperature [46]. The χ_M of the dopant *p*-CyDTV radical crystal follows a Curie-Weiss law ($\theta = -3.8$K) [46]. The χ_M of each doped crystal ($x = 0 \sim 0.07$) shows a broad maximum at the temperature of T_{max}, as shown in Fig. 7(b). The value of T_{max} decreases as the concentration (x) of *p*-CyDTV increases. The SP transition indicated by a drop of χ_M near 13 ~ 15K was observed in the samples with $x < 0.02$. The values of T_{max} and T_{SP} are listed in Table 3.

The temperature dependence of $\chi_M(x, T)$'s above T_{SP} can be simply described by the sum of the two contributions; one is from the 1D-Heisenberg AFM linear-chain system ($\chi_{1D\text{-}HAF}(T)$) with corrected T_{max}^{Corr} values and the other is a paramagnetic contribution ($\chi_{Curie}(T)$).

$$\chi_M(x, T) = C_{1D\text{-}HAF}(x)\chi_{1D\text{-}HAF}(T) + C_{Curie}(x)\chi_{Curie}(T) \quad (T > T_{SP}) \tag{4}$$

where $C_{1D\text{-}HAF}(x) + C_{Curie}(x) = 1$, and $\chi_{Curie}(T)$ is given by $\chi_{Curie}(T) = N_0 g^2 \mu_B^2 S(S+1)/3k_B T$ with $S = 1/2$. The T_{max}^{Corr}, $2J/k_B$, $C_{1D\text{-}HAF}(x)$, and $C_{Curie}(x)$ values obtained are listed in Table 3.

On the other hand, the $\chi_M(x, T)$'s of the doped radical crystals with $x = 0.01 \sim 0.07$ below T_{SP} were described by three terms [43, 46].

Table 3.
Magnetic properties of the doped-radical crystals, $(p\text{-CyDOV})_{1-x}(p\text{-CyDTV})_x$ ($x = 0$–0.07).

x	T_{max} (K)	T_{max}^{Corr} (K)	$2J/k_B$ (K)	T_{SP} (K)	T_{SP}' (K)	T_N (K)	C_{Curie}	C_{SP} ($T<T_{SP}$)	$C_{1D\text{-}HAF}$ ($T<T_{SP}$)	$C_{1D\text{-}HAF}$ ($T>T_{SP}$)
0	54	54	−84.2	15.0	5.6	0.135	0.004	0.996	0	0.996
0.01	54	55	−85.8	13.5	5.3	0.290	0.011	0.448	0.541	0.989
0.02	52	55	−85.8	13.0			0.012	0.406	0.582	0.988
0.03	50	53	−82.7				0.019		0.981	0.981
0.05	45	52	−81.1				0.024		0.976	0.976
0.07	44	52	−81.1 (−72.0) [a]	–	–	0.164	0.025		0.975	0.975

a) The value determined by the heat capacity measurement.

$$\chi_M(x, T) = C_{SP}(x)\chi_{SP}(T) + C_{1D\text{-}HAF}(x)\chi_{1D\text{-}HAF}(T) + C_{Curie}(x)\chi_{Curie}(T) \quad (T < T_{SP}) \quad (5)$$

where $C_{SP}(x) + C_{1D\text{-}HAF}(x) + C_{Curie}(x) = 1$, and $\chi_{Curie}(T)$ in Eq. (5) is the same as that in Eq. (4). The second term represents the contribution from 1D-Heisenberg AFM linear-chain (1D-HAF) subsystem. The susceptibility for the SP state ($\chi_{SP}(T)$) was calculated from the $\chi_M(x, T)$ for the pure *p*-CyDOV ($x = 0$) below T_{SP} by using Eq. (5), by assuming that the fraction $C_{1D\text{-}HAF}(0) = 0$ for $x = 0$. The $C_{SP}(x)$, $C_{1D\text{-}HAF}(x)$, and C_{Curie} values obtained are listed in Table 3.

Recently, the measurements of the field dependence of magnetization ($H = 0 \sim 55$ T) were performed for the doped verdazyl radical crystals, $(p\text{-CyDOV})_{1-x}(p\text{-CyDTV})_x$ ($x = 0$, 0.01, 0.02, 0.03, 0.05, and 0.07) [47]. The high-field magnetization $M(x, H)$ curves of the doped organic SP system below $T_{SP} = 15.0$K can be well explained by the sum of the magnetization (M) of the spin-Peierls state ($C_{SP}(x)M_{SP}(H)$), 1D Heisenberg antiferromagnet ($C_{1D\text{-}HAF}(x)M_{1D\text{-}HAF}(H)$), and Curie impurity ($C_{Curie}(x)M_{Curie}(H)$),

$$M(x, H) = C_{SP}(x)\, M_{SP}(H) + C_{1D\text{-}HAF}(x)\, M_{1D\text{-}HAF}(H) + C_{Curie}(x)\, M_{Curie}(H) \quad (6)$$

where $C_i(x)$ is the fraction of the *i* th contribution. The result suggests the coexistence of SP state and AFM long-range order in the doped *p*-CyDOV radical system at low temperature.

Heat capacity measurements ($T = 0.10 \sim 32$K) were performed for the doped verdazyl radical crystals, $(\text{p-CyDOV})_{1-x}(\text{p-CyDTV})_x$ ($x = 0$, 0.01, and 0.07) [48]. The AFM transitions (TN = 0.135, 0.290, and 0.164K) were observed for the doped crystals

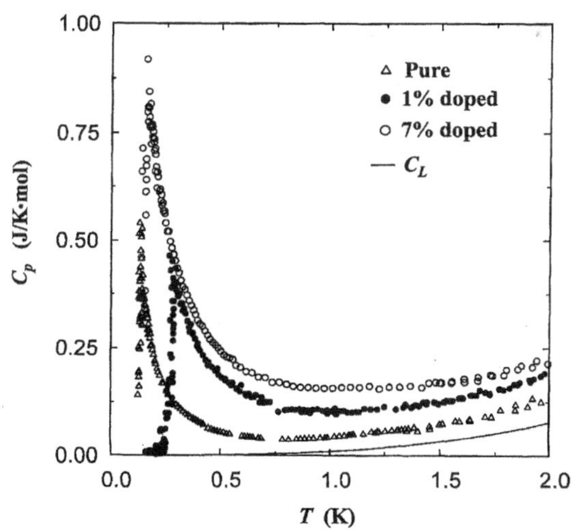

Fig. 8. Heat capacity (C_P) of the doped *p*-CyDOV radical system, $(p\text{-CyDOV})_{1-x}(p\text{-CyDTV})_x$ ($x = 0$, 0.01 and 0.07), in the temperature range of 0.10–2.0K. The solid line is the estimated lattice contribution C_L.

(x = 0, 0.01, and 0.07), respectively, indicating i) the coexistence of dimerized antiferromagnetic order (D-AF) and spin-Peierls state in the doped SP system with $x = 0$ and 0.01, and ii) the appearance of uniform antiferromagnetic order (U-AF) in the doped system with $x = 0.07$ at low temperatures (see Fig. 8). The doping effects for the organic SP system (*p*-CyDOV), which include the temperature (T: T_{SP} and T_N) - impurity (x) phase diagram (see Fig. 4 in Ref. 48), are similar to those for the inorganic SP system ($CuGeO_3$) [44, 45], although the one-dimensionality, i.e. $|zJ'/J| \cong 6\times10^{-4}$, in the organic *p*-CyDOV system is much higher than that ($\sim 10^{-1}$) in the inorganic $CuGeO_3$ system, where zJ' and J are inter- and intra-chain exchange interactions. This work is the first report of the heat capacity study of the doping effect on the organic SP transition system.

7 Spin Frustration in Verdazyl Radical Crystals

Spin frustration is one of the interesting magnetic behavior. However, the examples of the frustrated systems which have been experimentally studied so far are almost limited to inorganic compounds. The examples of real organic spin frustration compounds are very limited [49–52]. Awaga et al. [49] studied the structural and magnetic properties of the $(m\text{-MPYNN})^+(ClO_4^-)_x(I^-)_{1-x}\cdot 1/3(\text{acetone})$ $(0 \le x \le 1)$ (*m*-MPYNN = *m*-N-methylpyridinium α-nitronyl nitroxide) system, and found that this system can be characterized as a spin-1 *kagome* antiferromagnet. The example of the spin frustration compound was also reported for the BDPA·benzene radical crystal (BDPA = 1,3-bisdiphenylene-2-phenyl-allyl), showing the possibility of the isosceles triangle type spin frustration in the system [50]. Further, the example of an isosceles triangular three spin system with negative J's that would show spin frustration is reported for the organic nitroxide triradical [51]. A new spin-frustrated system consisting of $[Mn(hfac)_2]$ (manganese (II) bis(hexafluoroacetylacetone)) and *bnn* (bis(nitronyl nitroxide)) has been reported by Tanaka et al [52]. This heterospin system displays butterfly-type spin frustration between two organic radical molecules ($S = 1/2$) and two transition metal ions ($S = 5/2$), as shown in Fig. 10(c). In this four spin system the large AFM interaction (J_{13}) was made ineffective by the butterfly spin configuration. Although J_{13} is more negative than J, the latter wing-body AFM interaction (J) is dominated, and the two $S = 1/2$ spins of the *bnn* moiety are obliged to align in parallel. The butterfly configuration and two $S = 5/2$ spins of the manganese ions at the wing sites therefore play an important role indicating this spin configuration.

Anomalous magnetic properties have been observed for three kinds of 3-(aryl-substituted)-1,5-diphenyl-verdazyl radical crystals (*p*-FPDV, *p*-PyDV, and *m*-PyDV) (see Fig. 9(a)) [22]. The magnetic susceptibilities of these radicals show a broad maximum at T_{max} = 200, 272, and 157K, respectively. The values of T_{max} observed are one or two orders of magnitude larger than those for usual organic neutral radical crystals, suggesting strong intermolecular AFM interactions (J) in the crystals. The temperature dependence of the susceptibilities cannot be explained by a simple model, such as 1D Heisenberg AFM linear-chain and S-T one (see Fig. 9(b)). The χ_M's of the *p*-FPDV, *p*-PyDV, and *m*-PyDV are about 42, 76, and 35% larger than the corresponding those calculated by a 1D Heisenberg AFM linear-chain model at the temperature region (30–300K), respectively. The results of the X-ray structure analysis and unpaired spin density calculation of the *p*-FPDV radical indicate that *p*-FPDV radical molecules form a 2D magnetic network which includes a strong AFM intra-chain interaction (J_{I-I}) and two kinds of FM inter-chain interactions (J_{I-II} and J_{I-III}), forming butterfly-type spin frustration system (Fig. 10(a), (b)). As shown in Fig. 10(b), the exchange pathway observed for the *p*-FPDV system seems to be similar to that for the [Mn(hfac)$_2$](*bnn*) system, although all the spins are S =1/2, wing-body interactions (J_{I-II} and J_{I-III}) are ferromagnetic, and butterfly-type frustration spreads out the whole 2D network sheet in the former. If the FM inter-chain interactions (J_{I-II} and J_{I-III}) are not weak and comparable to AFM intra-chain interaction (J_{I-I}), all the spins in the butterfly spin system will prefer to take a parallel conformation. As a result, the magnetic moments are enhanced and the susceptibility will increase by about 42% at all the temperature region. The ratio ($\chi_M^{Obsd}/\chi_M^{1D}$) will depend on both the ratios ($|J_{I-II}|$ / $|J_{I-I}|$ and $|J_{I-III}|$ / $|J_{I-I}|$) in the crystals. *p*- and *m*-PyDV showed magnetic properties similar to that of *p*-FPDV. The unpaired spin distributions of *p*- and *m*-PyDV radical molecules are similar to that of the *p*-FPDV, as the results of McLachlan MO calculation indicate. Consequently, we can expect similar magnetic structure, that is, the existence of spin frustration in these radical crystals.

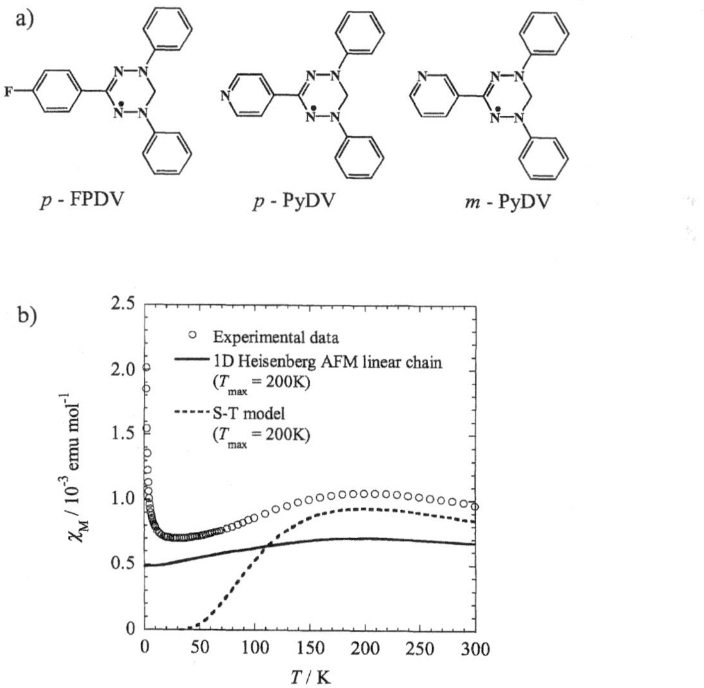

Fig. 9. (a) Molecular structure of the *p*-FPDV, *p*-PyDV, and *m*-PyDV radicals. (b) Temperature dependence of the χ_M (○) of the *p*-FPDV radical crystal.

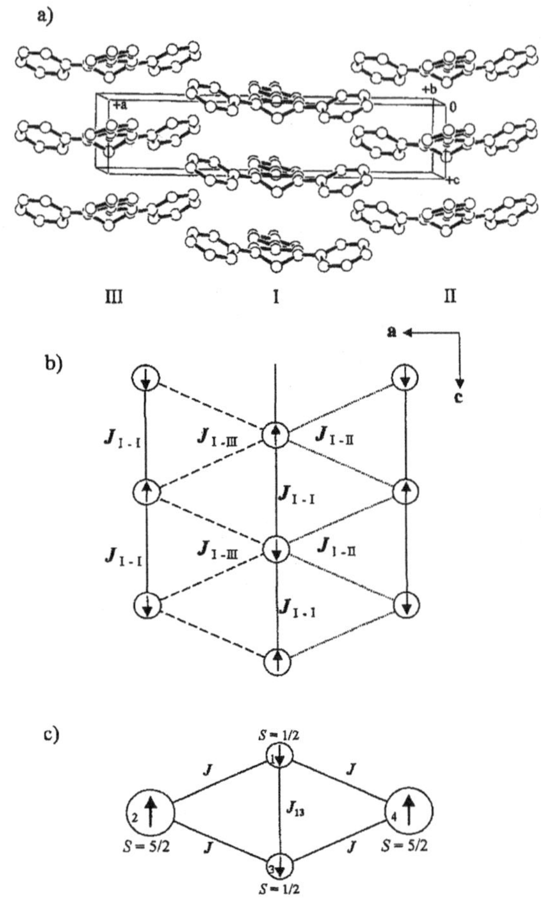

Fig. 10. (a) Columnar packing of the *p*-FPDV radical molecules showing the formation of 1D linear-chain along the *c*-axis. (b) The exchange pathways including spin frustration interaction and spin alignment for the ground state of *p*-FPDV. (c) Spin alignment for the ground state of $[Mn(hfac)_2]_2(bnn)]$ complex having butterfly-type spin frustration interaction.

8 Magnetic Property of Verdazyl Radical Alloy

The magnetic susceptibilities of 12 kinds of verdazyl radical alloy, $(p\text{-CDpOV})_{1-x}$ $(p\text{-BDpOV})_x$ ($x = 0 \sim 1$) (see Fig. 11(a)), were measured to study the doping effects of magnetic impurities on the quasi-1D Heisenberg ferromagnet (*p*-CDpOV) and antiferromagnet (*p*-BDpOV) with S = 1/2, which have similar van der Waals radius to each other [53]. Figure 11(b) shows a plot of $\chi_M T$ versus T for the radical alloys, where χ_M is the magnetic susceptibility per 1-mol of *p*-CDpOV and *p*-BDpOV radicals as a total. The χ_M's of the radical alloy show only a weak x dependence at the low concentration region of *p*-BDpOV ($0 \leq x \leq 0.50$). However, we can see that the χ_M's above $x = 0.50$ drastically change with doping.

The χ_M of the pure *p*-CDpOV radical crystal ($x = 0$) follows the Curie Weiss law with a positive Weiss constant (θ) of +3.5K in the temperature region 20 ~ 300K [13, 23]. As described in Section 3, the *p*-CDpOV radical crystal is a bulk ferromagnet with T_C = 0.21K and behaves as a quasi-1D Heisenberg ferromagnet with intra- and inter-chain exchange interactions ($2J$ and $z'J'$) of $2J/k_B$ = +8.2K and $z'J'/k_B$ = 0.03K, respectively, above T_C [13]. *p*-CDpOV has high one-dimensionality, because the ratio $J'/J = 3.7\times10^{-3}$ assuming $z' = 2$.

The χ_M's of the radical alloys $(p\text{-CDpOV})_{1-x}(p\text{-BDpOV})_x$ (x = 0.03 ~ 0.50) show behavior similar to that of the pure *p*-CDpOV radical crystal ($x = 0$). The χ_M's follow the Curie-Weiss law with positive Weiss constants of +3.5 ~ +4.7K in the temperature region 20 ~ 300K, as listed in Table 4. In the molecular field theory, the Weiss constant

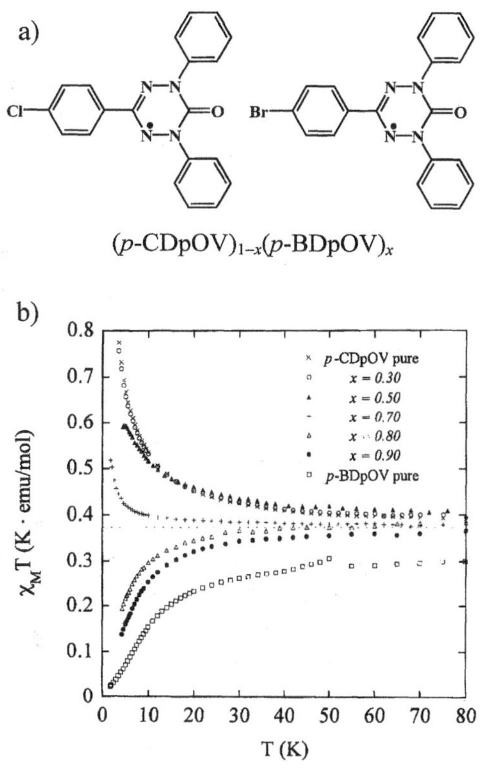

Fig. 11. (a) Molecular structures of *p*-CDpOV and *p*-BDpOV. (b) Temperature dependence of the product $\chi_M T$ for radical alloys, $(p\text{-CDpOV})_{1-x}(p\text{-BDpDO})_x$ ($x = 0 \sim 1$).

Table 4.
Magnetic properties of the radical alloys, $(p\text{-CDpOV})_{1-x}(p\text{-BDpOV})_x$.

x	θ/K	$2J/k_B$ /K
0	+ 3.5	+ 8.2
0.03	+ 4.3	+ 9.6
0.05	+ 4.2	+10.0
0.10	+ 3.9	+ 9.0
0.20	+ 4.7	+ 11.0
0.30	+ 3.6	+ 8.2
0.50	+ 4.4	+ 7.6
0.70	+ 0.65	+ 1.24
0.80	– 0.21	-----
0.90	– 2.7	-----
0.95	– 4.8	-----
1	– 7.7	– 14.8

(θ) is given as $\theta = 2zJS(S + 1)/3k_B$. Assuming $z = 2$ and substituting $S = 1/2$, we can obtain the relation, $\theta = J/k_B$. In fact, θ's observed for the radical alloys show good accordance with corresponding J/k_B's estimated by using Eq. (3) (see Table 4). The results indicate that the radical alloys also have a character as quasi-1D Heisenberg ferromagnet above T_C.

On the other hand, the χ_M of pure *p*-BDpOV ($x = 1$) follows the Curie-Weiss law with a negative Weiss constant (θ) of –7.7K in the temperature region 30 ~ 300K. The susceptibility of *p*-BDpOV shows a broad maximum at $T_{max} = 9.5$K. The χ_M's of *p*-BDpOV can be well explained by a 1D AFM Heisenberg linear chain model with $2J/k_B = -14.8$K. The χ_M's for $x = 0.80$, 0.90, and 0.95 show no maximum and monotonically increase as temperature is lowered below 300K, following the Curie-Weiss law with negative Weiss constants of $\theta = -0.21$, –2.7, and –4.8K, respectively. The result indicates that *p*-CDpOV molecules homogeneously distribute in the AFM chains in each sample, and the value of $|\theta|$ rapidly decreases by a small amount of *p*-CDpOV substitution.

As listed in Table 4, the value of θ remains constant ($\theta = +4.1 \pm 0.6$K) at the range of $x = 0 \sim 0.50$ and changes from positive to negative at $x = 0.70 \sim 0.80$. The value decreases rapidly with increasing x. The FM interaction indicated by a positive Weiss constant (θ) was observed in the samples with $x \leq 0.70$.

The effect of spin dilution in a FM chain has been studied in a previous work. Using the molecular FM compound, decamethylferrocenium tetracyanoethanide, $[FeCp_2^*][TCNE]$ as the parent compound, Narayan et al [54]. have determined the effects of nonmagnetic impurity in a low-dimensional magnetic system. The replacement of some $S = 1/2$ $[FeCp_2^*]^+$ sites by $S = 0$ $[CoCp_2^*]^+$ sites in a chain causes substantial changes in the magnetic properties. The diluted crystals, $[FeCp_2^*]_{1-x}[CoCp_2^*]_x[TCNE]$, with compositions $x = 0$, 0.045, 0.077, 0.085, and 0.145 form an isostructural dilution series. They found that the three-dimensional ordering temperature T_C decreases rapidly with the concentration (x) of spinless sites. The rate of decrease of T_C with x is in the order of the exchange spacial anisotropy (i.e., $J_{interchain}/J_{intrachain}$) of the parent system. The experimental curve of T_C (x) obtained for $[FeCp_2^*]_{1-x}[CoCp_2^*]_x[TCNQ]$ follows the trend predicted for diluted quasi-1D magnets. The results confirm the long-standing theoretical predictions concerning effects of spin dilution in quasi-1D molecular ferromagnets [54]. On the other hand, detailed information about the x dependence of the intrachain exchange interaction ($2J_{intrachain}$) has not been obtained from the analyses of the magnetic susceptibilities of the above diluted crystals.

To study the effect of magnetic impurity for quasi-1D Heisenberg ferromagnet (*p*-CDpOV), *p*-BDpOV radical was chosen as the best dopant [53]. First, *p*-BDpOV is expected to be substituted easily for *p*-CDpOV, because the molecular size, that is, the van der Waals radius is similar to each, except for the difference of 2% in the lengths of long axes of the *p*-CDpOV (15.0 Å) and *p*-BDpOV (15.3 Å) radical molecules. In fact, the X-ray diffraction patterns of the radical alloys ($x = 0.03 \sim 0.30$) are quite similar to that of pure *p*-CDpOV crystal ($x = 0$) (see Fig. 6 in Ref. 53). The result indicates that the *p*-BDpOV radical molecules homogeneously distribute in the quasi-1D FM chains of the *p*-CDpOV in each sample and the crystal structure of radical alloys ($x = 0.03 \sim 0.30$) is almost the same as that of the pure *p*-CDpOV.

Second, both the molecules are magnetic and have spin $S = 1/2$. Third, we can expect that *p*-BDpOV has electronic structure similar to that of *p*-CDpOV. In fact, the results of the McLachlan MO calculations indicate that the *p*-BDpOV molecule has unpaired electron distribution very similar to that of the *p*-CDpOV (see Fig. 1 in ref. 53), and both the radicals show a large spin polarization effect in the molecules. Therefore, we can expect similar magnetic properties for the radical alloys with $x = 0 \sim 0.30$.

In fact, at the region of $x = 0 \sim 0.30$, both the Weiss constant (θ) and the intrachain exchange interaction ($2J/k_B$) show constant values of about $+ 4.0 \pm 0.7$ K and 9.4 ± 1.6 K, respectively. The result suggests that in this system the intermolecular exchange interaction ($2J_{Cl\text{-}Br}$) between the neighboring *p*-CDpOV and *p*-BDpOV radical molecules is almost identical with the $2J_{Cl\text{-}Cl}$ between the neighboring *p*-CDpOV molecules and does not affect the propagation of spin correlations. Further, the exchange interaction $2J_{Br\text{-}Br}$ between the neighboring *p*-BDpOV radicals is also thought to be similar to $2J_{Cl\text{-}Cl}$, because the occurrence of the *p*-BDpOV-*p*-BDpOV bonding in the system with x is about x^2; in the case for $x = 0.30$ it is about 0.09. The existence of such a bond doesn't affect the magnetic property of the mixed crystal. The result suggests that the substituted *p*-BDpOV molecules play a role as *p*-CDpOV ones in the radical alloys with $x = 0 \sim 0.30$.

On the other hand, the X-ray diffraction pattern of radical alloy with $x = 0.50$ consists of those of *p*-CDpOV and *p*-BDpOV, although the contribution of the former is more remarkable than that of the latter (see Fig. 6 in Ref. 53). However, the θ and $2J$ for $x = 0.50$ are almost the same with corresponding those for the pure *p*-CDpOV. The probabilities of the occurrence of the *p*-CDpOV-*p*-CDpOV, *p*-CDpOV-*p*-BDpOV, and *p*-BDpOV-*p*-BDpOV bondings in the system with $x = 0.50$ are 0.25, 0.50 and 0.25, respectively. The results suggest that all the intrachain exchange interactions $2J_{Cl\text{-}Cl}$, $2J_{Cl\text{-}Br}$, and $2J_{Br\text{-}Br}$ have similar values to each other, although the X-ray diffraction pattern of the radical alloy with $x = 0.50$ shows the weak contribution of the *p*-BDpOV.

Similarly, the X-ray diffraction patterns of the radical alloys at high concentration region ($x = 0.80 \sim 0.95$) of *p*-BDpOV are quite similar to that of pure *p*-BDpOV crystal, indicating that *p*-CDpOV molecules homogeneously distribute in the 1D chain consisting of the *p*-BDpOV molecules in each sample. Therefore, we can expect similar magnetic properties for the samples with $x = 0.80 \sim 1$. However, the χ_M's of mixed radical crystals show notable x dependence at the high concentration region of *p*-BDpOV ($0.80 \leq x \leq 1$), and the θ's drastically change with doping. The reason of such a drastic change in the Weiss constant is not clear at present. This work is the first report of the effects of magnetic impurities on the quasi-1D ferromagnetic chain system.

9 Photo-Excited High Spin States on π-Conjugated Verdazyl Radicals

Intramolecular spin alignment and exchange interaction through π conjugation in purely organic spin systems are important in the field of molecule-based magnetism. However, most studies are limited to the ground state systems. The photo-induced spin alignment through π-conjugation is one of the key processes for photo-control of the molecule-based magnetism. Teki et al [55]. observed the photo-excited quartet ($S = 3/2$) and quintet ($S = 2$) states on purely organic π-conjugated radical-triplet pair systems, which were generated by a spin alignment through π-conjugation between dangling imino-nitroxide radicals ($S = 1/2$) and the excited triplet ($S = 1$) state of a phenyl- or diphenylanthracene. π-Conjugated spin systems arising from the aromatic hydrocarbons and the stable radicals are ideal systems on studying the photo-induced intramolecular spin alignment through π-conjugation.

Similar excited quartet and quintet states were observed for the 9-anthranene-(4-phenylverdazyl) radical (1) and 9,10-anthracene-bis(4-phenylverdazyl) diradical (2) (see Fig. 12) by time resolved electron spin resonance (TRESR) [56]. The TRESR spectrum was analyzed by the ordinary spin Hamiltonian with the Zeeman and fine structure terms. For the quartet state of (1), the g value, fine structure splitting, and relative population of the Ms sublevels have been determined to be $g = 2.0035$, $D = 0.0230$ cm^{-1}, $E = 0.0$, $P_{1/2'} = P_{-1/2'} = 0.5$ and $P_{3/2'} = P_{-3/2'} = 0.0$, respectively, by spectral simulation. The spin Hamiltonian parameters of the quintet state of (2) were determined to be $g = 2.0035$, $D = 0.0128$ cm^{-1}, $E = 0.0$, $P_{2'} = P_{-2'} = 0.0$, $P_{1'} = P_{-1'} = 0.37$ and $P_{0'} = 0.26$, respectively. Direct observation of the excited high spin state showed that photoinduced intramolecular spin alignment is realized between the excited triplet state ($S = 1$) of the phenyl- or diphenylanthracene moiety and the doublet spin ($S = 1/2$) of the dangling verdazyl radicals. *Ab initio* MO calculations (DFT) were carried out in order to clarify the mechanism of the photoinduced spin alignment. Similar measurement was also performed for the pyrene-verdazyl radical system, in which excited high-spin quartet ($S = 3/2$) state was observed [57].

Studies on the spin alignment between the metastable excited triplet state and the stable radicals will yield very important information on novel spin alignment. Further, this leads to a new strategy for the photoinduced/switching magnetic spin systems and the photocontrol of molecule-based magnetism.

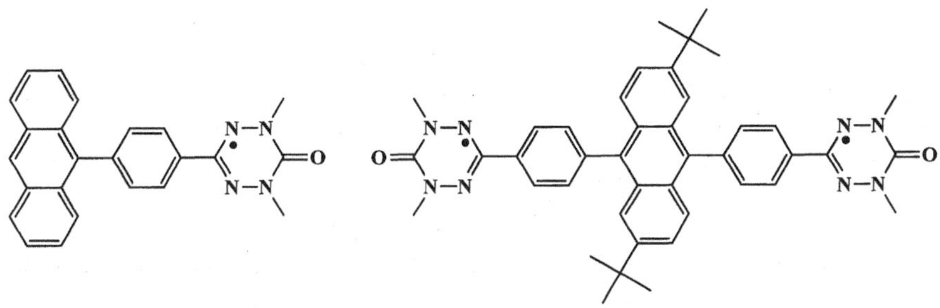

Fig. 12. Molecular structures of π-conjugated verdazyl radicals.

10 Transition Metal Complex of Verdazyl Radicals

The design and synthesis of hybrid magnetic materials consisting of transition metal ions and open-shell free-radical ligands are a major focus of molecular materials research. Among the most extensively studied paramagnetic ligands in this context are nitroxide radicals, including nitronyl nitroxide and imino nitroxides [58, 59]. Such systems not only provide a probe for exchange interactions between unpaired electrons but also in some instances have shown novel long range ordering effects. The stable verdazyl radical with four nitrogen atoms in a molecule represents an attractive alternative ligand system. However, there are currently very few examples of metal-verdazyl complexes [60, 61].

Brook et al. [60] reported the ESR and magnetic properties of the diradical 1,1'-5,5'-tetramethyl-6,6'-dioxobis(verdazyl) (BVD) and observed that the molecule has a singlet ground state with a thermally populated triplet lying 760 cm^{-1} above it. The diradical can chelate metals with relatively little geometric perturbation and consequently valuable information about the metal-radical and metal-metal interactions should be obtainable. BVD was reacted with copper (I) halides in acetonitrile and copper (II) halides in methanol to give copper (I) coordination polymers of composition $[Cu_2X_2(BVD)]_x$ [60]. Magnetic susceptibility measurements taken on the three polymers indicate that the spins couple in one-dimensional chains with alternating exchange parameters: J_1 = -190 cm^{-1}, J_2 = -116 cm^{-1} (X = Cl); J_1 = -200 cm^{-1}, J_2 = -110 cm^{-1} (X = Br); and J_1 = -271 cm^{-1}, J_2 = -200 cm^{-1} (X = I). However, all of these complexes are based on diamagnetic Cu(I) ion. The absence of metal-based unpaired spins in these systems precludes the possibility of verdazyl-metal magnetic exchange interactions.

Hicks et al. [61] reported the preparation of the first verdayl radical complexes of open-shell ions, namely Ni(II) and Mn(II). Reactions of pydv (see Fig. 13) with $M(hfac)_2{\cdot}2H_2O$ afforded (1) (M = Ni) and (2) (M = Mn) as air-stable complexes. Preliminary X-ray crystallographic data confirm the mononuclear, pseudo-octahedral nature of complexes (1) and (2), both of which contain a fully planar, chelating pydv

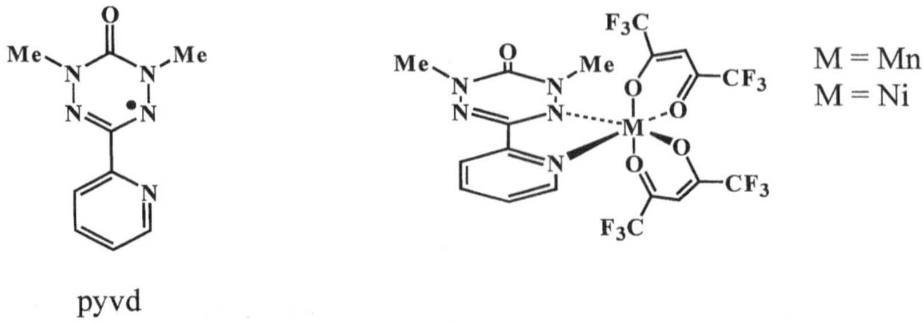

Fig. 13. Molecular structure of pyvd and its complexes with $M(hfac)_2$ (M = Mn and Ni).

ligand as shown in Fig. 13. The susceptibility of Mn complex (1) can be explained by a model in which the verdazyl and manganese spins are antiferromagnetically coupled with the following parameters: g = 1.987, θ = –0.8K, J_{AB} = –45 cm^{-1}. The magnetic behavior of Ni complex (2) is consistent with that of a ferromagnetically coupled $S_A = 1/2$, $S_B = 1$ pair with the following parameters: g = 2.1, θ = –2.0K, J_{AB} = +240 cm^{-1}. The verdayl-metal coupling is strong and ferromagnetic in the case of Ni(II). These results indicate the potential of creating new magnetic materials based on coordinated verdayl radicals and paramagnetic transition metal ions. Further, Hicks et al. found that the verdazyl radical forms the complex with hydroquinone molecule [61].

11 Magnetic Conductors Based on Verdazyl Radicals

Development of the molecular ferromagnet and that of the superconductor are two challenging targets in material science, and many studies have been performed with great success in recent years. Superconductivity and magnetism have long been considered incompatible to each other, because Cooper pairs are destroyed by an external field (or by the internal field generated in ferromagnets). Consequently, it is very interesting to establish superconductivity in a composite system containing localized magnetic moments. The first example of an organic/inorganic hybrid system in which paramagnetism and superconductivity coexist has been reported for $[BEDT\text{-}TTF]_4(H_2O)[Fe(C_2O_4)_3](C_6H_5CN)$ (T_C = 7K) [62]. Coexistence of ferromagnetism and metallic conductivity has been found for a similar system, $[BEDT\text{-}TTF]_3[MnCr(C_2O_4)_3]$ [63]. Further, it has been reported that $[\kappa\text{-}BETS]_2FeBr_4$ undergoes the transition from an AFM metal phase to superconducting phase at 1.1K [64]. Matsumoto et al. have recently reported an example of semiconducting ferrimagnet based on (2:1) charge-transfer salt of ethylenedithiotetrathiafulvalenoquinone-1,3-dithiolemethide with $FeBr_4^-$ ion [65]. However, the examples of the organic/inorganic hybrid system in which the magnetism and conductivity coexist are very limited, because of the difficulty in molecular design of the complexes. Further, the examples of the magnetic conductor consisting of purely organic compounds have not been reported.

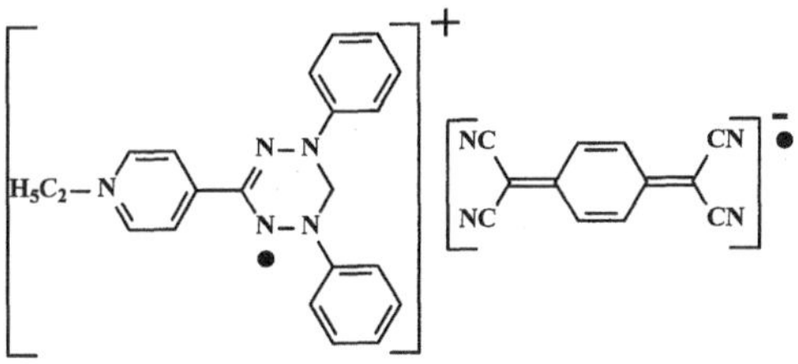

[*p*-EtPyDV]$^+$[TCNQ]$^-$ (1)

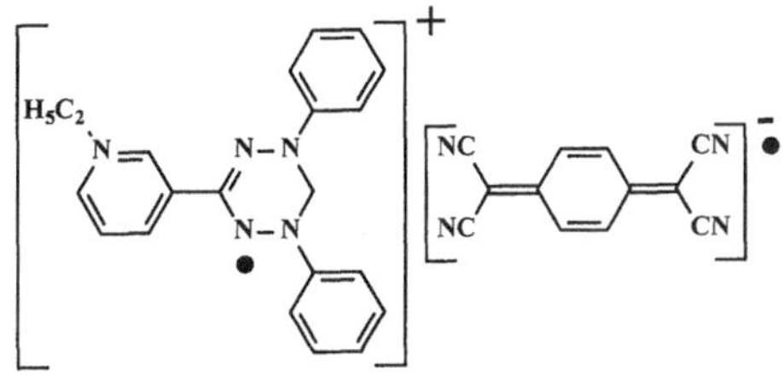

[*m*-EtPyDV]$^+$[TCNQ]$^-$ (2)

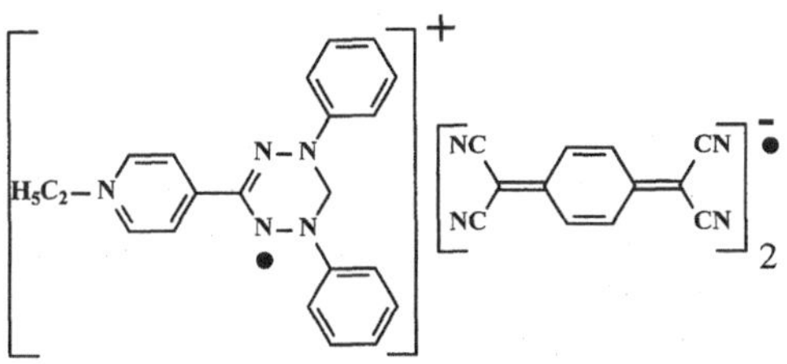

[*p*-EtPyDV]$^+$[TCNQ]$_2^-$ (3)

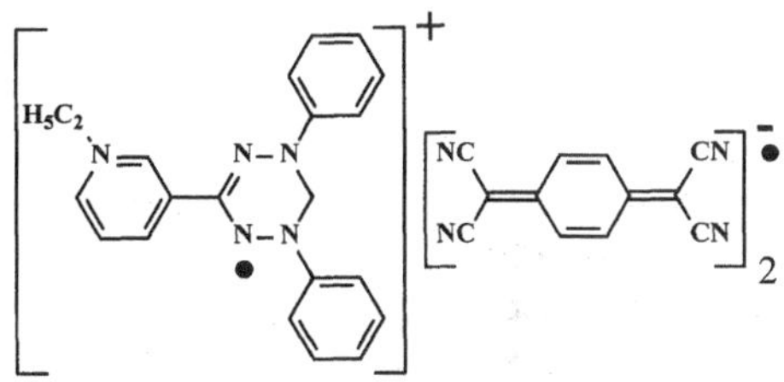

[*m*-EtPyDV]$^+$[TCNQ]$_2^-$ (4)

Fig. 14. Molecular structures of the (1:1) and (1:2) salts of verdazyl radical cations with TCNQ anion.

11.1 Pure organic Magnetic Semiconductors Consisting of Verdazyl Cation and TCNQ Anion

The magnetic properties of the charge-transfer (CT) complexes of 2-(4-alkyl-pyridinium)-imidazolin-1-oxyls (alkyl = Me, Et, *n*-Pr, and *n*-Bu) and 2-(3- and 4-methyl-pyridinium)-nitronylnitroxide with three kinds of electron acceptors, i.e., TCNQ, $TCNQF_4$, and hexacyanobutadienide (HCBD) have been studied by Sugimoto et al. [66]. Nakatsuji et al. [67] prepared the CT complexes of the verdazyl radicals with $TCNQF_4$ and DDQ, and studied the magnetic properties of the complexes. Recently, four kinds of (1:1) and (1:2) salts of 3-(4- and 3-ethyl-pyridinium)-1,5-diphenylverdazyl radical cations with TCNQ anion ((1) – (4), see Fig. 14) have been prepared, and the magnetic and electric properties and crystal structure of these radical salts have been studied [68]. The values of $\chi_M T$ at 300K are 0.437, 0.383, 0.620, and 0.482K emu/mol for the salts (1), (2), (3) and (4), respectively, and decrease by lowering the temperature (Fig. 15). The susceptibility of the salt (1) showed anomalous magnetic behavior that cannot be explained by a simple model. The χ_M of the salt (2) follows the Curie-Weiss law with a Curie constant of 0.374K emu/mol and a negative Weiss constant of –2.4K in the temperature region 1.8–300K, indicating disappearance of a net magnetic moment due to the formation of strong spin pairs in TCNQ anion molecules.

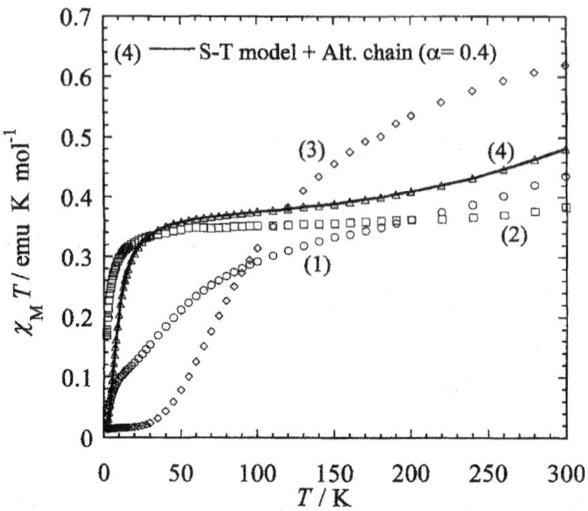

Fig. 15. Temperature dependence of the product $\chi_M T$ for salts (1), (2), (3), and (4).

The χ_M of the (1:2) salt (3) shows a broad maximum at T_{max} = 118K, and can be well explained by a S-T model (Eq. (2)) with $2J/k_B$ = -190K. The result shows that both the spins of verdazyl cation and TCNQ anion contribute to the magnetism of the salt (3). If the verdazyl cation and the TCNQ anion molecules form segregated columns in the crystal, we can expect two kinds of exchange interactions (J_{Cation} and J_{Anion}). The present result suggests that the cation and anion molecules form one-dimensional column in the crystal. The χ_M of the (1:2) salt (4) shows a broad maximum at T_{max}= 9.3K. The χ_M of salt (4) can be explained by the three term contributions (Eq. (7)).

$$\chi_M = C_{1D\text{-}Alt}\chi_{1D\text{-}Alt} + C_{S\text{-}T}\chi_{S\text{-}T} + C_{Curie}\chi_{Curie} \tag{7}$$

where $C_{1D\text{-}Alt} + C_{S\text{-}T} + C_{Curie} = 2$, and the first, second and third terms represent the contributions from 1D Heisenberg AFM alternating-chain system [26], S-T equilibrium system (Eq. (2)), and Curie impurity, respectively. At low temperature region (T < 100K), the susceptibility for the salt (4) can be well reproduced by the two term contributions ($\chi_{1D\text{-}Alt}$ and χ_{Curie}) with the value of $C_{1D\text{-}Alt}$ = 0.977 and C_{Curie} = 0.045 and with $2J_1/k_B$ = −15.2K (alternation parameter $\alpha = J_2/J_1$ = 0.4). Above ca. 100K, however, $\chi_M T$ increases with increasing temperature, as shown in Fig. 15. This increase indicates that thermal magnetic excitation occurs in addition to the alternating-chain spins and the Curie spins. The experimental data are interpreted by the S-T model with the value of $2J/k_B$ = −1.61 T_{max} = −741K (T_{max} = 460K) and $C_{S\text{-}T}$ = 0.978. As shown in Fig. 15, the experimental curve can be well reproduced by Eq. (7).

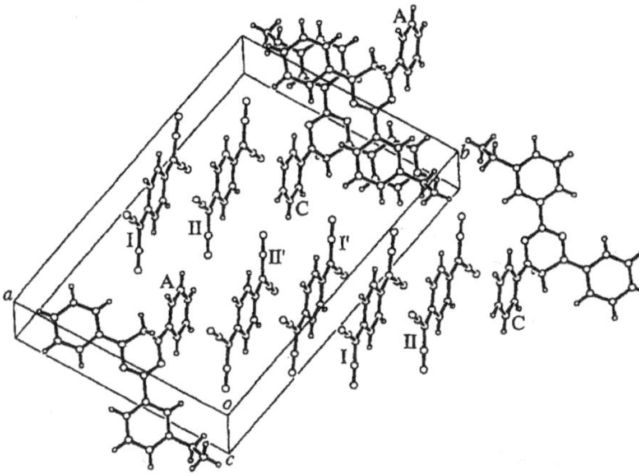

Fig. 16. Molecular packing of salt (4) showing sections of the TCNQ layer and their relation to the verdazyl cations.

The crystal structure could be determined for the (1:2) salt (4). As shown in Fig. 16, there are two crystallographically-independent TCNQ molecules, $TCNQ_I$ and $TCNQ_{II}$, in the crystal of $[m\text{-EtPyDV}]^+[TCNQ]_2^-$ (4), while they have little difference in structure. By comparing the bond lengths of $TCNQ_I$ and $TCNQ_{II}$ with those of neutral TCNQ, monoanion $TCNQ^-$, and $TCNQ^{-1/2}$ (see Table 2 in ref. 68), $TCNQ_I$ and $TCNQ_{II}$ are considered to exist as $TCNQ_I^{-1/2}$ and $TCNQ_{II}^{-1/2}$, respectively, in salt (4) [68]. As shown in Fig. 16, the four $TCNQ^{-1/2}$ molecules form two kinds of dimers, $TCNQ_I^{-1/2}$ - $TCNQ_{II}^{-1/2}$ and $TCNQ_{II'}^{-1/2}$ - $TCNQ_{I'}^{-1/2}$, having short intermolecular distances. The two dimer units are related by an inversion center. The dimers are stacked in parallel to the axis connecting the two verdazyl radical cations ($Verdazyl_A$ and $Verdazyl_C$). The interplanar distances of alternating $TCNQ^{-1/2}$ molecules are 3.36 Å for $TCNQ_I$ - $TCNQ_{II}$ and 3.51 Å for $TCNQ_I$ - $TCNQ_{I'}$. Shorter interplanar spacings in a pair indicate stronger spin interaction, so that the magnitude of interaction is supposed to be stronger in I - II than in I – I'. The large AFM exchange interaction ($2J/k_B = -741$K) observed for the salt (4) may be explained by the short interplanar distances of alternating $TCNQ^{-1/2}$ molecules in the crystal.

The verdazyl cation molecules ($[m\text{-EtPyDV}]^+$) in salt (4) form 1D columns along the *c*-axis with moderately short N···N contacts between the nitrogen atoms in the central hydrazidinyl moiety N1-N2-C1-N3-N4, having the large unpaired spin densities (see Fig. 4 in ref. 68) [68]. The column is built up by two kinds of radical pairs, $Verdazyl_A$ – $Verdazyl_B$ and $Verdazyl_A$ – $Verdazyl_C$, suggesting 1D alternating magnetic properties. In fact, the susceptibility below the temperature 30K can be well described by the 1D alternating-chain model with $2J_1/k_B = -15.2$K and $2J_2/k_B = -6.0$K.

The result of the measurement of electric conductivity shows that the two (1:1) salts (1) and (2) are insulators ($\sigma_{RT} = 2.0\times10^{-6}$ S cm^{-1} for (1) and $\sigma_{RT} = 1.3\times10^{-7}$ S·cm^{-1} for (2)) as expected for the (1:1) complexes. On the other hand, the two (1:2) salts (3) and (4) show semiconductive behavior ($\sigma_{RT} = 6.9\times10^{-2}$ S·cm^{-1} and $E_A = 0.072$ eV for (3) and

σ_{RT} = 1.5×10^{-3} $S\cdot cm^{-1}$ and E_A = 0.25 eV for (4)). As a conclusion, the present result shows that the two (1:2) salts, [*p*-EtPyDV]$^+$[TCNQ]$_2^-$ (3) and [*m*-EtPyDV]$^+$[TCNQ]$_2^-$ (4), are new genuine organic magnetic semiconductors. At lower temperature, we can expect AFM ordering for these salts.

The similar results were obtained for the (1:1) and (1:2) salts of *p*- and *m*-methyl-pyridinium-substituted verdazyl radicals with TCNQ. The two kinds of the (1:2) salts obtained were paramagnetic semiconductors [69].

11.2 Magnetic Semiconductors Consisting of Verdazyl Cation and Ni(dmit)$_2$ Anion

Extensive studies have been performed for the assembled metal complexes (Cation)[M(dmit)$_2$]$_2$ (M = Ni and Pd and dmit = 1,3-dithiol-2-thione- 4,5-dithiolate) of a closed shell cation with an M(dmit)$_2$ anion, exhibiting insulating, metallic and superconducting phases at low temperature under high pressure [70]. Two examples of molecular paramagnetic semiconductors consisting of an open-shell radical cation donor and a metal complex anion acceptor have been reported for the (1:1) salts of (i) the *p*-EPYNN (*p*-N-ethylpyridinium α-nitronyl nitroxide) and (ii) the Me_3N^+-TEMPO (=N,N,N-trimethyl(1-oxyl-2,2,6,6-tetramethylpiperidine-4-yl)ammonium) radical cations with a Ni(dmit)$_2$ anion [71].

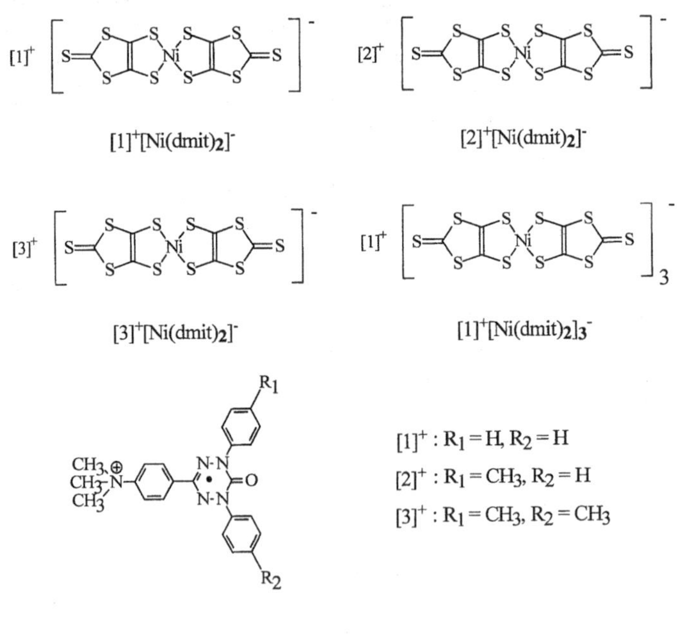

Fig. 17. Molecular structures of the (1:1) and (1:3) salts of verdazyl radical cations with Ni(dmit)$_2$ anion.

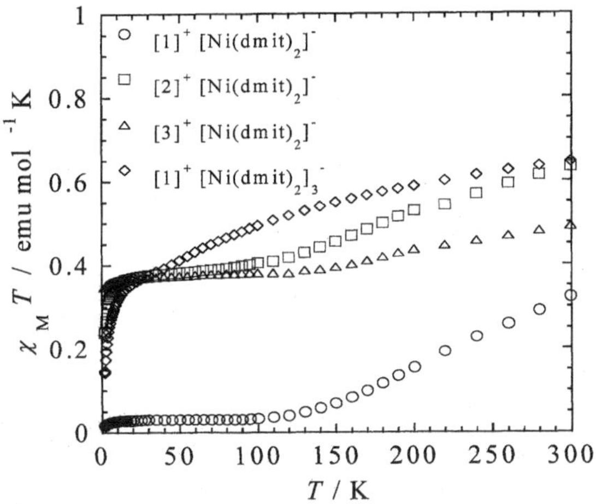

Fig. 18. Temperature dependence of the product $\chi_M T$ for salts (4), (5), (6), and (7).

Recently, four kinds of (1:1) and (1:3) salts of 3-[4-(trimethylammonio)phenyl]-1,5-diphenyl-6-oxoverdazyl radical cation $[1]^+$ and its mono- and di-methyl derivatives ($[2]^+$ and $[3]^+$) with $Ni(dmit)_2$ anions ($[1]^+[Ni(dmit)_2]^-$ (4), $[2]^+[Ni(dmit)_2]^-$ (5), $[3]^+[Ni(dmit)_2]^-$ (6), and $[1]^+[Ni(dmit)_2]_3^-$ (7)) (see Fig. 17) have been prepared, and the structural, magnetic and electric properties of these radical salts have been studied [72]. Figure 18 shows a plot of $\chi_M T$ versus T for three kinds of (1:1) salts (4), (5) and (6) and a kind of (1:3) salt (7). The values of $\chi_M T$ for (4), (5), and (6) at 300K are 0.323 (0.391 at 350K), 0.650, and 0.540K emu/mol, respectively, as listed in Table 5. These values are larger than that (0.376 K emu/mol) for $S = 1/2$ free spin system, suggesting that both spins of the verdazyl cation ($S = 1/2$) and $Ni(dmit)_2$ anion ($S = 1/2$) contribute to the magnetism of salts (4) - (6). The susceptibility of (1:3) salt (7) showed a behavior similar to that of the (1:1) salt (5).

The χ_M values of salts (5) and (7) can be well reproduced by the sum of the contributions from: (i) a Curie-Weiss system with a Curie constant of 0.376K emu/mol and negative Weiss constants (θ) of –0.4 and –1.7K and (ii) a dimer system with strong negative exchange interactions of $2J/k_B$ = –354 and –258K, respectively. The dimer formations in $Ni(dmit)_2$ anions have been ascertained by the crystal structure analyses of salts (4) - (6). In salts (4) and (6), $Ni(dmit)_2$ dimer molecules are sandwiched between two verdazyl cations, indicating the formation of a linear tetramer in (4) and (6). The molecular packing of salt (4) is shown in Fig. 19. The $Ni(dmit)_2$ anion molecules in salt (4) form a dimer ($Ni(dmit)_2$ (A) - $Ni(dmit)_2$ (B)) with short intradimer contacts of ca. 3.6–3.7 Å, suggesting strong AFM exchange interaction. However, the χ_M of (4) cannot be explained by a dimer model. In salt (4), the dimer molecules are sandwiched between two verdazyl cation $[1]^+$ molecules, forming linear tetramer ($[1]^+$ – $[Ni(dmit)_2]^-$ (A) – $[Ni(dmit)_2]^-$ (B) – $[1]^+$) (see Fig. 19). We can expect strong exchange interaction between the $Ni(dmit)_2$ anions and the verdazyl cations $[1]^+$, because there are many short contacts between the Ni, S(i) and C(i) atoms in the $Ni(dmit)_2$ anions and the N(*j*) atoms (*j* = 1, 2, 3, and 4) in the hydrazidinyl moiety (N1-N2-C7-N3-N4-C8), having large unpaired spin densities, of verdazyl cations $[1]^+$ [22, 53].

Table 5.
Conductivity (σ_{RT}), activation energy (E_A), $\chi_M T$ value (at 300K), and magnetism of salts (4) – (7).

Salts	σ_{RT} S cm^{-1}	$\chi_M T$ K emu/mol	Magnetism
$[1]^+[Ni(dmit)_2]^-$ 1/2CH_3CN (4)	1.9×10^{-7} [a, c]	0.323 (0.391 at 350K)	Linear tetramer ($T_{max}\sim$ 360K, $2J_1/k_B$ = –600K, $2J_2/k_B$ = –280K) + Curie impurity (3.0%)
$[2]^+[Ni(dmit)_2]^-$ (5)	1.8×10^{-7} [a, c]	0.650	Curie-Weiss (θ = –0.4K) + S-T ($2J/k_B$ = –354 K)
$[3]^+[Ni(dmit)_2]^-$ (6)	4.7×10^{-5} [b, d]	0.540	Linear tetramer ($2J_1/k_B$ = –30K, $2J_2/k_B$ = –580K)
$[1]^+[Ni(dmit)_2]_3^-$ (7)	1.0×10^{-1} [b, d] (E_A=0.10eV)	0.642	Curie-Weiss (θ = –1.7K) + S-T ($2J/k_B$ = –258K)

[a]Single crystal. [b]Pressed pellet. [c]Two-probe ac method. [d]Four-probe ac method.

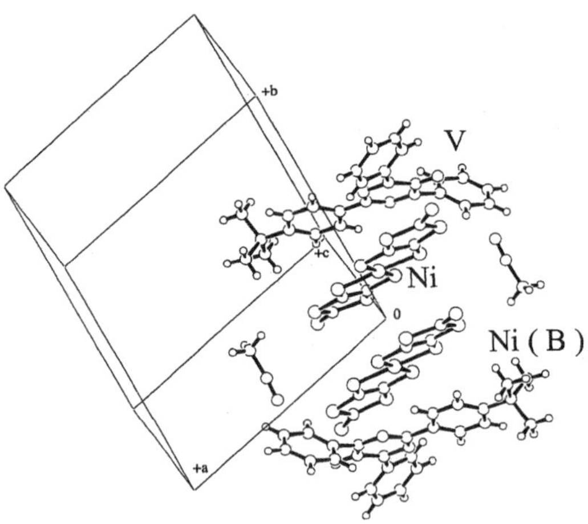

Fig. 19. Molecular packing in salt (4), showing the formation of a linear tetramer [verdazyl cation-$Ni(dmit)_2$ anion (A)-$Ni(dmit)_2$ anion (B)-verdazyl cation] along the *b*-axis.

The magnetic susceptibility data for salts (4) and (6) have been fitted to a linear tetramer model using an end exchange interaction of $2J_1/k_B$ = –600K and a central interaction $2J_2/k_B$ = –280K for (4), and $2J_1/k_B$ = –30K and $2J_2/k_B$ = –580K for (6), respectively. The results of the temperature dependence of the $g(T)$-value in salts (4) - (6) obtained by ESR measurement also support the above analyses. The (1:1) salts (4), (5) and (6) are insulators. On the other hand, the conductivity of (1:3) salt (7) at 20°C was σ = 0.10 S·cm^{-1} with an activation energy E_A = 0.099 eV, showing the semiconductor property. Salt (7) is a new molecular paramagnetic semiconductor.

The similar results were obtained for five kinds of (1:1), (1:3), and (2:1) salts of 3-[4-(diethylmethylammonio)phenyl]-1,5-diphenyl-6-oxoverdazyl radical cation $[V]^+$ with $M(dmit)_2$ anions (M = Ni, Zn, Pd, and Pt) ($[V]^+[Ni(dmit)_2]^-$ (8), $[V]^+[Ni(dmit)_2]_3^-$ (9), $[V]^+_2[Zn(dmit)_2]^{2-}$ (10), $[V]^+_2[Pd(dmit)_2]^{2-}$ (11), and $[V]^+_2[Pt(dmit)_2]^{2-}$ (12)) [73]. The salts (9) and (11) were paramagnetic semiconductors. The examples of the organic/inorganic hybrid system which shows the magnetism and conductivity are very limited, because of the difficulty in the molecular design of the complexes. The present work provides several examples of new molecular paramagnetic semiconductors consisting of open-shell verdazyl radical cations and metal complex anions.

References

1. M. Tamura, Y. Nakazawa, D. Shiomi, K. Nozawa, Y. Hosokoshi, M. Ishikawa, and M. Takahashi, M. Kinoshita, Chem. Phys. Lett. 186 (1991) 401.
2. Y. Nakazawa, M. Tamura, N. Shirakawa, D. Shiomi, M. Takahashi, M. Kinoshita, and M. Ishikawa, Phys. Rev. B46 (1992) 8906.
3. M. Kinoshita, Jpn. J. Appl. Phys. 33 (1994) 5718.
4. P.M. Lahti (Ed.) "Magnetic Properties of Organic Materials" (1999) Marcel Dekker, New York.
5. K. Itoh and M. Kinoshita (Ed.) "Molecular Magnetism: New Magnetic Materials", (2000) Amsterdam, Gordon & Breach (Kodansha as copublisher, Tokyo).
6. R. Kuhn, F.A. Neugebauer, and H. Trischmann, Angew. Chem. 76 (1964) 691.
7. K. Mukai, N. Azuma, and K. Ishizu, Bull. Chem. Soc. Jpn. 43 (1970) 3618.
8. N. Azuma, K. Ishizu, and K. Mukai, J. Chem. Phys. 61 (1974) 2294.
9. N. Azuma, Bull. Chem. Soc. Jpn. 55 (1982) 1357.
10. F.A. Neugebauer, H. Fisher, and R. Siegel, Chem. Ber. 121 (1988) 815.
11. F.A. Neugebauer, H. Fisher, and C. Krieger, J. Chem. Soc. Perkin Trans. 2 (1993) 535.
12. K. Takeda, K. Konishi, K. Nedachi, and K. Mukai, Phys. Rev. Lett. 74 (1995) 1673.
13. K. Takeda, T. Hamano, T. Kawae, M. Hidaka, M. Takahashi, S. Kawasaki, and K. Mukai, J. Phys. Soc. Jpn. 64 (1995) 2343.
14. K. Mukai, K. Konishi, K. Nedachi, and K. Takeda, J. Phys. Chem. 100 (1996) 9658.
15. M. Mito, K. Takeda, K. Mukai, N. Azuma, M.R. Gleiter, C. Krieger, and F.A. Neugebauer, J. Phys. Chem. 101 (1997) 9517.
16. S. Tomiyoshi, T. Yano, N. Azuma, M. Shoga, K. Yamada, and J. Yamauchi, Phys. Rev. B49 (1994) 16031.
17. R.K. Kremer, B. Kanellakopulos, P. Bele, H. Brunner, and F.A. Neugebauer, Chem. Phys. Lett. 230 (1994) 255.
18. M. Mito, H. Nakano, T. Kawae, M. Hitaka, S. Takagi, H. Deguchi, K. Suzuki, K. Mukai, and K. Takeda, J. Phys. Soc. Jpn. 66 (1997) 2147.

19. K. Mukai, M. Nuwa, T. Morishita, T. Muramatsu, T.C. Kobayashi, and K. Amaya, Chem. Phys. Lett. 272 (1997) 501.
20. K. Mukai, N. Wada, J.B. Jamali, N. Achiwa, Y. Narumi, K. Kindo, T. Kobayashi, and K. Amaya, Chem. Phys. Lett. 257 (1996) 538.
21. T. Hamamoto, Y. Narumi, K. Kindo, K. Mukai, Y. Shimobe, T.C. Kobayashi, T. Muramatsu, and K. Amaya, Physica B 246-247 (1998) 36.
22. K. Mukai, M. Matsubara, H. Hisatou, Y. Hosokoshi, K. Inoue, and N. Azuma, J. Phys. Chem. B 106 (2002) 8632.
23. K. Mukai, S, Kawasaki, J.B. Jamali, and N. Achiwa, Chem. Phys. Lett. 241 (1995) 618.
24. K. Mukai, M. Nuwa, K. Suzuki, S. Nagaoka, N. Achiwa, and J.B. Jamali, J. Phys. Chem. B 102 (1998) 782.
25. J.C. Bonner and M.E. Fisher, Phys. Rev. A135 (1964) 640.
26. J.C. Bonner, H.W.J. Blote, J.W. Bray and I.S. Jacobs, J. Appl. Phys. 50 (1979) 1810.
27. K. Awaga, T. Sugano, and M. Kinoshita, Chem. Phys. Lett. 141 (1987) 540.
28. K. Awaga and Y. Maruyama, J. Chem. Phys. 91 (1989) 2743.
29. T. Nogami, T. Ishida, M. Yasui, M. Iwasaki, N. Takeda, M. Ishikawa, T. Kawasaki, and K. Yamaguchi, Bull. Chem. Soc. Jpn. 69 (1996) 1841.
30. J. Yamauchi, K. Adachi, and Y. Deguchi, J. Phys. Soc. Jpn. 35 (1973) 443.
31. P.-M. Allemand, G. Srdanov, and F. Wudl, J. Am. Chem. Soc. 112 (1990) 9391.
32. T.C. Kobayashi, M. Takiguchi, C. Hong, A. Okada, K. Amaya, A. Kajiwara, A. Harada, and M. Kamachi, J. Phys. Soc. Jpn. 65 (1996) 1427.
33. A.J. Banister, N. Brickleband, I. Lavender, J.M. Rawson, C.I. Gregory, B.K. Tanner, W. Clegg, M.R.J. Elsegood, and F. Palacio, Angew. Chem. Int. Ed. Engl. 35 (1996) 2533.
34. F. Palacio, G. Antorrena, M. Castro, R. Burriel, J. Rawson, J.N.B. Smith, N. Bricklebank, J. Novoa, and C. Ritter, Phys. Rev. Lett. 79 (1997) 2336.
35. T. Sugimoto, M. Tsujii, H. Matsuura, and N. Hosoito, Chem. Phys. Lett. 235 (1995) 183.
36. T. Moriya, in: "Magnetism I," eds. G.T. Rado and H. Suhl, (1963) New York, Academic Press, , pp.85-125.
37. J.B. Jamali, N. Achiwa, K. Mukai, K. Suzuki, Y. Ajiro, K. Matsuda, and H. Iwamura, J. Mag. Mag. Mater. 177-181 (1998) 789.
38. L.J. De Jongh and A.R. Miedema, Adv. Phys. 23 (1974) 1.
39. I.S. Jacobs, J.W. Jr. Bray, H.R. Hart, L.V. Interrante, J.S. Kasper, G.D. Watkins, D.E. Prober, and J.C. Bonner, Phys. Rev. B14 (1976) 3036.
40. S. Huizinga, J. Kommandeur, G.A. Sawatzky, B.T. Thole, K. Kopinga, W.J.M. de Jonge, and J. Roos, Phys. Rev. B19 (1979) 4723.
41. M. Hase, I. Terasaki, and K. Uchinokura, Phys. Rev. Lett. 70 (1993) 3651.
42. S.B. Oseroff, S.-W. Cheong, B. Aktas, M.F. Hundley, Z. Fisk, and Jr.L.W. Rupp, Phys. Rev. Lett. 74 (1995) 1450.
43. Y. Ajiro, T. Asano, F. Masui, M. Mekata, H. Aruga-Katori, T. Goto, and H. Kikuchi, Phys. Rev. B 51 (1995) 9399.
44. Y. Masuda, A. Fujioka, Y. Uchiyama, I. Tsukada, and K. Uchinokura, Phys. Rev. Lett. 80 (1998) 4566, and references are cited therein.
45. B. Grenier, J.-P. Renard, P. Veillet, C. Paulsen, R. Calemczuk, G. Dhalenne, and A. Revcolevschi, Phys. Rev. B57 (1998) 3444.
46. a) K. Mukai, M. Yanagimoto, Y. Shimobe, K. Inoue, and Y. Hosokoshi, Chem. Phys. Lett. 311 (1999) 446; b) K. Mukai, Y. Shimobe, J.B. Jamali, and N. Achiwa, J. Phys. Chem. 103 (1999) 10876.

47. a) K. Mukai, M. Yanagimoto, H. Maruyama, Y. Narumi, and K. Kindo, J. Phys. Soc. Jpn. 71 (2002) 2539; b) K. Mukai, M. Yanagimoto, Y. Shimobe, K. Kindo, and T. Hamamoto, J. Phys. Chem. B106 (2002) 3687.
48. K. Mukai, M. Yanagimoto, S. Tanaka, M. Mito, T. Kawae, and K. Takeda, J. Phys. Soc. Jpn. 72 (2003) 2312.
49. K. Awaga, T. Okuno, A. Yamaguchi, M. Hasegawa, T. Inabe, Y. Maruyama, and N. Wada, Phys. Rev. B49 (1994) 3975.
50. N. Azuma, T. Ozawa, and J. Yamauchi, Bull. Chem. Soc. Jpn. 67 (1994) 31.
51. J. Fujita, M. Tanaka, H. Suemune, N. Koga, K. Matsuda, and H. Iwamura, J. Am. Chem. Soc. 118 (1996) 9347.
52. M. Tanaka, K. Matsuda, T. Itoh, and H. Iwamura, Angew. Chem. Int. Ed. 37 (1998) 810.
53. K. Mukai, K. Suzuki, K. Ohara, J. B. Jamali, and N. Achiwa, J. Phys. Soc. Jpn. 68 (1999) 3078.
54. K.S. Narayan, B.G. Morin, J.S. Miller, and A.J. Epstein, Phys. Rev. B46 (1992) 6195, and references cited therein.
55. a) Y. Teki, S. Miyamoto, K. Iimura, M. Nakatsuji, and Y. Miura, J. Am. Chem. Soc. 122 (2000) 984; b) Y. Teki, S. Miyamoto, M. Nakatsuji, and Y. Miura, J. Am. Chem. Soc. 123 (2000) 294.
56. Y. Teki, M. Nakatsuji, and Y. Miura, Mol. Phys. 100 (2002) 1385.
57. Y. Teki, M. Kimura, S. Narimatsu, K. Ohara, and K. Mukai, Bull. Chem. Soc. Jpn. 77 (2004) 95.
58. A. Caneschi, D. Gatteschi, and R. Sessoli, and P. Rey, Acc. Chem. Res. 22 (1989) 392.
59. K. Inoue, T. Hayamizu, H. Iwamura, D. Hashizume, and Y. Ohashi, J. Am. Chem. Soc. 118 (1996) 1803.
60. a) D.J.R. Brook, H.H. Fox, V. Lynch, and M.A. Fox, J. Phys. Chem. 100 (1996) 2066; b) D.J.R. Brook, V. Lynch, B. Conklin, and M.A. Fox, J. Am. Chem. Soc. 119 (1997) 5155.
61. a) R.G. Hicks, M.T. Lemaire, L.K. Thompson, and T.M. Barclay, J. Am. Chem. Soc. 122 (2000) 8077; b) R.G. Hicks, M.T. Lemaire, L. Ohrstrom, J.F. Richardson, L.K. Thompson, and Z. Xu, J. Am. Chem. Soc. 123 (2001) 7154.
62. M. Kurmoo, A.W. Graham, P. Day, S.J. Coles, M.B. Hursthouse, J.L. Caulfield, J. Singleton, F.L. Pratt, W. Hayes, L. Ducasse, and P. Guionneau, J. Am. Chem. Soc. 117 (1995) 12209.
63. E. Coronado, J.R. Galan-Mascaros, C.J. Gomez-Garcia, and V. Laukhin, Nature 408 (2000) 447.
64. H. Fujiwara, E. Fujiwara, Y. Nakazawa, B.Z, Narymbetov, K. Kato, H. Kobayashi, A. Kobayashi, M. Tokumoto, and P. Cassoux, J. Am. Chem. Soc. 123 (2001) 306, and references are cited therein.
65. T. Matsumoto, T. Kominami, K. Ueda, T. Sugimoto, T. Tada, S. Noguchi, H. Yoshino, K. Murata, M. Shiro, E. Negishi, N. Toyoda, S. Endo, and K. Takahashi, Inorg. Chem. 41 (2002) 4763, and references are cited therein.
66. T. Sugimoto, M. Tsujii, T. Suga, N. Hosoito, M. Ishikawa, N. Takeda, and M. Shiro, Mol. Cryst. Liq. Cryst. 272 (1995) 183, and reference are cited therein.
67. S. Nakatsuji, A. Kitamura, A. Takai, K. Nishikawa, Y. Morimoto, N. Yasuoka, H. Kawamura, and H. Anzai, Z. Naturforsch. 53b (1998) 495.
68. K. Mukai, S. Jinno, Y. Shimobe, N. Azuma, Y. Hosokoshi, K. Inoue, M. Taniguchi, Y. Misaki, and K. Tanaka, Polyhedron 20 (2001) 1537.
69. K. Mukai, S. Jinno, Y. Shimobe, N. Azuma, M. Taniguchi, Y. Misaki, K. Tanaka, K. Inoue, and Y. Hosokoshi, J. Mater. Chem. 13 (2003) 1614.

70. a) P. Cassoux, L. Valade, H. Kobayashi, A. Kobayashi, R.A. Clark, and A.E. Underhill, Coord. Chem. Rev. 110 (1991) 115; b) A.E. Pullen, and R.-M. Olk, Coord. Chem. Rev. 188 (1999) 211.
71. a) H. Imai, H. Otsuka, T. Naito, K. Awaga, and T. Inabe, J. Am. Chem. Soc. 121 (1999) 8098; b) S. Aonuma, H. Casellas, C. Faulmann, B. Garreau de Bonneval, I. Malfant, P. Cassoux, P.G. Lacroix, Y. Hosokoshi, and K. Inoue, J. Mater. Chem. 11 (2001) 337.
72. K. Mukai, N. Senba, T. Hatanaka, H. Minakuchi, K. Ohara, M. Taniguchi, Y. Misaki, Y. Hosokoshi, K. Inoue, and N. Azuma, Inorg. Chem. 43 (2004) 566.
73. K. Mukai, T. Hatanaka, N. Senba, T. Nakayashiki, Y. Misaki, K. Tanaka, K. Ueda, T. Sugimoto, and N. Azuma, Inorg. Chem. 41 (2002) 5066.

5

Quantum Spin Magnetism in Nitroxide-based Compounds

Y. Hosokoshi and K. Inoue

Osaka Prefecture University and Hiroshima University, Japan

1 Introduction

In the last decades, the magnetism of molecule-based material has drawn much interest. Since the first genuine organic ferromagnet was discovered in 1991 [1], more and more interest is given to organic magnetic materials which yield various spin networks. One of the characteristics of organic radicals consisting only of light elements, is the isotropic nature of electron spins because of the quenching of spin-orbital couplings. Another feature in organic crystals is low dimensionality due to the stacking structure of planar π-conjugated molecules. Low dimensional (Low-D) Heisenberg spin systems have attracted much attention because of their marvelous quantum effects. This chapter describes the quantum spin systems composed of organic radicals.

The interest for quantum spin systems originates from the Haldane's conjecture in 1983, in which different ground states are expected for the antiferromagnetic chain systems of integer and half-odd-integer spins [2]. It means the breaking of the universal rule where the ground state is independent of the spin quantum number. Intermediate between the $S = 1/2$ and $S = 1$ was studied by the $S = 1/2$ alternating chain with ferromagnetic and antiferromagnetic interactions [3]. Such a system was obtained using organic monoradical, MeNN [4]. As another kind of spin systems with both ferromagnetic and antiferromagnetic interactions, $S = 1/2$ distorted honeycomb lattices with ferromagnetic and antiferromagnetic interactions are also studied by organic monoradicals [5].

Intermolecular magnetic interactions originate from the molecular overlap. The molecular packings relevant to ferro- or antiferromagnetic interactions are classified in Ref. [6–8]. The magnitude of intermolecular antiferromagnetic interactions depends on the overlap between the molecular orbitals of unpaired electrons (SOMO's). Appearance of intermolecular ferromagnetic interactions, needs to minimize the overlap between the SOMO's [9]. It must be mentioned that it is difficult to obtain strong ferromagnetic interactions as intermolecular one.

In order to get strong ferromagnetic interactions, utilization of organic polyradical is efficient. An organic biradical, 1,3,-bis(*N*-*t*-butylaminoxyl)benzene, abbreviated as BNO, is known to have strong intramocelular ferromagnetic interactions and behaves as an $S = 1$ at room temperature [10]. There is a well-established guideline for the sign (ferromagnetic or antiferromagnetic) of the intramolecular exchange interactions between the unpaired electrons [11]. The topological arrangement of radical moieties governs the sign. In our study, stable nitroxides are employed as spin sources which are shown in Fig. 1. Another example of a biradical which behaves as an $S = 1$ at room temperature is PNNNO [12]. Systematic study on the PNNNO family revealed that the magnitude of the intramolecular interactions depends on the π-conjugation of the molecules [13]. Antiferromagnetic spin networks of the organic polyradicals give various quantum spin networks.

In this chapter, our recent study is presented on quantum spin systems by nitroxide based compounds. Section 2 deals with an $S = 1/2$ system. This is an example of molecular flexibility playing important role in the spin structure. In Section 3, the essence of our polyradical approach is presented. Section 4 describes the spin ladder compounds not only of $S = 1/2$ but also of $S = 1$. The mixed spin systems with $S = 1/2$ and $S = 1$ appear in Section 5 and 6. The first organic ferrimagnet is described in Section 5 and spin frustration due to molecular flexibility in Section 6. Section 7 deals with pressure effects on 'soft' molecular magnetic materials. Field induced phenomena appear in Section 8. In this chapter, the magnetic exchange coupling between $\boldsymbol{S}_i$ and $\boldsymbol{S}_j$ is written as $-2J\boldsymbol{S}_i \cdot \boldsymbol{S}_j$.

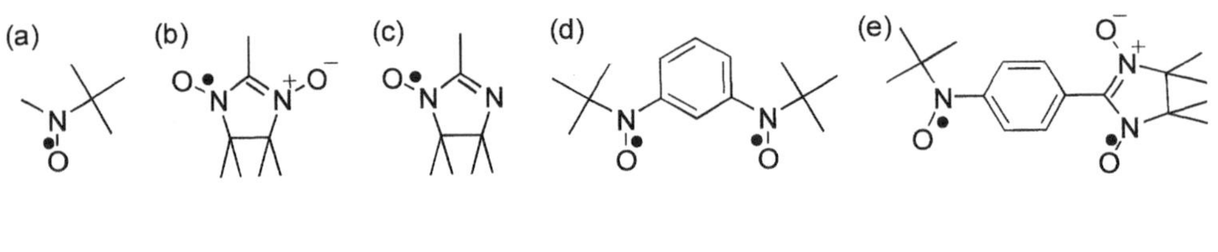

Fig. 1. Molecular structure of organic radicals appeared in this chapter. (a) *N-tert*-butyl nitroxide (abbreviated as *t*BuNO), (b) nitronyl nitroxide (= 4,4,5,5-tetra-metyl-4,5-dihydro-1*H*-imidazol-1-oxyl 3 oxide, abbreviated as NN), (c) imino nitroxide (= 4,4,5,5-tetramethyl-4,5-dihydro-1*H*-imidazol-1-oxyl, abbreviated as IM), (d) BNO, (e) PNNNO.

2 A Zigzag Spin Chain, F_2PIMNH [14]

2.1 A Zigzag Spin Chain with the Next-nearest-neighbouring Interactions

There are many examples of one-dimensional (1D) compounds, but not so many of double-spin chains. Recently, double-spin chain has drawn much attention because it is intermediate between one- and two-dimensional (2D) systems. Figure 2(a) shows a simple model of an 1D chain with the exchange coupling J_1 between neighbouring spins. If each spin has the next-nearest-neighbour interactions (abbreviated as the NNN interactions, J_2) as shown in Fig. 2(b), the model is re-drawn to a zigzag uniform chain with J_1 having the NNN interactions, J_2, like Fig. 2(c). If the interaction within a zigzag chain is alternately changed with J_1 and J_3, the model becomes the one drawn as Fig. 2(d). In the extreme case of J_3=0, the model is reduced to a two-leg spin ladder (Fig. 2(e)). The synthetic strategy for a spin ladder will appear in Section 4. In this section, we describe the zigzag spin chain with the NNN interactions of Fig. 2(c).

The model of Fig. 2(c) is also called as a railroad trestle model. The Hamiltonian can be written as

$$H = -2J_1 \sum_{i=1}^{N} \mathbf{S}_i \bullet \mathbf{S}_{i+1} - 2J_2 \sum_{i=1}^{N} \mathbf{S}_i \bullet \mathbf{S}_{i+2} . \tag{1}$$

In the case of J_1, $J_2 < 0$, spin-frustration is expected. In spite of the physical interests, experimental study has long prevented by the absence of the model compound. We succeeded in synthesizing the compound of a zigzag chain with the NNN interactions and observed an energy gap for the first time [14].

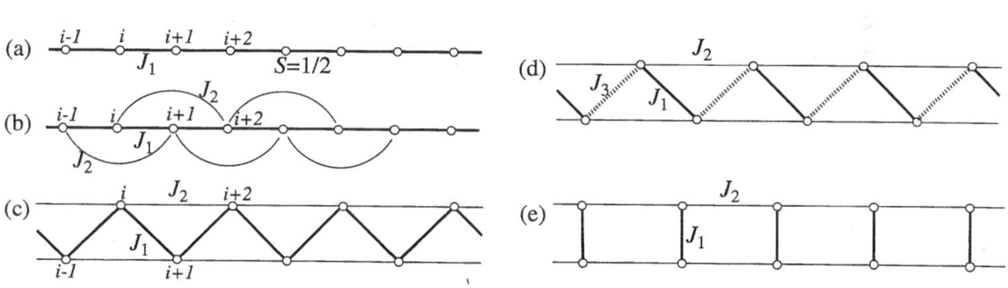

Fig. 2. Schematic illustration of S = 1/2 chain models. (a) A simple uniform chain with the nearest neighbour coupling J_1. (b) A uniform chain coupled with J_1 and the next-nearest-neighbour (NNN) coupling, J_2. (c) A zigzag uniform chain with the NNN interactions which is the same model with (b). (d) A zigzag alternating chain with J_1 and J_3, having NNN coupling J_2. This model is reduced to (c) when $J_1 = J_3$. (e) Two-leg spin ladder which corresponds to the model (d) with $J_3 = 0$.

2.2 Materials and Crystal Structures

We synthesize two related compounds of F_2PIMNH and Cl_2PIMNH. Molecular structure of F_2PIMNH is shown in Figure 3 as an inset. Replacement of the fluorine atoms by chlorine atoms yields Cl_2PIMNH. Both compounds crystallize to form zigzag chains by the hydrogen bondings between the O atom in the NO group and the H atom in the NH group. The interatomic distances are O...N 3.136(8) Å (O...H 2.254 Å) for F_2PIMNH and O...N 3.032(4) Å (O...H 2.128 Å) for Cl_2PIMNH. The chain structure is similar to each other. Figure 3(a) shows the chain structure of F_2PIMNH. The difference between the two compounds lies in the second contact between the next-nearest-neighbour (NNN) molecules within a chain. The interatomic distance between the N atom of the NH group and O atom of the NO group is 5.372(9) Å in F_2PIMNH. This kind of interatomic distance is longer in Cl_2PIMNH (6.236(4) Å). The different NNN contact comes from the difference in the molecular distortion. The dihedral angle between the imino nitroxide (IM) plane and the phenylene ring in F_2PIMNH and Cl_2PIMNH

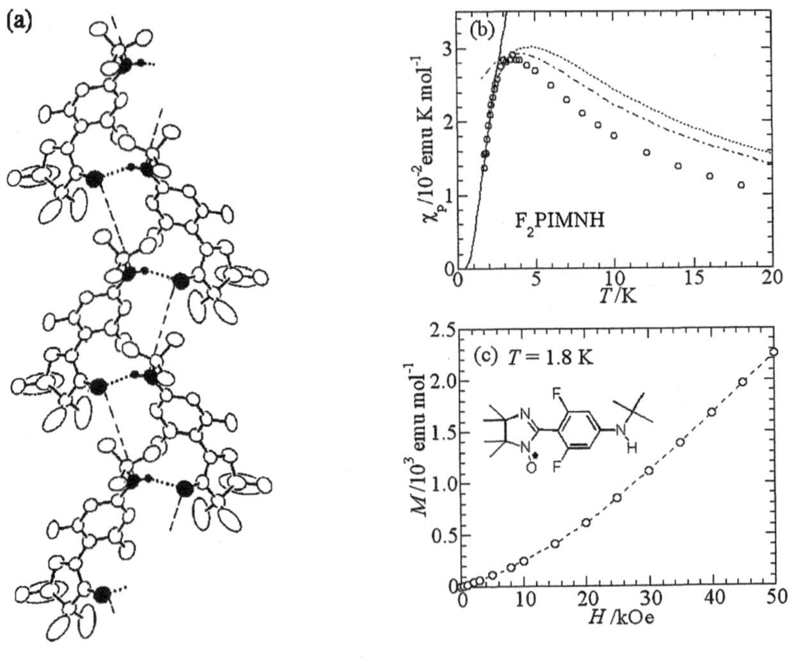

Fig. 3. (a) Crystal structure of F_2PIMNH viewed along the *b* axis. *Pca*2_1, a = 12.220(2) Å, b = 13.819(2) Å, c = 10.673(3) Å, V = 802.3(5) Å^3, Z = 4. The O atom of the NO group and the N and H atoms in the NH group are represented by solid ellipsoids. Dotted lines between the O and H atoms correspond

is 64° and 79°, respectively. The large dihedral angle is related to the electrostatic repulsion between the halogen atom on the phenylene ring and oxygen atom in the IM unit. It is strong contrast that many phenyl nitronyl nittroxides have small dihedral angle *ca.*20-30° by the weak hydrogen bonding between the O atom of the NO group and the H atom of the phenyl ring [6, 7]. The dihedral angle plays important role in the molecular packing in the crystals [1b, 13, 15]. The multi-dimensional network is always formed by the molecules having the large dihedral angles.

2.3 Magnetic Properties

The temperature dependences of the paramagnetic susceptibility (χ_p) of Cl_2PIMNH shows a round peak at around 2.0K which is well reproduced by the Heisenberg uniform chain model [16] with $2J_1/k_B = -3.2$K. Therefore, the magnetism of Cl_2PIMNH is understandable only by considering the nearest-neighbour exchange coupling along the hydrogen bonding.

On the other hand, the values of χ_p in F_2PIMNH take a maximum at around 3.4K, and below this temperature, decrease rapidly as decreasing temperature. As shown in Fig. 3(b), the value of χ_p goes exponentially to zero as $T \rightarrow 0$. We roughly estimate the magnitude of the gap (Δ) by fitting the temperature dependence o f χ_p below 2.4K by the following relation, $\chi \propto \exp(-\Delta/T)$. The solid curve in Fig. 3(b) represents the calculation for $\Delta = 3.1$K. The magnetization isotherm at the lowest temperature (1.8K) is shown in Fig. 3(c). The concave curve also suggests the existence of an energy gap above the singlet ground state. The Δ value estimated corresponds to 23 kOe, at which we can find the bend of the magnetization isotherm.

The gapless ground state in the $S = 1/2$ antiferromagnetic uniform chain is exactly elucidated. For the explanation of the gap observed in F_2PIMNH, we must introduce other interactions to the system. From the crystal structure, the NNN contacts are considered to play a significant role in the ground state with a gap. The magnetic model of F_2PIMNH is the one shown in Fig. 2(c).

There are several theoretical works concerning the properties of the $S = 1/2$ antiferromagnetic Heisenberg chain with the NNN interactions. In the case of $J_2/J_1 = 1/2$, the exact ground state is obtained [17]. The ground state when J_2/J_1 is a dimer state and there is an energy gap between the ground state and the first excited state [18]. According to the numerical study by Okamoto and Nomura [19], this model has an energy gap in the case of $J_2/J_1 > 0.2411$.

We cannot estimate the exchange couplings of J_1 and J_2 in F_2PIMNH, because there are no calculated susceptibilities on this model which we can use for the analysis [20]. The treatment of the NNN interactions by the mean-field approximation to the antiferromagnetic uniform chain cannot simulate the experimental results. This means that the NNN interactions (J_2) in this compound are of considerable magnitude compared with the nearest-neighbor interactions (J_1).

Another possible origin of the energy gap in F_2PIMNH is the breaking of symmetry at low temperatures due to a structural change. The crystals of F_2PIMNH involve carbon atoms in a *t*-butyl group with large thermal parameters. In such a case, there is a possibility of the symmetry breaking at the freezing point of the thermal vibration [6].

This kind of structural change always occurs around 100K. Thus, we performed an X-ray study of F_2PIMNH down to 80K and could not detect any breaking of symmetry in the crystal structure. In addition to this fact, the observed susceptibility data cannot be explained by the antiferromagnetic alternating chain model [21]. Therefore, we can conclude that the NNN interaction is the origin of the excitation energy gap in F_2PIMNH.

In summary, hydrogen bonded spin chains are presented. We have observed the singlet ground state and the existence of the finite excitation energy gap in F_2PIMNH. Although Cl_2PIMNH has a crystal structure similar to that of F_2PIMNH, it has a gapless ground state. The difference is related to the difference in magnitude of the NNN interactions. The large dihedral angle between the radical and phenylene planes, 64°, yields significant NNN interactions in F_2PIMNH. This is the first observation of the energy gap in a zigzag spin chain with the NNN interactions. We believe that this finding will contribute toward the progress of the study of the ground state properties of double chain systems.

3 Polyradical Approach

3.1 Magnetic Exchange Couplings

Although the wide variety of spin network can be constructed by the use of the variety of organic molecules, full control of molecular packings in crystals is difficult so far. As for the intramolecular magnetic interactions, both the size and the magnitude are controllable. We can design the spin cluster within which radical moieties are connected by intramolecular magnetic interactions. The advantage of the polyradical approach is to get robust high spin species.

The intramolecular magnetic exchange in an organic biradical BNO is exceptionally large. It is ten times as large as the one in the *meta*phenylen likage of NN's [22]. The small exchange coupling of NN is probably due to the fact that central carbon atom in the NN is a node of the SOMO. There are a few examples of organic biradicals which behave as $S = 1$ species at room temperature. Another example of a robust $S = 1$ species having strong intramolecular ferromagnetic exchange coupling is PNNNO [12]. In both compounds, the ferromagnetic exchange coupling between the radical moieties within a molecule is understood by the topological rule [11]. The PNNNO family is stable and single crystals are easily obtained. We have studied the series compounds of the PNNNO family [13] and the relation between the magnitude of the exchange coupling and the molecular structure are listed in Table 1.

The magnitude of the exchange couplings depends not only on the separation of the radicals but also on the dihedral angle between the radical planes. The reduction of the exchange coupling in PIMNO supports the canonical contribution of the quinoidal structure in PNNNO [12]. The magnitude of the magnetic exchange coupling is related to the delocalization of the unpaired π electrons. The delocalization of the unpaired electron is studied in UV/VIS absorption spectra. The bathochromic shift corresponds to greater delocalization of the unpaired π electrons. Observed absorption maxima exhibit

Table 1.
Magnetic exchange couplings in the PNNNO family.

Compounds		$2J/k_B$	θ_1	θ_2	(bip)[a]
PNNNO	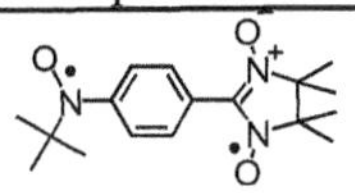	638K	22°	28°	
PIMNO	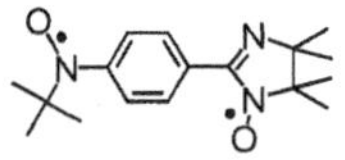	216K	37°	25°	
F_2PNNNO	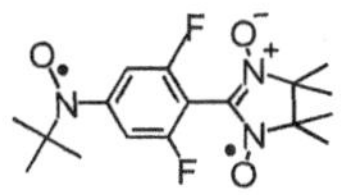	407K	2°	53°	
BIPNNNO	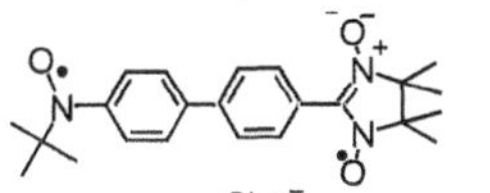	100K	30°	21°	23°
Cl_2PNNNO	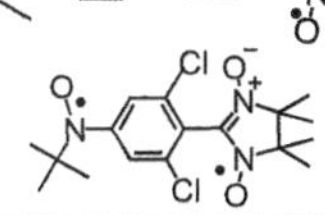	160K			

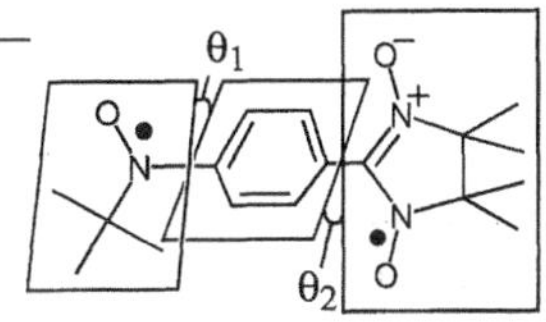

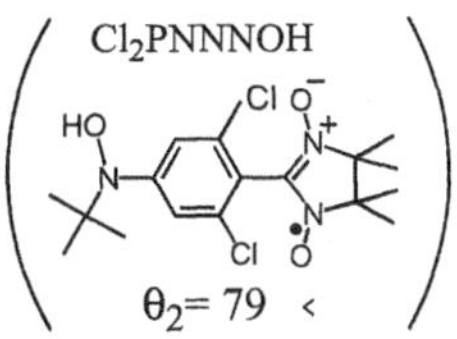

a) Dihedral angle between the two phenylene rings.
b) Single crystals of Cl_2PNNNO are not obtained due to the stability problem.

the same tendency to the exchange couplings. The exchange coupling in Table 1 was estimated by fitting the susceptibility (χ_p) data using the spin model based on the crystal structure. The measurements of χ_p were also done for isolated molecules dispersed in polyvinylchloride (PVC) matrix, and the exchange couplings estimated agree well with the ones for single crystals.

3.2 Molecular Designing

As mentioned above, both the sign and the magnitude of intramolecular magnetic exchange couplings are controllable. We can design spin clusters made of organic polyradicals having several radical units within a molecule. We must mention that the electrostatic energy plays substantial role in controlling the molecular packing even in the neutral organic crystal [7, 8, 14]. The partial charges on the O and N atoms on the NO group prefer the intermolecular contacts between them, which yields the antiferromagnetic exchange interactions [8]. In our polyradical approach using nitroxide-based compounds, first, a spin cluster is made by using intramolecular ferro- and antiferromagnetic exchange couplings. Next, the spin clusters are antiferromagnetically coupled by the intermolecular contacts between the NO groups.

The spin cluster including a BNO moiety as an $S = 1$ species is designed. The spin pair of BNO is named as BIP-TENO. The exchange coupling scheme within a BIP-TENO molecule is shown in Fig. 4(b) [23]. The related biradical, BIP-BNO is

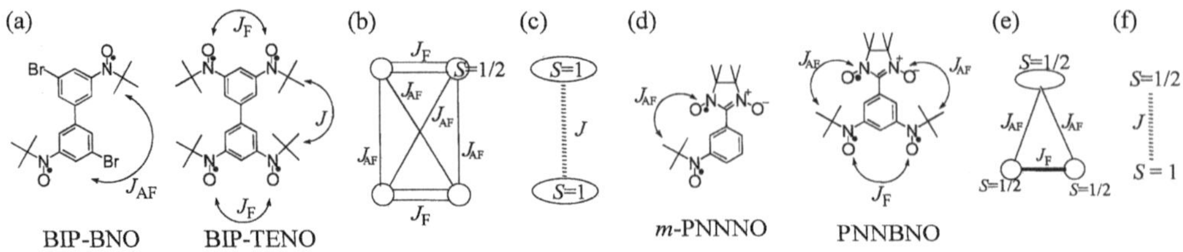

Fig. 4. (a) Molecular structures of BIP-BNO and BIP-TENO. (b) Schematic diagram of the magnetic model of a BIP-TENO molecule. A circle, a double bond and a single bond represent an $S = 1/2$ spin, a ferromagnetic exchange coupling, and an antiferromagnetic exchange coupling, respectively. (c) Extreme limit of the model (b) when $J_F \rightarrow \infty$. (d) Molecular structures of *m*-PNNNO and PNNBNO. (e) Schematic illustration of the magnetic model of a PNNBNO molecule. (f) Extreme limit of the model (e) when $J_F \rightarrow \infty$.

also studied and the exchange coupling within a molecule is estimated to $2J_{AF}/k_B = -67$K [24]. The measurements of χ_p for isolated BIP-TENO molecules was done and the temperature dependence obeys the isolated dimer model for $S = 1$ with the exchange coupling of $2J/k_B = -20$K [24]. The analysis based on the 4-spin cluster model in Fig. 4 (b) gives the ferromagnetic exchange coupling in a BNO unit to be $2J_F/k_B \geq 590$K.

The spin pair of an $S = 1$ and $S = 1/2$ is made by the combination of BNO and *m*-PNNNO. The antiferromagnetic exchange through the *m*-phenylene linkage of NN and *t*BuNO is proved by the study of *m*-PNNNO. In the crystals, each molecule is considerably separated to each other, and the intermolecular interaction is considered to be negligibly small. The temperature dependence of χ_p are well reproduced by the isolated dimer model with $2J_{AF}/k_B = -136$K [25]. The dihedral angles are $\theta_1 = 34.5°$ and $\theta_2 = 32°$ in the same definition in Table 1. The scheme of the exchange coupling in a PNNBNO molecule is shown in Fig. 4(e). The temperature dependence of χ_p of isolate PNNBNO molecules are analysed based on the 3-spin model and magnetic exchange couplings are estimated to $2J_F/k_B \geq 860$K and $2J_{AF}/k_B = -160$K [26]. At low temperature, a PNNBNO molecule behaves as the pair of an $S = 1/2$ and $S = 1$.

The intermolecular contacts between the NO groups are favorable from the electrostatic point of view. Stacking of these dimeric molecules in one-dimension results in the formation of the spin ladder.

4 Spin Ladder

4.1 Designing of Spin Ladder

There has been considerable current interest in the study of a low-dimensional quantum-spin system with an energy gap. Among them, the antiferromagnetic ladder system has been extensively investigated because it is related to the Haldane state and high-T_C superconductivity. Daggoto *et al.* [27]. and Rice *et al.* [28]. suggested that the ladder system is in a spin liquid state and has a spin gap. They also implied that when holes are lightly doped, the spin gap would remain and superconductivity possibly appears. The existence of the spin gap and superconductivity has been experimentally confirmed [29]. Experimental studies on spin ladder systems have been devoted mainly to the $S = 1/2$ systems, but $S = 1$ system is also attractive to study. For the ground state of the $S = 1$ ladder with antiferromagnetic legs ($J_{\parallel}$) and rungs ($J_{\perp}$), the Haldane state is expected in the extreme limit of $J_{\perp} \rightarrow 0$, and the dimer state in $J_{\parallel} \rightarrow 0$. Its ground state phase diagram is quite interesting. Our success in preparing an $S = 1$ ladder has stimulated the theoretical study. The ground state properties of an $S = 1$ antiferromagnetic ladder is investigated by the quantum Monte Carlo simulation, which revealed that there is no quantum phase transition between the Haldane state and the $S = 1$ dimer state [30].

We employed *t*BuNO as a spin source of $S = 1/2$, focusing on the stability and controllability of the crystal structure. Antiferromagnetic rungs and legs are composed of intra- and intermolecular antiferromagnetic interactions, respectively. In a BIP-BNO molecule, two spins of $S = 1/2$ are coupled antiferromagnetically each other. As an $S = 1$ species, BNO is used. A BIP-TENO molecule includes two BNO moieties coupled antiferromagnetically. The 1D stacking of the molecules results in a spin ladder. The positive and negative partial charges on the N and O atoms of the NO group are preferable to the intermolecular contacts between them. The intermolecular interactions between the NO groups work as the antiferromagnetic legs.

4.1 An S=1/2 Ladder, BIP-BNO

BIP-BNO molecules crystallize to a monoclinic system, *C*2/*c*, a = 14.333(3) Å, b = 12.583(3) Å, c = 11.567(2) Å, β = 91.74(2)°, V = 2085.2(6) Å^3, Z = 4, and by the *c*-glide reflection symmetry, a uniform chain structure is formed [23]. The intermolecular contact is N···O 4.654(6) Å, which is doubled by the two-fold symmetry within a molecule. Two NO groups in a BIP-BNO molecule are crystallographically equivalent. Thus, the system is regarded as a two-leg spin ladder system. As is shown in Fig. 5(a), the temperature dependence of χ_p exhibits a broad maximum at 45K. The χ_p value decreases as T decreases and goes exponentially to zero. We fit the observed data using the formula, $\chi \propto \exp(\Delta/T)/\sqrt{T}$, given by Troyer *et al.* [31], and estimate the excitation gap Δ = 47K [23]. In order to estimate the exchange coupling, we perform exact diagonalizations of a ladder chain with even length up to 16 spins and calculate the susceptibility [24]. The Hamiltonian used is

$$H = -2J_{\perp}\sum_i \boldsymbol{S}_{2i-1}\cdot\boldsymbol{S}_{2i} - 2J_{//}\sum_i (S_{2i-1}\cdot S_{2i+1} + \boldsymbol{S}_{2i}\cdot\boldsymbol{S}_{2i+2})\,. \quad (2)$$

The exchange couplings of $2J_{\parallel}/k_B$ = –25K and $2J_{\perp}/k_B$ = –67K ($J_{\parallel}/J_{\perp}$ = 0.38) well reproduce the experiments. The calculation using this parameter set is represented in Fig. 5 (a) as the dotted curve.

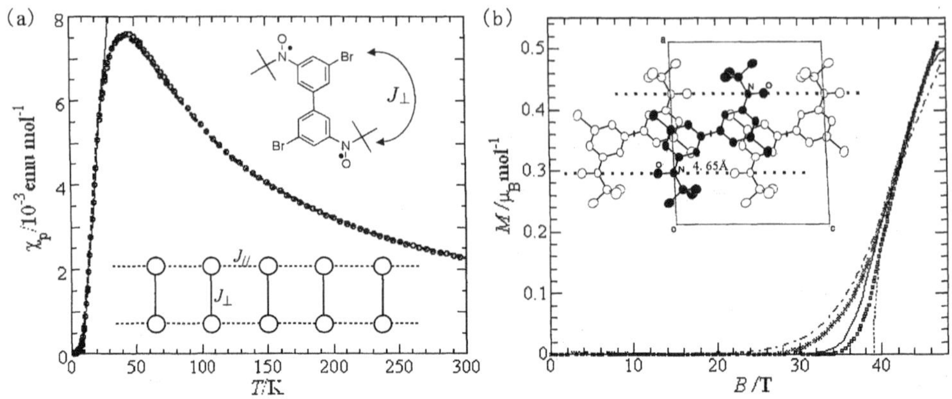

Fig. 5. (a) Temperature dependence of χ_p of BIP-BNO. Open circles represent observed data. The solid curve is the estimation the spin gap (Δ = 47K). The dotted curve is the calculation with $2J_{\parallel}/k_B = -25$K and $2J_{\perp}/k_B = -67$K ($J_{\parallel}/J_{\perp} = 0.38$). (b) Magnetization curves of BIP-BNO at 1.6 (circle) and 4.2K (diamond). Solid and dotted broken curves represent the calculations for 1.6 and 4.2K, respectively. Dotted curve is the estimation of the spin gap, 38.8 T (Δ= 52K). Inset: Crystal structure of BIP-BNO.

Figure 5 (b) shows the magnetization isotherms of BIP-BNO measured at 1.6 and 4.2K using a induction method in pulsed magnetic fields up to about 50 T with a time duration of 15 ms produced by a wire-wounded pulse magnet. The finite magnetization is not observed below 35 T, above which it begins to grow. By the same method used in the literature [32], we estimate the critical field of 38.8 T, which corresponds to the excitation gap of Δ = 52K. This value is consistent with the one estimated from the temperature dependence of the susceptibility. The magnetic exchange couplings are again estimated by comparing with our calculations. The exchange couplings of $2J_{\parallel}/k_B = -28$K and $2J_{\perp}/k_B = -73.5$K ($J_{\parallel}/J_{\perp} = 0.38$) well reproduce the experiments. Solid and dotted broken curves in Fig. 5(b) represent calculations. The estimated values of magnetic interactions are consistent between the magnetizations and susceptibilities [24].

4.2 An S=1 ladder, BIP-TENO

The BIP-TENO molecules crystallize to an orthorhombic system, *Pbcn*, a = 24.517(3) Å, b = 12.163(6) Å, c = 9.234(3) Å, V = 2753(2) Å^3, Z = 4 [23]. A BIP-TENO molecule includes a twofold symmetry, therefore two BNO units are crystallographically equivalent to each other. The molecules stack themselves with their molecular plane parallel one another. By the c-glide reflection symmetry, uniform chain structure with the N···O contacts is formed. Interatomic distances are O···N; 3.705(8) Å and O···O; 3.850(9) Å. These contacts are doubled by a twofold symmetry. The crystal

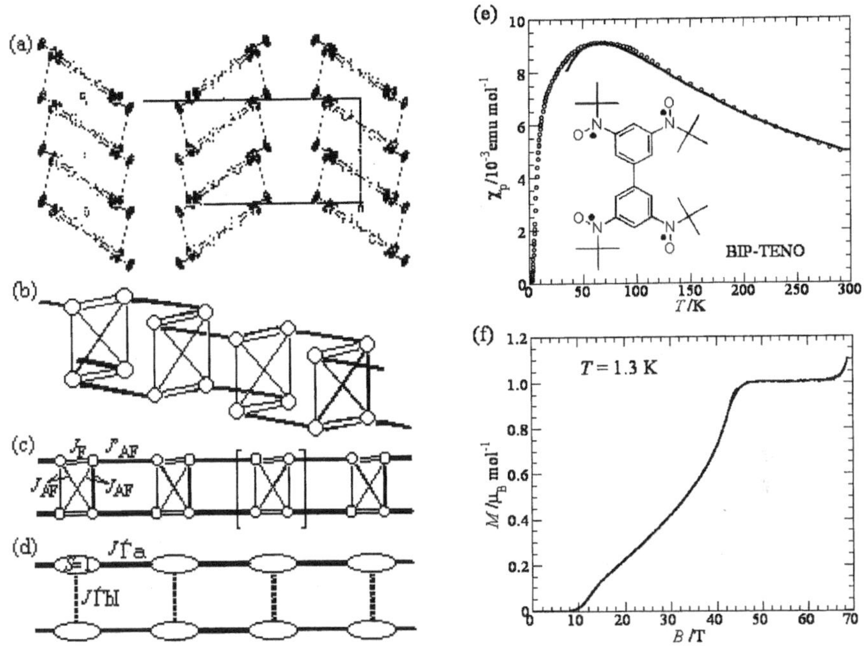

Fig. 6. (a) Crystal structure of BIP-TENO projected onto the *ac* plane. The NO groups are represented by solid ellipsoids. Thin lines correspond to the nearest-neighbor contacts of O···N (3.705(8) Å). (b) Schematic illustration of the chain structure of BIP-TENO along the *c* axis. A BIP-TENO molecule is represented by the same definition used in Fig. 4(b). A bold line represents the nearest-neighbor coupling between the molecules. (c) Magnetic model of the crystals of BIP-TENO. The definitions of the symbols are the same as those in (b). (d) The extreme limit of the model (c) when $J_F \to \infty$. (e) Temperature dependence of χ_p of BIP-TENO. Solid curve is the calculation for the $S = 1$ chain model with $2J_{\parallel}/k_B = -50$K adopted interchain interactions of $2J_{\perp}/k_B = -42$K by the mean-field treatment. (f) Magnetization curve of BIP-TENO at 1.3K.

sructure of BIP-TENO viewed along the *b* axis is shown in Fig. 6(a). The schematic illustration of the chain structure and the scheme of magnetic exchange is drawn in Figs. 6(b) and (c), respectively. An open circle means the NO group as an $S = 1/2$. A double bond and a single bond correspond to the intramolecular ferro- and antiferromagnetic exchange coupling, respectively. A bold line represents an intermolecular contact between the NO groups. A molecule is shown by a bracket. In the extreme limit of $J_F \to \infty$, this can be regarded as a ladder system with $S = 1$ shown in Fig. 6(d).

The temperature dependence of χ_p is characterised by the very broad peak at around 67K and the shoulder at around 15K as is shown in Fig. 6(e). The value of χ_p goes down to zero as $T \to 0$. The singlet ground state of this material is suggested. We have not yet succeeded in analyzing the whole temperature dependence of χ_p. The data above 50K is reproducible by some models [24]. The solid curve in Fig. 6(e) is the calculation with the S=1 antiferromagnetic chain of $2J\ J_{||}/k_B$ = –50K having interchain interactions of $2J_{\perp}/k_B$ = –42K (adopted by the mean-field treatment). This approximation is too rough to reproduce the shoulder at around 15K, and the estimated values of antiferromagnetic exchange couplings include ambiguity. Nevertheless, we are confident that at least two types of antiferromagnetic exchange couplings of $2J/k_B \sim$ 50K exist.

In Fig. 6(f), the magnetization curves of BIP-TENO crystals measured at 1.3K in pulsed high magnetic fields are shown [33]. The magnetization below 9 T is almost zero, and the singlet ground state is suggested. On assumption of the excited triplet state, the excitation gap is estimated to be Δ = 15.6K with B_{C1} = 11.6 T. Above a critical field B_{C1}, the magnetization increases rapidly. The increase seems to hold at B_{C2} = 44.8 T. The value of magnetization above B_{C2} is just a quarter of the complete saturation value (Ms = 4 μ_B/f.u.). This indicates that the 1/4 plateau appears just above B_{C2}. The plateau continues up to B_{C3} = 67 T, and a spin gap appears again above B_{C3}.

There are some spin systems having magnetization plateau. In some cases such as the alternating antiferromagnetic chain of S = 1 [34] and a 2D antiferromagnetic lattice of ferromagnetic dimmers[13], the local structure of Hamiltonian is the origin of the plateau. However, the local structure in BIP-TENO from which the 1/4 plateau originates, has not been observed, although the 1/2 plateau can appear in BIP-TENO. The mechanism of the occurrence of the 1/4 plateau is still open to study. At the same energy scale with the excitation gap, an anomaly is observed in the temperature dependence of χ_p as the shoulder at about 15K. Some theoretical works concerning the mechanism of 1/4 plateau of this compound were reported [35]. The possible appearance of the cusp in magnetization is pointed out [33]. The possibility of the structural change at low temperature and under magnetic fields is the problem to be examined also.

5 Ferrimagnetism

5.1 Single-component Strategy

In most organic radical crystals, the antiparallel spin alignment is much stable than the parallel one. Ferrimagnetism has long been considered as an effective strategy to give organic materials spontaneous magnetizations by the antiferromagnetic exchange coupling between two kinds of organic radicals having different spin-multiplicities [36]. There are many ferrimagnets reported among inorganic-organic hybrid systems. However, it is difficult to get a genuine organic ferrimagnet by mixed crystals of two radical components.

In order to achieve this subject of an organic ferrimagnet from a different viewpoint, we have proposed a single-component strategy: utilizing a triradical including an S = 1 and an S = 1/2 units within a molecule and connecting the S = 1 and S = 1/2 units by intra- and intermolecular antiferromagnetic interactions. According to this strategy, we have succeeded in achieving the first example of organic ferrimagnets. The organic triradical, PNNBNO forms a ferrimagnetic spin ladder and inter-ladder interaction is also ferrimagnetic. Magnetic ordering in three-dimension is proved by the heat capacity measurements.

5.2 The First Organic Ferrimagnet, PNNBNO [26]

The crystals of PNNBNO belong to the orthorhombic system, *Pnma*, a = 11.652(1) Å, b = 9.6117(9) Å, c = 20.605(2) Å, V = 2307.6(3) Å^3, and Z = 4. The molecule lies on the mirror plane and is completely planar. The PNNBNO molecules stack along the b axis in the 'head-to-tail' fashion: the S = 1 unit of BNO and the S = 1/2 unit of NN is overlapped to each other as is shown in Fig. 7(a). The interatomic distance between the O atom of the NN and the N atom of the *t*BuNO is 4.91 Å, which is indicated by a broken line in Fig. 7(a). This contact affords the intermolecular antiferromagnetic exchange coupling. The molecules are related to the two-fold screw symmetry and the scheme of the magnetic exchange couplings along the b axis is drawn in Fig. 7(b), where J_F and J_{AF} denote the intramolecular ferro- and antiferromagnetic exchange couplings, respectively, and J'_{AF} is intermolecular antiferromagnetic exchange coupling. This can be regarded as a ferrimagnetic ladder with rungs of intramolecular antiferromagnetic interactions and legs of intermolecular antiferromagnetic interactions between the S = 1/2 and the S = 1. The model of the extreme limit of $J_F \rightarrow \infty$ is drawn in Fig. 7(c). The important point of the crystal structure of PNNBNO is the interladder interactions. As is shown in Fig. 7 (d), each ladder is surrounded by four neighboring ladders with the contacts between NN (S = 1/2) and *t*BuNO's (S = 1) of 5.5–5.9 Å. Therefore, alternant arrangement between S = 1 and S = 1/2 in three-dimension is realized in PNNBNO crystals.

Figure 7(e) shows the temperature dependence of the product of χ_p and temperature, $\chi_p T$ (μ_{eff}^2, square of the effective magnetic moment). The values of $\chi_p T$ decrease with decreasing T, through a broad minimum with the temperature range of 20–50K, and then increase below 20K. The stationary value within 20–50K (*ca.* 0.375 emu·K·mol^{-1}) agrees well with the expectation of the formation of 1 mol of S = 1/2 species. It is evident that intermolecular antiferromagnetic exchange coupling between the S = 1/2 and S = 1 unit exists, because without intermolecular interaction, the values of χ_pT should approach the constant value with decreasing T. The increase of χ_pT below 20K means the development of the ferrimagnetic correlation between the S = 1/2 and the S = 1 species. The observed susceptibility is analyzed using the numerical calculations for the ferrimagnetic ladder model (Inset of Fig. 7(e)) with even length up to 12 spins of S = 1/2. The solid curve in Fig. 7(e) represents the calculation using the parameter set of 2JF/kB = 860K, 2JAF/kB = –216K, and 2J'AF/kB = –0.6K, which well reproduces the observed data. The temperature dependence of the heat capacity of PNNBNO is shown in Fig. 7(f). The thermal relaxation technique was used for a single crystal. The λ-shaped peak is observed at 0.28 K. The total magnetic entropy gain (ΔS) is estimated by the integration of C/T with T. The simple assumption based on the spin-wave model is used for the extrapolation of low temperature part. The estimated value of ΔS at 3.3K reaches 91% of NAkBln2, which is the theoretical expectation for the total magnetic entropy of 1 mol of S = 1/2 species. It is evident that the effective S = 1/2 species (ferrimagnetic spins) undergo a magnetic phase transition at 0.28K. The three-dimensional (3D) nature of this material in the even lower-temperature region is proved.

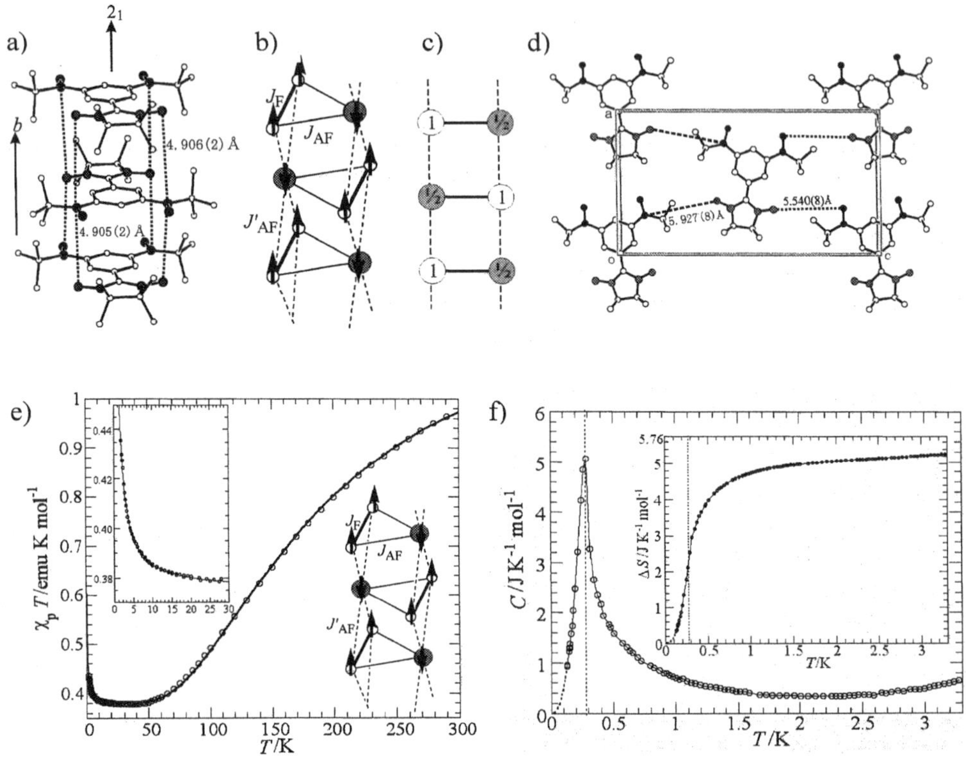

Fig. 7. (a) Uniform chain in the crystals of PNNBNO. (b) Schematic illustration and magnetic exchange coupling in PNNBNO along the b axis. (c) Ferrimagnetic ladder of $S = 1$ and $S = 1/2$ corresponding to the model (b) when $J_F \rightarrow \infty$. (d) Crystal structure of PNNBNO in the ac plane at $y = 1/4$. (e) Temperature dependence of $\chi_p T$ of PNNBNO crystals. (e) Temperature dependence of heat capacity of PNNBNO. Dotted line indicates transition temperature ($T_C = 0.28$K). Inset: Estimated entropy change.

6 Spin Frustration

6.1 A Mixed-spin System with S=1 and S=1/2, BIPNNBNO

Frustration effects in low-D antiferromagnetic quantum spin systems have attracted much attention because of their peculiar behaviour in low-energy physics. The quantum effect and the geometrical frustration cooperatively induce a strong fluctuation, which often gives rise to various non-magnetic ground states and quantum phase transitions. A typical example of such frustrated low-dimensional magnets is the $S = 1/2$ antiferromagnetic zigzag chain, which is described in Section 2. In this section, a novel frustrated spin system in 2D is presented.

The novel triradical, BIPNNBNO is regarded as the antiferromagnetic spin pair of $S = 1$ and $S = 1/2$. Replacement of the benzene ring in PNNBNO in previous section by biphenyl ring, gives the BIPNNBNO molecule. The measurements of χ_p of isolated molecules in PVC matrix, were done and the exchange couplings in a BIPNNBNO molecule are estimated to $2J_F/k_B = 860K$ and $2J_{AF}/k_B = -26K$. The magnitude of J_{AF} is reduced by 1/7. The reduction ratio is almost the same as the one for J_F between PNNNO and BIPNNNO described in Section 3. The BIPNNBNO molecule can be regarded as the spin pair of $S = 1$ and $S = 1/2$ because the magnitude of J_F is 30 times as large as the one of J_{AF}, whereas PNNBNO is described 3-spin model of $S = 1/2$ rather than the spin pair of $S = 1$ and $S = 1/2$. The molecules of BIPNNBNO also form ferrimagnetic chains. The flexibility of the biphenyl ring results in the interactions between the next-nearest-neighbouring (NNN) chains adding to that between the nearest neighboring chains, which gives rise to the spin frustration.

6.2 Magnetic Properties of BIPNNBNO

The crystal structure of BIPNNBNO is orthorhombic (*Pbcn*, $a = 9.0126(7)$ Å, $b = 17.578(1)$ Å, $c = 33.924(3)$ Å, $V = 5374.3(6)$ Å^3, $Z = 8$). Along the *b* axis, a uniform chain is formed by the contact between the NN and *t*BuNO. The interatomic distance between the O atoms is 3.951 Å. This contact induces antiferromagnetic exchange coupling between the $S = 1/2$ (NN) and $S = 1$ units (BNO). The alternant alignment between the $S = 1$ and $S = 1/2$ by the intra- and intermlecular antierromagnetic interactions construct a ferrimagnetic chain along the *b* axis. Between chains, we can find two kinds of intermolecular contacts along the *a* axis. The one is seen between the nearest neighbouring chains by the contact between the NN's. The interatomic distances between the O and N atoms are 4.42 and 4.74 Å. This contact yields antiferromagnetic exchange coupling between the two spins of $S = 1/2$. The other is the contact between the NNN chains. The interatomic distance between the O atoms of *t*BuNO's is 4.66 Å. By this contact, antiferromagnetic exchange coupling is given between the $S = 1$ units (BNO). Schematic illustration of the crystal structure is shown in Fig. 8(a). All of the three kinds of intermolecular interactions are antiferromagnetic. On the assumption of the ferrimagnetic spin alignment along the *b* axis, the antiferromagnetic interactions between the nearest neighbouring chains contradict the antiferromagnetic interactions between the NNN chains. That is, spin frustration occurs.

The value of χ_p of BIPNNBNO increases monotonically with increasing temperature and exhibits a broad peak at around 16K. Below 10K, the value again increases, reaches a maximum at 4K, and then abruptly decreases toward zero. The ground state of this compound is singlet with a spin gap. The behaviour of χ_p above 40K is identical between the crystalline state and the isolated molecules dispersed in PVC matrix. Therefore, the curious behaviour below about 20K is attributed to the effect of intermolecular interactions. Similar temperature dependence of double peaks was reported in another spin-frustrated system of an $S = 1/2$ antiferromagnetic Kagome lattice [37].

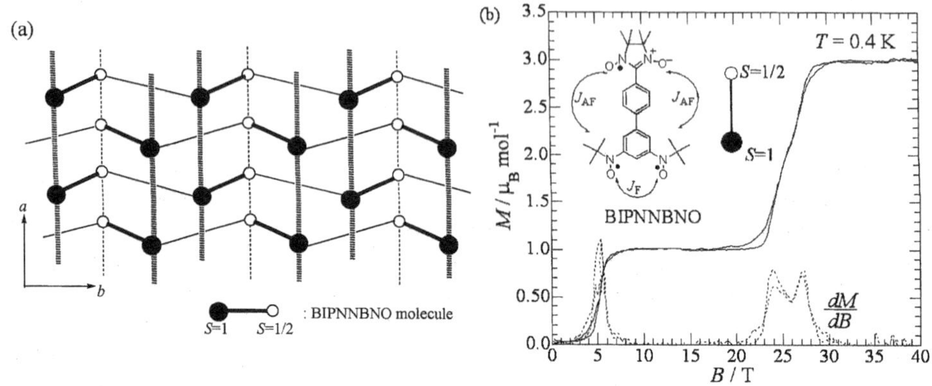

Fig. 8. (a) Schematic illustration of crystal structure of BIPNNBNO. (b) Magnetization curve of BIPNNBNO at 0.4K.

Figure 8(b) shows the complete magnetization process of BIPNNBNO. The magnetization in nearly zero below 4.5 T, increases rapidly above 4.5 T and shows a plateau of 1 μ_B, which corresponds to the 1/3 magnetization plateau. Full saturation moment is 3 μ_B. We consider that in the plateau region each molecule has an $S = 1/2$ spin aligned along the field direction. Above 23 T, it increases again and saturates to be 3 μ_B at 29 T. It should be noted that a very narrow magnetization plateau of 2/3 appears around 26 T. This may be related to the spin frustration in this quasi-2D spin system.

7 Pressure Effects

The light and soft organic crystals exhibit large pressure effects. There are many reports concerning the enhancement of the antiferromagnetic interactions under pressure. Compression of the molecular packing under pressure provides the larger molecular overlaps and stronger antiferromagnetic interactions. As another type of pressure effects, we discovered the suppression of the structural change under pressure. Moreover, we observed the first example of the reduction of the antiferromagnetic interactions under pressure.

In this section, crossover between the alternating and uniform chains is first described. This phenomenon has relation to the suppression of the structural change under pressure. Next, pressure effects in $S = 1$ systems are presented. Pressure dependence of the transition temperature is compared with the one in simple antiferromagnets with $S = 1/2$. Lastly, the dimensional crossover between 2D and 1D is reported. Experiments on the two compounds having similar chain structures are compared and the reduction of antiferromagnetic interactions is observed under pressure.

For the magnetic measurements under pressure, a small pressure clamp cell made of TiCu alloy was used. The use of the Quantum Design SQUID magnetometer equipped a pressure cell enables the magnetic mesurements with high accuracy. The detail of the experimental procedure is described in the literature [38, 39].

7.1 Crossover between the Alternationg and Uniform Chain, F_5PNN

An organic monoradical, F_5PNN crystallizes to a monoclinic system, *C2/c*, $a = 10.82(1)$ Å, $b = 11.70(4)$ Å, $c = 11.04(2)$ Å, $\beta = 102.98(7)°$, $V = 1362.4$ Å^3, $Z = 4$ at 180K. A molecule has the two-fold symmetry along the molecular long axis. In the crystals, noticeable is a chain structure along the *a+c* direction by the contacts between the NO groups. In spite of the uniform chain structure at high temperature, the magnetic properties at low temperature obey the antiferromagnetic alternating chain model with $2J_1/k_B = -5.6$K and the alternating ratio $\alpha = J_1/J_2 = 0.4$ [4, 8]. The structural change at low temperature is suggested. The law-temperature properties are well described by an $S = 1/2$ antiferromagnetic alternating chain. The temperature dependence of χ_P down to 0.5K [4], magnetization curves [4], and heat capacity [40] are consistently analysed.

We performed the X-ray crystal structure analysis down to 7K [41]. Slope of the temperature dependences of the lattice parameters change at around 70K. The crystal system at 7K is determined as *Cc*, in which the uniform chain structure is still conserved. The loss of the two-fold symmetry of a molecule occurs at around 70K. A small peak independent of the external magnetic field was observed at 5K in the heat capacity measurement by the thermal relaxation technique [42]. The structural change below 7K is quite rare. In a spin-Peierls transition, the change from a uniform chain ($\alpha = 1$) to a dimer state ($\alpha = 0$) occurs. In the present case, observed is the change to the alternating chain with $\alpha = 0.4$. Another mechanism of the transition is suspected. The crystal lattice of F_5PNN is so soft to be easily affected by external condition. The magnetostriction is observed below 4K [43]. When the magnetization isotherm measured at 0.5K is compared with the one calculated, the change of α from 0.4 to 0.5 is noticeable. The change of α is related to the change of the volume under magnetic fields.

We apply pressure on this compound and observed the gapless ground state under pressure. The temperature dependence of χ_P at 5 kbar obeys the antiferromagnetic uniform chain model with $2J_{AF}/k_B = -5.6$K [38]. The gradual change from the alternating to uniform chains is observed [40]. In our opinion, the mechanism of the pressure induced crossover from an alternating to a uniform chain in F_5PNN is the suppression of the structural change under pressure. In our study on an organic radical, galvinoxyl, pressure effects of the suppression of the structural change is observed. The mechanism is explained by the enhancement of the potential barrier of the structural transition [38, 39]. A galvinoxyl molecule has two-fold symmetry, which is lost at the transition temperature [44]. The common feature between F_5PNN and galvinoxyl is the moleuclar two-fold symmetry in spite of having thermally vibrating methyl or *t*-butyl groups, respectively. The driving force of the structural transition is considered as the freezing of the thermal vibration. The mechanism of the pressure effects on F_5PNN is still controversial, but this is the first observation of the pressure-induced crossover from an alternating to a uniform chain.

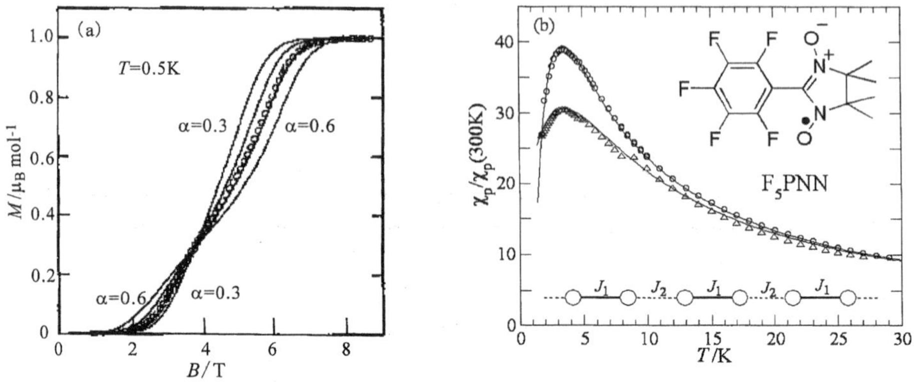

Fig. 9. (a) Magnetization process of F_5PNN at 0.5K. Solid curves are the calculations for the alternating chains with alternating ratio, $\alpha = 0.3$, 0.4, 0.5, and 0.6. (b) Temperature dependence of χ_p of F_5PNN at ambient pressure (circle), and at 5 kbar (triangle). Each experiment is fitted by the $S = 1/2$ alternating chain model which is indicated by a solid curve with $2J_1/k_B = -5.6$K and $\alpha = J_1/J_2$=0.4, and $2J_1/k_B = -5.6$K and $\alpha = 1$, respectively

7.2 Dimensional Crossover from 2D to 1D in an S = 1 System, F_2PNNNO [45]

7.2.1 A Ferromagnetic and Antiferromagnetic Alternating Chain, PIMNO

The crystal structure of PIMNO has uniform chains with the contacts between the *t*BuNO and IM. The temperature dependence of χ_p is well reproduced by the $S = 1/2$ ferromagnetic and antiferromagnetic alternating chain model with $2J_1/k_B = 216$K and $2J_2/k_B = -12$K ($\alpha = J_1/J_2 = -18$). This compound undergoes an antiferromagnetic phase transition at 2.5K due to the weak interchain interactions [13]. We performed the magnetic measurements under pressure of this compound, and observed that the transition temperature becomes higher as increasing pressure. We analyse the enhancement of T_C as the enhancement of interchain interactions by mean-field treatment, that is

$$\chi = \frac{\chi_{\text{chain}}}{1 - (2zJ'/Ng^2\mu_B^2)\chi_{\text{chain}}}, \tag{3}$$

where J' and z are the interchain exchange coupling and the number of the nearest neighbours, respectively. The temperature dependences of χ_p under several pressures were fitted with eq. (3) on the assumption of the fixed intrachain interactions. For the data under 5 kbar, $2zJ'/k_B = -3$K is obtained as the interchain interactions.

The pressure effect of PIMNO with the gradient $a = d(T_C(P))/dP = 0.22$ kbar^{-1} is larger than the ones reported for $S = 1/2$ antiferromagnets ($a = 0.086$–0.15) [46]. The large pressure dependence observed in PIMNO can be related to the nature of $S = 1$ species.

7.2.2 Pressure Effects on PNNNO and F_2PNNNO

Both crystals of PNNNO and F2PNNNNO involve uniform chains along the b axes with the contacts between the NN units. The difference lies in the interchain interactions. The large dihedral angle between the phenyl and NN planes in a F2PNNNO molecule yields short contact between tBuNO's, O···O 3.88 Å. Thus, a distorted honeycomb structure is formed. On the other hand, the interchain contact in PNNNO is as long as 5 Å and the system is one dimensional. The magnetic models are shown in Fig. 10(b). The magnetic exchange couplings are estimated based on the crystal structure, 2JF/kB = 638K, 2JAF/kB = –14.5K, 2J'AF/kB = 0 for PNNNO and 2JF/kB = 407K, 2JAF/kB = –7.4K, 2J'AF/kB = –67K for F2PNNNO [13]. The different dimensionality appears obviously in their magnetization curves. The monotonous increase of magnetization is observed in PNNNO. On the other hand, F2PNNNO shows a spin gap with 9 T and two-fold saturation. (See Section 8.2).

Figure 10 shows the temperature dependence of χ_p of F_2PNNNO measured at ambient pressure and at 3 kbar. At ambient pressure, the value of χ_p takes a maximum at around 16K and decreases to zero as temperature decreases. The ground state at ambient pressure is singlet with an energy gap. At 3 kbar, the value of χ_p takes a maximum at 10K, and below 10K, decreases to ward a finite value as $T \rightarrow 0$. It is clear that the

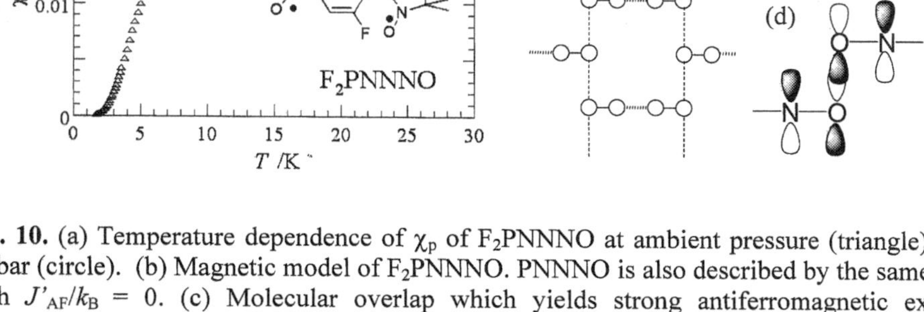

Fig. 10. (a) Temperature dependence of χ_p of F_2PNNNO at ambient pressure (triangle) and at 3 kbar (circle). (b) Magnetic model of F_2PNNNO. PNNNO is also described by the same model with $J'_{AF}/k_B = 0$. (c) Molecular overlap which yields strong antiferromagnetic exchange coupling. (d) Accidental orthogonality of the molecular orbitals having the negligibly small overlap integral.

energy gap disappears at 3 kbar. The behaviour of χ_p is almost the same under the pressure region of 3–7 kbar.

The pressure effects on PNNNO are also studied. The temperature dependence of χ_p is independent of the applied pressure up to 8 kbar. This fact means that the intermolecular interactions between the NN's within a chain is not influenced by pressure less than 8 kbar.

The gapless ground state of F_2PNNNO at 3 kbar means the weaker exchange coupling between chains, J'_{AF}, considering from the origin of the energy gap. Judging from the pressure independent behaviour in the chain compound of PNNNO, the change of the intrachain exchange coupling, J_{AF}, is negligibly small. The dimensional crossover from 2D to 1D in F_2PNNNO is induced by pressure.

The antiferromagnetic interactions are related to the overlap between the SOMO's. Shorter intermolecular contacts by applied pressure always result in the stronger antiferromagnetic exchange couplings. In the PNNNO family, the SOMO is localized on the *t*BuNO and NN. The origin of the strong interchain interactions in F_2PNNNO at ambient pressure is the short contact between the *t*BuNO's, whereas the one in PNNNO is considerably separated. We must mention that the SOMO has a node between the nitrogen and oxygen atoms. Thus, the magnitude of the overlap between the SOMO is affected by the relative orientation between the two *t*BuNO's as is shown in Fig. 10(c). If the relative orientation of NO groups is changed by applied pressure, the antiferromagnetic exchange coupling can become smaller. That is the case of the F_2PNNNO. The orthogonal condition between the SOMO's can yield even ferromagnetic exchange coupling [6]. This is the first example of the reduction of the antiferromagnetic interactions under pressure.

8 Field Induced Magnetic Ordering

8.1 An Alternating Antiferromagnetic Chain, F_5PNN

The magnetic properties of quantum spin systems under magnetic fields have attracted much attention both from theoretical and experimental points of view. Spin-gapped systems such as Haldane systems, spin-Peierls systems, two-leg ladder systems, and alternating chain systems have a finite energy gap (Δ) between the nonmagnetic ground state and the lowest excited state because of notable quantum effects. When a magnetic field is applied to these systems, the lowest branch of the first excited state goes down due to the Zeeman effect and intersects the ground state at the lower critical field $B_{C1} = \Delta/g\mu_B$, indicating that the gap closes. For the higher field, the magnetization starts to grow and saturates at the upper critical field B_{C1} with a spin-polarized state. For $B_{C1} \leq B \leq B_{C2}$, the system has the gapless excitation on the magnetic ground state and is described by the universality class of 1D quantum systems called the

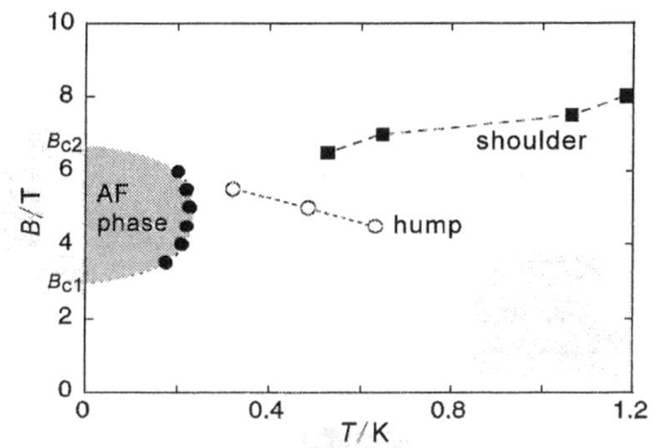

Fig. 11. Magnetic field versus temperature phase diagram for F_5PNN. Closed circles represent the transition temperatures, and the field induced magnetic ordered phase is marked by shadow. Open circles and Squares denote the temperature at which hump and shoulder appear, respectively. (See text.)

Tomonaga-Luttinger liquid (TLL), in which heuristic features, such as the power-law temperature dependence of the NMR relaxation rate and the linear-T dependence of the low-temperature specific heat are theoretically predicted [47]. Experimental probing of the TLL state is also reported, but the study covering the whole field region below B_{C1} and above B_{C2}, is quite limited. Chaboussant *et al.* reported the observation of the TLL state in an S = 1/2 two-leg ladder $Cu_2C_5H_{12}N_2Cl_4$ (CuHpCl) in the NMR study in magnetic fields up to above [48].

As is described in Section 7.1, F_5PNN is an antiferromagnetic alternating chain with $2J_1/k_B = -5.6$K and the alternating ratio $\alpha = J_2/J_1 = 0.4$. In F_5PNN, B_{C1} and B_{C2} are estimated to 3 and 6.5 T, respectively. It should be noted here that these two critical fields are much smaller than those of other spin-gapped compounds and can be easily attained with a conventional superconducting magnet, which enables us to perform the detailed specific heat measurements on F_5PNN in the three different phases, the gapped, gapless, and spin-polarized phases.

Under the field between B_{C1} and B_{C2}, antiferromagnetic order was observed below 0.3K. The H-T phase diagram of F_5PNN is shown in Fig. 11. Outside the field induced magnetic ordered phase, we observed a hump at around 5 T. This hump is intrinsic to the one-dimensionality of this material. In the field region just above B_{C2}, shoulder appears. The shoulder develops with increasing magnetic fields. It is noted that the heat capacity does not seem to approach zero monotonously down to 0K at 7 T.

The temperature and field dependence of the specific heat are qualitatively in good agreement with the theoretical calculation on an $S = 1/2$ two-leg ladder [49]. This agreement suggests that the observed behaviors are related with the low-energy excitation in the Tomonaga-Luttinger liquid [50].

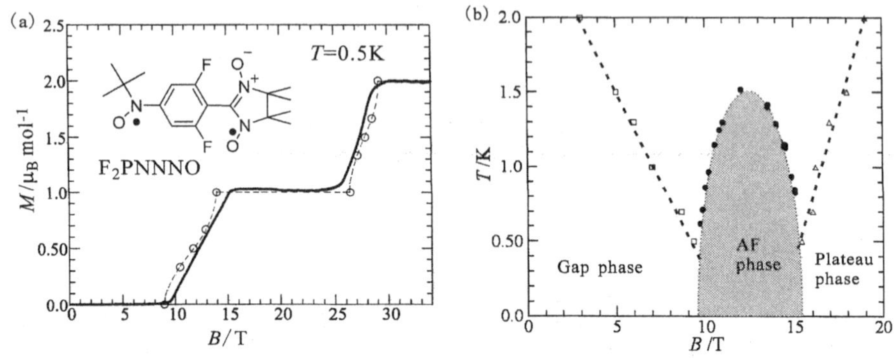

Fig. 12. (a) Magnetization curve of F_2PNNNO at 0.5K. Open circles represent the calculations. (See ref.13) (b) Magnetic phase diagram of F_2PNNNO.

8.2 A Distorted Honeycomb Lattice with S = 1, F_2PNNNO

An $S = 1$ distorted honeycomb lattice compound, F_2PNNNO has a singlet ground states with the spin gap of $B_{C1} = 9$ T. The magnetization saturates with two steps. At 0.5K, the finite magnetization begins to grow above B_{C1}(= 9 T). Magnetization reaches half the value of the full saturation magnetization, 1 μ_B, at $B_{C2} = 15$ T and keep the value up to $B_{C3} = 25$ T. Above B_{C3}, it again increases and reaches a full saturation, 2 μ_B, at $B_{C4} = 29$ T. The linear field dependence of magnetization was observed above B_{C1}, where as a typical low-dimensional quantum spin system shows the shape of inverse-S-type. The linear field dependence observed between B_{C1} and B_{C2}, suggests the antiferromagnetic ordering in magnetic fields.

We have investigated the field induced magnetic ordering by the measurements of the temperature dependence of magnetization under various magnetic fields up to 15 T. We have observed the antiferromagnetic phase transition as a bent in the temperature dependence of the magnetizations [51]. The phase diagram of F_2PNNNO is shown in Fig.12.

The relation between the field-induced magnetic ordering and the Bose-Einstein condensation has drawn current interests [52]. The estimation of the critical exponent of the critical field and transition temperature is important to understand the field-induced magnetic ordering. The determination of the transition temperature in our magnetization measurements includes some ambiguity. The exact determination of the critical exponent will be done by the method of ^{19}F-NMR study [53] and specific heat measurements [54] which are now in progress.

Acknowledgements

The authors are greatly indebted to Prof. T. Goto and Dr. H. Nakano for their kind collaboration and fruitful discussions. We would like to express sincere gratitude to all the collaborators for their help in accomplishing this work.

The authors are grateful for financial support by a Grant-in-Aid for Scientific Research from the Ministry of Education, Culture, Sports, Science and Technology of Japan. Financial supports from Mitsubishi Chemical Corporation Fund, The Murata Science Foundation, The Asahi Glass Foundation, The Mitsubishi Foundation, Saneyoshi Scholarship Foundation, Hayashi memorial Foundation for Female natural Scientists, and Yamada Science Foundation are also acknowledged.

References

1. (a) M. Tamura, Y. Nakazawa, D. Shiomi, K. Nozawa Y. Hosokoshi, M. Ishikawa, M. Takahashi, and M. Kinoshita, Chem. Phys. Lett. 186 (1991) 401; (b) Y. Nakazawa, M. Tamura, M. Shirakawa, D. Shiomi, M. Takahashi, M. Kinoshita, and M. Ishikawa, Phys. Rev.B46, (1992) 8906.
2. (a) F.D.M. Haldane, Phys. Rev. Lett. 50 (1983) 1153; (b) F.D.M. Haldane, Phys. Lett. 93A (1983) 464.
3. K. Hida, Phys. Rev. B45, (1992) 2207.
4. M. Takahashi, Y. Hosokoshi, H. Nakano, T. Goto, M. Takahashi, and M. Kinoshita, Mol. Cryst. Liq. Cryst. 306 (1997) 111.
5. (a) Y. Hosokoshi, M. Takahashi, T. Goto, and K. Inoue, J. Mag. Mag. Mater. 177-181 (1998) 713; (b) H. Nakano, Y. Hosokoshi, and M. Takahashi, J. Mag. Mag. Mater. 177-181 (1998) 717.
6. Y. Hosokoshi, M. Tamura, M. Kinoshita, H. Sawa, R. Kato, Y. Fujiwara, and Y. Ueda, J. Mater. Chem. 4 (1994) 1219.
7. Y. Hosokoshi, M. Tamura, H. Sawa, R. Kato, and M. Kinoshita, J. Mater. Chem. 5 (1995) 41.
8. Y. Hosokoshi, M. Tamura, D. Shiomi, N. Iwasawa, K. Nozawa, M. Kinoshita, H. Aruga Katori, and T. Goto, Physica B 201 (1994) 497.
9. Y. Hosokoshi, M. Tamura, K. Nozawa, S. Suzuki, H. Sawa, R. Kato, and M. Kinoshita, Mol. Cryst. Liq. Cryst. 271 (1995) 115
10. K. Mukai, H. Nagai, and K. Ishizu, Bull. Chem. Soc. Jpn. 48 (1975) 2381.
11. H. C. Longuet-Higgins, J. Chem. Phys. 18 (1950) 265.
12. K. Inoue and H. Iwamura, Angew. Chem. Int. Ed. Engl. 34 (1995) 927.
13. Y. Hosokoshi, Y. Nakazawa, K. Inoue, K. Takizawa, H. Nakano, M. Takahashi, and T. Goto, Phys. Rev. B60 (1999) 12924.
14. Y. Hosokoshi, K. Katoh, K. Inoue, and T. Goto, J. Phys. Soc. Jpn. 68 (1999) 2910.
15. M. Tamura, Y. Hosokoshi, D. Shiomi, M. Kinoshita, Y. Nakasawa, M. Ishikawa, H. Sawa, T. Kitazawa, A. Eguchi, Y. Nishio and K. Kajita, J. Phys. Soc. Jpn. 72 (2003) 1735.
16. W.E. Hatfield, R.R. Weller, and J.W. Hall, Inorg. Chem. 19 (1980) 3825.
17. (a) C.K. Majumdar, J. Phys. C 3 (1970) 911; (b) C.K. Majumdar and D.K. Ghosh, J. Math. Phys. 10 (1969) 1399; (c) B.S. Shastry and B. Sutherland, Phys. Rev. Lett. 47 (1981) 964.
18. I. Affleck, T. Kennedy, E.H. Liebe and H. Tasaki, Commun. Math. Phys. 477 (1988) 115.
19. K. Okamoto and K. Nomura, Phys. Lett. A 169 (1992) 433.
20. The susceptibilities for the special cases are presented in refs. (a) T. Nakamura, Phys. Rev. B.57 (1998) R3197; (b) T. Nakamura and K. Okamoto, Phys. Rev. B58 (1998) 2411.
21. J.W. Hall, W.E. Marsh, R.R. Weller, W.E. Hatfield, Inorg. Chem. 20 (1981) 1033.
22. D. Shiomi, M. Tamura, H. Sawa, R. Kato, and M. Kinoshita, J. Phys. Soc. Jpn. 62 (1993) 289.
23. K. Katoh, Y. Hosokoshi, K. Inoue, and T. Goto, J. Phys. Soc. Jpn. 69 (2000) 1008.

24. K. Katoh, Y. Hosokoshi, K. Inoue, M.I. Bartashevich, H. Nakano, and T. Goto, J. Phys. Chem. Solids, 63 (2002) 1277.
25. Y. Hosokoshi, K. Katoh, and K. Inoue, Synth. Metals 133 (2003) 527.
26. Y. Hosokoshi, K. Katoh, Y. Nakazawa, H. Nakano, and K. Inoue, J. Am. Chem. Soc. 123 (2001) 7921.
27. E. Dagotto, J. Riera, and D. Scalapino, Phys. Rev. B45 (1992) 5744.
28. T.M. Rice, S. Gopalan, and M. Sigrist, Europhys. Lett. 23 (1993) 445.
29. (a) Z. Hiroi, M. Azuma, M. Takano, K. Ishida, and Y. Kitaoka, Phys. Rev. Lett. 73 (1994) 3463; (b) G. Chaboussant, P.A. Crowell, L.P. Levy, O. Piovesana, A. Madouri, and D. Mailly, Phys. Rev. B55 (1997) 3046; (c) Z. Honda, Y. Nonomura, K. Katsumata, J. Phys. Soc. Jpn. 66 (1997) 3689; (d) M. Uehara, T. Nagata, J. Akimitsu, H. Takahashi, N. Mori, and K. Kinoshita, J. Phys. Soc. Jpn. 65 (1998) 2764.
30. S. Todo, M. Matsumoto, C. Yasuda, and H. Takayama, Phys. Rev. B64 (2001) 224412.
31. M. Troyer, H. Tsunetsugu, and D. Würtz, Phys. Rev. B50 (1994) 13515.
32. T. Goto, H. Aruga Katori, and Y. Ajiro, J. Phys. Soc. Jpn. 61 (1992) 4155.
33. T. Sakai, K. Okamoto, K. Okunishi, K. Kindo, Y. Narumi, Y. Hosokoshi, K. Katoh, K. Inoue, and T. Goto, Physica B 346-347 (2004) 34.
34. (a) T. Tonegawa, T. Nakao, and M. Kaburagi, J. Phys. Soc. Jpn. 65 (1996) 3317; (b) M. Hagiwara, Y. Narumi, K. Kindo, M. Kohno, H. Nakano, R. Sato, and M. Takahashi, Phys. Rev. Lett. 80 (1998) 1312.
35. (a) K. Okamoto, N. Okazaki, T. and Sakai, J. Phys. Soc. Jpn. 70 (2001) 636; (b) T. Sakai, N. Okazaki, K. Okamoto, K. Kinodo, Y. Narumi, Y. Hosokoshi, K. Katoh, K. Inoue, and T. Goto, Phys. Stat. Sol.(b) 236 (2003) 429.
36. A.L. Buchachenko, Dokl. Akad. Nauk. Engl. Ed. 244 (1979) 107.
37. N. Wada, T. Kobayashi, H. Yano, T. Okuno, A. Yamaguchi, and K. Awaga, J. Phys. Soc. Jpn. 66 (1997) 961.
38. Y. Hosokoshi, M. Tamura, and M. Kinoshita, Mol. Cryst. Liq. Cryst. 306 (1997) 423.
39. Y. Hosokoshi, M. Mito, M. Tamura, K. Takeda, K. Inoue, and M. Kinoshita, Rev. High Pressure Sci. Tech. 7 (1998) 620.
40. M. Mito, T. Kawae, Y. Hosokoshi, K. Inoue, M. Kinoshita, and K. Takeda, Solid State Commun. 111 (1999) 607.
41. A. Sato, A. Kobayashi, Y. Hosokoshi, and K. Inoue, unpublished data.
42. Y. Nishio, M. Tamura, K. Kajita, Y. Hosokoshi, K. Inoue, and M. Kinoshita, unpublished data.
43. M. Takahashi, Y. Hosokoshi, and T. Goto, K. Inoue, unpublushed data.
44. H. Sawa, Y. Hosokoshi, and M. Kinoshita, unpublished data.
45. Y. Hosokoshi and K. Inoue, Synth. Metals, 103 (1999) 2323.
46. K. Takeda, K. Konishi, M. Tamura, and M. Kinoshita, Phys. Rev. B53 (1996) 3374.
47. F.D.M. Haldane, Phys. Rev. Lett. 45 (1980) 1358.
48. G. Chaboussant *et al.*, Phys. Rev. Lett. 80 (1998) 2713.
49. X. Wang and L. Yu, Phys. Rev. Lett. 84 (2000) 5399.
50. Y. Yoshida, N. Tateiwa, M. Mito, T. Kawae, K. Takeda, Y. Hosokoshi, and K. Inoue, Phys. Rev. Lett. 94 (2005) 037203.
51. H. Mitamura, Y. Hosokoshi, T. Goto, and K. Inoue, to be submitted.
52. (a) A. Oosawa, M. Ishii and H. Tanaka, J. Phys. Cond. Matt. 11 (1999) 265; (b) T. Nikuni, M. Oshikawa, A. Oosawa, and H. Tanaka, Phys. Rev. Lett. 84 (2000) 5868.
53. S. Matsubara, M. Takigawa, Y. Hosokoshi, and K. Inoue, to be submitted.
54. H. Tsujii, Y. Takano, Y. Hosokoshi, and K. Inoue, unpublished data.

Carbon-Based Magnetism
T. Makarova and F. Palacio (Editors)

6

The Magnetism of π-Orbitals under Pressure

K. Takeda[1] and M. Mito[2]

[1]*Department of Applied Quantum Physics, Faculty of Engineering, Kyushu University, Fukuoka 812-8581, Japan*
[2]*Faculty of Engineering, Kyushu Institute of Technology, Kitakyushu 804-8550, Japan*

1 Introduction

The artificial control of the overlapping of π-orbitals under the pressure gives one of the effective ways to investigate the electronic state and the mechanism of intermolecular interactions. One can trace the variation of the physical properties of a certain model compound continuously as a function of the applied pressure to look into the intrinsic parameters that characterize the properties. This physical experimental approach contrasts with the chemical way, which controls the overlapping of the wave functions by substituting different moieties or chemical parts in the original molecule. Such a substituting procedure sometimes gives a too discrete change to identify what might be the origin of the variation. In comparison with inorganic materials, organic compounds are mechanically soft and then pressurization provides larger changes to the values of the transfer integrals of the molecular orbital and of the Coulomb potentials in a wider range. The pressure-induced ferro-to-antiferromagnetic transition in organic compounds, for instance, occurs under a pressure range that is one order magnitude smaller than the counterpart of inorganic compounds as will be shown later. In this chapter, various typical pressure effects on genuine organic magnetic compounds are discussed, particularly the relation between the magnetic transition temperature and intermolecular interactions under hydrostatic pressures up to 100 kbar (10 GPa).

2 Experiments under Pressure

Prior to the detailed description of pressure effects on organic magnetic materials, it will be helpful to outline high pressure experimental methods. Depending on the pressure range and the physical parameters to be investigated, a suitable type of pressure cell is used. When roll played by weak intermolecular interactions is important, low temperature cryogenic techniques are required in addition to the pressurization technique.

For the pressure range up to 20 kbar (2GPa), the magnetic-metal free CuBe clamp cells are normally used for the measurements of heat capacity, magnetic susceptibility and magnetization, as well as for the structural analysis under the pressure. In Fig. 1 is shown an experimental setup, with which one can perform the simultaneous measurements of thermal and magnetic properties above 0.5K [1, 2].

The value of the hydrostatic pressure at low temperatures is obtained from the pressure dependence of the superconducting transition temperature of a superconducting metal, such as Sn, or Al, which is placed in the pressure cell, and its transition temperature is detected by the coil system in the magnetic susceptibility measurements. When lower temperatures are required, the ^{3}He pot of a ^{3}He cryostat in Fig. 1 is replaced with a mixing chamber of the ^{3}He-^{4}He dilution refrigerator.

In order to obtain the absolute value of the heat capacity of the sample, one has to know the pressure dependence of the heat capacity of the pressure transmission medium, e.g. an Apiezon grease, which amounts a substantial quantity in some cases.

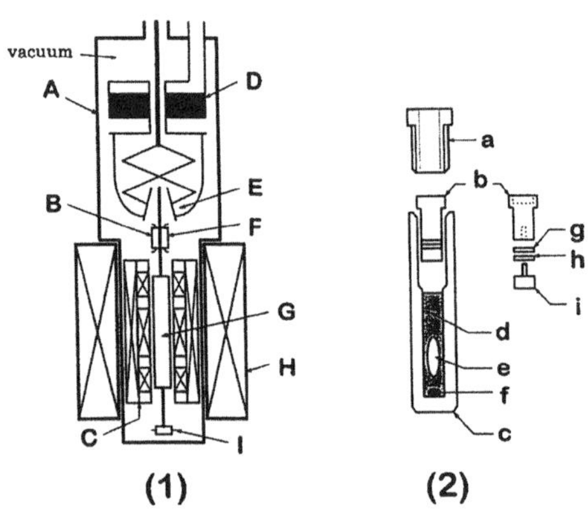

Fig. 1. (1) Schematic structure of the cryostat for the simultaneous measurement of magnetic susceptibility and heat capacity under the hydrostatic pressure. A; adiabatic tube, B; thermometer 1, C; coil system for ac susceptibility, D; liquid ^{3}He pot, E; mechanical heat switch, F; thermometer 2, G; pressure clamp cell, H; superconducting magnet, I; heater and subheater. (2) Pressure clamp cell. a; locking nut, b; piston, c; cylinder, d; pressure transmitting oil, e; sample, f; Sn, g; Cu seal, h; Teflon seal, i; mushoom plug.

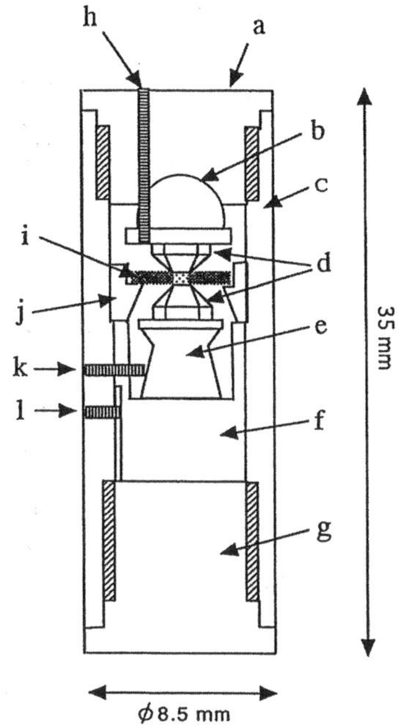

Fig. 2. Overall cross-sectional view of the miniature diamond anvil cell for the SQUID magnetometer. The main parts are (a) grand nut, (b) half-sphere seat for the upper diamond, (c) cylinder, (d) diamonds, (e) anvil plate for the lower diamond, (f) piston, (g) nut for applying pressure, (h) screw for tilt adjustments, (i) gasket, (j) supporter for the gasket, (k) screw for X–Y adjustments and (l) piston guide screw.

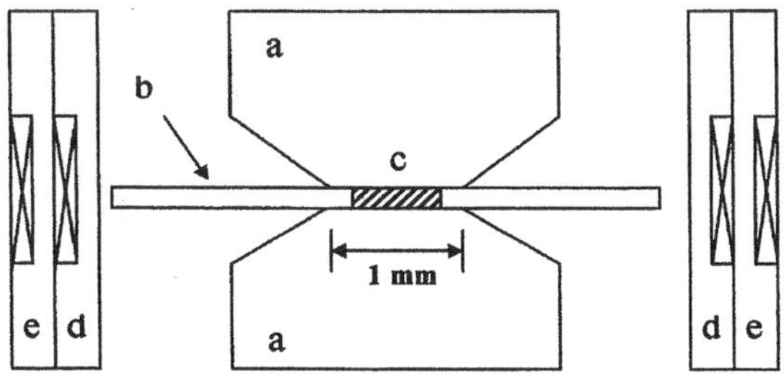

Fig. 3. A schematic view of the high pressure system for the ac susceptibility measurement: (a) diamond anvil, (b) gasket, (c) sample space, (d) primary coil, (e) secondary coil.

Although the sample contribution is normally a small fraction of the total heat capacity, one can get relatively large contributions when the entropy of the sample is consumed at low temperatures. The relative contribution to the total heat capacity and overall experimental check points are given in Refs. 3 and 4, together with some results for real compounds.

For the precise magnetic measurements up to 300 kbar (30 GPa), a miniature diamond anvil cell (DAC) has been developed which can be installed in a commercial SQUID magnetometer as shown in Fig. 2 [5]. With this type of pressure cell, the disappearance of bulk magnetization of metallic Fe and Gd at high pressures has been detected [6, 7].

Another type of diamond anvil cell is given in Fig. 3 which is attached to the ^{3}He-^{4}He dilution refrigerator. With this cell, the disappearance of the ferromagnetism in organic compounds has been observed at low temperatures [8, 9].

Hereafter, some prominent pressure effects are given for typical genuine organic compounds, such as antiferromagnets, ferromagnets and weak-ferromagnets, studied by the use of these pressure techniques.

3 Pressure Effects on Genuine Organic Antiferromagnets

Almost all of the organic compounds are diamagnetic, or less commonly antiferromagnetic,when they contain unpaired electrons or consist in free radicals. In the later case, the SOMO-SOMO overlapping between molecules is favoured by nature. Pressurization of an antiferromagnet generally tends to enhance the overlapping, and hence the intermolecular interaction increases under pressures as well as its transition temperature does.

Let us show as an example the pressure dependence of the Néel temperature $T_N(P)$ of the five antiferromagnetic compounds illustrated in Fig. 4.

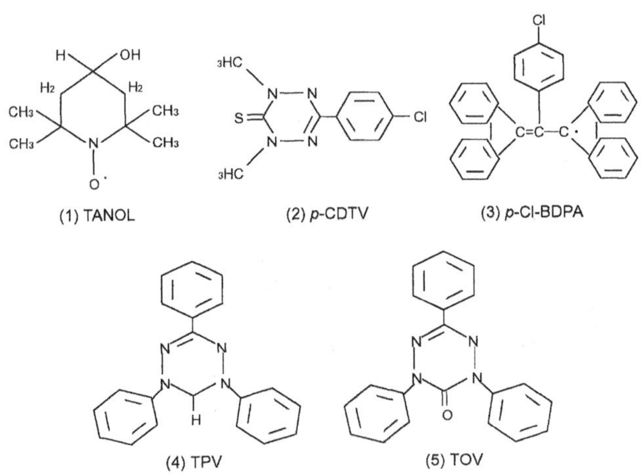

Fig. 4. Molecular structure of five antiferromagnetic compounds: (1) 2,2,6,6-tetramethyl-4-piperidinol-1-oxyl (TANOL), (2) 3-(4-chlorophenyl)-1,5-dimethyl 6-thioxoverdazyl (*p*-CDTV), (3) 1,3-bisdiphenylene-2-pchlorophenyl allyl (*p*-Cl-BDPA), (4) triphenyl verdazyl (TPV), (5) 1,3,5-triphenyl-6-oxoverdazyl (TOV).

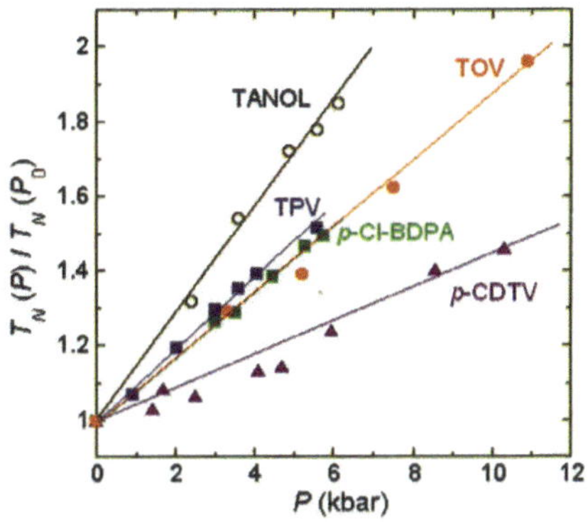

Fig. 5. Pressure dependence of $T_N(P) / T_N(P_0)$ for the five antiferromagnetic compounds in Fig. 4.

Table. 1.
The values of $T_N(P_0)$ and the initial slope a of the genuine organic antiferromagnets.

	Compound	$T_N(P_0)$	a	Ref.
1	TANOL	0.49	0.15	10, 11, 12
2	*p*-CDTV	0.67	0.044	13, 14, 15
3	*p*-Cl-BDPA	3.25	0.086	12
4	TPV	1.70	0.093	11, 12
5	TOV	5.0	0.086	16

The shift of their Néel temperatures under pressure P is detected by heat capacity, magnetization or magnetic susceptibility measurements and for small P it is generally expressed by the linear relation,

$$T_N(P)/ T_N(P_0) = 1 + a\,P \quad (1)$$

The value of the initial slope a and the experimental results of $T_N(P)$ are given in Table 1 and Fig. 5, respectively, for the five compounds [10–16].

In the magnets without itinerant electrons, the magnetic transition temperature is proportional to the superexchange interaction $J(r)$. The 10-th power law on the magnetic Grüneisen constant $n = -\,d(\ln J)/d(\ln r) = 10$ is applied to some metallic oxides, where r is the interatomic distance[17]. When $J(r) \propto r^{-n}$, which can not be distinguished from $J(r) \propto \exp(-b\,r)$ for the small change of r, we have

$$T_N(P)/ T_N(P_0) = 1 + k\,n\,P \quad (2)$$

where $k = -(d\,r/d\,p)/r$. Equation (2) qualitatively explains the experimental linear relation between $T_N(P)$ and P. For the quantitative analysis, we need to know the low temperature compressibility k along each pathway of the exchange interactions.

Figure 6 shows the magnetic heat capacity of TANOL under pressure, which is approximated as a quasi-one-dimensional Heisenberg antiferromagnet. In the mean field theory, $T_N(P)$ is given by the intrachain, $J(P)$, and interchain, $J'(P)$, interactions by

$$k_B T_N(P) = S^2 \{2 z J(P) J'(P)\}^{1/2} \tag{3}$$

where S is the spin value and z is the number of neighbouring chains. From the experimental values of $T_N(P)$ and $J(P)$, which are estimated from the broad heat

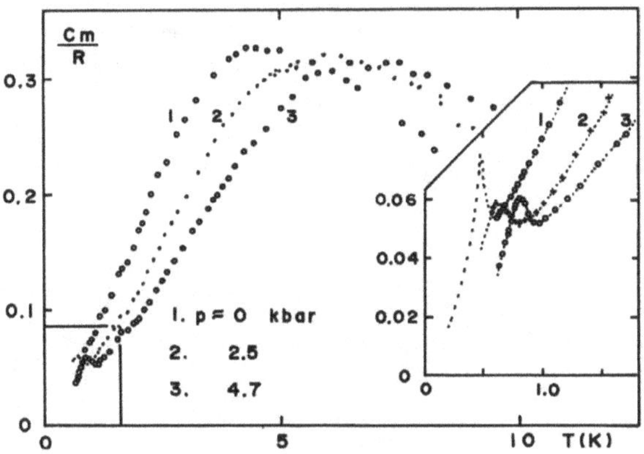

Fig. 6. Magnetic heat capacity of TANOL at some representative pressures.

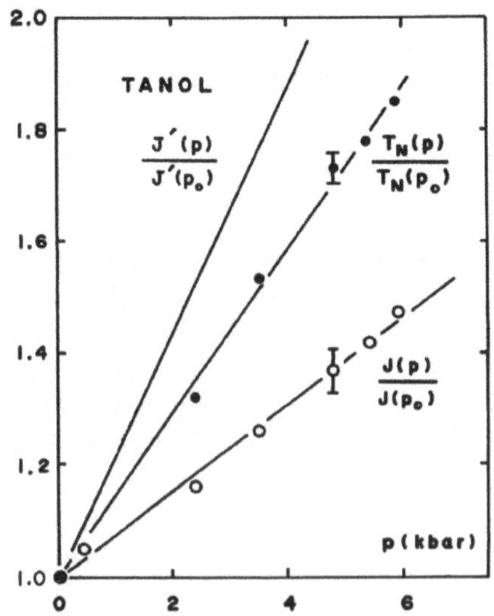

Fig. 7. Pressure dependence of the transition temperature $T_N(P)$, intrachain interaction $J(P)$ and interchain interaction $J'(P)$ of TANOL.

capacity maximum, we get the value of $J'(P)$ from Eq. 3. The relative pressure dependence of these quantities is given in Fig. 7 [10].

The crystallized *p*-CDTV is also a quasi-one-dimensional Heisenberg magnet with intrachain ferromagnetic interactions loosely coupled antiferromagnetically between chains [18]. The value of a = 0.044 (kbar^{-1}) is the smallest of the five. As will be mentioned in the next section, pressurization normally gives a negative effect on the strength of ferromagnetic interaction in organic systems. The intrachain ferromagnetic interaction of *p*-CDTV is suppressed relatively strongly under pressure, while in TANOL the intrachain antiferromagnetic interaction is enhanced, as well as the interchain interaction, to lead to a larger value of a = 0.15 (kbar^{-1}) than that of *p*-CDTV.

As it is noted in Table 1 the other three compounds have nearly the same value of a. This feature points out higher magnetic lattice dimensionality [11, 19, 20] in these compounds than in the previous two radicals. The unpaired electrons are expected to delocalize over a wider space within the larger molecules and the overlap of the π-orbitals or the transfer integrals are considered to be more susceptible under the pressure.

4 Pressure Effects on Genuine Organic Ferromagnets

There have been several well-defined genuine organic ferromagnets after the first realization of the β-phase *p*-NPNN in 1991 [21]. Here we will discuss the pressure dependence of the interesting four ferromagnets whose chemical composition is described in Fig. 8, where the first two compounds exhibit the pressure-induced ferro- to antiferromagnetic transition. Some of the pressure effects are given in Table.2,

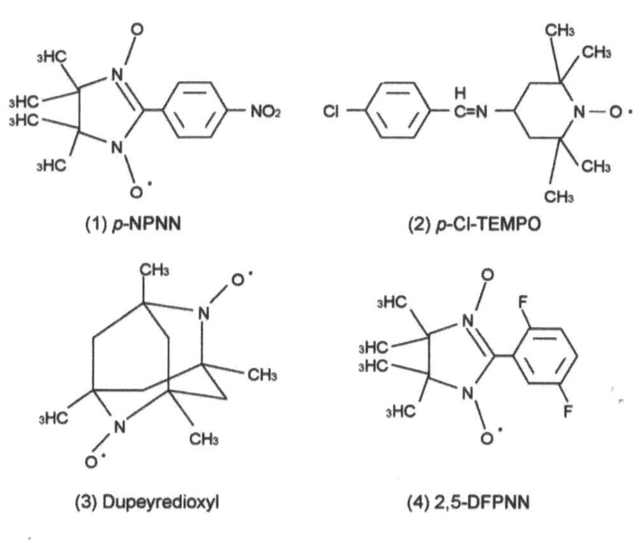

Fig. 8. Molecular structure of four representative organic ferromagnets.

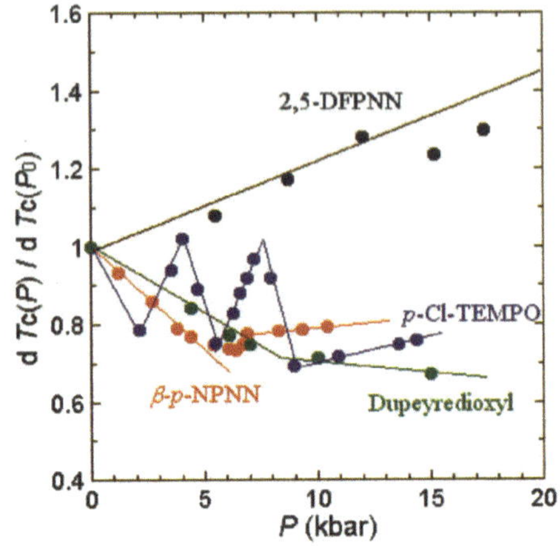

Fig. 9. Pressure dependence of $T_C(P)/ T_C(P_0)$ for the four ferromagnets of Fig. 8.

and in Fig. 9 [1, 2, 8, 9, 22–28]. It is noted that only 2,5-DFPNN gives a positive value for *a*, implying that the ferromagnetic interaction is rare and likely to be unstable in organic systems under the pressure [8, 9].

4.1 Pressure-induced Ferro-to-antiferromagnetic Transitions in β-p-NPNN

The ferromagnetism of *β-p*-NPNN is well established [21]. An example of the simultaneous measurement of its magnetic susceptibility and heat capacity is given in Fig. 10(a). Magnetic susceptibility rapidly increases at $T_C(P_0)$ = 0.61K with the decreasing temperature, where the heat capacity shows a sharp peak at $T_C(P_0)$. The pressurization makes the heat capacity peak shift towards lower temperatures, and at the pressure of P = 7.2 kbar the heat capacity shows a broad maximum just above the sharp peak as in Fig.10(b), indicating the reduction of magnetic lattice dimensionality [1, 22]: We can fit the broad maximum to the theory for two-dimensional Heisenberg system with S = 1/2 and a ferromagnetic exchange interaction J_f/k_B = 0.40K.

The pressure dependence of the magnetic susceptibility of *β-p*-NPNN is shown in Fig. 11 [23–25]. It is clear that the ferromagnetic feature is suppressed under high pressures, leading to the appearance of a cusp at $T_C(P)$ at pressures higher than 6.4 kbar. Above this pressure the cusp shifts towards higher temperatures as the pressure is increased, as seen in the inset of Fig. 11. At P = 6.9 kbar, the application of an external magnetic field has the effect to shift the cusp towards lower temperatures as observed for normal antiferromagnets. This upturn of $T_C(P)$ is indicative of a pressure-induced ferro-to-antiferromagnetic transition for P > 6.9 kbar. We can estimate the averaged value of antiferromagnetic exchange interactions to be zJ_{af}/k_B = –0.03K at P = 6.9 kbar from the estimated critical field Hc = 430 Oe at T = 0K with the relation $Hc = 2\times 2zJ_{af} \langle S\rangle/g\mu_B$. This value of J_{af} is about one tenth of J_f at P = 7.2 kbar. The heat capacity

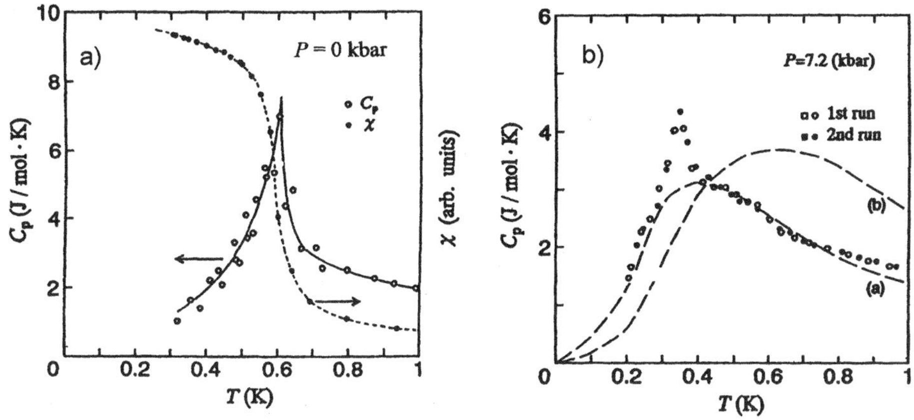

Fig. 10. (a) Results of the simultaneous measurements of magnetic heat capacity and magnetic susceptibility of *β-p*-NPNN at ambient pressure; (b) Magnetic heat capacity of *β-p*-NPNN under the pressure of $P = 7.2$ kbar. The curves correspond to the theoretical estimates of the heat capacity for the two-dimensional Heisenberg (a) ferromagnet and (b) antiferromagnet ($S = 1/2$, $J/k_B = 0.4$K).

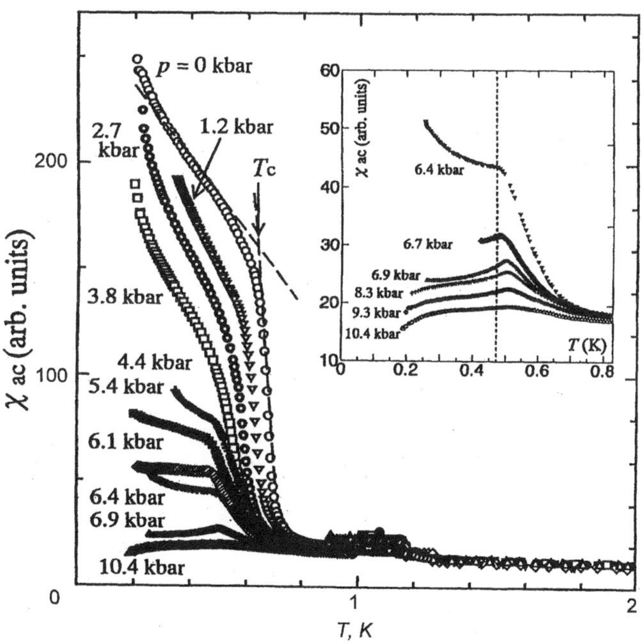

Fig. 11. Pressure Dependence of the ac-magnetic susceptibility of *β-p*-NPNN under pressure up to $P = 10.4$ kbar around the transition temperature.

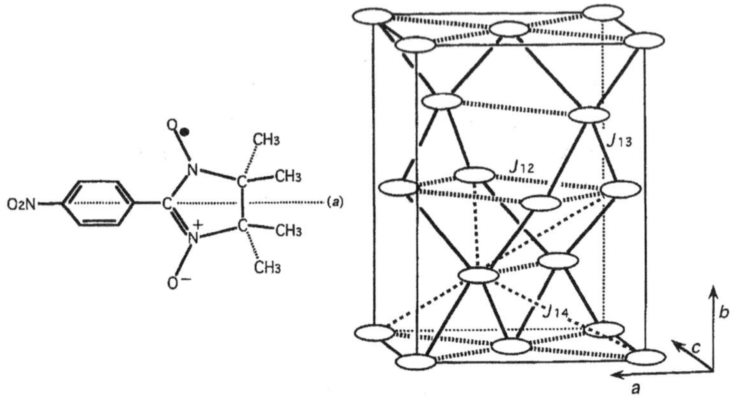

Fig. 12. Molecular structure of *β-p*-NPNN and a schematic diamond-like crystal structure of *β-p*-NPNN. The J_{ij} ($i = 2, 3, 4$) are main paths for exchange interaction to be considered.

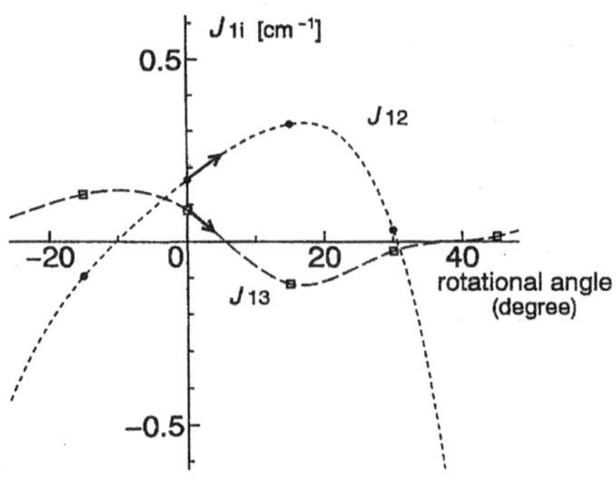

Fig. 13. Calculated dependence of J_{12} and J_{13} on the rotational angle indicated as in Fig. 14. The arrows indicate the expected direction of the change in *β-p*-NPNN.

curve in Fig. 10(b) reflects the short-range ferromagnetic ordering in two dimensions for $T > 0.4$K, followed by the three-dimensional long-range ordered state triggered by interlayer antiferromagnetic interaction. The antiferromagnetic feature for $P > 6.9$ kbar is also detected by the magnetization curve with a spin-flop-like behaviour at $T < T_c(P)$ [23–25].

From the overall experimental and theoretical analyses for this compound, the dominant three exchange interactions at ambient pressure are evaluated to be $J_{12}/k_B = 0.35$K, $J_{13}/k_B = 0.14$K, and $J_{14}/k_B = -0.04$K for the exchange pathways illustrated in Fig. 12 [29, 30]. It is reasonable to consider that the initial intralayer

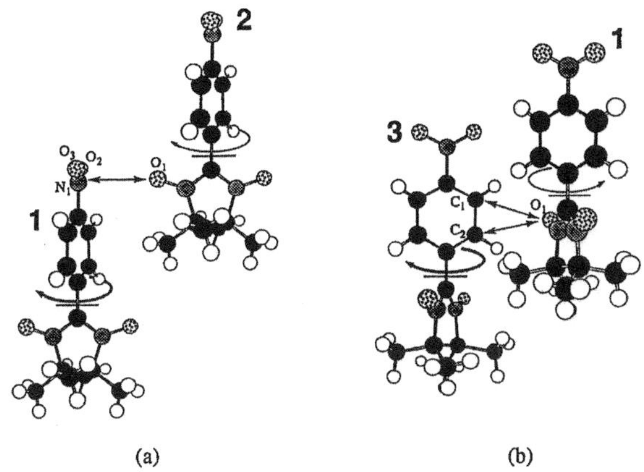

Fig. 14. Schematic illustrations of the radical rotation in *β-p*-NPNN. The positive direction for the rotational angle is indicated by the arrows.

interaction J_{12}/k_B = 0.35K increases on the *ab*-plane, and becomes J_f/k_B = 0.40K at P = 7 kbar, while the interlayer J_{13} and J_{14} ones are averaged to reduce and even change its sign to J_{af} for pressures higher than 6.9 kbar. Such a pressure dependence of the exchange interactions is consistent with theoretical results, where the relative rotation of the molecules is regarded to have the primary importance than the uniform sharing or shrinkage to meet the experimental results [29, 30]. Figure 13 shows the ab-initio calculation of J_{12} and J_{13} as a function of the rotational angle taken as indicated in Fig. 14 [29]. The rotation about 5 degrees seems to agree with the experiments. The structural analysis of this compound under pressures has been carried out at room temperature, however, it gives a variation of lattice parameters enough to decrease J_{13} to have negative value and at the same time to uprise J_{12} as shown in Fig. 13 [23–25].

4.2 Pressure Effects on the Ferromagnetic Interaction Intermediated by the β-hydrogen of a Methyl Group

The pressure dependence of $T_C(P)$ of *p*-Cl-TEMPO exhibits a negative initial slope followed by a zig-zag behaviour with increasing pressure as shown in Fig. 9 [26, 27]. In Fig. 15 the results of the systematic observation of magnetic susceptibility are shown for various pressures up to 15 kbar. It is noted that the peak value of the susceptibility decreases in accordance with increasing pressure, while its position on the temperature scale makes the zig-zag change for $P < 9$ kbar. For higher pressures than 9 kbar, the $T_C(P)$ shows a linear increase for P, as in the case of *β-p*-NPNN. The magnetization experiment of *p*-Cl-TEMPO below 0.23K, indicates the onset of antiferromagnetic behaviour at P = 13.6 kbar. For understanding these results, it is instructive to look into the spin polarization along the exchange pathway in the molecular space. As shown in Fig.16, the nitroxide oxygen atom in the TEMPO-based ferromagnets is always located near methyl- and/or methylen-hydrogen atoms at a *β*-position from an nitroxide group

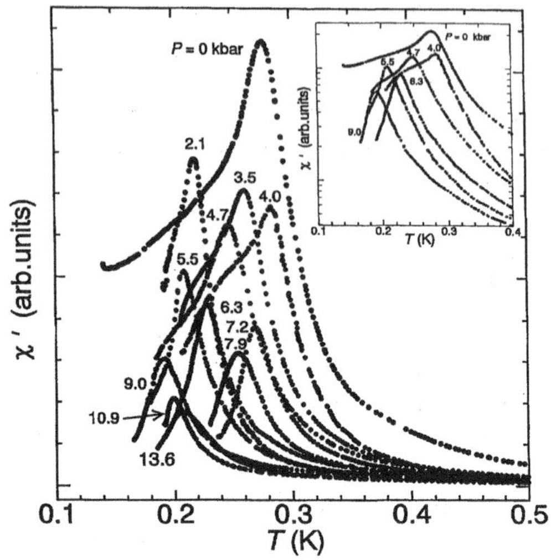

Fig. 15. Results of χ'_{ac} for *p*-Cl-TEMPO in the pressure region up to 13.6 kbar. The ordinate of the inset figure is plotted in a logarithmic scale.

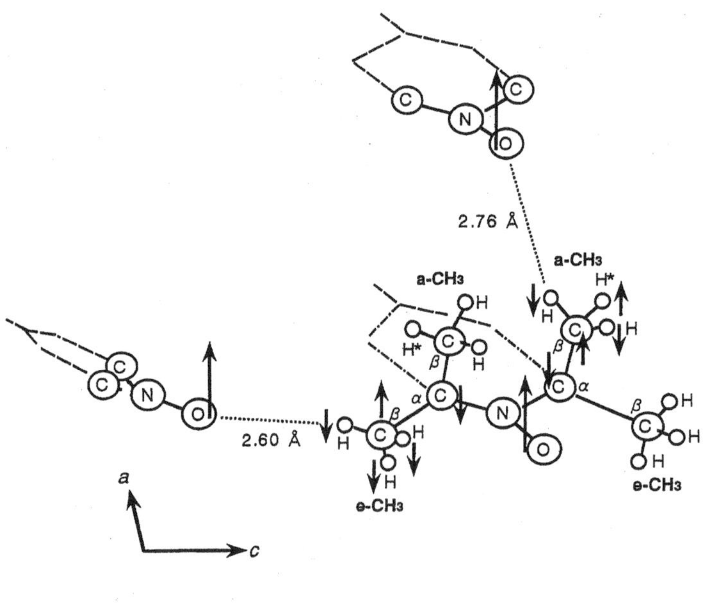

Fig. 16. Space alignment of principal spin polarizations of *p*-Cl-TEMPO molecules. The carbon adjacent to the nitroxide is called as α-carbon, and the carbon of the methyl moiety is called as β-carbon. Four methyl moieties linked to the TEMPO ring are classified to two equatorial methyl (e-CH_3) groups and two axial methyl (a-CH_3) ones.

in adjacent molecules, and the interatomic distance (2.5–2.9 Å) is close to the van der Waals radii of H and O atoms [27]. The spin polarization of NO moiety is propagated to the adjacent NO radical via two carbon atoms and hydrogen atoms of equatorial and axial methyl as illustrated with arrows in Fig. 16. The zig-zag behaviour of $T_C(P)$ is considered to be related to the rotation of these methyls with hydrogen atoms with positive or negative spin polarization under pressures.

4.3 Pressure Effects on Dupeyredioxyl

Dupeyredioxyl is a biradical compound which has the highest Curie temperature $T_C(P_0)$ = 1.48K of the four. Here we only show the pressure dependence of magnetic susceptibility measured with two kinds of pressure cells [28]. Fig. 17 shows the magnetic susceptibility results obtained with a CuBe clamp cell. As the pressure increases, the cusp at $T_C(P)$ shifts to lower temperatures and the ferromagnetic feature becomes less evident. It is noted that the value of $T_C(P_0)$ is not completely reversible when it is measured after the measurements at 7 kbar, the pressure at which the original ferromagnetic contribution seems to be affected. The results measured with a diamond anvil cell indicate that the ferromagnetism of Dupeyredioxyl is almost destroyed at 15 kbar as is seen in Fig. 18.

As for the microscopic mechanism of the reduction of $T_C(P)$, the most notorious is the reduction of orthogonality of the SOMO's associated with the two NO moieties within a molecule, which gives the effective value of S = 1 at ambient pressure. The relative rotation of these NO axes by pressurization may cause the decrease of intramolecular ferromagnetic coupling (170K) between two S = 1/2. In any case the pressure dependence of T_C in this bi-radical compound indicates an effective reduction of the ferromagnetic interaction within a molecule and a pressure-induced reduction of the intermolecular ferromagnetic interactions as well.

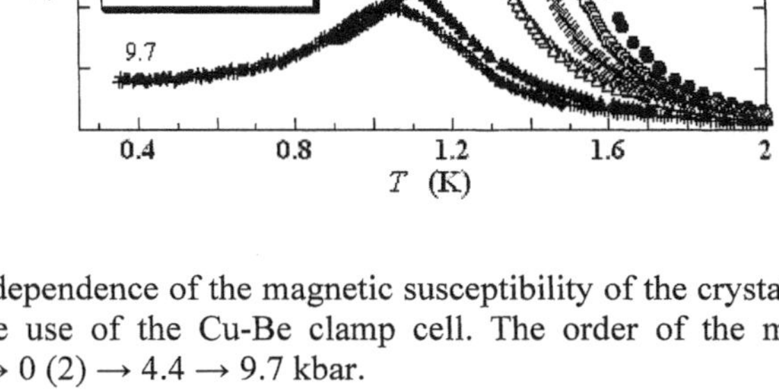

Fig. 17. Pressure dependence of the magnetic susceptibility of the crystalline dupeyredioxyl up to 9.7 kbar by the use of the Cu-Be clamp cell. The order of the measurement is 0 kbar (1) → 6.1 → 7.0 → 0 (2) → 4.4 → 9.7 kbar.

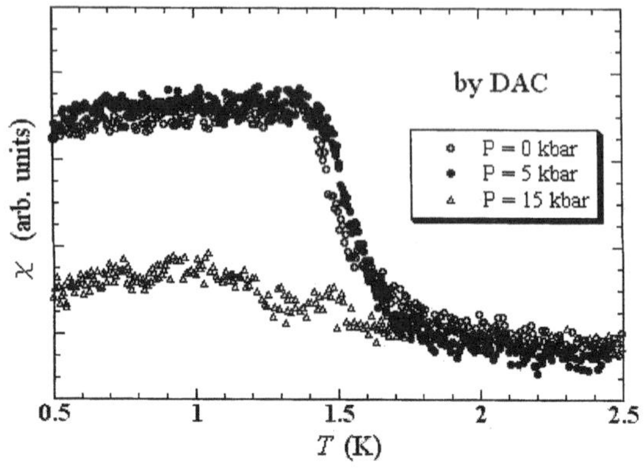

Fig. 18. Magnetic susceptibility of the crystalline dupeyredioxyl observed by the use of DAC.

4.4 Pressure Effects on 2,5-DFPNN

The pressure effects on this compound are investigated up to 10 GPa (100kbar), where there appears short interatomic distances of neighbouring atoms almost comparable to the sum of their respective van der Waals radii, and the bulk ferromagnetic feature disappears [8, 9]. The coefficient of the initial slope a of $T_C(P)$ is positive as indicated in Table 2 and in Fig. 9. The pressure dependence of the heat capacity and the magnetic susceptibility measured under hydrostatic pressures with a CuBe clamp cell are shown, respectively, in Figs. 19 and 20. Figure 21 gives the results of the susceptibility for higher pressures measured with a DAC. The Bragg peaks in the powdered X-ray diffraction patterns measured at different pressures shift toward higher angles without changing their shape up to 4.7 GPa. Their analyses provide the pressure dependence of the lattice parameters of the orthrohmbic crystal structure of this quite a sharp heat capacity peak appears at T = 0.49K. The long tail of the heat capacity for $T > T_C(P_0)$ indicates the quasi-one-dimensional (along a- axis) feature of the magnetic interactions as expected from the structural analysis. The application of pressure uprises $T_C(P)$ and the heat capacity tail grows to approach to the theoretical curve for the two-dimensional Heisenberg ferromagnet with J/k_B = 0.70K for $P >$ 1.2 GPa. This increase of the magnetic lattice dimensionality is thought to be due to the larger shrinkage of the c-axis rather than that of the other two axes in this pressure region. The ferromagnetic feature, probably on the ac-plane, is clearly seen in the susceptibility curves shown in Fig. 20. At higher pressures, however, the ferromagnetism is suppressed as shown in Fig. 21. The hump of the susceptibility at P = 3.9 GPa indicated in the inset of Fig. 21 seems to indicate the appearing of compound, space group *P*bca, at ambient temperature as shown in Fig. 22. At P = 0 GPa, antiferromagnetic ordering as appeared in the previous section. When this is the case, the ferromagnetic layers couple antiferromagnetically in this pressure region. The X-ray structural analysis experiment at room temperature seems to show a saturation of the shrinkage for $P >$ 5 GPa, where the interatomic distances are almost coincident with thesum of the van der Waals radii. The bulk ferromagnetism disappears for P >5 GPa up to 10 GPa as shown in Fig. 21(b).

Table 2.
The values of $T_C(P_0)$ and the initial slope a of the four genuine organic ferromagnets.

	Compound	$T_C(P_0)$	a (initial slope)	Ref.
1	*β-p*–NPNN	0.61	–0.048	1, 2, 22–25
2	*p*–Cl–TEMPO	0.28	–0.099	26, 27
3	Dupeyredioxyl	1.48	–0.034	28
4	2,5–DFPNN	0.45	0.079	8, 9

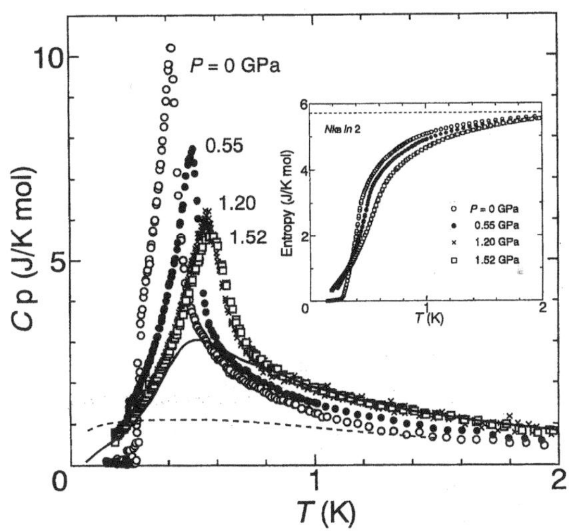

Fig. 19. Pressure dependence of the magnetic heat capacity (Cp), and the magnetic entropy of 2,5-DFPNN in the hydrostatic pressure region up to 1.52 GPa (inset). The dotted and solid curves express the theoretical curves of S = 1/2 one-dimensional Heisenberg ferromagnetic system with intrachain interaction J/k_B = 0.70K and S = 1/2 two-dimensional (square-lattice) one with intraplane interaction J_{2D}/k_B = 0.70K, respectively. The dotted line in the inset represents the total magnetic entropy Nk_B ln(2S+1) for S = 1/2.

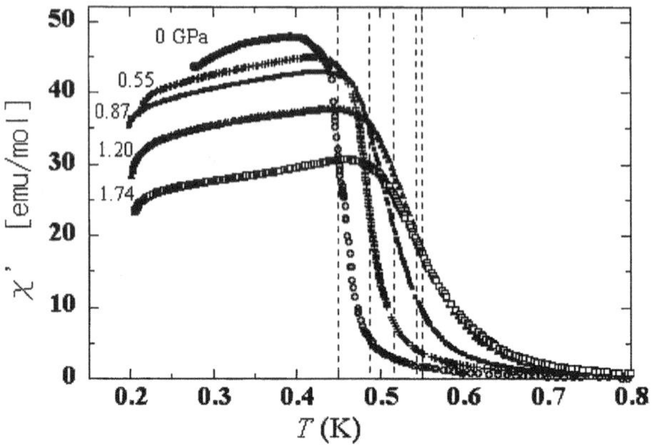

Fig. 20. Pressure dependences of in-phase component of ac magnetic susceptibility for 2,5-DFPNN under hydrostatic pressure up to P =1.74 GPa. The transition temperature (Tc) to the ferromagnetic long-range order at each pressure is shown by dotted line, shifting toward high temperatures with increasing pressure.

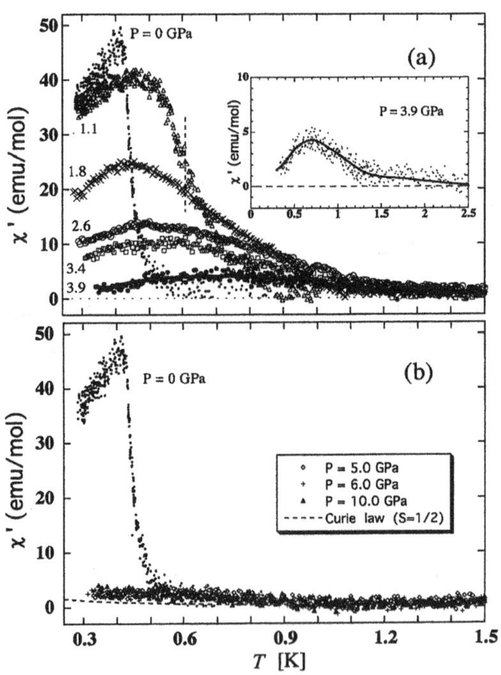

Fig. 21. Pressure dependence of in-phase ac magnetic susceptibility of 2,5-DFPNN under nonhydrostatic pressure: (a) for $P < 3.9$ GPa, and (b) for $P = 5.0$, 6.0, and 10.0 GPa. The dotted line in (a) shows the estimated Tc at P =1.1 GPa. The dotted curve in (b) expresses the Curie's law for $S = 1/2$.

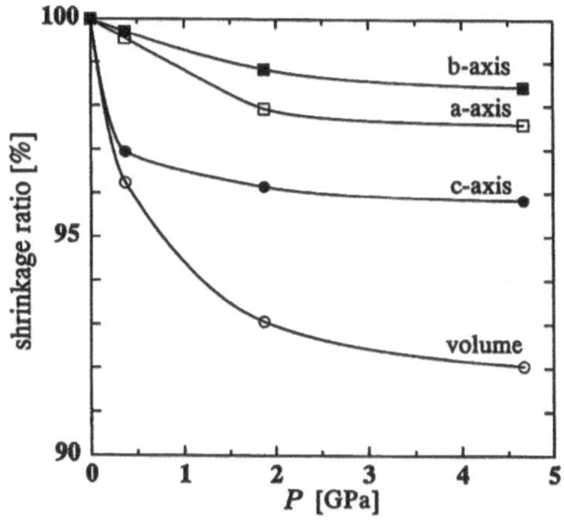

Fig. 22. Pressure dependences of the lattice constants (*a*, *b*, and *c*) and the volume of unit cell (*V*) of 2,5-DFPNN. The solid curves are the guides for the eye.

5 Pressure Effects on Sulfur-based Organic Magnets with Ferromagnetic Moments

Recently tiazyl-based organic compounds with ferromagnetic moments have been reported, as mentioned in other chapters of this book. Here we will briefly describe the pressure effects of two derivatives of the dithiadiazolyl radical, the *p*-NCC_6F_4CNSSN (weak-ferromagnet with $T_N = 36K$) [31, 32] and the *p*-$O_2NC_6F_4CNSSN$ (ferromagnet with $T_C = 1.3K$) [33] (Fig. 23). The spin densities are mainly localized on the S and N atoms of the five-membered ring in these compounds.

5.1 Pressure-induced Enhancement of the Transition Temperature of p-NCC_6F_4CNSSN from 36K to 72K

The pressure dependence of the magnetization of this compound is shown in Fig. 24. It is noted that the transition temperature increases with pressure from 36K at ambient pressure up to 72K at $P = 16.4$ bar [31, 32]. This value of the transition temperature is the highest of genuine organic compounds with net magnetic moments. As shown in Fig. 25, the transition temperature extrapolates to the liquid nitrogen temperature 77K at around $P = 20$ kbar. On the other hand, the absolute value of magnetization decreases with pressure. These features are reasonable for the weakferromagnetic systems with the Dzyalosinsky-Moriya interaction or systems with nonequivalent values of the *g*-tensor. In these systems, the pressurization normally has the effect of increasing the antiferrom agnetic interaction, which results in a enhancement of the transition

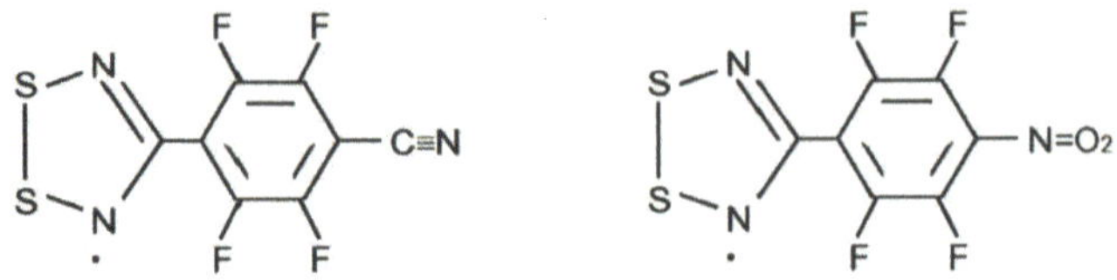

(1) p-NC.C6F4.CNSSN (2) p-O2N.C6F4.CNSSN

a) b)

Fig. 23. Molecular structure of p-NCC_6F_4CNSSN (a) and p-$O_2NC_6F_4CNSSN$ (b).

temperature as mentioned in section 3. At the same time, the enlarged antiferromagnetic exchange interaction makes the anti-parallel spin arrangement stronger as compared to the canting tendency originated from the anisotropic effects, thus resulting in a decrease of the net moment. Then, the product of magnetization and the antiferromagnetic exchange interaction (or the transition temperature) gives a constant value when the two anisotropic effects are pressure independent. In practice, this relation is found to be held for this compound in the pressure region $P < 10$ kbar.

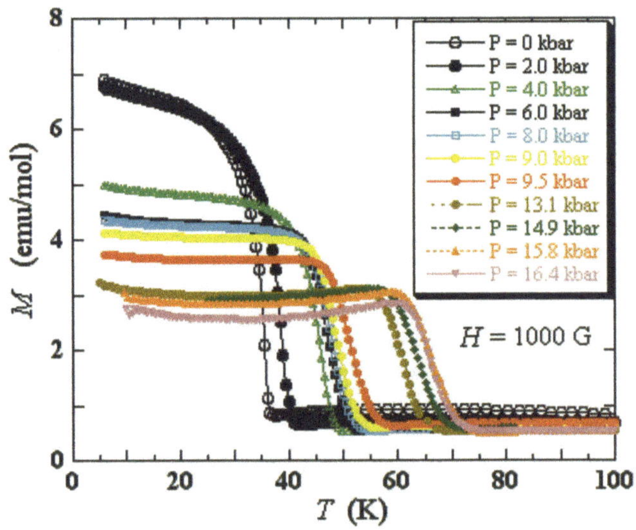

Fig. 24. Temperature dependence of the magnetization for p-NCC_6F_4CNSSN under pressure up to $P = 16.4$ kbar.

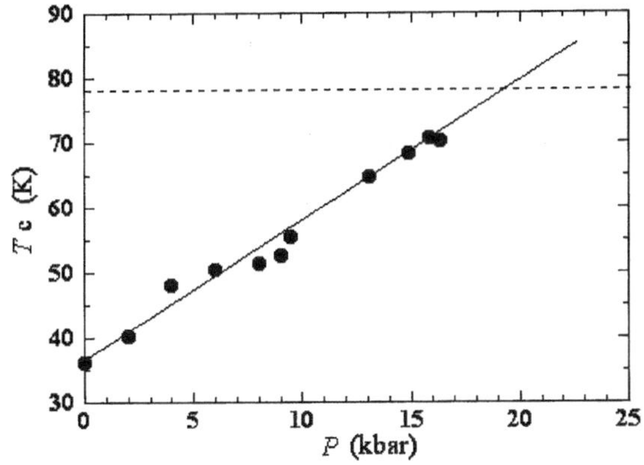

Fig. 25. Pressure dependence of weak-ferromagnetic transition temperature for *p*-NCC_6F_4CNSSN.

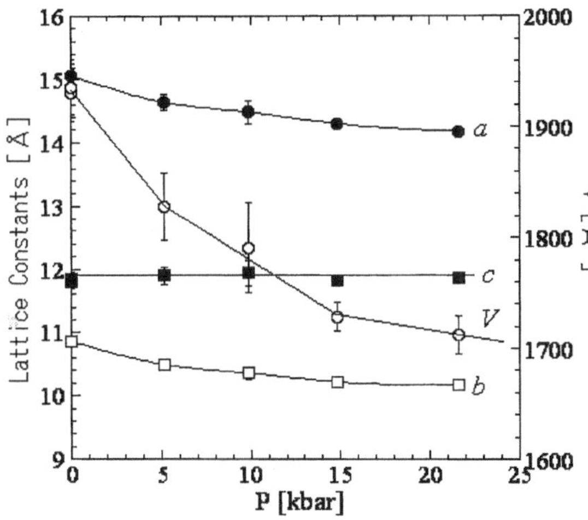

Fig. 26. Pressure dependence of the lattice parameters for *p*-NCC_6F_4CNSSN (F_{dd2}).

The change of the lattice parameters of the F_{dd2} space group of this compound is obtained from the X-ray analysis at room temperature as shown in Fig. 26. The lattice constant for the *c*-axis, which is parallel to the molecular axis, seems to be pressure independent, while those for *a*- and *b*-axes are highly pressure dependent and show relatively the same behavior. The pressure effects of the present compound are detailed somewhere [32].

5.2 Relation between $T_C(P)$ and the Averaged Intermolecular Exchange Interaction $J(P)$ in the Genuine Organic Ferromagnet p-$O_2NC_6F_4CNSSN$

This sulfur-based radical compound exhibits ferromagnetism below $T_C(P_0) = 1.30$K, as mentioned in another chapter of this book. Here we give a detailed analysis of the pressure dependence of the magnetic heat capacity, and the magnetization up to $P = 13.4$ kbar [33]. Fig. 27 shows the raw data of heat capacity obtained with a CuBe clamp cell (46.443 g) containing a sample of *p*-$O_2NC_6F_4CNSSN$ (0.0829 g) and Apiezon-J grease (0.0482 g). The sample of this small amount gives relatively large contribution at low temperatures. By the subtraction of the contribution of the addenda, the intrinsic heat capacity is obtained as shown in Fig. 28 from which $T_C(P)$ is obtained

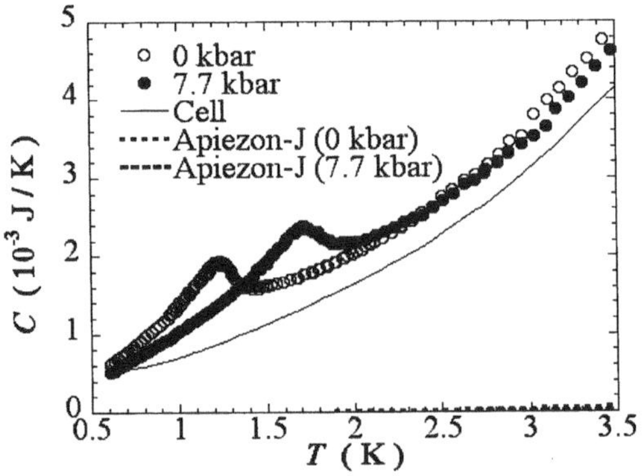

Fig. 27. An example of raw data of heat capacity measurements under pressures $P = 0$, and 7.7 kbar for the sample of *p*-$O_2NC_6F_4CNSSN$ (0.0829 g), CuBe pressure cell (46.443 g) and Apieson-J oil(0.0482 g).

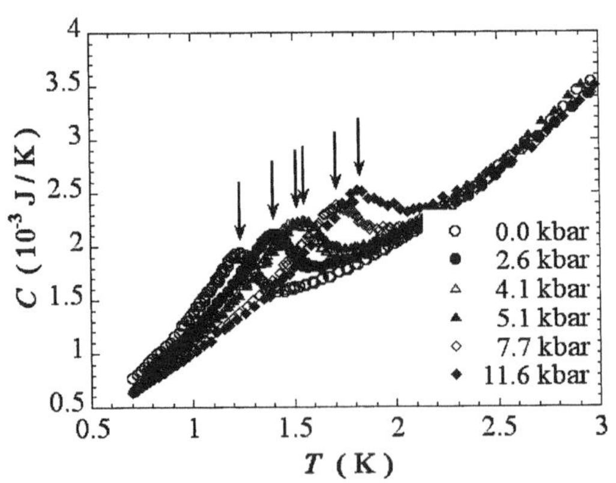

Fig. 28. Pressure dependence of the heat capacity of *p*-$O_2NC_6F_4CNSSN$ near the Curie temperature.

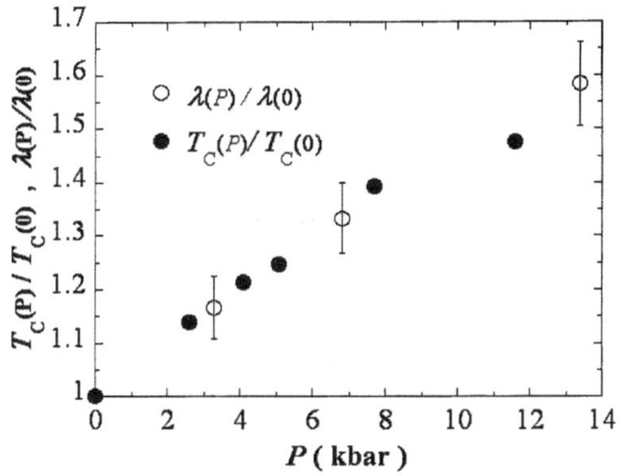

Fig. 29. Pressure dependence of $T_C(P)$ and $\lambda(P)$ ($=2zJ(P)/N(g\mu_B)^2$) of of p-$O_2NC_6F_4CNSSN$ normalized at the respective values $T_C(P_0)$, and $\lambda(P_0)$.

as illustrated in Fig. 29. The initial slope a = +0.045 ($kbar^{-1}$) seems to continue up to P = 10 kbar. Figure 30 shows six heat capacity measurements in the pressure range from 0 to 12 kbar, each normalized by the peak coordinates (Tc, C_{max}). It is noted that almost all data points fall on a universal curve around the critical temperature giving a law of corresponding state in this pressure region. This implies that the magnetic parameters are uniformly modified as the pressure on the compound increases.

Investigation of the pressure dependence of the magnetization allows to determine the averaged variation of the intermolecular interactions and, hence, that of $T_C(P)$. Fig. 31 shows the results of magnetization at P = 13.6 kbar at various constant

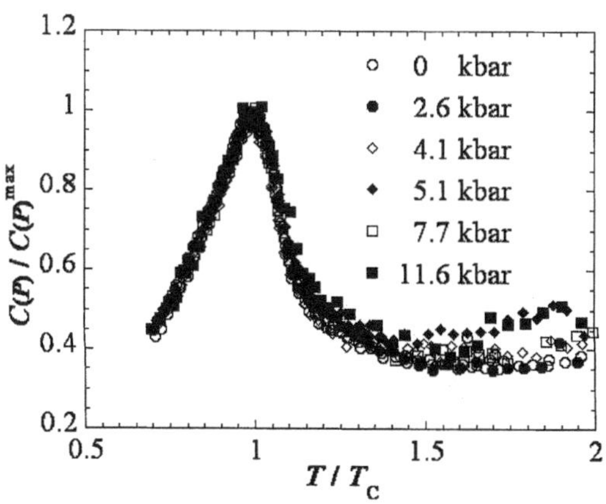

Fig. 30. Pressure dependence of the heat capacity of p-$O_2NC_6F_4CNSSN$ normalized by the peak value at the respective pressure versus temperature normalized by the corresponding $T_C(P)$.

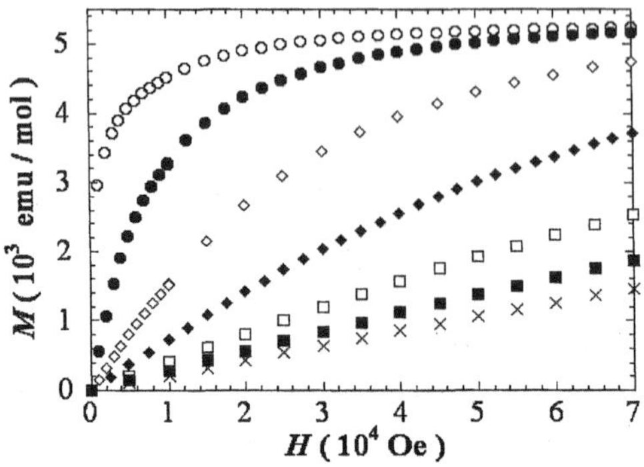

Fig. 31. Magnetization of p-$O_2NC_6F_4CNSSN$ under P =13.6 kbar above $T_C(P)$ at T = 1.9, 3.0, 5.0, 8.0, 12.0, 16.0, and 20.0K (from up to down).

temperatures above $T_C(P_0)$. In the applied field H, a spin feels the molecular field λM from the neighbour spins through exchange interactions J_i, in addition to H. Then the internal field becomes $H + \lambda M$, where

$$\lambda = \Sigma_{(i=1)} 2z_i\, J_j / N (g\,\mu_B)^2 = 2z\, J / N (g\,\mu_B)^2 \quad (4)$$

The value of λ is determined so that the experimental data of M give the best fit to the Brillouin function for $S = 1/2$ when plotted against $g\,\mu_B\,(H+\lambda M)/k_BT$. At P = 13.6 kbar, we have the best fitting value of λ = 9.5 as in Fig. 32, and this corresponds to $2zJ/k_B$ = 14.2K. In the same way, the pressure dependence of these

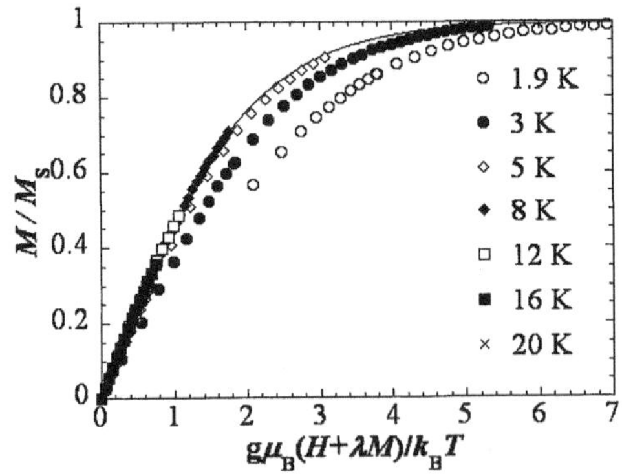

Fig. 32. Magnetization of p-$O_2NC_6F_4CNSSN$ under P = 13.6 kbar plotted agaist $g\mu_B(H+\lambda M)/k_BT$. Except temperatures just above $T_C(P)$ (1.9 and 3.0K), almost all of the data points fall on the Brillouin function for S = 1/2 with $\lambda(P)$ = 9.5.

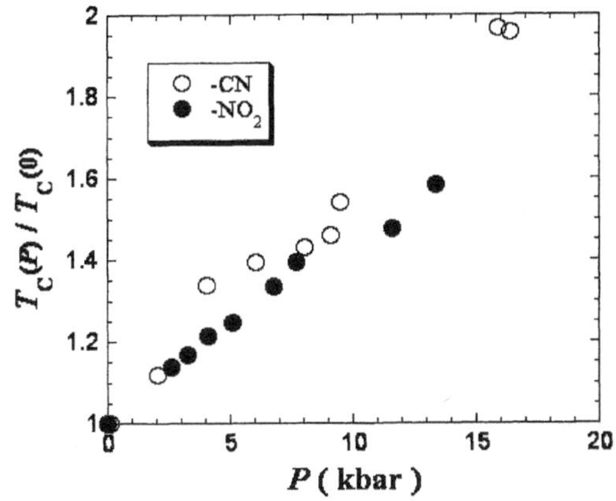

Fig. 33. Comparison of $T_C(P)$ for the two sulfur-based magnets p-NCC_6F_4CNSSN (open circle) and p-$O_2NC_6F_4CNSSN$ (closed circle) normalized at the respective $T_C(P_0)$.

values of λ and zJ/k_B are obtained, including $\lambda = 6.0$ and $2zJ/k_B = 9.0$K at ambient pressure. Each of these values is linearly proportional to the transition temperature. The pressure dependence of λ normalized by the value at ambient pressure is also given in Fig. 29.

It is interesting to compare $T_C(P)$ of p-$O_2NC_6F_4CNSSN$ with that of p-NCC_6F_4CNSSN on the common scale as in Fig. 33, from which the transition temperatures and magnetic interactions of these sulfur-based compounds show nearly the same pressure effects beyond antiferromagnetic and ferromagnetic features.

6 Other Interesting Pressure Effects on Thermally and Structurally Unstable Compounds

The galvinoxyl radical and pentafluorophenyl nitronyl nitroxide radical (F_5PNN) show interesting pressure effects different from those mentioned above, in the sense that pressure works to maintain the room temperature magnetic and structural properties.

6.1 Galvinoxyl

This compound has played an important role in the research of genuine organic ferromagnets. Its high and positive Curie-Weiss temperature of +11K [34, 35] gave an important thrust to the investigation of ferromagnetic interactions, although its ferromagnetic behaviour disappears below 85K accompanied by a structural change as seen in the magnetic susceptibility in Fig. 34. The recent study reveals that the ferromagnetic feature is preserved under pressure down to ^{4}He temperatures [36]. The magnetic heat capacity measured at P = 6.4K in a external magnetic field exhibits clear ferromagnetism down to 0.7K [37], being consistent with the theoretical work for a

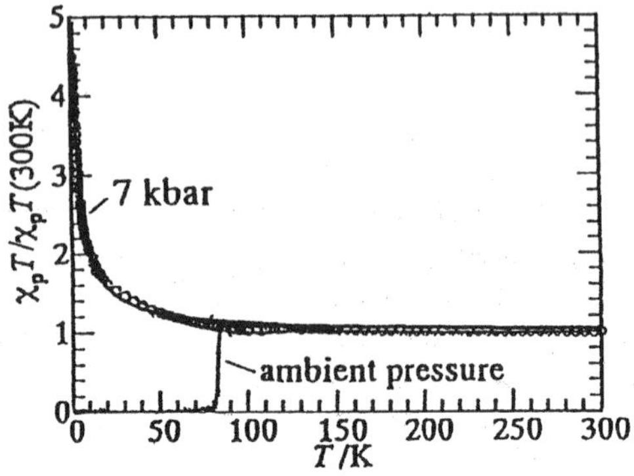

Fig. 34. Temperature dependence of ac-susceptibility of galvinoxyl at ambient pressure and $P = 7$ kbar. The solid curve expresses the theory of $S = 1/2$ one-dimensional Heisenberg ferromagnetic chain with $2J/k_B = 25K$.

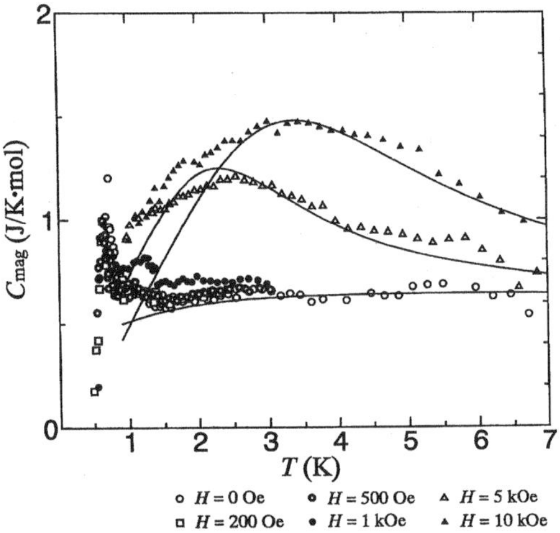

Fig. 35. Temperature dependence of magnetic heat capacity of galvinoxyl under external fields at $P = 6.4$ kbar. Each solid curve expresses the theory of $S = 1/2$ one-dimensional Heisenberg ferromagnetic chain with $2J/k_B = 18.0K$ in such magnetic fields as $H = 0$, 5 and 10 kG.

quasi-one-dimensional spin system with the intrachain interaction of $2J/k_B = +18K$ and the interchain one $zJ'/k_B = -0.2K$, as shown in Fig. 35 [37]. Below 0.7K, antiferromagnetic ordering arises [37].

6.2 F_5PNN

At room temperature, the crystal symmetry of this compound suggests the uniform magnetic chain structure [36, 38]. The magnetic properties at low temperatures, however, show the features of the alternating chain system in various magnetic and thermal observations. Recently, field induced magnetic ordering has been found above the spin gap energy in this system [39, 40]. Fig. 36 shows the pressure effects on the temperature dependence of the magnetic susceptibility. At ambient pressure, the susceptibility follows that of an alternating spin system with energy gap at the lowest temperature, where the alternating exchange ratio a = (J_2/J_1) = 0.4; $(J_1/k_B = -5.6\text{K})$ reproduces the experiments. This contradicts the uniform structure at room temperature.

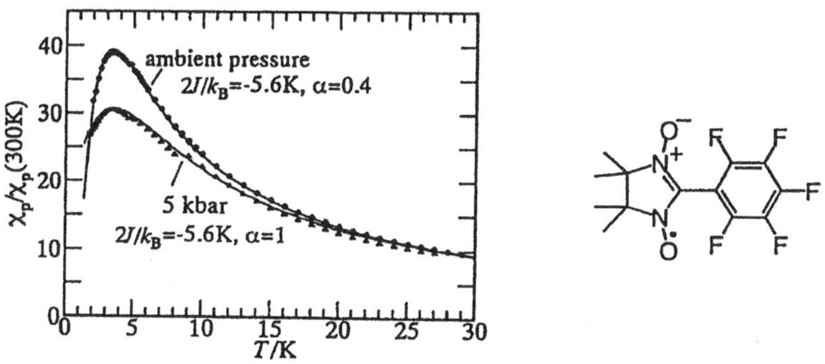

Fig. 36. Temperature dependence of ac-susceptibility of F_5PNN at ambient pressure and $P = 5$ kbar. Solid curves express the theoretical curves for $S = 1/2$ Heisenberg antiferromagnetic alternating chain system.

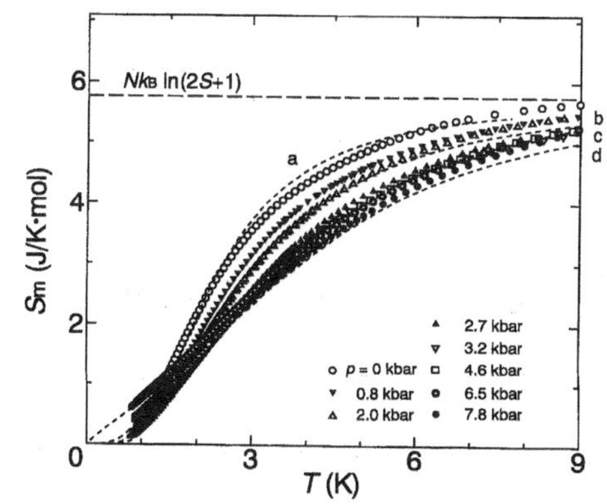

Fig. 37. Temperature dependence of magnetic entropy of F_5PNN for some pressures. The broken straight line shows the total magnetic entropy of $Nk_B \ln(2S+1)$. Four dotted curves (a-d) show the Duffy *et al.*'s theoretical curves with such two parameters as α and J/k_B ; (a) $\alpha = 0$, $J/k_B = -3.1$K, (b) $\alpha = 0.6$, $J/k_B = -3.9$K, (c) $\alpha = 0.8$, $J/k_B = -4.4$K, and (d) $\alpha = 1.0$, $J/k_B = -4.5$K.

The structure at low temperature has not been solved yet. It is noted that the results at $P = 5$ kbar fit to the theory for an uniform antiferromagnetic chain with the same value of $J_1/k_B = -5.6K$ [41]. The pressure dependence of the heat capacity of this compound has been investigated under several pressures below 10 kbar, giving the same crossover feature from the alternating to the uniform magnetic interactions [41]: At ambient pressure, the heat capacity exhibits an exponential increase indicating the spin gap state at low temperatures. As the pressure approaches to 6.5 kbar, the magnetic heat capacity becomes linearly dependent with the temperature indicating the uniform chain system.

For $P > 6.5$ kbar, the uniformity seems to be kept with the increasing values of J_1 and the Debye temperature. This pressure dependence of the heat capacity is quantitatively explained with the above mentioned theories [42]. The magnetic entropy also shows a clear cross-over feature under pressure, from the exponential to the linear dependence on temperature as indicated in Fig. 37.

7 Conclusion

By the artificial control of the overlapping of π-orbitals under the pressure, the electronic state and the mechanism of intra- and intermolecular interactions have been investigated for various prototypes of organic magnets, such as antiferromagnets, ferromagnets, weak-ferromagnets and so on. The enhancement and/or reduction of the magnetic transition temperatures and pressure-induced magnetic and structural transitions have been quantitatively analyzed based on the thermal, magnetic, and structural measurements under the pressure, and interpreted on the words of molecular magnetism. For the further development of this physical method under the pressure, it is indispensable to trace the precise topological change of all of the constituent atoms, especially at low temperatures. On the other hand, pressure effects on low-dimensional field-induced spin ordering systems are expected to bring about a heuristic quantum excitations associated with the Tomonaga-Luttinger liquid [40].

References

1. K. Takeda, K. Konishi, M. Tamura, and M. Kinoshita, Phys. Rev. B53 (1996) 3374.
2. K. Takeda, K. Konishi, M. Tamura, and M. Kinoshita, Thermo. Acta, 266 (1996) 175.
3. K. Takeda, M. Wada, M. Inoue, and T. Haseda, Jpn. J. Appl. Phys. 26 (1987) 947.
4. M. Wada, K. Takeda, A. Ohtani, A. Onodera, and T. Haseda, J. Phys. Soc. Jpn. 52 (1983) 3188.
5. M. Mito M. Hitaka, T. Kawae, K. Takeda, T. Kitai, and N. Toyoshima, Jpn. J. Appl. Phys. 40 (2001) 6641.
6. K. Takeda and M. Mito, J. Phys. Soc. Jpn. 71 (2002) 729.
7. T. Iwamoto, M. Mito, M. Hidaka, T. Kawae, and K. Takeda, Physica B 329-333 (2003) 665.
8. M. Mito, T. Tanimoto, T. Kawae, M. Hitaka, K. Takeda, S. Nakatsuji, H. Morimoto, and H. Anzai, Polyhedron 20 (2001) 1505.
9. M.Mito, H. Deguchi, T. Tanimoto, T. Kawae, S. Nakatsuji, H. Morimoto, H. Anzai, H. Nakao, Y. Murakami, and K.Takeda, Phys. Rev. B67 (2003) 24427.
10. K. Takeda, N. Uryu, M. Inoue, and J. Yamauchi, J. Phys. Soc. Jpn. 56 (1987) 736.

11. K. Takeda, H. Deguchi, T. Hoshiko, K. Konishi, K. Takahashi and J. Yamauchi, J. Phys. Soc. Jpn. 58 (1989) 3361.
12. J. Yamauchi, K. Takeda, and K. Konishi, Mol. Cryst. Liq. Cryst. 234 (1993) 97.
13. M. Mito, H. Nakano, M. Ito, M. Hitaka, T. Kawae, K. Konishi, K. Mukai, and K. Takeda, Czech. J. Phys. 46. Suppl S4 (1996) 2107.
14. K. Takeda, M. Mito, M. Ito, T. Kawae, M. Hitaka, M. Matsubara, and K. Mukai, Mol. Cryst. Liq. Cryst. 306 (1997) 177.
15. K. Takeda, M. Mito, T. Kawae, H. Nakano, M. Hitaka, K. Suzuki, K. Mukai, J. Magn. Magn. Mater. 177-181 (1998) 817.
16. M. Mito, M. Hitaka, T. Kawae, K. Takeda, K. Suzuki, and K. Mukai, Mol. Cryst. Liq. Cryst. 334 (1999) 369.
17. D. Bloch, J. Phys. Chem. Solids 27 (1966) 881.
18. K. Takeda, K. Konishi, K. Nedachi, and K. Mukai, Phys. Rev. Lett. 74 (1996) 1673.
19. M. Mito, H. Nakano, T. Kawae, M. Hitaka, S. Takagi, H. Deguchi, K. Suzuki, K. Mukai and K. Takeda, J. Phya. Soc. Jpn. 66 (1997) 2147.
20. K. Takeda, M. Mito, H. Nakano, T. Kawae, M. Hitaka, S. Takagi, H. Deguchi, S. Kawasaki and K. Mukai, Mol. Cryst. Liq. Cryst., 306 (1997) 431.
21. M. Kinoshita, P. Turek, M. Tamura, K. Nozawa, D. Shiomi, Y. Nakazawa, M. Ishikawa, M. Takahashi, K. Awaga, T. Inabe, and Y. Maruyama, Chem. Lett. (1911) 1225.
22. K. Konishi, K. Takeda, M. Tamura, and M. Kinoshita, Mol. Cryst. Liq. Cryst. 279 (1996) 115.
23. M. Mito, T. Kawae, M. Takumi, K. Nagata, M. Tamura, M. Kinoshita, and K. Takeda, Phys. Rev. B56 (1997) R14255.
24. M. Mito, T. Kawae, M. Takumi, K. Nagata, M. Tamura, M. Kinoshita, and K. Takeda, J. Magn. Magn. Mater. 177-181 (1998) 801.
25. K. Takeda, M. Mito, T. Kawae, M. Takumi, K. Nagata, M. Tamura, and M. Kinoshita, J. Phys. Chem. B102 (1998) 671.
26. M. Mito, T. Kawae, K. Takeda, T. Ishida, and T. Nogami, Mol. Cryst. Liq. Cryst. 334 (1999) 323.
27. M. Mito, T. Kawae, M. Hitaka, K. Takeda, T. Ishida, and T. Nogami, Chem.Phys.Lett. 333 (2001) 69.
28. K. Takeda, M. Mito, K. Kinoshita, M.A. Novak, J.L. Tholenece, and A. Rassat, Polyhedron 22 (2003) 2287.
29. K. Takeda, M. Mito, T. Kawae, M. Takumi, H. Deguchi, S. Takagi, M. Okumura, T. Kawakami, K. Yamaguchi, and M. Kinoshita, Chem. Phys. Lett. 308 (1999) 181.
30. M. Okumura, T. Kawakami, A. Oda, K. Yamaguchi, M. Mito, and K. Takeda, Mol. Cryst. Liq. Cryst. 335 (1999) 623.
31. M. Mito, T. Kawae, K. Takeda, S. Takagi, Y. Matsushita, H. Deguchi, J.M. Rawson, and F. Palacio, Polyhedron 20 (2001) 1509.
32. M. Mito, S. Takagi, H. Deguchi, J. M. Rawson, F. Palacio and K. Takeda, in preparation for publication.
33. K. Takeda, Y. Yoshida, Y. Inagaki, T. Kawae, J. M. Rawson, F. Palacio, and M. Mito, in preparation for publication.
34. K. Mukai, H. Nishiguchi, and Y. Deguchi, J. Phys. Soc. Jpn. 23 (1967) 125.
35. K. Mukai, Bull. Chem. Soc. Jpn. 42 (1969) 40.
36. Y. Hosokoshi, M Tamura, and M. Kinoshita, Mol. Cryst. Liq. Cryst. 306 (1997) 423.
37. Masaki Mito, Dr. Thesis, Kyushu University, Fukuoka (1998).
38. M. Takahashi, Y. Hosokoshi, H. Nakano, T. Goto, M. Takahashi, and M. Kinoshita, Mol. Cryst. Liq. Cryst. 306 (1997) 111.

39. Y. Yoshida, K. Yurue, M. Mito, T. Kawae, Y. Hosokoshi, K. Inoue, M. Kinoshita, and K. Takeda, Physica B 329-333 (2003) 979.
40. Y. Yoshida, N. Tateiwa, M. Mito, T. Kawae, K. Takeda, Y. Hosokoshi, and K. Inoue, Phys. Rev. Lett. 94 (2005) 037203.
41. M. Mito, T. Kawae, Y.Hosokosh, K. Inoue, M. Kinoshita, and K. Takeda, Solid Sate Commun. 111 (1999) 607.
42. W. Duffy. Jr and K.P. Barr, Phys. Rev. 165 (1968) 647.

7

Spin Density Distribution and Interaction Mechanisms in Thiazyl-based Magnets

J. Campo[1], J. Luzón[1], F. Palacio[1], and J. Rawson[2]

[1]*Instituto de Ciencia de Materiales de Aragon, CSIC-Universidad de Zaragoza, E-50009, Zaragoza, Spain*
[2]*Department of Chemistry, The University of Cambridge, Lensfield Road, Cambridge CB2 1EW, UK*

1 Introduction

This chapter presents a study of the magnetic properties and magnetic interaction mechanisms in a series of thiazyl radicals, which are of particular interest in the field of molecular magnetism.

The chapter is organized as follows. First, there is an introduction to organic molecular magnetism and thiazyl radicals. The importance of an accurate knowledge of the spin density distribution in order to understand the mechanisms of the magnetic interactions is discussed. This is highlighted with complimentary experimental and theoretical *ab initio* determinations of the spin density distribution in the dithiadiazolyl radical, *p*-$O_2NC_6F_4CNSSN^•$. An *ab-initio* computation of the magnetic interactions in this radical is presented in relation to its spin density distribution and magnetic behaviour. This approach is then extended to other dithiadiazolyl and thiazyl radicals. Finally, the conclusions revealed from this study are summarized.

1.1 Organic Magnets

Since the discovery of the *β*-phase of *p*-nitrophenyl nitronyl nitroxide (*p*-NPNN) as the first organic ferromagnet free of metallic ions [1] the design of new purely organic ferromagnets has been the subject of a considerable amount of research.

The presence of unpaired electrons are the basis of magnetic phenomena. In the case of organic molecules they can be obtained in two different ways. One is based on charge-transfer processes between electron donor and electron acceptor molecules to produce open-shell cations and anions. TDAE-C_{60} [tetrakis (dimethylamino) ethylene-C_{60}], which exhibits long-range magnetic ordering at 16.1K, is the best representative example of this approach [2].

The second approach, which is most commonly followed, is the generation of open-shell molecules, the so-called free-radicals. In the design of free-radical-based magnets there are several difficulties that have to be overcome.

Almost all the free radicals are unstable species due to their preference of organic systems to form closed-shell configurations via oxidation, reduction or covalent bond formation. Several strategies are used to increase the kinetic stability of the free radicals. Steric shielding of the atoms containing the unpaired electrons is often employed to inhibit covalent bond formation *via* a free radical mechanism. A second strategy is the use of Coulombic electrostatic repulsions for keeping apart the reactive centres. The tendency of organic systems to form closed-shell configurations via bond formation is typically reflected in a propensity to dimerize in the solid state, which renders them diamagnetic. In order to provide thermodynamic stabilisation of the radical centre it is desireable to provide a singly occupied molecular orbital which is low-lying. This is typically achieved through the introduction of electronegative heteroatoms and/or π-delocalization of the unpaired electrons and decreases its reactivity. The reader interested in a more extended revision of the strategies for stabilizing free radicals is referred to ref. [3].

The interaction between unpaired electrons from different radicals is a necessary requirement for obtaining long-range magnetic order in the solid. Therefore, these strategies for stabilizing free radicals based on the isolation of the unpaired electrons must be employed in moderation. In other words, the kinetic stability and the strength of the magnetic interaction of the free radicals are two opposing requirements that must be balanced in the design of purely organic systems that show bulk magnetic behaviour. A consequence of this balance is the low ordering temperatures of these kinds of materials, almost all of them quite below liquid helium temperature (4.2K), which greatly limits the applications of such systems.

In organic molecular systems the two main sources of magnetic anisotropy are magnetic dipolar interactions and spin-orbit coupling. Both sources usually have a small magnitude and purely organic magnets are considered as isotropic Heisenberg systems. The weakness of the dipolar interactions is due to the delocalization of the unpaired electrons. On the other hand, the unpaired electrons are located in molecular orbitals composed of *p* atomic orbitals with no orbital moment. Therefore, the spin-orbit coupling anisotropy comes from the spin-orbit coupling of the ground state with excited states with orbital angular momentum, and is a second-order term. This low magnetic anisotropy introduces another difficulty in the design of purely organic magnets, since in Heisenberg systems long-range magnetic ordering is only possible if the magnetic interactions are propagated in a three-dimensional network.

In summary, there are only three requirements for obtaining long-range magnetic ordering in purely organic solids: the existence of magnetic centres, the magnetic interaction among these magnetic centres and the propagation of these magnetic interactions throughout the solid. In practice it proves extremely difficult to achieve all these goals in any one single system. In particular, the last two requirements imply a crucial role of the unpaired electron distribution in the free radicals and of the packing of these free radicals in the crystal.

Among the systems showing long-range magnetic order, those which exhibit a spontaneous magnetization are of particular interest, since they can potentially lead to applications, *e.g.* magnetic sensors and magneto-optic applications. The different approaches for the design of materials with spontaneous magnetization are described in the next section.

1.2 Strategies for Obtaining Spontaneous Magnetization

As mentioned above, a desired property of the molecular magnetic materials is a long-range spontaneous magnetization. Several strategies have been proposed in order to achieve this spontaneous magnetization.

1.2.1 Orthogonal Orbitals

In the Anderson formalism for the magnetic exchange interaction the kinetic term is usually dominant and the magnetic interaction is antiferromagnetic. However, when the net overlap between the natural magnetic orbitals is zero, i.e. the orbitals are orthogonal, the kinetic term vanishes and the magnetic interaction is ferromagnetic. This mechanism, for instance, has been used in order to explain the ferromagnetism below 90K of the Prussian-blue structured $CsNi^{II}[Cr^{III}(CN)_6]\cdot 2H_2O$ [4].

1.2.2 Spin Polarization (McConnell I Mechanism)

Within the Unrestricted Hartree-Fock (UHF) formalism, the spin polarization phenomenon arises from the different exchange interaction between the unpaired electron in the singly occupied molecular orbital (SOMO) with the 'spin up' (α) and 'spin down' (β) electrons in the fully occupied molecular orbitals (FOMO). In the UHF formalism the electron-electron repulsion between the unpaired electron of α spin of the SOMO with α-spins in the FOMOs will be less than the repulsion between the SOMO α-spin and the corresponding β-spins in the FOMOs. In order to minimise this electron-electron repulsion the FOMOs' β-spins try and adopt regions of space not occupied by the SOMO α-spin. As a consequence the α and β spins have slightly different spatial distributions. The total spin density distribution is the difference between the total α- and β-spin distributions and often leads to negative spin density accumulating in regions where the SOMO α-spin has no contribution. The correlation among the electrons, not considered in the Hartree-Fock formalism, also influences the spin polarization effect.

The basis of the MacConnell I mechanism is that when a region of negative spin density in one molecule overlaps with a region of positive spin density in another molecule, a ferromagnetic coupling should be expected. In order to support this mechanism McConnell proposed a new Hamiltonian version for the magnetic interaction between two molecules, A and B [5]

$$H = -2\vec{S}_A\vec{S}_B\sum_{ij} J_{ij}\rho^i_A\rho^j_B$$

Here $\boldsymbol{S}_A$ and $\boldsymbol{S}_B$ are the total spins on A and B. J_{ij} is a magnetic coupling constant between the *i* and *j* atoms located in the A and B molecules respectively. ρ^i_A and ρ^j_B are the spin populations on the *i* and *j* atoms. Since the magnetic coupling constants are usually negative, a ferromagnetic interaction between the two molecules can be achieved if the spin densities of the interacting atoms have opposite signs.

The above expression has been shown to lack theoretical rigour and cannot be naively applied by considering only the closest atoms with high spin populations between the two molecules [5, 6]. Whilst it is frequently invoked to explain the magnetic exchange between free radicals, it should be applied with some caution.

1.2.3 McConnell II Mechanism

This mechanism was proposed for charge-transfer salts composed of chains of alternating electron-donor and electron-acceptor molecules in which the SOMO orbitals of the donor in the neutral state are degenerate [7]. The triplet state of the neutral donor, according to the Hund's rule for degenerate states, induces a triplet state in the charge-transfer state through a configuration interaction. Such a triplet state is propagated along the donor-acceptor chain. Several modifications of this mechanism have been proposed, where the triplet state of the charge-transfer state is stabilized through configuration interaction with different excited triplet states [8, 9, 10].

The McConnell II mechanism was proposed to explain the bulk ferromagnetic behaviour of the $[Fe(Cp_2^*)]^+[TCNE]^-$ compound below 4.8K [11]. However, a polarized neutron diffraction experiment suggests the McConnell I mechanism may be more adequate to explain the ferromagnetic interaction between the donor and the acceptor molecules [12].

1.2.4 Ferrimagnetism

Up to now, all the above mechanisms aiming to design materials with long-range spontaneous magnetization are based on the achievement of ferromagnetic interactions. On the other hand, the last two strategies profit from the tendency of magnetic interaction to be antiferromagnetic. The so-called "ferrimagnetic approach" is based in the construction of chains containing alternating magnetic centres with unequal spin values. This is the strategy used in the bimetallic approach for inorganic molecular magnets. Although a similar strategy has been proposed for organic magnets, it is only very recently that the first example has been discovered [13].

1.2.5 Weak Ferromagnetism

There are compounds whose magnetic behaviour at high temperature can be ascribed to an antiferromagnetic material but which show a weak spontaneous magnetization below a transition temperature. This phenomenon, known as weak ferromagnetism, was explained by Dzyaloshinskii [14] and Moriya [15] as an antiferromagnetic structure in which there exists a canting angle between the magnetic moments of different sub-lattices with opposite signs of the magnetic moments. This non-collinearity of the magnetic moments results in a net magnetization.

There are two main mechanisms that can produce the non-colinearity of the magnetic moments. In the first one, the non-colinearity of the magnetic moments reflects the non-colinearity of the direction of the local magnetic anisotropy. The second mechanism is due to the combination of the spin-orbit coupling and the magnetic interaction between two magnetic centres. This mechanism can be accounted by a new term in the magnetic interaction Hamiltonian:

$$H_{DM} = -\vec{D} \cdot (\vec{S}_A \times \vec{S}_B)$$

The conditions allowing canting between two magnetic moments are very strict. In relation to the crystal field, the nuclear and the magnetic unit cells must be the same and the sites whose magnetic moments are non-colinear must not be related by an inversion centre. In other words, weak ferromagnetism is favoured in molecular magnetism due to the higher propensity for low-symmetry molecules to adopt structures without inversion centre. The organic magnets with the highest transition temperatures are weak ferromagnets, in particular the 1,3,5-Triphenyl-6-oxoverdazyl (TOV) compound [16] and the β phase of the dithiadiazolyl radical *NC*-C_6F_4-CNSSN• [17, 18], whose transition temperatures are 4.8K and 36K respectively. The magnitude of the spin-canting is related to the magnetic anisotropy which, as we have already seen (section 1.1), is very small for organic free radicals. As a consequence, organic free radicals tend to exhibit very low canting angles and, consequently, very low values of the magnetization, e.g., 1.5×10^{-3} μ_B/mol at 0K are observed. (*cf* 1.0 μ_B/mol for organic ferromagnets built upon an equivalent S = 1/2 spin system).

1.2.6 Application to Purely Organic Systems

The most obvious structure presenting spontaneous magnetization is the ferromagnetic one. However, only a few compounds have been reported as ferromagnetic. The first discovered examples were the already mentioned *p*-NPNN radical [1, 19 and references therein], the TDAE-C_{60} charge-transfer salt [20] and the diazaadamantane dinitroxyl biradical [21]. Recently, the discovery of two S-based ferromagnets have been reported: the dithiadiazolyl derivative *p*-$O_2NC_6F_4$CNSSN• [22] and the organic cation radical salt [BBDTA][$GaCl_4$].CH_3CN [23] with ferromagnetic transition temperatures of 1.3K and 6.7K respectively. The former radical belongs to the family of compounds studied in this chapter. The reason for the low number of ferromagnetic free-radical crystals is that antiferromagnetic interactions are usually more favourable. Whilst the magnetic interaction between radicals may be strongly ferromagnetic in two dimensions, a net antiferromagnetic coupling in the third dimension will lead to an antiferromagnetic ground state, i.e. very strict conditions must be met to obtain ferromagnetic interactions.

1.3 Dithiadiazolyl Radicals

Almost all the attempts in the design of purely organic systems showing spontaneous magnetization have involved nitroxide NO• radical derivatives as building blocks (see ref. [24] and contributions therein). In these radicals the spin delocalization over the NO• groups provides high stability. Whilst the spin-bearing oxygen atoms are available at the molecular periphery, the electrostatically repulsive nature of the $O^{\delta-}...O^{\delta-}$ contacts simultaneously disfavours close approach of regions of spin density. A systematic search of the Crystallographic database reveal no O...O contacts less than the sum of the van der Waals radii [6]. To date the highest ferromagnetic transition temperature of a nitroxide radical derivative is 1.48K in the diazaadamantane dinitroxyl biradical [21].

One approach to raise the magnetic ordering temperature utilises the more diffuse nature of the *3p* sulphur orbitals compared to the *2p* ones of the oxygen atoms. Sulphur-based radicals should in principle favour stronger molecular interactions than oxygen-based ones [25, 26, 27], since the more radially diffuse *3p* orbitals should facilitate interactions at greater distances (*cf* van der Waals radii of O and S at 1.50 and 1.74 Å). In addition S/N based radicals favour electrostatically attractive $S^{\delta+}...N^{\delta-}$ contacts and exhibit a multitude of S...N contacts less than the sum of the van der Waals radii.

The heterocyclic dithiadiazolyl radicals $RCN_2S_2^{\bullet}$ have provided promising results in the framework of this sulphur-based approach for obtaining purely organic ferromagnets, as it will be shown later. Of the two known isomers, the 1,3,2,4-dithiadiazolyl radical, undergoes rearrangement to the thermodynamically more stable 1,2,3,5-dithiadiazolyl radical [28].The latter have been the object of extensive studies [29, 30], particularly in relation to their potential application in the design of organic conductors via regular stacks of the heterocyclic rings. Rawson *et al.* [31] have reviewed the chemistry of the dithiadiazolyl radicals. This Chapter reviews the magnetic properties of these radicals.

The SOMO orbital in the 1,2,3,5-dithiadiazolyl radical is an antibonding π* orbital localized on the heterocyclic ring with a node on the carbon atom due to the a_2 symmetry of the orbital (see Fig. 1) [32]. A very important consequence of the node at the carbon atom is that the electronic properties of this radical, e.g., hyperfine coupling constants, solution reduction and oxidation potentials, and gas phase ionization potentials show very little dependence on the *R* substituent at the carbon atom. Therefore, these *R* substituents can be varied in order to tune the crystal packing of the radicals for material applications without significantly changing the electronic properties of the dithiadiazolyl ring. *Ab initio* calculations have indicated a significant

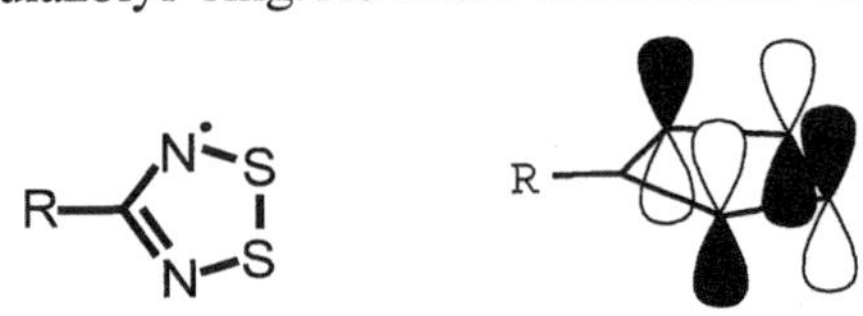

a) b)

Fig. 1. a) Scheme of the 1,2,3,5-dithiadiazolyl radical. b) SOMO orbital for the 1,2,3,5-dithiadiazolyl radical

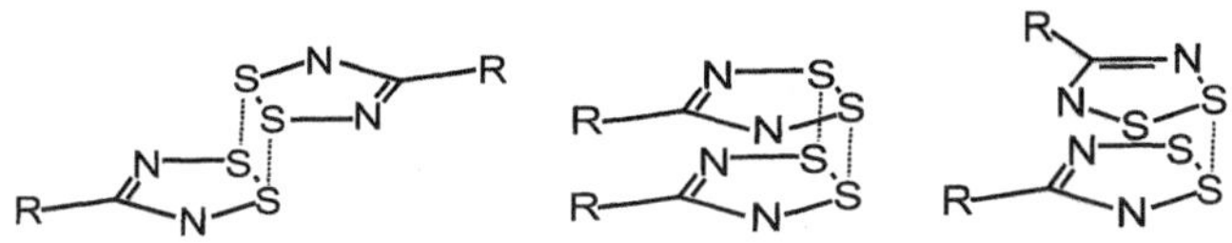

Fig. 2. Dimerization geometries of the dithiadiazolyl radicals.

polarity of the S-N bond, in the sense $S^{\delta+}$—$N^{\delta-}$. Consequently, there will be a strong tendency for these radicals to pack in such a way as to maximise the intermolecular $S^{\delta+}$—$N^{\delta-}$ interactions. Since the S and N atoms bear the majority of the spin density (see Fig. 1b and Section 2) then these close contacts should favour strong magnetic interactions. In addition the inclusion of sulphur generates additional spin-orbit coupling leading to a greater anisotropy than nitroxide radicals [33].

The main difficulty of utilising these radicals in the design of molecular magnetic materials is the lack of sufficient steric protection to inhibit dimerisation. Solution EPR studies on dithiadiazolyl radical derivatives in solution have determined a favourable dimerization enthalpy of ~ 35 kJ/mol [31]. Such a high dimerisation energy is due to the strong π-π bonding interaction between the SOMO orbitals with close S...S and/or N...S and N...N contacts. The significant overlap between the SOMO orbitals from each radical in the dimer produces a spin-paired singlet ground state and therefore diamagnetic behaviour. Three typical dimerisation geometries of the dithiadiazolyl radicals are shown in Fig. 2.

Perfluoraryl groups have been proposed as the R substituents in order to avoid the dimerisation [31, 34] since i) the intramolecular N...F repulsions lead to a large twist angle between perfluorophenyl and dithiadiazolyl ring planes, making the phenyl group more sterically demanding; ii) the electrostatic repulsion between the perfluorophenyl rings may inhibit dimerisation; iii) the inclusion of strongly structure-directing groups. These may lead to structures in which the structure-directing group becomes the dominant intermolecular interaction and crystal growth utilising these functional groups may occur at the expense of the dimerisation process. This appears particularly successful when either the heterocyclic N or S atoms are themselves part of the structure-directing interaction, either through formation of favourable $S^{\delta+}...X^{\delta-}$ or $N^{\delta-}...X^{\delta+}$ interactions.

To date, five perfluoroaryl-dithiadiazolyl radical crystals in which the dimerization tendency has been inhibited have been synthesized (see Fig. 3) [17, 22, 35, 36, 37, 38]. These five structures exhibit very different magnetic behaviour. One of the most interesting is the *p*-NCC_6F_4CNSSN• (**1**) which is polymorphic and crystallises in two different phases, α and β. The β phase, **1**β, exhibits the highest transition temperature to a magnetic ordered phase which displays a spontaneous magnetization [18]. In this compound the magnetic interactions are antiferromagnetic but the magnetic space group allows the canting of the spins, which results in weak ferromagnetism (see Section 1.2.5). Both the α phase of the same dithiadiazolyl radical [35] and *p*-BrC_6F_4CNSSN• (**2**) present paramagnetic behaviour with a negative value of the Weiss constant [39]. These negative Weiss constants indicate the existence of antiferromagnetic interactions between the radicals but an absence of long range order. Conversely the *p–*$O_2NC_6F_4CNSSN\bullet$ radical (**3**) shows ferromagnetic interactions and a ferromagnetic transition at 1.32K [22, 40]. The fourth compound, the *p-* $NCC_6F_4C_6F_4CNSSN\bullet$ (**4**) exhibits a structure-directing CN···S interaction as observed in $\mathbf{1}\alpha$ and $\mathbf{1}\beta$ but exhibits near perfect Curie paramagnet ($|\theta| < 1K$) consistent with large distances between heterocyclic rings beyond 6Å [38]. The magnetic behaviour of the first four radicals will be explained in more detail later in this chapter. It should be emphasised that these dithiadiazolyl radicals are not the only dithiadiazolyl radicals to exhibit solid state paramagnetism (see also refs. [41, 42, 43]) but, as we shall see, their strong electronic and structural similarities provide an excellent group of molecules from which to draw meaningful conclusions.

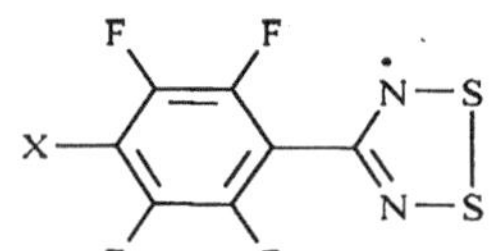

1 X = CN
2 X = Br
3 X = O_2N
4 X = NCC_6F_4

Fig. 3 Paramagnetic dithiodiazyl derivatives

As seen in this general introduction to organic magnets, the unpaired electron distribution in the free radicals and the crystal packing play a crucial role in the magnetic behaviour of the free-radical solids. In fact, the important parameter is the spin density distribution, which not only reflects the unpaired electron distribution but also the polarization effects of these unpaired electrons on the doubly occupied molecular orbitals. A favourable packing to achieve long-range magnetic order must allow the proximity of the spin density of different free radicals and the propagation of the magnetic interactions through the solid. In addition, the nature of the magnetic interaction depends strongly on the relative geometry between the interacting magnetic orbitals which is also determined by the crystal packing. In other words, the different magnetic behaviour of the monomeric dithiadiazolyl radicals, even in two polymorphic crystals of the same radical, is due to the differences in their crystal packing. In the sections which follow we utilise radicals **1** – **3** to exemplify the sensitivity of the magnetic behaviour to variations in structure whilst retaining a constant spin density distribution.

2 Spin Densities in the Dithiadiazolyl Ring

The magnetic behaviour of several dithiadiazolyl radicals, including both polymorphs of **1** and the bromo derivative **2**, have been studied in detail by Antorrena together with other thiazyl radicals [44]. The work also investigated the spin density in the perfluoroaryl derivatives of the dithiadiazolyl radical.

The spin density of dithiadiazolyl radicals has been undertaken from a theoretical perspective with semi-empirical and *ab initio* calculations and from an experimental point of view with EPR and Electron Spin Echo Envelope Modulation (ESEEM) spectroscopic techniques and with polarized neutron diffraction experiments.

The spectroscopic methods have several pecularities with respect to the neutron polarized diffraction experiments: i) they only provide estimates of how much of the spin density is associated with each spectroscopically active nucleus but does not provide information on the spatial density distribution about each nucleus; ii) EPR can only act as a reporter of the spin density distribution at those nuclei with non-zero nuclear spin e.g. ^{14}N and ^{19}F; iii) the signal assignment is not always simple; iv) in EPR and ESEEM the sign of the spin density is not trivially determined; v) in EPR and ESSEM the experiments are typically performed in solution and reflect the spin density distribution in magnetically isolated ions not on the actual crystal structure (in the solid state dipolar line broadening often gives rise to line broadening which inhibits observation of the hyperfine interactions which reflect the spin density distribution); vi) antiferromagnetic compounds can be studied using EPR spectroscopy.

Different approaches to the experimental determination of the spin density in dithiadiazolyl derivatives using continuous-wave EPR (CW-EPR) and ESEEM spectroscopies have been carried out [31, 45). Only the local spin population at the heterocyclic nitrogen atom and the aromatic fluorine atoms could be investigated with these techniques. In addition to these experimental results, there are also *ab-initio* calculations of the Mulliken spin populations [18, 44]. All these studies suggest that the spin density is located basically on the dithiadiazolyl ring and it is almost independent of the *R* substituent. Hence, the spin density distribution in the dithiadiazolyl ring for the different derivatives should be very similar to each other. Therefore, the spin density distribution obtained experimentally from one compound can be extrapolated to the other ones.

Unlike resonance techniques, polarized neutron diffraction (PND) provides a direct probe of the actual spatial distribution of the spin density in the solid. This powerful method has already been applied to other purely organic compounds [46, 47, 48], where it has contributed significantly to the understanding of their magnetic behaviour.

3 Spin Density Studies on the *p*-nitro-tetrafluorophenyl- dithiadiazolyl radical (3)

The ideal situation for a PND experiment is to have all the spins parallel. Hence **1**β is not suitable for this kind of experiments because of its antiferromagnetic phase transition at 36K. The magnetic field necessary to polarize significantly the spins at low temperature is much higher than the accessible laboratory magnetic field (about 10 Tesla). Among the other dithiadiazolyl radicals, **3** was chosen for the PND experiment [49] for several reasons: i) single crystals up to 1–2 mm^3 were available; ii) the other radicals present antiferromagnetic interactions, which hinders the induced spin polarization; iii) this radical is intrinsically interesting due to its ferromagnetic transition at 1.32K. In the

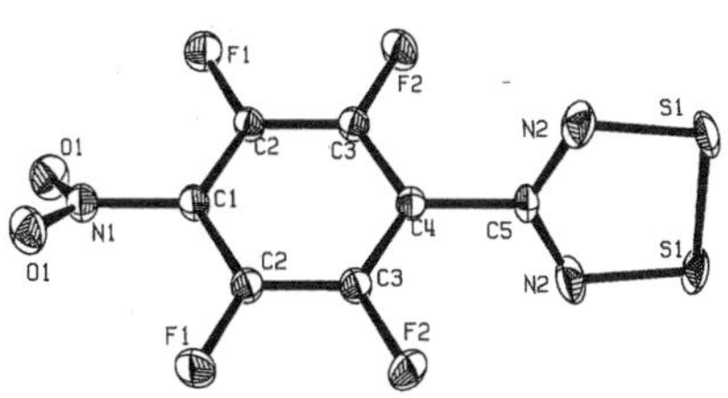

Fig. 4. View of the *p*–$O_2NC_6F_4$CNSSN• molecule determined by single-crystal neutron diffraction experiments at 20K. The thermal ellipsoids are at the 99 % probability level.

following sections the crystallographic and magnetic behaviour of this compound will be described with some detail before discussing the PND results.

3.1 Crystal Structure and Magnetic behaviour of the p-$O_2NC_6F_4CNSSN$• Radical

The crystal structure of **3** has been determined by X-ray and neutron diffraction at, respectively, 180 and 20K (Fig. 4) [22, 49]. A 180K, this radical crystallizes in a non-centro symmetric tetragonal structure, space group $P4_12_12$, with four molecules per unit cell and with the following cell parameters: $a = 8.138$ Å, $c = 15.049$ Å (Fig. 5a). Molecules are arranged in sheets perpendicular to the *c* axis rotated 90 degrees with respect to each other. In these sheets, the molecules form chains through electrostatic $S^{\delta+}...O^{\delta-}$ interactions and these chains are linked via intermolecular O...F contacts in the *ab* plane (Fig. 5b). These sheets are related by a 4-fold screw axis parallel to the *c* axis, with electrostatically favourable inter-layer $S^{\delta+}...N^{\delta-}$ contacts. There is no crystallographic transition is observed in the whole temperature range from room temperature to 1.5K. The cell parameters at 20K reflect a slight contraction of the cell upon cooling ($a = 8.1125(6)$ Å, $c = 14.7646(9)$ Å).

With the exception of BBDTA-$GaCl_4$-CH_3CN [23], **3** is the only S-based free radical derivative that has been described as a pure ferromagnet [22]. The high temperature magnetic susceptibility shows a Curie-Weiss behaviour with $\theta_{CW} = 1.6 \pm 0.1$K. The positive value of the Curie-Weiss constant indicates the presence of ferromagnetic interactions between the 1/2 spin centres. Indeed, the low temperature *ac* magnetic susceptibility on a single crystal exhibits a ferromagnetic transition at 1.32 ± 0.02K (Fig. 6). In the magnetic measurements small anisotropy is observed as expected for a compound where the only possible source of magnetic anisotropy is the magnetic dipolar interactions between the radicals.

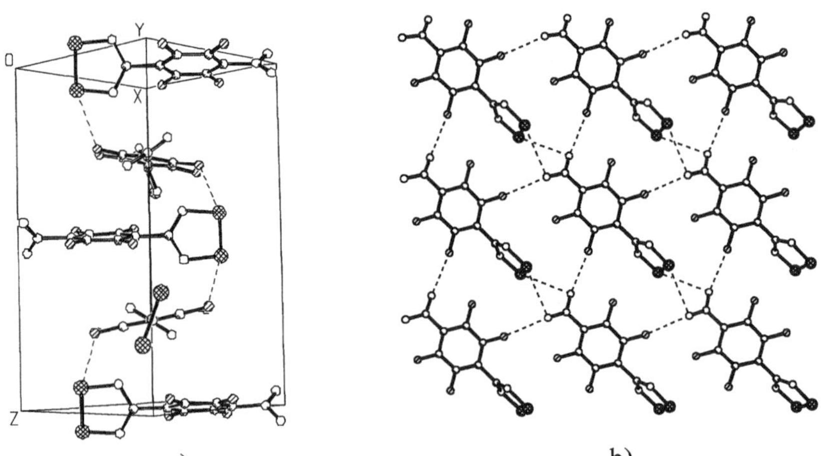

a) b)

Fig. 5. a) The four molecules in the unit cell of **3**. b) A sheet of ***3*** in the *ab* plane.

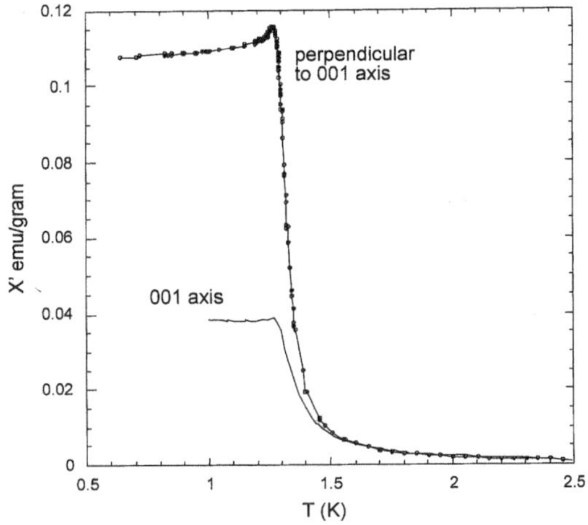

Fig. 6. Single crystal magnetic *ac* susceptibilities (ν = 1.11 Hz) *vs* temperature in **3** for the (001) and perpendicular to the (001) orientations of the crystal with respect to the *ac* field.

3.2 Polarized Neutron Diffraction Experiments on the p-$O_2NC_6F_4CNSSN$• Radical

The same crystal of **3** that was used in the structure determination at 20K was used for the polarized neutron diffraction experiments. The magnetic saturation of the sample was achieved by the application of magnetic field of 9 Tesla at 1.5K [49].

In a non-centrosymmetric crystal structure the nuclear and magnetic structure factors are complex numbers and the magnetic structure factors must be calculated from analytical modelling of the spin density distribution in order to fit the experimental flipping ratios. The wavefuntion [50, 51] and the multipolar expansion [46] are the two different approaches commonly used for modelling the spin density distribution. Whereas the wavefunction approach provides a straightforward interpretation of the physical origin of the spin density distribution, the multipole approach confers more flexibility to the model, enlighting details which cannot be deduced from the wavefunction approach in a simple manner. For complementarity, results from both approaches will be discussed in this chapter in order to provide an accurate picture of the spin density distribution in the radical.

Several models considering different number of parameters and different constraints were used to calculate the magnetic form factors within each approach. In the case of the wavefunction approach the model providing the best results included the p_z orbitals of the carbon, nitrogen and sulphur atoms of the heterocyclic ring together with the p_x and p_y orbitals of the nitrogen and sulphur atoms and a coefficient for the 2*s* nitrogen orbital to account for the sp^2 hybridation of the nitrogen atoms. The $2p_z$ orbital was the only orbital considered for the carbon atom because preliminary Density Functional Theory (DFT) calculations [44] indicated that its spin population is one order of magnitude lower than the sulphur and nitrogen ones. More simplified models, like ignoring the 2*s* orbital in the nitrogen atoms, give poorer agreement factors. Instead, the multipolar expansion approach seems to be less sensitive and a simple model assuming that the plane of the heterocyclic ring is a symmetry plane, hence considering only even multipoles in z, provided the best results with the minimum number of parameters. Calculations using more complex models including odd multipoles in z and the carbon atoms in the aromatic ring provided essentially the same results.

The main results from both approaches are compared in Table 1, where the total spin density population on the atoms of the dithiadiazolyl ring is normalised to 1 μ_B per radical, the sum of the populations before the normalisation is represented by $m(\mu_B)$ which corresponds to the magnetisation of the radical, the number of parameters of the model is represented by n_v and χ^2 is the agreement factor of the fit.

Figure 7 and 8 give several projections of the spin density distribution in relation to the atoms of the dithiadiazolyl ring as calculated following the wavefunction and the multipolar expansion approach, respectively. The results from both approaches indicate that the spin density is almost entirely localized in p_z orbitals of the nitrogen and sulphur atoms of the dithidiazolyl radical. This is consistent with the SOMO nature: an anti-bonding π orbital extended on the sulphur and nitrogen atoms. Beside this spin density coming from the SOMO orbital, there exists some p_z spin density on the carbon atom induced by spin polarization. As for the spin density populations, they are in agreement with previous values from the resonance experiments and *ab-initio* calculations.

The negative spin density at the carbon atom is due to a polarisation effect produced by the SOMO orbital on some of the doubly occupied orbitals. In the dithiadiazolyl ring there are seven electrons in π orbitals, six in three doubly occupied molecular orbitals and one in the SOMO orbital. Among the three doubly occupied orbitals there are two with b symmetry, and therefore without a node at the carbon atom. In these orbitals the β orbital (spin down) will be more localized in the carbon atom than the α orbital (spin up) in order to avoid the electrostatic repulsion with the electron in the SOMO orbital. In the α orbitals the electrostatic repulsion is in certain way compensated by the exchange interaction between electrons with parallel spins. This spatial difference between the α and β orbitals is the origin of the negative spin density at the carbon atom.

In the wavefunction approach a canting can be clearly observed in the projection of the sulphur and nitrogen spin density distribution on the *yz* plane (Fig. 7). Besides the possibility that the wavefunction approach is not completely well adapted to describe the experimental spin density due to its constraints, the physical origin of this canting can be attributed to the proximity of the fluorine atom in the *ortho* position of the per-fluoraryl ring. Due to the torsion between the two rings, this fluorine atom is not in the plane of the heterocyclic ring but closer to one of the lobes of the nitrogen $3p_z$ orbital The polarisation of the electron density away from nitrogen by the electronegative

Table 1
Comparative results of the spin density population on the atoms of the dithiadiazolyl ring, magnetisation of the radical, $m(\mu_B)$, number of parameters used, n_v, and agreement factor for the two approaches followed to analyse the PND results.

Approach	Population (μ_B)			$m(\mu_B)$	n_v	χ^2
	S	N	C			
Wavefunction	0.301(1)	0.222(1)	-0.046(1)	0.894(2)	9	2.33
Multipolar expansion	0.284(1)	0.242(1)	-0.053(2)	0.923(3)	21	1.29

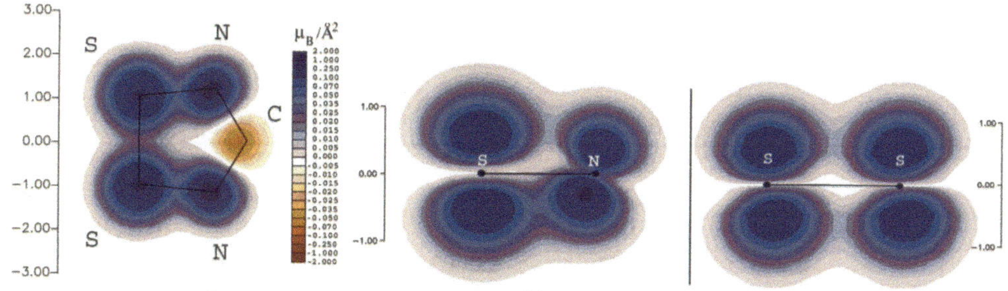

a) b) c)

Fig. 7. Projection of the calculated spin density distribution following the wavefunction approach: a) onto the plane containing the dithiadiazolyl ring; b) in a plane perpendicular to the dithiadiazolyl ring containing S and N; c) in a plane perpendicular to the dithiadiazolyl ring containing S and S.

fluorine may well be the origin of the canting and is supported by EPR studies and DFT which reflect some unpaired spin density at the ortho-fluorine atoms.

The multipolar expansion approach provides two important features of the spin density distribution that cannot be present in the wavefunction approach because of its constraints. The first is the displacement of the sulphur spin density away from the sulphur and nitrogen neighbouring atoms. This effect is clearly observed in Fig. 8, where the projection of the sulphur spin density is not centred on the sulphur atom but is displaced away from the centre of the ring. This displacement of the spin density is consistent with the anti-bonding nature of the SOMO orbital: while in the bonding molecular orbitals there is a concentration of electronic density in the bond, in the anti-bonding one the electronic density is away from the bonds. The second feature in these models is the negative spin density between the sulphur atoms and between the sulphur and nitrogen atoms. This negative spin density could be due to polarization effects produced by the SOMO orbital on the other double occupied molecular orbitals, although it could also be a spurious effect of the models trying to fit correctly the spin density displacement in the sulphur atom. A similar spurious effect has been noted in the PND spin density distribution in a nitronyl nitroxide radical [52].

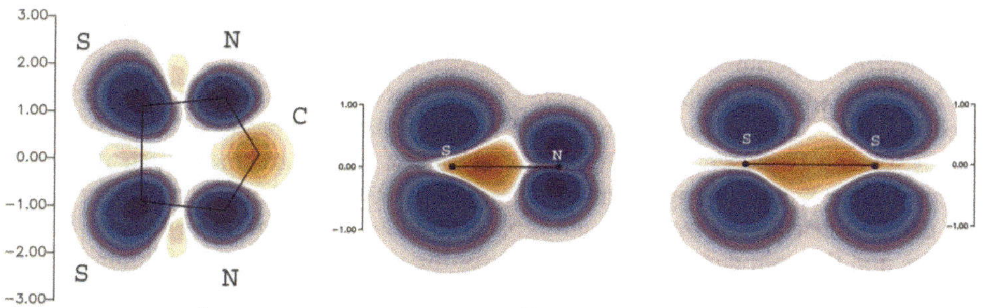

a) b) c)

Fig. 8. Projection of the calculated spin density distribution following the multipolar expansion approach: a) onto the plane containing the dithiadiazolyl ring, b) onto a plane perpendicular containing both S and N atoms and c) containing both S atoms.

Table 2.
DFT Mulliken spin populations in **3**.

Atom	DMOL3			Gaussian98	
	Periodic	Radical	PW91	B3LYP	UHF
S1	0.320	0.311	0.2910	0.2787	0.1390
N2	0.213	0.224	0.2455	0.2868	0.6623
C5	–0.063	–0.065	–0.0667	–0.1252	–0.6238
C4	0.005	0.005	0.0003	0.0078	–0.0446
C3	–0.003	–0.004	–0.0030	–0.0069	0.0840
C2	0.002	0.001	0.0014	0.0036	–0.0993
C1	–0.003	–0.003	–0.0033	–0.0067	0.0974
F2	0.000	0.000	0.0003	0.0000	0.0014
F1	0.000	0.000	0.0001	0.0002	–0.0007
N1	0.001	0.000	0.0002	0.0005	–0.0130
O1	0.000	0.000	–0.0005	–0.0007	0.0053

The fact that calculations including the carbon atoms in the aromatic ring provided essentially the same results indicate that there is no significant spin population away from the dithiadiazolyl ring. The population in these carbon atoms is, indeed, one order of magnitude less than the population in the carbon atom of the dithiadiazolyl ring.

In conclusion, the PND experiments indicate that the spin density is almost entirely located in the sulphur and nitrogen atoms of the dithiadiazolyl ring in a p_z-orbital-type distribution with the z axis perpendicular to the ring. This spin density arises from the anti-bonding nature of the SOMO orbital and is displaced from the centre of the atoms outside the bonds This displacement is most pronounced at the sulphur atoms, which will favour through space magnetic interactions with other radicals.

3.3 Theoretical Spin Densities

PND results have been complemented with theoretical computation of the spin density distribution using two different *ab initio* packages, the DMOL3 package and the Gaussian98 package [53]. The geometry obtained from the low-temperature non-polarized neutron experiment was used in all the calculations [54].

The Mulliken spin populations calculated for all the atoms of **3** using both packages are listed in Table 2. (Note that the molecule sits on a two-fold rotation axis. The atom numbering scheme is presented in Fig. 4). They are in good agreement with the PND results, confirming that all the spin density is located on the dithiadiazolyl ring. Moreover, the spin density distribution from the *ab-initio* calculation confirms the p_z nature of the spin density and the spin density displacement for the sulphur atoms. In addition there is also a negative spin density distribution in the plane of the dithiadiazolyl ring, as can be observed in Fig. 9. The value of this negative spin density is however much smaller than the negative spin density observed in the PND experiment.

4 Magnetic Interactions in Dithiadiazolyl Radicals

The magnetic interactions of the closely-related group of monomeric dithiadiazolyl radicals, *p*-$XC_6F_4CNSSN\cdot$, where X = CN (**1**), Br (**2**), O_2N (**3**) are discussed in this section. Since all the molecules differ only in the group linked to the tetrafluorophenyl ring, where there is no measurable spin density, they are all magnetically similar. When these molecules are assembled in the solid state, they exhibit very different intermolecular contacts. As a consequence their magnetic properties turned to be very different from one another. Theoretical calculations show that the dominant exchange interactions between these thiazyl radical groups are through direct exchange processes and emphasise the extreme influence that the molecular packing has on the bulk magnetic properties of the compound [55].

In recent years DFT has proved to be an extremely powerful approach to the determination of the exchange interaction in both metal clusters [56] and through space interactions in organic radicals. Indeed systematic studies by Novoa and Yamaguchi (see chapter in this book and references therein) have shown that the bulk magnetic behaviour of organic solids can be determined through an evaluation of the local nearest-neighbour exchange interactions between radicals [6]. In particular, Novoa's studies on aryl-substituted nitronyl nitroxide radicals clearly showed that whilst the phenyl substituent itself bore negligible spin density, it contributed significantly to the magnetic exchange pathway.

The general approach to determine the magnetic exchange interaction between two nearest neighbour radicals is to compute the energies of the triplet (E_T) and broken symmetry singlet (E_{BS}) states. The broken symmetry singlet state is a DFT state composed of pure magnetic states in which all the α-magnetic orbitals are located on one magnetic center and all the β-magnetic orbitals are located on the other magnetic center [57, 58, 59].

Within the context of the Hamiltonian $H = -2\boldsymbol{J}\mathbf{S}_1\cdot\mathbf{S}_2$, where $\boldsymbol{J}$ is the exchange interaction between spins S_1 and S_2, the energy between the singlet and triplet is denoted by $2\boldsymbol{J}$. Yamaguchi proposed [60] that the exchange interaction could be estimated from the energies of the broken symmetry singlet and triplet states and their expectation values $<S^2>$, according to the expression:

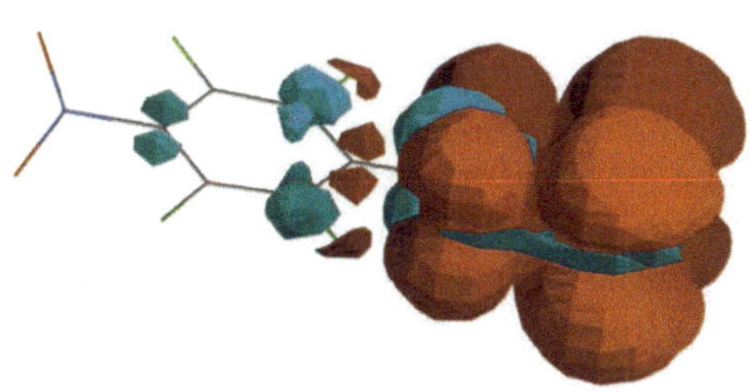

Fig. 9. Two spin density isosurfaces, from the molecular DMOL3 Perdew-Wang 91 calculation, for a value of 2×10^{-4} μ_B Å^3 in modulus. Red and blue areas refer to regions of positive and negative spin density respectively.

$$J = -\frac{E_{HS} - E_{BS}}{\langle S^2 \rangle_{HS} - \langle S^2 \rangle_{BS}}$$

Density functional calculations were carried out using the hybrid exchange correlation functional B3LYP which has proved to correct the tendency of the local and GGA exchange correlation functionals to overestimate the stability of the singlet state with respect to the triplet [61]. Two different basis sets were employed: the polarized split-valence double-zeta (6-31G**) and triple-zeta (6-311G**) Pople basis sets [62] in order to test the convergence of the calculation with the size of the basis set. Values of $|J|$ were considered significant when $|J| > 1$ cm^{-1}.

In the following of the section the geometry of all the nearest neighbour interactions within the structures of the three above mentioned *p*-tetrafluorophenyl derivatives will be discussed together with their respective magnetic properties. Since the magnetic interaction depends on the distance *r* between the magnetic centres as *exp(-r)* [63], possible magnetic pathways are selected within two cut-off distances. As almost all the spin density is located on the dithiadiazolyl ring, selected radical pairs are with an intermolecular distance shorter than 7 Å between atoms in both dithiadiazolyl rings. Moreover, since it has been demonstrated that magnetic interactions can be transmitted through atoms with very low spin populations [64], radical pairs with an atom of the dithiadiazolyl ring in one radical and an atom in a neighbour radical at a distance shorter than 4 Å are also considered. This methodology for determining all the possible magnetic pathways is similar to the Deumal-Novoa methodology [65].

4.1Exchange Interactions in p-$O_2NC_6F_4CNSSN$•

As explained in section 2.1, radical **3** is one of the rare examples of a bulk organic ferromagnet with a Curie temperature of 1.32K at ambient pressure. Its behaviour under applied pressure is also unusual for most of the organic ferromagnets investigated exhibit a decrease in Tc as pressure increases whereas the opposite occurs in antiferromagnets and weak ferromagnets [66]. In the case of **3** the ordering temperature raises up to 1.8K under 11.6 kbar [67]. The structure of **3** has been described in section 2.1. Four magnetic interaction pathways can be identified within the rules explained above. They are represented in the scheme of Fig. 10:

- J_1: This pathway connects a radical with four other radicals, two in the upper *ab* plane and two in the lower *ab* plane through heterocyclic S···N contacs of 3.658 Å, the shortest between atoms of neighbour dithiadiazolyl rings.
- J_2: The J_2 pathway presents a similar structure of the interaction propagation to J_1, although in this case the shortest contact between atoms of neighbour dithiadiazolyl rings is very large, 6.989 Å. However, there are several intermolecular contacts between other atoms and the dithiadiazoyl ring shorter than 4 Å that can potentially contribute to propagate magnetic interactions.
- J_3: This pathway links the radical with its four nearest neighbours along the *a* and *b* axes, where the most relevant contact is between the sulphur of the dithiadiazolyl ring and a fluorine atom in meta position at 3.907 Å, just below the cut-off limit.

- J_4: Like J_3, this pathway is in the *ab* plane and it propagates magnetic interactions through the chains formed by electrostatic interactions between the NO_2 group and the sulphur atoms with O···S distances of 3.186 Å and 3.323 Å.

The only exchange pathway connecting neighbouring dithiadiazolyl rings is J_1. Thus, J_1 should be much stronger than the other three magnetic interactions. This interaction propagates in a three-dimensional network, thus allowing a magnetic transition without the contribution of other magnetic interactions. Since this compound undergoes a ferromagnetic transition at 1.32K, the interaction must be ferromagnetic.

A theoretical analysis of the exchange interactions indicates that the strongest exchange coupling is via this S...N contact and is weakly ferromagnetic ($\boldsymbol{J}_1 = +1.14\ cm^{-1}$). All other interactions including those formed via the NO_2...S contact ($\boldsymbol{J}_4 = -0.03\ cm^{-1}$) are more than an order of magnitude smaller (see Table 3 for more detailed results). A simple analysis of these interactions predict Weiss constants of +1.80K and +1.64K for the 6–31G** and 6–311G** basis sets respectively. Within the mean-field approximation, this leads to an upper limit for the magnetic ordering temperature of 1.80–1.64K, in excellent agreement with the observed value of 1.3K [22].

It is interesting to notice that since the spin populations on the sulphur and nitrogen atoms in both dithiadiazolyl rings are positive, the J_1 interaction should be antiferromagnetic on the basis of the McConnell I mechanism. However, the McConnell I mechanism is a simplified model based on phenomenological observations and is not theoretically rigorous [5]. This model only considers the spin density populations, neglecting other factors such as the geometrical orientation of the magnetic orbitals. In order to understand the possible ferromagnetic nature of the J_1 pathway, we recall some points about the mechanisms of the magnetic interactions. As explained in Section 1.2, the magnetic interaction between electrons in two magnetic orbitals can be decomposed into two principal terms, one antiferromagnetic and the other ferromagnetic. In general, the antiferromagnetic term dominates, resulting in an antiferromagnetic interaction. This antiferromagnetic component depends linearly on the overlap between the magnetic orbitals. Consequently, if the magnetic orbitals are orthogonal the antiferromagnetic component vanishes and the resulting magnetic interaction is ferromagnetic. This "orthogonality method" is one of the methods used in molecular magnetism in order to obtain ferromagnetic interactions.

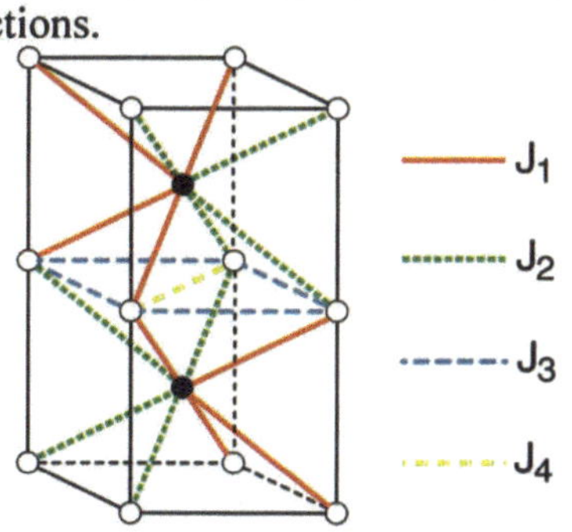

Fig. 10. Scheme of the possible magnetic pathways of **3** in the unit cell. The radical rings are represented by spheres. For clarity, the interactions in the *ab* plane are only described in the plane *z* = 0.5.

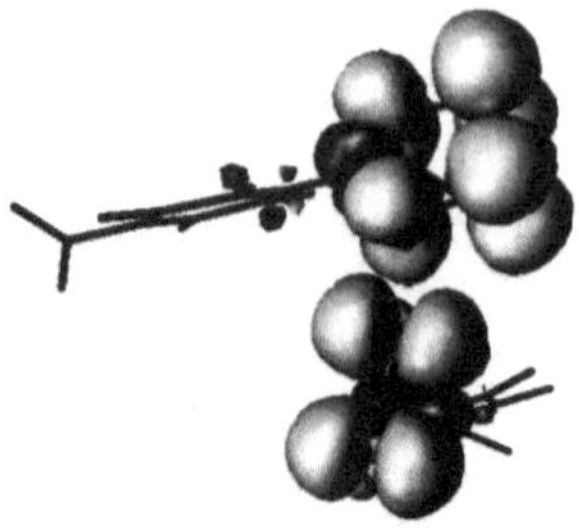

Fig. 11. Spin density isosurfaces for the J_1 radical dimer computed by DFT. Light grey isosurface is at 0.001 μ_B Å^3 and the black isosurface is at –0.001 μ_B Å^3.

Table 3.
Broken Symmetry DFT magnetic coupling constants for **3**.

Pathway	Contact	d (Å)	J(cm^{-1}) (6-31G**)	J(cm^{-1}) (6-311G**)
J_1	S···N	3.658	+1.26	+1.15
J_2	S···N C_1···N_2	6.989 3.503	+0.05	+0.04
J_3	S···F	3.907	–0.03	–0.03
J_4		3.186, 3.323	–0.03	–0.04

Exchange interactions are based upon the crystal data at 160K.

Both radicals involved in the J_1 pathway are represented in Fig. 11. Although the SOMOs of both radicals are not strictly orthogonal by symmetry, their overlap should be very low because of their relative geometry. In the closest intermolecular contact N2–S1, one of the $2p_z$ lobes of the nitrogen atom is pointing to the sulphur atom almost perpendicular to its $3p_z$ orbital. This near orthogonality between the p_z orbitals taking part in the magnetic interaction pathway favours the ferromagnetic part against the antiferromagnetic one in the magnetic interaction. Furthermore, the negative spin density in the plane of the ring would also reinforce the ferromagnetism due to the McConnell I mechanism. However, since this negative density is very low, according to the DFT calculations, its contribution to the ferromagnetic interaction will be of second order.

4.2 Exchange Interactions and Magnetic behaviour of α-p-NCC_6F_4CNSSN•

The α-phase of *p*-NCC_6F_4CNSSN (**1α**) was the first reported dithiadiazolyl radical to retain its monomeric nature in the solid state [35]. Preliminary magnetic measurements on a Faraday Balance exhibit a broad maximum in χ around 8K indicative of short range antiferromagnetic interactions. Above 100K, the susceptibility follows a Curie-Weiss law with $\theta = -25$K.

The structure of **1α** is triclinic, space group *P-1*, with the two molecules per unit cell related by an inversion centre [35]. The low symmetry of the triclinic crystal system leads to a more varied range of close intermolecular contacts. The shortest of these are $CN^{\delta-}$ ...$S^{\delta+}$ interactions which link radicals together into molecular chains ($d_{CN\cdots S}$ = 3.068 Å and 3.105 Å). An additional web of S...S contacts links chains in an antiparallel fashion in the *ab* plane (d_1 = 3.601, d_2 = 3.668 Å). Along the *c*-axis a pair of close-heterocyclic S···N contacts ($d_{S...N}$ = 4.186 and 4.269Å) form a centrosymmetric dimer. These loosely associated dimers are linked via N···N contacts (5.064 Å) about a crystallographic inversion centre. In Fig. 12 2×2×2 unit cells of **1**α are viewed along the *a* axis.

An analysis of all nearest neighbour exchange interactions in **1**α reveals two significant interactions which alternate in a linear chain and comprise competing anti-ferromagnetic (J_1) and ferromagnetic (J_2) interactions of similar magnitude. Interestingly enough, a third interaction through the shorter of the two intermolecular S···S contacts in the *ab* plane (J_3) propagates a negligibly small exchange coupling. These three shortest contacts, J_1, J_2 and J_3, are pictured for one radical in Fig. 12. The data presented in Table 4 indicate that the determination of the exchange interactions has not fully converged at the 6-311G** level thus leaving some uncertainty in their respective magnitudes, although it is clear that J_1 and J_2 are the dominant interactions and propagate throughout the crystal lattice to generate a one-dimensional chain structure with alternating ferromagnetic and antiferromagnetic interactions. Nevertheless, the alternating signs of two interactions along the chain produces an expected upper limits

Table 4.
Broken Symmetry DFT magnetic coupling constants for **1**α.

Pathway	Contact	*d* (Å)	*J*(cm^{-1}) (6-31G**)	*J*(cm^{-1}) (6-311G**)
J_1	S···S	3.682	−6.02	−8.80
J_2	S···N S···N	4.186 4.269	+9.15	+7.58
J_3	S···S	3.664	−1.00	+0.04

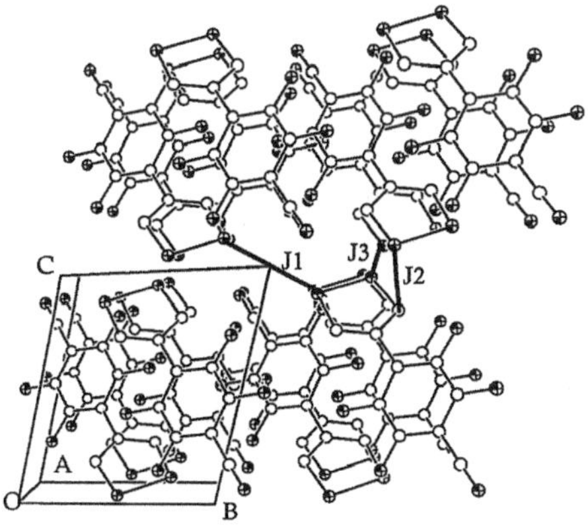

Fig. 12. 2×2×2 unit cells of the *α-NC*-C_6F_4-CNSSN• radical viewed along to the *a* axis. The J_1, J_2 and J_3 pathways are represented for one radical.

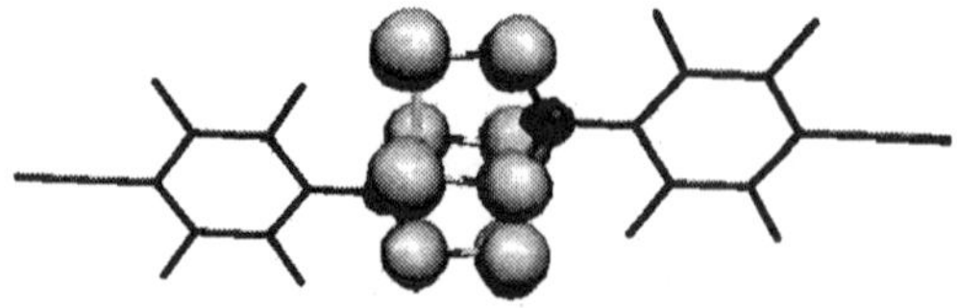

Fig. 13. Spin density isosurfaces for the J_1 radical dimer of the *α-p-NC*C_6F_4CNSSN• radical. Light grey isosurface is at 0.01 μ_B Å^3 and the black isosurface is at –0.01 μ_B Å^3. The plane of the view is parallel the dithiadiazolyl ring of the nearest radical.

for the short-range maximum in the susceptibility of a few Kelvin, somewhat less than the experimental value (*ca.* 8K). This inconsistency merits further experimental studies.

The J_2 interaction, which is ferromagnetic and stronger than the one that develops ferromagnetism in **3**, deserves some discussion. . Due to the symmetry the SOMO orbitals in the J_1 pair are almost orthogonal, thus favouring a ferromagnetic interaction. Moreover, there is an overlap between the positive spin density of a sulphur atom and the negative spin density of the carbon atom in the dithiadiazolyl ring, which also favours a ferromagnetic interaction. The spin density isosurfaces are pictured in Fig. 13.

4.3 Exchange Interactions and Magnetic behaviour of β-NC-C_6F_4-CNSSN•

The *β* phase of the *p-NC*C_6F_4CNSSN•, **1**β., is one of the most extensively studied organic magnets. In this phase molecules of **1** crystallize in the non-centrosymmetric group *Fdd2* with eight molecules per unit cell. As in the α phase, the crystal packing is governed by the electrostatic interactions between the sulphur atoms of the heterocyclic ring and the nitrogen atom of the cyano group producing a chain-like motif along the *c* axis with short $CN^{\delta-}\cdots S^{\delta+}$ contacts (2.986 Å). Contrary to the α phase, the chains are packed in a parallel way with favourable electrostatic interchain S···N interactions.

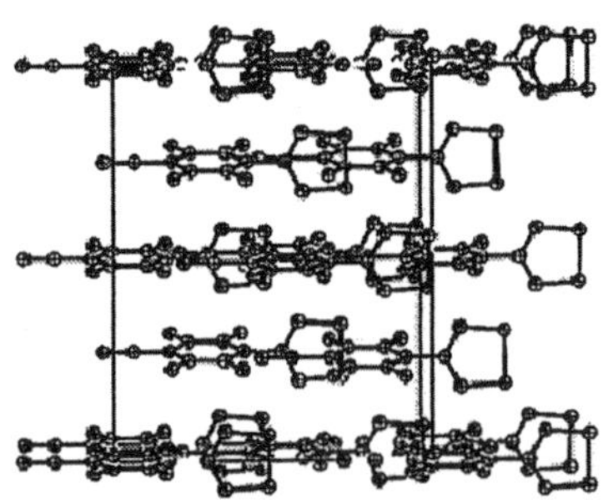

Fig. 14. Molecular packing diagram of the *β-NC*-C_6F_4-CNSSN• radical viewed perpendicular to the *b* axis.

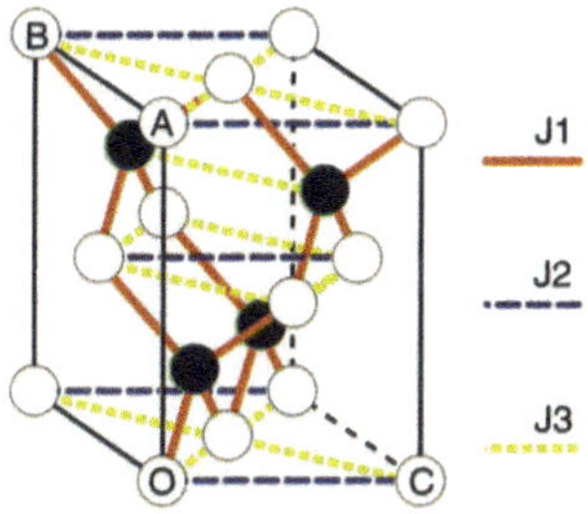

Fig. 15. Scheme of the possible magnetic pathways in the unit cell of the *β*-*NC*-C_6F_4-CNSSN• radical. The radicals are represented by spheres. The J_4 interaction has been omitted for clarity.

The crystal packing viewed perpendicular to the *b* and *c* axis is represented in Fig. 14.

At 36K this compound exhibits a transition to an antiferromagnetic phase. Since the two sub-networks with antiparallel spins are not related by a centre of symmetry, the Dzyaloshinskii-Moriya term in the spin Hamiltonian is non-zero and allows canting of the spins from colinearity. In turn this results in a weak spontaneous magnetization [17, 18]. Extrapolation of the spontaneous magnetisation to 0K yields a canting angle of 0.26(2)° with respect to colinearity. This magnetic phase transition has been characterised by magnetic and heat capacity techniques. The average antiferromagnetic structure has been determined from neutron diffraction techniques [18] and it has been further investigated by μ-SR [68] and single crystal EPR experiments [40, 69]. Furthermore, the pressure dependence of the magnetic ordering of this compound shows that Tc raises up to 72K at 16.4 kbar [66, 70].

The three possible magnetic exchange pathways obtained from the distance cut-off criteria are represented in Fig. 15 and the calculated exchange interactions presented in Table 4. It has proved possible to probe the magnetic pathway via the closest intermolecular $CN^{\delta-}\cdots S^{\delta+}$ contact between the cyano group and the dithiadiazolyl ring experimentally and has been found to be negligible [71] in agreement with the calculated exchange interaction. Of the remaining two pathways, only one antiferromagnetic interaction, $\boldsymbol{J_1}$, is significant and propagates through the lattice in a diamond-like network. This exchange interaction is associated with a short S···N contact of 3.488 Å (see Table 4) between heterocyclic rings. The mean-field approximation based upon the 6-311G** magnetic coupling constant $\boldsymbol{J_1}$, predicts an antiferromagnetic transition at 45.15K, in good agreement with the experimental value of 36K.

Table 5.
Broken Symmetry DFT magnetic coupling constants for 1α.

Pathway	Contact	*d* (Å)	J(cm^{-1}) (6-31G**)	J(cm^{-1}) (6-311G**)
J_1	S···N	3.488	–32.58	–31.38
J_2	CN···S	2.986	–0.03	–0.04
J_3	F···S		–0.006	–0.007

Exchange interactions are based upon the crystal data at 160K

4.4 Exchange Interactions and Magnetic behaviour of p-BrC$_6$F$_4$CNSSN•

Like **1**β and **3**, this radical, **2**, also crystallizes in a polar space group, *Aba2*, [39]. In this radical the difference of electronegativity between the bromine and the sulphur atoms is not enough to force a chain-like motif in the structure as in **1**α and **1**β. Instead, as shown in Fig. 16, the radicals are connected along the crystallographic *c* axis by short S···N contacts of 3.175 Å and along the *b*-axis by N...Br contacts. In addition the rings are π-stacked in columns parallel to the crystallographic *a* axis by means of *out-of-plane* electrostatic interactions between the sulphur atoms of the heterocyclic rings leading to S···N and S···S contacts in the range 3.675–3.999 Å (compared with the sum of van der Waals radii perpendicular to the ring plane (3.2 and 4.0 Å for S···N and S···S, respectively).

The magnetic susceptibility of **2** above 60K follows a Curie-Weiss law with θ = −27K. From this value, either an antiferromagnetic transition at 27K (within the mean-field approximation) a broad peak around 27K due to short range order interactions would be expected. However, the experimental magnetic susceptibility down to 1.8K shows neither an antiferromagnetic transition nor a broad peak but increases continually with decreasing temperature [39]. Amongst the several tentative explanations, the existence of competing ferromagnetic and antiferromagnetic interactions has also been proposed as the possible origin of this magnetic behaviour.

The analysis of the magnetic pathways provides seven possible ones within the distance cut-off criteria. However, only three are significant (Table 5). J_1 and J_2 connect molecules along the *a*-axis and alternate in sign. The third exchange pathway, J_3, connects the radicals in the *c* direction and is antiferromagnetic.

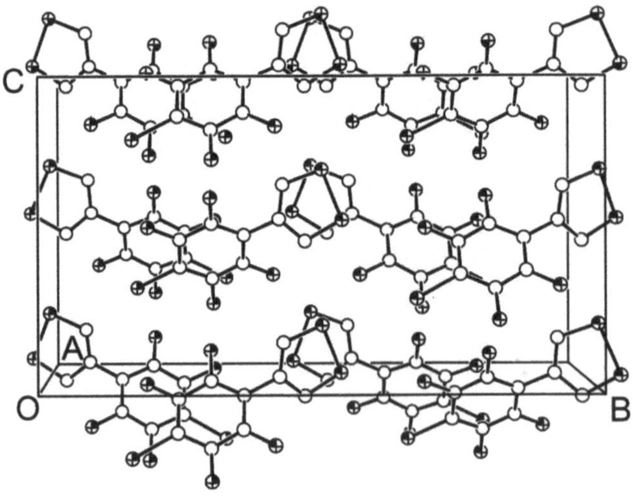

Fig. 16. Molecular packing diagram of *p-Br*C$_6$F$_4$CNSSN viewed along the *a* axis.

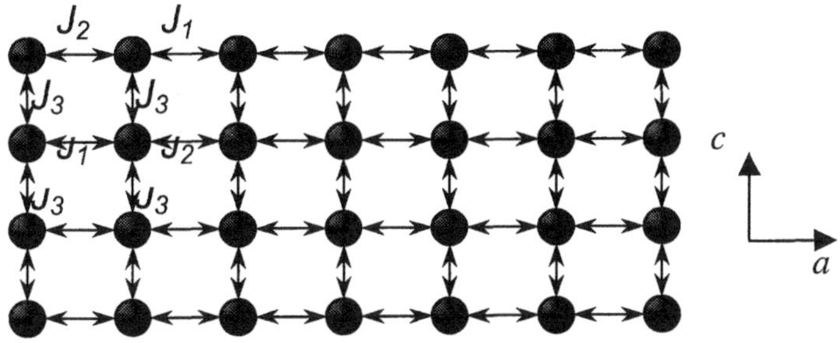

Fig. 17. Exchange pathway in *p-Br*C_6F_4CNSSN with relative orientations of the crystallographic axes

The magnetic motif generated by these three interactions is a two-dimensional grid (Fig. 17) where each molecule simultaneously interacts ferro (J_2) and antiferromagnetically (J_1 and two J_3). Within the mean-field approximation these interactions lead to an expected Weiss constant of –26K (6-31G**) or –24K (6-311G**) in excellent agreement with the experimental value.

The absence of long-range magnetic order can be directly attributed to the combined effects of the very small magnetic anisotropy characteristic of π-type orbitals and the two-dimensional nature of the interactions network (Fig. 17). In addition, the competing signs of the exchange interactions leads to frustration in the magnetic system and to the suppression of short-range order.

Table 6.
Broken Symmetry DFT magnetic coupling constants for **2**.

Pathway	Contact	d (Å)	J (cm^{-1}) (6-31G**)	J (cm^{-1}) (6-311G**)
J_1	N(1)...S(1a)	3.634		
	N(1)...S(2a)	4.182		
	S(1)...S(1a)	3.865	–9.861	–7.618
	S(2)...C(1a)	4.196		
	N(2)...N(2a)	4.377		
J_2	S(2)...S(2a)	3.675		
	S(2)...N(2a)	3.738	+8.513	+7.519
	S(2)...S(1a)	4.371		
J_3	S...N	3.175	–8.383	–8.250

Exchange interactions are based upon the crystal data at 160K.

5 Other Thiazyl Radicals

The methodology described in the previous sections, in which magnetic measurements are combined with studies of the spin density distribution in the organic molecule and with *ab initio* calculations of the exchange pathways, has not been extensively applied to other S-based radicals. There are, however, some interesting candidates whose magnetic properties have been investigated although their spin density distribution have not been studied experimentally. Such is the case of some dithiazolyl radicals, such as the methylbenzodithiazolyl, MBDTA, (**5**), the benzo[1,2:4,5-d']bis[1,3,2]dithiazole, BBDTA, (**6**) and the trithiatriazapentalenyl, TTTA, (**7**). Here we extend this approach and show it is equally applicable to other S-based free radicals. With the exception of TTTA (**7**) whose properties are described in detail elsewhere in this book [72], no calculations of the strength of the exchange interaction have been reported. We have shown that these calculations are important in determining not only the sign and magnitude of the intermolecular interactions, but also the lattice dimensionality.

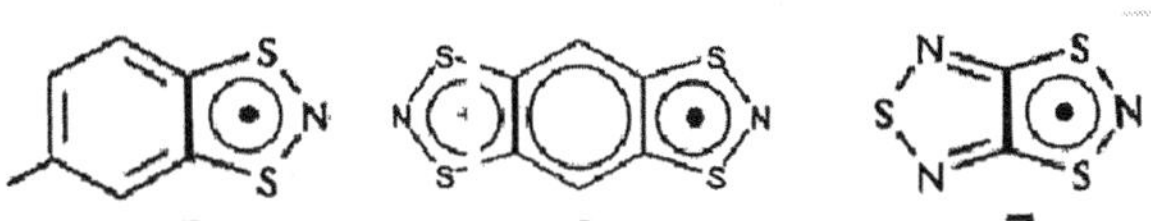

5 6 7

5.1 *The Methylbenzodithiazolyl Radical (MBDTA)*

The asymmetric unit of MBDTA (**5**) contains a single molecule of unexceptional geometry with the fused ring essentially planar. Molecules of MBDTA form a herringbone array with π-stacking along the crystallographic *c*-axis. The distance between equivalent atoms in neighbouring rings in the stack coincides with the length of the *c*-axis, but the molecular plane is inclined at 60.7° to the *ab* plane and the closest contact along the stacking direction is 3.742 Å [73]. A view of MBDTA in the *ab* plane is shown in Fig. 18a. In addition to the intra-stack contacts, there are a series of inter-stack S···S contacts in the range 3.71–3.82 Å. These intermolecular contacts give rise to a two-dimensional sheet of interactions in the crystallographic *bc* plane.

The magnetic behaviour of (**5**) exhibits a broad maximum at 140K (Fig. 18b). The absence of any field dependent magnetisation below 140K indicates that the maximum in χ is associated with the onset of low dimensional antiferromagnetic order. Below ca. 25K the susceptibility curve increases rather sharply most likely due to the contribution from a small mole fraction of non correlated paramagnetic centres in the sample [73].

The low dimensionality of the magnetic structure is consistent with the two-dimensional nature of the dithiazolyl interactions in the crystal structure and the compound was analysed as a two dimensional Heisenberg system, using high temperature series expansions for the square planar Heisenberg model [74]. A good agreement between experimental data and this theoretical model was obtained with an exchange term, $J \sim -72$K.

Whilst the fused nature of the BDTA and MBDTA ring systems provides an opportunity for π-delocalisation of the spin density away from the heterocyclic ring, theoretical calculations have shown that the majority of the spin density is still localised on the S and N atoms. Consequently the main pathway for magnetic exchange can be considered to be via the intermolecular S···N and S···S interactions already described. This gives rise to a two-dimensional sheet structure in the *bc* plane; propagation along

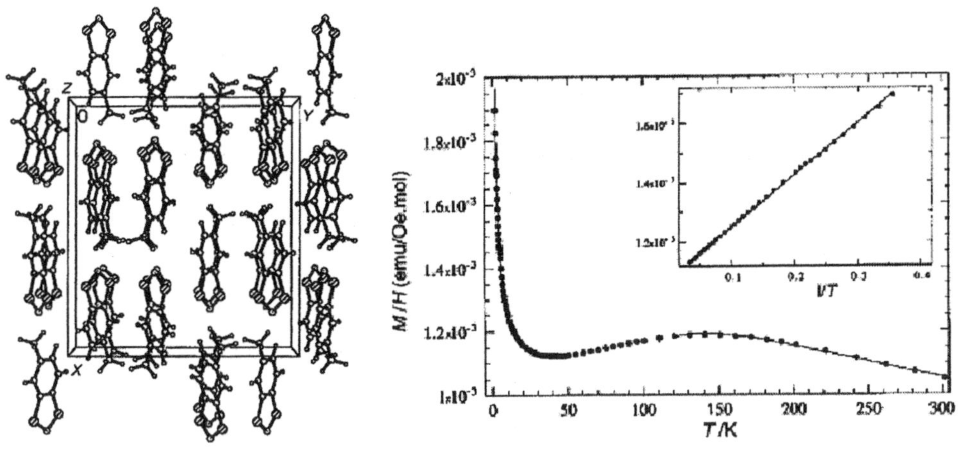

(a) (b)

Fig. 18. a) View of MBDTA in the *ab* plane; b) Temperature dependence of MBDTA.

the crystallographic *a* direction relies on N···H interactions. A determination of the calculated exchange interactions reveal three dominant interactions propagated via close contacts between heterocyclic rings. The dominant interaction is antiferromagnetic ($\boldsymbol{J}_1$ = –92.3K at the 6-311G** level) and propagates along the crystallographic *c*-axis, whilst the other two are ferromagnetic ($\boldsymbol{J}_2$ = +1.9 and $\boldsymbol{J}_3$ = 7.8K) and propagate along the *b*-axis [54].

5.2 The benzo[1,2:4,5-d']bis[1,3,2]dithiazole (BBDTA)

This compound, which preparaton was reported some time ago [10, 75, 76], is diamagnetic despite the fact that its structure consists in discrete, unassociated molecules [77]. The interest of this compound stays in its radical cation salts [BBDTA][$Fe^{III}Cl_4$]·CH_3CN, (**6a**) and [BBDTA][$GaCl_4$]·CH_3CN (**6b**). In these crystals the BBDTA molecules form a ladder-type structure through which the [XCl_4] anions are weakly linked. Mössbauer experiments in (**6a**) indicate antiferromagnetic ordering below 6.6K [78]. Instead (**6b**) is diamagnetic. However, removal of the crystal solvent results in a drastic change from diamagnetic to paramagnetic. The bulk desolvated solid orders as a ferromagnet below 6.7K despites the fact that the [$GaCl_4$] is a non-magnetic anion [23]. It is certain that the removal of solvent molecules induces a packing modification in the BBDTA radical cations that triggers dominant ferromagnetic interactions.

5.3 The Trithiatriazapentalenyl, (TTTA)

This compound (7), first reported by [79], has been found to exhibit a first-order structural phase transition with a wide thermal hysteresis loop in the vicinity of room temperature that strongly affects its magnetism [80, 81, 82]. Since it is discussed in detail in another chapter of this book [72] only a brief account of its magneto-structural behaviour will be given here.

DFT calculations using the molecular geometry determined from the single crystal crystal studies reported by Wolmershauser [79] indicate that the unpaired electron resides in a π^* orbital which is delocalised over the entire molecule [81]. However, the spin density distribution is asymmetric with 84% based on the SNS fragment and just 22% on the NSN fragment. The slight excess of spin density on the heteroatoms, N and S, is compensated by a small (3%) negative spin density at C. The high temperature (HT) phase, which structure consists of a regular polar stack of molecules along the *b*-axis, exhibits an effective magnetic moment of $1\mu_B$ at room temperature, significantly less than that anticipated for a S = 1/2 paramagnet. The magnetic data for this phase (225–350K) was modeled as a one-dimensional Heisenberg chain of S = 1/2 ions and yielded an intra-stack coupling of $\boldsymbol{J}$ = –320K, but required an additional inter-stack coupling of $\boldsymbol{J'}$ = –60K [80]. Although the one-dimensional chain model reproduces the experimental data, the magnitude of $\boldsymbol{J'/J}$ (~0.2) indicates that this is not a good one-dimensional system. On cooling below 225K and by 200K the paramagnetism is essentially quenched, consistent with the strong dimerisation along the stacking direction induced by the structural change leading to the low temperature (LT) phase.

On warming the LT phase to room temperature, the susceptibility is essentially constant to 310K but then increases abruptly, reaching the susceptibility of the original HT phase at 330K. The observation that the sample is essentially diamagnetic to 310K is consistent with a strongly antiferromagnetically coupled regime. Using the Bleaney-Bowers equation for an exchange-coupled dimer of S = 1/2 ions [83], the singlet-triplet separation must be in excess of $2\boldsymbol{J}$ = -2071K. This is in agreement with the value reported by Awaga ($\boldsymbol{J}$ = –1300K) [80]. DFT calculations on the LT, dimeric, phase indicated that the ground state electronic configuration is an open shell singlet with a singlet-triplet separation of $2\boldsymbol{J}$ = –2657K. This is in contrast to many dithiadiazolyl radicals [84] and other dithiazolyl radicals [85, 86] in which a closed-shell singlet (spin-paired dimer) is the electronic ground state.

A key question inherent to TTTA is its bistability. In order to achieve bistability, there must be two polymorphs of comparable energy. Polymorphism is not uncommon, particularly in molecular materials in which the forces between molecules are often weak, and a number of thiazyl-based radicals have been shown to exhibit polymorphism [87, 88, 89] but do not exhibit bistability. The bistability arises because the energy pathway to interconversion must not be too prohibitive. In the case of **7** the differences in intra-stack exchange energy favour the LT phase whereas electrostatic interactions seem to favour the HT phase [81].

6 Conclusions

This chapter has overviewed a series of S-based molecular magnetic materials of interest because of their magnetic ordering and magnetic interactions. It emphasises the importance of the spin density distribution determinations and *ab initio* calculations of the exchange interactions to complement the information gained from magnetic measurements. This approach allows a deep understanding of the interaction mechanisms governing the magnetism of this type of compounds and the way interactions propagate.

It has been shown that in the case of the dithiadiazolyl radicals the spin density is almost entirely located on the sulphur and nitrogen atoms of the dithiadiazolyl ring reflecting the nature of the SOMO orbital. The magnetism of four compounds of this series presenting very different magnetic behaviour, ferromagnetism, antiferromagnetism, paramagnetism and spin frustration have been discussed.

The presence of ferromagnetic interactions in three of the five radical structures indicates that dithiadiazolyl rings are good candidates in the quest to synthesize purely organic magnets with higher transition temperatures. An adequate molecular design could lead to suitable crystal packing with relatively strong ferromagnetic interactions propagating in the three dimensions. The magnitude of the ferromagnetic exchange couplings determined here would give an upper limit to ferromagnetic ordering temperatures of ca. 10^1K. Conversely the antiferromagnetic interactions are substantially stronger and could lead to ordering temperatures up to 10^2 K. Approaches to generate high T_C magnets might therefore be to focus on strategies which favour weak ferromagnetism, as in the case of **1**β, or develop novel two-spin systems to form ferromagnetic materials.

It would be of interest to extend these studies to other series of S-based radicals, in particular to the dithiazolyl-based compounds which magnetism seems to be as rich as in the case of the dithiadiazoly ones.

References

1. M. Tamura, Y. Nakazawa, D. Shiomi, K. Nozawa, Y. Hosokoshi, M. Ishikawa, M. Takahashi, and M. Kinoshita, Chem. Phys. Lett. 186 (1991) 401.
2. P.M. Allemand, K.C. Khemani, A. Koch, F. Wudl, K. Holczer, S. Donovan, G. Gruner, and J.D. Thompson, Science 253 (1991) 301.
3. J. Veciana, in: Molecular Magnetism: From Molecular Assemblies to the Devices, Eds. E. Coronado, P. Delhaes, D. Gatteschi, and J. S. Miller, NATO ASI series, 321 (1996) 425, Kluwer Acad. Publ.
4. V. Gadet, T. Mallah, I. Castro, P. Veillet, and M. Verdaguer, J. Am. Chem. Soc. 114 (1992) 9213.
5. M. Deumal, J.J. Novoa, M.J. Bearpark, P. Celani, M. Olivucci, and M.A. Robb, J. Phys. Chem. A 102 (1998) 8404.
6. J.J. Novoa and M. Deumal, Struct. and Bonding 100 (2001) 33.
7. H. M. McConnell, Proc. Robert A. Welch Found. Conf. Chem. Res. 11 (1967) 144.
8. J. Thomaides, P. Maslak, and R. Breslow, J. Am Chem. Soc. 110 (1988) 3970.
9. J.B. Torrance, I. Bagus, A.I. Nazzal, and S.P. Parkin, J. Appl. Phys. 63 (1988) 2962.

10. E. Dorman, M.J. Nowak, K. Williams, R.G. Angus, and F.J. Wudl, J. Am. Chem. Soc. 109 (1987) 2594.
11. J.S. Miller, J.C. Calabrese, H. Rommelmann, S.R. Chittapedi, R.W. Zhang, W.M. Reiff, and A.J. Epstein, J. Am. Chem. Soc. 109 (1987) 769.
12. J. Schweizer, S. Golhen, E. Lelievre-Berna, L. Ouahab, Y. Pontillon, and E. Ressouche, Physica B 297 (2001) 213.
13. C. Herring, Magnetism IIB, Direct exchange between well-separated atoms, 5, Eds. G.T. Rado and H. Suhl (1966) Academic Press, New York.
14. I. Dzyaloshinsky, J. Phys. Chem. Solids 4 (1958) 241.
15. T. Moriya, Phys. Rev. 120 (1960) 91.
16. R.K. Kremer, B. Kanellakopulos, P. Bele, H. Brunner, and F.A. Neugebauer, Chem. Phys. Lett. 230 (1994) 255.
17. A.J. Banister, N. Bricklebank, I. Lavender, J.M. Rawson, C.I. Gregory, B.K. Tanner, W. Clegg, M.R.J. Elsegood, and F. Palacio, Angew. Chem. Int. Ed. Engl. 35 (1996) 21, 2533.
18. F. Palacio, G. Antorrena, M. Castro, R. Burriel, J.M. Rawson, J.N.B. Smith, N. Bricklebank, J.J. Novoa, and C. Ritter, Phys. Rev. Lett. 79 (1997) 2336.
19. M. Kinoshita, 2001, π Electron Magnetism: from Molecules to Magnetic Materials, Ed. J. Veciana, Springer Verlag, Berlin, p. 1
20. P.M. Allemand, K. C. Khemani, A. Koch, F. Wudl, K. Holczer, S. Donovan, G. Gruner and J. D. Thompson, Science 253 (1991) 301.
21. R. Chiarelli, M.A. Novak, A. Rassat, and J. L. Tholence, Nature 363 (1993) 147.
22. A. Alberola, R.J. Less, C.M. Pask, J.M. Rawson, F. Palacio, P. Oliete, C. Paulsen, A. Yamaguchi, and R. D. Farley, Angew. Chem. Int. Ed. Engl. 42 (2003) 4782.
23. W. Fujita and K. Awaga, Chem. Phys. Lett. 357 (2002) 385.
24. P.M. Lahti, Magnetic Properties of Organic Materials, Editor, (1999) Marcel Dekker Inc. (NY).
25. G. Antorrena, J.E. Davies, M. Hartley, F. Palacio, J.M. Rawson, J.N. B. Smith, and A. Steiner, Chem. Comm. 15 (1999) 1393.
26. F. Palacio and J.M. Rawson, in: Magnetic Properties of Organic Materials. Ed. by P.M. Lahti (1999) 703, Marcel Dekker Inc. (NY).
27. J.M. Rawson and G.D. McManus, Coord. Chem. Rev. 189 (1999) 135.
28. N. Burford, J. Passmore, and M.J. Schriver, J. Chem. Soc, Chem. Commun. 140 (1986).
29. A.W. Cordes, R.C. Haddon, and R.T. Oakley, The Chemistry of Inorganic Ring Systems, 295 (1992) Eds. Steudel, Elsevier, Amsterdam, The Netherlands.
30. J.M. Rawson and A.J. Banister, The Chemistry of Inorganic Ring Systems, Eds. R. Steudel, 323 (1992) Elsevier, Amsterdam, The Netherlands.
31. J.M. Rawson, A.J. Banister, and I. Lavender, Adv. Heterocycl. Chem. 62 (1995) 137.
32. J. Campbell, D. Klapstein, P.F. Bernath, W.M. Davis, R.T. Oakley, and J.D. Goddard, Inorg. Chem. 35 (1996) 4264.
33. J.M. Rawson and F. Palacio, π Electron Magnetism: from Molecules to Magnetic Materials, Ed. J. Veciana, (2001) 93, Springer Verlag, Berlin.
34. J.M. Rawson, R.J. Lessand J.N.B. Smith, F. Palacio, and G. Antorrena, Mol. Cryst. Liq. Cryst. 334 (1999) 275.
35. A.J. Banister , N. Bricklebank, W. Clegg, M.R.J. Elsegood, C.I. Gregory, I. Lavender, J.M. Rawson, and B.K. Tanner, J. Chem. Soc. Chem. Comm. 6 (1995) 679.
36. G. Antorrena, F. Palacio, J.M. Rawson, and J.N.B. Smith, in: Supramolecular Engineering in Synthetic Metallic Materials. Ed. by J. Veciana, C. Rovira, and D.B. Amabilino, NATO ASI, C518 (1999) 217, Kluwer Publ.
37. C.M. Pask, R.J. Less, J.M. Rawson, F. Palacio, and P. Oliete, Phosphorus Sulfur. 168 (2001) 457.

38. A. Alberola, R.J.Less, F. Palacio, C.M. Pask and J.M. Rawson, Molecules, 9 (2004) 771.
39. G. Antorrena, J. E. Davies, M. Hartley, F. Palacio, J. M. Rawson, J. N. B. Smith and A. Steiner, Chem. Comm., 15 (1999) 1393.
40. A. Alberola, C. M. Pask, J. M. Rawson, E.J.L. McInnes, J. Wolowska, H. El-Mkami and G.M. Smith, J. Phys. Chem. B, 107 (2003) 14158.
41. C.D. Bryan C.D., A.W. Cordes, R.M. Fleming, N.A. George, S.H. Glarum, R.C. Haddon, C.D. MacKinnon, R.T. Oakley, T.T.M. Palstra and A.S. Perel, J. Am. Chem. Soc., 117 (1995) 6880.
42. R.A. Beekman, R.T. Boere, K.H. Moock, and M. Parvez, Can J. Chem., 76 (1998) 85.
43. T.M. Barclay, A.W. Cordes, N.A. George, R.T. Oakley, R.C. Haddon, and M.E. Itkis, Chem. Commun. (1999) 2269.
44. G. Antorrena, Estudio de Materiales Magnéticos Moleculares: Radicales S-N Y Polímeros de Coordinación., (1998) Thesis, University of Zaragoza, Zaragoza.
45. P.J. Alonso, G. Antorrena, J.I. Martinez, J.J. Novoa, F. Palacio, J.M. Rawson, and J.N.B. Smith, Appl. Magn. Reson. 20 (2001) 231.
46. P.J. Brown, A. Capiomont, B. Gillon, and J. Schweizer, J. Magn. Magn. Mater. 14 (1979) 289.
47. P.J. Brown, A. Capiomont, B. Gillon, and J. Schweizer, Mol. Phys. 48 (1983) 4.
48. A. Zheludev, V. Barone, M. Bonnet, B. Delley, A. Grand, E. Ressouche, P. Rey, R. Subra, and J. Schweizer, J. Am. Chem. Soc. 116 (1994) 2019.
49. J. Luzón, J. Campo, F. Palacio, G.J. McIntyre, J.M. Rawson, A. Alberola, and A. Goeta, 2005, (submitted to publication).
50. W.J. Hehre, R.F. Stewart, and J.A. Pople, J. Chem. Phys. 51 (1969) 2657.
51. W.J. Hehre, R. Ditchfield, R.F. Stewart, and J.A. Pople, J. Chem. Phys. 52 (1970) 2769.
52. E. Ressouche, Densité de Spin Dans Les Radicaux Libres Nitroxides et Leurs Complexes Cuivriques, Thesis of Université Joseph Fourier (1991) Grenoble.
53. Gaussian 98, Revision A.11.3, M.J. Frisch, G.W. Trucks, H.B. Schlegel, G.E. Scuseria, M.A. Robb, J.R. Cheeseman, V.G. Zakrzewski, J.A. Montgomery, Jr. , R.E. Stratmann, J.C. Burant, S. Dapprich, J.M. Millam, A.D. Daniels, K.N. Kudin, M.C. Strain, O. Farkas, J. Tomasi, V. Barone, M. Cossi, R. Cammi, B. Mennucci, C. Pomelli, C. Adamo, S. Clifford, J. Ochterski, G.A. Petersson, P.Y. Ayala, Q. Cui, K. Morokuma, N. Rega, P. Salvador, J.J. Dannenberg, D.K. Malick, A.D. Rabuck, K. Raghavachari, J.B. Foresman, J. Cioslowski, J.V. Ortiz, A.G. Baboul, B.B. Stefanov, G. Liu, A. Liashenko, P. Piskorz, I. Komaromi, R. Gomperts, R.L. Martin, D.J. Fox, T. Keith, M.A. Al-Laham, C.Y. Peng, A. Nanayakkara, M. Challacombe, P.M.W. Gill, B. Johnson, W. Chen, M.W. Wong, J.L. Andres, C. Gonzalez, M. Head-Gordon, E.S. Replogle, and J.A. Pople, Gaussian, Inc. , Pittsburgh PA, (2002).
54. J. Luzón, Magnetic Interaction Mechanisms in Molecular Materials, (2004) Thesis, University of Zaragoza, Zaragoza.
55. J.M. Rawson , J. Luzón, and F. Palacio, Coord. Chem. Rev. (2005) in press.
56. M. Deumal, J. Ribas-Arino, M.A. Robb, J. Ribas, and J.J. Novoa, Molecules 9 (2004) 757.
57. L. Noodleman and J.G. Norman, J. Chem. Phys. 70 (1979) 4903.
58. L. Noodleman, J. Chem. Phys. 74 (1981) 5737.
59. L. Noodleman and E.R. Davidson, Chem. Phys. 109 (1985) 131.
60. H. Nagao, M. Nishino, Y. Shigeta, T. Soda, Y. Kitagawa, T. Onishi, Y. Yoshioka, and K. Yamaguchi, Coord. Chem. Rev. 198 (2000) 265.
61. E. Ruiz, J. Cano, S. Alvarez and P. Alemany, J. Chem. Comp. 20 (1999) 1391.
62. R. Krishnan, J.S. Binkley, R. Seeger, and J.A. Pople, J. Chem. Phys. 72 (1980) 650.

63. C. Herring, Magnetism IIB, Direct exchange between well-separated atoms, 5, Eds. G.T. Rado and H. Suhl (1966) Academic Press, New York.
64. M. Deumal and J.J. Novoa, J. Mol. Struc. 506 (2000) 287.
65. M. Deumal, J.J. Novoa, M.J. Bearpark, P. Celani, M. Olivucci, and M.A. Robb, J. Phys. Chem. A 106 (2001) 1299.
66. K. Takeda and M. Mito, in: Carbon-based magnetism. Ed. by T. Makarova and F. Palacio. Elsevier. p 131.
67. K. Takeda, Y. Yoshida, Y. Inagaki, T. Kawae, J.M. Rawson, F. Palacio, and M. Mito, in preparation for publication.
68. F.L. Pratt, A.E. Goeta, F. Palacio, J.M. Rawson, and J.N.B. Smith, Physica B289 (2000) 119.
69. J.M. Rawson, A. Alberola, H. El-Mkami, and G.M. Smith, J. Phys. Chem. Sol. 65 (2004) 727.
70. M. Mito, T. Kawae, K. Takeda, S. Takagi, Y. Matsushita, H. Deguchi, J. M. Rawson, and F. Palacio, Polyhedron 20 (2001) 1509.
71. P.J. Langley, J.M. Rawson, J.N.B. Smith, M. Schuler, R. Bachmann, A. Schweiger, F. Palacio, G. Antorrena, G. Gescheidt, and A. Quintel, J. Mater. Chem. 9 (1999) 1431.
72. K. Awaga, W. Fujita, T. Tanaka, H. Matsuzaki, and H. Okamoto, in: Carbon-based magnetism. Ed. by Makarova, T. and F. Palacio. Elsevier (2006). p. 189.
73. G.D. McManus, J.M. Rawson, N. Feeder, F. Palacio, and P. Oliete, J. Mater. Chem. 10 (2000) 2001.
74. R. Navarro, in: Magnetic Properties of Layered Transition Metal Compounds. Ed. L.J. de Jongh. (1990) Kluwer Acad.
75. G. Wolmershäuser, M. Schnauber, T. Wilhelm, and L.H.Sutcliffe, Synth. Met. 14 (1986) 239.
76. K.A. Williams, M.J. Nowak, E. Dormann, and F. Wudl, Synth. Met. 14 (1986) 233.
77. T.M. Barclay, A.W. Cordes, R.H. de Laat, J.D. Goddard, R.C. Haddon, D.Y. Jeter, R.C. Mawhinney, R.T. Oakley, T.T.M. Palstra, G.W. Patenaude, R.W. Reed, and N.P.C. Westwood, J. Am. Chem. Soc. 119 (1997) 2633.
78. G. Wolmershäuser, G. Wortman, and M. Schnauber, J. Chem. Research (S) (1988) 358.
79. G. Wolmershäuser and R. Johann, Angew. Chem., Int. Ed. , Engl. 28 (1989) 920.
80. W. Fujita and K. Awaga, Science 286 (1999) 261.
81. G.D. McManus, J.M. Rawson, N. Feeder, J. van Duijn, E.J.L McInnes, J.J. Novoa, R. Burriel, F. Palacio, and P. Oliete, J. Mater. Chem. 11 (2001) 1992.
82. W. Fujita, K. Awaga, H. Matsuzaki, and H. Okamoto, Phys. Rev. B65 (2002) 064434.
83. B. Bleaney and K.D. Bowers, Proc. Roy. Soc. (London) Ser. A. 214 (1952) 451.
84. A.W. Cordes, C.D. Bryan, W.M. Davis, R.H. de Laat, S.H. Glarum, J.D. Goddard, R.C. Haddon, R.G. Hicks, D.K. Kennepohl, R.T. Oakley, S.R. Scott, and N.P.C. Westwood, J. Am. Chem. Soc. 115 (1993) 7232.
85. E.G. Awere, N. Burford, R.C. Haddon, S. Parsons, J. Passmore, J.V. Waszczak, and P.S. White, Inorg. Chem. 29 (1990) 4821.
86. E.G. Awere, N. Burford, C. Mailer, J. Passmore, M.J. Schriver, P.S. White, A.J. Banister, H. Oberhammer, and L.H. Sutcliffe, J. Chem. Soc. Chem. Commun. (1987) 66.
87. A.W. Cordes, R.C. Haddon, R.G. Hicks, R.T. Oakley, and T.T.M. Palstra, Inorg. Chem. 31 (1992) 1802.
88. A.W. Cordes, R.C. Haddon, R.G. Hicks, R.T. Oakley, T.T.M. Palstra, L.F. Schneemeyer, and J.V. Waszczak, J. Am. Chem. Soc. 114 (1992) 1729.
89. C.D. Bryan, A.W. Cordes, R.C. Haddon, R.G. Hicks, D.K. Kennepohl, C.D. McKinnon, R.T. Oakley, T.T.M. Palstra, A.S. Perel, S.R. Scott, L.F. Schneemeyer, and J.V. Waszczak, J. Am. Chem. Soc. 116 (1994) 1205.

Carbon-Based Magnetism
T. Makarova and F. Palacio (Editors)

8

Magnetic Bistability in Molecule-based Magnetic Materials

K. Awaga[1], W. Fujita[2], T. Tanaka[1], H. Matsuzaki[3], and H. Okamoto[3]

[1]*Department of Chemistry, Graduate School of Sience, Nagoya University, Chikusa-ku, Nagoya, 464-8602, Japan*
[2]*Research Center of Materials Science, Nagoya University, Chikusa-ku, Nagoya, 464-8602, Japan*
[3]*Department of Advanced Material Science, Tokyo University, Kashiwa, Chiba 2778561 Japan*

1 Introduction

There has been considerable interest in the bistability of molecule-based materials, which means a property of molecular assembly existing under two stable (or metastable) electronic states in a given range of external parameters (temperature, pressure, etc.), for the purpose of its application to electronic devices, such as thermal sensors, switching units, information storage media and so on [1–3]. This is recognized as a key to develop next-generation electronic devices. In the recent studies of molecule-based materials, bistable materials have been discovered in the material range from molecules to liquid crystals, to solid state materials, and in the property range from optical, to electrical, to magnetic.

The molecule-based magnetic materials have been studied extensively in these three decades, during which a variety of magnetic behavior were revealed in addition to a remarkable development of molecular ferromagnetism [4, 5]. One of the most important characters of these materials is a strong spin-lattice interaction; a small structural modification brings about a significant change in magnetic property. Most of these materials exhibit the Curie-Weiss paramagnetism at high temperatures, because their magnetic exchange interactions are weak in general. When the temperature is lowered, the exchange interactions become effective, and some of them make magnetic ordering, just like inorganic magnetic materials. However they are not a majority of the molecule-based magnetic materials; some other materials make a structural phase transition toward a less paramagnetic low-temperature phase, reflecting their strong spin-lattice interactions. Since these transitions are structural, they are always accompanied by thermal hysteresis upon cooling and heating. When the hysteresis loop width is large enough, they can be regarded as magnetic bistable materials. The most spectacular examples include the specific Fe(II) spin-crossover complexes, in which the transitions occur between a high-spin ($S = 2$) high-temperature phase and a low-spin ($S = 0$) low-temperature phase with a wide thermal hysteresis loop [1, 6–19].

The molecule-based materials exhibit various types of bistability besides magnetic one. The electrical bistability has been found in several organic materials; they exhibited a transition from a high-impedance state and a low-impedance state as a function of applied voltage [3, 20–22]. Such electrical bistability is attracting recent interest, because of possible application in organic electronic memory. Many organic molecules exhibit photochromism even in solid state, so that they can be regarded as optical bistable materials [23]. The molecule-based materials have flexibility in various aspects: molecular structure, intermolecular packing, spin state, electronic state, charge distribution, etc. Such flexibility probably allows manifestations of bistability.

2 Magnetic Bistability in Organic Radical Solids

The magnetic properties of organic radical solids have been studied extensively [4, 5], and the magnetic bistability has been discovered in several compounds. In this section we will describe their behavior, focusing on the studies of a thiazyl radical, 1,3,5-trithia-2,4,6-triazapentalenyl (abbreviated as TTTA, see the inset of Fig. 1). The heterocyclic thiazyl radicals possess unique chemical and physical properties [24, 25]. They are chemically stable in contrast to the instability of the most organic radicals, so that they do not need the protection groups on their molecular skeletons. This brings about a close packing and a strong intermolecular interaction in solid state. Their molecular skeletons include a large electronic polarization, leaving positive polarized charges on sulfur atoms and negative ones on nitrogen atoms. This polarization makes short intermolecular, interatomic S•••N contacts. The thiazyl radical solids often have a multi-dimensional network that consists of face-to-face π-π overlaps and side-by-side S•••N contacts. They have a strong spin-lattice interaction, reflecting the fact that the magnetic molecular orbital, or singly occupied molecular orbital, always exhibits a phase change between bonded sulfur and nitrogen. These interesting characters are believed to realize unique solid-state physical properties, such as high-temperature magnetic ordering, magnetic bisitability, etc. [26–35]

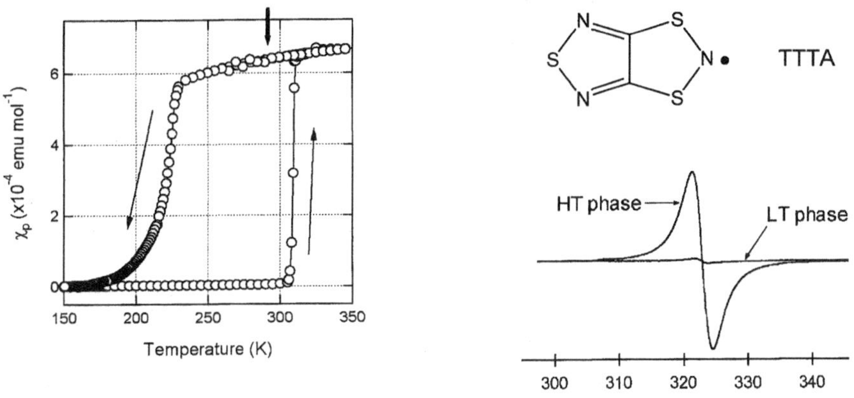

a) b)

Fig. 1. (a) Temperature dependence of the paramagnetic susceptibility χ_p for a polycrystalline sample of TTTA. The bold arrow indicates the χ_p value for the virgin sample just after sublimation. (b) X-band EPR spectra of the HT and the LT phase of TTTA at room temperature.

2.1 Magnetic Properties

TTTA was synthesized by Wolmershäuser and Johann in 1989 [36]. It exhibits a first-order phase transition with a surprisingly-wide thermal hysteresis loop in the vicinity of room temperature [37–39]. Figure 1(a) depicts the temperature dependence of the paramagnetic susceptibilities χ_p for a polycrystalline sample of TTTA. The bold arrow in this figure indicates the value of χ_p for the virgin sample just after the sublimation. As the sample was cooled from the room temperature, χ_p showed a slight decrease. At 230K χ_p began to quickly decrease, becoming zero at 170K. TTTA was intrinsically diamagnetic at low temperatures. When the sample was heated from a low temperature, a diamagnetic-to-paramagnetic transition was found not around 230K but at 305K. The magnetic measurements indicated a first-order phase transition with $T_{c}\downarrow$=230K and $T_{c}\uparrow$=305K. Since the hysteresis loop includes the room temperature (290 K), the material exhibits magnetic bistability at room temperature. There was little change in the shape of the loop even after repeats of this thermal cycle. Such a paramagnetic-diamagnetic phase transition is not very rare in molecule-based magnetic materials, but it is characteristic of TTTA to exhibit stable room-temperature bistability. To prove that the hysteresis loop observed is intrinsic, the following experiments were carried out. The high-temperature (HT) phase was slowly cooled from room temperature to 230K ($T_{c}\downarrow$). Just during the phase transition to the diamagnetic low-temperature (LT) phase, the sample cooling was stopped and then the sample was heated, following the magnetic response. The results for four runs with different starting points are depicted in Fig. 2 [38]. In every run the plots of χ_p clearly indicate that there is little change in the ratio between the HT and LT phases in the temperature range within the loop. This indicates that the two phases can stably coexist in this range, and,

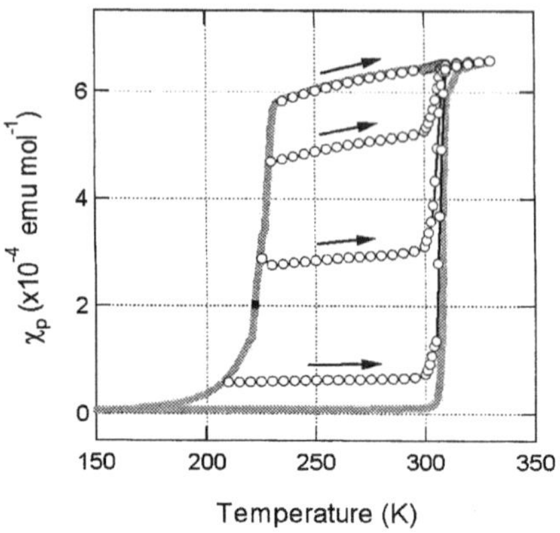

Fig. 2. The magnetic responses of TTTA at $T_{c\downarrow}$ to increases in temperature. The four kinds of plots represent the results on the samples with different starting points. The bold gray curves depict the hysteresis loop shown in Fig. 1(a).

therefore, is a firm evidence that the hysteresis is intrinsic at least in the time scale of laboratory.

Organic radicals are usually EPR active, so that we can obtain the precise spin information by means of this high-sensitivity measurement. The temperature-variable X-band EPR experiments indicated the phase transition of TTTA at *ca.* 200 and 310K upon cooling and heating, respectively [38, 39]. Figure 1(b) shows a comparison between the single-crystal EPR signals for the two phases at 300K. The spectrum of the HT phase consists of an intense absorption, in contrast to the weak signal of the LT phase. The g-factor of the HT phase signal is g = 2.0043 that is typical for the thiazyl radicals. Since EPR is highly sensitive to the spin state of organic radicals, the two phases of TTTA can be very easily distinguished by EPR rather than SQUID measurements.

2.2 Crystal Structures

There is a significant difference between the crystal structures of the HT and LT phases of TTTA [37–39]. The structure of the HT phase consists of a polar 1D stacking column, in which the molecules are related by translational relations with a constant interval. This stacking column is surrounded by six neighboring columns with very short S•••N and S•••S contacts. In contrast, TTTA exhibits a strong dimerization along the stacking direction in the LT phase. This radical dimerization must be responsible for the diamagnetic property of the LT phase. Figure 3 shows a schematic comparison between the nearest neighbor intermolecular arrangements in the two phases. It is

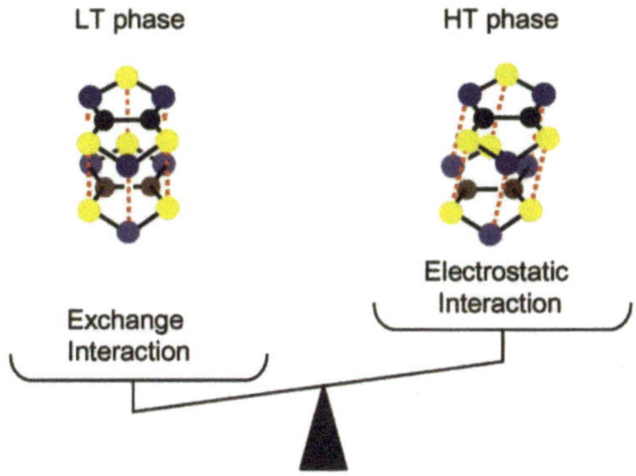

Fig. 3. Nearest neighbor intermolecular overlaps of TTTA in the HT and LT phases. The electrostatic interaction prefers a shifted overlap, while the exchange interaction does an eclipsed one.

notable that the molecules have an eclipsed overlap in the LT phase, in contrast to a shifted overlap in the HT phase. Exactly speaking, the molecular planes in the LT-phase dimer are not parallel; the distance between the -S-N-S- moieties is shorter than that between the -N-S-N- moieties, probably reflecting a bonding interaction between the unpaired electrons which are concentrated on the -S-N-S- moieties. This suggests that the bonding formation would be a driving force of this phase transition.

We believe that there is a competition between the exchange and electrostatic interactions. The exchange interaction prefers an eclipsed overlap, such as found in the LT phase, because this structure maximizes the intermolecular overlap between the magnetic molecular orbitals. However this is the most disagreeable structure from the viewpoint of electrostatic energy, because it includes the intermolecular, interatomic contacts between the polarized charges of the same sigh, namely $S^{\delta+}\cdots S^{\delta+}$ and $N^{\delta-}\cdots N^{\delta-}$. This competition presumably causes the drastic phase transitions in the thiazyl radical family.

2.3 Thermodynamic Properties

The phase transition of TTTA was examined by means of DSC [38]. In the cooling and the heating process, an exothermic and endothermic transitions were observed at 234 and 315K (on-set temperatures), respectively. The transition enthalpy ΔH at $T_{c\downarrow}$ and $T_{c\uparrow}$ were obtained to be 2210 and 2340 J·mol^{-1}, respectively. The difference must be due to the fact that the HT and LT phases have different heat capacities. The transition entropies ΔS were estimated as $\Delta S_{\downarrow} = \Delta H_{\downarrow}/T_{c\downarrow} = 9.44$ J·K^{-1}·mol^{-1} and $\Delta S_{\uparrow} = \Delta H_{\uparrow}/T_{c\uparrow} = 7.43$ J·K^{-1} mol^{-1} upon cooling and heating, respectively. The maximum estimation of the magnetic contribution is $R\ln 2$ (= 5.76 J·K^{-1} mol^{-1}). Since the HT phase involves rather strong antiferromagnetic interactions, the real contribution would be smaller than $R\ln 2$. In spite that $R\ln 2$ is the maximum estimation, the ΔS values observed are larger than this value. The presence of such excess entropy indicates a cooperation of the lattice system in this phase transition.

2.4 High-pressure Effects

The effects of a quasi-hydrostatic pressure on TTTA were studied to control the magnetic bistability. In general, the pressure effects on molecule-based materials are useful and important, because of their large compressibilities [40–43]; the physical properties are significantly affected by lattice modulations induced by pressure. Despite extensive studies of high-pressure effects on molecular crystals, non-equilibrium phenomena, namely hysteresis behavior, have still been left unexplored in great detail [18, 19, 44].

The temperature dependence of χ_p was examined under the pressures of 0.38, 0.75, and 1.5 GPa [45]. The diamagnetic-paramagnetic transition of TTTA was found to occur sharply even under pressure, with a significant shift of the transition temperatures. The pressure dependence of $T_{c\uparrow}$ and $T_{c\downarrow}$ are shown in Fig. 4. The values of $T_{c\uparrow}$ and $T_{c\downarrow}$ exhibited significant increases with increased pressure and thus the bistable range was shifted toward higher temperature. Although room temperature (290K) is just below $T_{c\uparrow}$ at ambient pressure, it falls in the center of the hysteresis loop at 0.75 GPa. That is, room-temperature bistability can be stabilized by pressure. The two solid curves in Fig. 4 are only for eye guides, but the gradients at the ambient pressure are roughly estimated to be $dT_{c\downarrow}/dp = 45 \pm 20$ K·GPa^{-1} and $dT_{c\uparrow}/dp = 20 \pm 10$ K·GPa^{-1}. The theoretical value of dT_c/dp is given by the Clapeyron equation, $dp/dT_c = \Delta S/\Delta V = \Delta H/(T_c\Delta V)$, where ΔH, ΔS and ΔV are the enthalpy, entropy and volume changes, respectively, though this equation is valid only under thermal equilibrium. Enthalpy changes for the transitions in TTTA were obtained by DSC measurements, as described above. The difference between the unit cell volumes of the

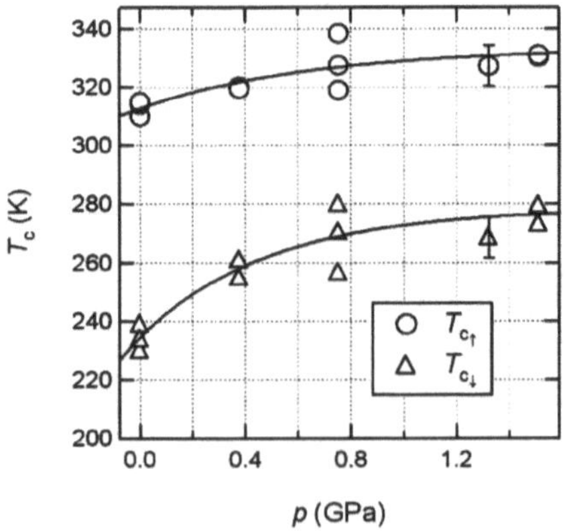

Fig. 4. Pressure dependence of the transition temperatures, $T_{c\uparrow}$ and $T_{c\downarrow}$, for TTTA. Solid curves are for eye guides.

two phases is 3.4 Å^3 at room temperature [38]. Since the unit cell includes four molecules, the volume difference is $\Delta V_{RT} = 3.4N_A/4$ Å^3 mol^{-1}, where N_A is the Avogadro constant. The theoretical value for dT_c/dp at p=0 GPa can be estimated to be 60 K·GPa^{-1} under the assumption that $\Delta V = \Delta V_{RT}$. The agreement between the theoretical and experimental values is fair (the same order of magnitude), and the equation can explain the tendency of a shift to higher transition temperatures with increasing pressure.

2.5 Photo-induced Phase Transition

A phase control by photoirradiation was successfully achieved on TTTA, taking advantage of the bistability around room temperature [47]. Figures 5(a) and 5(b) shows polarized microscope images of the crystal surfaces of the HT and LT phases, respectively. There was very clear chromism between red-purple in the HT phase and yellow-green in the LT phase, which made it possible to observe the photo-induced phase transition visually. Figure 5(c) depicts an image of the crystal in the LT phase after a single-shot (6 ns pulse) irradiation at 2.64 eV. This crystal exhibited a permanent color change from yellow-green to red-purple in the radiated area. This photo-induced transition from the diamagnetic LT phase to the paramagnetic HT phase was found both inside (296K) and outside (11 K) the hysteresis loop. The excitation photon density dependence of the LT-to-HT conversion efficiency was examined, and a clear threshold was concluded in it. This strongly indicates that the observed is not a thermal effect, but a photo-induced phase transition.

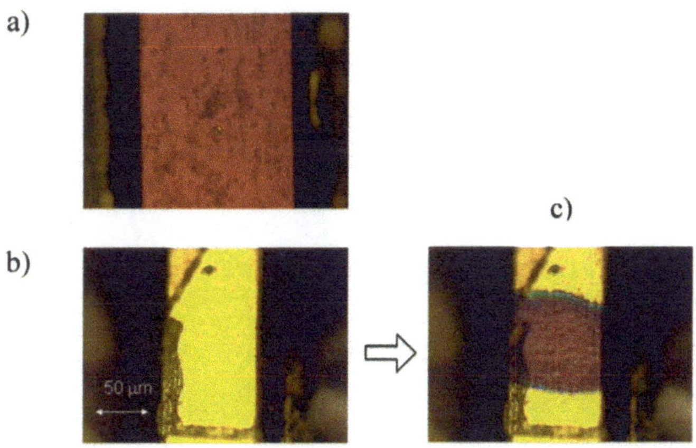

Fig. 5. The images of polarized microscope for TTTA; (a) HT phase; (b) LT phase; (c) after radiation of laser pulse on the sample surface of the LT phase.

2.6 Comparison with Spin-crossover Transitions

The spin-crossover phenomena have been studied extensively [47, 48], and photo-induced transitions [49, 50] and magnetic bistability [1, 6–19] have been discovered in them before TTTA. These photo-induced phenomena are called LIEEST (Light-Induced Excited Spin-State Trapping). It is interesting to compare the magnetic bistability in TTTA with those in the spin-crossover complexes.

The magnetic data for TTTA in Fig. 1(a) are re-plotted as $\chi_p T$ vs. T in Fig. 6(a) to compare with those of a Fe(II) spin-crossover complex, $[Fe(Htrz)_{3-3x}(4\text{-}NH_2trz)_{3x}](ClO_4)_2$ [1], whose magnetic behavior is schematically depicted in Fig. 6(b). The values of $\chi_p T$ for the HT phase of TTTA are smaller than those for the HT phase of the spin-crossover complex by one order of magnitude. The Fe(II) complex exhibits a change in spin multiplicity between $S = 0$ and $S = 2$, in contrast that TTTA has not a change in molecular spin multiplicity ($S = 1/2$) but intermolecular interaction. While the values of $\chi_p T$ for the HT phase of the Fe(II) complex are temperature independent, those of the HT phase of TTTA show a significant decrease with a decrease in temperature because of the antiferromagnetic intermolecular interactions. The change in magnetic susceptibility at the phase transition of TTTA is much smaller than that of the spin crossover complex, but the spin states in TTTA can be sensitively distinguished by EPR in contrast to the EPR silence of the Fe(II) spin crossover complex. The loop width of TTTA is more than 70K, which is much larger than that of the complex, while $[Fe(4\text{-}NH_2trz)_3](NO_3)_{1.7}(BF_4)_{0.3}$ preencapsulated in a polymer was reported to exhibit the loop width of *ca.* 60K [1].

$[Fe(phy)_2](BF_4)_2$ (phy=1,10-phenanthroline-2-carbaldehydephenylhydrazone) exhibits the spin-crossover transition between $S = 0$ and $S = 2$, with a large thermal

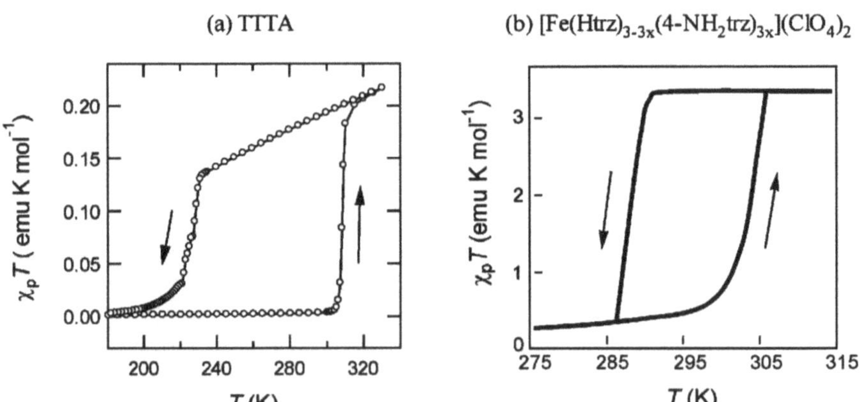

Fig. 6. Comparison between the magnetic data for TTTA (a) and $[Fe(Htrz)_{3-3x}(4\text{-}NH_2trz)_{3x}](ClO_4)_2$ (b).

hysteresis [19]. The pressure effects on this compound were examined up to 0.4 GPa, and a shift of the bistable range to higher temperatures was observed with $dT_c/dp = 30\ K{\cdot}GPa^{-1}$ (the average of $dT_{c\uparrow}/dp$ and $dT_{c\downarrow}/dp$). We now compare the pressure effects on TTTA with those on this spin-crossover complex. The dT_c/dp value for the Fe(II) complex is similar to that of TTTA. In $[Fe(phy)_2](BF_4)_2$ the volume change between the two phases is $15N_A\ Å^3{\cdot}mol^{-1}$, while that of TTTA is $3.4N_A/4\ Å^3{\cdot}mol^{-1}$. Therefore, by the Clapeyron equation the volume term leads to a greater increase in dT_c/dp for $[Fe(phy)_2](BF_4)_2$ compared to TTTA. The entropy change of the spin-crossover transition is also moderate, $85\ J{\cdot}mol^{-1}{\cdot}K^{-1}$, because it includes the spin contribution, $R\ln5$, caused by the change from $S = 0$ to 2. In contrast, the spin contribution to the transition entropy of TTTA is smaller than $R\ln2$ (= $5.76\ J{\cdot}K^{-1}{\cdot}mol^{-1}$), because the transition in TTTA occurs between a diamagnetic phase and a 'suppressed' paramagnetic phase due to antiferromagnetic intermolecular interactions. As such, it is considered that the small spin contribution is responsible for a small TTTA entropy change, namely $8.5\ J{\cdot}K^{-1}{\cdot}mol^{-1}$ (the average of $\Delta S_\uparrow$ and $\Delta S_\downarrow$), which likely causes a high sensitivity to pressure. The transitions in TTTA and $[Fe(phy)_2](BF_4)_2$ exhibit similar shifts to high temperature with increasing pressure, but the basic origins are probably very different.

3 Molecule-based Magnetic Bistable Materials

In the previous section, we described the magnetic bistability in TTTA and compared it with those of spin-crossover compounds. Such magnetic bistability has been found one after another in the family of heterocyclic thiazyl radicals [50–55]. Their magnetic and structural behaviors accompanied by the transitions are similar to those of TTTA.

Potassium 2,3,5,6-tetrachloro-4-oxido-1-phenoxyl (known as potassium, chloranil) exhibits a nearly diamagnetic-paramagnetic phase transition with a remarkable thermal hysteresis loop at $T_{c\uparrow} = 265K$ and $T_{c\downarrow} = 220K$ [56, 57]. The magnetic data for the LT phase can be reproduced by the antiferromagnetic dimer model, while the data for the HT phase can be fit to 1D antiferromagnetic Heisenberg model. Although the chloranil radicals form an ideal 1D stacking chain, the potassium ions bridges the distances between the radical columns.

The spirobiphenalenyl neutral radicals are reported to simultaneously exhibit bistability in three physical channels: electrical, optical, and magnetic [58]. The HT phase is paramagnetic and transparent in IR region, while the LT phase is diamagnetic and opaque with radical dimerization. Against all expectation, the conductivity increases by two orders of magnitude in the LT phase.

Magnetic bistability has also been found in several transition metal compounds besides the spin-crossover complexes. The cobalt-iron Prussian blue analogues exhibits an electron transfer transition between a Fe(II)($S = 0$)-CN-Co(III)(S=0) LT phase and a Fe(III)(S=1/2)-CN-Co(II)($S = 3/2$) HT phase with a thermal hysteresis of 40K [59]. $[Co(III)Cp_2]^+[Fe(III)I_4]^-$ is reported to exhibit a 30K thermal hysteresis in magnetic susceptibility that is attributed to a structural phase transition [60].

There is no clear guiding principle to obtain molecule-based bistability at this stage, but it would be much easier to produce room-temperature magnetic bistabily in organic materials than to synthesize room-temperature organic ferromagnets. Various types of molecule-based bistability will be certainly developed as a characteristic function of the molecule-based materials.

4 Summary

We described the room-temperature bistability in an organic radical, TTTA, and compared it with those in the other materials, such as spin-crossover complexes. The comparison produced semi-quantitative interpretations on their magnetic bistabilities. The research on bistability has become a new branch in materials science on molecule-based materials.

References

1. O. Kahn and C.J. Martinez, Science 279 (1998) 44.
2. O. Kahn, Chemtronics 3 (1988) 140.
3. "Molecular Electronic Devices", F.L. Carter ed., (1982) Marcel Dekker, New York.
4. "Molecular Magnetism – New Magnetic Materials", K. Itoh, M. Kinoshita eds., (2000) Gordon and Breach, Japan.
5. S.J. Blundell, F.L. Pratt, J. Phys.: Condens. Matter, 16 (2004) R771.
6. J. Kröber, E. Codjvie, O. Kahn, F. Groliere, and C. Jay, J. Am. Chem. Soc. 115 (1993) 9810.
7. J. Kröber, J.P. Audiere, R. Claude, E. Codjvie, O. Kahn, J.G. Haasnoot, F. Groliere, C. Jay, A. Bousseksou, J. Linares, F. Varret, and A. Gonthiervassal, Chem. Mater. 6 (1994) 1404.
8. T. Nakamoto, A. Bhattacharjee, and M. Sorai, Bull. Chem. Soc. Jpn. 77 (2004) 921.
9. Z.J. Zhong, J.-Q. Tao, Z. Yu, C.-Y. Dun, Y.-J. Liu, and X.-X. You, J. Chem. Soc., Dalton Trans. (1998) 327.
10. Y. Murakami, T. Komatsu, and N. Kojima, Synth. Met. 103 (1999) 2157.
11. R. Jinek, J. Spiering, P. Gütlich, and A. Hauser, Chem. Eur. J. 2 (1996) 1435.
12. H. Daubric, C. Cantin, C. Thomas, J. Kliava, J.-F.-Létard, and O. Kahn, Chem. Phys. 244 (1999) 75.
13. D.L. Reger, J.D. Elgin, M.D. Smith, F. Grandjean, L. Rebbouh, and G.J. Long, Eur. J. Inorg. Chem. (2004) 3345.
14. P. Guionneau, J.-F. Létard, D.S. Yufit, D. Chasseau, G. Bravic, A.E. Goeta, J.A.K. Howard, and O. Kahn, J. Mater. Chem. 9 (1999) 985.
15. H. Oshio, K. Kitazaki, J. Mishiro, N. Kato, Y. Maeda, and Y. Takashima, J. Chem. Soc., Dalton Trans., (1987) 1341.
16. V. Niel, M.C. Muñoz, A.B. Gaspar, A. Galet, G. Levchenko, and J.A. Real, Chem. Eur. J. 8 (2002) 2446.
17. V. Niel, A.L. Thompson, M.C. Muñoz, A. Galet, A.E. Goeta, J.A. Real, Angew. Chem. Int. Ed. Engl. 42 (2003) 3760.
18. V. Ksenofontov, G. Levchenko, H. Spiering, P. Gütlich, J.-F. Létard, Y. Bouhedja, and O. Kahn, Chem. Phys. Lett. 294 (1998) 545.

19. V. Ksenofontv, H. Spiering, A. Schreiner, G. Levchenko, H.A. Goodwin, and P. Gültlich, J. Phys. Chem. Solids 60 (1999) 393.
20. L.P. Ma, W.J. Yang, Z.Q. Xue, and S.J. Pang, Appl. Phys. Lett. 73 (1998) 850.
21. Y. Iwasa, T. Koda, Y. Tokura, S. Koshihara, N. Iwasawa, and G. Saito, Appl. Phys. Lett. 55 (1989) 2111.
22. L. Ma, J. Liu, S. Pyo, and Y. Yang, Appl. Phys. Lett. 80 (2002) 362.
23. For review, see M. Irie, Chem. Rev. 100 (2000) 1685.
24. J.M. Rawson and G.D. McManus, Coor. Chem. Rev. 189 (1999) 135.
25. J.M. Rawson, A.J. Banister, and I. Lavender, Adv. Heterocyc. Chem. 62 (1995) 137.
26. J. Banister, N. Bricklebank, I. Lavender, J.M. Rawson, C.I. Gregory, B.K. Tanner, W. Clegg, M.R. Elsegood, and F. Palacio, Angew. Chem. Int. Ed. Engl. 35 (1996) 2533.
27. C.D. Bryan, A.W. Cordes, R.M. Fleming, N.A. George, S.H. Glarum, R.C. Haddon, R.T. Oakley, T.T. Palstra, A.S. Perel, L.F. Schneemeyer, and J.V. Waszczak, Nature 365 (1993) 821.
28. C.D. Bryan, A.W. Cordes, R.M. Fleming, N.A. George, S.H. Glarum, R.C. Haddon, C.D. MacKinnon, R.T. Oakley, T.T. Palstra, and A.S. Perel, J. Am. Chem. Soc. 117 (1994) 6880.
29. Y. Teki, K. Itoh, A. Okada, H. Yamakage, T. Kobayashi, K. Amaya, S. Kurokawa, S. Ueno, and Y. Miura, Chem. Phys. Lett. 270 (1997) 573.
30. T.M. Barclay, A.W. Cordes, R.H. de Laat, J.D. Goddard, R.C. Haddon, J.D. Jeter, R.C. Mawhinney, R.T. Oakley, T.M.M. Palstra, G.W. Patenaude, R.W. Reed, and N.P.C. Westwood, J. Am. Chem. Soc. 119 (1997) 2633.
31. E. Dormann, N.J. Norwak, K.A. Williams, R.O. Angus, and Jr., F. Wudl, J. Am. Chem. Soc. 109 (1987) 2594.
32. G. Wolmershäuser, G. Wortmann, and M. Schnauber, J. Chem. Res. Synop. (1988) 358.
33. W. Fujita and K. Awaga, J. Am. Chem. Soc. 123 (2001) 3601.
34. W. Fujita and K. Awaga, Chem. Phys. Lett. 357 (2002) 385.
35. W. Fujita, K. Awaga, M. Takahashi, M. Takeda, and T. Yamazaki, Chem. Phys. Lett. 362 (2002) 97.
36. G. Wolmershäuser, and R. Johann, Angew. Chem. Int. Ed. Engl. 28 (1989) 920.
37. W. Fujita and K. Awaga, Science 286 (1999) 261.
38. W. Fujita, K. Awaga, H. Matsuzaki, and H. Okamoto, Phys. Rev. B 65 (2002) 064434.
39. G.D. McManus, J.M. Rawson, N. Feeder, J. van Duijn, E.J.L McInnes, J.J. Novoa, R. Burriel, F. Palacio, P. Oliete, J. Mater. Chem. 11 (2001) 1992.
40. K. Awaga, Y. Maruyama, Phys. Rev. B 44 (1991) 2589.
41. K. Awaga, T. Sekine, M. Okawa, W. Fujita, S.M. Holmes, G.S. Gregory, Chem. Phys. Lett. 293 (1998) 352.
42. K. Takeda, K. Konishi, M. Tamura, and M. Kinoshita, Phys. Rev. B 53 (1996) 3374.
43. D. Jerome, A. Mazaud, M. Ribault, and K. Bechgaard, J. Phys. Lett. 41 (1980) 95.
44. Y. Galcia, V. Ksenofontv, G. Levchenko, and P. Gültlich, J. Mater. Chem. 10 (2000) 2274.
45. T. Tanaka, W. Fujita, and K. Awaga, Chem. Phys. Lett. 393 (2004) 150.
46. H. Matsuzaki, W. Fujita, K. Awaga, and H. Okamoto, Phys. Rev. Lett. 91 (2003) 017403.
47. P. Gütlich, Struct. Bonding (Berlin) 44 (1981) 83.
48. P. Gütlich and A. Hauser, Coor. Chem. Rev. 91 (1990) 1.
49. S. Decurtins, P. Gütlich, C.P. Köhler, H. Spiering, and A. Hauser, Chem. Phys. Lett. 105 (1984) 1.
50. S. Decurtins, P. Gütlich, K.M. Hasselbach, A. Hauser, and H. Spiering, Inorg. Chem. 24 (1985) 2174.

51. J.L. Brusso, O.P. Clements, R.C. Haddon, M.E. Itkis, A.A. Leitch, R.T. Oakley, R.W. Reed, and J.F. Richardson, J. Am. Chem. Soc. 126 (2004) 14692.
52. J.L. Brusso, O.P. Clements, R.C. Haddon, M.E. Itkis, A.A. Leitch, R.T. Oakley, R.W. Reed, and J.F. Richardson, J. Am. Chem. Soc. 126 (2004) 8256.
53. H. Du, R.C. Haddon, I. Krossing, J. Passmore, J.M. Rawsson, and M.J. Schriver, Chem. Commun. (2002) 1836.
54. J. Banister, N. Bricklebank, I. Lavender, J.M. Rawson, C.I. Gregory, B.K. Tanner, W. Clegg, M.R.J. Elsegood, and F. Palacio, Angew. Chem. Int. Ed. Engl. 35 (1996) 2533.
55. W. Fujita, K. Awaga, Y. Nakazawa, K. Saito and K.M. Sorai, Chem. Phys. Lett. 352 (2002) 348.
56. T. Sugano, Polyhedron 20 (2001) 1285.
57. J.J. Andre and G. Weill, Chem. Phys. Lett. 9 (1971) 27.
58. M.E. Itkis, X. Chi, A.W. Cordes, and R.C. Haddon, Science 296 (2002) 1443.
59. N. Shimamoto, S. Ohkoshi, O. Sato, and K. Hashimoto, Inorg. Chem. 41 (2002) 678.
60. J.M. Migliori, W.M. Reiff, A.M. Arif, and J.S. Miller, Inorg. Chem. 43 (2004) 6875.

Carbon-Based Magnetism
T. Makarova and F. Palacio (Editors)

9

The Nature of Effective Exchange Interactions

K. Yamaguchi, S. Yamanaka, and Y. Kitagawa

Department of Chemistry, Graduate School of Science, Osaka University, Toyonaka, Osaka 560-0043, Japan

1 Introduction

In this chapter, the nature of effective exchange interactions is elucidated on the basis of fundamental theories of quantum chemistry. About 30 years ago, Hay, Thibeault and Roffmann (HTR) [1] presented a general expression for the effective exchange integrals (J_{ab}) on the basis of the perturbation theory by the use of the extended Hückel molecular orbitals (EHMO) for transition metal complexes such as [Cu(II)O_2Cu(II)]L. Katriel and Paunz [2] investigated the less screening rule (LSR) for relative stability between singlet (S) and triplet (T) states of carbon atom and related species from the viewpoint of the open-shell restricted Hartree-Fock (ORHF) and related symmetry-adapted (SA) methods. We performed MO-theoretical description of organic diradical and antiaromatic molecules on the basis of the broken-symmetry (BS) unrestricted Hartree-Fock (UHF) approximation [3, 4]. The spin vector model (SVM) [5] followed by the magnetic group theory is also derived to diagnose relative stability between the low-spin (LS) and high-spin (HS) states of these species. McConnell [6] pointed out that EHMO perturbation theory favors the Hund-type explanation of ferromagnetic coupling of anti-stacking dimer of benzyl radical, while the BS MO theory supports the McConnell-type and SVM pictures for it. Theoretical explanations of the mechanism of ferromagnetic interaction are different in these models. Now, reliable and accurate wavefunctions are available to examine the validity and applicability of these three approaches to molecular magnetism. Here, conceptual and physical basis of them is discussed in detail on the basis of computational results for several selected atoms, molecules and clusters. The SVM picture is found to be important for the prediction of intermolecular ferromagnetic spin alignments, though simple Hund rule is often breakdown.

2 The Nature of Chemical Bonds and Magnetic Interactions

2.1 Chemical Bond of H_2^+ and Virial Theorem

The exact solution is desirable to elucidate the nature of effective exchange interaction in molecular magnetism [7]. The analytical and very accurate solutions of Schrödinger equation for hydrogen molecular ion (H_2^+) are now available for elucidation of the origin of chemical bond of two-center one-electron system. These solutions should satisfy the Hellmann-Feynman and virial theorems. The virial theorem is given for a component atom [7] as

$$2T + V = 0, \tag{1}$$

where T and V are the kinetic and potential energy terms, respectively. Similarly the virial theorem for diatomic molecules is given under the Born-Oppenheimer approximation as

$$2T + V + R\left(\frac{dE}{dR}\right) = 0, \tag{2}$$

where R denotes the interatomic distance. The third term in eq.(2) disappears at the equilibrium geometry (R_e). Then the energy difference between diatomic molecule and separated atoms should have the following relation.

$$\Delta E = -\Delta T = \frac{1}{2}\Delta V. \tag{3}$$

This indicates that the binding energy for diatomic molecule arises from the lowering of potential energy term (V). The relation (3) is satisfied by the exact solution and even by the SCF solution.

Here, let us consider the simplest molecular orbitals (MO) of H_2^+ and H_2 as

$$\phi(b) = (\varphi_A + \varphi_B)/\sqrt{2(1+S)}, \tag{4a}$$

$$\phi(a) = (\varphi_A - \varphi_B)/\sqrt{2(1-S)}, \tag{4b}$$

where b and a denote the bonding and antibonding orbitals, respectively, and φ_c ($c = A, B$) and S are Slater-type atomic orbital (AO) of c and overlap integral.

$$\varphi_C = \left(\pi\alpha^3\right)^{1/2} \exp\left(-\frac{r}{\alpha}\right), \; S = \langle\varphi_A|\varphi_B\rangle. \tag{4c}$$

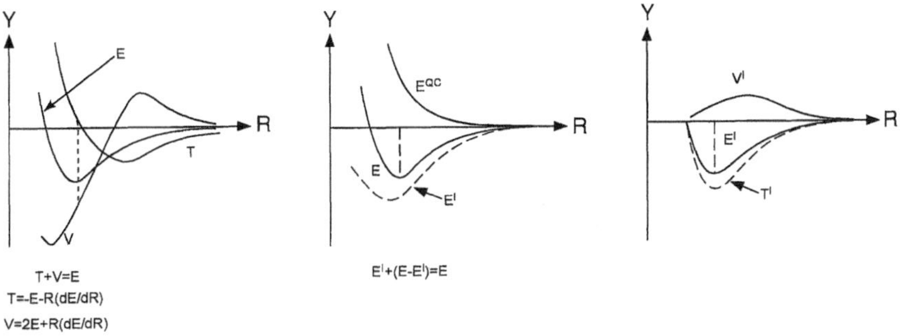

a) b) c)

Fig. 1. Variations (Y) of (a) E, T and V, (b) E^{QC} and E^{I} and (c) T^{I} and V^{I} with interatomic distance (R) by the exact solution of H_2^+.

The optimized exponent α for H_2^+ is $(1.2380)^{-1}$, though it is 1.0 for hydrogen atom: note that $T(H_2^+) = 0.5865$ ($= -E$), $V(H_2^+) = -1.1730$ (at $R = R_{eq}$) and $E(H) = -0.5$ (a.u.). Clearly the energy difference satisfies the relation in eq.(3) at the equilibrium geometry.

Next, the R-dependence of each term is examined by using the relation $E = T + V$ and eq.(2), to provide the following equations:

$$T = -E - R\left(\frac{dE}{dR}\right), \tag{5a}$$

$$V = 2E + R\left(\frac{dE}{dR}\right). \tag{5b}$$

Figure 1 shows variations of T, V and E with the change of interatomic distance (R). The kinetic energy decreases in the separated atom region ($R >> R_{eq}$), but it rather increases in the strong bonding region ($R \approx R_{eq}$) because of contraction of atomic orbital ($\alpha < 1.0$) for chemical bonding. The reverse tendency is recognized for the potential energy term (V) and the binding energy is attributable to the rapid decrease of V from the viewpoint of the virial theoretm. This type of explanation via the virial theorem is often referred to as physicist's view of chemical bonding [8].

2.2 Quantum Interference and Kinetic Exchange Interaction

In this section, we examine the chemist's view that molecule(s) consists of atomic components. To this end, the electron density given by eq. (4) is divided into quasi-classical (QC) and quantum interference (I) terms, respectively [8],

$$\rho(b) = (\phi(b))^2 = \rho^{QC} + \rho^{I}(b), \tag{6a}$$

$$\rho(a) = (\phi(a))^2 = \rho^{QC} + \rho^{I}(a), \tag{6b}$$

where

$$\rho_A = \varphi_A^2,\ \rho_B = \varphi_B^2,\ \rho^{QC} = (\rho_A + \rho_B)/2, \tag{6c}$$

$$\rho^I(b) = \rho(i)/(1+S),\ \rho^I(a) = -\rho(i)/(1-S), \tag{6d}$$

$$\rho(i) = \left\{\varphi_A\varphi_B - \frac{1}{2}S\left(\varphi_A^2 + \varphi_B^2\right)\right\}. \tag{6e}$$

From the above equations, the main difference between bonding and antibonding orbitals lies in the sign of the interference density in Eq.(6e). Actually the electron density accumulates in the interatomic (bonding) region in the case of the bonding (b) MO, while it decreases in the region for the antibonding (a) MO.

The total energy of H_2^+ by the bonding MO is divided into the QC and I terms as shown in Fig. 1(b): $E = E^{QC} + E^I$. The quantum interference energy E^{I} plays an essential role for chemical bonding in the whole region of H_2^+. Furthermore, the E^{I} term is expressed by the kinetic (T^{I}) and potential (V^{I}) energy terms, respectively, as shown in Fig. 1(c). The T^{I} term plays an important role for chemical bonding of H_2^+. This leads to the chemist view [8] that chemical bonding arises from the quantum interference of atomic orbitals. The bonding parameter β is defined by using this term as [9]

$$\beta(-t) = \left\langle \rho(i)\hat{H} \right\rangle \tag{7a}$$

$$= \langle \varphi_A | \hat{H} | \varphi_B \rangle - \frac{1}{2}S\left\{\langle \varphi_A | \hat{H} | \varphi_A \rangle + \langle \varphi_B | \hat{H} | \varphi_B \rangle\right\} \tag{7b}$$

$$\cong S \cdot const. \tag{7c}$$

This is the origin of the kinetic exchange term in molecular magnetism.

2.3 Chemical Bond of H_2 and Electron Repulsion Effect

The nature of chemical bond of H_2^+ [8] is not altered for H_2 [7]. By adding one electron to H_2^+, hydrogen molecule H_2 is formed and its wavefunction is therefore expressed by the superposition of the ionic and covalent terms as

$$\phi(1,2) = \left(\frac{N}{2}\right)^{1/2}\left[\cos\gamma\left\{\varphi_A(1)\varphi_A(2) + \varphi_B(1)\varphi_B(2)\right\} + \sin\gamma\left\{\varphi_A(1)\varphi_B(2) + \varphi_B(1)\varphi_A(2)\right\}\right], \tag{8a}$$

where N is the normalizing constant

$$N = \left(1 + S^2 + 2S\sin 2\gamma\right)^{-1}. \tag{8b}$$

The Weinbaum-type wavefunction is given by the optimized values : tan $\gamma = 3.9$ and $\alpha = (1.193)^{-1} < 1.0$ [7]. Therefore, the weight of the covalent term (second term in Eq.(8a)) is much larger than that of the ionic (first) term. The on-site Coulomb repulsion term U newly appears because of contribution of the ionic term for H_2. The pair density Π_c is introduced to evaluate the Coulomb repulsion as

$$\Pi_C = \left(\frac{N}{2}\right)\cos^2\gamma\{\rho_A(1)\rho_A(2)+\rho_B(1)\rho_B(2)\}, \tag{9a}$$

$$\langle r_{12}^{-1}\rangle = \left\langle\frac{\Pi_C}{r_{12}}\right\rangle = N\cos^2\gamma\, U, \tag{9b}$$

$$U = \int\rho_C(1)(r_{12}^{-1})\,\rho_C(1)\,d\tau \;\; (C = A, B). \tag{9c}$$

The Hubbard model is defined by both kinetic term in Eq.(7) and on-site repulsion U in Eq.(9).

$$\hat{H} = \sum_{a,b} -t_{ab}\hat{\chi}_a^\dagger\hat{\chi}_b + \sum_c U_c\hat{\Pi}_c, \tag{10}$$

where $\hat{\chi}_c^\dagger$ ($\hat{\chi}_c$) denotes the creation (annihilation) operators.

The virial theorem for H_2^+ is equally applicable to $H_2\{E = -T = \frac{1}{2}(V_{en} + V_{ee} + U_{nn})\}$, showing that the electron-nuclear attraction energy (V_{en}) plays a crucial role of chemical bonding at R_{eq} even if the electron-electron repulsion energy (V_{ee}) term is added. However, it is noteworthy that the *T*-term is mainly contributable to the chemical bonding of H_2 if the Heitler - London (HL) wavefunction (sin $\gamma = 1$ in Eq.(8a)) is employed under the assumption of the fixed exponent ($\alpha = 1$) even for H_2 molecule. Inflexible wavefunctions like HL with the fixed α often provide different explanation of chemical bonding of H_2 from that of the accurate one satisfying the virial theorem (see below). Thus, the right answer of chemical bonding by the right wavefunction is the contradiction of electron cloud in H_2 instead of the HL perturbation picture. The HL valence-bond (VB) picture has been extended to molecular magnetic materials by Ovchinnikov [12] and Klein [13]. Here, the VB theory is not repeated.

2.4 Hartree-Fock Calculations of H_2^+ and H_2

Several variational theories are feasible for molecular magnetism. Broken-symmetry (BS) methods such as the unrestricted Hartree-Fock (UHF) [3, 4] are often utilized conveniently to estimate the effective exchange interactions between open-shell species. In order to elucidate reliability of the BS approach, the UHF/6-311G** calculations of H_2^+ have been carried out, and its total energy is divided into the *T*- and V_{en} and V_{nn} (nuclear repulsion) terms. Figure 2 illustrates variations of each term with *R*. The UHF

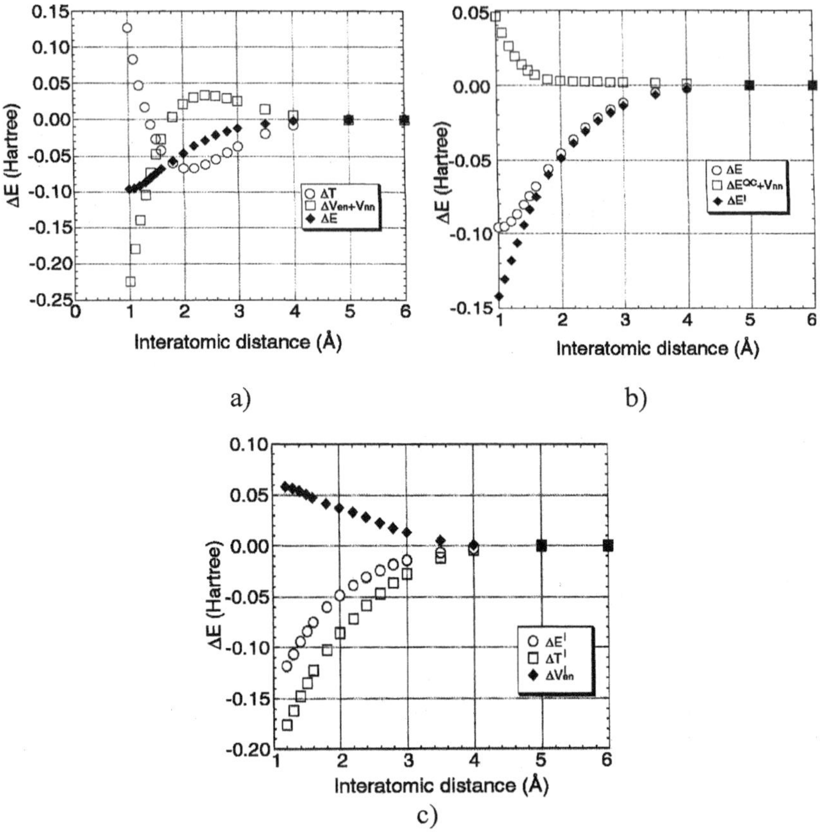

Fig. 2. Variations of (a) ΔE, ΔT, ΔV_{en} and ΔV_{nn}, (b) ΔE^{QC}, ΔE^{I} and (c) ΔT^{I} and ΔV^{I}_{en} with R by UHF / 6-311G**.

calculations can reproduce characteristic features of the exact energy decompositions in Fig. 1: (i) V_{en} determines the binding energy (physical picture), (ii) ΔE^{I} is its origin (chemical picture) and (iii) ΔT^{I} is the main component of ΔE^{I} (kinetic exchange picture).

The BS UHF calculations of H_2 are performed by using both minimal basis set (STO-3G) and extended one (6-311G**). Figure 3 illustrates variations of T, V_{en}, V_{ee} and E. From Fig. 3(a), the kinetic energy term (ΔT) plays a main contribution to chemical bonding throughout the dissociation (association) when the STO-3G basis set is employed. This situation is the same for the HL wavefunction with the fixed α. On the other hand, the electron-nuclear attraction (V_{en}) plays an essential role of the bonding in the case of the 6-311G** basis set. The contraction (or orbital relaxation) is crucial to obtain the same picture as that of the exact solution. Thus use of the flexible

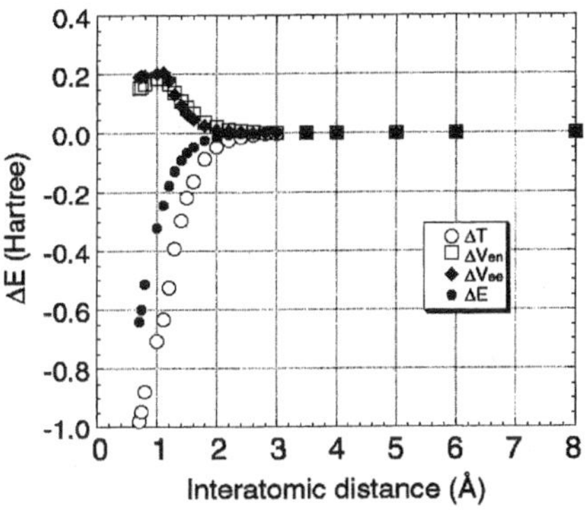

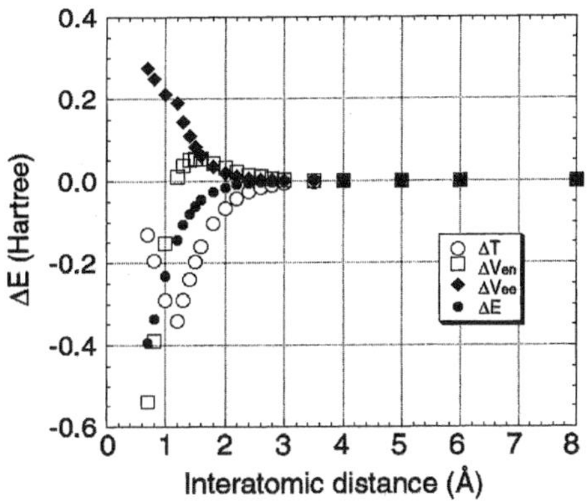

Fig. 3. Variations of ΔT, ΔV_{en} and ΔV_{ee} with R by (a) UHF / STO-3G and (b) UHF/6-311G**.

basis set is crucial for the correct description of the chemical bonding like the wavefunction of the optimized α in Eq. (4). On the other hand, in the VB CI language in Eq. 8, many ionic configurations expressed by using extended basis sets are crucial in this regards. UHF with a flexible basis set can grasp the key behaviors of T, V_{en}, V_{ee}, though it breaks the spin symmetry.

If we may focus attention to the magnetic region ($R >> R_{eq}$), the kinetic energy term (ΔT) plays a dominant role for chemical bonding even if the 6-311G** basis sets are employed. This in turn indicates that the traditional kinetic exchange picture is reliable enough for theoretical understanding of the direct effective exchange interactions between localized spins [14]

$$\hat{H}(HB) = -\sum J_{ab}\hat{\mathbf{S}}_a\,\hat{\mathbf{S}}_b, \tag{11}$$

where $\mathbf{S}_c$ (c = a, b) denotes the local spin operator and J_{ab} denotes the effective exchange integral. However, it is noteworthy that localized spins in molecules often interact indirectly via polarization of π-networks. As shown later, BS methods are often used for *ab initio* computation of J_{ab} values in both cases.

2.5 Instability of HF and DFT Solutions and Broken-symmetry Solutions

The closed-shell MOs in eq.(4a) bifurcates into the broken-symmetry (BS) MOs for the up- and down-spins, respectively, if the interatomic distance (R) exceeds through a certain limit [3, 4]. These spin-polarized MOs are given by using the bonding and antibonding MOs as

$$\psi^{+}(\uparrow) = \cos\theta\phi(b) + \sin\theta\phi(a), \tag{12a}$$

$$\psi^{-}(\downarrow) = \cos\theta\phi(b) - \sin\theta\phi(a), \tag{12b}$$

where θ is the orbital mixing parameter which is determined by the BS methods. The localized MO (LMO) are defined by the orbital mixing limit [10] as

$$\chi_A = \psi^{+}\left(\theta = \frac{\pi}{4}\right) = \frac{1}{\sqrt{2}}\{\phi(b)+\phi(a)\} \cong \varphi_A - S\varphi_B, \tag{13a}$$

$$\chi_B = \psi^{-}\left(\theta = \frac{\pi}{4}\right) = \frac{1}{\sqrt{2}}\{\phi(b)-\phi(a)\} \cong \varphi_B - S\varphi_A, \tag{13b}$$

where χ_A and χ_B have small tails for the orthogonality condition. The BS MOs in Eq. (12a) are rewritten by LMO as

$$\psi^{+} = \cos\omega\chi_A + \sin\omega\chi_B, \tag{14a}$$

$$\psi^{-} = \cos\omega\chi_B + \sin\omega\chi_A, \tag{14b}$$

where $\omega = \theta + \pi/4$ and $\beta = \langle\chi_A|\hat{H}|\chi_B\rangle$.

Generally, the BS MOs are variable from the closed-shell MO limit ($\theta = 0$, $\omega = \pi/4$) to the valence-bond (VB) limit ($\theta = \pi/4$, $\omega = \pi/2$) with the change of the bonding parameter: $x = \beta/U$ as illustrated in Fig. 4a. If the Hubbard model in Eq.(10) is employed, the bifurcation from the closed-shell MO, i.e. the restricted Hartree-Fock (RHF) solution, to the BS UHF solution occurs at $x = \beta/U(HF) = 1/2$ as shown in Fig. 4(b). On the other hand, the on-site Coulomb repulsion (U) is considerably reduced under the hybrid density functional theory (HDFT) as

$$U(HDFT) = f(k)U(HF), f(k) < 1, \tag{15}$$

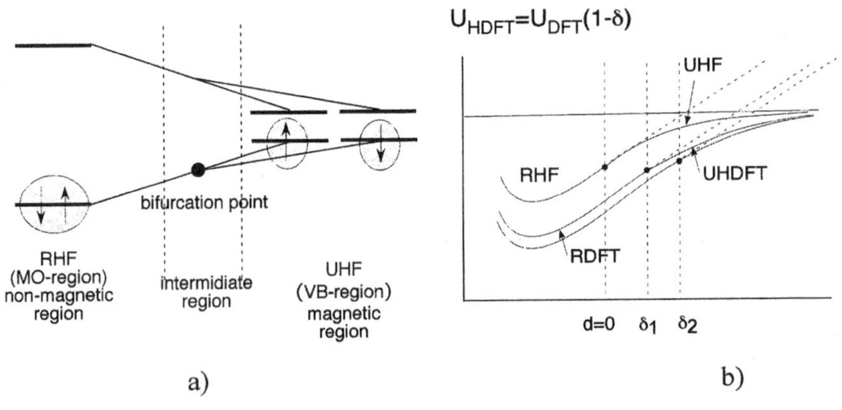

Fig. 4. (a) Bifurcation from RHF to UHF in the course of dissociation of covalent bond and (b) Potential energy cuves for HF, DFT and HDFT.

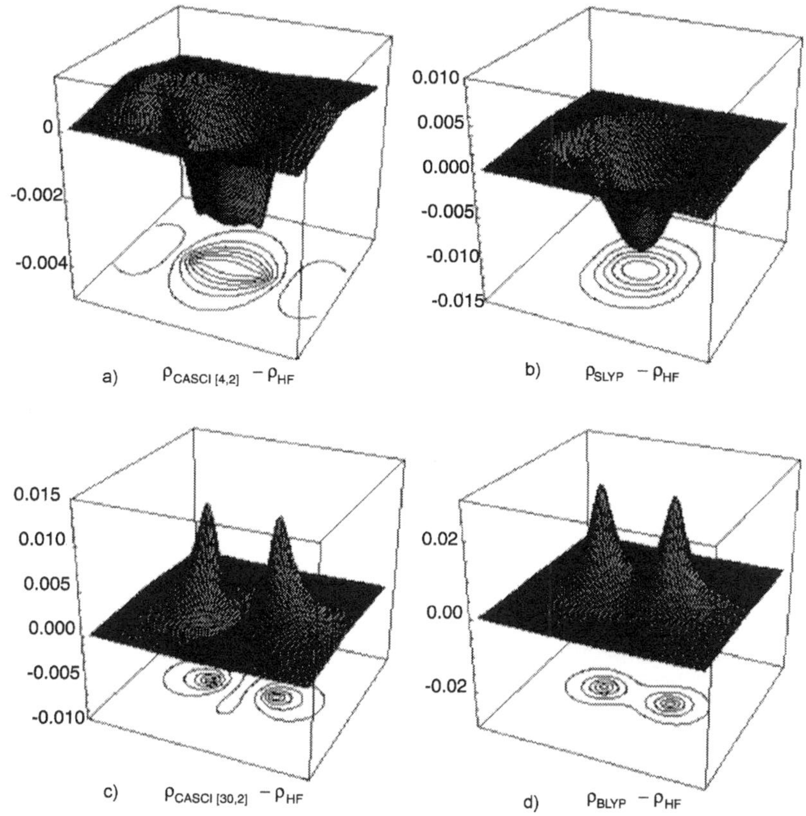

Fig. 5. Three-dimensional maps of the difference densities obtained by $\rho_X - \rho_{HF}$ (X = CASCI[4,2], CASCI [30,2], SLYP and BLYP)

where k is the mixing parameter of the HF exchange potential. Then the bifurcation point is also reduced to $x' = \beta'/U(\mathrm{HDFT}) = \dfrac{f(k)\beta(HF)}{f(k)U(HF)} = 1/2$ ($\beta'(\mathrm{HDFT}) = f(k)\beta(\mathrm{HF})$). In fact, UHF often overestimates the magnetic (bifurcating) region, while HDFT provides reasonable results (see below), though the determination of k-value is usually empirical like B3LYP (k = 20%).

The above scaling of U by HDFT is closely related to screening of electron repulsion by the electron correlation effect [16]. The difference of total electron density between post HF and HF (RHF) can be used as a useful measure to analyze the electron correlation effects. Figure 5 illustrates the difference densities for H_2 at R_{eq} by complete active space (CAS) configuration interaction (CI), DFT and HDFT methods, where the cc-pVTZ basis set is employed [17]. The CAS CI by the use of four active orbital two active electron [4,2] indicates the decrease of electron density in the bonding region (removal of the ionic term in Eq.(8)), showing the nondynamical correlation effect.

While the CASCI by the use of the 30 active MO 2-electron [30,2] shows the contraction of electron on atoms, which is responsible for dynamical correlation (decrease of α in Eq.(8)). The electron correlation effect is therefore divided into nondynamical and dynamical parts in general. The difference density map by the Hartree-Fock-Slater (HFS) method is rather similar to that of small CASCI [4,2]. On the other hand, the large-scale CASCI [30,2] result can be well reproduced by the HDFT methods such as UB3LYP. This implies that HDFT takes into account for both nondynamical and dynamical correlation effects through variations of electron density determined by the exchange correlation (XC) functions. Therefore the energy gap between the lowest (LS) and highest (HS) spin states is determined by subtle balance of T, V_{en} and V_{ee} terms in the case of HDFT.

2.6 Chemical Indices and Cluster Models for Effective Exchange Interaction

The natural orbital (NO) analyses [3, 4] of BS solutions is one of useful methods to extract necessary information for chemical bonds.

$$\rho(1,1') = \sum_i n_i \, \phi_i(1)\phi_i(1'), \tag{16}$$

where n_i denotes the occupation number. For example, it is given by UHF and spin-polarized DFT (UDFT) as

$$\rho = (\psi^+)^2 + (\psi^-)^2 = 2\cos^2\theta(\phi(b))^2 + 2\sin^2\theta(\phi(a))^2, \tag{17a}$$

where $n_b = 2\cos^2\theta$ and $n_a = 2\sin^2\theta$. Then the effective bond order (b_i) is defined [18] by

$$b_i = \frac{n_b - n_a}{2} = \cos 2\theta, \tag{18}$$

where $b_i = 1$ for the closed-shell MO (RHF and RDFT), while $0 \le b_i < 1$ for the non-closed shell MO by BS approach. This implies that the antibonding orbital is partially occupied in the BS solution because of the pseudo excitation via electron correlation. The first-order density is also defined by LNO(see Eq.(13)) as

$$\rho = \chi_A^2 + \chi_B^2 + 2\chi_A\chi_B \sin 2\omega\,. \tag{17b}$$

Then the effective bond order is directly given by the interference term: $b_i = 2\sin 2\omega/2 = \sin 2\omega$, which is equivalent to the orbital overlap $T = \langle \psi^+ | \psi^- \rangle$ (see Eq.(14)).

The up- and down-spin densities appear even in the singlet state by the HF solution because of electron localization via correlation. The spin density (Q) is defined [3, 4] by

$$Q=\left(\psi^{+}\right)^{2}-\left(\psi^{-}\right)^{2}=\cos 2\omega\left(\chi_{A}^{2}-\chi_{B}^{2}\right), \tag{19}$$

where cos2ω = 0 for RHF (RDFT) and cos2ω ≠ 0 for BS. Then the magnitude of unpaired electron density is given by [19]

$$D=(\cos 2\omega)^{2}=1-T^{2}, \tag{20}$$

where T is the orbital overlap $\left\langle\psi^{+}\middle|\psi^{-}\right\rangle$.

The information entropy is a common concept in the field of science and technology [20]. The information entropy (I_i) for chemical bond is defined by

$$I_{i}=-\sum_{j} n_{i}\ln n_{i}, \tag{21a}$$

where I_c = –2ln2 for RHF (RDFT) and I_0 = 0 for the dissociation limit. Then the normalized entropy is given by

$$I_{n}=(I_{c}-I_{i})/I_{c}, \tag{21b}$$

where $I_n = 1$ for RHF (RDFT) and $0 \le I_n < 1$ for UHF (UDFT). The I_n-value is useful even for post HF methods such as QCISD and CASCI with fractional occupation numbers defined by Eq.(16).

The BS solutions are closely related to the symmetry-adapted (SA) solutions via Eq.(12) [18] . The low-spin (LS) BS solutions involve higher-spin components. The pure spin states can be generally obtained by the spin projection method. For example, the spin-projected UHF (PUHF) solution for H_2 is given by the 2×2 MO CI form as

$$\Phi(PUHF)=N_{AP}\left\{C^{+}\left|\phi(b)\bar{\phi}(b)\right|-C^{-}\left|\phi(a)\bar{\phi}(a)\right|\right\}, \tag{22a}$$

where

$$C^{\pm}=(1\pm T) \text{ and } N_{AP}=\left\{2\left(1+T^{2}\right)\right\}^{1/2}. \tag{22b}$$

The PUHF solution is rewritten by LNO as

$$\Phi(PUHF)=\left(\frac{N}{2}\right)\left[\sin 2\omega\left\{\left|\chi_{A}\bar{\chi}_{A}\right|+\left|\chi_{B}\bar{\chi}_{B}\right|\right\}+\cos 2\omega\left\{\left|\chi_{A}\bar{\chi}_{B}\right|+\left|\bar{\chi}_{A}\chi_{B}\right|\right\}\right]. \tag{22c}$$

This VB-CI form is equivalent to the Weinbaum-type wavefunction in Eq.(8), though atomic orbitals are replaced by the orthogonalized LNO. Therefore the CASCI by the use of LNO can be utilized for polyradicals instead of VB CI [10]. The LNO CASCI is very useful for investigation of charge-transfer or ionic effects via heteroatoms and others [11]. The biradical character (y_i) is given by the weight (W_D) of double excitation configuration in the 2×2 MO CI form as

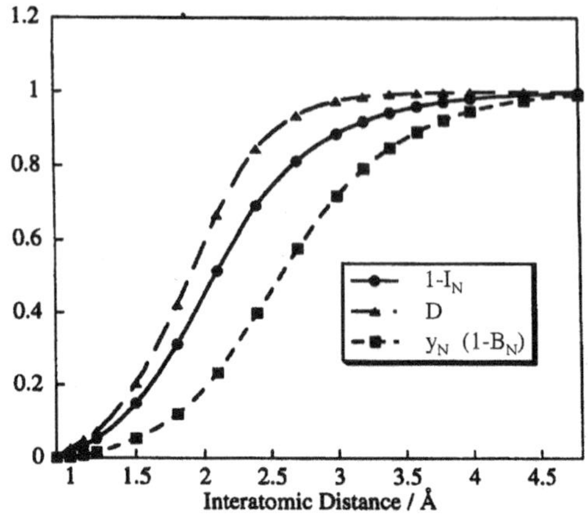

Fig. 6. Variations of chemical indices of the $(H)_{50}$ radical clusters with the interatomic distance.

$$y_i = 2W_D = 1 - \frac{2T_i}{1+T_i^2}, \tag{23}$$

where i denotes the i-th bonding-antibonding pair. The effective bond order (B_i) after spin projection (P) is given by

$$B_i = \frac{n_b(P) - n_a(P)}{2} = \frac{2T_i}{1+T_i^2} = 1 - y_i. \tag{18b}$$

Then the increase of biradical character is nothing but the decrease of effective bond order in the projected BS approximation [20].

The merit of the BS approach is that it can be applicable to relatively large systems, for which symmetry-adapted (SA) CASCI and CASSCF approach are not feasible. The chemical bond indices for $2N$-site mesoscopic radical clusters are given by the average form as

$$X_N = \frac{1}{N}\sum_{i=1}^{N} X_i \quad (X = b, B, y, I \text{ and } D), \tag{23}$$

where i denotes the i-th pair. As an example, UB3LYP calculations of the linear hydrogen atoms with $N = 50$ have been carried out to elucidate variations of information entropy, unpaired electron density and effective bond order with interatomic distance (R) as shown in Fig. 6 [20]. The HDFT method with and without spin projection is useful for theoretical investigation of chemical bonds in mesoscopic radical clusters. It is noteworthy that CASCI [50,50] and CASSCF [50,50] are impossible at the present time.

3 Less Screening Rule and Hund Rule

3.1 Less Screening Rule for Carbon Atom and Related Species

The Hund rule considering the Coulombic exchange interaction (K) has been employed for simple and lucid explanation of the high-spin ground state of atoms and molecules [2]. However, it is often replaced by the less screening rule if the virial theorem is required for wavefunctions employed. The carbon atom (C) and its isoelectronic atom ions (N^+, O^{2+} and F^{3+}) have the orthogonal p-atomic orbitals as illustrated in Fig. 7(a), showing the triplet ground state in conformity with the Hund rule. The electron-electron interaction (V_{ee}) between the open-shell (radical) two electrons of these species is given for the ground triplet (^{3}P) and excited singlet (^{1}D) states as

$$^1V_{ee}(^1\mathrm{D}) = \mathrm{G} + \mathrm{K},\ ^3V_{ee}(^3\mathrm{P}) = \mathrm{G} - \mathrm{K}, \tag{24a}$$

$$\Delta V_{ee} = {}^1V_{ee}(^1\mathrm{D}) - {}^3V_{ee}(^3\mathrm{P}) = 2\mathrm{K}\ (= 2J), \tag{24b}$$

where G and K are Coulomb and exchange integrals, respectively.

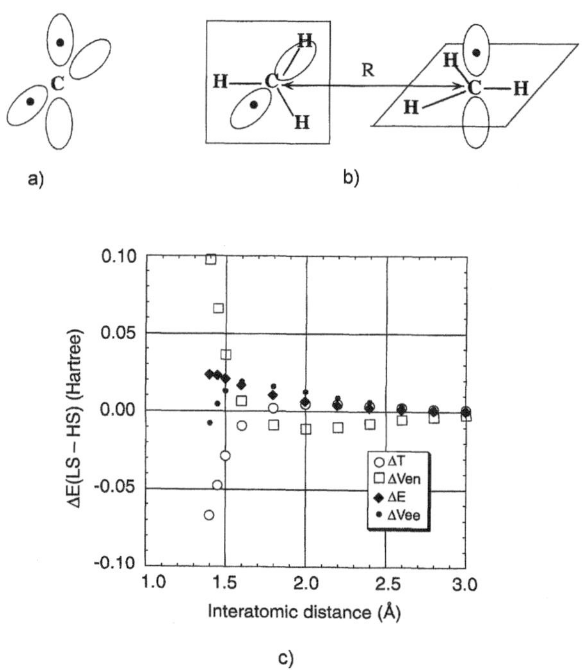

Fig. 7. Orbital configurations of (a) carbon atom, (b) methyl radical dimer $(\mathbf{1})_2$ and (c) energy partitioning of $(\mathbf{1})_2$ by UHF / 6-311G**.

$$G = \int 2p_x(1)2p_y(2)r_{12}^{-1}2p_x(1)2p_y(2)d\tau_1 d\tau_2, \quad (25a)$$

$$K = \int 2p_x(1)2p_y(2)r_{12}^{-1}2p_x(2)2p_y(1)d\tau_1 d\tau_2. \quad (25b)$$

The ^{3}P state is more stable by 2K than the ^{1}D state. This classical Hund rule is chemically interpreted by the orthogonality of radical orbitals.

However, the Hund (2K) rule has been derived under the assumption that the same Slater orbitals, i.e., the same exponent in Eq.(4c), can be used for both spin states, and then the kinetic (T) and electron-nuclear (V_{en}) attraction terms are also the same: $\Delta T = \Delta V_{en} = 0$. On the other hand, the open-shell RHF (ORHF) calculations of both spin states by the use of different effective nuclear charges (Z_{eff}) for different spin states have been carried out so as to satisfy the virial theorem (see section II.1). The V_{ee} terms are calculated to be 12.861 > 12.849 > 12.824 (a.u.) for the ^{3}P, ^{1}D and ^{1}S states, respectively. The tendency : $V_{ee}(^3\mathrm{P}) > V_{ee}(^1\mathrm{D}) > V_{ee}(^1\mathrm{S})$ is different from the classical Hund rule. This in turn implies that G and K values should be different between spin states, for example ^{3}P and ^{1}D:

$$\Delta V_{ee} = (^1\mathrm{G} + {}^1\mathrm{K}) - (^3\mathrm{G} - {}^3\mathrm{K}) = (^1\mathrm{G} - {}^3\mathrm{G}) + (^1\mathrm{K} + {}^3\mathrm{K}) < 0. \quad (26)$$

This also indicates that Coulomb repulsion for ^{3}P is larger than that of ^{1}D since $^1\mathrm{K}+{}^3\mathrm{K} > 0$: $^3\mathrm{G} > {}^1\mathrm{G}$, in accord with the tendency for the optimized Z_{eff} values: 1.568 (^{3}P) > 1.536 (^{1}D) > 1.487 (^{1}S). Because of this orbital contractions, the electron-nuclear attraction (V_{en}) terms become –88.106, –87.963 and –87.746 (a.u.) for ^{3}P, ^{1}D and ^{1}S, respectively. The relative stability between ^{3}P and ^{1}D is determined by the V_{en} term, showing the breakdown of the classical Hund (2K) rule. The situation is the same for the isoelectronic series (N^+, O^{2+} and F^{3+}). Since the contraction of electron cloud in ^{3}P is more predominant than in ^{1}D, spin alignment via this effect is referred to as the less screening rule [21–22]. Therefore the less screening rule has a common physical foundation of chemical bonds of H_2^+ and H_2.

3.2 Less Screening Rule for Simple Atoms and Molecules

The breakdown of the classical Hund rule was first discovered for the He atom [23]. For example, the V_{ee} terms are 0.256 and 0.268 for ^{1}S and ^{3}S, respectively, while the V_{en} terms are –4.541 and –4.619 (a.u.), because of large Z_{eff} (1.41) of ^{3}S than that (1.20) of ^{1}S. Similarly, the V_{ee} (V_{en}) terms are 0.245 (–4.493) and 0.267 (–4.533) for ^{1}P and ^{3}P, respectively. Apparently the less screening rule [22] should be applied for these states.

The multi-reference (MR) CI calculations [24–27] of diatomic molecules and simple molecules have been carried out to elucidate the applicability of the less screening rule. Table 1 summarizes the calculated results. For two and multi-center systems, the nuclear-nuclear repulsion (V_{nn}) term is one of the important factors determining the singlet-triplet energy difference.

$$^1E - {}^3E = 2J = \Delta T + \Delta V_{en} + \Delta V_{ee} + \Delta V_{nn} . \quad (27)$$

Table1.
Energy gaps (a.u.) between singlet and triplet and contribution of each term to them caluclated by Multi-reference CI method.

Term	BH ($^{1,3}\pi$)	NH ($^{1}\Delta$,$^{3}\Sigma$)	C_2 ($^{1,3}\Pi$)	O_2 ($^{1}\Delta$,$^{3}\Sigma$)	CH_2 (^{1,3}B)	H_2CO (^{1,3}A")
ΔE	0.0688	0.0667	0.0397	0.0374	0.0625	0.0029
ΔT	–0.0380	–0.0821	–0.0387	–0.0950	–0.0151	–0.0605
ΔV_{ee}	–0.0779	–0.0282	–0.0932	–0.1991	0.0799	–0.3908
ΔV_{en}	0.2172	0.1694	0.2434	0.5186	–0.1425	0.9015
ΔV_{nn}	–0.0326	0.076	–0.0719	–0.1871	0.1401	–0.4473

From Table 1, the ΔT term is negative for all species, and the ΔV_{ee} term is also negative except for CH_2, indicating the general tendencies: $^{3}V_{ee} > {}^{1}V_{ee}$ and $^{3}T > {}^{1}T$. On the other hand, a general trend is not found for the V_{nn} term. Therefore, we may conclude that the V_{en} term plays a predominant role for determination of the S-T gaps of BH, NH, C_2, O_2 and H_2CO* [24–27]. The ΔV_{ee} term is positive ($^{1}V_{ee} > {}^{3}V_{ee}$) for CH_2 because of the cancellation of ΔV_{en} and ΔV_{nn}.
The above examples shows that the less screening rule is applicable for diatomic molecules and simple molecules where distances of radical electrons are short. This may in turn indicates that the classical Hund rule is applicable for well-separated localized spins. The energy partitioning has been examined by changing the intermolecular distance (R) between methyl radicals (**1**) as shown in Fig. 7(b) [17]. The geometry of CH_3 has been fixed for both singlet and triplet states. The V_{en} term plays a predominant role for determination of the S-T gap in the relatively shorter R region, supporting the less screening rule. On the other hand, the V_{en} term rather stabilizes the LS state in the large R region of $(\mathbf{1})_2$, showing that ΔT and ΔV_{ee} terms are significant for stabilization of the HS state. The electron-electron (V_{ee}) term is important in the direct exchange ($R >> R_{eq}$) region, supporting the classical Hund rule.

2.3 Less Screening and Hund Rules for Polyradicals

There are two different effective exchange interactions between localized spins in polyradical species: (i) direct exchange interaction (J_{direct}) and (ii) indirect effective exchange interaction ($J_{indirect}$) via chemical bonds such as π-bond. As an example, let us consider meta-(phenylene) bis methyl (**2**) diradical, which can be regarded as methyl radical dimer $(\mathbf{1})_2$ linked with phenyl group as illustrated in Fig. 8(a). We have performed UHF, UB2LYP, UB3LYP and UBLYP/6-31G* calculations of **2** [17]. Table 2 summarizes relative contributions of T, V_{en}, V_{ee}, and V_{nn} terms to the singlet-triplet (ST) energy gap (eV).

First, let us examine these terms at the average geometry ($\mathbf{2}_{ave}$) for both states. The ΔV_{ee} term is positive for HDFT except for UHF. This is quite different from one-center

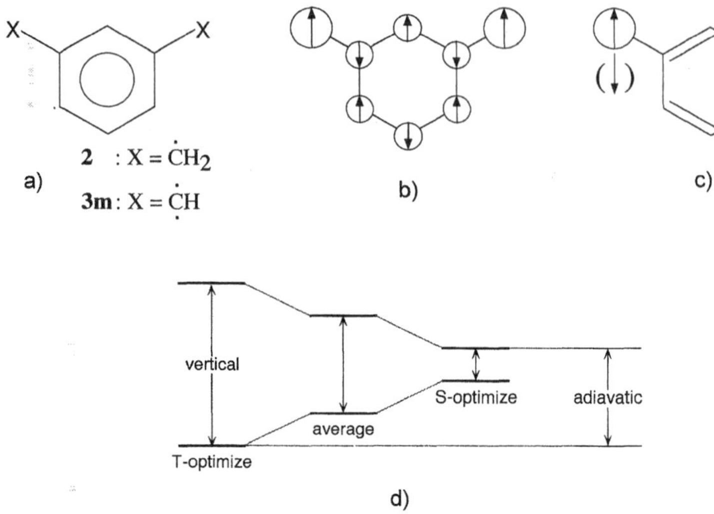

Fig. 8. (a) Structure of *m*-phenylene bis(methyl) (**2**) and bis(methylene) (**3**), (b) spin vector model of **2** (**3**) and (c) ST energy gaps for three geometries of **2**.

diradicals such as C and He^* and two-center diradicals such as C_2 and O_2 in Table 1. For UHF, spin polarization (SP) effect of the π-bond is abnormally large. Indeed, the magnitude of spin density on carbon atoms of phenyl ring is calculated to be about 1.0, showing an antiparallel spin alignment as illustrated in Fig. 8(b). Therefore the distance between unpaired spins is effectively very short (only 1.4 Å), though localized spins are really separated. Therefore the less screening rule (LSR) is applicable even for **2** under the UHF approximation. The contribution of the one-electron part $\Delta E_{one} = \Delta T + \Delta V_{en}$ (= 0.303 eV) is much larger than ΔV_{ee} (= 0.173 eV) even for UB2LYP. The LSR via the SP effect should be considered to be applicable under UB2LYP. On the other hand, the ΔE_{one} term by UB3LYP (UBLYP) is 0.066 (0.055), and the ΔV_{ee} term is 0.214 (0.125). Then the two electron repulsion term is predominant than the one-electron part by UB3LYP and UBLYP. Therefore, we may consider that the classical Hund rule is also applicable for **2**, from the comparison between ΔE_{one} and ΔV_{ee}.

However, the kinetic energy increase $|\Delta T|$ overweighs the electron-electron repulsion (V_{ee}) effect: $|\Delta T| > V_{ee}$ by both UHF and HDFT calculations of **2**. This in turn leads to another viewpoint that the V_{en} term can be regarded as the main origin of the S-T gap. This is nothing but the LSR for **2**. This type explanation is already presented for the excited P state of Ne^{8+}: $\Delta T = -0.262$, $\Delta V_{ee} = 0.187$ and $\Delta V_{en} = 0.337$ (a.u.) 2, giving the relations: ΔE_{one} (=0.072) < ΔV_{ee} but $|\Delta T| > V_{ee}$. The physicist's view emphasizes the LSR even for Ne^{8+}. Our chemist's view is that the Hund picture of the ST gap of **2** is rather responsible for direct effective exchange of localized spins and the LSR picture arises from the indirect effective exchange via the SP effect of π-electrons. Therefore evaluation of each contribution is inevitable for us as discussed in section IV.

Table 2
Energy gaps (eV) between singlet and triplet and contribution of each term to them of *m*-(phenylene) bis (methylene) calculated by UHF and UDFT method

structure	method	ΔE	ΔV_{nn}	ΔT	ΔV_{en}	ΔV_{ee}
$\mathbf{2}_{ave}$	UHF	1.1658	0.0000	−5.8991	8.1282	−1.0634
	UB2LYP	0.4756	0.0000	−1.2692	1.5720	0.1728
	UB3LYP	0.2799	0.0000	−0.4358	0.5018	0.2140
	UBLYP	0.1805	0.0000	−0.2176	0.2727	0.1254
$\mathbf{2}_{T}$	UHF	1.2860	0.0000	−6.4607	9.0705	−1.3238
	UB2LYP	0.5241	0.0000	−1.3878	1.8231	0.0888
	UB3LYP	0.3109	0.0000	−0.5307	0.6812	0.1604
$\mathbf{2}_{S}$	UHF	0.9698	0.0000	−5.5937	7.3607	−0.7972
	UB2LYP	0.4233	0.0000	−1.1078	1.2931	0.2380
	UB3LYP	0.2714	0.0000	−0.4200	0.4738	0.2177

The full geometry optimization of **2** has been carried out to elucidate the nuclear-nuclear repulsion (V_{nn}) contribution. Table 2 summarizes the relative contributions for the singlet (S)- and triplet (T)- optimized geometries of **2**. The ΔE_{one} and ΔV_{ee} terms are 0.435 (0.150) and 0.088 (0.160) by UB2LYP (UB3LYP) at the T- optimized geometry ($\mathbf{2}_T$), showing the enhancement of the LSR tendency expressed by Fig. 8(b). The π-bonds deform to stabilize the T-stable geometry via the SP effect. On the other hand, they are 0.185 (0.054) and 0.238 (0.218), respectively, for S-optimized geometry ($\mathbf{2}_S$). The relation: $|\Delta E_{one}|>V_{ee}$ is recognized for $\mathbf{2}_S$ even by UB2LYP, though UHF still indicates the LSR behavior. The SP effect of π-bond is weakened at $\mathbf{2}_S$ by HDFT, leading to the spin-paired representation of $\mathbf{2}_S$ by Fig. 8(d). The adiabatic ST gap of **2** is positive, indicating the ground triplet state as shown in Fig. 8(c), though the sign of J_{ab} is determined by the subtle balance of the ΔT, ΔV_{en}, ΔV_{ee}, and ΔV_{nn} terms.

4 Spin Polarization Rule and Related Spin Alignment Rules

4.1 CASCI and CASSCF for Direct Exchange Integrals

The Hückel molecular orbital (HMO) theory provide two degenerated nonbonding MOs (NBMOs) for $\mathbf{2}_{ave}$. The CASCI by the use of two-active NBMOs two-electrons [2,2] is theoretically equivalent to the CASCI [2,2] for the C atom and O_2 molecules. This means that the ST gap for $\mathbf{2}_{ave}$ is given by the Coulombic exchange integral K_{AB} where χ_A = NBMO1 and χ_B = NBMO2.

Table 3.
Energy gaps (eV) between singlet and triplet and contribution of each term to them of *m*-(phenylen) bis (methylen) calculated by CASSCF method

structure	method	ΔE	ΔV_{nn}	ΔT	ΔV_{en}	ΔV_{ee}
$\mathbf{2}_{ave}$	CASSCF[2,2]	0.1845	0.0000	0.9119	-1.2374	0.5100
	CASSCF[6,6]	0.5212	0.0000	0.4494	-0.4114	0.4833
$\mathbf{2}_{T}$	CASSCF[2,2]	0.1273	0.0000	0.6646	-0.9326	0.3952
$\mathbf{2}_{S}$	CASSCF[2,2]	0.1119	0.0000	0.5939	-0.8389	0.3569

$$^{1}E - {}^{3}E = 2K_{AB} \tag{28}$$

This is a well-accepted Hund rule for $\mathbf{2}_{ave}$ which is often refereed as the NBMO rule by Longett-Higgins.

However, the above relation is resulted under the assumption $\Delta T = \Delta V_{en} = \Delta V_{coul} = 0$; namely the fixed α (exponent) approximation both S and T state. The situation is altered if the orbital relaxation effects and electron-electron interactions via other nondynamical electrons are included explicitly. As an example, the CASSCF [*n*,*n*] ($n \geq 2$) / 6-31G* results are shown in Table 3. The ΔT and ΔV_{en} terms by CASSCF [2,2] are largely positive (0.912 eV) and negative (–1.237 eV), respectively, showing the large orbital relaxation effect. However, the one-electron term $|\Delta E_{one}|$ (=0.325) is less than the ΔV_{ee} term. The classical Hund rule for $\mathbf{2}_{ave}$ is therefore supported after inclusion of the orbital relaxation effect within CASSCF [2,2]. Interestingly, the ST gap by CASSCF [2,2] is similar to that of UBLYP. Next, two-bonding and two-antibonding π-orbitals have been added to CAS [2,2] to include the nondynamical electron correlation effects involving the SP effect. The ΔT and ΔV_{en} terms become 0.449 and –0.411 (eV), respectively, canceling each other, though the V_{ee} term remains almost constant. As a result, the ST gap becomes larger than that of UB2LYP. Moreover, the ΔV_{ee} term by CASSCF [6,6] is still larger than those of HDFT in Table 2. This indicates that more larger CASSCF [*n*,*n*] ($n >> 6$) is crucial for screening of the electron-electron repulsion term by inclusion of the dynamical correlation correction: for example a simple case of $\Delta E = -\Delta T = \Delta V_{en} = \Delta V_{ee} = 0.2$. As illustrated in Fig. 5, the difference density map between a computational method *X* and open-shell RHF (ORHF) would be useful for pictorial understanding of correlation corrections.

$$\Delta\rho(X) = \rho(X) - \rho(ORHF) \ (X = \text{CASSCF}[n,n], \text{HDFT, etc}) \tag{29}$$

The ST gap for $\mathbf{2}_{T}$ becomes smaller than that of $\mathbf{2}_{ave}$ after inclusion of the geometry relaxation because of the decrease of ΔV_{ee}. The same situation is also recognized for $\mathbf{2}_{S}$. In order to understand these interesting relations, we must remember that the one-electron part ($\hat{T} + \hat{V}_{en}$) and two electron part ($\hat{V}_{ee}$) is not commutable.

Table 4.
Effective exchange integrals (J_{ab}) (cm^{-1}) of *m*- (**3m**) and *p*- (**3p**) (phenylen) bis (methylen) calculated by several methods.

Methods	3m	3p
CASCI[b)](CASSCF)[4,4]	188(170)	–1992
CASCI[b)](CASSCF)[8,8]	507(519)	–1453(–1653)
CASPT2[4,4]	406	
UHF	2342	–3753
UCCSD(T)	402	–1895
UBLYP	418	–1981

a) 4-31G basis set was used, b) UNO was used.

$$\left[\hat{T}+\hat{V}_{en},\hat{V}_{ee}\right]\neq 0 \tag{30}$$

Therefore variations of these terms are not independent in general. In this sense, the HDFT solutions intend to simulate the energy partitionings of the exact solution by using the appropriate mixing of exchange correlation functions. This is the reason why pure DFT can provide reasonable ST gap even though it does not involve explicitly Coulombic exchange operator $\hat{K}$. While the CASSCF [*n*,*n*] approaches to the exact one by expanding active orbitals, together with use of flexible basis sets. Unfortunately we cannot obtain the very accurate solution satisfying both the virial theorem and full correlation corrections for **2** to determine exactly the relative weight of contributions of the Hund and LSR effects.

The reliable wavefunctions of meta (*m*) and para (*p*) - phenylene bis methylene (**3m** and **3p**) are available by our group [28, 29]. Table 4 summarizes the calculated J_{ab} values in the Heisenberg model (see Eq.(11)). For this purpose, we have used our general formula [30]

$$J_{ab}=\frac{{}^{LS}E(X)-{}^{HS}E(X)}{{}^{HS}\langle S^2\rangle(X)-{}^{LS}\langle S^2\rangle(X)} \tag{31a}$$

where ${}^{Y}E(X)$ and ${}^{Y}\langle S^2\rangle(X)$ denote, respectively, the total energy and total spin angular momentum obtained by the computational method *X* for the spin state *Y*. It is noteworthy that both BS and SA methods such as HDFT (UHF) and CASCI (CASSCF) can be used as *X* in our scheme.

The J_{ab} value for **3m** by CASCI (CASSCF) [4,4] is 188 (170) (cm^{-1}), which is responsible for direct effective exchange interaction between triplet methylene groups. It becomes 507 (519) after the inclusion of the spin polarization (SP) and nondynamical correlations by CASSI (CASSCF) [8,8]. The overestimation of the J_{ab} value is further improved by the addition of dynamical correlation to CASSCF[4,4] by the second-order perturbation (PT2) method. Indeed, it is reduced to 406 cm^{-1} by CASPT2. Interestingly, the J_{ab}-value by CASPT2 [4,4] can be reproduced by the BS UHF coupled cluster (CC)

SD(T) method: UCCSD(T), and also by UBLYP. Judging from the J_{ab}-values by CASSCF [4,4] and CASPT2 [4,4], both classical Hund and LSR rules are almost equally contributable to the effective exchange interaction of **3m**. On the other hand, the direct exchange interactions almost determine the magnitude of the J_{ab}-value of the para-isomer **3p**, where a closed-shell structure is available.

4.2 Intramolecular Spin Alignment Rules

Here, let us consider applicability of the Hund, NBMO and LSR rules on the basis of the numerical results in Table 2-4. The Coulombic exchange integral K_{AB} between NBMOs for alternate hydrocarbones is approximately given by

$$K_{AB} = \sum_j c_{Aj}^2 c_{Bj}^2 U_j \tag{32}$$

where χ_A and χ_B are expressed by the NBMO coefficients on carbon atom j,

$$\chi_A = \sum_j c_{Aj}\varphi_j, \quad \chi_B = \sum_j c_{Bj}\varphi_j . \tag{33}$$

Therefore the K_{AB} integral is zero if the product $c_{Aj}^2\, c_{Bj}^2$ for NBMO is zero at all the site j. For example, the product is indeed zero for 3,3'-biphenylene-bis(methylene) (**4**)

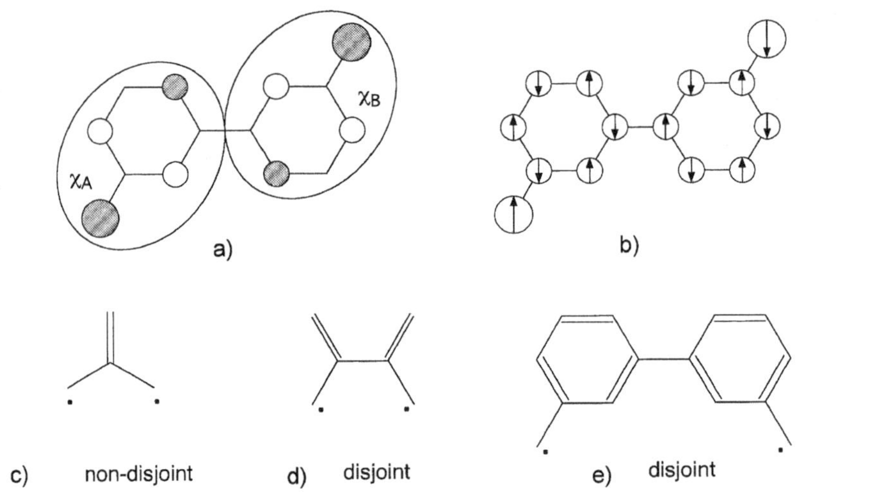

Fig. 9. (A) SOMO of 3,3'-biphenylene bis(methyl) (**4**), (B) Spin vector model of **4**, and (C) joint and disjoint orbital pairs.

because NBMO χ_A and χ_B are localized on the left and right benzyl radicals, respectively, as shown in Fig. 9(a). The ST gap of **4** is really zero by CASSCF [2,2], though the simple Hund rule and NBMO rule predict the triplet (high-spin) ground state. Therefore several methods [31-33] such as the joint-disjoint (JDJ) classification of the overlapping have been presented for further refinements of the Hund and NBMO rules as illustrated in Fig. 9(c-e).

The CASCI / 4-31G by the use of 8-active MO and 8-active electrons [8,8] of **4** indicated that the J_{ab} value for **4** is -254 cm^{-1}, showing the anti-ferromagnetic effective exchange interaction. For this calculation, we have used Eq.(31). In order to understand the LS (singlet) ground state of **4**, spin vector model has been introduced to express the SP plus EC effects evaluated by CASCI [8,8] as shown in Fig. 9(b). As shown in Table 2, our spin vector model is responsible for a simplified expression of the less screening rule. Our expression is different from the VB-type (*, non * representation) one in the physical content, though both predict similar spin alignment rules for alternant hydrocarbons. However, it is noteworthy that the VB model is hardly applicable to cation (or anion) radicals as well as radicals involving hetero atoms. In a previous review 11, we have examined a lot of polycarbon radicals from the viewpoints of both Hund (NBMO) and spin vector models.

4.3 Spin Fluctuation and Noncollinear Magnetism

Recently, spin fluctuation systems [5] have been accepted great interest from several fields of spin science. For example, let us consider triangular cluster of triplet methylene $(CH_2)_3$. The UHF and UDFT solutions with axial, i.e., one-dimensional (1D) spin modulation, exhibits a spin frustration since triplet (parallel)-type spin pair is formed in three degenerated 1D spin configurations in Fig. 10(a). Anderson 34 has discussed the resonating VB (RVB) representation of this spin frustration system. The triangular spin alignment with 2D spin configuration in Fig. 10(b) is feasible to maintain the singlet-type spin alignment (correlation) [5]. This spin structure can be described by general spin orbitals (GSO) defined [35] by

$$\eta_i = \eta_i^+ \alpha + \eta_i^- \beta = \begin{pmatrix} \eta_i^+ \\ \eta_i^- \end{pmatrix} \tag{34}$$

where $\eta_i^{\pm}$ denote the up (α)- and down (β)-spin components, respectively.

The J_{ab} values for the spin frustrated system can be obtained by using Eq.(31), where GSO HF and GSO HDFT solutions are utilized for LS. For example, the J_{ab} values for $(CH_2)_3$ are –746 (R = 1.7Å), –478 (R = 1.8Å) and –300 (R=1.9Å) (cm^{-1}) by GSO SVWN5 / 4-31G method. Similarly, they are –2828, –2112 and –761 (cm^{-1}) for $R_{C\text{-}N}$ = 1.7, 1.8 and 1.9 (Å), respectively, in the case of $N(CH_2)_3$ in Fig.10(c). The negative J_{ab}-values indicate that the LS doublet state is the ground state even for the D_{3h} geometry of these species. Of course, the quantum behaviors 34 of the spin system are

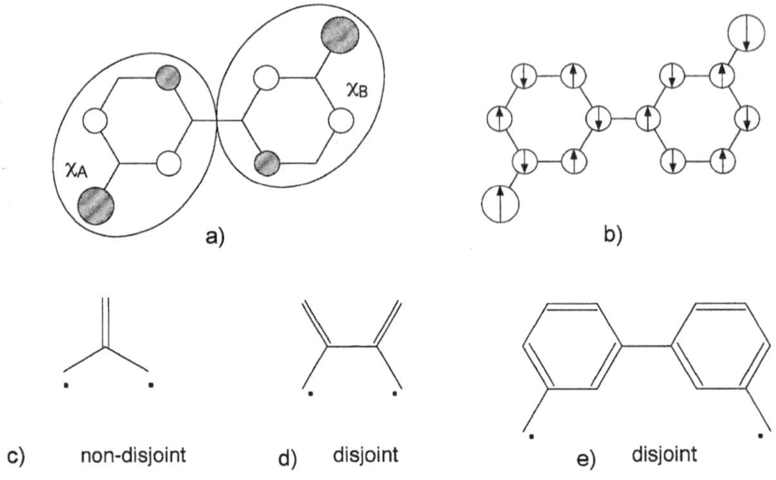

Fig. 10. Axial low-spin (A), helical (B, C) and axial high-spin (D) configuration of organic polyradicals.

examined by the exact dianonalization of the Heisenberg Hamiltonian involving the calculated J_{ab} values [36, 37]. On the other hand, the J_{ab}-value for $Ph(CH_2)_3$ in Fig.10D is calculated to be 1452 (cm^{-1}) by GSO SVWN5 / 4-31G, showing the HS (S = 3/2) ground state. Chemical synthesis of large triangular organic molecules is really desirable for molecular material science.

5 Intermolecular Spin Alignment Rules

5.1 Perturbation and Variational Theories

Intermolecular spin alignment rules are examined on the basis of both perturbation and variation theories in this section [6]. Here, let us consider the well-known syn- and anti-dimer of benzyl radical (**5**) or phenyl carbene (**6**) as illustrated in Fig. 11. Since the intermolecular distance (R) is long for $(\mathbf{5})_2$, the perturbation theory is a challenging approach. In fact, Yoshizawa and Hoffman [38] have performed the perturbation calculation of $(\mathbf{5})_2$ by using their own Eq.(35) as

$$J_{AB} = K_{AB} - \frac{(\varepsilon_a - \varepsilon_b)^2}{G_{AA} - G_{AB}} \tag{35}$$

where G_{AA} and G_{AB} denote the intramolecular and intermolecular Coulombic repulsion integrals, and ε_b and ε_a mean the orbital energies of the bonding and antibonding molecular orbitals of dimer by the extended Hückel MO (EHMO) method. They have concluded that the J_{AB}-value becomes positive (ferromagnetic) because the intermolecular Coulombic exchange integral (K_{AB}) is stronger than the antiferromagnetic orbital interaction term (second term in Eq.(35)). This implies that the intermolecular Hund rule is applicable to (**5**)$_2$ because the orbital splitting second term is weak.

We have performed the CASCI and CASSCF calculations of (**5**)$_2$ at three different R-values [39, 40]. Table 5 summarizes the calculated results. The CASCI [2,2] by the use of the UHF natural orbitals (UNO) (see Eq.(16)) [41] have shown that the J_{AB}-value becomes negative (antiferromagnetic). In the case of CASCI [2,2], the orbital relaxation term is not taken into account for both the spin states. Therefore the J_{AB} value is expressed by

$$J_{AB}(SOMO - SOMO) = K_{AB} - const \times (S_{AB})^2 \tag{36}$$

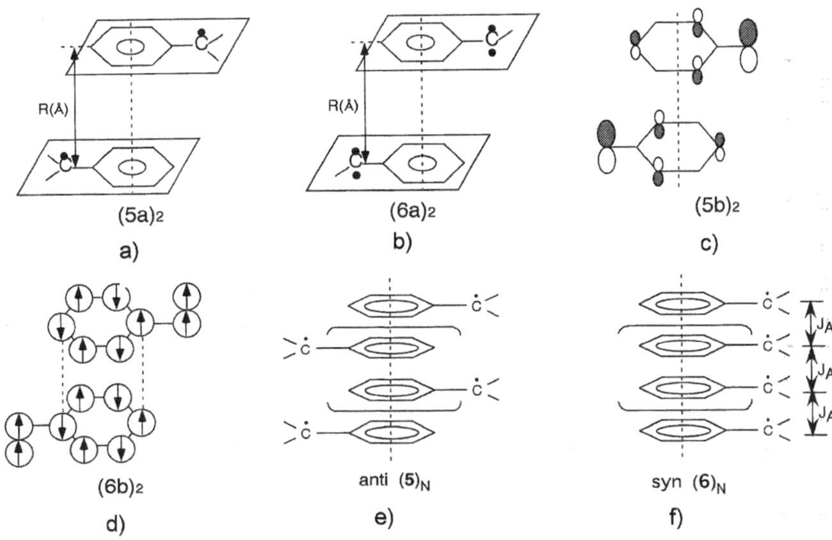

Fig. 11. Sandwich dimer of benzylradical (a), phenyl carbene (b), SOMO-SOMO interaction (c) and Spin vector model (d). Syn- and anti-dimer of **5** are shown by e and f, respectively.

Table 5
SOMO-SOMO effective exchange integrals (J_{ab}) (cm^{-1}) of benzyl radical (**5**) and phenyl carbene (**6**) dimers calculated by several methods$^{a)}$.

system	R(Å)	CASSCF[m,m]	UNO CASCI[m,m]	DNO CASCI[m,m]
	3.0	−75.0	−84.6	−53.3
(**5**)$_2$ $^{b)}$	3.4	−31.9	−41.5	−22.0
	3.8	−10.3	−17.6	−5.01
	3.0	−12.2	−16.7	−12.8
(**6**)$_2$ $^{c)}$	3.4	−5.33	−8.44	−5.82
	3.8	−1.66	−3.75	−1.98

a) anti-type stacking, b) m = 2, c) m = 4

where the second term denotes the kinetic exchange (KE) interaction term between singly occupied MO (SOMO) of **5** (see Eq.(7c)). Therefore the KE term is often referred to as the orbital (SOMO)-overlap (OO) interaction term [10, 11]. The CASCI [2,2] indicates that the SOMO-SOMO overlap term overweights the potential exchange (K_{AB}) term. In fact, the orbital overlap for $(\mathbf{5})_2$ is not zero as illustrated in Fig. 11(c). This situation is not altered even if the natural orbitals (DNO) [42] of UB3LYP are employed as shown in Table 5.

Next, in order to examine the contribution of orbital relaxation effects, the CASSCF [2,2] calculations of $(\mathbf{5})_2$ have been carried out. The sign of J_{AB} remains negative, though its magnitude is a little reduced as compared with the corresponding UNO CASCI [2,2] results. Thus, the orbital relaxation effects do not change the whole situation since the intermolecular distance (R) is long as in the case of methyl radical dimer (R > 2.0Å) in Fig. 7(c). The above discussions are wholly applicable to the dimer of phenyl carben $(\mathbf{6})_2$ in Fig. 11b, though the CASCI[4,4] and CASSCF[4,4] are responsible for the SOMO-SOMO interactions. So there is no need to repeat them. These example indicate that the orbital-overlap (OO) rule is predominant if the orbital overlap (S_{AB}) does not disappear because of symmetry.

5.2 Spin Polarization and Correlation Effects

The preceding CASCI and CASSCF results for $(\mathbf{5})_2$ and $(\mathbf{6})_2$ indicate that spin polarization (SP) and electron correlation (EC) effects should play substantial roles to bring about intermolecular ferromagnetic exchange interactions in these model clusters. Then the CASCI [10,10] have been carried out for $(\mathbf{5})_2$. The calculated JAB values are summarized in Table 6. The J_{ab} values by CASCI [10,10] are positive (ferromagnetic)

Table 6.
Effective exchange integrals (J_{ab}) (cm^{-1}) of benzyl radical (**5**) and phenyl carbene (**6**) dimers calculated by CASCI methods.

R(Å)	$(\mathbf{5})_2$ a)	$(\mathbf{6})_2$ b)
3.0	116.6	28.05
3.2	47.34	12.65
3.0	6.738	5.583
4.0	0.065	0.355

a) CASCI[10,10], b) CASCI[12,12]

Table 7.
Effective exchange integrals (J_{ab}) (cm^{-1}) of cluster spin systems.

N	anti-$(\mathbf{5})_N$ d)	syn-$(\mathbf{6})_N$ d)
2	103	−1034
6	142	−1160
8	167	−1242
10	197	−1340

after the SP and EC corrections. Similarly the CASCI [12,12] predicts the ferromagnetic interaction for phenyl carbene dimer (**6**)$_2$. These results indicate that the SP and EC effects for phenyl group play important roles for the intermolecular ferromagnetic coupling. The intermolecular spin vector model can be applicable to grasp essential features of the SP plus EC effects as shown in Fig.11(d). The post HF methods including these effects is necessary to obtain correct answer for the origin of the ferromagnetic couplings.

However, the CASCI method cannot be applicable to clusters of **5** or **6**. Therefore we have performed the UB2LYP calculations of syn- and anti-stacking clusters of **5** as shown in Fig. 11(e) and f [40]. To this end, Eq.(31) is extended to N-site clusters (each spin size = s)

$$J_{AB} = \left[{}^{LS}E(X) - {}^{HS}E(X)\right]/\Delta \tag{31b}$$

where

$$\Delta = {}^{HS}\langle s^2\rangle(X) - {}^{LS}\langle s^2\rangle(X) - s\frac{(N-2)^2}{N} \quad (N=2n) \tag{31c}$$

The calculated J_{AB}(N)-values are given in Table 7. They are positive (ferromagnetic) for anti-type clusters because of the SP plus EC effects. While they are largely negative (antiferromagnetic) for syn-type clusters because of the large SOMO-SOMO overlap. The general tendency is not altered by formation of clusters (**5**)$_N$. The situation is the same for (**6**)$_N$. Thus the HDFT method can reproduce CASCI results for dimer, being applicable to N-site clusters ($N \leq 50$) (see Fig. 8).

5.3 Spin Vector Model and McConnell Model

The post HF calculations show that carbon polyradical species belong to the intermediate correlation regime. This implies that both spin polarization (SP) and electron correlation (EC) effects play crucial roles to determine electronic structures of these species. Now we have obtained reasonable answer for McConnell's question6 for benzyl radical dimer (**5**)$_2$ in Fig.11(a): EHMO perturbation theory supports the classical Hund rule, while BS MO calculation is consistent with the McConnell model [43]. The CASCI and CASSCF calculations are not inconsistent with the BS MO model and McConnell model [44], when the latter model emphasize the SP effect. Electron correlation (EC) effect is also important for ferromagnetic spin coupling of (**5**)$_2$ even in qualitative sense. Further examination of our conclusion by using very accurate wavefunction is an interesting future problem.

The spin vector model in Fig. 11(d) is often utilized in our papers for pictorial understanding of spin alignment rules which are based on the SP and EC effects. It is a classical vector representation of the less screening rule (LSR) for correlated electron motions in polyradical species. In this review, we do not discuss the spin vector model (SVM) followed by the magnetic group [45]

$$M = H + T(G - H), \tag{32}$$

where M denotes the magnetic symmetry, and T is the time-reversal operation. A spatial symmetry is expressed by G, and H is a subgroup of G. The SVM is replaced by the spin correlation function (SCF) in the case of CAS-based and exact methods [46].

$$K_2(r_1,r_2) = \int \mathbf{S}(1)\cdot\mathbf{S}(2)\rho_2(r_1,r_2)d\tau_1 d\tau_2 \tag{33}$$

The SVM (SCF) 4 predicts that the ground spin structure tends to have fully-ordered magnetic symmetry (maximum magnetic symmetry rule (MMSR)). Several examples are shown in Ref. 11. We do not touch extra hole or electron doping problems concerning magnetic conductivity and superconductivity 11. Therefore the double exchange and related problems are not discussed here. Theoretical studies on magnetic conductivity or magnetic metals have been performed in our other papers [47].

6 Conclusion

The nature of effective exchange interactions in atom, molecule and cluster is elucidated on the basis of analytical, very accurate *ab initio* wavefunctions, which satisfy the Hellmann-Feynman and virial theorems. The reference density map for them classifies important roles of nondynamical and dynamical correlation corrections for the mean-field (RHF) solution. This analysis in turn clarifies physical foundation and applicability of density functional theory (DFT) and hybrid density functional theory (HDFT):

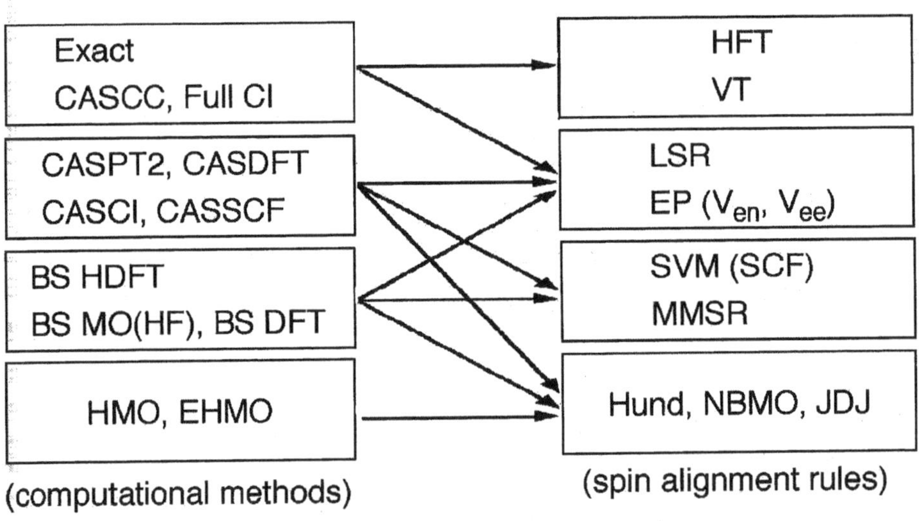

Fig. 12. Computational schemes and spin alignment rules (JDJ: Joint-Disjoint, HFT: Hellmann -Feynman theorem, VT: virial theorem, EP: energy partitioning, SVM: spin vector model, SCF: spin correlation function, MMSR: maximum magnetic symmetry rule, LSR: less screening rule).

note that pure DFT does not explicitly involve the exchange operator $\hat{K}$. Our computational scheme of the effective exchange integrals of Heisenberg model is applicable to symmetry-adapted (SA) CASCI, CASSCF and CASPT2 but also broken-symmetry (BS) HF, DFT and HDFT methods. Scope and limitation of several spin alignment rules such as Hund and less screening rules are examined on the basis of the sign and magnitude of J_{ab} determined by *ab initio* computational results. Carbon-based poly-radicals belong to the intermediate correlation regime, and therefore spin polarization (SP) and electron correlation (EC) effects play important roles for determination of their electronic structures. The energy partitioning of T, V_{en}, V_{ee} and V_{nn} by the post HF methods are performed to elucidate the origin of the effective exchange interactions (J_{ab}). These analyses support the less screening rule (LSR) [22] and maximum magnetic symmetry rule (MMSR) [4] in carbon-based polyradicals. They have also confirmed our previous proposals for intermolecular ferromagnetic couplings: (i) reduction of the SOMO-SOMO overlap (OO) term (see eq. (36) and Fig. 11C) to suppress antiferromagnetic interaction and (ii) enhancement of ferromagnetic spin density product (SPD) term (see Fig. 11D) via spin polarization (SP). Our spin alignment rules [52] are found to be consistent with the experiments and other theoretical results [6, 53-56]. Several examples are also presented in refs 10, 11 and 15. Figure 12 illustrates mutual relationships for computational methods and spin alignment rules. Theory of molecular magnetism of carbon-based polyradicals is closely related to the nature of labile chemical bonds. Therefore it is one of fundamental and substantial problems in the intersection area of chemistry and physics.

References

1. P.J. Hay, J.C. Thibeault, and R. Hoffmann, J. Am. Chem. Soc. 97 (1975) 4884.
2. J. Katriel and R. Pauntz, Adv. Quant. Chem. 10 (1975) 143.
3. K. Yamaguchi, Chem. Phys. Lett. 33 (1975) 330.
4. K. Yamaguchi, Chem. Phys. Lett. 35 (1975) 230.
5. K. Yamaguchi, Chem. Phys. Lett. 30 (1975) 288.
6. H.M. MacConnel, in: Magnetic properties of organic materials, Ed. P.M. Lahti (1999) 1, Marcel Dekker Inc, New York.
7. K. Rundenberg, Rev. Mod. Phys. 34 (1962) 326.
8. M.J. Feinberg, K. Rundenberg, and E.L. Mehler, Adv. Quant. Chem. 10 (1975) 27.
9. R.S. Mulliken, J. Chim. Phys. 46(1949) 497.
10. T. Kawakami et al., in: Molecular-based magnetic materials, Eds. M.M. Turnbell, T. Sugimoto, and L.K. Thompson, ACS Symp. Series 644 (1996) 30.
11. K. Yamaguchi et al., in: Molecular Magnetism, Eds. K. Itoh and M. Kinoshita (2000) 9, Kodansha Gordon and Breach Pub., Tokyo.
12. A.A. Ovchinnikov, Theoret. Chim. Acta. 47 (1978), 297.
13. D.J. Klein et al., Theoret. Chim. Acta. 69 (1986) 393.
14. W. Heisenberg, Z. Phizik 49 (1928) 619.
15. K. Yamaguchi et al., in: Functionality of Molecular Systems, Ed. S. Nagashima (1998) 67, Springer, Tokyo.
16. T. Kawakami et al., Polyhedron, 22 (2003) 2051.
17. S. Yamanaka et al., Int. J. Quant. Chem. in press.

18. K. Yamaguchi, Self-consistent-field, Eds. R. Carbo and M. Klobukowski (1990) 727, Elsevier, The Netherlands.
19. K. Takatsuka, K. Yamaguchi and T. Fueno, Theoret. Chim. Acta. 48 (1978) 175.
20. K. Yamaguchi et al., Int. J. Quant. Chem. 90 (2002) 370.
21. J. Katriel, Theoret. Chim. Acta. 26 (1972) 163.
22. R.J. Boyd, Nature 310 (1984) 480.
23. E.R. Devidson, J. Chem. Phys. 42 (1965) 4199.
24. K.V. Darvesh et al., J. Chem. Phys. 94 (1990) 3480.
25. K.V. Darvesh et al., Chem. Phys. 157 (1991) 99.
26. K.V. Darvesh and R.J. Boyd, J. Chem. Phys. 87 (1987) 5329.
27. K.V. Darvesh and R.J. Boyd, J. Chem. Phys. 90 (1989) 5638.
28. M. Mitani et al., J. Chem. Phys., 113 (2000) 4035.
29. M. Mitani et al., J. Chem. Phys. 113 (2000) 10486.
30. K. Yamaguchi, Y. Takahara, and T. Fueno, in: Applied Quantum Chemistry, Eds V.H. Smith et al. (1986) 155, D. Reidel, Lancaster.
31. W.T. Borden and E.R. Davidson, J. Am. Chem. Soc. 99 (1977) 4587.
32. W.T. Borden, H. Iwamura, and J.A. Berson, Acc. Chem. Res. 27 (1994) 109.
33. J.A. Berson, in: Magnetic properties of organic materials, Ed. P.M. Lahti (1999) 7, Marcel Dekker Inc, New York.
34. P.W. Anderson, Mat. Res. Bull 8 (1973) 153.
35. K. Yamaguchi and H. Fukutome, Prog. Theoret. Phys. 54 (1975) 1599.
36. M. Nishino et. al., Phys. Rev. B61 (2000) 4033.
37. M. Nishino et. al., Phys. Rev. B62 (2000) 9463.
38. K. Yoshizawa and R. Hoffmann, J. Am. Chem. Soc. 117 (1995) 6921.
39. R. Takeda et al., Synthetic Metals 133-134 (2003) 593.
40. Y. Kitagawa et al., Int. J. Quant. Chem. in press.
41. K. Yamaguchi, Int. J. Quant. Chem. in press.
42. K. Yamaguchi, Int. J. Quant. Chem. S14 (1980) 269.
43. K. Yamaguchi, Chem. Phys. Lett. 68 (1979) 269.
44. H. M. McConnell, J. Chem. Phys. 39 (1963) 1910.
45. K. Yamaguchi et al., Chem. Phys. Lett. 66 (1979) 395.
46. K. Yamaguchi, Chem. Phys. 29 (1978) 117.
47. K. Yamaguchi et al., Synth. Metals 19 (1987) 81.
48. K. Yamaguchi, H. Fukui and T. Fueno, Chem. Lett. (1986) 625.
49. K. Yamaguchi, et al. Chem. Lett. (1986) 629
50. K. Yamaguchi, et al. Synth. Metals 19 (1987) 87.
51. K. Yamaguchi, H. Namimoto and T. Fueno, Mol. Cryst. Liq. Cryst. 176 (1989) 151
52. K. Yamaguchi, Y. Toyoda and T. Fueno, 159 (1989) 459.
53. A. Izuoka et al. J. Am. Chem. Soc. 107 (1985) 1786
54. T. Sugawara et al. J. Am. Chem. Soc. 108 (1986) 4272
55. K. Awaga, T. Sugano and M. Kinoshita, 41 (1987) 540.
56. Experimental and theoretical results in this book.

Carbon-Based Magnetism
T. Makarova and F. Palacio (Editors)

10

Magnetism of Fullerene Charge-transfer Complexes

A. Omerzu[1] and M. Tokumoto[2]

[1]*Jozef Stefan Institute, Jamova 39, 1000 Ljubljana, Slovenia*
[2]*Nanotechnology Research Institute, Natl Inst. of Advanced Industrial Science and Technology (AIST), 1-1-1 Umezono, Tsukuba, Ibaraki 305-8568, Japan*

1 Introduction

The research on magnetic properties of fullerene charge-transfer (CT) complexes was sparked in 1991 when Fred Wudl's group in Santa Barbara, in their course of studying a reduction of fullerenes with strong organic donors, discovered a compound tetrakis(dimethylamino)ethylene-C_{60} (TDAE-C_{60}), which surprisingly showed a ferromagnetic transition at 16K [1]. That was a temperature, which exceeded Curie temperatures of any other pure organic material known so far by more then order of magnitude. Since 1991 a lot of experimental and theoretical work has been done towards understanding of the ferromagnetic ordering in TDAE-C_{60}.

The discovery of TDAE-C_{60} motivated researchers from several laboratories around the world to try to synthesize new organic, fullerene-based ferromagnets. Their approaches could be divided into four main groups: (1) reduction of higher fullerens with TDAE, (2) reduction of C_{60} with different organic or organometallic donors, (3) functionalization of C_{60} and subsequent doping and (4) complexing C_{60} with rare earth elements. Each of those approaches and their results will be presented in following sections.

2 TDAE-C_{60}

In the original work [1] the authors synthesized TDAE-C_{60} by adding TDAE (liquid) to a toluene solution of C_{60}. The result was black, microcrystalline powder, which precipitated from the solution almost instantaneously after adding of TDAE. The material was highly air sensitive. All manipulations had to be done in a glove box with an inert atmosphere. According to elemental analysis it was claimed that stoichiometry was 1:1.16, which was later proven to be false. TDAE and C_{60} react in a simple 1:1 ratio to form a charge transfer salt $TDAE^+C_{60}^-$. Measurements of temperature and field dependence of magnetisation demonstrated a clear transition to the ferromagnetic phase below T_c = 16K. The magnetisation increased abruptly below T_c and T_c increased with the measuring field. The field dependence of magnetisation below T_c displayed an *S* curve, characteristic to ferromagnets but without observable hysteresis. Magnetic susceptibility measurements at high temperatures ($T >$ 30K) revealed almost temperature-independent behavior, quite distinct from the Curie-Weiss behavior $\chi = C/(T-\theta)$, which is expected for systems of localised magnetic moments. Furthermore, the electric conductivity measured on a compressed pellet turned out to be quite high, 10^{-2} S·cm^{-1}. From those findings the authors concluded that TDAE-C_{60} might be an itinerant soft ferromagnet.

Soon after the discovery, the first X-ray diffraction study on TDAE-C_{60} was performed in Brookhaven National Laboratory [2]. It was demonstrated that TDAE and C_{60} crystallize in 1:1 stoichiometric ratio, indeed. The structure was determined to be *c*-centered monoclinic (space group *C2/m*). The intramolecular C_{60}-C_{60} separation is the shortest along the *c* axis (9.98 Å) and much greater in the *a-b* plane (10.25 Å). The two-fold axis of TDAE molecules also orients itself along *c* axis. From the structural parameters it could be concluded that TDAE-C_{60} has an anisotropic low-dimensional band structure, which could account for unusual electronic and magnetic properties.

Later, experiments of Tanaka's group showed a different picture [3]. From the *ESR* measurements they found that unpaired spins reside mostly on C_{60} molecules (the *g*-value 2.0003 of TDAE-C_{60} is much closer to 1.999 for electrochemically prepared C_{60}^- then 2.0036 for $TDAE^+$). They also predicted the Jahn-Teller distortion of C_{60}^- or, in other words, polaron formation in C_{60}. In contrast to original findings they observed the Curie-Weiss behavior of the magnetic susceptibility. From the shape of *M(H)* curves at 4.5K they calculated that magnetic moments at low temperatures form clusters with an average size of 170 spins per cluster. Evidently, TDAE-C_{60}, at least in its powder form, showed the properties of a superparamagnet.

At this point it was not yet clear if TDAE-C_{60} is a "proper" ferromagnet, but the following experiments cleared that issue. Suzuki et al. [4] and Dunsch et al. [5] found that TDAE-C_{60} could exhibit a hysteresis curve although with a very small coercive field, $H_c \sim$ 2 Oe and a remanent magnetic moment $M_r \sim 3\times10^{-4}$ emu. Suzuki et al. [4] also measured the *AC* susceptibility with a nonzero imaginary part, i.e. energy losses, related to the hysteresis. The final proof for the ferromagnetic state in TDAE-C_{60} came from a zero field μSR experiment by Lappas et al. [6]. The authors demonstrated the existence of an internal magnetic field of 68 Gauss with very broad distribution (48 Gauss) reflecting spatial inhomogenity.

The μSR experiment wasn't the only one which indicated inhomogeneities, structural or magnetic, in the system. Even in the early stage of the research some experimental results appeared to be in contradiction with the hypothesis of a long-range ferromagnetic ordering in TDAE-C_{60}. First of all, the temperature dependence of the *ESR* linewidth showed a relatively small line broadening and no frequency shift [7] with a non-exponential and very slow decay of the magnetization [8]. Both of these features are characteristic of random magnetic systems without a long-range order. Hence, it was suggested that TDAE-C_{60} could be a spin glass. The experiment by Mihailovic et al. [9] established a direct connection between orientational degrees of freedom of C_{60} molecules and magnetic interactions in the system. Thus, this hypothesis seemed to be even more plausible. This connection has been demonstrated also by theoretical calculations [10, 11]. By freezing C_{60} molecules in random orientations one can obtain a distribution of exchange interactions in the system, and consequently, magnetic disorder and frustration – two essential conditions for a spin glass. Later measurements of linear and non-linear susceptibilities [12] partly confirmed the spin glass hypothesis. The linear susceptibility χ_1 exhibited a broad peak centered at 10K and the non-linear susceptibilities χ_3, χ_5 and χ_7 diverged at the same temperature. The only feature which deviated from the spin-glass behavior was the absence of any shift of the peak position with frequency, which is characteristic for spin glasses. Obviously, TDAE-C_{60} was showing some characteristics of spin glasses and some of ferromagnets and it wasn't inconceivable that both phases coexist in a sample.

At that stage of research it was evident that lot of questions on the nature of TDAE-C_{60} (ferro)magnetism remained, which couldn't be answered by the experiments on powder samples. A reproducibility of physical properties for the powder samples was unsatisfactory even for the samples from the same group. Due to different sample purity, which was mainly affected by solvent inclusion into the crystal structure and oxygen contamination, as well as varying grain sizes in the powder samples, TDAE-C_{60} exhibited an inconsistent behaviour. The samples were changing their properties even by aging, usually by increasing their ferromagnetic signal. It was clear that availability of monocrystals was essential for making any progress in understanding of ferromagnetism in TDAE-C_{60}. The first attempt to grow single crystals was done by Suzuki et al. [13], but their crystals were rather small (0.3 mm in length, 0.05 mm in diameter) and of poor quality. So, it was impossible to determine a crystal structure. Finally, the crystals even didn't show the ferromagnetic transition.

A similar approach for the single crystal growth by diffusion method was adopted by one of the authors (A.O.). The improved method allowed one to reduce the diffusion rate by a smart design of the crystal-growing cell and a temperature control. The results were much better. First, the crystals showed a ferromagnetic transition at 16K. Second, it was possible to obtain high quality single crystals of millimeter size. That breakthrough paved the way for the further experiments.

Having macroscopic single crystals available, one of the first questions, which should be answered, is a mechanism of the electrical conductivity. Although microwave conductivity measured by Schilder et al. [14] and optical conductivity measured by Bommeli et al. [15] confirmed an insulating behavior of TDAE-C_{60} (Fig.1), their experiments were performed

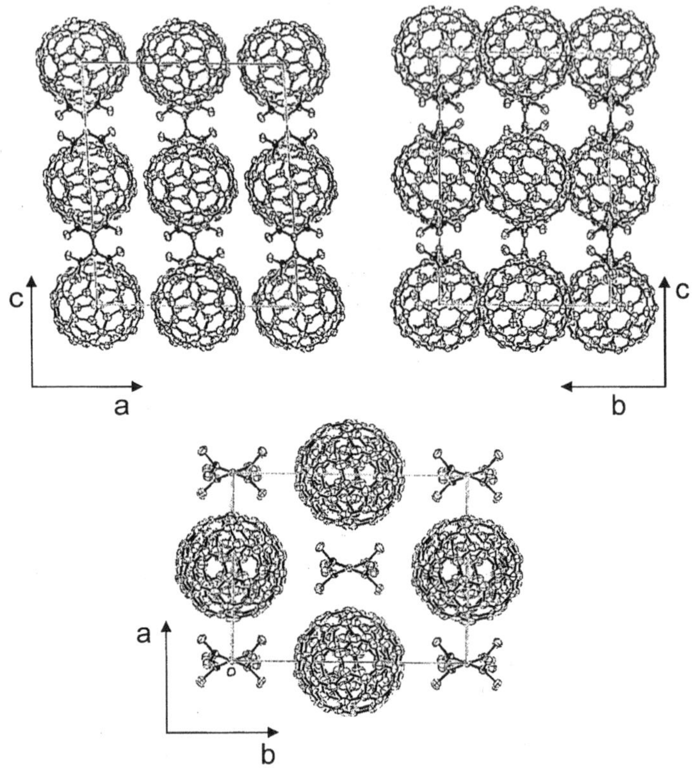

Fig. 1. Three different views on the TDAE-C_{60} crystal structure along *b*, *a* and *c* axis.

on microcrystalline samples with grains typically 10–100 nm in size. Large surface to volume ratio and air sensitivity of the material obviously complicated a clear discrimination between the insulating and the metallic intrinsic conducting state of TDAE-C_{60}. Omerzu et al. [16] circumvented that problem by measuring the *AC* and the *DC* conductivity on single crystals of TDAE-C_{60} with direct electrical contacts. They found that conductivity could be decomposed into two components: frequency-dependent, temperature-independent tunnelling and temperature-dependent phonon-assisted hopping. A dynamic, rotational disorder of C_{60} molecules plays a key role in the conductivity. The conductivity shows a crossover at $T_0 = 150$K (Fig. 2). It is a temperature, which separates the high-temperature orientationally disordered state from the low-temperature ordered state as it was demonstrated by ^{13}C NMR measurements (Fig. 3) [17]. The hopping mechanism prevails at $T > T_0$ where the hopping probabilities are higher, but at $T < T_0$ the tunnelling is a more efficient conducting channel.

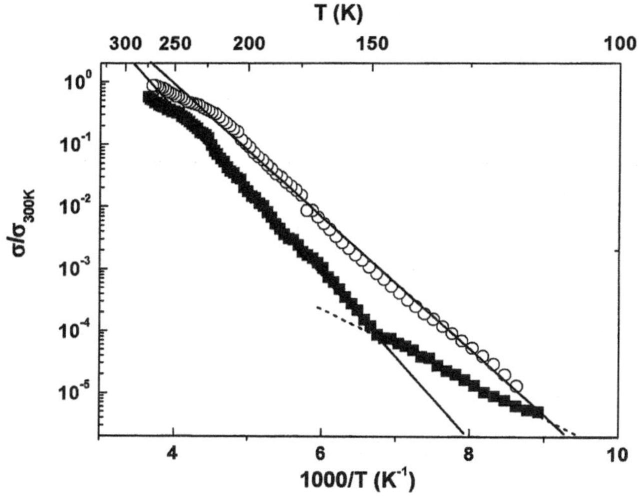

Fig. 2. The *DC* conductivity of TDAE-C_{60} single crystal as a function of temperature. The full squares were measured for cooling at rate 0.1 K/min, while the open circles were measured in near-quench conditions, 33 K/min.

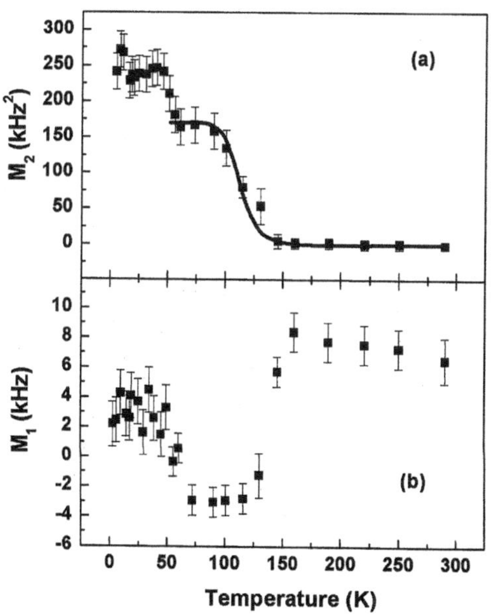

Fig. 3. Temperature dependence of the second (a) and the first (b) moment of the ^{13}C NMR spectra in powdered TDAE-C_{60}.

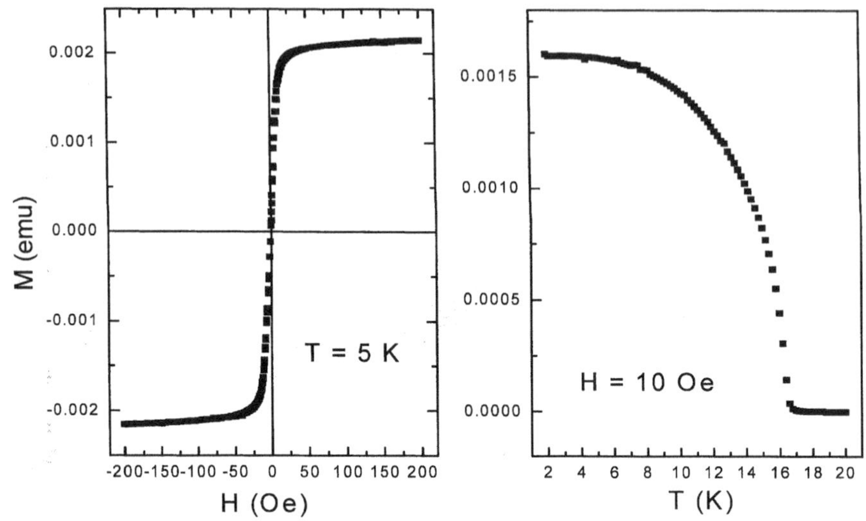

Fig. 4. The field dependence (left) and the temperature dependence (right) of magnetisation of α-TDAE-C_{60}.

Single crystals, in contrast to powder samples, have reproducible physical properties and offer a possibility for another intricate property of TDAE-C_{60} to be also explained. Namely, the powder samples frequently showed much lower magnetization as expected, and even worse, its value for a particular sample changed with aging. To resolve that intricacy, Mrzel et al. [18] chose TDAE-C_{60} crystals grown at 10°C. As-grown samples showed no ferromagnetic signal at low temperatures. Samples were treated in several heating cycles at temperatures between 50°C and 110°C. After each heating cycle the temperature dependence of the *ESR* signal was measured. A sharp increase in the intensity of the ferromagnetic signal was observed after the sample was treated at 70°C or higher. The ferromagnetic signal eventually disappears when the sample was heated above 100°C. From then on it has become clear that TDAE-C_{60} can exist in at least two crystal modifications: the usual or α-TDAE-C_{60} which has a ferromagnetic phase below 16K (Fig. 4) and newly discovered α'-TDAE-C_{60} modification without the ferromagnetic phase. The α' modification is the metastable one and can be irreversibly transformed into the stable α modification by thermal treatment.

When the existence of two different modifications of TDAE-C_{60} was firmly established, researchers started with experiments, which would determined the nature of the two TDAE-C_{60} modifications magnetic ground states. Arcon et al. [19] measured a ferromagnetic resonance in α-TDAE-C_{60}. By using the low-field *ESR* technique they showed a nonlinear variation of the resonance frequency ν with resonance field H and proved the existence of long range magnetic order. From the ν versus H dependence they were able to rule out an antiferromagnetic behavior as well as a paramagnetic or a spin-canted one. From an extremely low value of anisotropy field (29 Gauss) they concluded that α-TDAE-C_{60} is an example of an easy axis Heisenberg ferromagnet with the easy axis along crystallographic *c*-axis, the axis of the closest C_{60}^- approach.

Another insight to the nature of the ferromagnetic transition in α-TDAE-C_{60} offered measurements of the critical behavior near the ferromagnetic phase transition point by Omerzu et al. [20]. The authors presented the results of independent measurements of the static critical exponents for susceptibility $\chi(T) \sim (T/T_c-1)^{-\gamma}$, spontaneous magnetization $M_s \sim (1- T/T_c)^{\beta}$ and critical isotherm $H \sim M^{\delta}$ in the vicinity of the transition temperature T_c. The obtained results $\gamma = 1.22 \pm 0.02$, $\beta = 0.75 \pm 0.03$ and $\delta = 2.28 \pm 0.14$ differed significantly from those expected for a 3D Heisenberg ferromagnet, $\gamma = 1.38$, $\beta = 0.36$ and $\delta = 4.8$. In addition, the exponents didn't obey the scaling relation $\gamma = \beta(\delta - 1)$. The authors found an explanation for such a discrepancy in a reduced effective dimensionality of the system caused by additional degrees of freedom coming from C_{60} molecular rotation. Those induce an important degree of randomness into the system and alter the nature of the ferromagnetic transition.

The presence of intrinsic randomness in α-TDAE-C_{60} was clearly demonstrated in measurements of linear and non-linear *AC* susceptibilities by Omerzu et al. [21]. It is known that for ferromagnetic systems with a relatively low degree of disorder in magnetic interactions a re-entrant spin glass (RSG) transition follows the ferromagnetic transition at a lower temperature $T_{RSG} < T_{FM}$. Measurements of odd and even harmonics of *AC* magnetic response in TDAE-C_{60} revealed an additional broad peak centered at 7K,

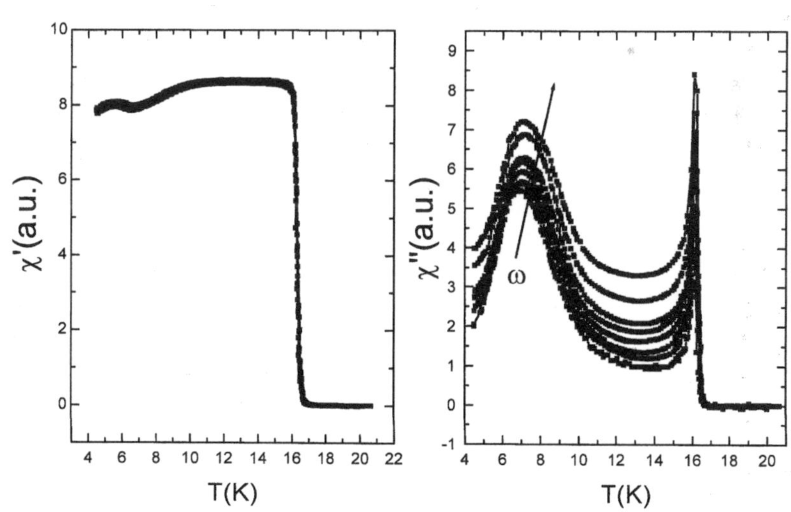

Fig. 5. The temperature dependence of the real (left) and the imaginary (left) parts of the linear *AC* susceptibility of α-TDAE-C_{60} measured at different frequencies between 33 Hz and 3 kHz.

but only for odd harmonics. The reason is that at the spin glass transition the time reversal symmetry is not broken in contrast to the ferromagnetic transition. Indeed, the measurements showed a divergence in both odd and even harmonics at $T = T_{FM}$. The frequency dependence of the peak in the imaginary part of the linear susceptibility at 7K gave an additional confirmation for the reentrant spin glass transition (Fig. 5). Thus, the riddle of the coexistence of the long-range ferromagnetic order and the short-range spin-glass disorder was resolved.

Magnetic properties of α'-TDAE-C_{60} are much simpler. By measuring the macroscopic magnetic properties i.e. the field (Fig. 6) and the temperature (Fig. 7) dependence of the magnetisation, Omerzu et al. [22] showed that α'TDAE-C_{60} is a paramagnet. The field-temperature dependence of the magnetisation exactly follows the Brillouin formula

$$M = N\mu \tanh(\mu H / k_B T),$$

where N is the number of spins in a sample and μ is the magnetic moment (in the case of $S = ½$, $\mu = \mu_B$ – the Bohr magneton). From the formula and the measured magnetisation and the mass of the sample it was possible to calculate an effective number of spins per formula unit N_{eff}. It turned out that N_{eff} equals the number of C_{60}^- ions in the sample. That notion immediately posed a question on missing contribution of $TDAE^+$ spins. Additional measurements at higher temperatures showed that N_{eff} increases from 1 to 2 per formula unit as temperature approaches 100K. A mechanism

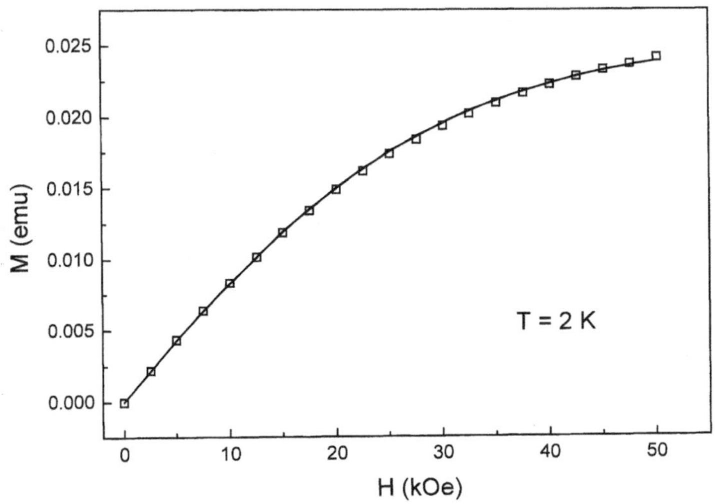

Fig. 6. Magnetisation of α'-TDAE-C_{60} as a function of an applied field at $T = 2$K. The solid line is the Brillouin function, $M = N\mu \tanh(\mu H / k_B T)$.

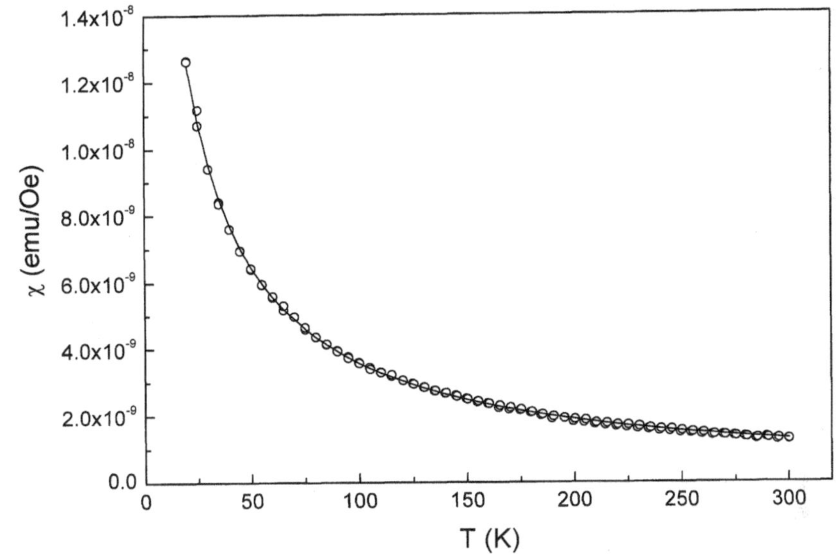

Fig. 7. Magnetisation of α'-TDAE-C_{60} as a function of temperature measured in an external field of 10 kOe. The solid line is the Curie-Weiss function, $\chi = C/(T-\theta)$.

which could account for such behavior might be an antiferromagnetic correlation among $TDAE^+$ spins, which causes the $TDAE^+$ subsystem of spins to "freeze out" from the bulk magnetization at low temperatures. It could also explain why antiferromagnetic correlations were frequently observed in measurements of high-temperature susceptibility. However, the role of the $TDAE^+$ spins in the TDAE-C_{60} magnetism remains provocative until now.

The fact that TDAE-C_{60} appears in two modifications with completely different magnetic properties was a clue for the microscopic understanding of its magnetism. An irreversible transition from the metastable, nonferromagnetic form α'-TDAE-C_{60} into the stable, ferromagnetic form α-TDAE-C_{60} can be performed in a controlled way. Usually, crystals of α'-TDAE-C_{60} are sealed into glass or quartz capillaries under He. The transformation takes place at 70°C. It needs 6 hours for completion and any excess heating can gradually degrade the samples. The whole procedure can be controlled by measuring magnetization curves at low temperatures before and after the annealing (Fig. 8).

Since the transformation occurs at mild conditions one would suppose only minor structural differences between α' and α TDAE-C_{60}. Narymbetov and co-workers [23] found the two modifications to be structurally indistinguishable at room temperature. Differences appeared at temperatures below 50K as additional diffuse lines in a diffraction

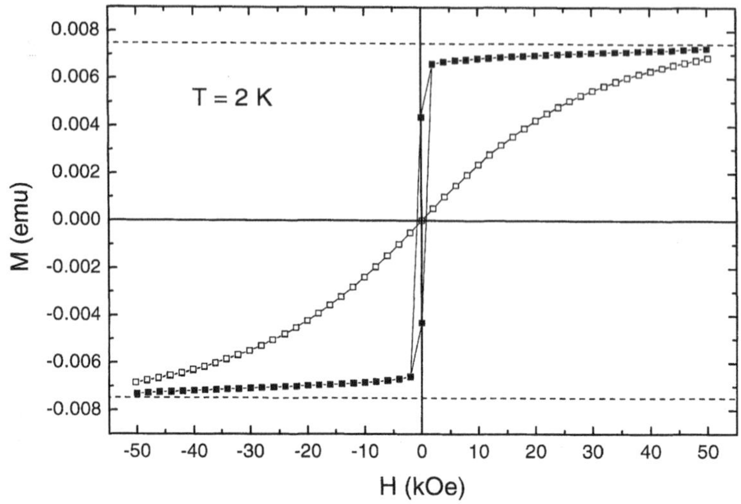

Fig. 8. The field dependence of the magnetisation of TDAE-C_{60} before (open squares) and after (filled squares) annealing. The measurements were performed on the same single crystal.

pattern. By further cooling down to 7K those lines disappeared in the case of paramagnetic (PM) α'-TDAE-C_{60} and in the case of ferromagnetic (FM) α-TDAE-C_{60} they coalesced into additional sharp diffraction spots. Those additional diffraction spots for α-TDAE-C_{60} correspond to a primitive unit cell, indicating that the crystal transformed from a *C*-centered structure to a primitive one. A refinement of the crystal structure was possible only after introducing additional degree of freedom - a relative rotation of C_{60} molecules around their three-fold axis by ±60° with 50% occupancy. In the PM sample, the relative C_{60} orientations are similar to those encountered in other C_{60} solids: the 6-6 double bond faces the center of the hexagon on the neighbouring molecule. In the FM sample on the other hand, a new orientation appears (±60°), which leads to three possible relative configurations of the C_{60}s. However, only configuration in which the neighbouring C_{60} are rotated relative to each other by ±60° is compatible with 50% occupancy of two rotations determined from the structural refinement. In that configuration, the double bond on one molecule faces the center of the pentagon of its neighbour, leading to C_{60} molecules ordered along the *c*-axis with alternating orientations (Fig. 9 and Fig. 10).

Kambe et al. [24] explored a temperature dependence of α-TDAE-C_{60} structure in more detail. They found that additional diffraction spots, which correspond to the primitive lattice, start to appear at 180K. They followed an increase of the new Bragg reflection as the samples temperature decreased. From the smooth increase of the intensity they concluded that the *C*-centered to primitive lattice structural transition in α-TDAE-C_{60} is of the second order.

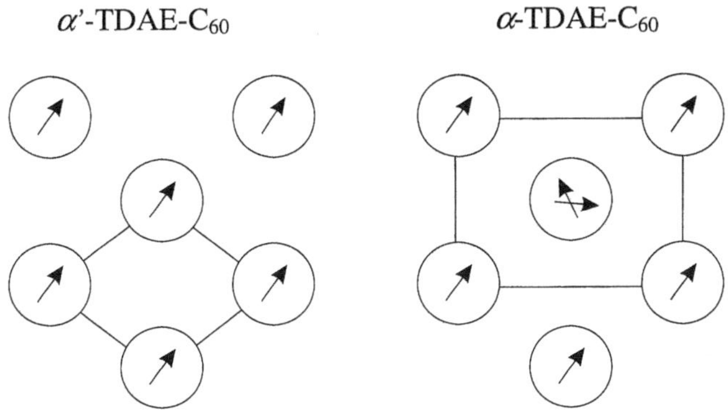

Fig. 9. A schematic diagram of the C_{60} molecular orientations in the *a-b* plane for the PM (left) and the FM (right) structures. The corresponding *C*-centered and primitive unit cells in the *a-b* plane are shown.

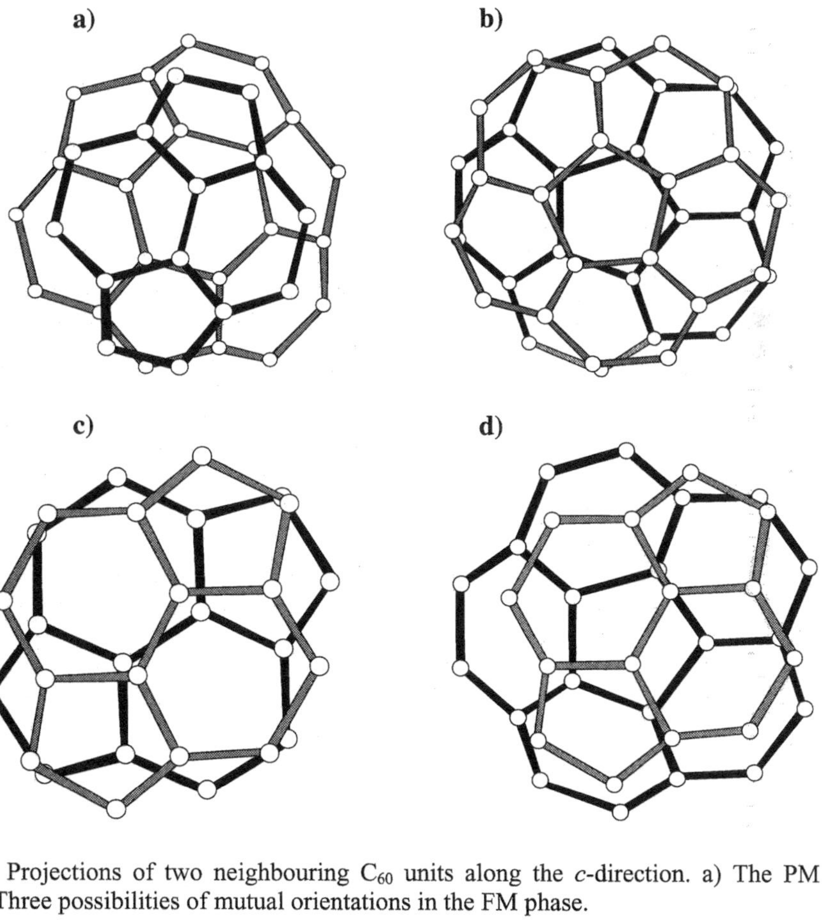

Fig. 10. Projections of two neighbouring C_{60} units along the *c*-direction. a) The PM phase. b) to d) Three possibilities of mutual orientations in the FM phase.

When it seemed that the relation between TDAE-C_{60} structure and its magnetic properties was satisfactory resolved a new discovery appeared. In their investigation of pressure effect in TDAE-C_{60} Mizoguchi et al. [25] found that (i) T_c of the ferromagnetic transition in α-TDAE-C_{60} decreased with increasing pressure and eventually disappeared at 9 kbar; and (ii) above 10 kbar at 300K the α-TDAE-C_{60} polymerised. The new, polymer β-TDAE-C_{60} phase consist of 1D C_{60} chains covalently interconnected by [2+2] cycloaddition in a similar way as in polymer o-Rb_1C_{60}. The new phase was stable even after pressure release. Garaj et al. [26] measured the temperature dependence of the β-TDAE-C_{60} *ESR* signal above 300K and found that β-TDAE-C_{60} depolymerised at 520K. This process was irreversible and the depolymerized samples showed magnetic properties similar to the ferromagnetic α-TDAE-C_{60}. There is another interesting property of β-TDAE-C_{60}: it is a paramagnet with the magnetic susceptibility showing the Curie-Weiss temperature dependence but with twice as many spins as in α-TDAE-C_{60}. From a shift in the *ESR* g-factor from 2.0005 in α-TDAE-C_{60} to 2.0028 in β-TDAE-C_{60}, which is much closer to 2.0036 in $TDAE^+$ cation radical, the authors concluded that the missing $TDAE^+$ spins revived in β-TDAE-C_{60}. $TDAE^+$ spins, which are mutually cancelled in α' and α-TDAE-C_{60} (the physical origin of that is still unknown) appeared to be localized and noninteracting in the polymer β-TDAE-C_{60}.

3 Higher Fullerenes Reduced with TDAE

In the early stage of research on TDAE-C_{60} magnetism it was interesting to compare it with higher fullerenes reduced with TDAE. It turned out that C_{70}, C_{84}, C_{90} and C_{96} readily form charge-transfer (CT) complexes with TDAE [27]. Their magnetic properties were characterised mainly with *ESR*. The g-factor and linewidth of the *ESR* line for all of TDAE-higher fullerene samples were almost temperature independent and the intensity of the *ESR* line, which is proportional to the spin susceptibility, followed the Curie low $I \sim C/T$. Hence, the TDAE-higher fullerenes CT complexes are simple paramagnets. From the g value the authors concluded that the unpaired spins reside mainly on fullerene units. Later, Oshima et al. [28] were able to crystallize TDAE-C_{70}-toluene complex and to obtain its crystal structure. In their samples C_{70} molecules formed singly bonded dimers. Magnetically, the crystals were paramagnets down to 1K. It was supposed that spins on C_{70} dimers form spin singlets, so that magnetic signal of TDAE-C_{70} could only originate from $TDAE^+$ cation radicals. That would suggest that also in other TDAE-higher fullerenes complexes $TDAE^+$ spins are not silent.

Tanaka et al. [29] succeeded to synthesise molecular alloys TDAE-$(C_{60})_{1-x}(C_{70})_x$ in a broad x range from 0.1 to 0.9. The low temperature magnetic properties of the alloys were monotonically changing from a ferromagnetic for TDAE-C_{60} to a paramagnetic for TDAE-C_{70}. Interestingly, the Curie temperature, T_c also linearly decreased as the content of C_{70} increased. This is consistent with the mean field result for T_c: $T_c = 2JzS(S+1)/3k_B$, where z is the effective number of nearest neighbours. This result is important because if the magnetic interactions in TDAE-C_{60} would be only along chains in the *c* direction, any amount of impurities (C_{70} substitutions) would break the ferromagnetic order.

4 Reduction of C_{60} with Different Organic or Organometallic Donors

C_{60} combines with many organic or organometallic donors to form charge-transfer salts. Here we will mention only those, which are relevant for magnetism of fullerene-based charge-transfer compounds.

Klos et al. [30] were stimulated by the discovery of ferromagnetism in TDAE-C_{60} to try reduction of C_{60} with other amines similar to tertiary amine TDAE. For that purpose they synthesised tertiary amines diazobicyclononene (DBN) and diazobicycloundecene (DBU). In contrast to TDAE, where its eight methyl groups sterically hinder direct reaction with C_{60}, DBN and DBU are not so well protected. DBN^+ and DBU^+ reacted with C_{60}^- forming covalent bonds and only a few percent of nominal spin survived. Nevertheless, those residual spins showed in the case of DBU-C_{60} a short range magnetic order, which evolves below 70K.

In 1994 Wang and Zhou [31] published the results of magnetic measurements on the charge transfer complex [1, 1', 3, 3'-tetramethyl-$\Delta^{2,2'}$-bi(imidazolidine)]$^+$-C_{60}^- (TMBI-C_{60}). The complex showed ferromagnetic behavior up to 140K with large hysteresis loop (coercive field 1000 Oe!). However, very soon Schilder et al. [32] showed that it was false. In fact, TMBI even hardly made a CT complex with C_{60} and resulting product was mainly diamagnetic (the *ESR* signal came only from impurities).

Otsuka et al. [33] used a variety of electron donors including aromatic amines, phenothiazines, phenazines, tetratianofulvalene derivatives and metallocenes to form charge-transfer complexes with C_{60}. Among them, only CT complexes with metalocenes: decamethyferrocene (Cp_2^*Fe), cobaltocene (Cp_2Co) and nicklocene (Cp_2Ni), showed ferromagnetic characteristics. All three complexes exhibited an *S*-shaped *M(H)* curve even at room temperature with a narrow hysteresis. Similarly, the charge transfer complex with 1,1'-biferrocene [34] showed the signs of ferromagnetism at 20K. Unfortunately, synthesis and magnetic properties of charge-transfer complexes of C_{60} with the metallocenes were not reproducible.

5 Charge Transfer Complexes of C_{60} Derivatives

Another approach in the synthesis of novel fullerene-based molecular magnets was based on functionalizing of C_{60} by covalently attaching different adducts to the fullerene cage. Subsequently, such derivatives of C_{60} would be combined with organic or organometallic donors to form charge-transfer complexes. The idea was to slightly alter the fullerene electronic properties, e.g. electronic affinity, as well as to hinder C_{60} rotational degrees of freedom.

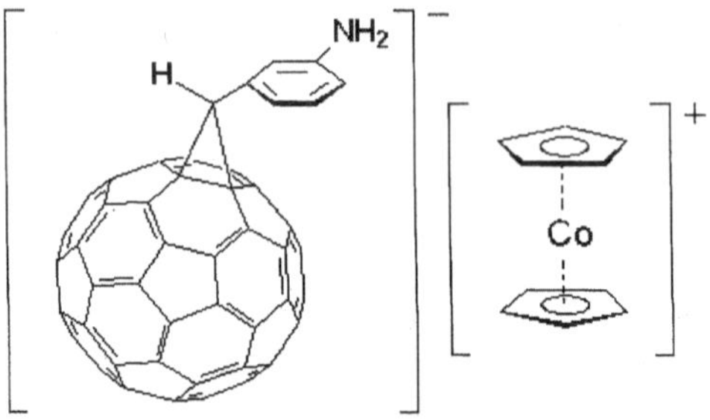

Fig. 11. Fullerene derivative 1-(3-aminophenyl)-1*H*-methanofullerene[C_{60}] (left) and cobaltocene Cp_2Co (right).

In 1994 Venturini et al. [35] presented the first successful synthesis of doped fullerene derivative ferromagnet. It was dinitro-spiromethanofullerene (C_{61}"No_2") doped with bicyclopentadiene cobalt (Cp_2Co or cobaltocene). It showed paramagnetic to ferromagnetic transition at 8K. Transition temperature was lower then in TDAE-C_{60} but it was encouraging sign for a further exploration. The second breakthrough happened in 1998 when Mrzel et al. [36] reported ferromagnetic transition in a cobaltocene doped C_{60} derivative at 19K - higher then in TDAE-C_{60}. The derivative was 1-(3-aminophenyl)-1*H*-methanofullerene[C_{60}] (Fig. 11). The important feature in both compounds was that cobaltocene in its oxidised state Cp_2Co^+ has spin $S = 0$ and by no means could contribute to the magnetic signal. Only spins $S = ½$ on fullerene moieties contributed to the magnetic ordering. That was an essential discovery having in mind that the role of $TDAE^+$ spin $S = ½$ in TDAE-C_{60} ferromagnetic ordering was unknown.

For cobaltocene-doped fullerene derivatives it was found that the temperature at which the doping was performed plays a crucial role in determining the low-temperature magnetic properties of these materials. A detailed study [37] revealed the optimum conditions, particularly the temperature for the synthesis of ferromagnetic material. The magnetisation of samples differed markedly both in a magnitude and in a critical temperature T_c. The magnetisation was highest when the synthesis was performed in the vicinity of 45°C and fell off rapidly on either side of that temperature. The critical temperatures ranged from 13K to 17K. The low-temperature magnetisation in a weak external field (the spontaneous magnetisation) could vary approximately by a factor of three among different samples. The samples in Ref. [36] showed also a hysteretic behavior in their magnetic curves below T_c with a coercive field H_c ~ 100 Oe and a remanent magnetisation M_r which is about 0.1 percent of the expected saturation magnetisation M_s. The magnetisation did not show saturation in fields up to 1 kOe.

6 Intercalation of Magnetic Ions

As soon as various atom and molecules can be intercalated into C_{60} crystal, it was expected that new magnetic C_{60} compounds could be synthesised where the magnetic moments would be carried by intercalants. For that purpose Eu was an obvious choice

[38]. Europium has a magnetic moment $7\mu_B$ in the divalent state, while it is nonmagnetic in the trivalent state. In a fulleride Eu_6C_{60} europium ions are in the divalent state and they order ferromagnetically below 12K [39]. Substitution of Eu with nonmagnetic Sr ions in $Eu_{6-x}Sr_xC_{60}$, (x = 1 –5) had little effect on the transition temperature, T_c. In addition, Eu_6C_{60} showed a huge negative magnetoresistance at and below T_c. Evidently, there exist a strong interaction between conduction carriers and localized magnetic moments; namely, the strong π-f interaction in Eu_6C_{60}. This fact indicates that the ferromagnetism in Eu_6C_{60} comes from the indirect exchange interaction via C_{60} molecules, which is quite in contrast with the case of magnetic semiconductor EuO.

Second fulleride with rare earth intercalated ions, which showed a magnetic behavior was Ce_xC_{60} [40]. Cerium has outermost electronic configuration $4f^1 5d^1 6s^2$. In the case of Ce_xC_{60} cerium ion is in a trivalent state Ce^{3+} with unpaired $4f^1$ electron. Magnetic properties of Ce_xC_{60} were rather controversial. When cooled in a zero magnetic field the system showed transition to the superconducting state below 13.5K. On the contrary, cooling even in a very low magnetic field (2 Oe) destroyed superconductivity and the system exhibited a ferromagnetic transition at 15K. The ferromagnetic state was also characterised by a hysteresis loop in a *M(H)* curve. Although proximity of two different ground states in Ce_xC_{60} hasn't been explained yet, we could roughly ascribe superconductivity to the doped C_{60} subsystem and ferromagnetism to the superexchange interactions between Ce^+ ions.

References

1. P.M. Allemand, K.C. Khemani, A. Koch, F. Wudl, K. Holczer, S. Donovan, G. Gruner, and J.D. Thompson, Science 253 (1991) 301.
2. P.W. Stephens, D. Cox, J.W. Lauher, L. Mihaly, J.B. Wiley, P.-M. Allemand, A. Hirsch, K. Holczer, Q. Li, J.D. Thompson, and F. Wudl, Nature 355 (1992) 331.
3. K. Tanaka, A.A. Zakhidov, K. Yoshizawa, K. Okahara, T. Yamabe, K. Yakushi, K. Kikuchi, S. Suzuki, I. Ikemoto, and Y. Achiba, Phys Rev. B47 (1993) 7554.
4. A. Suzuki, T. Suzuki, R.J. Whitehead, and Y. Maruyama, Chem. Phys. Lett. 223 (1994) 517.
5. L. Dunsch, D. Eckert, J. Frohner, A. Bartl, and K.-H. Muller, J. Appl. Phys. 81 (1997) 4611.
6. A. Lappas, K. Prassides, K. Vavekis, D. Arcon, R. Blinc, P. Cevc, A. Amato, R. Feyerherm, F. N. Gygax, and A. Schenck, Science 267 (1995) 1799.
7. P. Venturini, D. Mihailovic, R. Blinc, P. Cevc, J. Dolinsek, D. Abramic, B. Zalar, H. Oshio, P.-M. Allemand, A. Hirsch, and F. Wudl, Int. J. Mod. Phys. B 6 (1992) 3947.
8. R. Blinc, P. Cevc, D. Arcon, D. Mihailovic, and P. Venturini, Phys. Rev. B50 (1994) 1.
9. D. Mihailovic, D. Arcon, P. Venturini, R. Blinc, A. Omerzu, and P. Cevc, Science 268 (1995) 400.
10. K. Tanaka, Y. Asai, T. Sato, T. Kuga, T. Yamabe, and M. Tokumoto, Chem. Phys. Lett. 259 (1996) 574.
11. T. Sato, T. Saito, T. Yamabe, and K. Tanaka, Phys. Rev. B 55 (1997) 11052.
12. K. Tanaka, T. Sato, K. Yoshizawa, K. Okahara, T. Yamabe, and M. Tokumoto, Chem. Phys. Lett. 237 (1995) 123.
13. A. Suzuki, T. Suzuki, and Y. Maruyama, Solid State Commun. 96 (1995) 253.

14. A. Schilder, H. Klos, I. Rystau, W. Schultz, and B. Gotschy, Phys. Rev. Lett. 73 (1994) 1299.
15. F. Bommeli, L. Degiorgi, P. Wachter, D. Mihailovic, A. Hassanien, P. Venturini, M. Schreiber, and F. Diedrich, Phys. Rev B51 (1995) 1366.
16. A. Omerzu, D. Mihailovic, S. Tomic, O. Milat, and N. Biskup, Phys. Rev. Lett. 77 (1996) 2045.
17. D. Arcon, J. Dolinsek, and R. Blinc, Phys. Rev. B 53 (1996) 9137.
18. A. Mrzel, P. Cevc, A. Omerzu, and D. Mihailovic, Phys. Rev. B 53 (1996) R2922.
19. D. Arcon, P. Cevc, A. Omerzu, and R. Blinc, Phys. Rev. Lett. 80 (1998) 1529.
20. A. Omerzu, M. Tokumoto, B. Tadic, and D. Mihailovic, Phys. Rev. Lett. 87 (2001) 177205-1.
21. A. Omerzu, D. Mijatovic, and D. Mihailovic, Synth. Metals 121 (2001) 1155.
22. A. Omerzu, D. Mihailovic, and M. Tokumoto, Phys. Rev. B61 (2000) R11883.
23. B. Narymbetov, A. Omerzu, V.V. Kabanov, M. Tokumoto, H. Kobayashi, and D. Mihailovic, Nature 407 (2000) 883.
24. T. Kambe, Y. Nogami, and K. Oshima, Phys. Rev. B61 (2000) R862.
25. K. Mizoguchi, M. Machino, H. Sakamoto, T. Kawamoto, M. Tokumoto, A. Omerzu, and D. Mihailovic, Phys. Rev. B63 (2001) 140417-1.
26. S. Garaj, T. Kambe, L. Forro, A. Sienkiewicz, M. Fujiwara, and K. Oshima, Phys. Rev. B68 (2003) 144430-1.
27. K. Tanaka, A. A. Zakhidov, K. Yoshizawa, K. Okahara, T. Yamabe K. Kikuchi, S. Suzuki, I, Ikemoto, and Y. Achiba, Solid State Comunn. 85 (1993) 69.
28. K. Oshima, T. Kambe, M. Fujiwara, and Y. Nogami, Synth. Met. 133-134 (2003) 699.
29. K. Tanaka, T. Sato, T. Yamabe, K. Yoshizawa, K. Okahara, and A.A. Zakhidov, Phys. Rev. B51 (1994) 990.
30. H. Klos, I. Rystau, W. Schutz, B. Gotschy, A. Skiebe, and A. Hirsch, Chem. Phys. Lett. 224 (1994) 333.
31. H. Wang and D. Zhu, J. Phys. Chem. Solids 55 (1994) 437.
32. A. Schilder, B. Gotschy, A. Seidl, and R. Gompper, Chem. Phys. 193 (1995) 321.
33. A. Otsuka, T. Teramoto, Y. Sugita, T. Ban, and G. Saito, Synth. Met. 70 (1995) 1423.
34. W.-J. Wang, Y. L. Lay, C.-S. Chang, H.-S. Chiu, K.-S. Chuang, and B.-C. Wang, Synth. Met. 86 (1997) 2293.
35. P. Venturini, V. Krasevec, D. Mihailovic, M. Eiermann, G. Srdanov, N. S. Sariciftci, C. Li, and F. Wudl, Proceedings of International winterschool on Electronic Properties of Novel Materials, Singapore: World Scientific, 1994, pp. 514-517.
36. A. Mrzel, A. Omerzu, P. Umek, D. Mihailovic, Z, Jaglicic, Z. Trontelj, Chem. Phys. Lett. 298 (1998) 329.
37. P. Umek, A. Omerzu, D. Mihailovic, and ., Tokumoto Chem. Phys. 253 (2000) 361.
38. Y. Ksari-Habiles, D. Claves, G. Chouteau, P. Touzain, C. Jrandey, J. L. Oddou, and A.J. Stepanov. Phys. Chem. Solids 58 (1997) 1771.
39. K. Ishii, A. Fujiwara, H. Suematsu, and Y. Kubozono, Phys. Rev. B 65 (2002) 134431-1.
40. Y. Maruyama, S. Motohashi, N. Sakai, M. Watanabe, K. Suzuki, H. Ogata, and Y. Kubozono, Solid State Commun. 123 (2002) 229.

Carbon-Based Magnetism
T. Makarova and F. Palacio (Editors)

11

Diamagnetism of Diamond and Graphite

A.V. Nikolaev[a,b] and B. Verberck[a]

[a]*Department of Physics, University of Antwerp, Universiteitsplein 1, 2610 Antwerp, Belgium*
[b]*Institute of Physical Chemistry of the Russian Academy of Sciences, Leninskii prospekt 31, 117915 Moscow, Russia*

1 Introduction

A material is said to be diamagnetic when its magnetic susceptibility χ is negative, *i.e.*, when its induced magnetic moment $\vec{M} = \chi\vec{H}$ is opposite to the applied magnetic field $\vec{H}$. Diamagnetism has been reported for practically all carbon modifications: graphite [1], diamond [2], C_{60} and C_{70} fullerene [3–5], and carbon nanotubes [5]. The most recent study of the temperature and field dependence of the magnetic susceptibility of these carbon forms is that of Heremans *et al.* [5], for temperatures ranging from 4.5K to 310K and magnetic fields up to 55 kOe. The results of Heremans *et al.* [5] are summarized in Fig. 1. Here, we focus on the data for diamond and graphite since reviews of the magnetic properties of fullerenes and carbon nanotubes are the subject of other chapters of this book.

The magnetic moment of diamond has linear magnetic field dependence at all fields, and is temperature-independent ("+" symbols in Fig. 1). The susceptibility is found to be $\chi = (-4.5 \pm 0.2) \times 10^{-7}$ emu/g. Note that in Ref. [5], χ is defined as $\frac{dM}{dH}$ and not as $\frac{M}{H}$.

Graphite has an anisotropic diamagnetic susceptibility [1, 6, 5]. For fields perpendicular to the hexagonal carbon sheets, the susceptibility is very large: $\chi_{\perp} \approx -30 \times 10^{-6}$ emu/g [1, 5]. The in-plane susceptibility (fields parallel to the sheets) is about equal to the free-atom diamagnetic susceptibility: $\chi_{\parallel} \approx -5 \times 10^{-7}$ emu/g [7]. For

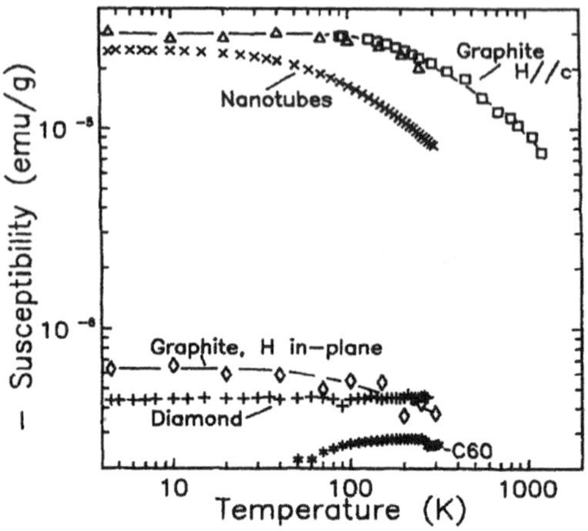

Fig. 1. Temperature dependence of the magnetic susceptibility χ of varuous carbon materials, after Ref. 5. The measurements on graphite by Heremans *et al.* [5] for fields perpendicular to the carbon sheets (at 200 Oe) are represented by "△" symbols and compared to the data of Ganduli and Krishnan [1], represented by "□" symbols. Graphite's in plane susceptibility is represented by "○" symbols [5]. The "+" symbols represent the susceptibility of diamond [5]. Note that the ordinate is - χ.

magnetic fields parallel to the carbon sheets, the susceptibility $\chi_{\parallel}$ does not depend on temperature and field, at least in the low-field approximation. (The data of Heremans *et al.* [5] may result from a small $\chi_{\perp}$ contribution.) At fields H ≥ 4 kOe, the moment is opposed to and varies linearly with the field; below 20K, $\chi_{\parallel} \approx -(5.5 \pm 1) \times 10^{-7}$ emu/g. At room temperature it decreases to $\chi_{\parallel} \approx -(4 \pm 1) \times 10^{-7}$emu/g. In Fig. 1, the slope $\frac{dH}{dM}$, measured at 4 kOe, is shown as a function of temperature ("◊" symbols). The measurements of Heremans *et al.* [5] of $\chi_{\perp}$ were performed on highly-oriented pyrolytic graphite (HOPG); they show good linearity of $M(H)$ below 20 kOe. The observed deviations from linearity at higher fields and $T \leq 20$K are the onset of de Haas–van Alphen oscillations [5]. In Fig. 1, the temperature dependence of $\chi_{\perp}$ measured at 200 Oe - a field at which de Haas-van Alphen oscillations are absent - is shown (△ symbols) and compared to early data of Ganguli and Krishnan [1] (□ symbols).

In the following sections, we give an overview of theoretical models designed to describe the diamagnetic behavior of diamond and graphite. We also briefly introduce the "classical" models of Van Vleck, Langevin, Landau and Pauli, and discuss the development of a theoretical approach to diamagnetism due to conduction electrons. The emphasis lies on the models' theoretical aspects and assumptions/approximations. For more details and extensive experimental information, we refer the reader to other reviews (Refs. [8] and [9]).

2 Theory of (Dia)magnetism

It is easily seen that within a classical theory, the magnetization of a system in thermal equilibrium must always vanish (Bohr–von Leeuwen theorem [10]). Indeed, the hamiltonian of a system in a magnetic field $\vec{H}$ is obtained from the system's hamiltonian in absence of a field by replacing the particle momenta $\vec{p}_i$ by $\vec{p}_i - \frac{q_i}{c}\vec{A}(\vec{r}_i)$, with q_i the charge of particle i and $\vec{A}$ the vector potential $(\vec{H} = \vec{\nabla} \times \vec{A})$. Hence, the partition function Z, involving integrations over all momenta $\vec{p}_i$, does not contain any dependence on the magnetic field since a change of variables $\vec{p}_i \rightarrow \vec{p}_i + \frac{q_i}{c}\vec{A}(\vec{r}_i)$ can be made. But if Z and the free energy $F \propto -\ln Z$ do not depend on $\vec{H}$, the magnetization $\vec{M}$, being proportional to $\frac{\partial F}{\partial H}$, must vanish. Therefore, in order to explain magnetic phenomena, a quantum theory is required from the very start.

2.1 Diamagnetism of Localized (Atomic) Electrons

In the presence of a uniform magnetic field $\vec{H}$, the hamiltonian of an electronic system - say, an atom - is modified in two ways [11]: (i) The momentum $\vec{p}_i$ of each electron (i indexes electrons) is replaced by $\vec{p}_i + \frac{e}{c}\vec{A}(\vec{r}_i)$, so that the total electronic kinetic energy reads

$$T = \frac{1}{2m_e}\sum_i \left[\vec{p}_i + \frac{e}{c}\vec{A}(\vec{r}_i) \right]^2 \tag{2.1}$$

with m_e the electron mass and $-e < 0$ the charge of an electron. For a uniform field, the vector potential $\vec{A}(\vec{r})$ is usually chosen to be

$$\vec{A}(\vec{r}) = -\frac{1}{2}\vec{r} \times \vec{H}. \tag{2.2}$$

(ii) All energies H_i due to the interaction of the field with the electron spins $\vec{s}_i$,

$$H_i = g_0 \mu_B s_{iz} H, \tag{2.3}$$

have to be added to the hamiltonian. Here, $\mu_B = \frac{e\hbar}{2m_e c}$ is the Bohr magneton and $g_0 = 2.0023 \approx 2$ is the electronic g-factor. The magnetic field is chosen to lie parallel to the z-axis: $\vec{H} = H\vec{e}_z$.

The total electronic kinetic energy operator T [Eq. (2.1)] then reads

$$T = \frac{1}{2m_e}\sum_i p_i^2 + \mu_B\, \vec{L}\cdot\vec{H} + \frac{e^2}{8m_ec^2}H^2\sum_i\ (x_i^2+y_i^2) \tag{2.4}$$

where $\vec{L}$ is the total electronic orbital angular momentum:

$$\vec{L} = \frac{1}{\hbar}\sum_i \vec{r}_i\times\vec{p}_i \tag{2.5}$$

The spin terms (2.3) combine with T [Eq. (2.4)] to give the following field-dependent hamiltonian terms:

$$\mathcal{H}_{\text{mag}} = -\vec{M}\cdot\vec{H} + \frac{e^2}{8m_ec^2}H^2\sum_i (x_i^2+y_i^2) \tag{2.6}$$

with $\vec{M} = -\mu_B(\vec{L}+g_0\sum_i \vec{s}_i)$.

For small fields, the magnetic contribution to the hamiltonian - Eq. (2.6) - can be considered as a small perturbation relative to the zero-field hamiltonian $\mathcal{H}_0$. Applying second-order perturbation theory yields the following energy spectrum:

$$\varepsilon_n = \varepsilon_n^0 + \Delta\varepsilon_n = \varepsilon_n^0 + A_nH - \frac{1}{2}(B_n+D_n)H^2, \tag{2.7a}$$

with

$$A_n = (M_z)_{nn}, \tag{2.7b}$$

$$B_n = 2\sum_{n'\neq n}\frac{|(M_z)_{nn'}|^2}{\varepsilon_{n'}^0-\varepsilon_n^0} \tag{2.7c}$$

$$D_n = -\frac{e^2}{4m_ec^2}\sum_i (x_i^2+y_i^2)_{nn}. \tag{2.7d}$$

Here, brackets () stand for matrix elements: $(f)_{nn'} \equiv \langle\psi_n|f|\psi_{n'}\rangle$. For calculating the Helmholtz free energy F, an approximation assuming $\Delta\varepsilon_n \ll T$ ($k_B \equiv 1$), valid for small magnetic fields, can be made by taking only the leading terms of the expansion of Z into powers of H into account [12]:

$$Z \equiv \sum_n e^{-\frac{\varepsilon_n}{T}} = \sum_n e^{-\frac{\varepsilon_n^0}{T}}\left[1+\frac{A_nH}{T}+\frac{A_n^2H^2}{2T^2}+\frac{(B_n+D_n)H^2}{2T}\right]$$
$$= Z^0\left[1+\frac{H^2}{2T}\left(\frac{\langle A^2\rangle}{T}+\langle B\rangle+\langle D\rangle\right)\right]. \tag{2.8}$$

Here, $\langle\ \rangle$ stands for thermal averaging (with unperturbed energy levels ε_n^0), and $Z^0 = \sum_n e^{-\frac{\varepsilon_n^0}{T}}$. In Eq. (2.8), the thermal average of A_n has vanished since absence of magnetic order is presumed:

$$\langle A\rangle = \frac{1}{Z^0}\sum_n (M_z)_{nn}\, e^{-\frac{\varepsilon_n^0}{T}} = 0 \qquad (2.9)$$

The free energy F is then obtained via

$$F = -T\ln Z. \qquad (2.10)$$

Finally, one can calculate the magnetic moment $\vec{M} = M\vec{e}_z$:

$$M = -\left(\frac{\partial F}{\partial H}\right)_{T,V} \equiv \chi H. \qquad (2.11)$$

The foregoing analysis yields

$$\chi \approx \frac{\langle A^2\rangle}{T} + \langle B\rangle + \langle D\rangle. \qquad (2.12)$$

The first term, $\frac{\langle A^2\rangle}{T}$, gives the well-known Curie law for the magnetic susceptibility, while the second term, $\langle B\rangle$ – called the Van Vleck term –, is temperature-independent. The first and second terms of the right-hand side of Eq. (2.12) are usually dominant. They lead to paramagnetism – a net nonvanishing magnetization $\vec{M}$ parallel to the applied field $\vec{H}$. Only the third term, $\langle D\rangle$, gives a negative contribution to χ. If it prevails, it results in a diamagnetic effect – a net nonvanishing magnetization antiparallel to the applied magnetic field.

The diamagnetic effect occurs in a special but important case of atoms with completely filled electronic shells. These atoms have a spherically symmetric electron density distribution as follows from the generalized Unsöld theorem [13], and $\langle L\rangle = \langle S\rangle = 0$. The Curie and Van Vleck contributions are then zero, and the magnetic susceptibility reduces to

$$\chi = -\frac{e^2}{6m_e c^2}\sum_i \langle r_i^2\rangle, \qquad (2.13)$$

because the spherical symmetry further implies

$$\langle x_i^2 \rangle = \langle y_i^2 \rangle = \frac{1}{3}\langle r_i^2 \rangle. \tag{2.14}$$

This case is known as Langevin or Larmor diamagnetism. Not only noble gases display this type of diamagnetism, but also ionic crystals like the alkali halides AX, since therein, the ions (approximately) have completely filled electronic shells. From the above discussion, it is clear that in principle, all core electron shells result in a diamagnetic contribution described by Eq. (2.13).

2.2 Diamagnetism of Free Electrons

We consider a degenerate electron gas, characterized by its thermodynamic variables T, V and μ (chemical potential). The thermodynamic potential Ω is then the key thermodynamic function. The magnetic moment can be derived from it [compare with Eq. (2.11)]:

$$M = -\left(\frac{\partial \Omega}{\partial H}\right)_{T,V,\mu} \equiv \chi H. \tag{2.15}$$

The energy levels of orbital motion of an electron in a magnetic field read [14, 15]

$$\varepsilon_n(p_z) = \frac{p_z^2}{2m_e} + (2n+1)\mu_B H, \tag{2.16}$$

with $n = 0, 1, 2, \ldots$ and p_z the electron's linear momentum in the direction of the field ($\vec{H} = H\vec{e}_z$), taking on continuous values from $-\infty$ to $+\infty$. For a given value of n, the number of states in the interval dp_z equals

$$2\frac{VeH}{(2\pi\hbar)^2 c} dp_z, \tag{2.17}$$

where spin degeneracy is taken into account. The thermodynamic potential can be rewritten as

$$\Omega(\mu) = 2\mu_B H \sum_{n=0}^{\infty} f(\mu - [2n+1]\mu_B H), \tag{2.18}$$

where the function f is defined as

$$f(\mu) = -\frac{Tm_e V}{2\pi^2\hbar^3}\int_{-\infty}^{+\infty}\partial p_z \ln\left(1+\exp\left(-\frac{\frac{p_z^2}{2m_e}-\mu}{T}\right)\right) \tag{2.19}$$

As before $k_B \equiv 1$. After some algebra one obtains the following expression for Ω.

$$\Omega(\mu) = \Omega_0(\mu) - \frac{1}{6}\mu_B^2 H^2 \frac{\partial^2\Omega_0(\mu)}{\partial\mu^2}, \tag{2.20}$$

with $\Omega_0(\mu)$ the thermodynamic potential in absence of a field. Recalling that $\frac{\partial\Omega_0(\mu)}{\partial\mu} = -N$, the magnetic susceptibility is found as [14, 12]

$$\chi \equiv \chi_{\text{Landau}} = \frac{\mu_B^2}{3V}\frac{\partial^2\Omega_0(\mu)}{\partial\mu^2} = -\frac{\mu_B^2}{3V}\left(\frac{\partial N}{\partial\mu}\right)_{T,V} < 0. \tag{2.21}$$

In the limit $T \to 0$, $\left(\frac{\partial N}{\partial\mu}\right)_{T,V}$ can be interpreted as the electronic density of states at the Fermi level. Diamagnetic effect (2.21), due to the quantization of orbital motion of electrons in a magnetic field, is called Landau diamagnetism.

For free electrons, there is another magnetic effect. It is of paramagnetic nature, and usually dominates the diamagnetic Landau susceptibility, so that the net susceptibility is again paramagnetic. Indeed, the spin energy of an electron, $\pm\mu_B H$, has to be taken into account as well (the + and - signs correspond to the two values $\pm\frac{1}{2}$ of the spin component along the field). Since ε enters the Fermi distribution together with the chemical potential μ in the form $\varepsilon - \mu$, adding $\pm\mu_B H$ to ε is equivalent to substituting μ by $\mu \mp \mu_B H$. The thermodynamic potential Ω is therefore given by

$$\Omega(\mu) = \frac{1}{2}\Omega_0(\mu + \mu_B H) + \frac{1}{2}\Omega_0(\mu - \mu_B H). \tag{2.22}$$

The factors $\frac{1}{2}$ respect the number of electron states for a fixed spin component. Expanding $\Omega(\mu)$ into powers of $\mu_B H$ yields

$$\Omega(\mu) \approx \Omega_0(\mu) + \frac{1}{2}\mu_B^2 H^2 \frac{\partial^2\Omega_0(\mu)}{\partial\mu^2}, \tag{2.23}$$

and finally,

$$\chi \equiv \chi_{\text{Pauli}} = -\frac{\mu_{\text{B}}^2}{V}\frac{\partial^2\Omega_0}{\partial\mu^2} = \frac{\mu_{\text{B}}^2}{V}\left(\frac{\partial N}{\partial\mu}\right)_{T,V} > 0. \tag{2.24}$$

This contribution to χ is known as the Pauli paramagnetic susceptibility [11]. In contrast to the susceptibility of paramagnetic ions obeying the $\frac{1}{T}$ Curie law, the Pauli susceptibility of conduction electrons is essentially independent of temperature. From this point of view it bears similarity to the Van Vleck term, Eq. (2.7c).

The total susceptibility χ induced by a magnetic field, experimentally determined via measurements of the induced bulk moment, is a combination of the Pauli paramagnetic susceptibility, the Landau diamagnetic susceptibility and the Langevin diamagnetic susceptibility of the closed-shell ion cores (see Sec. 2.1). Clearly, it is not straightforward to distinguish (experimentally) between particular contributions to the susceptibility.

Considering only contributions (2.21) and (2.24) of the free-electron model, one sees that

$$\chi_{\text{Landau}} = -\frac{1}{3}\chi_{\text{Pauli}}, \tag{2.25}$$

and that the resulting susceptibility therefore reads

$$\chi = \chi_{\text{Landau}} + \chi_{\text{Pauli}} = \frac{2}{3}\chi_{\text{Pauli}} > 0. \tag{2.26}$$

According to Eq. (2.26), a degenerate free-electron gas exhibits paramagnetism. In reality, however, electron-electron interactions and correlations destroy the validity of expressions (2.25) and (2.26). Doped semiconductors are examples of conducting materials in which conduction-electron diamagnetism can be substantially larger than paramagnetism [16]. Important for our discussion is that the same goes for both diamond and graphite.

2.3 Diamagnetism of Itinerant Electrons in Solids

In general, the diamagnetic susceptibility of a solid is small and temperature-independent. At very low temperatures, additional contributions displaying oscillatory dependence on the magnetic field - the well-known de Haas-Van Alphen effect [11] - arise. In the following, we limit ourselves to the steady diamagnetic susceptibility. We also neglect electronic correlations and do not consider very high magnetic fields.

As discussed in Sec. 2.2, Landau obtained a diamagnetic effect [Eq. (2.21)] for the motion of free electrons in a magnetic field [14, 12]. The quantity $\left(\frac{\partial N}{\partial\mu}\right)_{T,V}$, which enters Eq. (2.21), can be interpreted as the electronic density of states at the Fermi level in the limit $T \to 0$. Assuming that the temperature dependence of $\left(\frac{\partial N}{\partial \mu}\right)_{T,V}$ is negligible, we then compute it from the relation

$$N = V\frac{(2m_e\mu)^{3/2}}{3\pi^2\hbar^3}, \tag{2.27}$$

known from the theory of the degenerate electron gas [11]. The magnetic susceptibility per unit volume then reads

$$\chi = -\frac{e^2 k_{\mathrm{F}}}{12\pi^2 m_e c^2}. \tag{2.28}$$

Here, k_{F} is the wave number at the Fermi level.

The first calculation of the magnetic susceptibility of a solid due to conduction electrons has been performed by Peierls [17]. Essentially, it is a free-electron model, extended and modified for a periodic potential. Within a tight-binding approximation, Peierls arrived at a susceptibility consisting of three parts:

$$\chi = \chi_1 + \chi_2 + \chi_3. \tag{2.29}$$

Here, χ_1 is the susceptibility due to free (non-interacting) atoms (described in Sec. 2.1), χ_3 a renormalized Landau susceptibility [compare with Eq. (2.28)],

$$\chi_3 = -\frac{e^2 k_{\mathrm{F}}}{12\pi^2 m_e^* c^2}, \tag{2.30}$$

and m_e^* is an effective band mass. Finally, χ_2 in Eq. (2.29) is a complicated interference term, the magnitude and sign of which are uncertain. We observe that the free atom and Landau contributions to the magnetic susceptibility add up to the total susceptibility only if $\chi_2 = 0$.

Inspired by Peierls' result, some authors prefer to consider two different electron masses m_e and m_e^* for the Pauli and Landau terms. Although in this way it is possible to explain why in some materials χ_{Landau} can exceed χ_{Pauli} and result in a net diamagnetic effect, this approach is not really logical, since the same function $\left(\frac{\partial N}{\partial \mu}\right)_{T,V}$ enters both Eq. (2.21) and Eq. (2.24). If the effect of a periodic potential is simply the modification of the electronic density of states, then the main conclusion drawn from the free-electron model – expressed by Eqs. (2.25) and (2.26) - must remain.

After Peierls' pioneering work, many attempts to solve the problem of Landau diamagnetism and Pauli paramagnetism in a solid in a satisfactory and rigorous way were undertaken. Wilson [18] tackled the diamagnetism of Bloch electrons by studying the density matrix as a function of the magnetic field. Though he was not able to derive a complete formula via this approach, Hebborn and Sondheimer [19] continued along the same lines and obtained a complete expression for χ, albeit in a very complicated way. Much off the complexity of their method originates from the lack of periodicity of the magnetic perturbing terms of the hamiltonian giving rise to singular matrix elements in the partition function. To simplify things, Hebborn *et al.* [20] considered a space-dependent periodic external field; the result for a uniform field is then obtained by finally letting the period approach infinity.

Kjeldaas and Kohn [21] used a generalized effective-mass theory by introducing an expansion of $\left|\vec{k}-\vec{k}_F\right|$ up to fourth order. Unfortunately, their result has limited usefulness, since it applies only to the very special case of electrons confined to the vicinity of the top or the bottom of an energy band. They used their result to calculate the diamagnetic susceptibility of Li and Na, but – as they themselves pointed out – in the case of Li, the ratio of the second to the first term in the expansion in powers of $\left|\vec{k}-\vec{k}_F\right|$ equals 0.6, so higher-order terms may be important.

Equivalent methods have been presented by Roth [22], Blount [23] and Wannier and Upadhaya [24]. All considered the fact that the effect of a magnetic field on a band is twofold: one effect is the gradual transformation of the parameters of the band, the other effect is the breaking up of the band into a series of discrete states. The bands thereby become field-dependent.

Samoilovich and Rabinovich [25] calculated the diamagnetic susceptibility of conduction electrons in a nearly-free-electron approximation. Following a similar procedure, also within the nearly-free-electron approximation, Glasser [26] obtained an expression for the total magnetic susceptibility of metals. These results have a simpler form than others, but have the undesirable feature that expressions diverge when the Fermi surface touches the zone boundary. Furthermore, the approximation has been justified by the pseudopotential method in absence of a magnetic field, but it is not clear whether it is valid when a magnetic field is present.

Misra and Roth [27] generalized the pseudopotential method to the case of a solid in a magnetic field in order to calculate the diamagnetic susceptibility correctly and accurately. They obtained a general equation of motion, from which they first derived an expression for the core diamagnetism. Then they developed an expression for the diamagnetic susceptibility of Bloch electrons in terms of a pseudopotential. Their method has the advantage that the results can be derived relatively simply (as opposed to other aforementioned works). Using the pseudopotential technique and degenerate perturbation theory, Misra and Roth expressed the result in a form allowing easy computation of χ. In contrast to other models, they have found that the diamagnetic and paramagnetic susceptibilities are additive. The diamagnetic susceptibility of their model is given by

$$\chi=\chi_0\left[1+\sum_{\vec{G}} D_{\vec{G}}(a)\right], \tag{2.31}$$

where χ_0 is the diamagnetic contribution of free electrons. The sum runs over reciprocal lattice vectors $\vec{G}$; the quantities $D_{\vec{G}}$ are highly complicated and include a parameter a, related to the chemical potential μ: $a = \frac{8m_e\mu}{G^2}$. For the computation of χ for a face-centered lattice, Misra and Roth chose eight nearest neighbor reciprocal (111) $\vec{G}$ vectors. Their result for aluminum is shown in Fig. 2. For the nondegenerate case, the magnitude of $\sum_{\vec{G}} \vec{G}(a)$ approaches infinity for $a \to 1$ and $a_0 = \frac{8m_e\mu_0}{G^2} \approx 1.1$ (μ_0 stands for the chemical potential in absence of a magnetic field). The degenerate-perturbation-theory result remains finite even when $a \to 1$ ($a_0 \to 1.1$). Misra and Roth concluded that in many cases, the deviation from free-electron behavior is small because the parameter a does not get very close to one. Large deviations from the free-electron results can occur if the parameter a is close to one and if the interaction is sufficiently large. If this is the case, the deviation can be in either direction, as seen in Fig. 2.

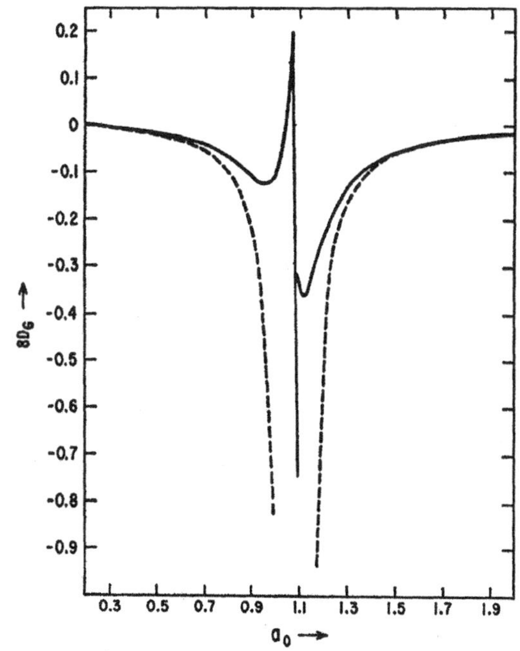

Fig. 2. Plot of $\sum_{\text{shell}} D_G(a_0)$ versus a_0 for Al, after Ref. [27]. The dashed curve corresponds to the second-order nondegenerate perturbation theory, the solid curve to the degenerate perturbation theory of Misra and Roth.

3 Diamagnetism of Diamond

For an understanding of the diamagnetism of tetrahedral semiconductors like diamond, Si and Ge, Hudgens *et al.* [2] proposed a simple model based on magnetic contributions of localized electrons: two Langevin diamagnetic terms χ_c and χ_v, arising from the core and valence electrons, respectively, and a Van Vleck paramagnetic susceptibility χ_p:

$$\chi = \chi_c + \chi_v + \chi_p. \tag{3.1}$$

Such terms have been discussed already in Sec. 2.1.

The electronic spectrum of diamond and other diamond-structured semiconductors exhibits a gap at the Fermi level; the gap separates the valence and the conduction band. Therefore, at the Fermi level, the electronic density of states $\frac{\partial N}{\partial \mu}$ vanishes, which is why Landau diamagnetism and Pauli paramagnetism are excluded in Eq. (3.1) – $\chi_{\text{Landau}} = 0$ [Eq. (2.22)] and $\chi_{\text{Pauli}} = 0$ [Eq. (2.24)]. The Van Vleck term, obtained via second order perturbation theory, Eq. (2.7c), can be considered as resulting from virtual magnetic dipole transitions between the valence band and the conduction band. In Ref. [2], Hudgens *et al.* approximated the Van Vleck term by a one-oscillator model with oscillator energy E_m. This energy is the average energy separation of valence and conduction band states connected by the magnetic dipole operator. Such an approximation was made earlier by Philips to describe the dielectric susceptibility [28]. It is based on the Penn isotropic two-band model approximating the low frequency dielectric constant by

$$\varepsilon(0) - 1 = \left(\frac{\hbar \omega_p}{E_g} \right)^2, \tag{3.2}$$

with $\hbar\omega_p$ the valence electron plasma energy and E_g a weighted energy separation of valence and conduction band states connected by the electric dipole operator. In general, E_m and E_g are different. However, if magnetic and electric dipole matrix elements vary slowly with energy, it is expected that $E_m = \beta E_g$, with the coefficient of proportionality $\beta \sim 1$. Under such assumptions χ_p becomes

$$\chi_p = \frac{B}{\beta E_g}, \tag{3.3}$$

where the parameter B is proportional to an average interband magnetic dipole matrix element. Hudgens *et al.* [2] argued that B depends mainly on the symmetry of the effective potential or of the charge density, and that it is zero for spherically or

Table 1.
Measured values of the magnetic susceptibility χ and extracted components χ_c, χ_v, χ_p for diamond, Si and Ge (in units 10^{-6} cm^3/mole). The measured values of $\frac{\partial x}{\partial t}$ are in 10^{-9} cm^3/mole·K. The parameter B (in units 10^{-4} eV cm^3/mole) and the effective radius of valence electrons r_v (in units Å) are described in the text, is the interatomic distance (in units Å). All data are taken from Hudgens *et al.* (Ref. [2]).

	diamond	Si	Ge
χ	– 5.9±0.1	–3.2±0.1	– 7.85±0.25
χ_c	– 0.15	–2.3	– 8.3
χ_v	–12.35±2	–19.65±3	–25.15±2
χ_p	6.6±2	18.75±3	25.6±2
$\frac{d\chi}{dT}$	– 0.007±0.003	0.75±0.03	1.8±0.05
B	0.9±0.3	0.9±0.15	1.1±0.1
r_v	1.04±0.15	1.32±0.1	1.48±0.06
d	1.548	2.346	2.450
$\frac{2r_v}{d}$	1.34±0.19	1.13±0.09	1.21±0.05

cylindrically symmetric electronic density distributions. Therefore, B should be approximately equal for all tetrahedral semiconductors.

The diamagnetic contribution χ_c due to the core electrons is given by Eq. (2.13),

$$\chi_c = -\frac{e^2}{6m_e c^2}\sum_{\text{core}}\langle r^2\rangle. \tag{3.4}$$

as is the diamagnetic response of the four valence electrons,

$$\chi_v = -\frac{4e^2}{6m_e c^2}\langle r^2\rangle_{\text{valence}}. \tag{3.5}$$

The latter expression implicitly assumes equivalence of the four valence electrons, because they form four hybrid sp^3 molecular orbitals. However, initial expression (2.13) is a result for spherical charge distributions (see Sec. 2.1). Clearly, this it is not the case for the four valence carbon (or Si or Ge) electrons, so Eq. (3.5) is a serious simplification. On the other hand, tetrahedral symmetry is relatively high (24 symmetry elements), so the anisotropy is relatively small, and in the end, relation (3.5) may be reasonable.

As remarked before, it is extremely difficult to distinguish different contributions to the total magnetic susceptibility experimentally. Hudgens *et al.* [2] managed to keep track of their various contributions [Eq. (3.1)] by studying the temperature dependence of χ and assuming that $\langle r^2 \rangle_{\text{valence}}$ has the same temperature coefficient of expansion as the square of the covalent bond length. The temperature behavior of χ is then described by

$$\frac{d\chi}{dT} = 2\alpha\chi_{\text{v}} - \frac{d \ln E_{\text{g}}}{dT} \chi_{\text{p}}, \tag{3.6}$$

with α the linear expansion coefficient. The temperature evolution of E_{g} is obtained from measurements of $\frac{d\varepsilon(0)}{dT}$.

The results of the measurements and calculations of Ref. [2] are summarized in Table 1. The data for diamond are compared with Si and Ge, sharing the tetrahedral structure. The parameter r_{v} stands for an effective radius, calculated from χ_{v} by means of

$$r_{\text{v}} = \sqrt{\sum_{\text{valence}} \langle r^2 \rangle} = 2\sqrt{\langle r^2 \rangle_{\text{valence}}} = \sqrt{\frac{-6m_e c^2 \chi_{\text{v}}}{e^2}}. \tag{3.7}$$

The ratio $\frac{2r_{\text{v}}}{d}$ (d being the interatomic separation) and the parameter B are almost the same for all materials. The fact that the results for diamond are comparable with those of Si and Ge is remarkable because diamond's band gap E_g (5.4 eV [29, 30]) is significantly larger than that of Si (1.11–1.17 eV [29, 30]) and Ge (0.66–0.744 eV [29, 30]). The small negativeness of $\frac{d\chi}{dT}$ for diamond is due to the first term in the right-hand side of Eq. (3.6). It is also interesting that χ_{v} and χ_{p} are large compared to χ_{c} and that these terms cancel one another in the case of Si and Ge.

4 Graphite: Electronic Band Structure

As shown by Peierls [17] and Wilson [18], the magnetic susceptibility χ of a solid, for a magnetic field $\vec{H}$ along the z axis ($\vec{H} = H\vec{e}_z$), involves (see Eq. (5.1) below) the quantity

$$\frac{\partial^2 \varepsilon}{\partial k_x^2}\frac{\partial^2 \varepsilon}{\partial k_y^2} - \left(\frac{\partial^2 \varepsilon}{\partial k_x \partial k_y}\right)^2, \tag{4.1}$$

with $\varepsilon(\vec{k})$ the electronic dispersion relation. We conclude that a thorough understanding of graphite's electronic band structure is necessary first. Therefore, as an introduction to the next section dealing with the susceptibility itself, in the following we briefly review the most important models addressing the electronic spectrum of graphite.

4.1 Tight-binding Model of Wallace

Pioneering theoretical work on graphite was done by Wallace [31]. He derived the band structure of graphite within a tight-binding formalism. His main concern was the explanation of graphite's electrical conductivity σ, but knowledge of the spectrum $\{\varepsilon(\vec{k})\}$ allowed him to make some statements about other physical properties, including the magnetic susceptibility χ. In the following, we sketch Wallace's theoretical approach [31].

4.1.1 Single Hexagonal Layer

A first step towards the calculation of the three-dimensional (3D) electronic band structure of graphite is the consideration of a single hexagonal layer (Fig. 3). Physically, this means neglecting the interaction between planes and assuming that conduction takes place in layers only. Since the bond lengths $l = 1.42$ Å within one graphite layer are small compared to the interlayer distance $\frac{c_0}{2} = 3.354$ Å, this makes a reasonable starting point. Wallace uses two carbon $2p_z$ functions as basis functions for his tight-binding approach [31]: one for each of the two carbon atoms in the two-dimensional (2D) primitive unit cell. The electrons of the carbon atoms forming co-planar bonds are excluded. This leads to two bands; the lowest one (valence band) is completely filled at $T = 0$, the upper one (conduction band) is then completely empty. As customary in the tight-binding formalism, overlap integrals of basis functions on different lattice

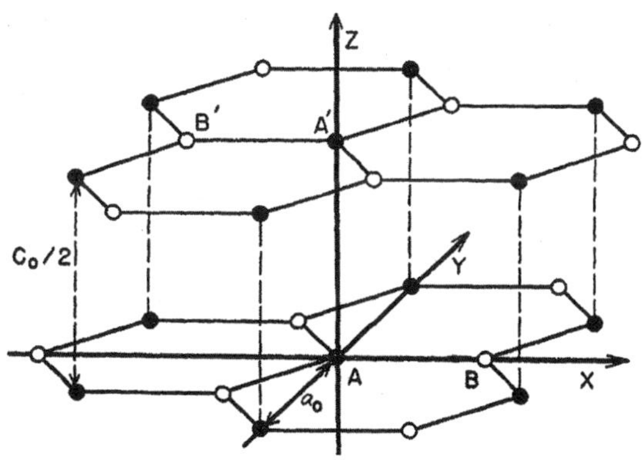

Fig. 3. Graphite consists of parallelly stacked hexagonal layers, spaced $\frac{c_0}{2} = 3.354$ Å apart. Within one layer, two inequivalent carbon atoms can be distinguished ('A' type and 'B' type). The distance between two nearest equivalent atoms is $a_0 = \sqrt{3}l = 2.46$ Å. Taken from Ref. [36].

sites, as well as exchange integrals over atoms more distant than nearest neighbors, are neglected. These approximations result in the following two-band spectrum:

$$\varepsilon_{\pm} = H'_{11} \pm |H'_{12}|, \tag{4.2a}$$

$$H'_{11} = \varepsilon_0 - 2\gamma'_0\left[\cos(2\pi k_y a_0) + 2\cos(\pi k_x a_0\sqrt{3})\cos(\pi k_y a_0)\right],$$
(4.2b)

$$H'_{12} = -\gamma_0\left[e^{-2\pi i k_x a_0/\sqrt{3}} + 2\cos(\pi k_y a_0)e^{2\pi i k_x a_0/2\sqrt{3}}\right].$$
(4.2c)

Here, ε_0 is the energy of a carbon $2p_z$ electron, $a_0 = \sqrt{3}l$, and γ'_0 and γ_0 are exchange integrals over nearest neighboring atoms of the same and different types, respectively. An investigation of the spectrum reveals that zero band gaps are located at the corners of the (hexagonal) first Brillouin zone. The anisotropy of the bands near the corner points of the first Brillouin zone is a consequence of the hexagonal symmetry of the graphite lattice [31].

4.1.2 Stacked Hexagonal Layers

Graphite consists of several parallelly stacked hexagonal layers of the kind just described. Since neighboring layers display a shift with respect to each other (see Fig. 3), the 3D primitive unit cell contains four, and not two, inequivalent carbon atoms. Therefore, the tight-binding method yields four bands, obtained from solving a 4×4 secular equation. It reads, taking only first and second neighbors [31],

$$\begin{vmatrix} \varepsilon - H_0 & -\gamma_0 S & \gamma_1\Gamma & \gamma'_1\Gamma S \\ -\gamma_0 S^* & \varepsilon - H_0 & \gamma'_1\Gamma S & \gamma'_1\Gamma S \\ \gamma_1\Gamma & \gamma'_1\Gamma S & \varepsilon - H_0 & -\gamma_0 S^* \\ \gamma'_1\Gamma S^* & \gamma'_1\Gamma S & -\gamma_0 S & \varepsilon - H_0 \end{vmatrix} = 0, \tag{4.3a}$$

with

$$H_0 = \varepsilon'_0 - 2\gamma'_0\left[\cos(2\pi k_y a_0) + 2\cos(\pi\sqrt{3}k_y a_0)\cos(\pi k_y a_0)\right], \tag{4.3b}$$

$$S = e^{-2\pi i k_x a_0/\sqrt{3}} + 2\cos(\pi k_y a_0)e^{2\pi i k_x a_0/2\sqrt{3}}, \tag{4.3c}$$

$$\Gamma = 2\cos(\pi k_z c_0). \tag{4.3d}$$

Here, ε_0' is a constant energy, and γ_0, γ_1 and γ_1' are exchange integrals between certain nearest neighboring atoms. At first, the approximative assumption $\gamma_1' = 0$ may be made. Indeed, $|\gamma_0| > |\gamma_1| > |\gamma_1'|$. The four solutions $\varepsilon_{\pm\pm}$ of Eq. (2.3a) then read

$$\varepsilon_{\pm\pm} = H_0 \pm \frac{1}{2}\gamma_1\Gamma \pm \sqrt{\frac{1}{4}\gamma_1^2\Gamma^2 + \gamma_0^2 |S|^2}. \tag{4.4}$$

The energy contours for planes of constant k_z look similar as in the 2D case. Zero band gaps occur at some points — the twelve corner points of the hexagonal prism constituting the first Brillouin zone (Fig. 4) – as in the single-layer case. This conclusion remains when γ_1' is properly taken into account. It is found that then, the energy $\varepsilon_{\pm\pm}$ is increased by an amount

$$\varepsilon = \gamma_0\gamma_1'\Gamma \frac{\gamma_0 |S|^2 (S + S^*) - 2\varepsilon_{\pm\pm}(S^2 + S^{*2})}{2\varepsilon_{\pm\pm}\gamma_1\Gamma \pm 4\gamma_0^2 |S|^2}. \tag{4.5}$$

Wallace [31] stated that the strong anisotropy of some of the properties of graphite, like the magnetic susceptibility, is caused by the high degree of anisotropy of the energy surfaces near the zone boundary. Indeed, as already noted, the magnetic susceptibility χ is directly related to the band structure via expressions containing quantities like (4.1). The (diamagnetic) susceptibility $\chi_{\parallel}$ across graphite layers is of the order of $9\pi^4 a_0^4 \frac{\gamma_0^4}{\gamma_1^2}$, whereas the (temperature-dependent) susceptibility within graphite layers is found to be approximately $3\pi^4 a_0^2 c_0^2 k_B T \frac{\gamma_0^2}{\gamma_1}$, or $k_B T \frac{\gamma_1}{\gamma_0^2}$ times $\chi_{\parallel}$. Finally, Wallace remarked that the anisotropy $\frac{\chi_{\perp}}{\chi_{\parallel}}$ is of the same order of magnitude as that of the conductivity, $\frac{\sigma_{\perp}}{\sigma_{\parallel}}$.

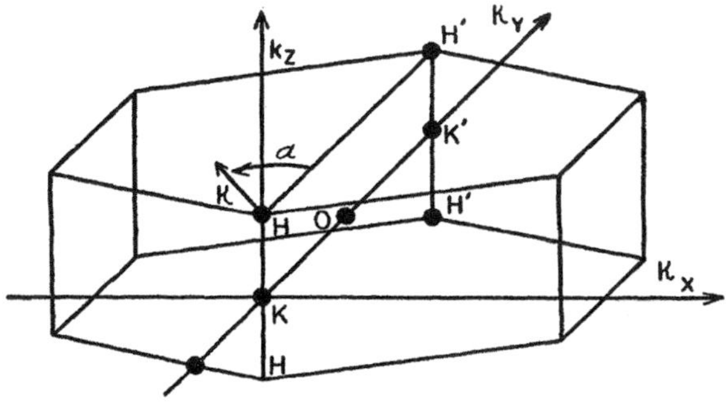

Fig. 4. First Brillouin zone of graphite. Trajectories of interest are *HKH* and *H'K'H'*. The origin of k_x and k_y has been chosen on *HKH*. Taken from Ref. [36].

4.2 Group-theoretical Model of Slonczewski and Weiss

Slonczewski and Weiss analyzed the electronic spectrum of graphite from a group-theoretical point of view [32]. They noted that the tight-binding approaches of Wallace [31], Hove [33] and McClure [34] did not allow for the presence of holes with effective masses different from those of the electrons and carried out a 3D treatment based on the symmetry of the graphite lattice, *i.e.* using concepts from group theory. The following outlines their results.

4.2.1 2D Lattice

Slonczewski and Weiss also started by considering a single layer of graphite [32]. They obtained the following expression for $\varepsilon_{\pm}(\vec{k})$ near the corner point $\vec{k}_0$ of the first Brillouin zone (see Fig. 4):

$$\varepsilon_{\pm}(\vec{k} = \vec{k}_0 + \vec{\kappa}) = E(\vec{k}_0) \pm \frac{\hbar p_0}{m_e}\kappa, \tag{4.6}$$

with $\kappa = \sqrt{\kappa_x^2 + \kappa_y^2}$, m_e (as in earlier sections) the electron mass, and p_0 a parameter to be determined from explicit wave functions (or experiments). Result (4.6) is independent of any approximations regarding wave functions and purely reflects the symmetry of an isolated graphite layer (neglecting spin-orbit coupling[1]). Eq. (4.6) confirms Wallace's result - Eqs. (4.2a) – (4.2c). One has

$$p_0 = \sqrt{\frac{3}{2}} a_0 \frac{m_e}{\hbar} \gamma_0. \tag{4.7}$$

4.2.2 3D Lattice

The 3D treatment of Slonczewski and Weiss [32] focuses on $\varepsilon(\vec{k})$ along and near HKH lines in $\vec{k}$-space connecting opposing corner points H and H of the first Brillouin zone - see Fig. 4. A general point on such a line of varying k_z is labeled S. Slonczewski and Weiss [32] present four wave functions obtained from an earlier group-theoretical analysis of the 3D graphite lattice [35]. These wave functions are approximate because they are constructed from 2D layer functions of π symmetry, assumed to be unaffected by neighboring layers. They were argued to be satisfactory. It is found that for the HKH path in reciprocal space, two of the four bands $\varepsilon_{i=1,\ldots,4}(\vec{k})$ coincide:

---

[1]The effect of spin-orbit coupling is a splitting of $E(\vec{k}_0)$, but by an amount less than 10^{-4} eV.

$$\varepsilon_1(k_z) = E_1^H + \delta_1 \cos\left(\frac{c_0 k_z}{2}\right), \tag{4.8a}$$

$$\varepsilon_2(k_z) = E_1^H - \delta_1 \cos\left(\frac{c_0 k_z}{2}\right), \tag{4.8b}$$

$$\varepsilon_3(k_z) = \varepsilon_4(k_z) = E_3^H + \delta_2 \left[1 - \sin^2\left(\frac{c_0 k_z}{2}\right)\right]. \tag{4.8c}$$

Here, $E_1^H = \varepsilon_1(\vec{k} = \vec{k}_H)$ and $E_3^H = \varepsilon_3(\vec{k} = \vec{k}_H)$. Note the degeneracy of ε_1 and ε_2 at the H-point, a consequence of the symmetry of the crystal. The difference of the energies E_1^H and E_3^H comes from the inequivalency of the two atoms of the 2D primitive unit cell, experiencing different crystal fields. The parameters δ_1 and δ_2 have to be calculated using explicit wave functions or derived from experimental data. The variation of ε_3 and ε_4 with k_z is reasoned to be much smaller than that of ε_1 and ε_2, since, respectively, second nearest and first nearest neighbor layer overlaps are involved: $|\delta_2| \ll |\delta_1|$. The actual values of $E_1^H - E_3^H$, δ_1 and δ_2 determine whether a band gap is present or not. Slonczewski and Weiss pictured two scenarios (Figs. 5 and 6 in Ref. [32]). The second one, exhibiting overlap between the conduction and the valence band, was found to be compatible with McClure's analysis of the de Haas–van Alphen effect [34]. Later, that picture was corrected by Dresselhaus [45] in respect to electron and hole states along the HKH line — see next subsection and Fig. 6. The same method leading to Eq. (4.6) of the previous subsection was applied to the full 3D case to investigate the bands near the HKH paths. The outcome reads

$$\varepsilon_1(\kappa) = \frac{E_1^S + E_3^S}{2} + \sqrt{\frac{(E_1^S - E_3^S)^2}{4} + \frac{\hbar^2 p_0^2}{m_e^2}\kappa^2}, \tag{4.9a}$$

$$\varepsilon_2(\kappa) = \frac{E_2^S + E_3^S}{2} + \sqrt{\frac{(E_2^S - E_3^S)^2}{4} + \frac{\hbar^2 p_0^2}{m_e^2}\kappa^2}, \tag{4.9b}$$

$$\varepsilon_3(\kappa) = \frac{E_1^S + E_3^S}{2} - \sqrt{\frac{(E_1^S - E_3^S)^2}{4} + \frac{\hbar^2 p_0^2}{m_e^2}\kappa^2}, \tag{4.9c}$$

$$\varepsilon_4(\kappa) = \frac{E_2^S + E_3^S}{2} - \sqrt{\frac{(E_2^S - E_3^S)^2}{4} + \frac{\hbar^2 p_0^2}{m_e^2}\kappa^2}, \tag{4.9d}$$

with $E_i^S = \varepsilon_i(\vec{k} = \vec{k}_s)$. Some terms have been neglected to get solutions (4.9a) – (4.9d). Hence, the spectrum consists of four hyperbolas of revolution; they never lead to overlap for fixed k_z. Expressions (4.9a) – (4.9d) may be used to determine the effective mass of electrons and holes in planes of constant k_z.

4.3 Slonczewski–Weiss–McClure Hamiltonian

The Slonczewski–Weiss hamiltonian was rewritten by McClure (Refs. [36] and [37]) by keeping only the first nonvanishing k_z-dependent term in each matrix element. The hamiltonian matrix reads (in absence of a magnetic field):

$$H = \begin{pmatrix} E_1 & 0 & H_{13} & H_{13}^* \\ 0 & E_2 & H_{23} & -H_{23}^* \\ H_{13}^* & H_{23}^* & E_3 & H_{33} \\ H_{13} & -H_{23} & H_{33}^* & E_3 \end{pmatrix}, \tag{4.10}$$

with

$$E_1 = \Delta + \gamma_1 \Gamma + \frac{1}{2}\gamma_5 \Gamma^2, \tag{4.11a}$$

$$E_2 = \Delta - \gamma_1 \Gamma + \frac{1}{2}\gamma_5 \Gamma^2, \tag{4.11b}$$

$$E_3 = \frac{1}{2}\gamma_2 \Gamma^2, \tag{4.11c}$$

$$H_{13} = \frac{1}{2}\sqrt{\frac{3}{2}}\left(\gamma_4 \Gamma - \gamma_0\right) a_0 K e^{i\alpha}, \tag{4.11d}$$

$$H_{23} = \frac{1}{2}\sqrt{\frac{3}{2}}\left(\gamma_4 \Gamma + \gamma_0\right) a_0 K e^{i\alpha}, \tag{4.11e}$$

$$H_{33} = \frac{\sqrt{3}}{2}\gamma_3 \Gamma a_0 K e^{i\alpha}, \tag{4.11f}$$

where

$$K = \sqrt{k_x^2 + k_y^2}, \tag{4.11g}$$

$$\Gamma = 2\cos\left(\frac{c_0 k_z}{2}\right). \tag{4.11h}$$

(The quantity Γ coincides with *Gamma* introduced by Wallace [31], Eq. (4.3d), but Wallace's definition of reciprocal space differs by factors 2π, which is why expressions (4.11h) and (4.3d) differ by a factor 2π.) In Eqs. (4.11a) – (4.11h), α is the polar angle about the zone edge and k_x and k_y have their origin at the zone edge (see Fig. 4). The form of H as written in Eq. (4.10) applies to the zone edge HKH; the complex conjugate H^* applies to $H'K'H'$ (see Fig. 4). Concerning conduction properties, there are four bands of interest: two almost unoccupied and two almost completely occupied. All four are near in energy in the vicinity of the six *HKH* edges of the hexagonal Brillouin zone. In the literature, the model hamiltonian (4.10) is known as the Slonczewski–Weiss–McClure (SWMcC) hamiltonian. In total, there are seven band parameters in the SWMcC hamiltonian: γ_0, γ_1, γ_2, γ_3, γ_4, γ_5 and Δ. Their magnitude and physical significance are described in Table 2. Each of the seven parameters of the SWMcC model can be identified with overlap and transfer integrals within the framework of the tight-binding approach, but in practice, they are evaluated experimentally (see Table 2). The SWMcC model was generalized by Johnson and Dresselhaus [46], adding the dispersion relation for the π-bands.

The eigenvalues of the SWMcC hamiltonian correspond to the energy dispersion relations $\left\{\varepsilon\left(\vec{k}\right)\right\}$ schematically illustrated in Fig. 5 (reproduced from Ref. [45]). Along the Brillouin zone edge *HKH*, two of the four solutions are doubly degenerate (labeled E_3). The remaining two solutions are nondegenerate (denoted by E_1 and E_2).

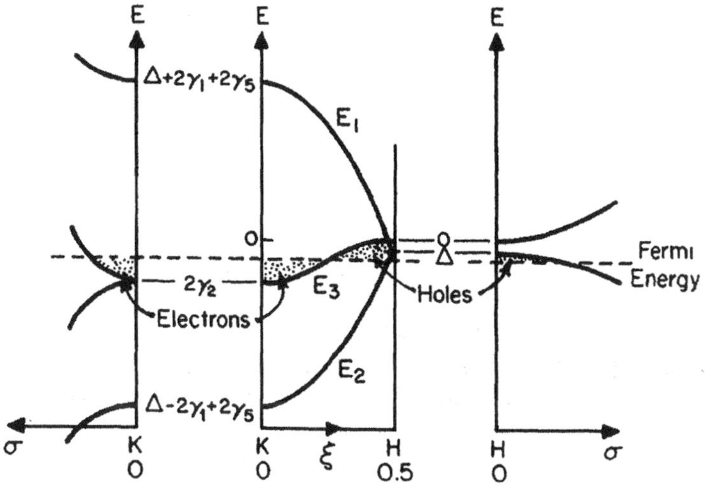

Fig. 5. Electronic energy bands near the *HK* axis in 3D graphite as obtained from SWMcC band model $\left(\xi = \frac{c_0 k_z}{2\pi}\right)$. After Ref. [45].

The degeneracy of the two E_3 levels is lifted when moving away from the zone edge; this is shown on the left-hand panel for the plane $k_z = 0$. At the H point ($k_z = \frac{\pi}{c_0}$), the levels E_1 and E_2 are degenerate; this degeneracy is conserved on the planes $k_z = \pm\frac{\pi}{c_0}$, as shown on the right-hand panel of Fig. 5. Note that Fig. 5 predicts the electron Fermi surface to lie around the K point, as opposed to the picture assumed in Refs. [32] and [36] (see Fig. 6).

It is worth noting that whereas 2D graphite is a zero-gap semiconductor, 3D graphite is semimetallic with a band overlap of $2\gamma_2$ (≈ 0.040 eV) and a bandwidth along the Brillouin zone edge of $4\gamma_1$ ($\approx$ 1.56 eV). Therefore, in 3D graphite, the density of states has a finite value at $E = 0$ (Ref. [45]). In the 2D limit, the only nonvanishing band parameter is γ_0, and the energy dispersion relation becomes linear in K.

Table 2.
Slonczewski–Weiss–McClure (SWMcC) band parameters for graphite. After Refs. [38], [8] and [9]. References are for the experimental determination of the band parameter values.

Band parameter	Value (in units eV)	Physical origin	Ref.
γ_0	3.16 ± 0.05	Overlap of neighboring atoms in a single layer plane. Determines 2D properties	[39]
γ_1	0.39 ± 0.01	Overlap of orbitals associated with carbon atoms located one above the other in adjacent layer planes. Width of π-bands at K equals $4\gamma_1$.	[40]
γ_2	-0.020 ± 0.002	Interactions between atoms in the next nearest layers and from coupling between π- and σ-bands. Band overlap equals $2\gamma_2$. The Fermi surface is most sensitive to γ_2.	[41,43]
γ_3	0.315 ± 0.015	Coupling of the two E_3 bands by a momentum matrix element. Trigonal warping of the Fermi energy is determined by γ_3.	[42]
γ_4	$\sim 0.044 \pm 0.024$	Coupling of E_3 bands. Determines inequality of K-point effective masses in valence and conduction bands.	[43]
γ_5	0.038 ± 0.005	Interactions between second nearest layer planes. Introduced in E_1 and E_2 to be consistent with E_3 in the order of the Fourier expansion of $E(k)$.	[40]
Δ	-0.008 ± 0.002	Difference in crystal fields experienced by inequivalent carbon sites in layer planes. Volume of minority hole carrier pocket is sensitive to Δ.	[39]
E_F	-0.024 ± 0.002	The Fermi level is measured with respect to the H-point extremum. It is fixed by the condition that the electron concentration must be equal to the hole concentration.	[44,43]

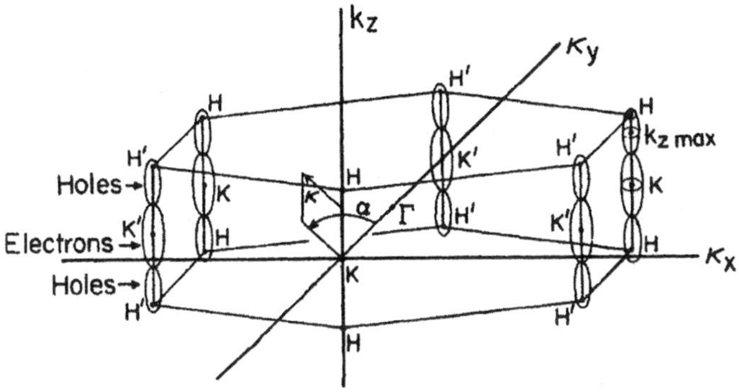

Fig. 6. Graphite's first Brillouin zone showing several high symmetry points and a schematic version of electron and hole Fermi surfaces located along the *HKH* and *H'K'H'* axes. After Ref. [45].

The SWMcC model has been applied to interpret a large variety of experiments related to the electronic structure and the Fermi surface of graphite [8, 38]. First-principles 3D calculations of the electronic band structure of graphite are in good agreement with the widely used SWMcC model near the Fermi level [47].

5 Graphite: Magnetic Susceptibility

As already noted, knowledge of the electronic band structure $\{\varepsilon(\vec{k})\}$ of graphite (Sec. 4) allows the calculation of its magnetic susceptibility χ. Such a calculation is of interest not only for graphite in particular, but also for the general theory of diamagnetism due to conduction electrons. Indeed, experiments have revealed a considerable diamagnetic susceptibility for graphite, yet from theory it is known that the 2D graphite (a single graphite sheet) is a zero-gap semiconductor. For a finite gap semiconductor $\frac{\partial N}{\partial \mu} = 0$ and the direct application of the Landau formula, Eq. (2.21), is not possible. Since graphite is close to this case, a more refined calculation of χ is required.

It should be mentioned that, in contrast to diamond (Sec. 3), for graphite, only one contribution to χ is considered in all theoretical approaches, namely the diamagnetic term due to the conduction electrons. This has become a common starting point, although a rigorous theoretical justification is missing.

5.1 Hove's Single Band Model

The first attempts to describe the diamagnetism of graphite were based on the Landau theory [48]: taking a 2D approximation for the band structure (like in Wallace's original paper, Ref. [31]), the magnetic susceptibility could be explained in a qualitative way. However, difficulties arose when making quantitative comparisons with experiments. For this reason, Hove made a more detailed study using a 3D band structure [33]. He applied the nondegenerate perturbation theory of Peierls [17] and Wilson [18], according to which the conduction electron contribution to the magnetic susceptibility (per unit volume) reads

$$\chi = \frac{\mu_{\mathrm{B}}^2 m_e^2}{12\pi^2 h^4} \int d\vec{k} \left[\frac{\partial^2 \varepsilon}{\partial k_x^2} \frac{\partial^2 \varepsilon}{\partial k_y^2} - \left(\frac{\partial^2 \varepsilon}{\partial k_x \partial k_y} \right)^2 \right] \frac{\partial f_0}{\partial \varepsilon}, \tag{5.1}$$

for a uniform magnetic field $\vec{H} = H\vec{e}_z$. In Eq. (5.1), the integration runs over the whole $\vec{k}$ space and f_0 is the Fermi-Dirac distribution function. Hove [33] employed the 3D Wallace electron energy band model exhibiting two touching bands (see Ref. [31] and Sec. 4), having the following energy values:

$$\varepsilon = -\gamma_1 \cos\left(\frac{c_0 k_z}{2} \right) \pm \sqrt{\gamma_1^2 \cos^2\left(\frac{c_0 k_z}{2} \right) + \frac{3}{4} \gamma_0^2 a_0^2 \kappa_{xy}^2}\,. \tag{5.2}$$

Here, it is understood that the origin of $\vec{k}$ is located at a corner of the (hexagonal) first Brillouin zone and that the lower band (–) is completely filled while the upper one (+) is completely empty. In Eq. (5.2), $\kappa_{xy}^2 = \kappa_x^2 + \kappa_y^2$. The overlap integrals γ_0 and γ_1 may be regarded as constants to be determined by experiment. Using Eq. (5.2), one can rewrite Eq. (5.1) into the following form:

$$\chi = \frac{\mu_{\mathrm{B}}^2 m_e^2}{4\pi^2 h^4 k_{\mathrm{B}} T} \frac{a_0^2}{c_0} \frac{\gamma_0^2}{\gamma_1} \int_{-\infty}^{\infty} \partial\varepsilon \cosh^{-2}\left(\frac{\varepsilon - \Delta\mu}{2k_{\mathrm{B}}T} \right) h\left(\frac{\varepsilon}{\gamma_1} \right), \tag{5.3}$$

where $\Delta\mu$ is the Fermi energy measured from the top of the filled band. The even function $h(x)$ is rather complicated [33].

The variation of χ with temperature has been reported in a number of publications [1, 49, 50]. The best fit of expression (5.3) to experimental data is obtained for γ_1 between 0.5 eV and 0.8 eV, with 0.5 eV preferable. This is about five times larger than the value $\gamma_1 = 0.09$ eV estimated from the theoretical calculation of Coulson and Taylor [51]. The fitting procedure further gives a value of 10.5 eV for γ_0, which is very different from estimated values of γ_0 between 1 and 3 eV [51]. Therefore, the value thus obtained for the actual magnitude of the magnetic susceptibility is lower than that observed by a factor of forty. Hove then concluded that the use of single band theory for graphite is satisfactory in explaining all relative variations of the susceptibility but fails in predicting the magnitude [33].

5.2 McClure's 2D Calculation

As seen in Sec. (4), two bands are near in energy in the vicinity of the H-point in $\vec{k}$-space. In such cases, band-to-band transitions caused by the magnetic field should be taken into account. In his first work on graphite, Ref. [34], McClure calculated the energy levels in the presence of a magnetic field employing the 2D Wallace band structure (Ref. [31]). He used the formalism of Luttinger and Kohn [52], specially designed to treat the effects of perturbations on degenerate and nearly-degenerate bands. The hamiltonian is first written down in terms of base functions, being the unperturbed solutions at a point of degeneracy in $\vec{k}$-space, times plane-wave factors $e^{i\vec{k}\cdot\vec{r}}$ [52]. The coupling between degenerate states and all others is removed to first order by a canonical transformation, yielding a set of coupled differential equations [52, 53]. McClure used a slightly generalized version of the equations [34]:

$$\sum_{j'}\left[\left(k_\alpha-\delta_{x\alpha}is\frac{\partial}{\partial k_y}\right)D_{jj'}^{\alpha\beta}\left(k_\beta-\delta_{\beta x}is\frac{\partial}{\partial k_y}\right)+\frac{\hbar}{m_e}\left(k_\alpha-\delta_{x\alpha}is\frac{\partial}{\partial k_y}\right)p_{jj'}^{\alpha}\right]B_{j'}(\vec{k})$$

$$=\varepsilon B_j(\vec{k}). \qquad (5.4)$$

Here, $s=\frac{eH}{\hbar c}$, ε is an energy eigenvalue in the presence of a field, $B_j(\vec{k})$ is the wave function in $\vec{k}$-space, indices j and j' refer to the degenerate states, $p_{jj'}^{\alpha}$ is the α-th cartesian component of a momentum matrix element and the numbers $D_{jj'}^{\alpha\beta}$ are analogous to effective-mass components. The repeated indices α and β stand for summations over x, y and z. The functions $B_j(\vec{k})$ are required to be periodic in $\vec{k}$-space. As before, Pauli spin paramagnetism has been neglected.

In order to use Eq. (5.4), McClure first calculated the momentum matrix elements $p_{jj'}^{\alpha}$. Then, neglecting all $D_{jj'}^{\alpha\beta}$, he managed to rewrite the system of Eqs. (5.4) as

$$\hbar v(-ik_x-s\frac{\partial}{\partial k_y}+k_y)B_2=\varepsilon B_1, \qquad (5.5a)$$

$$\hbar v(ik_x+s\frac{\partial}{\partial k_y}+k_y)B_1=\varepsilon B_2, \qquad (5.5b)$$

with v the common magnitude of the velocities of the charge carriers near the zone edge.

McClure solved the problem of finding the degeneracies of the levels by reasoning as follows. The Fourier transforms of the B-functions are factors in the direct-space wave function. The complex exponential factor in the B-function causes the center of the direct-space wave function to lie at $y' = \frac{k'_x}{s}$. Confining y' to a normalized volume then limits k'_x, and the counting of states proceeds as in the free-electron case. Finally, the magnetic susceptibility is obtained by calculating the Helmholtz free energy and taking the appropriate derivative. Assuming that the level splitting is small compared to the thermal energy (high-temperature limit), McClure obtained

$$\chi = -0.0014 \frac{\gamma_0^2}{T} \cosh^{-2}\left(\frac{\mu}{2k_B T} \right), \tag{5.6}$$

in units emu/g if γ_0 is expressed in units eV and T in kelvin. (μ is the energy of the Fermi level.) The experimentally determined anisotropic susceptibility at high temperatures is $-\frac{0.01}{T}$ emu/g [1]. The experimental data can be reproduced for $\gamma_0 = 2.6$ eV, which is within the range of the theoretical estimates of Coulson and Taylor [51]. When the energy level spacing is larger than the thermal energy, the calculation of McClure does give the de Haas–van Alphen effect, but the results do not fit the observed data. McClure concluded that a more accurate band structure is needed to account for low-temperature experiments.

5.3 McClure's 3D Calculation

McClure presented his calculation of the conduction electron diamagnetism based on graphite's 3D band structure as Ref. [37]. A magnetic field $\vec{H}$ parallel to the z-axis is introduced via the vector potential $\vec{A} = -Hy\vec{e}_x$. Correspondingly, one replaces k_x by $k_x + \frac{ieH}{\hbar c}\frac{\partial}{\partial k_y}$ in the Luttinger–Kohn formalism. By a canonical transformation of the initial Slonczewski–Weiss–McClure hamiltonian [Eq. (4.10)], McClure obtained an effective, field-dependent hamiltonian. The secular equation derived from this effective hamiltonian reads

$$B^2 n(n+1) - \left(n + \frac{1}{2}\right) B \left(\frac{E - E_1}{(1-\nu)^2} + \frac{E - E_2}{(1+\nu)^2} \right)(E - E_3) + \frac{(E - E_1)(E - E_2)(E - E_3)^2}{(1-\nu^2)^2} = 0, \tag{5.7}$$

where $\nu = \Gamma \frac{\gamma_4}{\gamma_0}$ and $B = \frac{\gamma_0^2 3a_0^2 eH}{2\hbar c}$; n labels Landau energy levels. Obtaining the energy eigenvalues requires the solution of Eq. (5.7), a quartic equation. However, in many cases, Eq. (5.7) can be reduced to a quadratic equation.

As for the 2D case [34], the magnetic susceptibility follows from the Helmholtz free energy, $\chi = -\frac{1}{H}\frac{\partial F}{\partial H}$. The free energy is given by

$$F = N\mu + \mu_B \sum_i \sum_n g(E_{i,n}), \tag{5.8a}$$

$$g(E) = -k_B T \ln\left(1 + e^{\frac{\mu - E}{k_B T}}\right). \tag{5.8b}$$

Here, i is the band branch index which specifies one of the four solutions for a given n. The final result reads:

$$\chi_0 = N_0 \gamma_0^2 \int_{-\pi}^{\pi} d\xi (D + I), \tag{5.9}$$

with

$$D = -\frac{1}{12}(\omega_1[f(E_1) - f(E_3)] + \omega_2[f(E_2) - f(E_1)]), \tag{5.10a}$$

$$I = \int_{E_3}^{E_1} dE\left(-\frac{\partial f}{\partial E}\right)\phi_1 - \int_{E_3}^{E_2} dE\left(-\frac{\partial f}{\partial E}\right)\phi_2. \tag{5.10b}$$

Here, $f(E)$ is the Fermi-Dirac distribution function ($f = \frac{\partial g}{\partial E}$), and

$$\xi = c_0 k_z, \tag{5.11a}$$

$$N_0 = \left(\frac{e}{\hbar c}\right)^2 \frac{3a_0^2}{2\pi^2 c_0}, \tag{5.11b}$$

$$\omega_1 = \frac{(1-\nu)^2}{E_1 - E_3}, \tag{5.11c}$$

$$\omega_2 = \frac{(1+\nu)^2}{E_2 - E_3}, \tag{5.11d}$$

$$\phi_1 = \frac{1}{8\Omega_0} \ln \frac{(E - E_3)(E_1 - E_2)}{\Omega(E_1 - E_3)(1+\nu)^2}, \tag{5.11e}$$

$$\phi_2 = \frac{1}{8\Omega_0} \ln \frac{(E - E_3)(E_1 - E_2)}{\Omega(E_2 - E_3)(1-\nu)^2}, \tag{5.11f}$$

$$\Omega = -\frac{E - E_1}{(1-\nu)^2} + \frac{E - E_2}{(1+\nu)^2}. \tag{5.11g}$$

If $E_1 = E_2$, which is the case for the top and the bottom of the Brillouin zone, and $\Delta \to 0$ (we recall that Δ is one of the parameters of the SWMcC tight-binding model, Sect. 4.3), χ_0 reduces to the 2D result [34]: a susceptibility proportional to the derivative of the Fermi function evaluated at the degeneracy energy. For very high temperatures ($k_B T \gg |E_1 - E_3|, |E_2 - E_3|$), the 2D result holds for the entire susceptibility. At low temperatures, the integral term I is more important than the difference term D. The latter always contributes paramagnetically. In the high-temperature limit, the integral term gives a contribution equal to 1.5 of the total, and the difference term cancels the extra 0.5.

McClure argued that his result explains the anisotropy of the charge carrier diamagnetism. Indeed, in his calculation, a magnetic field parallel to the c_0-axis is considered. If the magnetic field is perpendicular to the c_0-axis, the interband interaction is not important and the Landau formula is valid. For the same γ_0 and γ_1, it gives about one seventh of the $\vec{H} \parallel \vec{e}_z$ result.

Calculations have been made for a variety of band parameter values; the best fit with experimental data [1] is achieved for $\gamma_0 = 2.8$, $\gamma_1 = 0.27$, $\gamma_2 = 0.02$, $\gamma_4 = -0.03$, $\gamma_5 = 0.02$, and $\Delta = 0.025$, all in units eV. (The value of γ_3 is irrelevant, as will follow from the next subsection.) The values of γ_0 and γ_1 were chosen to fit the experimental curve at high temperatures. It was found that the only parameter having an appreciable effect on the susceptibility above 300K is γ_1. At 100K, the parameter Δ is the most effective. The quoted parameters also give the correct de Haas–van Alphen periods, with electron and hole masses of $0.03\, m_e$ and $0.06\, m_e$, respectively, for a Fermi level of 0.029 eV. This Fermi level (0.029 eV) corresponds to impure (doped) graphite. For pure graphite and the band parameters used it is 0.027 eV.

5.4 Calculation of Sharma, Johnson, and McClure

General formulas for calculating diamagnetism due to band electrons are always complicated. In his calculations, McClure [34, 37] considered it easier to solve the Landau levels directly before evaluating the partition sum. Sharma, Johnson and McClure [54] performed a new calculation of graphite's diamagnetism, using the general formula for diamagnetism due to band electrons derived by Fukuyama [55]:

$$\chi = \frac{e^4}{(\hbar c)^2 \pi^3} k_B T \sum_{n=-\infty}^{+\infty} \int d\vec{k} Tr\left[\gamma^x g \gamma^y g \gamma^x g \gamma^y g\right]. \tag{5.12}$$

Here, γ^x and γ^y are components of the momentum matrix and g is the Matsubara Green's function [56]:

$$g = (z_n + \mu - \hat{H})^{-1}, \tag{5.13}$$

with $z_n = (2n+1) i\pi k_B T$, $\hat{H}$ the hamiltonian in absence of a magnetic field and μ the Fermi energy. In Eq. (5.21), the integration over $\vec{k}$ runs over the first Brillouin zone, and the trace is taken with respect to the band indices. The susceptibility is in units emu/cm^3; spin degeneracy is included. Fukuyama has shown that formula (5.12) is equivalent to the previous — more complicated — formulas when the Matsubara sum over n is performed before carrying out the matrix algebra and the integrations. The approach of Sharma *et al.* takes the effect of trigonal warping of the Fermi surface into account, while before [34, 37], McClure assumed that the Fermi surface has rotational symmetry about the magnetic field direction [57].

Using the $\vec{k}\cdot\vec{p}$ approximation, Sharma *et al.* applied Eq. (5.12) to the Slonczewski-Weiss (SW) hamiltonian and obtained

$$\chi = \chi_0 + \delta\chi, \tag{5.14}$$

where χ_0 is independent of γ_3 and $\delta\chi$ is the correction up to second-order in γ_3. The expression for χ_0 is equivalent to Eq. (5.9), previously obtained by McClure [37]. The extra term $\delta\chi$ in Eq. (5.14) is given by

$$\delta\chi = -\frac{N_0\gamma_3^2}{4}\int_{-\pi}^{\pi} d\xi\Gamma^2 \frac{\partial}{\partial E_3}\left[\int_{E_2}^{E_1} dE\frac{G(E)-G(E_3)}{E-E_3} + G(E_3)\ln\left(\frac{1-\nu}{1+\nu}\right)^2 + G_1(E_3) - \frac{8}{3}f(E_3)\right], \tag{5.15}$$

where

$$G(E) = f(E)\frac{x_1+x_2}{(x_1-x_2)^3}\left(\frac{7}{2}x_1^2 + 24x_1x_2 + \frac{7}{2}x_2^2\right), \tag{5.16a}$$

$$G_1(E) = \frac{31}{2}f(E)\left(\frac{x_1+x_2}{x_1-x_2}\right)^2. \tag{5.16b}$$

Here

$$x_1 = (1+\nu)^2(E-E_1), \tag{5.17a}$$

$$x_2 = (1-\nu)^2(E-E_2). \tag{5.17b}$$

Up to second order in γ_4/γ_0, the high-temperature limits of χ_0 and $\delta\chi$ read

$$\chi_0 = -\frac{\pi N_0\gamma_0^2}{6k_B T}\left[1-\frac{3}{5}\left(\frac{\gamma_4}{\gamma_0}\right)^2\right], \tag{5.18a}$$

$$\delta\chi = \frac{\pi N_0 \gamma_3^2}{6k_B T}\left[1 - \frac{24}{5}\left(\frac{\gamma_4}{\gamma_0}\right)^2\right]. \tag{5.18b}$$

The corrections due to γ_4 are small — typically, $\left(\frac{\gamma_4}{\gamma_0}\right)^2 \approx 0.003$ — while γ_3 results in a correction to χ of 1% at high temperatures. At low temperatures, χ_0 is approximately proportional to $\frac{\gamma_0^2}{\gamma_1}$ and $\delta\chi$ to $\frac{\gamma_3^2}{\gamma_2}$. The fractional correction $\frac{\delta\chi}{\chi_0}$ is thus significantly larger for this limit.

To obtain χ as a function of both temperature and Fermi level, the integrations over E and ξ must be done numerically. The zero-temperature susceptibility as a function of the Fermi level is shown in Fig. 7. Calculations show that the effect of taking $\gamma_3 = 0.3$ eV (while keeping the other SW parameters, see caption of Fig. 7) is an increase of the diamagnetism by 13% at low temperature and a reduction by 1% at high temperature.

To calculate χ at finite temperature, Sharma *et al.* used the relation

$$\chi(\mu,T) = -\int_{-\infty}^{+\infty} \partial E\, \chi(E,0) \frac{\partial f(E-\mu)}{\partial E}. \tag{5.19}$$

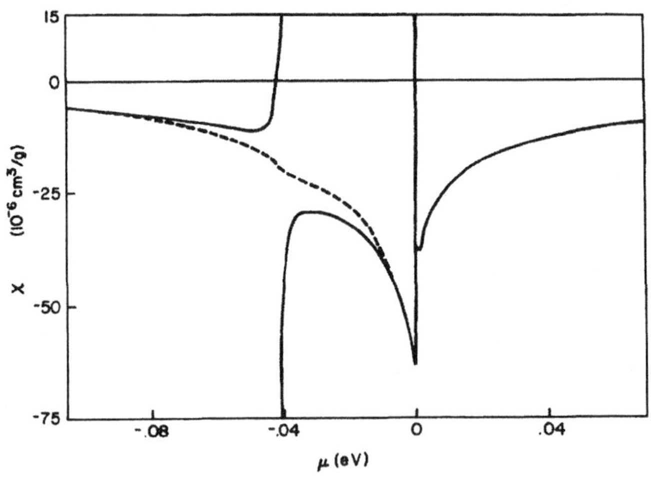

Fig. 7. Magnetic susceptibility of graphite at 0K as a function of the Fermi level; taken from Ref. [54]. The parameters of the Slonczewski-Weiss model are (in units eV): $\gamma_0 = 3.11$, $\gamma_1 = 0.376$, $\gamma_2 = -0.0208$, $\gamma_4 = 0.197$, $\gamma_5 = -0.01$ and $\Delta = -0.012$. The solid curve is for $\gamma_3 = 0.29$ eV, the broken curve for $\gamma_3 = 0$. Note that χ changes sign at $\mu = 0$ and -0.4 eV (for $\gamma_3 = 0.29$ eV).

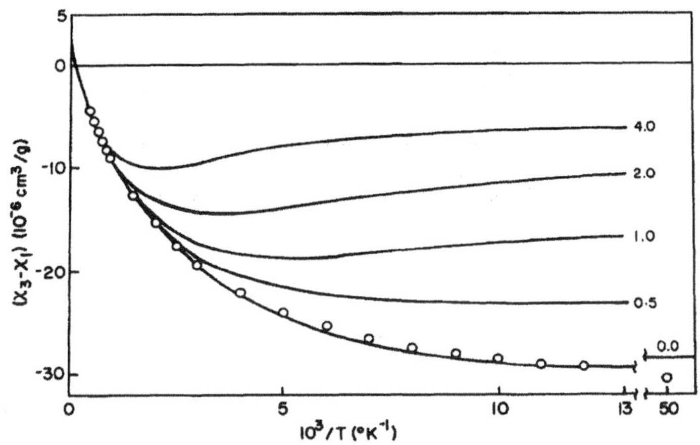

Fig. 8. Magnetic anisotropy of graphite as a function of inverse temperature, taken from Ref. [54]. The circles represent the experimental data of Ref. [6]. The numbers on the curves represent the difference of free holes and electrons per carbon atom.

Experimentally, the magnetic susceptibility of pure graphite as a function of temperature has been determined by the so-called torque method [6, 1]. The experiment yields the magnetic anisotropy $\chi_3 - \chi_1$, where χ_3 is the magnetic susceptibility along the c_0-axis and χ_1 the susceptibility perpendicular to the c_0-axis. From theory, it is known that

$$\chi_3 - \chi_1 = \chi + \chi_B, \tag{5.20}$$

where χ_B is independent of the temperature and the Fermi level. Formula (5.20) has been used to obtain a good fit to the experimental data, resulting in $\chi_B = 2.0 \times 10^{-6}\,\text{cm}^3/\text{g}$. The comparison between experiment [6] and theoretical calculations for the set of parameters given in the caption of Fig. 7 is shown in Fig. 8. Sharma *et al.* concluded that that parameter set agrees with the diamagnetism, the de Haas–van Alphen effect and the magnetoreflection at the H and K points. However, there is a discrepancy with the value of γ_3 found from cyclotron resonance measurements [58, 59].

7 Concluding Remarks

While there is only one theoretical calculation on the diamagnetism of diamond [2], the electronic band structure of graphite and its magnetic susceptibility χ have been the subject of intense study in the past, which has resulted in a well-established theoretical model [9, 8].

Based on the tight-binding electronic band model of graphite formulated by Slonczewski and Weiss [32], and McClure [36, 37] (SWMcC hamiltonian), Sharma, Johnson and McClure [54] have calculated of the diamagnetic susceptibility of graphite, which accounts for the measured data [1, 6, 5, 45, 38] to a high degree of accuracy. The values of band parameters extracted from susceptibility measurements agree very well with other experiments (optical and magneto-optical, Shubnikov–de Haas and de Haas–van Alphen effects). We conclude that correlation effects are not important for graphite and that the one-electron approach has been proved to be adequate.

It has been shown that the knowledge of χ of graphite and diamond can be used further to characterize the degree of graphitization of a carbon material [61]. For highly-ordered carbons, measurement of the anisotropy of χ provides a sensitive characterization technique. On the other hand, Flandrois and coworkers [9, 38, 60] have demonstrated that from magnetic susceptibility experiments one can obtain band parameters of various types of graphite (KISH, HOPG, PGCCL, etc.), which in turn give a very detailed knowledge of its electronic band structure.

In this review we have considered only the hexagonal form of graphite. The magnetic properties of the rhombohedral structure have been studied experimentally and theoretically, for a review see Ref. [62] and references therein.

Finally, theoretical models established for graphite have been of much use for the understanding of electronic properties of other carbon materials — fullerenes and nanotubes [63].

Acknowledgements

The authors thank T.L. Makarova for inviting them to write this contribution. This work has been financially supported by the Bijzonder Onderzoeksfonds, Universiteit Antwerpen (BOF-NOI) and by the Fonds voor Wetenschappelijk Onderzoek, Vlaanderen. B.V. is a research assistant of the Fonds voor Wetenschappelijk Onderzoek - Vlaanderen.

References

1. N. Ganguli and K.S. Krishnan, Proc. Roy. Soc. (London) A177 (1941) 168.
2. S. Hudgens, M. Kastner, and H. Fritzsche, Phys. Rev. Lett. 33 (1974) 1552.
3. R.C. Haddon, L.F. Schneemeyer, J.V. Waszczak, S.H. Glarum, R. Tycko, G. Dabbagh, A.R. Kortan, A.J. Muller, A.M. Musjsce, M.J. Rosseinsky, S.M. Zahurak, A.V. Makhija, F.A. Thiel, K. Raghavachari, E. Cockayne, and V. Elser, Nature (London) 350 (1991) 46.
4. R.S. Ruoff, D. Beach, J. Cuomo, T. McGuire, R.L. Whetten, and F. Diedrich, J. Phys. Chem. 95 (1991) 3457.
5. J. Heremans, C.H. Olk, and D.T. Morelli, Phys. Rev. B49 (1994) 15122.
6. E. Poquet, N. Lumbroso, J. Hoarau, A. Marchand, A. Pacault, and D.E. Soule, J. Chim. Phys. 57 (1960) 866.
7. K.S. Krishnan, Nature (London) 133 (1934) 174.

8. M.S. Dresselhaus, G. Dresselhaus, K. Sugihara, I.L. Spain, and H.A. Goldberg, in: Graphite Fibers and Filaments, Springer Series in Materials Science 5 (1988):, Springer, Berlin.
9. S. Flandrois, in: Graphite and Precursors. World of Carbon (2001) ed. by P. Delhaes, Gordon and Breach, Netherlands.
10. D.C. Mattis, in: The Theory of Magnetism I. Statics and Dynamics (1988) 21, Springer, Berlin.
11. N.W. Ashcroft and N.D. Mermin, Solid State Physics (1976), Holt, Rinehart and Winston.
12. L.D. Landau and E.M. Lifshitz, Course of theoretical physics 5 (1997): Statistical Physics, Pergamon, Oxford.
13. M. Tinkham, Group Theory and Quantum Mechanics (1964) McGraw-Hill, New York.
14. L. Landau, Z. Physik 64 (1930) 629.
15. L.D. Landau and E.M. Lifshitz, Course of theoretical physics 3 (1997): Quantum Mechanics: non-relativistic theory, Pergamon, Oxford.
16. R. Bowers, J. Phys. Chem. Solids 8 (1959) 206.
17. R.E. Peierls and Z. Physik 80 (1933) 763; R.E. Peierls, Quantum Theory of Solids (1974) Clarendon, Oxford.
18. A.H. Wilson, Theory of Metals, (1936) University Press, Cambridge; A.H. Wilson, Proc. Cambridge Phil. Soc. 49 (1953) 292.
19. J.E. Hebborn and E.H. Sondheimer, J. Phys. Chem. Solids 13 (1960) 105.
20. J.E. Hebborn, J.M. Juttinger, E.H. Sondheimer, and P.L. Stiles, J. Phys. Chem. Solids 25 (1964) 741.
21. T. Kjeldaas and W. Kohn, Phys. Rev. 105 (1957) 806.
22. L.M. Roth, J. Phys. Chem. Solids 23 (1962) 433.
23. E.I. Blount, Phys. Rev. 126 (1962) 1636.
24. G.H. Wannier and U.N. Upadhaya, Phys. Rev. A136 (1964) 803
25. A.G. Samoilovich and E. Y. Rabinovich, Soviet Fiz. Tverd. Tela 5 (1963) 567.
26. M.L. Glasser, Phys. Rev. A 134 (1964) 1296.
27. P.K. Misra and L.M. Roth, Phys. Rev. 177 (1969) 1089.
28. J.C. Philips, Rev. Mod. Phys. 42 (1970) 317.
29. C. Kittel, in: Quantum Theory of Solids (1996) 201, 7th edition, John Willey & Sons, New York.
30. Landolt-Börnstein New Series III, 22a (1987) in: Zahlenwerte und Funktionen aus Naturwissenschaften und Technik, ed. by O. Madelung and M. Schulz, Springer, New York.
31. P.R. Wallace, Phys. Rev. 71 (1947) 622.
32. J.C. Slonczewski and P.R. Weiss, Phys. Rev. 109 (1958) 272.
33. J.E. Hove, Phys. Rev. 100 (1955) 645.
34. J.W. McClure, Phys. Rev. 104 (1956) 666.
35. J.L. Carter, Thesis (Cornell University, 1953); C. Herring, J. Franklin Inst. 233 (1942) 525.
36. J.W. McClure, Phys. Rev. 108 (1957) 612.
37. J.W. McClure, Phys. Rev. 119 (1960) 606.
38. A. Maaroufi, S. Flandrois, C. Coulon, and J.C. Rouillon, J. Phys. Chem. Solids 43 (1982) 1103.
39. W.W. Toy, M.S. Dresselhaus, and G. Dresselhaus, Phys. Rev. B15 (1977) 4077.
40. A. Misu, E. Mendez, and M.S. Dresselhaus, J. Phys. Soc. Japan 47 (1979) 199.
41. D.E. Soule, J.W. McClure, and L.B. Smith, Phys. Rev. A134 (1964) 453.

42. R.E. Doezema, W.R. Datars, H. Schaber, and A. Van Schyndel, Phys. Rev. B19 (1979) 4224.
43. E. Mendez, A. Misu, and M.S. Dresselhaus, Phys. Rev. B21 (1980) 827.
44. R.O. Dillon, I.L. Spain, and J.W. McClure, J. Phys. Chem. Solids, 38 (1977) 635.
45. M.S. Dresselhaus, G. Dresselhaus, and J.E. Fischer, Phys. Rev. B15 (1977) 3180.
46. L.G. Johnson and G. Dresselhaus, Phys. Rev. B7 (1973) 2275.
47. H. Nagayoshi, M. Tsukada, K. Nakao, and Y. Uemura, J. Phys. Soc. Japan 35 (1973) 396; A. Zunger, Phys. Rev. B17 (1978) 626.
48. R. Smoluchowski, Rev. Modern Phys. 25 (1953) 178; W.P. Eatherly, see comments in discussion following R. Smoluchowski.
49. M. Goldsmith, J. Chem. Phys. 18 (1950) 523.
50. M. Owen, Ann. Physik 37 (1912) 657.
51. C.A. Coulson and R. Taylor, Proc. Phys. Soc. (London) A65 (1952) 815.
52. J.M. Luttinger and W. Kohn, Phys. Rev. 97 (1955) 869.
53. J.M. Luttinger, Phys. Rev. 102 (1956) 1030.
54. M.P. Sharma, L.G. Johnson, and J.W. McClure, Phys. Rev. B9 (1974) 2467.
55. H. Fukuyama, Prog. Theor. Phys. 45 (1971) 704.
56. A.A. Abrikosov, L.P. Gorkov, and I.E. Dzyaloshinski, in: Methods of Quantum Field Theory in Statistical Physics (1963) Prentice-Hall, New York.
57. Y. Yafet, Phys. Rev. 115 (1959) 1172.
58. P.R. Schroeder, M.S. Dresselhaus, and A. Javan, J. Phys. Chem. Solids Suppl. (1971) 139.
59. H. Ushio, T. Uda, and Y. Uemura, J. Phys. Soc. Jap. 32 (1972) 1551.
60. P. Biensan, J.C. Roux, H. Saadaoui, and S. Flandrois, Microsc. Microanal. Microstruct. 33 (1991) 465.
61. A. Pacault, A. Marchand, P. Bothorel, J. Zanchetta, J.V. Boy Cherville, and M.J. Oberlin, J. Chim. Phys. 57 (1960) 892.
62. S. Chehab, K. Guérin, J. Amiell, and S. Flandrois, Eur. Phys. J. B13 (2000) 235.
63. M.S. Dresselhaus, G. Dresselhaus, and P.C. Eklund, in: Science of Fullerenes and Carbon Nanotubes, (1996) Academic, San Diego.

Carbon-Based Magnetism
T. Makarova and F. Palacio (Editors)

12

Electronic and Magnetic Properties of Nanographites

K. Wakabayashi

Institut für Theoretische Physik, ETH-Hönggerberg, CH-8093 Zürich, Switzerland, and Department of Quantum Matter Science, AdSM, Hiroshima University, 739-8530 Higashi-Hiroshima, Japan

1 Introduction

The discovery of fullerene molecules and carbon nanotubes [1, 2] has triggered the intensive research on various nanometer-sized carbon materials. In these systems, the geometry of sp^2 carbon networks crucially affects their electronic states near the Fermi level [3, 4]. Studies with scanning tunneling microscopy and spectroscopy have confirmed the connection between the electronic states of single wall carbon nanotubes (SWCN) and their geometry [5, 6]. Besides the closed carbon molecules, nanometer-sized systems with open boundaries called "nanographites", also display unusual features, where the presence of edges of various shapes determines the electronic states [7]. There are two typical shapes in graphite edges, armchair and zigzag. Zigzag edges possess localized edge states near the Fermi level, while such localized states are completely absent for armchair edges. These edge states correspond to the non-bonding molecular orbitals. They generate a large peak in the density of states and give rise to pronounced effects in electronic, magnetic and transport properties in nanometer-sized graphite systems[7– 12]. Recent experiments for nanographites report direct observation of the edge states [13], the paramagnetic response at low-temperature [14, 15], the ferromagnetic response at the graphite edges [16, 17] and the fabrication of nanographite ribbons[18] and nanographite discs [19]. The purpose of this chapter is to elucidate the role of the edge states in the low-energy physical properties of nanographites.

2 Electronic States of Nanographite Ribbons

There are two typical shapes of a graphite edge, called *armchair* and *zigzag* (Fig. 1(a)). The two edges have 30 degrees difference in their cutting direction. Here I briefly discuss the way that the graphite edges drastically change the π electronic structures. Especially, a zigzag edge provides the localized edge state, while an armchair edge does not show such localized states. We hereafter call these localized states "edge state" [7].

A simple and useful model to study the edge and size effect is one of the graphite ribbon models as shown in Figs. 1(b) and (c). We define the width of graphite ribbons as N, where N stands for the number of the dimer (two carbon sites) lines for the armchair ribbon and by the number of the zigzag lines for the zigzag ribbon, respectively. It is assumed that all dangling bonds at graphite edges are terminated by hydrogen atoms, and thus give no contribution to the electronic states near the Fermi level. Throughout this chapter, it is also assumed that graphite ribbons have the translational invariance, and the unit cells can be defined as shown in Figs. 1(b) and (c). We employ a single-orbital tight binding model for the π electron network. The model has successfully been used for the calculation of electronic states of fullerene molecules, carbon nanotubes and other carbon-related materials [3, 20]. The Hamiltonian is written as,

$$H = \sum_{\langle i,j \rangle} t_{ij} c_i^\dagger c_j, \tag{1}$$

where the operator $c_i^\dagger$ creates an electron on the site i. $\langle i, j \rangle$ denotes the summation over the nearest neighbor sites. t the transfer integrals between all the nearest neighbor sites are set, for simplicity. This is sufficient to show the intrinsic difference in the electronic states originating from the topological nature of each system. The value of t is considered to be about 3.0 eV in graphite [21].

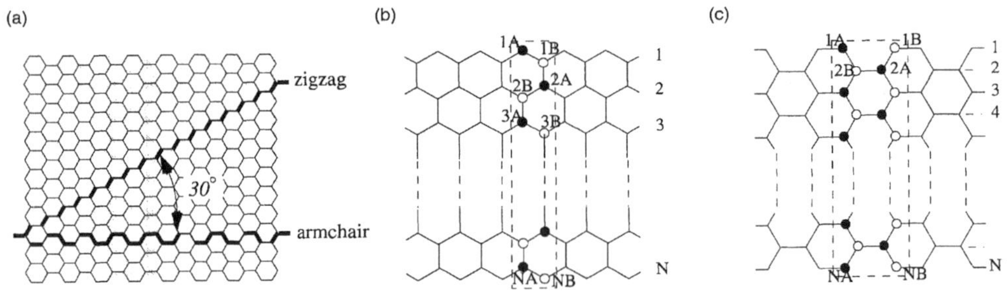

Fig. 1. (a) The typical graphite edge, *armchair* and *zigzag*. The structure of graphite ribbons with (b) zigzag edges and (c) armchair edges. The rectangle with the dashed line is the unit cell.

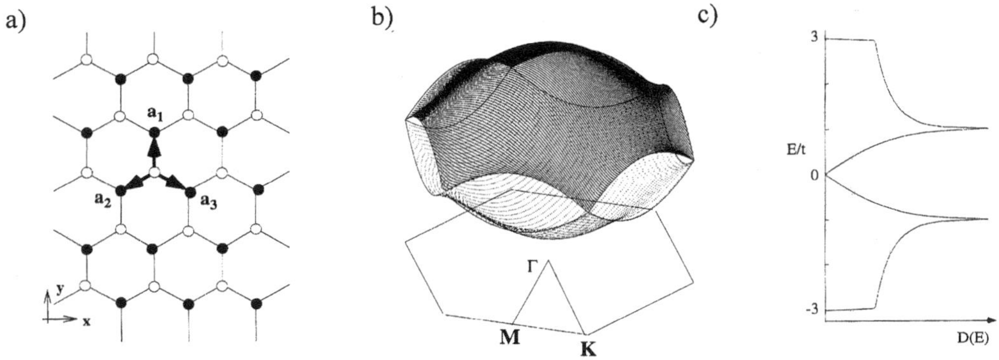

Fig. 2. (a) Graphite sheet in real space, where the black (white) circles mean the A(B)-sublattice site. (b) The π band structure and (c) the density of states of graphite sheet. The valence and conduction bands make contact at the degeneracy point K.

Prior to the discussion of the π-electronic states of graphite ribbon systems, we shall briefly review the π-band structure of a graphite sheet [20]. To diagonalize the Hamiltonian for a graphite sheet, we use a basis of two-component spinor, $c_{\mathbf{k}}^{\dagger} = (c_{A\mathbf{k}}^{\dagger},\ c_{B\mathbf{k}}^{\dagger})$, the Fourier transform of $(c_{i\in A}^{\dagger},\ c_{i\in B}^{\dagger})$. Let a_1, a_2, a_3 be the displacement vectors from a B site to its three nearest-neighbor A sites, defined so that $\hat{z}\cdot a_1 \times a_2$ is positive (Fig. 2(a)). $\hat{z}$ is the normal vector to the graphite sheet. In this representation, the Hamiltonian is written as $H = \sum_{\mathbf{k}} \mathbf{c}_{\mathbf{k}}^{\dagger} H_{\mathbf{k}} \mathbf{c}_{\mathbf{k}}$ and

$$H_{\mathbf{k}} = -t\sum_{i=1}^{3}\left(\cos(k\cdot a_i)\hat{\sigma}_x + \sin(k\cdot a_i)\hat{\sigma}_y\right), \tag{2}$$

where $\hat{\sigma} = (\hat{\sigma}_x, \hat{\sigma}_y, \hat{\sigma}_z)$ are the Pauli matrices. Then, the energy eigenvalues are $E_{\mathbf{k}}^{\pm} = \pm t\,|\sum_{i=1}^{3}\exp(k\cdot a_i)|$. Since one carbon site has one π-electron on average, only $E_{\mathbf{k}}^{-}$-band is completely occupied. Hereafter, we call $E_{\mathbf{k}}^{-}$ [$E_{\mathbf{k}}^{+}$] valence [conduction] band.

In Fig. 2(b) and (c), the energy dispersion of π-bands in the 1st Brillouin Zone (BZ) and the density of states are depicted, respectively. Near the Γ point, both valence and conduction bands have the quadratic form of k_x and k_y, *i.e.* $E_k = \pm(3 - 3|k|^2/4)$. At the M points, the middle points of sides of the hexagonal BZ, the saddle point of energy dispersion appears and the density of states diverges logarithmically. Near the K point

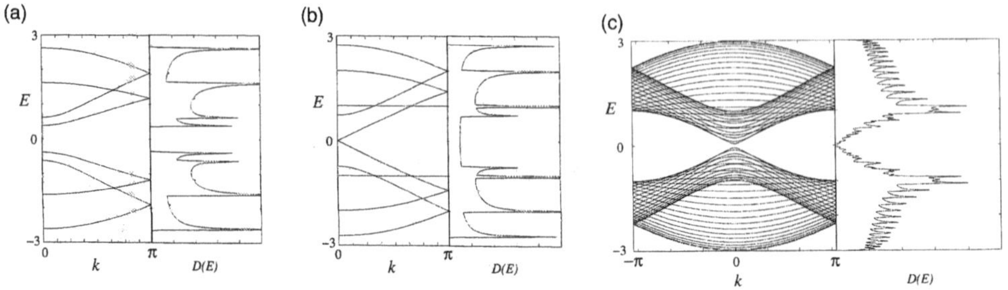

Fig. 3. Energy band structure $E(k)$ and density of states $D(E)$ of armchair ribbons of various widths [(a) $N=4$, (b) 5 and (c) 30].

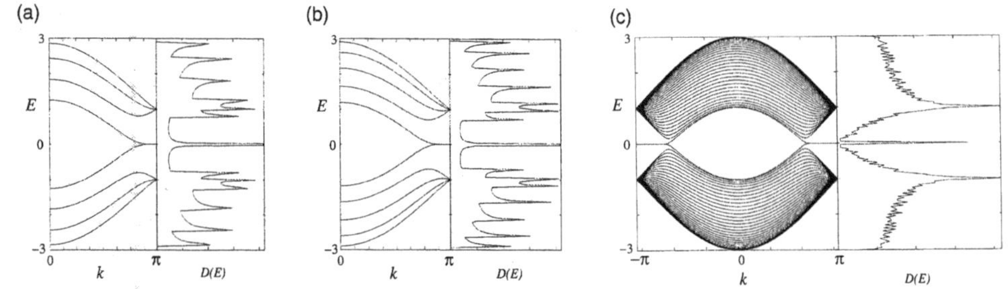

Fig. 4. Energy band structure $E(k)$ and density of states $D(E)$ of zigzag ribbons of various widths [(a) $N=4$, (b) 5 and (c) 30].

of the corner of hexagonal 1st BZ, the energy dispersion is linear in the magnitude of wave vector, $E_k = \pm 3ta|k|/2$, where the density of states linearly depends on energy. Here $a(=|a_i|(i=1,2,3))$. The Fermi energy is located at the K points and there is no energy gap at these points, since E_k vanishes at these points by the hexagonal symmetry. Thus, the graphite sheet is a zero-gap semi-conductor on this 2D model.

The calculated band structures of armchair ribbons are shown in Figs. 3(a)-(c), for three different ribbon widths, together with the density of states. The wave number k is normalized by the length of the primitive translation vector of each graphite ribbon, and the energy E is scaled by the transfer integral t. The top of the valence band and the bottom of the conduction band are located at $k=0$. It should be noted that the ribbon width decides whether the system is metallic or semiconducting . As shown in Fig. 3(b), the system is metallic when $N=3M-1$, where M is an integer. For the semiconducting ribbons, the direct gap decreases with increasing ribbon width and tends to zero in the limit of very large N.

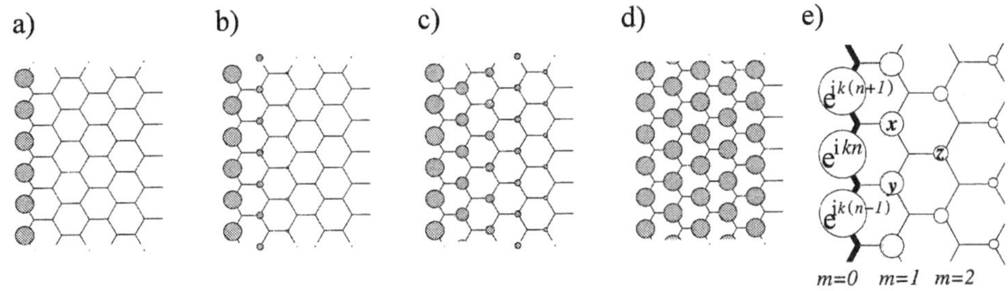

Fig. 5. Charge density plot for analytic solution of the edge states in a semi-infinite graphite, when (a) $k=\pi$, (b) $8\pi/9$, (c) $7\pi/9$ and (d) $2\pi/3$. (e) An analytic form of the edge state for a semi-infinite graphite sheet with a zigzag edge, emphasized by bold lines. Each carbon site is specified by a location index n on the zigzag chain and by a chain order index m from the edge. The magnitude of the charge density at each site, such as x, y and z, is obtained analytically (see text). The radius of each circle is proportional to the charge density on each site, and the drawing is made for $k = 7\pi/9$.

For zigzag ribbons, however, a remarkable new feature arises in the band structure, as shown in Figs. 4(a)-(c). We see that the highest valence band and lowest conduction band are always degenerate at $k=\pi$. It is found that the degeneracy of the center bands at $k=\pi$ does not originate from the intrinsic band structure of 2D graphite. These two special center bands get flatter with increasing ribbon width. A pair of almost flat bands appears within the region of $2\pi/3 \leq |k| \leq \pi$, where the bands sit in the vicinity of the Fermi level. No such flat band is expected for the band structure of 2D graphite.

The electronic state in the partial flat bands of the zigzag ribbons is found to be characterized by the localized state near the zigzag edge via examining the charge density distribution [7–9, 22, 23]. Here we show that the puzzle for the emergence of the edge state can be solved by considering a semi-infinite graphite sheet with a zigzag edge. First to show the analytic form, we depict the distribution of charge density in the flat band states for some wave numbers in Fig. 5(a)-(d), where the amplitude is proportional to the radius. The wave function has non-bonding character, *i.e.* finite amplitudes only on one of the two sublattices which includes the edge sites. It is completely localized at the edge site when $k=\pi$, and starts to gradually penetrate into the inner sites as k deviates from π reaching the extend state at $k=2\pi/3$.

Considering the translational symmetry, we can start constructing the analytic solution for the edge state by letting the Bloch components of the linear combination of atomic orbitals (LCAO) wavefunction be ..., $e^{ik(n-1)}$, e^{ikn}, $e^{ik(n+1)}$,... on successive edge sites, where n denotes a site location on the edge. Then the mathematical condition necessary for the wave function to be exact for $E=0$ is that the total sum of the components of the complex wave function over the nearest-neighbor sites should vanish. In Fig. 5(e), the above condition is $e^{ik(n+1)}+e^{ikn}+x=0$, $e^{ikn}+e^{ik(n-1)}+y=0$ and $x+y+z=0$. Therefore, the wave function components x, y and z are found to be

$D_k e^{ik(n+1/2)}$, $D_k e^{ik(n-1/2)}$, $D_k^2 e^{ikn}$, respectively. Here $D_k = -2\cos(k/2)$. We can thus see that the charge density is proportional to $D_k^{2(m-1)}$ at each non-nodal site of the m-th zigzag chain from the edge. Then the convergence condition of $|D_k| \leq 1$ is required, for otherwise the wave function would diverge in a semi-infinite graphite sheet. This convergence condition defines the region $2\pi/3 \leq |k| \leq \pi$ where the flat band appears.

From the analytic expression of the edge state, we can derive the analytic expression for density of states (DOS) near Fermi energy. As we have seen, the edge state penetrate to inner sites when the wave number changes from π to $2\pi/3$. If we consider the graphite ribbons with width N, two edge states which come from both sides of edge will overlap with each other and develop the bonding and anti-bonding interaction. Since the magnitude of the overlap becomes larger when the wave number approaches $2\pi/3$, the band gap between the bonding and anti-bonding state formed by the two edge states gets larger toward $k = 2\pi/3$. Therefore the partly flat bands acquire a slight dispersion which depends on the ribbon width N. The energy dispersion is calculated by the overlapping of two edge states. The amplitude of the edge state which penetrates from the first zigzag line is given by $\psi_n = D_k^{n-1} \equiv \psi_A$, which is located only on the A-sublattice. On the other hand,the amplitude of the edge state which penetrate from N^{th} zigzag line, is given by $\psi_{N-n} = D_k^{n-1} \equiv \psi_B$, which is located only on the B-sublattice. By using the tight binding Hamiltonian, the overlapping of two edge states is easily calculated,

$$\langle \psi_A | H | \psi_B \rangle = -2tND_k^{N-1}(1 + D_k/2) \equiv \eta_k. \tag{3}$$

Therefore, the energy spectrum of the edge states in lowest order perturbation in the overlap is given by the following eigenvalue problem, *i.e.*

$$\begin{pmatrix} 0 & \eta_k \\ \eta_k & 0 \end{pmatrix} \begin{pmatrix} \psi_A \\ \psi_B \end{pmatrix} = \varepsilon \begin{pmatrix} \psi_A \\ \psi_B \end{pmatrix}. \tag{4}$$

The energy spectrum is $\varepsilon = \pm\eta_k$, which gives the form of $\varepsilon \sim k^N$ around $k = \pi$. Therefore, the DOS related to the edge states has the form,

$$\rho(\varepsilon) = \frac{\partial k}{\partial \varepsilon} \sim \frac{1}{N}\varepsilon^a, \tag{5}$$

where $\alpha = \frac{1}{N} - 1$. Note that this DOS has a power-law dependence, which is different from the ordinary van Hove singularity of $\rho \sim 1/\sqrt{E}$ observed in one-dimensional system. It is also found that the renormalized DOS is inversely proportional to the ribbon width, which has been confirmed by numerical calculation [8].

Next we analytically show that both the energy gap Δ_a at $k=0$ of armchair ribbons and the energy gap Δ_z at $k=2\pi/3$ of zigzag ribbons are inversely proportional to the width of graphite ribbon. This result supports that the physical quantities related to the energy gap can be scaled by the ribbon width.

We examine the energy gap Δ_a at $k=0$ of armchair ribbons. It is easy to find that at $k=0$ the Hamiltonian for the armchair ribbon can be rewritten as

$$H=-t\sum_{j=1}^{N}\left[\sum_{\mu=1}^{2}(a^{\dagger}_{j,\mu}a_{j+1,\mu}+h.c.)+a^{\dagger}_{j,1}a_{j,2}+h.c.\right], \tag{6}$$

which is equivalent to the tight binding model for the 2-leg ladder lattice having N rungs[7,9]. The site indices (j,1) and (j,2) correspond to the j A(B) and j B(A) sites, respectively, when j is even(odd). The eigenvalues are evaluated as $\varepsilon^{\pm}=-2t\cos\left(\frac{n\pi}{N+1}\right)\pm t$ $(n=1,2,\ldots,N)$, the corresponding eigenfunctions are $\Psi_{j\mu}=(\mp)^{\mu+1}B\sin\left(\frac{nj\pi}{N+1}\right)$, where $\Psi_{j\mu}$ is the wavefunction at (j,μ)-site and B is the normalization factor. The system is metallic only when $N=3m-1$, because ε^{+} and ε^{-} become zero for $n=m$ and $2m$, respectively. Note that every third rung has a zero node in the wavefunction when $N=3m-1$. Thus, Δ_a s are given as,

$$\Delta_a=\begin{cases}0 & N=3m-1\\ 2\left[2t\cos\left(\dfrac{m}{3m+1}\pi\right)-t\right] & N=3m\\ 2\left[2t\cos\left(\dfrac{m+1}{3m+2}\pi\right)-t\right] & N=3m+1.\end{cases} \tag{7}$$

After expressing N in terms of $W=(N-3)\frac{\sqrt{3}}{2}+\sqrt{3}$, which is the ribbon width in the unit of the lattice constant a, and the Taylor expansion under the condition of $1/W\ll 1$, we can obtain the following results.

$$\Delta_a\sim\begin{cases}0 & N=3m-1\\ \dfrac{\pi}{W+\frac{\sqrt{3}}{2}} & N=3m\\ \dfrac{\pi}{W} & N=3m+1\end{cases} \tag{8}$$

Thus the Δ_a is inversely proportional to the ribbon width.

Since in the zigzag ribbons two states are degenerate at $k=\pi$, the gap of the zigzag ribbons is always zero and the system is always metallic. However, the energy bands have a gap at $k=2\pi/3$, because the bonding and anti-bonding configurations between two edge states coming from both edges is developed toward $k = 2\pi/3$ from $k = \pi$. According to the projection of the band structure of a graphite sheet onto the 1st BZ of zigzag ribbons, the degenerate points of valence and conduction band of a graphite sheet ties at $k = 2\pi/3$. Thus in the limit of infinite ribbon width the energy gap Δ_z at $k = 2\pi/3$ has to be zero. The Hamiltonian of zigzag ribbons at $k = 2\pi/3$ is rewritten as

$$H = -t\sum_{i=1}^{2N}(a_i^\dagger a_{i+1} + h.c.), \tag{9}$$

which is equivalent to the tight binding model for the one-dimensional lattice having $2N$ sites. The site index i corresponds to iA, if i is an odd number, and to iB, if i is an even number. The eigenvalues are evaluated as $\varepsilon = -2t\cos\left(\frac{n\pi}{2N+1}\right)$ ($n = 1,2,\ldots,2N$) and the corresponding eigenfunction at the jth site, Ψ_j, is evaluated as $\Psi_j = B\sin\left(\frac{nj\pi}{2N+1}\right)$, where B is the normalization factor. Therefore, Δ_z is given as,

$$\Delta_z = 4t\cos\left(\frac{N}{2N+1}\pi\right). \tag{10}$$

Taylor expansion under the condition of $1/N \ll 1$, we can obtain the following results.

$$\Delta_z \sim \frac{\pi}{N} = \frac{3\pi}{2}\frac{1}{W+1} \tag{11}$$

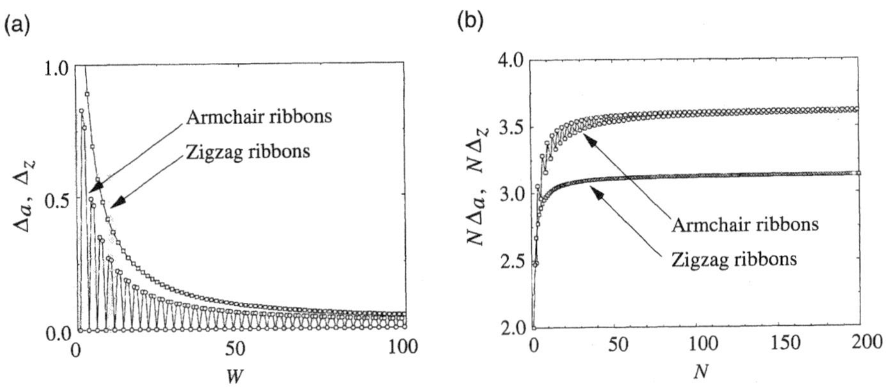

Fig. 6. (a) The width dependence of the energy gap of armchair ribbons at $k = 0$ (Δ_a) and zigzag ribbons at $k = 2\pi/3$ (Δ_z); (b) The plot of $N\Delta_a$ ($N\Delta_z$) versus N.

Thus the Δ_a is also inversely proportional to the ribbon width. Here $W = 3N/2 - 1$ is the ribbon width in the unit of the lattice constant.

In Fig. 6(a), the width dependence of the energy gap is shown for both armchair and zigzag ribbons. Figure 6(b) shows the plot of *N* times the gap Δ, $N\Delta$, versus *N*. The $N\Delta$ becomes constant for values more than about $N = 30$ ($N = 60$) for zigzag (armchair) ribbons.

3 Electronic States in a Magnetic Field

In this section we consider the electronic states of nanographites in a magnetic filed. The magnetic field B perpendicular to the graphite plane is incorporated in the transfer integral t_{ij} by means of the Peierls phase defined as $t_{ij} \to t_{ij}\exp\left[\mathrm{i}2\pi(e/ch)\int_i^j \mathrm{d}l \cdot A\right]$, where A is the vector potential and the integral is along the hopping path. We define the magnitude of the magnetic flux passing through a single hexagon ring of graphite sheet in the unit of the quantum flux ($\phi_0 = ch/e$) as ϕ. Since the single hexagon ring has the area $S_{\text{hex}} = 3\sqrt{3}a^2/2$, where $a = 1.42$ Å is the lattice constant of the graphite, ϕ is given as $\phi = BS_{\text{hex}}/\phi_0$. Here $B = |B|$, and $\phi = 1$ corresponds to 7.9×10^4 T. In this chapter, we treat the weak magnetic field limit. The cyclotron radius (magnetic length), l, is given by $l = \sqrt{c\hbar/eB}$. We use Landau gauge with $A = (0, \mathrm{B}x, 0)$, where we define the translational invariant direction of each ribbon as the y-axis, and the x-axis lies perpendicular to y-axis.

It should be noted that the same number N for both zigzag and armchair ribbons does not give the same ribbon width, when the ribbons are measured by the same unit of length. Therefore, when we compare physical quantities of zigzag and armchair ribbons with the same width W, we will use the following definition

$$W = \begin{cases} \frac{3}{2}Na - a \equiv W_z & \text{zigzag ribbons} \\ (N-1)\frac{\sqrt{3}}{2}a \equiv W_a & \text{armchair ribbons} \end{cases} \tag{12}$$

where a is the lattice constant.

In ribbon-shaped systems, the ratio between the width of a ribbon and cyclotron diameter effectively characterizes the electronic states of ribbon in a magnetic field. Therefore, we define the ratio between the ribbon width and the cyclotron diameter as the effective magnetic field $\tilde{B}$ for a convenience, which is given by $\tilde{B} = (W/d)^2$, where $d = 2l$. In the case of $\tilde{B} < 1$ since the cyclotron diameter is larger than the ribbon width the cyclotron motion of electrons is impeded, and the Landau levels are not formed. Secondly in the case of $\tilde{B} > 1$, since the cyclotron diameter becomes smaller than the ribbon width the cyclotron motion of electrons is not disturbed except in the vicinity of ribbon edges, and the Landau levels appear. Finally in the case of $\tilde{B} \gg 1$, where the cyclotron diameter is much smaller than the ribbon width and becomes the order of the lattice constant, the topology of the lattice becomes important.

Now we show that the band dispersions near $E=0$ can be scaled by the ribbon width W for sufficiently wide ribbons. In Fig. 7, the scaled band structures are shown, when (a) $\tilde{B}=0$, (b) $\tilde{B}=4$ and (c) $\tilde{B}=16$. Here the energy and the wave number k' are scaled by W, and k' means the wave number measured from $k=2\pi/3$. In the figures, the bold lines are dispersions of zigzag ribbons, and the dotted lines are the corresponding Landau levels of graphite sheet for comparison. Similarly, the band

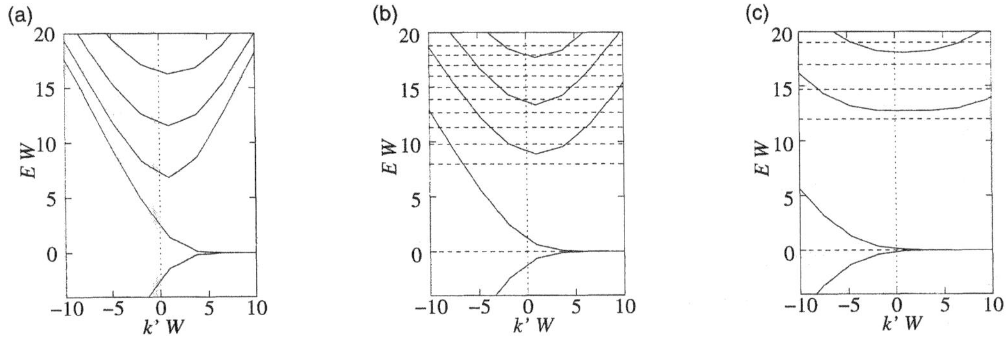

Fig. 7. The scaled energy band structures of zigzag ribbons near $k=2\pi/3$ for (a) $\tilde{B}=0$, (b) $\tilde{B}=4$ and (c) $\tilde{B}=16$. Bold lines are dispersions of zigzag ribbons. Here k' means the wave number measured from $k=2\pi/3$. Dotted lines are the corresponding Landau levels of graphite sheet for comparison.

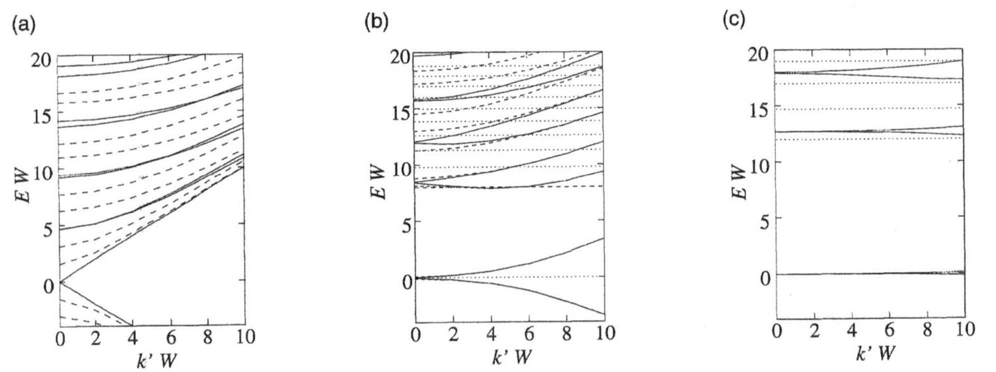

Fig. 8. The scaled energy band structures of armchair ribbons near $k=0$ for (a) $\tilde{B}=0$, (b) $\tilde{B}=4$ and (c) $\tilde{B}=16$. Bold lines are metallic armchair ribbons ($N=3m-1$) and dashed lines are semiconducting armchair ribbons ($N\neq 3m-1$). Dotted lines are the corresponding Landau levels of graphite sheet for comparison.

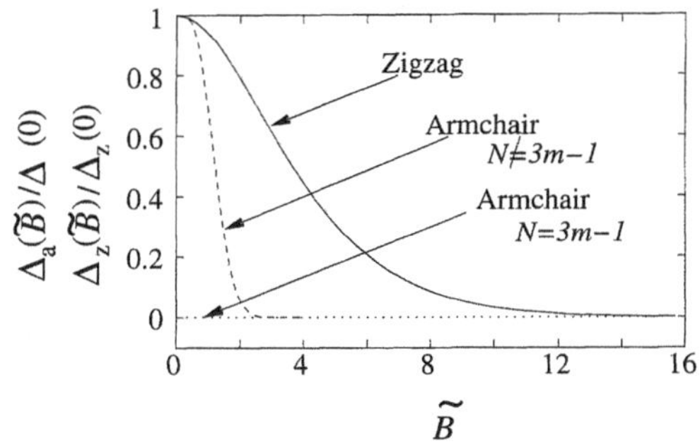

Fig. 9. Magnetic field dependence of the energy gap of armchair (zigzag) ribbons at $k = 0$ ($k = 2\pi/3$), $\Delta_a(\tilde{B})$ ($\Delta_z(\tilde{B})$). The energy gaps are normalized by the energy gap, $\Delta_a(0)$ ($\Delta_z(0)$), of $\tilde{B} = 0$.

dispersions of the armchair ribbons can be scaled. In Figure 8, the scaled band structures are shown, when (a) $\tilde{B} = 0$, (b) $\tilde{B} = 4$ and (c) $\tilde{B} = 16$. Bold lines are metallic armchair ribbons ($N = 3m - 1$) and dashed lines are semiconducting armchair ribbons ($N \neq 3m - 1$). The dotted lines are the corresponding Landau levels of graphite sheet for comparison.

In Fig. 9, the magnetic field dependence of the energy gap for armchair and zigzag ribbons is shown. The energy gap of semiconducting armchair ribbons becomes negligible when the effective magnetic field $\tilde{B}$ is much larger than 2. This condition is same as the case of an usual quantum wire. However, we need stronger magnetic field to collapse the energy gap (at $k = 2\pi/3$) of zigzag ribbons. This is due to the non-bonding character of the edge states. In the zero-field limit, $(1 - \Delta_{a(z)}(\tilde{B})/\Delta_{a(z)}(0))$ depends on $\tilde{B}^4$ for armchair ribbons, and $\tilde{B}^2$ for zigzag ribbons.

4 Bearded Edge and Cove Edge

In this section, we discuss two other important edge shapes having translational symmetry of zigzag axis, called *bearded* and *cove*. Although these two edges look rather artificial than the zigzag edge, they are interesting because they also show the non-bonding edge localization.

A Bearded edge is a zigzag edge with additional π-electron hopping bonds as shown in Fig. 10(a). This type of edge was firstly studied by Klein [24]. In Fig. 10(b), the band structure of a semi-infinite graphite sheet with a bearded edge is shown. Interestingly, a partial flat band appears in the region of $|k| \leq 2\pi/3$, which is the opposite condition in the semi-infinite graphite sheet with a zigzag edge. The analytic form of this edge state can be derived in the similar manner described in the previous section for the derivation of the edge state of zigzag edges. If we consider the case of infinite N (Fig. 10(a)), the analytic solution can be written as,

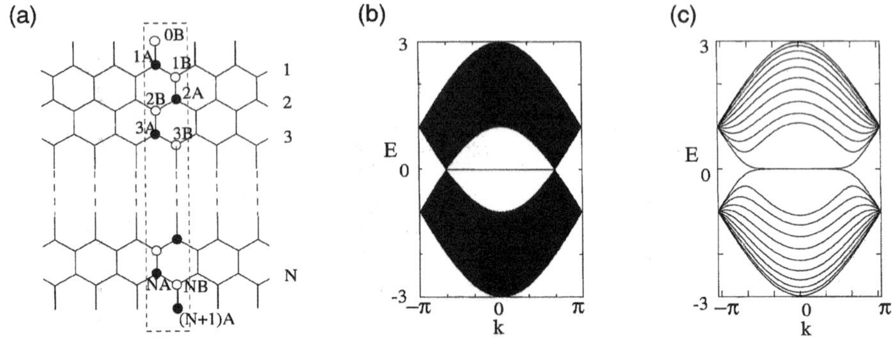

Fig. 10. (a) The structure of a graphite ribbon with two bearded edges. (b) The band structure of a semi-infinite graphite sheet with a bearded edge. (c) The band structure of bearded ribbon for $N = 10$.

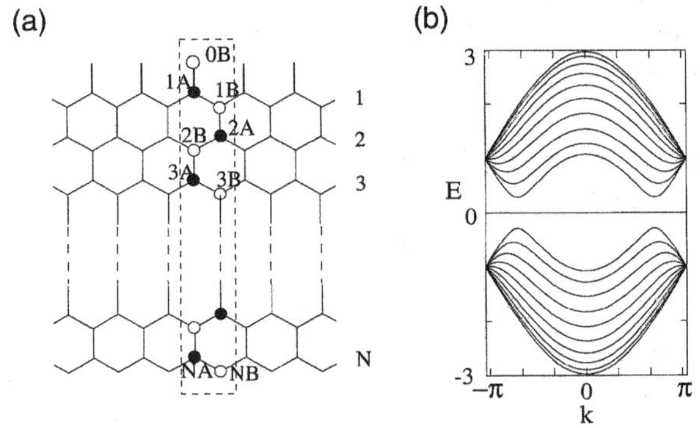

Fig. 11. (a) The structure of a graphite ribbon with a zigzag and a bearded edge and (b) the band structure for $N = 10$.

$$\Psi_{nB} = (1/D_k)^n \quad \text{and} \quad \Psi_{nA} = 0, \tag{13}$$

where $D_k = -2\cos(k/2)$. The convergence condition of the wave function, $|1/D_k| \le 1$, gives the region of edge state as $|k| \le 2\pi/3$. Since the penetration depth of the edge states get longer toward $k = \pm 2\pi/3$, the small energy gaps appear near $k = \pm 2\pi/3$ in the case of finite width ribbons due to the bonding and anti-bonding interaction between two edge states.

Next we show the ribbons having zigzag and bearded edge as shown in Fig. 11(a). Since in this ribbon $|N_{\mathrm{A}} - N_{\mathrm{B}}| = 1$, where $N_{\mathrm{A}}(N_{\mathrm{B}})$ means the number of sites belonging to the A(B)-sublattice, there is a flat band at $E = 0$ all over the 1st BZ, as shown in Fig. 11(b).

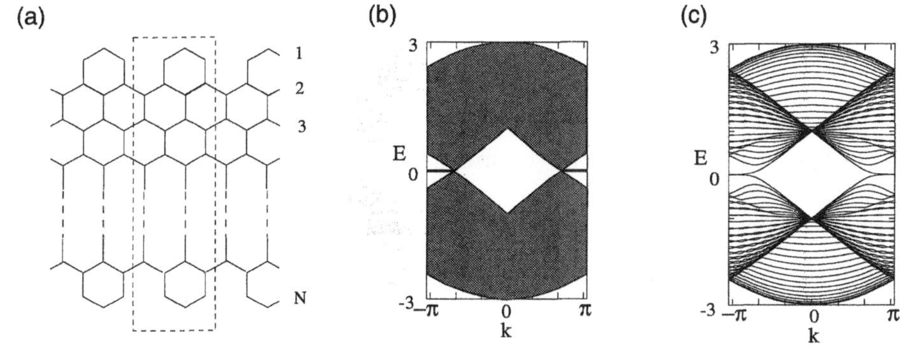

Fig. 12. (a) The structure of a graphite ribbon with two cove edges. (b) The band structure of a semi-infinite graphite sheet with a cove edge. (c) The band structure of bearded ribbon for $N = 10$.

The analytic solution of this flat band can be easily understood by the combination of two edge states for zigzag and bearded edges. In the region of $|k| < 2\pi/3$, the electrons are localized at the bearded edge, and in the region of $|k| > 2\pi/3$, the electrons are localized at the zigzag edge. At $k = \pm 2\pi/3$, the electrons are delocalized. It should be noted that this ribbon is insulating because the flat band does not carry currents and there are energy gaps between the flat band and next subbands.

Cove edge is a zigzag edge with attached additional hexagon rings. The graphite ribbons with two cove edges are shown in Fig. 12(a). In Fig. 12(b), the band structure of a semi-infinite graphite sheet with a cove edge is shown. This case also provides a partly flat band in the region of $|k| \leq 2\pi/3$. The analytic solution is not simple such as in the cases of zigzag and bearded edges, because the unit cell is the double size of the other two ribbons.

The important conclusion in this section is that an edge which is not parallel to an armchair axis provides edge states. Actually, graphite ribbons with mixed armchair edges and zigzag edges also show localized states [8]. Thus it is considered that the edges whose cutting direction is not parallel to the armchair axis show edge localized states. The reason why an armchair edge does not provide an edge state will be clarified in connection with a surface bound state of anisotropic superconductivity in the next section.

5 Connection with Anisotropic Superconductivity

After the discovery of the high temperature superconductor (HTSC), considerable effort has been devoted to the studies on the mechanism causing superconductivity. Today, a great deal of experimental data gives the convincing evidence that the pair wavefunction has $d_{x^2-y^2}$-wave symmetry rather than s-wave symmetry. This unconventional pairing symmetry induces much interest in the physical properties, one of them is the Andreev bound states at [110]-surface of $d_{x^2-y^2}$-wave superconductor, which is observed as a zero-bias conductance peak in the current-voltage characteristics [25–27]. In this section, we show that the mechanism of the surface bound states at the [110]-surface of $d_{x^2-y^2}$-wave superconductor is closely related to the edge states of graphite edges[10]. The reason why the armchair edge does not have the edge states will be clarified.

Let us consider how the electron (or hole) behaves at the surface of $d_{x^2-y^2}$- wave superconductor (SC). In Figure 13(a), the classical trajectories of electrons and holes in the region of the [100]-surface are shown. The change of the momentum of these trajectories by the reflection at the surface always connects same signs of the pairing potential. However, at the [110]-surface, as shown in Fig. 13(b), the change of the momentum of these trajectories by the reflection at the surface always connect different signs of the pairing potential, and lead to a phase shift π. This π-phase shift is the origin of the Andreev bound states at the [110]-surface of $d_{x^2-y^2}$-wave SC.

We start with the tight binding Hamiltonian of a graphite sheet given in the section 2, and derive the Bogolubov-de Gennes (BdG)-type equation. The tight binding equation of the graphite sheet in the k-space is written as $(\varepsilon_k \hat{\sigma}_- \varepsilon_k^* \hat{\sigma}_+)\hat{\Psi} = E\hat{\Psi}$, where $\varepsilon_k = \sum_i \exp(ik \cdot a_i)$, ${}^t\hat{\Psi} = (\Psi_{Ak}, \Psi_{Bk})$ and $\hat{\sigma}_\pm = (\hat{\sigma}_x \pm \hat{\sigma}_y)/2$. Applying the transformation, *i.e.*, $\Psi_p = (\Psi_A + \Psi_B)/\sqrt{2}$ and $\Psi_h = (\Psi_A - \Psi_B)/\sqrt{2}i$, we will have

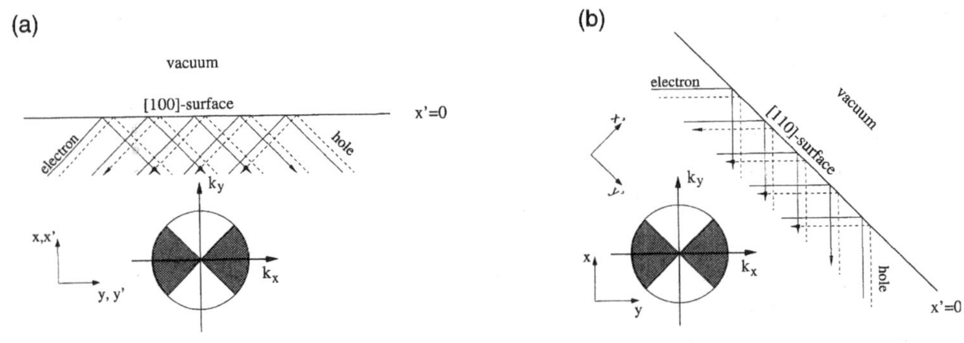

Fig. 13. Schematic view of Andreev reflection in the surface region of $d_{x^2-y^2}$-wave superconductor. The trajectories at the [100]-surface (a) does not connect gap regions of opposite sign, however at the [110]-surface (b) the trajectories connect gap regions of opposite sign which lead to a phase shift π and zero-energy bound states. Shading means the sign of the pair potential.

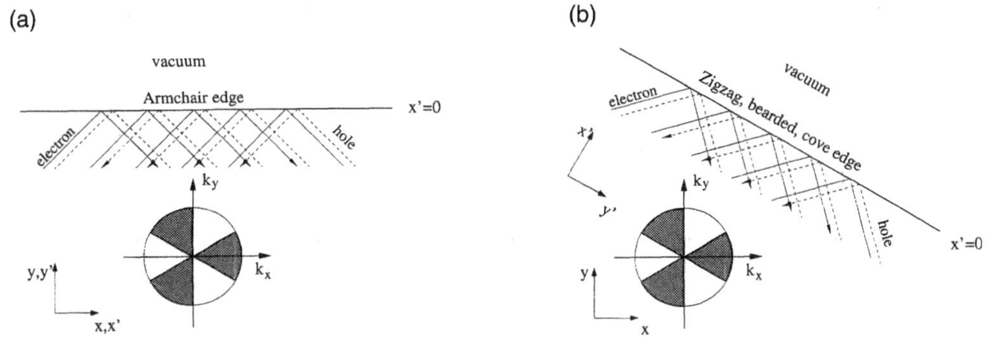

Fig. 14. Schematic view of Andreev reflection in the surface region of graphite (f-wave superconductor) surface. The trajectories at the armchair edge (a) does not connect gap regions of opposite sign, however at the zigzag, beaded and cove edge (b) the trajectories connect gap regions of opposite sign which lead to a phase shift π and zero-energy bound states.

$(\varepsilon_k^+ \hat{\sigma}_z - i\varepsilon_k^- \hat{\sigma}_x)\hat{\Phi}_k = E\hat{\Phi}_k$, where $\varepsilon_k^{\pm} = (\varepsilon_k \pm \varepsilon_k^*)/2$. In this equation, the diagonal terms formally correspond to the kinetic terms of particles and holes in the BdG equation, and the off-diagonal terms can be considered as the pair-potential of the superconductor. In the continuum limit, we can rewrite the above equation as follows, $(\hat{\xi}_k \hat{\sigma}_z + \Delta_k \hat{\sigma}_x)\hat{\Phi}_k = E\hat{\Phi}_k$, where $\hat{\xi}_k = -(\hbar^2/2m)\nabla^2 - \mu$. Here μ is the chemical potential and Δ_k represents the pair-potential of the superconductor, $\Delta_k = \left(\varepsilon_k - \varepsilon_k^*\right)/2i$. More simply, this pair potential can be rewritten as $\Delta = |\Delta| \cos(3\theta)$, close to the Fermi level, where θ is the angle of the Fermi vector measured from the k_x-axis. Interestingly, this pair potential means an odd-parity superconductivity, precisely f-wave superconductivity and deduce the following results: since the armchair (zigzag) axis corresponds to k_x (k_y), the trajectories at the armchair edge does not connect gap regions of opposite sign, however at the zigzag, beaded and cove edge the trajectories connect the gap regions of opposite sign which lead to a phase shift π and zero-energy bound states. Thus, the armchair edge does not show the edge localized states. Similar discussion can be found in ref. [30].

6 Orbital Diamagnetism and Pauli Paramagnetism

In the following two sections, the magnetic properties of nanographites are discussed in the absence and presence of the electron-electron interactions. It is well known that bulk graphite shows a large anisotropic diamagnetic susceptibility, while aromatic molecules have only weak diamagnetism. This means that the orbital diamagnetic susceptibility is sensitive to the size of graphite fragments. On the other hand, the sharp peak in the density of states due to the edge states gives rather strong Pauli paramagnetic response, so that the competition between these two components occurs in nanographite systems. Since the edge states also give the possibility of spin polarization at low-temperature due to the electron-electron interaction, we discuss the possibility of appearance of localized spins at the edges.

6.1 Orbital Magnetization and Susceptibility

The observed magnetic susceptibility χ is the sum of four components: (1) localized spin susceptibility $\chi_{\rm spin}$, (2) diamagnetic susceptibility due to the core electrons $\chi_{\rm core}$, (3) Pauli paramagnetic susceptibility $\chi_{\rm P}$ and (4) orbital diamagnetic susceptibility $\chi_{\rm orb.}$ due to the cyclotron motion of the itinerant electrons. Since we neglect electron-electron interaction for the moment, $\chi_{\rm spin}$ can be neglected. Furthermore, $\chi_{\rm core}$ is unimportant for us, because it is small and basically temperature independent. On the other hand, the Pauli paramagnetic susceptibility is related to the DOS at the Fermi level, which represents an important component in zigzag nanographite ribbons where an enhanced density of states appears at the Fermi level. Note that $\chi_{\rm P}$ is negligible in armchair ribbons, aromatic molecules and graphite sheets, because their DOS is suppressed at the Fermi level. We will see below that since the DOS introduced by the edge states is sharply peaked at the Fermi energy, $\chi_{\rm P}$ introduces a very pronounced temperature dependence which is nearly Curie-like. The diamagnetic contribution to the susceptibility is very familiar from the magnetic properties of graphite sheets. It is due to the orbital cyclotron motion of electrons in a field with a finite component perpendicular to the plane. Naturally, this diamagnetic response is very anisotropic and only weakly temperature dependent. From this we can conclude that in nanographite ribbons with zigzag edges the susceptibility should consist mainly of these two competing contributions, $\chi_{\rm P}$ and $\chi_{\rm orb.}$. Hence, a crossover occurs from a high-temperature diamagnetic to a low-temperature paramagnetic regime, where the characteristic temperature depends on the width of the ribbon and of the orientation of the external field. It is worth noting that the field direction is an important tool to distinguish the magnitude of the two components.

In this section, we briefly summarize the way to calculate the orbital diamagnetic susceptibility $\chi_{\rm orb}$ for graphite ribbons. We use the tight binding model for the calculation of the orbital diamagnetic susceptibility $\chi_{\rm orb}$. The free energy $F(H,T)$ including the magnetic field is given by,

$$F(H,T) = \mu N - \frac{1}{\beta\pi}\int_{\rm BZ} {\rm d}k \sum_{n} \ln\left(1 + {\rm e}^{-\beta\left(\varepsilon_{k,n}(H)-\mu\right)}\right) \tag{14}$$

where $\beta = 1/k_{\rm B}T$ and μ is the chemical potential and $\varepsilon_{k,n}(H)$ (n is the band index) is the energy spectrum of the graphite ribbons in the magnetic field calculated based on the tight binding model. Then the magnetic moment $M(H)$ and the magnetic susceptibility $\chi(H)$ per site for finite temperature and arbitrary magnetic field H are given by the 1st and 2nd derivative of the free energy with respect to H, respectively.

$$M(H) = -\frac{1}{N_e}\frac{\partial F}{\partial H}, \quad \text{and} \quad \chi(H) = \frac{1}{N_e}\frac{\partial M}{\partial H}. \tag{15}$$

Before we show the results on the graphite ribbons, let us briefly review the large orbital diamagnetism in graphite. The origin of the large diamagnetism in graphites is due to the appearance of the Landau level at $E = 0$. This was first shown by McClure [32]. When the energy dispersion is linear versus k near the K or K' points, the Landau levels E_n in the 1st BZ of the graphite sheet is expressed as,

$$E_n = \pm\frac{\sqrt{3}}{2}ta\sqrt{2nS} \quad (n = 0, 1, 2, \cdots), \tag{16}$$

where n is an index of the Landau levels and S is given by $S = eH/\hbar c$. t is the nearest neighbor transfer integral, and considered to be about 3.0 eV. Each Landau level has a characteristic feature and is very different from the Landau levels of ordinary 3D free electron gas, as seen in Fig. 15. It is noted that the zero-th Landau level is always located in the zero gap at the K point. The zero-th Landau level does not shift by a magnetic field. We can easily show that all the Landau levels with zero and negative Landau indices which are occupied by the valence electrons act to increase the free energy in a magnetic field and thus the orbital diamagnetism appears. When the Fermi energy is located in the zero energy gap, the oscillation of the free energy, which is known to cause de Haas-van Alphen effect at low temperatures, has a cusp at $E_F = 0$

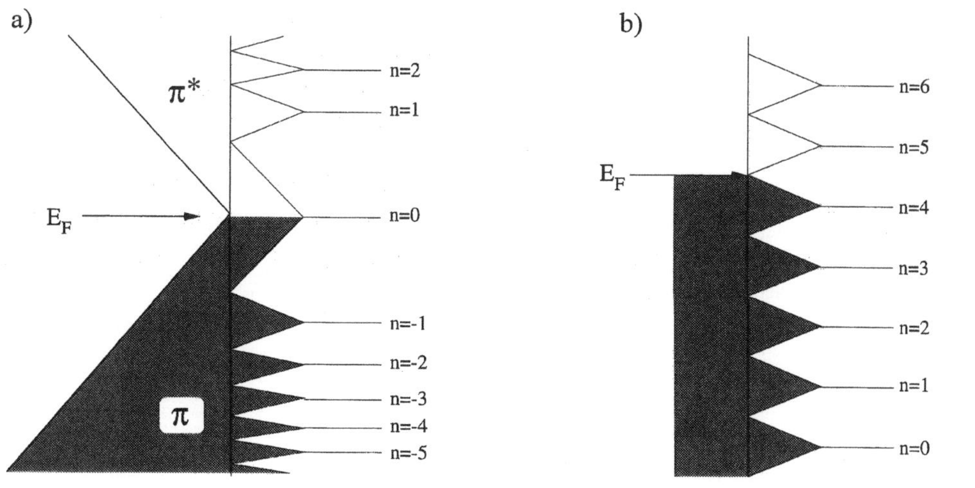

Fig. 15. (a) Landau levels of graphite sheet near K point (b) Landau levels of 3D metal.

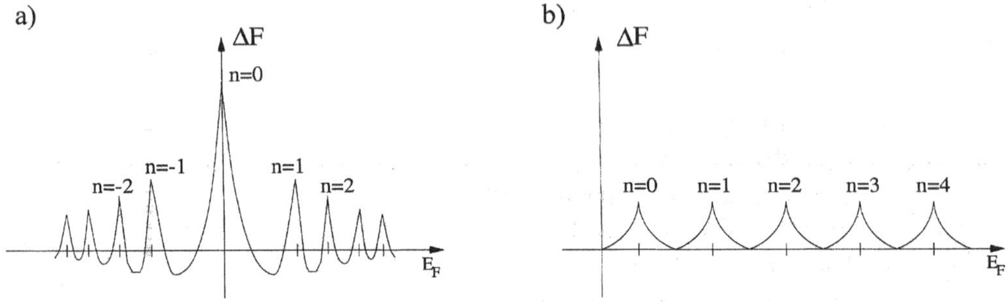

Fig. 16. (a) Oscillation of free energy of graphite sheet as a function of E_F. When the Fermi energy is located at zero, this oscillation has sharp peak, however (b) the simple 3D metal does not show such sharp peaks. The existence of this peak is the origin of large diamagnetism in graphites.

and yields large orbital diamagnetism as shown in Fig. 16(a). In Fig. 16(b), we also show the free energy as a function of E_F in the case of 3D electron gas. As seen in this figure, there are no sharp peaks in the oscillation in contrast to the results in Fig. 16(a) for graphite sheet. The expression of orbital susceptibility of graphite sheet at finite temperatures has been derived by McClure [32] based on the $k \cdot p$ approximation is

$$\chi_{\mathrm{MC}} = -0.044\left(\frac{4}{\pi C_{\mathrm{L}}}\right)^{3/2}(ta)^2\left(\frac{e}{\hbar c}\right)^2(k_B T \rho)\,\mathrm{sech}^2\left(\frac{\mu}{2k_{\mathrm{B}}T}\right)[\mathrm{emu/g}], \qquad (17)$$

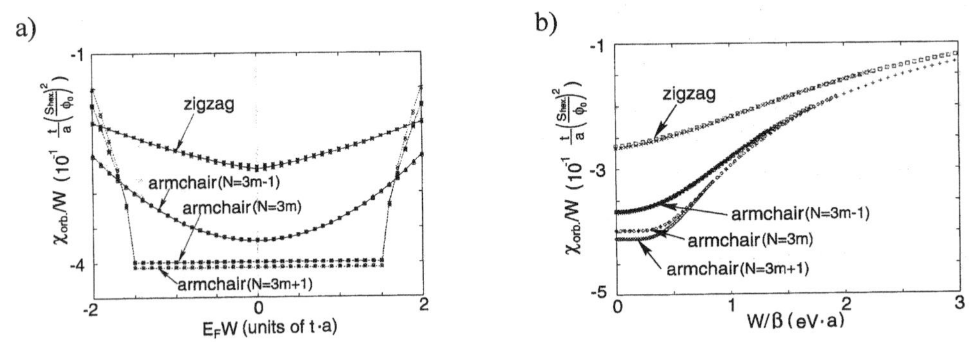

Fig. 17. (a) The Fermi energy dependence of the orbital magnetic moments χ_{orb} of graphite ribbons at T = 0. (b) The temperature dependence of χ_{orb}, where χ_{orb} is scaled by $1/W$ and β is scaled by W.

where μ is the Fermi energy and ρ is the density of carbon atoms in the unit volume of graphite. The typical value of $\chi_{\rm MC}$ for the graphite at room temperature is 21.0×10^{-6} emu/g.

The orbital magnetism is influenced by the system geometry and size. Here we show the Fermi energy dependence of $\chi_{\rm orb}$. Actually in real graphite materials, a small change in the carrier density from the half-filling is possible and can even be controlled by substrate properties. The calculated Fermi energy dependence is shown in Fig. 17(a), where it is found that $\chi_{\rm orb}/W$ is a universal function of μW. We normalize $\chi_{\rm orb}$ by dividing it by W, since it is proportional to W. Furthermore we multiply E_F by W, because the direct gap at $k=0$ is proportional to $1/W$ at $k=0$ for armchair ribbons and at $k=2\pi/3$ for zigzag ribbons for large W.

In Fig. 17(b), the temperature dependence of $\chi_{\rm orb}$ is shown. It is important from the viewpoint of experiments on nanographites. In all cases the magnitude of $\chi_{\rm orb}$ decreases with increasing temperature. It is also found that the temperature dependence of $\chi_{\rm orb}/W$ scales as a function of βW, because the energy gap is proportional to $1/W$. Our calculation also demonstrates that the edge effect becomes more significant at lower temperature. Similar scaling properties can also be found in carbon nanotubes [33, 34].

6.2 Pauli Paramagnetism

In the previous section, we have seen that the orbital diamagnetic susceptibility depends on the edge shape in nanographite ribbons. Here we discuss another important component of the magnetic susceptibility, Pauli paramagnetic susceptibility $\chi_{\rm P}$, because zigzag ribbons have a sharp peak of DOS at the Fermi level. The width of this peak has the order of meV, which is comparable to the temperature scale at room temperature. Therefore, we expect that the Pauli susceptibility of zigzag ribbons is sensitive to temperature, although the Pauli susceptibility of usual metals is temperature independent. On the other hand, since the DOS of armchair ribbons at $\varepsilon=0$ is zero or very tiny, we can neglect the effect of the Pauli paramagnetism in armchair ribbons.

The magnetic moment due to the Zeeman effect is $M=\mu_{\rm B}\left(n_{\uparrow}-n_{\downarrow}\right)$, where $\mu_{\rm B}$ is the Bohr magneton and $n_{\uparrow}$ ($n_{\downarrow}$) denotes the electron density with up-spin (down-spin). The electron density at arbitrary temperature for each spin is given by

$$n_\sigma=\frac{1}{\pi}\int_{1st\rm BZ}\mathrm{d}k\sum_n\frac{1}{1+\mathrm{e}^{\beta\left(\varepsilon_{n,k}-\sigma\mu_{\rm B}H\right)}},\tag{18}$$

where $\sigma(=\uparrow,\downarrow)$ means spin index. Therefore, the Pauli susceptibility $\chi_{\rm P}$ per site is given by

$$\chi_{\mathrm{P}} = \lim_{T\to 0}\frac{\partial M}{\partial H} = \frac{\beta\mu_{\mathrm{B}}^2}{\pi N_e}\sum_n \int \mathrm{d}k \frac{1}{\cosh\left(\beta\varepsilon_{n,k}\right)}. \tag{19}$$

We numerically calculate the finite temperature Pauli susceptibility of graphite ribbons using this equation up to room temperature.

It is possible to separate the contribution of the edge states to χ_{P}. As we have seen in Sec.2 of this chapter, the DOS due to the edge states is given by Eq.(5). After the substitution of Eq.(5) into Eq.(24), we replace the k-integration by the energy integration. Then we can obtain the χ_{P} contribution due to the edge states as,

$$\chi_p = \frac{1}{N_e N \beta^{\alpha}}\int dx \frac{x^{\alpha}}{\cosh x + 1} \sim \frac{1}{N} T^{\alpha} \tag{20}$$

where $x = \beta\varepsilon_k$ and $\alpha = \frac{1}{N} - 1$. Interestingly, χ_{P} has the Curie-like temperature dependence. The exponent of χ_{P} depends on the ribbon width through α. When N becomes infinite, the exponent α approaches -1 and χ_{P} shows the Curie-law. However, in this limit, the contribution of χ_{P} is diminished by a factor $1/N$ in Eq.(25). Numerical results of the Pauli susceptibility χ_{P} of zigzag ribbons up to room temperature are shown in Fig. 18(a) for various values of N. As expected, because of the edge states, χ_{P} shows Curie-like temperature dependence.

The observed susceptibility χ is essentially the sum of the orbital $\chi_{\mathrm{orb.}}$ and the Pauli susceptibility χ_{P}. The temperature dependence of the total susceptibility χ is shown in

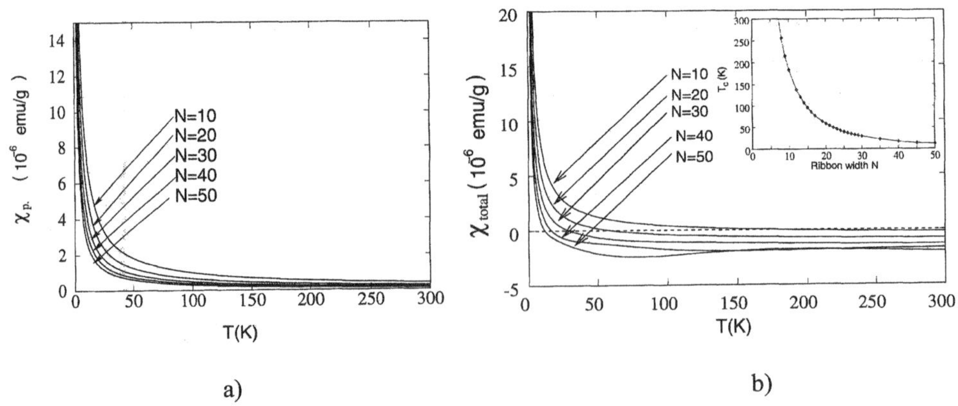

Fig. 18. (a)The temperature dependence of χ_{P} for N= 10, 20, .. , 50 up to room temperature. (b) The temperature dependence of total susceptibility χ, which is $\chi_{\mathrm{orb.}} + \chi_{\mathrm{P}}$ is shown for N= 10, 20, .. , 50. In the inset, the width dependence of crossing temperature, T_0, where $\chi = 0$.

Fig. 18(b). The total susceptibility χ shows the diamagnetic behavior in the high temperature regime and paramagnetic behavior in the low temperature regime. In the inset, the width dependence of the crossing temperature, *i.e.* $\chi = 0$, is plotted, which is approximately proportional to $1/N$.

Here we should remind that both aromatic molecules and bulk graphite show diamagnetic behavior, however, nanographite with zigzag edges have a remarkable paramagnetic behavior because of the edge states. If this paramagnetic behavior is experimentally detected, it will be an indirect evidence of the existence of the edge states.

7 Magnetic Instability

The presence of the sharp peak in the density of states should induce the lattice distortion by the electron-phonon interaction and/or the magnetic polarization by the electron-electron interaction. Because of the non-bonding character of the edge states, the lattice distortion in the vicinity of zigzag edges is unlikely with the realistic strength of the electron-phonon coupling [35]. The absence of the lattice distortion is also confirmed in terms of the density functional approach [36]. Thus we examine the effect of the electron-electron interaction by using the Hubbard model with unrestricted Hartree-Fock approximation. Here we find a possibility of spontaneous magnetic ordering near the edge, peculiar to the nanometer scale fragments of graphite.

In order to study the magnetic instability in nano-graphite ribbons, we use the Hubbard model with the unrestricted Hartree-Fock (HF) approximation. The Hamiltonian is written as

$$H_{HF} = -t \sum_{\langle i,j \rangle, s} c_{i,s}^{\dagger} c_{j,s} + U \sum_{i,s} \left(\langle n_{i,-s} \rangle - \frac{1}{2} \right) n_{i,s}, \tag{21}$$

where the operator $c_{i,s}^{\dagger}$ creates an electron with spin s on the site i and $n_{i,s} = c_{i,s}^{\dagger} c_{i,s}$. The indices of the sites in graphite ribbons are shown in Fig. 1(b) and (c). The parameters t and U are the nearest-neighbor transfer integral and on-site Coulomb repulsion, respectively. $\langle \ldots \rangle$ denotes the expectation value in the HF state. We solve the

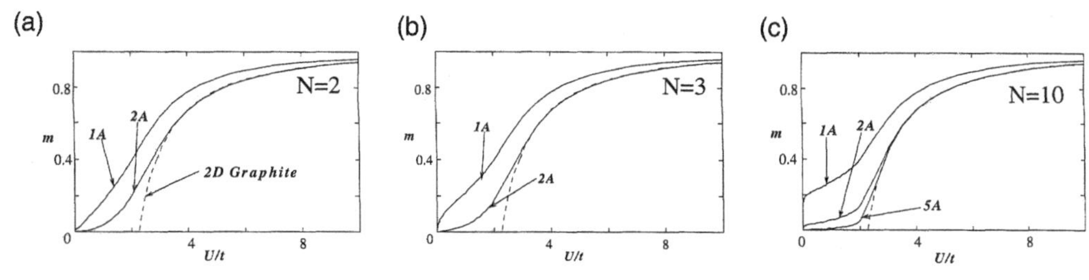

Fig. 19. The U dependence of the magnetization m for the zigzag ribbons of $N =$ (a)2, (b)3 and (c)10. The dashed lines mean the mean field solutions for the graphite sheet.

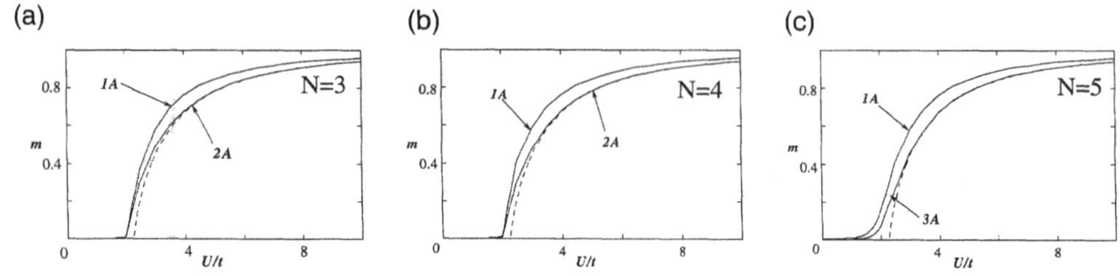

Fig. 20. The U dependence of the magnetization m for the the armchair ribbons of $N=$(a)3, (b)4 and (c)5. The dashed lines mean the mean field solutions for the graphite sheet.

unrestricted Hartree-Fock (HF) Hamiltonian with the self-consistence conditions, *i.e.* $m_i = \sum_{i,s} s\langle n_{i,s}\rangle$, and m_i means the magnetic moment at site i in the unit of Bohr magneton.

In Fig. 19, the U dependence of magnetization m for the zigzag ribbon of (a) $N=2$, (b) $N=3$ and (c) $N=10$ are shown. The dashed lines are HF solutions for a 2D graphite sheet. Similarly, we show the cases for armchair ribbons of (a) $N=3$, (b) $N=4$ and (c) $N=5$ in Fig. 20. First, we can clearly see a peculiar feature for the zigzag ribbons, *i.e.*, large magnetic moments emerge on the edge carbons even for the weak U, which is explained as follows: Since the 2D graphite is a zero-gap semiconductor whose DOS is zero at the Fermi level, the broken line stands up at a finite value $U(=U_C)$. This is consistent with the fact that graphite is non-magnetic, where U is expected to be much smaller than t. On the other hand, the zigzag ribbon has a large density of states at the Fermi level originating from the edge states. Thus, non-zero magnetic solutions can emerge for infinitesimally small U region as indicated in the present mean field result. However, special emphasis should be put on the behavior of the magnetization at the edge site 1A. As shown in Fig. 19, the magnetization at the site $1A$ rapidly rises up and reaches about 0.2 even at small $U(\approx 0.1)$, when the width of ribbon is increased. We note that the armchair ribbon does not show such singular magnetic behavior as shown in Figs. 20(a)-(c).

Next, we should also stress the local ferrimagnetic structure for the zigzag ribbons. We exhibit the magnetic structure of the ribbon with $N=10$ at $U/t=0.1$ in Fig. 21(a), where spin alignment is visible at both edge sites. The origin of this structure is also explained from the nature of the edge states, which are responsible for the magnetization. Since the amplitude of the edge state is non-zero only on one of the two sublattices at an edge and damps inwards, the magnetic moment selectively grows on this sublattice forming local ferrimagnetic spin configuration, which is getting smaller and promptly on inner sites. The opposite edge sites, however, belong to the different sublattices, the total magnetization of the zigzag graphite ribbon is zero, although this vanishing total spin for the ground state is consistent with the exact statement of the half-filled Hubbard model [37]. In Fig. 21(b), the ribbon width, N, dependence of the

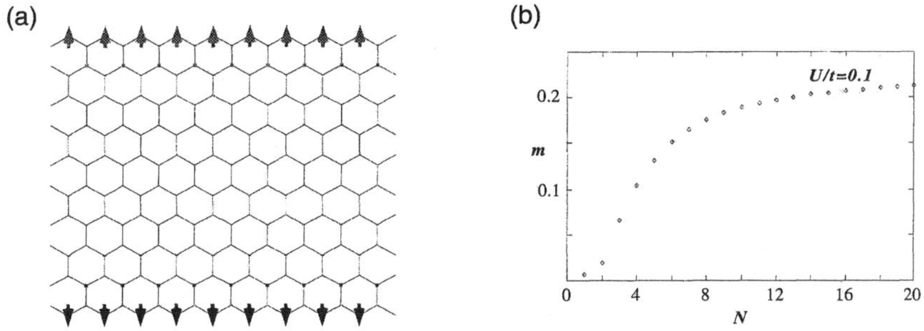

Fig. 21. (a) The schematic magnetic structure for the zigzag ribbon of $N = 10$ at $U/t = 0.1$. (b) The ribbon width, N, dependence of the magnetization at the most outer site (1A). The magnetization at site NB has same magnitude but opposite sign.

magnetization at the most outer site (1A or NB). The magnitude of the magnetization rapidly increases as ribbons getting wider and then saturates around $N = 10$.

The ferrimagnetic spin polarizations along the zigzag edges are interesting in view of the magnetic properties of nanographites. Nevertheless, the long-range order derived from the mean-field calculation is spurious, because no finite-momentum long-range spin order is expected in an one-dimensional system with full spin-rotation symmetry [38]. Even we may argue that quasi-long-range order, similar to the spin-1/2 Heisenberg chain, is not realized in zigzag ribbons of any finite width for the following reason: The unit cell of the ribbons contains an even number of sites such that Haldane's conjecture applies, *i.e.* the system should exhibit a spin gap [39]. This is very analogous to the case of the ladder systems with even number of legs, which display a resonating valence bond (RVB) ground state, *i.e.*, a short range correlated spin liquid state. With increasing the ribbon width, however, the analysis by the random phase approximation (RPA) shows that the spin gap Δ_s decreases exponentially due to the diminished overlap between two edges [40]. This fact means that the zigzag edges favors toward the spin polarization with ferromagnetic alignment. The systematic analysis of topological network in nanographites gives a good indication for designing the new magnetic carbon materials[41–44].

8 Summary

In this chapter, we investigated the electronic and the magnetic properties of nanographite systems. It is found that the electronic states are strongly influenced by the existence and the shape of graphite edge. Zigzag edges produce the localized edge states, which give a sharp peak in the density of states at the Fermi energy.

The orbital magnetic susceptibility χ_{orb} of nanographites has the intermediate values between those of aromatic molecules and bulk graphites. The χ_{orb} can be scaled as a function of the Fermi energy, temperature and ribbon width. Zigzag ribbons show the large Pauli paramagnetic response at low-temperature due to the existence of the sharp peak in the density of states at the Fermi energy, while in armchair ribbons the Pauli paramagnetic susceptibility, χ_{Pauli}, is negligible. Since the width of the peak is the order of meV, the χ_{Pauli} is sensitive to the temperature, resulting in the Curie-like behavior. In nanographite systems, this paramagnetic response competes with the diamagnetic responses, resulting in the crossover from high-temperature diamagnetic response to low-temperature paramagnetic response.

The edge states are not stable even for small onsite Coulomb interaction, which yields localized spins of about $0.2\mu_B$ at the edge sites. In zigzag ribbons, the spins ferrimagnetically align due to the nature of non-bonding edge states. The two magnetic states which polarized at the zigzag edge couple antiferromagnetically, resulting in the total magnetization being zero which is consistent with the Lieb's argument. Since the zigzag ribbons have even sites in the unit cell, the ground state of the system is considered as a resonating valence bond state, which induce a gap for the spin excitation. However, since the coupling between two edges rapidly decreases with increasing width, the spin gap is negligible small when the ribbon width has a nanometer-scale.

The argument we discussed here is applicable to the other edge shapes having the non-bonding edge states, *i.e.* bearded and cove edges. Based on the discussion of the previous section, the non-bonding edge states are expected at the edges which are not parallel to the armchair edge. Thus, there is a class of nanographites which show the crossover from the paramagnetic response at low temperature and the diamagnetic response at high temperature can be assigned as the characteristic feature of the nanographite systems.

In the activated carbon fiber (ACF) and graphitized nanodiamonds, the dangling bonds are terminated by other elements. Nevertheless, their magnetic susceptibility shows Curie-Weiss behavior originating from localized spins [14, 15]. Since in the ACF and graphitized nanodiamond systems, the dangling bonds are terminated, the presence of dangling bond spins can be ruled out. Our results show that the non-bonding edge states give strong Pauli paramagnetic response even in the absence of electron-electron interactions. Thus, we conclude that the localized spin originating from the non-bonding edge states is one of the strong candidates to explain the behavior of magnetic response in nanographite systems.

Up to now, it has been considered that the graphite-related materials are quite silent for the magnetic response. However, our results suggest that in nanographite systems the paramagnetic response due to the non-bonding edge states are crucial, resulting in that the nonographites is assigned as a completely different systems from bulk graphites and aromatic molecules in the viewpoint of the magnetic properties.

References

1. H.W. Kroto, J.R. Heath, S.C. O'Brien, R.F. Curl, and R.E. Smalley, Nature **318** (1985) 162.
2. S. Iijima, Nature **354** (1991) 56.

3. R. Saito, M. Fujita, G. Dresselhaus, and M. S. Dresselhaus, Phys. Rev. B46 (1992) 1804.
4. N. Hamada, S. Sawada, and A. Oshiyama, Phys. Rev. Lett. 68 (1992) 1579.
5. J.W.G. Wildöer, L.C. Venema, A.G. Rinzler, R.E. Smalley, and C. Dekker, Nature 391 (1998) 59.
6. T. W. Odom, J. Huang, P. Kim, and C. M. Lieber, Nature, 391 (1998) 62.
7. M. Fujita, K. Wakabayashi, K. Nakada, and K. Kusakabe, J. Phys. Soc. Jpn. 65 (1996) 1920.
8. K. Nakada, M. Fujita, G. Dresselhaus, and M.S. Dresselhaus, Phys. Rev. B54 (1996) 17954.
9. K. Wakabayashi, M. Fujita, H. Ajiki, and M. Sigrist, Phys. Rev. B59 (1999) 8271.
10. K. Wakabayashi, Ph.D Thesis, Tsukuba University (2000).
11. K. Wakabayashi and M. Sigrist, Phys. Rev. Lett. 84 (2000) 3390.
12. K. Wakabayashi, Phys. Rev. B64 (2001) 125428.
13. Y. Niimi, T. Matsui, H. Kambara, K. Tagami, M. Tsukada, and H. Fukuyama, Appl. Surf. Sci. 241 (2005) 43..
14. Y. Shibayama, H. Sato, T. Enoki, and M. Endo, Phys. Rev. Lett. 84 (2000) 1744.
15. O.E. Andersson, B.L. V. Prasad, H. Sato, T. Enoki, Y. Hishiyama, Y. Kaburagi, M. Yoshikawa, and S. Bandow, Phys. Rev. B58 (1998) 16387.
16. K. Han, D. Spemann, P. Esquinazi, R. Höhne, V. Riede, and T. Butz, Adv. Mater. 15 (2003) 1719.
17. P. Esquinazi, D. Spemann, R. Höohne, A. Setzer, K.-H. Han, and T. Butz, Phys. Rev. Lett. 91 (2003) 227201.
18. L.G. Cançado, M.A. Pimenta, B.R.A. Neves, G. Medeiros-Ribeiro, T. Enoki, Y. Kobayashi, K. Takai, K. Fukui, M.S. Dresselhaus, R. Saito, and A. Jorio, Phys. Rev. Lett. 93 (2004) 047403.
19. E. Dujardin, T. Thio, H. Lezec, and T. Ebbesen, Appl Phys. Lett. 79 (2001) 2474.
20. P.R. Wallace, Phys. Rev. 71 (1947) 622.
21. M.S. Dresselhaus, G. Dresselhaus, K. Sugihara, I.L. Spain, and H.A. Goldberg in: Graphite Fibers and Filaments (1988) Springer-Verlag.
22. K. Tanaka, S. Yamashita, H. Yamabe, and T. Yamabe, Synth. Met. 17 (1987) 143.
23. K. Kobayashi, Phys. Rev. B47 (1993) 1757.
24. D.J. Klein, Chem. Phys. Lett. 217 (1994) 261.
25. C.R. Hu, Phys. Rev. Lett. 72 (1994) 1526.
26. Y. Tanaka and S. Kashiwaya, Phys. Rev. Lett. 74 (1995) 3451.
27. M. Sigrist, Prog. Theor. Phys. 99 (1998) 899.
28. P.G. de Gennes, in: Superconductivity of Metals and Alloys, Chapter 5, (1966) Benjamin, New York.
29. A.F. Andreev, Sov. Phys-JETP 19 (1964) 1228.
30. S. Ryu and Y. Hatsugai, Phys. Rev. Lett. 89 (2002) 077002.
31. R.M. White, Quantum Theory of Magnetism, Springer-Verlag.
32. J.W. McClure, Phys. Rev. 104 (1956) 666.
33. J.P. Lu, Phys. Rev. Lett. 74 (1995) 1123.
34. H. Ajiki and T. Ando, J. Phys. Soc. Jpn. 62 (1993) 2470.
35. M. Fujita, M. Igami, and K. Nakada, J. Phys. Soc. Jpn. 66 (1997) 1864.
36. Y. Miyamoto, K. Nakada, and M. Fujita, Phys. Rev. B59 (1999) 9858.
37. E.H. Lieb, Phys. Rev. Lett. 62 (1989) 1201.
38. L. Pitaevskii and S. Stringari, J. Low. Temp. Phys. 85 (1991) 377.
39. E. Dagotto and T.M. Rice, Science 271 (1996) 618.

40. K. Wakabayashi, M. Sigrist, and M. Fujita, J. Phys. Soc. Jpn. **67** (1998) 2089, see also H. Yoshioka, J. Phys. Soc. Jpn. **72** (2003) 2145.
41. K. Kusakabe and M. Maruyama, Phys. Rev. B**67** (2003) 092406.
42. T.L. Makarova, Semiconductors, **38** (2004) 615.
43. K. Wakabayashi and K. Harigaya, J. Phys. Soc. Jpn. **72** (2003) 998.
44. S. Okada and A. Oshiyama, Phys. Rev. Lett. **87** (2001) 146803.

Carbon-Based Magnetism
T. Makarova and F. Palacio (Editors)

13

Flat-band Ferromagnetism in Organic Crystals

K. Kusakabe

Graduate School of Engineering Science, Osaka University, 1-3 Machikaneyama-cho, Toyonaka, Osaka 560-8531, Japan

1 Introduction

Magnetism is one of the most attractive phenomena in condensed matter physics. This cooperative phenomenon occurs even above the room temperature in some magnetic materials. In carbon-based materials, after a seminal finding of the first pure organic ferromagnet, *i.e.* para-nitro-phenyl-nitronyl-nitroxide (p-NPNN) [1], great progress in development of organic magnets has been made. Due to these discoveries, the spin-polarization phenomenon in π-electron systems was established [2, 3].

Newly found carbon materials, including fullerenes, are on the list of organic magnets [4]. To add novel organic magnets to the list, an artificial material design work plays an important role. The design may start from molecules which have degeneracy in the highest occupied molecular orbital (HOMO). If one can find ferromagnetic coupling among electrons in these orbitals, and if the high-spin ground state appears, this molecule may be used as a building block. In the case of C_{60}, for example, the symmetry of the molecule creates 3-fold degeneracy in HOMO and also in the lowest unoccupied molecular orbitals (LUMO). However, the number of active degenerate states does not increases as it is. If one can synthesize a polymeric system, which has high degeneracy in the electronic state and has spin moments of an order of the size of the polymer, one might obtain bulky magnetic substance [5, 6]. If one can create a molecular crystal in which inter-molecular exchange interaction is also ferromagnetic, one may have bulk magnets. These approaches to design magnetic materials are synthetic methods utilizing addition of magnetic or non-magnetic elements [2, 3].

We may reach the final goal by using another method than the constructive approach. For the materials design, we should also consider a well-known material, as well as newly synthesized materials. If we discovered ferromagnetism or ferrimagnetism as an intrinsic nature of graphite, *i.e.* a pure π electron system, the achievement would be astonishing and valuable. The finding would deepen understanding of magnetism in π electrons. The knowledge could open future application of the soft magnetic materials for our technology and environment. One might immediately deny this naive idea, because it is hopeless to expect localized magnetism for the bulk graphite, which is the most typical diamagnetic material. However, the very mechanism acting in known organic magnets has a room to work, if the graphitic structure becomes small or modified in nanometer-scale.

The goal of this chapter is to find theoretical evidences on spin polarization in graphitic structures. Keywords are "edges" and "defects". We focus our attention mainly on an effect due to edges of graphitic structures with a finite system size. This is because we can show a connection between historically established knowledge on organic ferromagnets and recent achievements on spin polarization of nanometer-sized graphitic systems, *i.e.* magnetism of "nano-graphite".

A key finding in this field is the edge states of nano-graphite. Nano-graphite was defined as graphitic systems in the nanometer-scale [7]. This category of graphitic structures includes fragment of graphene [8], nanometer-sized activated carbon fibers [9] and graphitized diamond nano-particles [10]. The nano-graphite is characterized by dependence on edge structures. The electronic structure around the Fermi level may be highly degenerated, when shape of the edge is the so-called zigzag edge. This degeneracy can cause magnetic instability. Basic magnetic mechanism is spin polarization in these highly degenerate orbitals or in a flat band.

However, we have to specify atomic configuration of the whole system to design real materials. Several theoretical predictions on magnetic structures have been done using the first-principles electronic structure calculations [11–13] . Some magnetic carbon structures relate to the magnetic zigzag edge. In this structure, there is high degeneracy in electronic states. In the terminology of the band theory, this phenomena corresponds to appearance of a flat-band. Thus magnetic nano-graphite may be a realization of flat-band ferromagnetism known in a class of the Hubbard models. We will give possible argument on this phenomenon both from the physics of the Hubbard model and the band theory. We will also comment on other magnetic effects found in various defects of graphitic structures and consider similarity to and difference from the graphite-edge effect.

2 Effective Models for Nano-graphite

Graphite is known to be a typical semimetal, which shows large diamagnetism in the normal state due to its special electronic states. Global band structure around the Fermi level E_F consists of σ-, π-, π^*- and σ^*-bands. These bands are in sp^2-hybridized bands. E_F lies at the touching points of π- and π^*-bands around the K- and K'-points in the hexagonal Brillouin zone. Conventional band-structure calculations

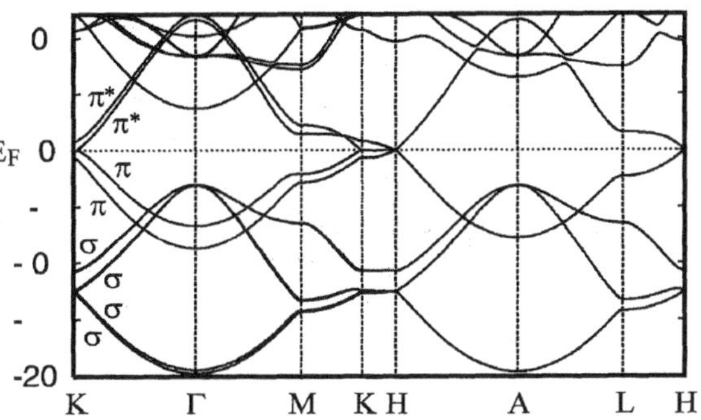

Fig. 1. The electronic band structure of bulk hexagonal graphite. The calculation was done using the local-density-approximation in the density functional theory. σ-, π- and π^*-bands are indicated.

conclude that the bulk graphite has a band structure with small electron pockets and hole pockets [14]. We display the electronic band-structure of graphite obtained by the local-density-approximation (LDA) in Fig. 1. This band structure suggests that to consider only π- and π^* bands is allowed for discussion of low energy physics.

We should comment on some details of this calculation here. For the electronic structure calculation, we rely on the density-functional theory [15]. Kohn and Sham showed that an effective single-particle system described by the next equation was given [16].

$$\left\{-\frac{\hbar^2}{2m}\Delta + \int d\mathbf{r}' \frac{e^2\rho(\mathbf{r}')}{|\mathbf{r}-\mathbf{r}'|} + \frac{\delta E_{xc}}{\delta\rho} + V_{ext}(\mathbf{r})\right\}\phi_{n,\sigma}(\mathbf{r}) = \varepsilon_n \phi_{n,\sigma}(\mathbf{r}). \quad (1)$$

Here, $\phi_{n,\sigma}(\mathbf{r})$ is a Kohn-Sham orbital specified by an orbital index n and a spin index σ. Constants e and m are the charge and the mass of the electron. A function $\rho(\mathbf{r})$ is the electron density and $V_{ext}(\mathbf{r})$ denotes the external potential mainly due to the ionic cores. To perform realistic calculation using this Kohn-Sham equation, we need to determine the exchange-correlation energy functional, E_{xc}. Among tractable energy functionals, we chose LDA given by Perdew and Wang for the calculation [17]. This is simply because another widely-used energy functional, *i.e.* the generalized-gradient-expansion approximation (GGA), does not reproduce lattice constants of graphite, especially the length of the c-axis. While LDA is known to give lattice constants with an error less than a few % difference from the experimental observation. This could be accidental. Indeed, a long-range non-local effect is essential for the description of graphite inter-layer interaction. The recently developed van-der-Waals density functional theory concludes that combination of the non-local energy functional determined by the screened polarization of non-uniform electron gas and the local energy functional given by GGA gives rather accurate mechanical properties even for the *c*-axis modification [18]. At present, what we can do best is to choose an energy functional. Once the functional is chosen, our first-principles calculation becomes a model calculation. We may regard LDA (or other model energy functional) as an extrapolation of the energy functional from a reference point like the uniform density limit.

To solve the Kohn-Sham equation, both Kohn-Sham orbitals and the electron charge density are expanded in a proper basis. The plane-wave basis is a useful one for the carbon systems, where we can utilize the ultra-soft pseudo-potentials [19]. Then, the first-principles calculation is possible for various nano-scale carbon structures including nano-graphite and carbon nanotubes. Taking enough cut-off energy of the plane-wave expansion, the result reproduces covalent bondings accurately. Essentially the same valence band structure is obtained, if we replace Perdew-Wang's functional by other LDA functionals or if we utilize other expansion techniques without the pseudo potentials.

To simplify the description of π-electrons in graphite, we sometimes utilize a model derived by an approximation called the linear combination of atomic orbitals, or the tight-binding description. To construct a tight-binding model of π electrons, we consider a 2-dimensional graphene sheet, instead of graphite. This is because inter-layer interaction between graphene sheets causes hybridization between a p_x orbital and orbitals having the other symmetry including s, p_x or p_y orbitals. In case that we need only properties at very low temperatures for the bulk graphite, we may be allowed to use renormalized π bands. Using the second- and higher-order perturbations, we may project other states than π states out from the model, and may derive effective electron transfer integrals between two π-orbitals on neighboring graphene sheets. However, we have to note that the whole π system takes part in the magnetism, as shown in the following sections. Thus the inter-layer electron transfer has to be carefully determined. Besides, we do not know whether the DFT-LDA band-structure calculation is accurate enough to determine inter-layer electron hopping. The problem of the inter-layer interaction should be directly related to the recent observation of curious transport phenomena along the c-axis [20, 21]. Different from the bulk graphite, pure π bands made only from p_z orbitals exist in a graphene sheet.

Thus in this section, we consider a tight-binding model for a 2-dimensional graphene sheet. We denote energy of π bands as $\varepsilon_{p,\mathbf{k}}$ (p $=\pm$). Here $\varepsilon_{+,\mathbf{k}}$ and $\varepsilon_{-,\mathbf{k}}$ denote π^*- and π-bands. The Wannier functions $\psi_{l,\mathbf{R}}$ of π-orbitals ($l = A, B$) at a unit cell at $\mathbf{R}$, may be constructed from the Bloch states $\psi_{p,\mathbf{k}}$ as

$$\psi_{A,\mathbf{R}} = \frac{1}{\sqrt{N}} \sum_{\mathbf{k}} \exp(-i\mathbf{k}\mathbf{R})(U_{A,+,\mathbf{k}}\psi_{+,\mathbf{k}} + U_{A,-,\mathbf{k}}\psi_{-,\mathbf{k}}),$$

$$\psi_{B,\mathbf{R}} = \frac{1}{\sqrt{N}} \sum_{\mathbf{k}} \exp(-i\mathbf{k}\mathbf{R})(U_{B,+,\mathbf{k}}\psi_{+,\mathbf{k}} + U_{B,-,\mathbf{k}}\psi_{-,\mathbf{k}}).$$

Here $U_{l,\pm,\mathbf{k}}$ is a (2×2) unitary matrix for each $\mathbf{k}$. ($l = A, B$). This matrix represents a gauge of the Wannier function. We can choose arbitrary $U_{l,\pm,\mathbf{k}}$ which changes center and spread of the Wannier function [22]. The Hamiltonian of π bands in the second quantizatiôn may be written as,

$$H_{LDA} = \sum_{p,\mathbf{k},\sigma} \varepsilon_{p,\mathbf{k}} c^{\dagger}_{p,\mathbf{k},\sigma} c_{p,\mathbf{k},\sigma} . \tag{2}$$

Here, $c^{\dagger}_{p,\mathbf{k},\sigma}$ and $c_{p,\mathbf{k},\sigma}$ denote creation and annihilation operators of an electron in $\psi_{p,\mathbf{k}}$ with spin σ.

Changing the basis from the Bloch function to the Wannier function does not change the spectrum. Let us choose

$$\begin{pmatrix} U_{A,+,\mathbf{k}} & U_{A,-,\mathbf{k}} \\ U_{B,+,\mathbf{k}} & U_{B,-,\mathbf{k}} \end{pmatrix} = \begin{pmatrix} e^{-i\theta(\mathbf{k})/2}\sqrt{2} & -e^{-i\theta(\mathbf{k})/2}\sqrt{2} \\ e^{i\theta(\mathbf{k})/2}\sqrt{2} & e^{i\theta(\mathbf{k})/2}\sqrt{2} \end{pmatrix}, \tag{3}$$

where $\theta(\mathbf{k}) = \sum_{j=1}^{3} e^{i\mathbf{k}\mathbf{a}_j}$ with $\mathbf{a}_j = (\sqrt{3}a/3, 0), (-\sqrt{3}a/6, a/2), (-\sqrt{3}a/6, -a/2)$. Here a is the lattice constant of graphene. By the inverse Fourier transformation, we immediately obtain,

$$H_{TBM} = -\sum_{<i,j>} \sum_{\sigma} t_{i,j} (c^{\dagger}_{i,\sigma} c_{j,\sigma} + \text{H.c.}) \tag{4}$$

Here, $c^{\dagger}_{i,\sigma}$ and $c_{i,\sigma}$ are creation and annihilation operators of an electron in a p_z orbital at an atomic site i with a spin σ. The index i is a unified notation representing $(l, \mathbf{R})$. $(l = A, B)$.

To reproduce the DFT-LDA result, we are required to include long range transfer integrals $t_{i,j}$ representing the electron hopping from the *i*-th site to the *j*-th site in the tight-binding model. However, this model is not convenient for the discussion of the magnetism. Thus we ignore $t_{i,j}$ except for those connecting the nearest-neighbor atoms. Then, the lattice structure or the graph given by non-zero $t_{i,j}$ becomes a so-called bipartite lattice or a bipartite graph. In the bipartite graph, vertices are grouped into A and B subgroups. Each edge connecting two vertices corresponds to a finite $t_{i,j}$. Here $t_{i,j} = 0$, if both of indexes i and j belongs to one of the sub group.

If this approximation is adopted, the π band becomes invariant with respect to the electron-hole transformation. Let's introduce a unitary transformation.

$$\begin{aligned} c^{\dagger}_{i,\sigma} &\longrightarrow d_{i,\sigma} \text{ for } i \in A, \\ c^{\dagger}_{i,\sigma} &\longrightarrow -d_{i,\sigma} \text{ for } i \in B, \end{aligned} \tag{5}$$

We can see that the transformation does not change H_{TBM} for the case of the bipartite lattice. Thus, the dispersion relation of electrons and that of holes are identical, which concludes that the band structure is symmetric with respect to the band center. Note that this symmetry does not exist in the original DFT model. However, this simplification is essential for the discussion of the Hubbard model in the next sections.

To consider magnetism, we need to introduce spin dependence in the model. This is often done by one of two complimentary approaches: the effective single-body description and the effective many-body description. The original Coulomb problem for the electron system may be or should be treated by one of the above descriptions. As a former approach, the local-spin-density approximation (LSDA) will be considered in the later sections. (See Section 5.1.) In this method, exchange interaction and correlation effects are estimated by extrapolating the exchange-correlation energy obtained for some solvable systems. Accurate calculations for spin-polarized electron-gas systems and small molecules are helpful cardinal points. Effective field thus determined is spin-dependent and spin-polarized solution in the single-body description may be obtained.

However, to treat magnetism appearing as a result of correlation effects, we should investigate the many-body description as well. Here, the correlation effects are defined as a physical effect which can be described only by a multi-Slater determinant Ψ and cannot be explained by the single-particle description or the ordinal Hartree-Fock approximation. Even for ferromagnetism, we faces at the correlation effect, if we consider relative stability of the ferromagnetic ground state against all low-energy excitations. The correlation effect will be detected as growth of a correlation function in low energy states. This growth of the fluctuation will be seen in relevant orbitals $\phi_i(\mathbf{r})$ which are partly occupied by electrons. Energetically, this fluctuation will occur only around the Fermi energy.

If we start from the density functional theory, we may construct an effective theory by introducing explicitly correlation terms in the energy functional [23]. By formulating as minimization of a functional $\bar{G}$ of a many-body wavefunction Ψ, we have an effective self-consistent equations. Here, using the electron field operator $\psi_\sigma(\mathbf{r})$, $\bar{G}[\Psi]$ is given by,

$$\bar{G}[\Psi] = \left\langle \Psi \,|\, \hat{T} + \hat{V}_{\mathrm{red}} \,|\, \Psi \right\rangle + \frac{1}{2}\int d^3r d^3r' \frac{e^2}{|\mathbf{r}-\mathbf{r}'|} n_\Psi(\mathbf{r}) n_\Psi(\mathbf{r}') + E_{\mathrm{rxc}}[n_\Psi]$$

$$+ \int d^3r v_{\mathrm{ext}}(\mathbf{r}) n_\Psi(\mathbf{r}),$$

$$E_{\mathrm{rxc}}[n_\Psi] = \min_{\Psi' \to n_\Psi} \left\langle \Psi' \,|\, \hat{T} + \hat{V} \,|\, \Psi' \right\rangle - \min_{\Psi' \to n_\Psi} \left\langle \Psi' \,|\, \hat{T} + \hat{V}_{\mathrm{red}} \,|\, \Psi' \right\rangle$$

$$-\frac{1}{2}\int d^3r d^3r' \frac{e^2}{|\mathbf{r}-\mathbf{r}'|} n_\Psi(\mathbf{r}) n_\Psi(\mathbf{r}'),$$

$$\hat{T} = -\frac{\hbar^2}{2m}\int d^3r \sum_{\sigma} \lim_{\mathbf{r}'\to\mathbf{r}} \psi_\sigma^\dagger(\mathbf{r}')\Delta_{\mathbf{r}}\psi_\sigma(\mathbf{r}),$$

$$\hat{V} = \frac{1}{2}\int d^3r d^3r' \frac{e^2}{|\mathbf{r}-\mathbf{r}'|}\sum_{\sigma,\sigma'}\psi_\sigma^\dagger(\mathbf{r})\psi_{\sigma'}^\dagger(\mathbf{r}')\psi_{\sigma'}(\mathbf{r}')\psi_\sigma(\mathbf{r}),$$

$$n_\Psi = \left\langle \Psi \,|\, \textstyle\sum_\sigma \psi_\sigma^\dagger(\mathbf{r})\psi_\sigma(\mathbf{r}) \,|\, \Psi \right\rangle.$$

The correlation term $\hat{V}_{\mathrm{red}}$ may be written as a density-density correlation on a localized orbital. Actually, if we let,

$$\hat{V}_{red} = \frac{1}{2}\int d^3r d^3r' g(|\mathbf{r}-\mathbf{r}'|)\,|\phi_i(\mathbf{r})|^2|\phi_i(\mathbf{r}')|^2$$

$$\left\langle \Psi \,\Big|\left\{\sum_{\sigma=\uparrow,\downarrow}\left(c_{i,\sigma}^\dagger c_{i,\sigma} - \left\langle c_{i,\sigma}^\dagger c_{i,\sigma}\right\rangle\right)\right\}\left\{\sum_{\sigma=\uparrow,\downarrow}\left(c_{i,\sigma}^\dagger c_{i,\sigma} - \left\langle c_{i,\sigma}^\dagger c_{i,\sigma}\right\rangle\right)\right\}\Big|\, \Psi\right\rangle$$

then we have a Hubbard-like model. Here $g(|\mathbf{r}-\mathbf{r}'|)$ is a coupling constant of the screened Coulomb interaction. By adopting a proper approximation for $E_{\mathrm{rxc}}[n_\Psi]$, an effective two-body interaction term is introduced in a tight-binding Hamiltonian [23].

After these modification, we reach an effective tight-binding model called the Hubbard model. The single band Hubbard model is described by the following Hamiltonian.

$$H = -\sum_{\langle i,j\rangle}\sum_{\sigma} t_{ij}(c_{i,\sigma}^\dagger c_{j,\sigma} + \mathrm{H.c.}) + \sum_i U_i n_{i,\uparrow} n_{i,\downarrow}\,. \tag{6}$$

Here we use $U = \int d^3r d^3r' g(|\mathbf{r}-\mathbf{r}'|)\,|\phi_i(\mathbf{r})|^2|\phi_i(\mathbf{r}')|^2$.

3 Nano-graphite and Edge States

In the chapter of "Electronic and Magnetic Properties of Nano-graphite", a comprehensive discussion was done on the physical properties of nano-graphite and peculiar edge states using the tight-binding model. Here, we summarize important characteristics of the edge states of nano-graphite, which are essential for the following discussion. Let us consider graphene ribbons shown in Fig. 2. In these figures, each circle represents a π-orbital. Electron hopping integral t_{ij} is assumed to be non zero only between two atoms connected by a line.

The edge of Fig. 2(a) is the zigzag edge, where edge-carbon atoms, *e.g.* those denoted by $x = 0$ and $x = N$, are assumed to have a hydrogen atom per each. Thus every carbon atom has a π electron in this structure. Another edge of Fig. 2(b) is called the Klein edge. There is an extra atomic site with a π electron attached to each edge- carbon

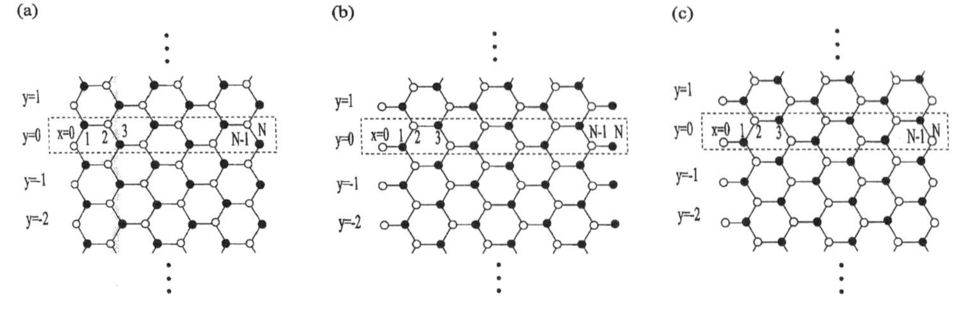

Fig. 2. Three typical graphene ribbons: (a) the zigzag ribbon, (b) the Klein ribbon and (c) the bearded ribbon. Circles represent positions of π-orbitals. In each structure, dotted squares represents a unit cell.

atom. Therefore, we assume that, on this extra site, there is a π-like orbital. Tentatively we assume that orbital energy of this extra π-orbital is equal to that of other π-orbitals. The Klein edge and its physical properties were studied by D.J. Klein and coworkers [24–26]. The third ribbon in Fig. 2(c) has two types of edges. This spatially asymmetric atomic configuration, which we call a bearded edge, has a band structure symmetric against reflection with respect to E_F as shown below. In all of these ribbons, π bands are assumed to be half-filled, where an electron occupies each π-orbital. Below, we call structures in Fig. 2(a), (b) and (c) the zigzag ribbon, the Klein ribbon and the bearded ribbon, respectively.

Properties of the tight-binding band of these structures are known. In contrast to the band structure of bulk graphite, flat bands appears in these bands at E_F. For the zigzag ribbon, we see two flat bands, which is nearly dispersionless in 1/3 of the whole one-dimensional Brillouin zone, $2\pi/3 \leq |k| \leq \pi$. For the Klein ribbon, the flat bands appears when $0 \leq |k| \leq 2\pi/3$. Appearance of the flat bands is understood by explicitly constructing eigen functions $\psi_k(x,y)$ of the tight-binding model with semi-infinite graphene with a zigzag edge or a Klein edge. Consider ribbons in Fig. 2s(a) and (b) with $N = \infty$. Let us determine wave functions $\psi_k(x,y)$ with energy 0. Here (x,y) denotes a lattice point at x in a y-th unit cell and k is the wave number in y-direction. For the zigzag edge, we have $\psi_k(x=2n,y) = (D_k)^n e^{iky}$, $\psi_{x=2n+1,y} = 0$ with $D_k = -2\cos(k/2)$ except for a normalization constant. We can check that this wave function is an eigen state of the semi-infinite zigzag ribbon by directly applying Eq. 4. For the Klein edge, we can find that $\psi_k(x=2n,y) = (D_k)^n e^{iky}$, $\psi_{x=2n+1,y} = 0$ with $D_k = -1/(2\cos(k/2))$. These wave functions are damping waves when $2\pi/3 \leq |k| \leq \pi$ for the zigzag edge and when $0 \leq |k| \leq 2\pi/3$ for the Klein edge. (We implicitly assume that $0^0 \equiv \lim_{\varepsilon\to 0+} \varepsilon^0 = 1$.) In addition, these solutions are non-bonding orbitals that have zero amplitude at every

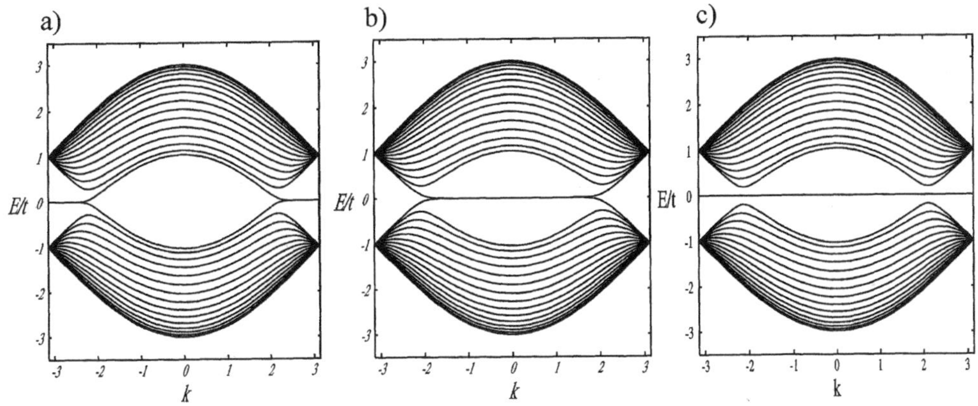

Fig. 3. The tight-binding band structure of graphene ribbons. (a) The Band structure of the zigzag ribbon, (b) that of the Klein ribbon, and (c) that of the bearded ribbon. For each band structure, flat bands appears (a) when $2\pi/3 \leq |k| \leq \pi$, (b) when $0 \leq |k| \leq 2\pi/3$, and (c) everywhere in the Brillouin zone.

other site. We call the non-bonding damping solution the edge state. The non-bonding nature fixes the band of the edge states at the center of the π band.

We should note that stability of the edge state against various deformation of the atomic structure. The Su-Shrieffer-Heeger model was adopted to explore this problem [27]. One might guess that there appears the Peierls instability similar to the case of poly-acetylene, since the zigzag ribbon with $N=1$ is nothing but the trans-poly-acetylene. However, this is not the case at least $N>3$. The result of Igami *et al.* suggests that the lattice deformation occurs not to create Peierls distortion along the ribbon but to expand the ribbon width. Near the zigzag edge, the bond length along the x-axis in Fig. 2 becomes longer than that of bulk graphite. This tendency was also confirmed by the first-principles calculation, where Peierls distortion was not seen within the numerical error but the elongation of the ribbon width for $N>3$ [28]. Similar tendency was also observed in the structure (c) of Fig. 2, which is important for the discussion in the section 5.2. This structural property is positive for magnetism originated by edge effects introduced in the following sections.

4 Flat-band Ferromagnetism in Organic Materials

For a well prepared materials design, we need to draw skeleton of the structure. Here, we start from the Hubbard model. The model is given by a natural simplification of the π electron systems as shown before. On the other hand, the Hubbard model is a prototype model for magnetism. Various magnetic mechanisms including the kinetic exchange, the super exchange, the double exchange and the *p-d* exchange, can be derived, if we consider a multi-orbital Hubbard model with inter-orbital interactions.

After a brief introduction of the Hubbard model, we will summarize known results on magnetism of the Hubbard model closely related to organic materials.

4.1 Flat-band Ferromagnetism in the Hubbard Model

Ground-state properties of the Hubbard model are depending on parameters of the model. The parameters may be categorized into three classes. The number of local orbitals N and the number of electrons N_e is the first one. We may use a fraction, N_e/N, which gives the electron occupation or the band filling. The case of $N_e/N=1$ is sometimes called the half-filling. This is because $N_e/N=1$ is just the half of the maximum value $N_e/N=2$ which is a fully filled band. A term, "site", in place of an orbital may be used, if the number of orbitals in concern is one at an atomic site. The second is the single-electron band structure given by the one-body part of the Hamiltonian, which is parametrized by $t_{i,j}$. The third is interaction strength in the two-body interaction terms (or the two-body part of the Hamiltonian). Sometimes, only the on-site Hubbard repulsion $U_i n_{i,\uparrow} n_{i,\downarrow}$ is considered for the discussion of magnetism. However, off-site terms and inter-orbital terms, *e.g.* the off-site repulsion $V_{i,j} n_i n_j$ and inter-orbital exchange $J_{i,j} \vec{S}_i \vec{S}_j$, may be considered to be essential for some kind of magnetic mechanism. Here, $n_{i,\sigma} = c^{\dagger}_{i,\sigma} c_{i,\sigma}$, $n_i = \sum_{\sigma} n_{i,\sigma}$ and $\vec{S}_i = \sum_{\sigma,\sigma'} c^{\dagger}_{i,\sigma} \vec{\sigma}_{\sigma,\sigma'} c_{i,\sigma'}$ with $\vec{\sigma}$ being the Pauli matrices σ_l ($l = x, y, z$). In this section we consider only the U term, *i.e.* the on-site repulsive interaction. We sometimes change $t_{i,j}$ and U_i arbitrary. This may be partly realized in real materials by applying mechanical (or chemical) pressure or chemical modification. Fixing a typical transfer integral, *e.g.* t for the nearest-neighbor pair of π-orbitals as a unit of energy, and assuming that $U_i = U$ is a constant, we may draw a magnetic phase-diagram as a function of U. The electron occupation is relevant to determine the magnetism. Even change in N_e by one may cause crucial difference in the Hubbard model as typically shown in the Nagaoka theorem, whose contents are beyond the scope of the present discussion [29].

Optimizing these parameters, we may have ferromagnetic phase. There are several interesting parameter sets, for which the ferromagnetic ground state is exactly shown [30]. As the exact solutions of the Hubbard model, we should mention the flat-band ferromagnetism. The name of the flat-band ferromagnetism is a kind of generic term. There are several proofs on the ferromagnetism relating to the Hubbard model with a flat band [30]. For each proof, there is a set of required conditions of parameters. Due to the requirement on $t_{i,j}$, a flat band appears in the tight-binding band. For the majority of proofs, the required filling factor N_e/N corresponds to a half-filling of the flat band. For example, Mielke-Tasaki's proof utilizes a flat band located at the bottom (or the top) of the whole band structure. Since the system has inevitably a multi-band structure, N_e/N is usually not 1. Here, for our purpose, the theorem by Lieb is relevant [31]. The proof is applicable for any bipartite structure with $N_e/N = 1$. We may say that alternant hydrocarbon has a π network, which is bipartite and filled by π electrons with $N_e = N$. Interestingly, the Lieb theorem connects two famous conclusions on design of the organic magnets and thus include the old knowledge. This Lieb's model will be discussed in Section 4.3.

4.2 The Longuet-Higgins Rule and The Ovchinnikov Rule

As discussed in Section 2, we have two approaches for the magnetism. Historically, there appeared the first discussion of magnetic organic molecules using the molecular orbital theory. Starting from the effective single particle description, we can explain this Longuet-Higgins discussion on the alternant hydrocarbon as follows [32]. The π-electron system may be described by a tight-binding model on a bipartite lattice. Using the idea of conjugation, each π-orbital is assumed to form a pair with neighboring orbital, if possible. This pair gives a double bond on the network. In some structures, there remain several unpaired non-bonding orbitals, even if double bonds are maximally created. By counting the number of double bonds, and by letting the number T, we have the number of non-bonding orbitals $N-2T$. Here N is the total number of π sites. For the bipartite lattice, this number is shown to be not smaller than $|N_A - N_B|$. If spins in these non-bonding orbitals are aligned following to the Hund's coupling, the system shows the high-spin ground state.

On the other hand, we may start from the effective many-body description. If we consider Hubbard repulsion U on each π orbital and derive an effective Hamiltonian in the limit of $U >> t$, the $S = 1/2$ isotropic Heisenberg model is obtained. For this model on a bipartite lattice, the Marshall-Lieb-Mattis theorem is known to hold. We denote numbers of sublattice sites N_A and N_B, respectively. This theoretical conclusion states that the ground state has the total spin $S = |N_A - N_B|/2$ Ovchinnikov applied the conclusion to determination of the ground-state spin of the Heisenberg system which models an organic molecule [33].

4.3 The Lieb Theorem in the Hubbard Model

One should refer to the original paper for the proof of the Lieb theorem [31]. In this section, we summarize sufficient conditions for the theorem and discuss physical meanings that we can draw out. The Lieb theorem consists of two theorems. One for the attractive Hubbard model with negative U_i and the other is for the repulsive Hubbard model with positive U_i. The former is essential and the latter is derived from the former using a unitary transformation called the partial electron-hole transformation. Here, we concentrate on the latter theorem, which make it possible to conclude ferrimagnetism in the Hubbard model.

Consider a bipartite lattice (or a bipartite graph) with two sets of sublattice sites, *A* and *B*. The numbers of sites in each sets are N_A and N_B, respectively. We introduce the

Hubbard model derived by identifying a vertex as a site and an edge as an electron path by $t_{i,j}$ in the graph. Assume that $U_i = U > 0$ *i.e.* a positive constant and that the band is half-filled, $N_e = N_s$. If these conditions are met, the ground state is unique apart from the trivial ($2S+1$)-fold degeneracy and has spin $S = |N_A - N_B|/2$.

Here we have to note the following. If one of these conditions is not met, there is no proof on the spin of the ground state. However, we may hope existence of a finite magnetic phase around the solvable model. If $|N_A - N_B| > 0$, and if the graph is identical to a lattice structure with periodicity, there appears a flat band in the band structure of the tight-binding model in which $U = 0$. This fact corresponds to the large degeneracy coming from the non-bonding orbitals discussed by Longuet-Higgins's argument. When U becomes infinitely large, the system may be regarded as a $S = 1/2$ Heisenberg model. In the model, the total spin is known to be $S = |N_A - N_B|/2$.

We have another strong conclusion for the repulsive Hubbard model on the bipartite lattice. Shen *et al.* have succeeded in proving existence of ferromagnetic order in the Hubbard model [34].

$$\vec{S}_i \cdot \vec{S}_j > 0 \quad \text{for } i \in A \text{ and } j \in A \text{ or } i \in B \text{ and } j \in B,$$

$$\vec{S}_i \cdot \vec{S}_j < 0 \quad \text{for } i \in A \text{ and } j \in B \text{ or } i \in B \text{ and } j \in A. \tag{7}$$

Interestingly, the proof is only possible for the Hubbard model.

For the purpose of the materials design, to consider a lattice structure shown in Fig. 2(c) is interesting, since it is a bipartite lattice with $|N_A - N_B| > 0$ and is extensive for a infinitely long ribbon. Following the conclusion of Lieb's theorem, we may draw a spin configuration as shown in Fig. 4. This figure is obtained by assuming the polarization occurs only on the flat band. In other word, the condition corresponds to the weak-coupling limit. It is interesting to compare the figure to Fig. 7(b), which is obtained by the DFT-LSDA model.

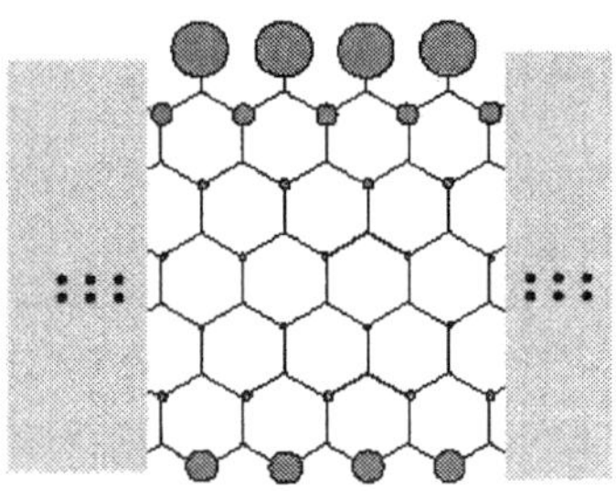

Fig. 4. Spin configuration of a model bearded ribbon. The spin moments are assumed to be made only from the spin-polarized flat band. Thus we have ferromagnetic component, only. At the Klein edge (upper edge) and at the zigzag edge (lower edge), $2\mu_B/3$ and $\pi_B/3$ of the moment appears.

5 Materials Design of Magnetic Carbon

Even if we notice that magnetic mechanism inducing ferrimagnetism or flat-band ferromagnetism is expected in π-electron systems, the knowledge by itself cannot answer whether we have magnetic carbon at hand or not. At the moment, there is no example of magnetic carbon by the edge effect, which is experimentally proven to appear. Our next purpose is to show a kind of theoretical counterparts possibly realized by magnetic mechanism given in the last section. Of course people know that theoretical methods are too weak to give the final answer. However, we believe that numerical data tells a right way.

This problem is a kind of materials design. It is known to be an inverse problem. To solve the inverse problem, one has to find an artificial material with desired properties. It is required to show a convincing clue on stability of the material, too. This is in contrast to usual demands for theory in the condensed matter physics, where explanation on phenomena is its goal. For the present purpose, one has to go beyond interpretation of phenomena and has to demonstrate realization of abstract idea. Here, we could refer to the history of organic magnets again. People design high-spin molecules. Some of them were really synthesized according to the design by chemists. Thus, we should follow the strategy of materials science, which are successfully applied to organic molecules and several solids.

5.1 Magnetism in the DFT-LSDA Model

For the present purpose of the materials design, we utilize the density-functional theory again and its resulting models. The so-called local-spin-density approximation (LSDA) offers us a tractable model of many-electron systems with non-uniform electron- and spin- density. If the spin density remains finite, one may conclude that magnetism can occur in a corresponding real material. The best justification of LSDA and its extension is to show examples indicating correspondence between their results and the real materials.

We consider the Kohn-Sham equation, eq. (1), again. If we introduce a model exchange-correlation energy functional of the electron density and of the spin density $E_{xc}[\rho_\uparrow, \rho_\downarrow]$, we obtain a set of self-consistent equations that consist of

$$\rho_\sigma = \sum_{\varepsilon_n \le \varepsilon_F} |\phi_{n,\sigma}(\mathbf{r})|^2 , \qquad (8)$$

and Eq. 1. This strategy may be understood as an extension of Levy's constrained search in order-parameter-specified constrained spaces [35, 36]. By looking for the most stable solution for a given explicit functional by varying the density and the spin density, which are order parameters, we can reach the true ground state with possible magnetism. The exact functional is, however, not known.

We utilize the local-spin density approximation as,

$$E_{xc}[\rho_\uparrow,\rho_\downarrow]=\int d^3r\varepsilon_{xc}(\rho_\uparrow(\mathbf{r}),\rho_\downarrow(\mathbf{r}))\,. \tag{9}$$

Then the equation becomes numerically tractable by expanding the Kohn-Sham orbital and the density using some proper basis sets for both the wavefunctions and the density.

Since the Kohn-Sham scheme utilizes a single-particle description, we can never hope that every correlation function in the original many-body electron system is reproduced. In this sense, there is no room to have correlated many-body states like the Mott gapped state, the spin gapped state and the Kondo singlet state starting from DFT-LSDA. To overcome the difficulty on the correlation effects, there exists an effective many-body description for DFT [23], which is developing.

However, an order parameter given by expectation value of a single particle operator could be similar to the true answer. For example, in known LSDA or GGA, the order parameter, *i.e.* the spin density given by the effective model, is adjusted to the best approximation for the Coulomb system around the uniform density limit. Thus, it is not surprising, if results of LSDA for some real materials are similar to reality. The single-particle description is naturally obtained when one perform a mean-field approximation. Mathematically, the present DFT-LSDA scheme is a kind of Hamiltonian approach, where the energy functional is modified from the Hartree approximation by the additional exchange-correlation term E_{xc}. Since E_{xc} is given by values of the order parameters ρ_σ, this scheme can be regarded as a generalized mean-field approach.

When some magnetic solution is found in a DFT-LSDA result, we can often find an analogy in a corresponding mean-field solution for a simplified many-body system *i.e.* the Hubbard model. For example, if we apply the Stoner approximation for the flat-band Hubbard model, we will have a fully spin polarized flat band in the solution. Similar explanation is often seen to interpret DFT-LSDA results on magnetic mechanisms including the double exchange and the p-d exchange. In these arguments, we have a rather understandable picture using the local orbitals, occupation of relevant bands and spin polarization of these states. However, since the mean-field Hamiltonian for the Hubbard model and DFT-LSDA is basically different model, there is no guarantee nor necessity for expectation that two models for many electron systems conclude the same result. We should carefully interpret the result.

5.2 Magnetic Nano-graphite

We now start the discussion on the magnetic nano-graphite. Our purpose in this section is to find an atomic structure corresponding to the skeleton in Fig. 2(c). To do the materials design, we have to specify the atomic structure of the zigzag edge and the Klein edge. Once these edges are prepared, we can design a π network satisfying $|N_A-N_B|>0$ and hence being magnetic.

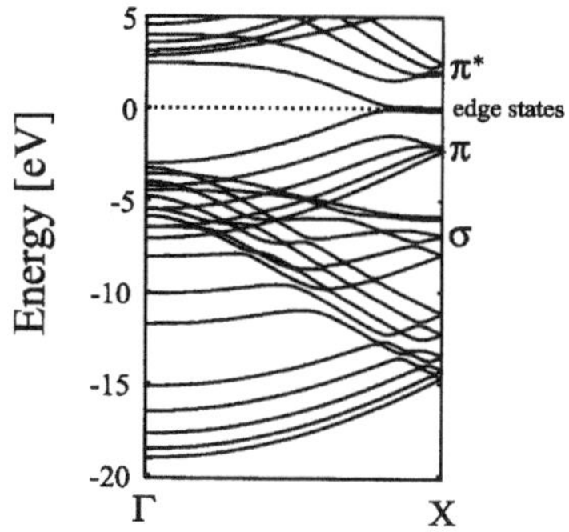

Fig. 5. The electronic band structure of a graphene ribbon with mono-hydrogenated zigzag edges. Two flat bands due to the edge states appears when $2\pi/3 \le |k| \le \pi$.

To create the zigzag edge, hydrogen termination is a way. Miyamoto *et al.* have shown that a graphene ribbon with hydrogen terminated zigzag edges has a DFT-LDA band structure, which is the same as the band structure of TBM as shown in Fig. 3(a) in essential characteristics [37]. Two center bands around E_F have an almost flat dispersion when $2\pi/3 \le |k| \le \pi$, and are the edge states (Fig. 5).

For the same zigzag ribbon, the DFT-LSDA calculation by Okada and Oshiyama concludes anti-ferromagnetic spin structure [12]. We may interpret that this corresponds to the mean-field solution of the Hubbard model as an effective model of the π-electron system on the ribbon [7]. The obtained spin density shows local ferrimagnetic spin structure at the edge and decays in the ribbon. Of course, this spin-density profile is a result of LSDA. True anti-ferromagnetic long-range order should not appear and the spin gap should open, if the real graphene ribbon behaves as a spin system or is well-described by the Hubbard model [7, 38]. However, the DFT-LSDA result suggests that local ferrimagnetic spin polarization would remain if the width of the ribbon becomes enough large.

Another type of the zigzag edge, *i.e.* the Klein edge was found by trial and error [11, 39]. One might think that it is not so difficult to put an additional π-orbital at the zigzag edge. The methylene group (-CH_2) could be a solution. However, due to geometrical restriction, the methylene group is attached perpendicular to the graphene plane. (Fig. 6(a)) Namely, each π-orbital on the methylene group turns to the perpendicular direction and does not hybridize with the host π system. The π network should be identified to the symmetric structure of Fig. 2(a) with extra one-dimensional π-orbitals at the right edge, but not to Fig. 2(c). The band structure is essentially the same as Fig. 5 with an additional one-dimensional π band. Thus the total spin is zero. If there is an interval between methylene groups at an edge, the plane of the methylene group may in the honeycomb plane of host graphene and π states may show magnetism due to a finite difference $|N_A - N_B|$. However, no magnetic solution has been obtained yet.

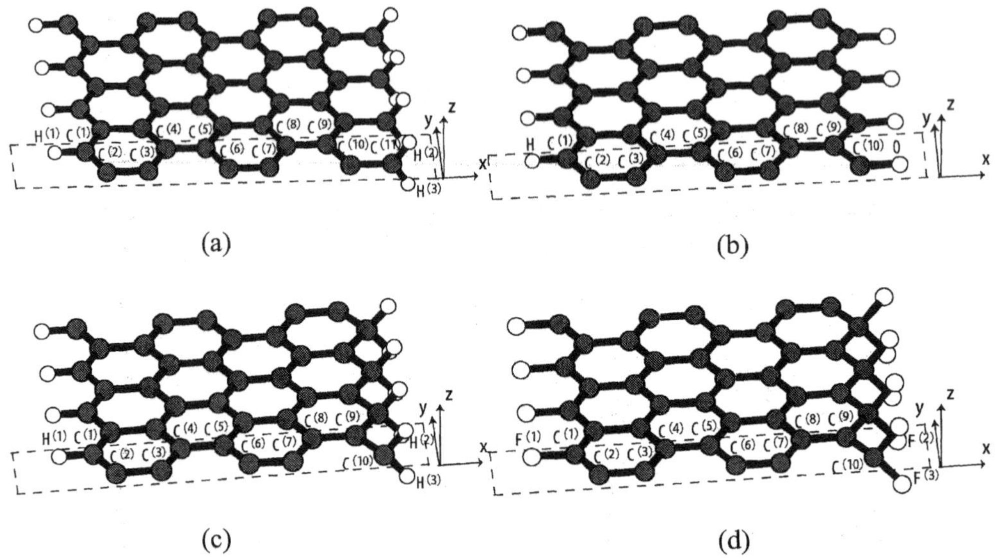

Fig. 6. Graphene ribbons with various types of edges. (a) A ribbon with a methylene substituted zigzag edge and a mono-hydrogenated zigzag edge. (b) A ribbon with a oxygen substituted zigzag edge and a mono-hydrogenated zigzag edge. (c) A ribbon with a di-hydrogenated zigzag edge and a mono-hydrogenated zigzag edge. (d) A ribbon with di-fluorinated zigzag edge and a mono-fluorinated zigzag edge. The structures of (b), (c) and (d) are magnetic nano-graphite, but (a) is not magnetic.

To avoid geometrical constraint, we next consider oxygen substitution. (Fig. 6(b)) Since an oxygen atom has two p states p_y and p_z, which are not fully occupied, there might appear π-orbital with spin moment. Unfortunately, DFT-LSDA calculation concludes that no spin moment appears at an oxygen-terminated zigzag edge. However, asymmetry is introduced in the ribbon and the hydrogen-terminated edge becomes spin polarized. We show spin density of the ribbon in Fig. 7(a). Ferrimagnetic spin structure is clearly seen at the hydrogenated zigzag edge. Electronic band structure shown in Fig. 8 indicates that a nearly flat band appears around E_F. The band splits at $2\pi/3 \leq |k| \leq \pi$, which corresponds to the spin moment at the mono-hydrogenated zigzag edge. There appears a cosine band whose band top is at the Γ point just above $E_F = 0$. This band comes from the p_y orbitals of oxygen. This example suggests that magnetism could appear at the hydrogenated zigzag edge, if the whole graphene system is structurally and electronically asymmetric.

Although we have a kind of magnetic carbon, the oxygen-terminated edge is not the Klein edge. Thus, substitution of a hydrogen by a function group with a π-orbital is not successful so far. Here, we seek for another solution. If a π-orbital disappears at the end carbon of a zigzag edge, the π network is modified and a Klein edge will be given. To

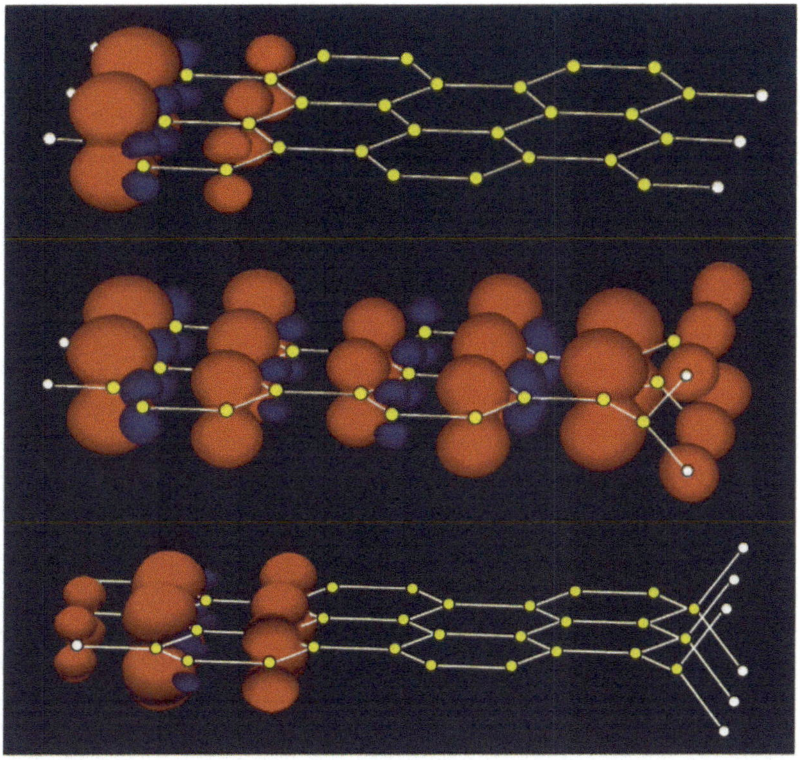

Fig 7. Spin density of magnetic graphene ribbons: (a) a ribbon with oxygen-substituted and mono-hydrogenated zigzag edges, (b) a ribbon with di-hydrogenated and mono-hydrogenated zigzag edges, (c) a ribbon with di-fluorinated and mono-fluorinated zigzag edges. Yellow circles represent carbon atoms, while white circles represent (a) hydrogen atoms for the left edge and oxygen atoms for the right edge, (b) hydrogen atoms and (c) fluorine atoms, respectively. Red and blue isosurfaces represent spin-up density and spin-down density, respectively.

remove π-orbital may be possible, if the end carbon becomes sp^3 hybridized. Addition of some function groups at the end carbon may reduce the number of π-orbitals. One possible method to realize this idea is to create di-hydrogenated carbon at the edge. In Fig. 2(c), the right edge is a di-hydrogenated zigzag edge. Each edge carbon has four covalent bondings. The bond length is d_{CC}= 1.460 Å for the C-C bonding and d_{CH}= 1.125 Å for the C-H bonding at the edge carbon atom. The H-C-H bond angle is close to 100.7°. Thus, at the edge carbon atom, the covalent character is almost that of the sp^3 bonding. This ribbon structure is an optimized structure by DFT-LSDA. The optimized structure means that the atomic configuration is optimized so that inter-atomic forces balance. Thus the di-hydrogenated structure is locally stable.

In Fig. 7(b), distribution of the spin density is shown. The red and blue isosurfaces represent spin-up and spin-down density maps, respectively. An integral of the spin-up density is 1 μ_B per a unit cell greater than that of the spin-down density. The up spins are placed at the A sites and the down spins are found at the B sites. Totally, ferrimagnetic structure is realized. We call this typical structure a magnetic nano-graphite ribbon. Carbon atoms located inside the ribbon have finite magnetic moments. This feature of the spin density suggests that the total π system is magnetically ordered. Interestingly, there appears spin density on hydrogen atoms at the di-hydrogenated zigzag edge in Fig. 7(b). This another characteristic tells us that hydrogen atoms may be a key for the magnetism. Thus this di-hydrogenated edge is a realization of Klein's edge. A DFT-LSDA-based proof of this statement is given in the followings.

To analyze origin of the spin density, we can refer to the band-structure of the magnetic nano-graphite. (Fig. 9) The most important characteristic of the band structure is appearance of a fully spin-polarized flat band at the center of the π bands. We easily recognize nine π bands. These π bands are spin-polarized. This fact corresponds to the magnetization map in Fig. 7(b).

The wavefunction ψ_k on the flat band has every characteristic of the edge state. It is a damping wave with non-bonding nature. The function ψ_k with $0 \le k < 2\pi/3$ has amplitude mainly on the di-hydrogenated edge, while ψ_k with $2\pi/3 < k < \pi$ has amplitude mainly on the mono-hydrogenated edge. If we identify the di-hydrogenated edge as the Klein's edge and the mono-hydrogenated edge as the zigzag edge, behavior of ψ_k is just what is expected in TBM. Furthermore, ψ_k has amplitude on two hydrogen atoms at the Klein's edge. The profile of ψ_k suggests that we should identify an anti-boding orbital made from two 1s orbitals as an extra π-orbital at a zigzag edge of the π network. Then, the π network is identical to the skeleton of Fig. 2(c) with $N = 11$. This interpretation can be regarded as a confirmation for the requirement of $|N_A - N_B| > 0$. We can see another π band in the band structure of the hydrogenated ribbon around the Y point (Fig. 9).

Here we consider the interpretation of the DFT-LSDA result by a localized electron model. The fact that the first π band is rather separated energetically from other lower π bands means that the orbital energy of hydrogen 1s orbital is rather different from the π states of carbons. This makes it difficult to adjust the tight-binding model to satisfy the requirement of Lieb's theorem. Thus, to make a connection between the DFT-LSDA result and the knowledge on the Hubbard model, it is required to show adiabatic connection between the model directly obtained from the DFT-LSDA result and a solvable point in which the Lieb's requirements are fulfilled. But, the spin texture tells us that the system is in a rather strong correlation regime.

Stability of magnetic solution is estimated by determining energy difference between magnetic solution and non-magnetic one. This difference $E_{NM} - E_M$ is 58 meV per a unit cell containing ten carbon atoms and three hydrogen atoms. Another estimation of the exchange energy is the band gap of the polarized flat band. This is approximately 0.2 eV. The exchange splitting itself is estimated to be 0.5 ~ 0.6 eV.

Creation of the di-hydrogenated zigzag edge is an endothermic reaction. The calculation suggests that $E_{G2}-(E_{G1}+E_{H_2})\simeq 0.3\,\mathrm{eV}$ for the ribbon. Here E_{G2}, E_{G1} and E_{H_2} are the total energy of the magnetic graphene ribbon per a unit cell, a graphene ribbon with mono-hydrogenated zigzag edges and a hydrogen molecule, respectively [11]. Therefor, to fabricate magnetic nano-graphite, hydrogenation is not necessarily the best way.

An exothermic reaction to synthesize magnetic nano-graphite is possible to design by considering fluorination. In Fig. 2(d), a corresponding fluorinated ribbon is shown. The DFT-LSDA calculation results in a magnetic solution with a spin-polarized mono-fluorinated zigzag edge. However, a di-fluorinated edge at the other side of the ribbon is non-magnetic. (Fig. 7(c)) This is because the fluorine atom prefers a closed shell structure. However, we see a finite spin moment at the fluorine atom on the other mono-fluorinated edge. These characteristic features are found their correspondences in the band structure of Fig. 10. Almost one-third of the center band, which is made of the edge states at the mono-fluorinated edge, is spin-polarized.

Fluorination is exothermic. The calculation suggests that $E_{G4}-(E_{G3}+E_{H_2})\simeq -1.74\,\mathrm{eV}$ for the ribbon. Here E_{G4} and E_{G3} are the total energy of the fluorinated graphene ribbon in Fig. 6(d) and a graphene ribbon with mono-fluorinated zigzag edges. The formation energy of the fluorinated ribbon from the mono-hydrogenated ribbon is given by $(E_{G4}+E_{H_2})-(E_{G1}+(3/2)E_{F_2})\simeq -5.33\,\mathrm{eV}$ [39]. Since fluorine atoms prefer to form bonding with carbon atoms, the reaction may be utilized to create magnetic nano-graphite.

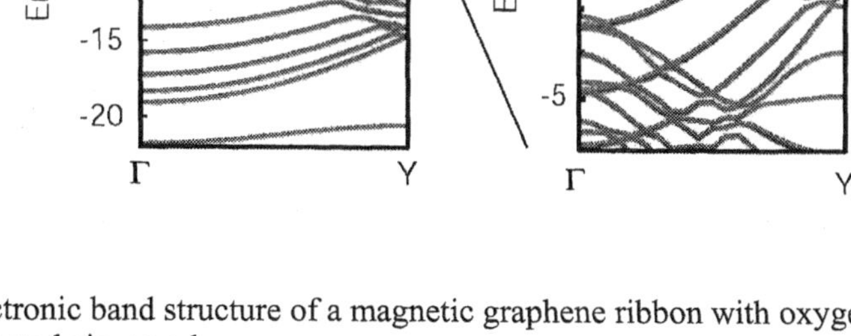

Fig. 8. The electronic band structure of a magnetic graphene ribbon with oxygen-substituted and mono-hydrogenated zigzag edges.

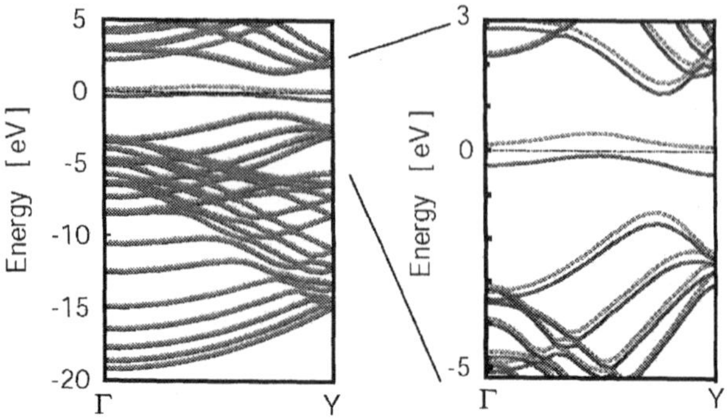

Fig. 9. The electronic band structure of a magnetic graphene ribbon with di-hydrogenated and mono-hydrogenated zigzag edges.

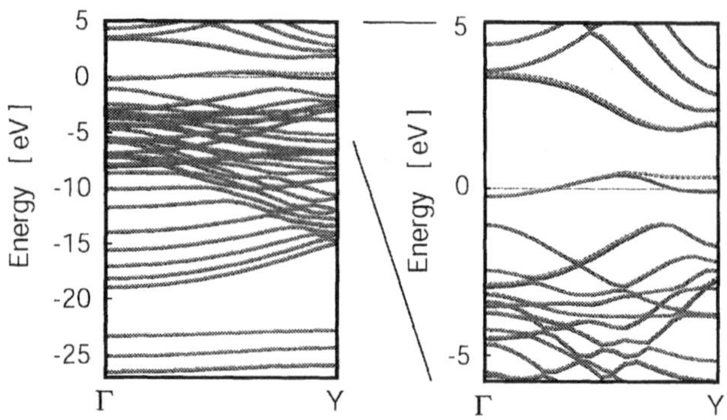

Fig. 10. The electronic band structure of a magnetic graphene ribbon with di-fluorinated and mono-fluorinated zigzag edges.

5.3 Magnetic Nanotubes by the Spin-polarized Edge States

The carbon nanotube is regarded as a rolled graphene sheet. If we can roll a magnetic graphene ribbon in a cylindrical shape, a magnetic nanotube can be obtained [11]. If very loosely rolled tubes can be created, or if the diameter of the tube is large enough, this expectation should hold. However, if the diameter of the obtained tube is small, we need to check the magnetism again for each specified tube. We can test the above idea and check whether magnetic nanotube is obtained as rolled magnetic nano-graphite.

The first report on magnetism in the finite nanotube studied by DFT-LSDA calculation was done by Okada and Oshiyama [12]. They considered finite zigzag nanotubes with two open mono-hydrogenated zigzag edges. They obtained magnetic solutions with the total spins of $S = 2$ and 3 for (7,0) and (10,0) finite nanotubes. These magnetic states are originated from spin polarization of the edge states. In Fig. 11(a), distribution of spin density is shown by isosurfaces.

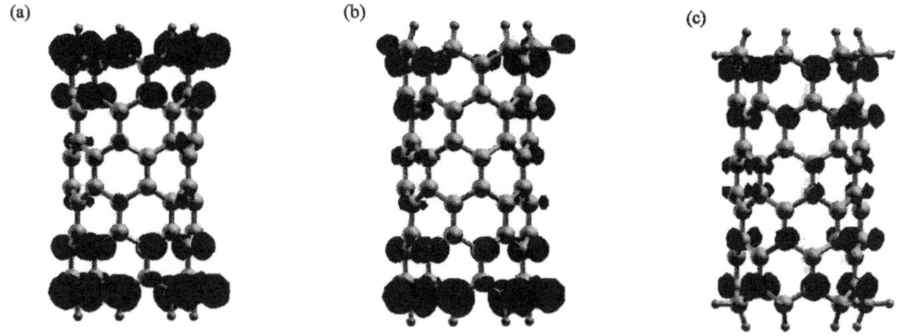

Fig. 11. Magnetic carbon nanotubes. (a) a (7,0) tube with mono-hydrogenated zigzag edges, (b) a (7,0) tube with a mono-hydrogenated zigzag edge and a di-hydrogenated zigzag edge, (c) a (7,0) tube with di-hydrogenated zigzag edges.

If we regard the π network on the nanotube as a rolled magnetic nano-graphite network, $|N_A - N_B|$ should be zero for these tubes. Thus, the counting rule does not hold. By the conventional kinetic exchange, we hope that the system has zero total moment. Thus, we see that another mechanism of spin polarization appears. when the diameter of tube is small. Interestingly, spin-polarization of degenerate edge states survives, which is concluded at least by the DFT-LSDA calculation. Although the counting rule of the total spin moment for planer graphene structure does not work, we have magnetic nanotubes.

Discrepancy between LSDA and the Hubbard model is also seen for tubes with other types of edges. For di-hydrogenated edges, the spin moment is again not large enough compared to the expected value from the counting rule. For example, $S = 3/2$ and $S = 1$ for tubes in Fig. 11(b) and (c), respectively. Since the degeneracy between edge states tend to be lifted by formation of di-hydrogenated edge, smaller magnetization is obtained for these tubes.

There are several reports on magnetism in finite nanotubes with open or closed ends [40, 41]. In some of the tubes, dangling bonds are the origin of the spin moments. Comparing to spins in the dangling bonds, the π electron spins couple with each other by the kinetic exchange, as suggested by the magnetic nano-graphite. Thus, we can conclude that tubes with the hydrogenated edges show stable magnetic structure.

5.4 Magnetic Graphite-diamond Hybrid Structures

Interface of the sp^2 network may be utilized to make a finite discrepancy between sublattice sites and let $|N_A - N_B| > 0$. The idea is exemplified by Okada and Oshiyama, who considered interfaces of graphic carbon structures and hexagonal BN in BCN compounds [42]. They succeeded in finding a magnetic solution. Similar consideration is possible by using an interface between sp^2 hybridized network and sp^3 hybridized network of carbon.

Fig. 12. Magnetic graphite-diamond hybrid structure. Graphene structures are attached to a diamond (100) surface. The isosurface of the spin density is shown.

In hydrogenated magnetic nano-graphite, the Klein edge was realized by π-like 1s orbitals of two hydrogen atoms. Consequently, we have not obtained an edge ended with sp^3 carbon atoms. To find π network modification by sp^3formation, we may consider an interface between graphite and diamond. This structure was considered by Baraban *et al.* who named the structures diamond-graphite hybrids [25]. However, their structures are unstable, because neighboring graphitic structures are too close with each other to keep the whole structure.

To stabilize the structure, we may consider graphitic structure created separately from others. (Fig. 12) In this structure, dangling bonds are eliminated by formation of dimers at the surface. If the diamond structure is bulky to keep the shape of the surface unit cell, the structure is stable. Magnetization profile reveals that the moments appear only on the graphite strip. Thus we call the structure a magnetic graphite-diamond hybrid structure. The π network appearing in the strip is identical to that of a magnetic organic polymer [43].

In the band structure, we can see appearance of a spin-polarized flat-band (Fig. 13). This result confirms that the design works well for this planer graphitic structure. Correspondence between LSDA and the Hubbard model is thus rather complete. If we

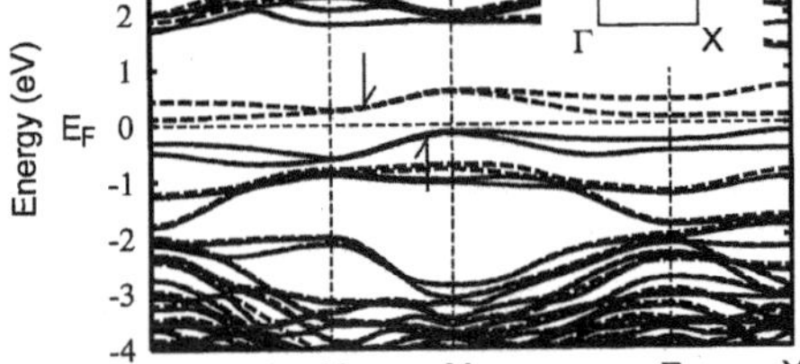

Fig. 13. The band structure of the graphite-diamond hybrid structure with a (3×1) unit cell of a modified diamond (100) surface. Spin polarized flat bands shown by arrows are seen at the Fermi level.

neglect difference in π-orbital energy among those on this comb structure, the application of the Shen theorem on the ferrimagnetic order becomes possible. Thus, the magnetic order at least in the ground state is expected for this one-dimensional structure from both the DFT-LSDA model and the Hubbard model.

6 Summary and Discussion

We have discussed possible magnetic nano-graphite structures. Starting from consideration on the models of the electronic system, we showed that combination of the DFT-LSDA model and the Hubbard-type model conclude a convincing conclusion for a planer nano-graphite structure. However, there remains uncertain discrepancy in results of these models for magnetic nanotubes. This point will be solved by development of the model of correlated electron systems.

We mainly discussed a mechanism for localized magnetism coming from the graphite edge states. This is because the resulting magnetic structure extends over the whole π network. The similar conclusion is known for another origin, *i.e.* topological defects made of odd-membered rings [13]. If we regard the edge as a topological defect, these conclusions suggests that π network can be intrinsically spin-polarized by the topological defects.

Lack of carbon sites and extra carbon atoms may be regarded as topological defects [44, 45]. We find similar magnetic effects in these defective structures. Especially, discussion on these defective structures [46] are regarded as essential for the explanation of interesting magnetic signals found in proton-irradiated graphite [20].

If the modified π network keeps bipartite structure, we may also conclude the magnetic order in graphitic systems. This can be an advantage for the design. However, to show universal behavior of defects in graphitic network would be demanded for the synthesis of magnetic nano-graphite, since the fabrication process may not be completely controlled. Although the story is on the way due to difficulty in describing the interacting electron system, we expect that valid theoretical findings will be experimentally proven to be true in near future.

Reference

1. M. Kinoshita, Mol. Cryst. Liq. Cryst. 233 (1993) 1.
2. K. Itoh and M. Kinoshita, "Molecular Magnetism: New Magnetic Materials" (2000) G & B Science Pub.
3. J.S. Miller and M. Drillon, 2001, "Magnetism: Molecules to Materials II: Molecule-Based Materials" (Wiley-VCH, Weinheim).
4. P.-M. Allemand, K.C. Khemani, A. Koch, F. Wudl, K. Holczer, S. Donovan, G. Gruner, and J.D. Thompson, Science 253 (1991) 301.
5. Y. Murakami and H. Suematsu, Pure Appl. Chem. 68 (1996) 1463.
6. T.L. Makarova, B. Sundqvist, R. Hohne, P. Esquinazi, Y. Kopelevich, P. Scharff, V.A. Davydov, L.S. Kashevarova, and A.V. Rakhmanina, Nature 413 (2001) 716.
7. M. Fujita, K. Wakabayashi, K. Nakada, and K. Kusakabe, J. Phys. Soc. Jpn., 65 (1996) 1920.
8. A. Nagashima, H. Itoh, T. Ichinokawa, C. Oshima, and S. Otani, Phys. Rev. B50 (1994) 4756.

9. A. Nakayama, K. Suzuki, T. Enoki, S.L. di Vittorio, M.S. Dresselhaus, K. Koga, M. Endo, and N. Shindo, Synth. Met. 57 (1993) 3736.
10. B.L.V. Prasad, H. Sato, T. Enoki, Y. Hishiyama, Y. Kaburagi, A.M. Rao, P.C. Eklund, K. Oshida, and M. Endo, Phys. Rev. B62 (2000) 11209.
11. K. Kusakabe and M. Maruyama, , Phys. Rev. B67 (2003) 092406.
12. S. Okada and A. Oshiyama, , J. Phys. Soc. Jpn. 72 (2003) 1510.
13. N. Park, M. Yoon, S. Berber, J. Ihm, E. Osawa, and D. Tománek, , Phys. Rev. Lett. 91 (2003) 237204.
14. J.-C. Charlier, J.-P. Michenaud, and X. Gonze, Phys. Rev. B46 (1992) 4531.
15. P. Hohenberg and W. Kohn, Phys. Rev. 136 (1964) B864.
16. W. Kohn and L.J. Sham, Phys. Rev. 140 (1965) A1133.
17. J.P. Perdew and Y. Wang, Phys. Rev. B45 (1992) 13244.
18. M. Dion, H. Rydberg, E. Schröder, D.C. Langreth, and B.I. Lundqvist, Phys. Rev. Lett. 92 (2004) 246401.
19. D. Vanderbilt, Phys. Rev. B41 (1990) 7892.
20. Y. Kopelevich, J.H.S. Torres, R.R. da Silva, F. Mrowka, H. Kempa, and P. Esquinazi, Phys. Rev. Lett. 90 (2003) 156402.
21. R. Ocaña, P. Esquinazi, H. Kempa, J. H. S. Torres, and Y. Kopelevich, Phys. Rev. B68 (2003) 165408.
22. N. Marzari and D. Vanderbilt, Phys. Rev. B56 (1997) 12847.
23. K. Kusakabe, J. Phys. Soc. Jpn. 70 (2001) 2038.
24. D.J. Klein, Chem. Phys. Lett. 217 (1994) 261.
25. A.T. Balaban, D.J. Klein and C.A. Folden, Chem. Phys. Lett. 217(1994) 266.
26. D.J. Klein and L. Bytautas, J. Phys. Chem. A 103 (1999) 5196.
27. M. Fujita, M. Igami and K. Nakada, J. Phys. Soc. Jpn. 66 (1997) 1864.
28. K. Nakada, M. Fujita, G. Dresselhaus and M.S. Dresselhaus, Phys. Rev. B54 (1996) 17954.
29. Y. Nagaoka, Phys. Rev. 147 (1966) 392.
30. H. Tasaki, Prog. Theor. Phys. 99 (1998) 489.
31. E.H. Lieb, Phys. Rev. Lett. 64 (1989) 1201.
32. H.C. Longuet-Higgins, J. Chem. Phys. 18 (1950) 265.
33. A.A. Ovchinnikov, Theor. Chem. Acta (1978) 297.
34. S.-Q. Shen, Z.-M. Qiu, and G.-S. Tian, Phys. Rev. Lett. 72 (1994) 1280.
35. M. Levy, Proc. Natl. Acad. Sci. (U.S.A.) 76 (1979) 6062.
36. K. Higuchi and M. Higuchi, Phys. Rev. B69 (2004) 035113; *ibid* 165118.
37. Y. Miyamoto, K. Nakada, and M. Fujita, Phys. Rev. B59 (1999) 9858.
38. H. Yoshioka, J. Phys. Soc. Jpn. 72 (2003) 2145.
39. M. Maruyama and K. Kusakabe, J. Phys. Soc. Jpn. 73 (2004) 656.
40. C. Kim, K. Seo, B. Kim, N. Park, Y.S. Choi, K.A. Park, and Y.H. Lee, Phys. Rev. B68 (2003) 115403.
41. Y.-H. Kim, J. Choi, K.J. Chang and D. Tománek, Phys. Rev. B68 (2003) 125420.
42. S. Okada and A. Oshiyama, Phys. Rev. Lett. 87 (2001) 146803.
43. D.J. Klein, C.J. Nelin, S. Alexander, and F.A. Matsen, J. Chem. Phys. 76 (1982) 3101.
44. Y. Ma, P.O. Lehtinen, A.S. Foster, and R.M. Nieminen, New J. Phys. 6 (2004) 68.
45. P.O. Lehtinen, A.S. Foster, A. Ayuela, A. Krasheninnikov, K. Nordlund, and R.M. Nieminen, Phys. Rev. Lett. 91 (2003) 17202.
46. P.O. Lehtinen, A.S. Foster, Y. Ma, A.V. Krasheninnikov, and R.M. Nieminen, Phys. Rev. Lett. 93 (2004) 187202.

Carbon-Based Magnetism
T. Makarova and F. Palacio (Editors)

14

Magnetism in Nanometer-scale Materials that Contain No Magnetic Elements

A. Oshiyama and S. Okada

Institute of Physics and Center for Computational Sciences, University of Tsukuba Tennodai, Tsukuba 305-8571, Japan

1 Preamble

Clarifying mechanisms of occurrence of magnetism has been a longstanding issue in condensed matter physics and in materials science. In localized electron systems, overlap of adjacent orbitals and geometric arrangements of lattice sites determine signs of exchange integrals and hereby induce a variety of magnetic ordering. In itinerant electron systems, Fermi-level density of states and electron correlation are two major factors that induce several magnetically ordered states. Each role, along with effects of orbital degeneracy and van-Hove singularity inherent to specific lattices, has been examined for more than a half century with several model Hamiltonians [1] or for real materials [2]. What is unequivocally clarified as mechanisms of the magnetic ordering are not many, however.

A lot of efforts in the past have clarified that balance of several factors, such as localized characters of electron states, exchange-correlation energies among electrons, peculiar characteristics of energy spectra occasionally due to arrangements of atomic positions, is decisive for appearance of magnetism. The balance in usual bulk materials is determined mainly by the choice of constituting elements from the periodic table: Each element has its own structure when assembled, and in the structures the balance possibly inducing magnetism is almost determined.

In nanometer-scale materials the above statement is not true. Shapes or boundaries in nano-structures play important roles in characteristics of electron states and hereby generates new balance among interactions. This opens a possibility to forge nano-scale magnets consisting solely of non-magnetic elements.

From technology viewpoints, spintronics that add new degrees of freedom, i.e., spin, to current electronics has recently attracted great attention. Candidate materials explored so far for spintronics are limited to conventional compound semiconductors doped with typical magnetic elements. Yet, if we are to follow and utilize highly developed fabrication technique in current semiconductor technology, magnetic elements are not easy to be treated. In the sense, exploration of nanometer-scale magnets containing no magnetic elements which are compatible with current technology is extremely important.

In this chapter, recent first-principle calculations which predict such unusual nano-scale magnets are presented.

2 Density Functional Theory

First-principle calculations which are described in this chapter rely on the density functional theory (DFT). Details are found in several reviews [3–5]. The DFT is based on a theorem which has been proved rigorously [6, 7]: i.e., the total energy E of the interacting electron system which is under the influence of nuclear potential $v_{\text{ext}}(\mathbf{r})$ is given as a functional,

$$E[n] = \int v_{\text{ext}}(\mathbf{r})\, n(\mathbf{r})\, dr + F[n] \tag{1}$$

of the single electron density $n(\mathbf{r})$ which, in turn, can be calculated, in principle, from the many-electron wavefunction. Here $F[n]$ is a universal functional of $n(\mathbf{r})$. This theorem implies that the many-electron wave function is not necessarily required to obtain the total energy but that the single electron density suffices. Then introducing a set of single electron wavefunction $\{\ \psi_i(\mathbf{r})\ \}$ to express $n(\mathbf{r})$ as,

$$n(r) = \sum_i |\psi_i(\mathbf{r})|^2 \tag{2}$$

we obtain

$$F[n] = -\frac{\hbar^2}{2m}\sum_i \int \psi_i{}^*(\mathbf{r})\,\nabla^2\,\psi_i(\mathbf{r})\,d\mathbf{r} + \frac{1}{2}\iint \frac{n(\mathbf{r})\ n(\mathbf{r}')}{|\mathbf{r}-\mathbf{r}'|} + E_{\text{XC}}[n] \tag{3}$$

where $E_{\text{XC}}[n]$ is the exchange-correlation energy coming from a fact that an electron is a fermion. Then by minimizing $E[n]$ with respect to $\{\ \psi_i(\mathbf{r})\ \}$, we obtain a set of variational equations named Kohn-Sham equation [8]:

$$\left\{-\frac{\hbar^2}{2m}\nabla^2 + v_{\text{ext}}(\mathbf{r}) + \int \frac{n(\mathbf{r}')}{|\mathbf{r}-\mathbf{r}'|}d\mathbf{r} + \frac{\delta E_{\text{XC}}[n]}{\delta n}\right\}\psi_i(\mathbf{r}) = \varepsilon_i\,\psi_i(\mathbf{r}) \tag{4}$$

Here $\delta E_{XC}[n]/\delta n$ is a functional derivative of the exchange correlation energy. A set of equations given as (2) and (4) should be solved selfconsistently. The $\psi_i(\mathbf{r})$ and ε_i are usually called Kohn-Sham (KS) orbital and Kohn-Sham energy, respectively. These may be regarded in usual cases as the energy and the wavefunction of the corresponding quasi-particle.

$v_{ext}(\mathbf{r})$ depends on nuclear coordinates. Hence we obtain forces by computing the derivatives of $E[n]$ with respect to nuclear coordinates, and then reach stable and metastable atomic geometries, or even explore reaction coordinates.

So far the theory is rigorous. We need an approximation for $E_{XC}[n]$ to proceed, however. The most easiest and successful approximation is the local density approximation (LDA) or the local spin density approximation (LSDA). These approximations are based on a notion that the single electron density $n(\mathbf{r})$ is relatively slowly varying. LDA and LSDA have been applied to a variety of materials and obtained great success, as documented in the reviews cited above. Sizable and important corrections to LDA and LSDA should be done in several cases. Several types of corrections are classified as generalized gradient approximation (GGA). In addition to the approximations, there are a lot of sophisticated techniques to obtain the solutions. For further details, reviews cited above or some literature [9–13] may be useful.

3 Edge States of Zigzag-shaped Graphite Ribbons

The first example is a graphite ribbon or a monolayer graphite strip. When a graphite strip has edges with zigzag shape, peculiar electron states located at Fermi energy E_F emerge. As in Fig. 1, two energy bands are degenerate at E_F in a part of the one-dimensional Brillouin zone for the wave number k along the ribbon direction: the dispersion-less flat bands [15–17]. It has been clarified that these states are extended along the edges but at the same time are localized at the edges for $k = \pm\pi/a$ (a: the periodicity along the edge), hereby called edge states. With decreasing k, the states at E_F gradually penetrate inside the ribbon. When the width of the ribbon is sufficiently wide, the two bands are degenerate in the region, $2\pi/3a < |k| < \pi/a$. With decreasing the width of the ribbon, the degeneracy is partly lifted and the two bands are degenerate in a smaller region, $k_0 < |k| < \pi/a$ ($k_0 > 2\pi/3a$). This flat-band nature is a consequence of delicate balance of electron transfer among π orbitals situated near edge atoms. It is noteworthy that such flat bands appear in the graphite ribbons only with zigzag shaped edges. In other ribbons with different edge-shapes such as armchair-shaped edges, flat bands are not induced. The shapes in nano-scale materials (nano-shapes) are crucial for electron states of the materials. It is also remarked that the edge states consist solely of π orbitals. No σ orbital is contained. This means that the existence of the edge states is generally unaffected by the adsorption of foreign atoms and molecules at the edge carbon atoms.

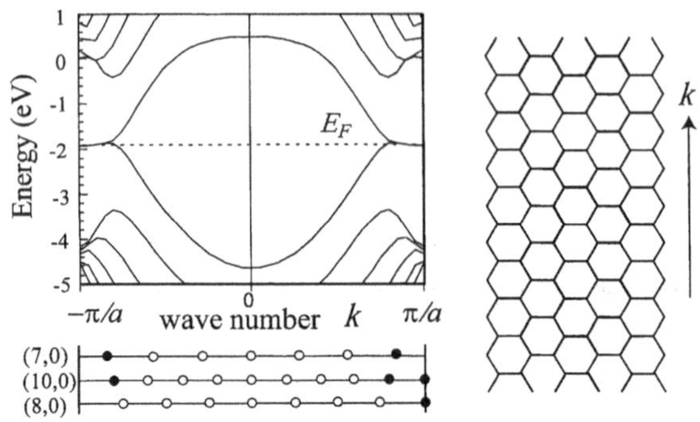

Fig. 1. Energy bands (left panel) of a zigzag-shaped graphite ribbon (right panel). Flat bands exist at E_F. Energy bands shown are obtained by LDA applied to the graphite ribbon with the width of $8b$ (b: the bond length) [14]. k_0 in this case (see text) is about $0.8\ \pi/a$ ($a = \sqrt{3}\,b$). Spin degrees of freedom are not considered here. The bottom part of the left panel illustrates additional boundary conditions for tube structures (see section 5 for details).

Flat bands in Fig. 1 indicate the existence of degenerate electron states at E_F. We then expect spin polarization in the vicinity of ribbon edges. Fig. 2 shows the spin density $n_\uparrow(\mathbf{r}) - n_\downarrow(\mathbf{r})$ of a zigzag-edged graphite ribbon calculated by LSDA [18]. The calculation clearly shows the existence of the magnetic ordering that is mainly located along the zigzag edges and decays gradually inside. The ferromagnetic spin ordering is observed along each edge. Yet the antiferro-magnetic ordering exists in the whole graphite ribbon.

There are two atomic sites in its hexagonal primitive cell of graphite and all the atomic sites are classified into two sublattices A and B (called bipartite lattice). The number of carbon atoms in each lattice of the ribbon is identical ($N_A = N_B$). It is clearly

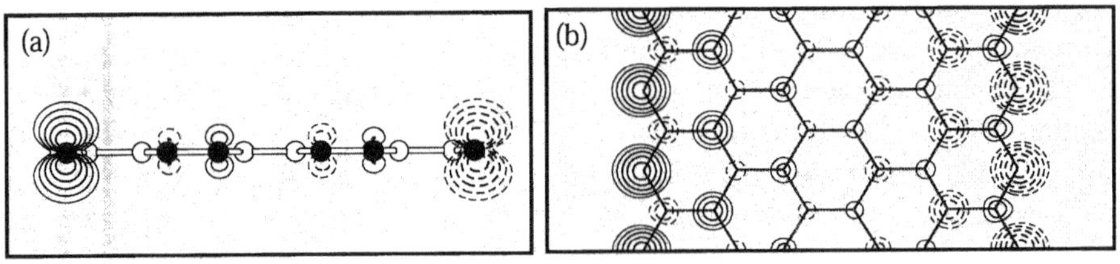

Fig. 2. Contour plots of spin density $n_\uparrow(\mathbf{r}) - n_\downarrow(\mathbf{r})$ on a plane perpendicular to a graphite ribbon with zigzag edges (a), and on a plane including the ribbon (b). In (a) the edges are perpendicular to the plane and C atoms on the plane are depicted by gray circles. Positive and negative values of the spin density are shown by solid and dashed lines, respectively. Each contour represents twice (or half) the density of the adjacent contour lines.

observed in Fig. 2 that the spin is polarized in one direction at one sublattice and it is in an opposite direction at the other. As a result the total spin S of the graphite ribbon vanishes. In the Hubbard model for a bipartite lattice, it is proved that the spin for the ground state is $S = (N_A - N_B) / 2$ for a half-filled case [19]. Although the proof requires vanishing transfer integrals beyond the nearest neighbor sites, the results from the present DFT calculations correspond to the statement obtained in the simple model.

Then it is expected that ferromagnetic or ferrimagnetic ordering appears in the whole ribbon when a modification of nano-scale shapes is introduced and the imbalance between N_A and N_B is generated. This point is further discussed below for other examples of nanomaterials.

4 Magnetic Ordering Related to Border States

Edges are regarded as borders between a material and vacuum. Hence borders in hetero-sheets are expected to induce peculiar electron states similar to the edge states. Spin polarization along borders is also expected.

4.1 Generalization of Edge States

The edge states are not peculiar to graphite ribbons. They could emerge whenever there are zigzag-shaped edges. Hexagonal boron nitride (*h*-BN) is another example of bipartite hexagonal lattice. Fig. 3 shows energy bands of the *h*-BN ribbon which has zigzag-shaped edges terminated with hydrogen atoms [20]. The ribbon is an insulator with the calculated energy gap of 4 eV. The highest occupied (HO) and the lowest unoccupied (LU) bands show extremely flat dispersion around the zone boundary X. The wavefunctions of the HO and LU bands clearly show that these states are edge states whose wavefunctions are localized at one of edges [Fig. 3(c)], showing similarity to the graphite ribbon. Interesting variety is further introduced in *h*-BN. Due to chemical difference of B and N, or due to atomic orbital-energy difference between the two elements, the LU edge state is localized at the edge of B atoms, whereas the HO edge state is localized at the edge of N atoms.

As stated above, termination of the edge atoms with foreign species, hydrogen atoms in this case, does not prevent the appearance of the edge states which consist of π orbitals near the edges (Fig. 3). These edge states are not expected from the projection analysis of energy bands of the *h*-BN monolayer. They emerge due to subtle balance of electron transfers among π orbitals situated in the vicinity of edge sites. The edge state is a common product in hexagonally bonded ribbons with zigzag-shaped edges.

More generally, even the edges are not required. Zigzag-shaped borders are enough to produce the peculiar electron states. Fig. 4 shows energy bands of BNC_4 and BNC_6 in which graphite strips and *h*-BN strips are gathered and become patchworks [20]. The border between a graphite strip and an *h*-BN strip is of zigzag form. We have found that this hexagonally bonded sheet is a semiconductor with a direct energy gap. When the width of the graphite strips is increased, the gap energy monotonically decreases. It is

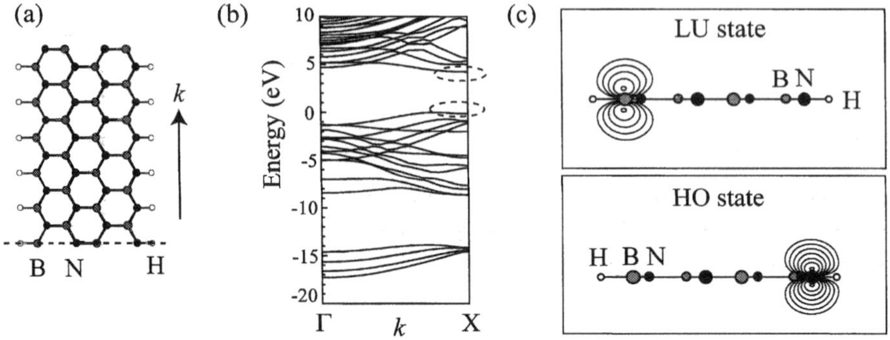

Fig. 3. (a) Top view of *h*-BN ribbon terminated with H. Black, dark and white circles denote N, B and H atoms, respectively. (b) Energy bands of the H-terminated *h*-BN ribbon. Electrons are accommodated in the states with minus energies. The flat band areas are emphasized by dashed-line circles. (c) Contour plots of the squared wavefunctions of the lowest unoccupied (LU) and the highest occupied (HO) states at X point represented on the cross section of the ribbon containing the dashed line in (a). Each contour represents twice (or half) the density of its adjacent lines. The highest values are 1.21 electron/cubic angstrom and 3.37 electron/cubic angstrom for the LU and the HO states, respectively.

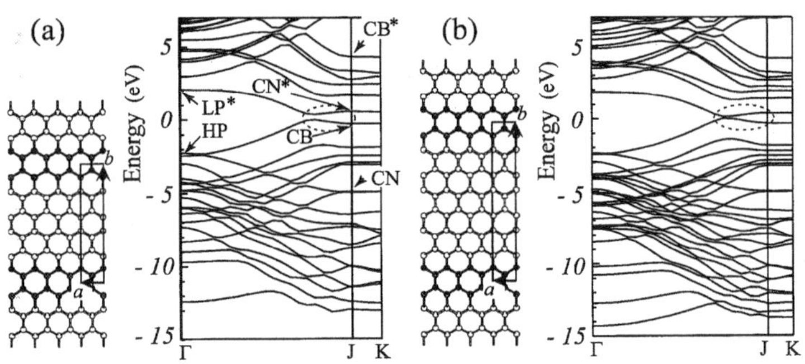

Fig. 4. Energy bands of hexagonally bonded sheets, BNC_4 (a) and BNC_6 (b). Atomic arrangements of a part of the sheet are shown on the left where white, black and gray circles denote C, N and B atoms, respectively. Periodicity in the sheet is represented by vectors *a* and *b*. The lines ΓJ and JK are parallel and perpendicular, respectively, to zigzag borders. The origin of the energy is the top of the occupied states. The flat-band area is marked by a dashed circle. See text for labels used for some bands at the zone boundary and the center.

shown in Fig. 4 that the highest occupied π and the lowest unoccupied π* bands show almost flat dispersion around the zone boundary J. This feature is similar to those of the energy bands of hexagonally bonded ribbons described above.

The wavefunctions of the flat band states in BNC_4 are shown in Fig. 5. It is found that the highest π state at J (labeled CB in Fig. 4(a)) is localized at C and B atoms which constitute one of borders of *h*-BN and graphite strips. It should be remarked that the wavefunction of CB is extended along the border. As in Fig. 5(a), the state has bonding π character with substantial hybridization of π orbitals of C and B atoms. It is then naturally expected that an antibonding state also exists. It is found that a state in the conduction bands has such character [CB* in Fig. 4(a)]. The wavefunction of CB* shown in Fig. 5(b) is located along the border and clearly manifests its antibonding character. The wavefunction of the lowest conduction band at J [CN* in Fig. 4(a)] is also shown in Fig. 5(c). It is localized at C and N atoms in this case which constitute the other border of *h*-BN and graphite strips. Again it extends along the border. The state has antibonding character. After exploration of the valence bands, the state labeled CN in Fig. 4(a) has been found. The wavefunction of CN is located along the border formed by C and N and is of bonding character, as is observed in Fig. 5(d). It is now clarified that zigzag borders of hexagonally bonded hetero-sheets induce a new class of electron states which are localized around atoms near borders and propagate along the borders. Then these states are called border states.

The border-state character is less prominent with decreasing the wavenumber along the border and is finally lost at the zone center. The wavefunctions of the states at Γ point [labeled HP and LP* in Fig. 4(a)] are shown in Figs. 5(e) and 5(f). It is found that those states are extended over the whole BNC_4 hetero-sheet. Consequently the total valence electron density is also extended over the sheet. The border states are therefore extremely unique and different from the usual interface states which maintain their localized character in whole BZ. One may rather notice that the property of the border state is similar to the edge state. Zigzag-shaped border in the hetero-sheet and the zigzag-shaped edge, i.e., the border of the material and vacuum, play identical role for electron states near E_F. Hence the border state may be regarded as a generalization of the edge state.

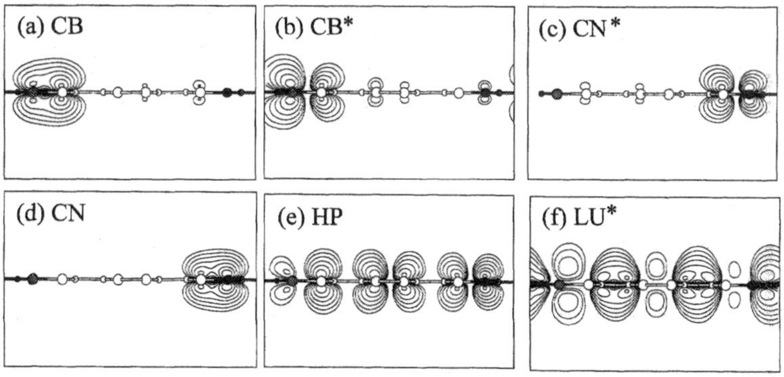

Fig. 5. Wavefunctions (Kohn-Sham orbitals) and charge density of BNC_4. Contour plots of squared wavefunctions of the flat-band states of BNC_4 at J, which are labelled as CB (a), CB* (b), CN* (c) and CN (d), respectively, in Fig. 4. Plots are on the vertical cross-section of the sheet which contains the vector *b* in Fig. 4(a). White, black, and shaded circles denote C, N and B atoms, respectively. Contour plots of squared wavefunctions of the highest occupied π (e) and the lowest unoccupied π^* (f) states at the zone center Γ are also shown. Plots are on the same cross-section as in (a).

It is emphasized that the states are peculiar to the zigzag borders. It is indeed found that the border states are absent in BNC_x sheets where the borders between graphite and *h*-BN strips are of armchair shape [20]. The LDA calculation shows that the strips with the with the armchair borders are semiconductors with a direct energy gap at Γ point and that the highest occupied and the lowest unoccupied states around the Γ point do not show the flat band nature. The characteristics of the band structure are completely different from those of the hetero-sheets with zigzag borders.

The electronic structures of BNC_x described above have been obtained after complete optimization of atomic geometries. The calculated bond length of C-N and B-C at the border is 1.37 and 1.50 Å, respectively. The length of C-C bonds is 1.40–1.42 Å and that of B-N is 1.40 and 1.43 Å. The bond lengths of C-C and B-N are similar to those in graphite and *h*-BN. The hetero-bonds, B-C and N-C, on the other hand, are less strong than the C-C or B-N bonds. Hence the phase separation leading to the striped structures of the *h*-BN and graphite discussed above is energetically favorable.

As stated above, the highest occupied π and the lowest unoccupied π* states of the hetero-sheet BNC_x with zigzag-shaped edges are the border states in their characters around the zone boundary J, whereas they are extended over the sheet around the zone center Γ. This implies potential applicability of BNC_x compounds. For instance, the striped BNC_x compound is capable of modulating the carrier flow by using the border states as channels of conduction. Carriers injected into the compound are first accommodated in either the highest π or the lowest π* bands at J. The injected carriers are thus of one-dimensional character along the borders. Then by exciting carriers or by increasing the number of carriers, the states around Γ become occupied. In the case, the conduction is isotropic reflecting the character of the wavefunctions.

Unusual distribution of the wavefunction is also utilized for nanocoils when BNC_x is formed in a tubular form. Upon relatively light carrier doping, the current is one-dimensional along the border of the graphite and the *h*-BN strips. Therefore by tuning the width ratio of the strips and by forming appropriate tubular structure, we can control chirality of the current along the BNC_x tube. This is a generalization of the chiral conductivity first proposed for the BNC_2 tube [21] where anisotropic conductivity of the BN monolayer is utilized. As a limiting case, the current flow in the BNC_x tube consisting of the *h*-BN and the graphite strips with its index (n,0) (zigzag tube) is completely circular around the tube, being regarded as a nano-solenoid.

4.2 Magnetic Ordering along Borders

The hetero-sheets which we have introduced in the previous subsection may look artifacts. The existence of the hetero-sheets in nature is likely, however. First, the honeycomb structures consisting of B, N and C atoms have been synthesized in forms of particles, tubes and flakes indeed [22–25]. Second, phase separation of graphite and BN regions leading to the striped structures above is energetically favorable. In fact, we have performed the total energy calculations for graphite, BN, BC, and NC hetero-sheets by DFT. The calculated bond energies of B-C and N-C are smaller than that of

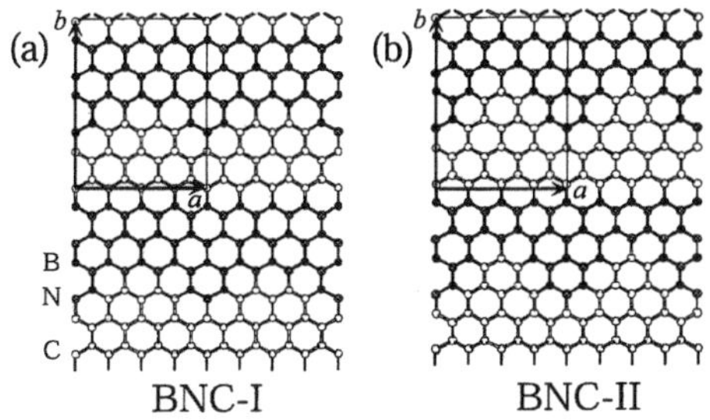

Fig. 6. Top views of BNC hetero-sheets with undulating zigzag borders, (a) BNC-I and (b) NC-II. White, shaded and black circles denote C, B, N atoms, respectively. The rectangle in each figure denotes the unit cell.

graphite by 1.52 eV and 0.81 eV, respectively. On the other hand, the bond energy of B-N is smaller than that of graphite only by 0.31 eV, inferring the formation of C-C bonds as well as B-N bonds.

Undulating borders rather than straight borders may be abundant in real hetero-sheets. This undulation leads to ferromagnetic, or more exactly, ferrimagnetic ordering along borders [18], as is shown below. We here consider honeycomb sheets consisting of B, N and C atoms shown in Figs. 6(a) and 6(b). In these BNC sheets, zigzag borders are undulating. This undulation causes the difference in the numbers of C atoms, N_A and N_B which denote the numbers belonging to sublattices *A* and *B*. In the hetero-sheet shown in Fig. 6(a) (labeled BNC-I hereafter), a unit cell contains 13 B, 14 N atoms, and 21 C atoms which are grouped into A-sublattice and B-sublattice carbon atoms ($N_A = 11$ and $N_B = 10$). Similarly, in the hetero-sheet shown in Fig. 6(b) (labeled BNC-II), a unit cell contains 11 B and 13 N atoms, and 24 C atoms which are grouped into $N_A = 13$ and $N_B = 11$ carbon atoms.

For those representatives of honeycomb hetero-sheets consisting of B, N, and C atoms, BNC-I and BNC-II, It is found that both ferrimagnetic and non-magnetic states as solutions of the Kohn-Sham (KS) equations in the LSDA. The ferrimagnetic state is lower in total energy than the non-magnetic state: The total energy differences per unit cell are 20 meV and 11 meV for BNC-I and BNC-II, respectively. This clearly indicates the occurrence of the ferrimagnetic ordering in the BNC hetero-sheets. The calculated values of the polarized spin per unit cell are one and two for BNC-I and BNC-II, respectively. Fig. 7 shows calculated energy bands (Kohn-Sham energy levels) of BNC-I and BNC-II in their ferromagnetic states. We observe a single flat-band state in BNC-I and two flat-band states in BNC-II. For these flat-band states, the dispersions even along the *a*-direction (parallel to the border direction) are relatively small, a few tenth of eV. It is of interest that the number of the flat-band states appearing in the BNC hetero-sheets corresponds to the difference in numbers, N_A - N_B, between *A*-sublattice and *B*-sublattice carbon atoms. The flat-band states for majority (MJ) and minority (MN) spins split by 0.3 - 0.4 eV for BNC-I and BNC-II, as is shown in Fig. 7. In the present *k*-point sampling and even in the doubly increased *k*-point sampling in the self-consistent-field (SCF) calculations, the flat-band states for the MJ spin are

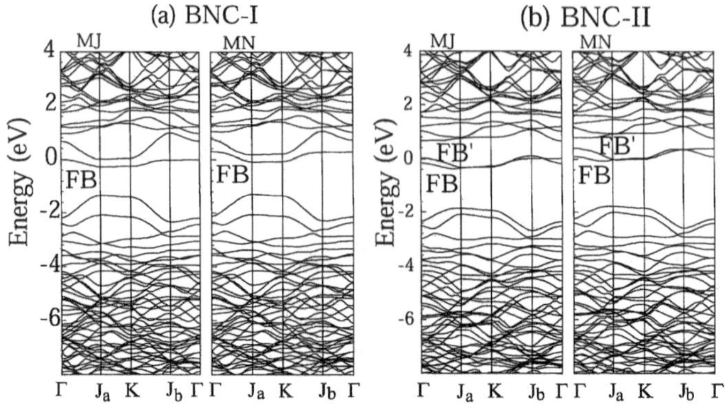

Fig. 7. Energy bands (Kohn-Sham energy levels) of (a) BNC-I and (b) BNC-II for majority (MJ) and minority (MN) spins along the symmetry lines. The directions Γ-J_a and K-J_b are parallel to *a* in Fig. 6, whereas the directions J_a-K and J_b-Γ are parallel to *b*. Flat band states are labeled by FB or FB'. The Fermi energy is located at $E_F = 0$. The MJ- and the MN-spin levels are split completely in the SCF calculation (see text), though they look overlapped along the symmetry lines.

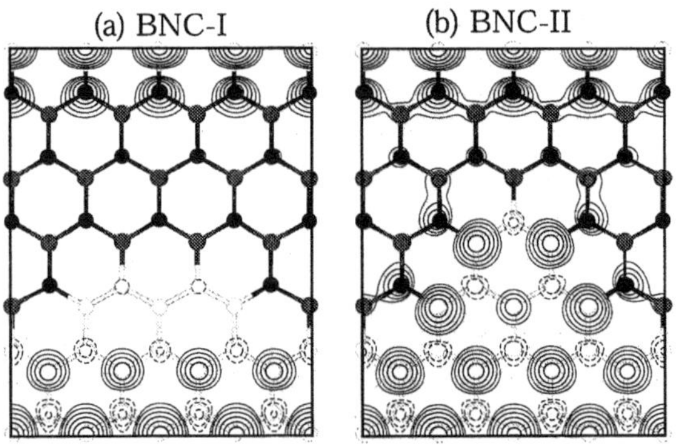

Fig. 8. Contour plot of the spin density $n_\uparrow(\mathbf{r}) - n_\downarrow(\mathbf{r})$ on the BNC hetero-sheet. (a) BNC-I and (b) BNC-II. Positive and negative values of the spin density are shown by solid and dashed lines, respectively. White, shaded and black circles denote positions of C, B, and N atoms, respectively. Each contour represents twice (or half) the density of its neighboring contour lines. The lowest contour represents 5.27×10^{-3} electron per cubic angstrom.

completely occupied, whereas those for the MN spin are empty. Semiconducting ferrimagnetic states have been thus obtained as ground states.

Analyses of KS orbitals of the flat-band states (FB and FB' in Fig. 7) clarify their characteristics. First they consist of π orbitals of constituent atoms. Second, they are localized in the border regions of graphite and BN, but at the same time extended along borders. Yet this peculiar nature is only for a part of BZ: The flat-band states have such characters only near BZ center, whereas the orbitals for FB and FB' near BZ boundaries are extended over the whole sheet.

Fig. 8 shows calculated spin density $n_{\uparrow}(\mathbf{r}) - n_{\downarrow}(\mathbf{r})$ (the upspin is the majority) of BNC-I and BNC-II. It is clear that the majority spin is distributed exclusively on the sublattice A of graphite, whereas the minority spin is on the sublattice B. Boron sites correspond to the A sublattice sites in the present honeycomb network so that the majority spin is also on the boron sites near the border (one eighth compared with the maximum density). The calculated total spin is identical to $S = (N_A - N_B)/2$ for both BNC-I and BNC-II. This indicates that there are ferrimagnetic ordering in flat bands in the bipartite lattice. BNC-I and BNC-II have periodic structures along the borders. This is unnecessary, however, to make the ferro- (ferri-) magnetic ordering exist. The zigzag borders between chemically different elements and the imbalance between the numbers of carbon sublattice sites suffice.

The calculated spin density in BNC-I and BNC-II is distributed on whole graphite region. This is because the graphite region is so small that the borders occupy a substantial portion of the region. When we prepare larger graphite regions, the spin density is distributed only near borders. On the other hand, the BN honeycomb region works as a separator between the spin densities. The hexagonal BN sheet has large energy gap. Hence the electron states around the Fermi level in the BNC sheets mainly consist of the π states of C atoms. This results in the spin density mainly located in the graphite region. When the BN region is wide enough, the spin density becomes one-dimensional so that the fluctuation may be substantial. In order to make the ferromagnetic ordering stable, there may be an optimum width of the BN honeycomb region. This situation opens a possibility of spin-density engineering in the BNC honeycomb sheet.

In order to further clarify the nature of the ferrimagnetic ordering in BNC-I and BNC-II, we consider another honeycomb structure BNC-III [Fig. 9(a)] where the BN region intervenes the graphite region of BNC-I and the resulting triangular graphite regions are separated from each other. In BNC-III, the LSDA calculations show that there is only a non-magnetic state. The energy bands (KS energy levels) shown in Fig. 9(b) clearly exhibit the existence of two flat bands. This corresponds to the difference in sublattice sites, $N_B - N_A = 2$ in this case. The energy dispersions of the flat bands are 0.2 eV, smaller than those in BNC-II, and the corresponding KS orbitals are distributed near borders. The situation in BNC-III is similar to that in BNC-I or BNC-II. The ferri-magnetic ordering does not exist, however. This is due to the weak coupling between polarized spins located in the triangular graphite regions. In this case the anti-ferromagnetic coupling between the spins dominates. In fact, further LSDA calculations for BNC-III using double-periodicity unit cells have revealed that the total energy for the anti-ferromagnetic state is lower than that of the non-magnetic state by 20 meV per unit cell. This finding in turn corroborates an argument that the ferromagnetic ordering in BNC-I and BNC-II is categorized into the flat-band ferromagnetism [26].

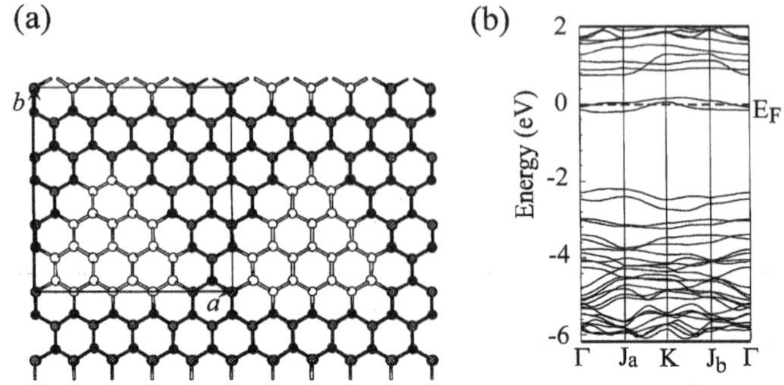

Fig. 9. (a) Top view of atomic arrangements and (b) energy-band structures of BNC-III hetero-sheet. White, shaded and black circles denote C, B, and N atoms, respectively. The Fermi level E_F is shown by the dashed line.

It is of importance to examine effects of carrier doping on the ferrimagnetic ordering in BNC-I and BNC-II. When we dope electrons in the flat bands which are empty in Fig. 7, the non-magnetic states become relatively favorable. The LSDA calculations show that the transition from the ferrimagnetic to the non-magnetic states occurs at 0.02 e and 0.05 e per unit cell for BNC-I and BNC-II, respectively. This is the quantification by DFT for the robustness of the flat-band ferromagnetism against carrier doping, which has been discussed only in Hubbard model previously [26, 27].

5 Magnetic Carbon Nanotubes

Lengths of carbon nanotubes (CNTs) are usually finite and open ends are indeed observed in transmission electron microscope images [28]. Finite-length CNTs are regarded as rolls of finite-width graphite ribbons. In such nanotubes with finite lengths, zigzag-shaped ends are naturally expected. Hence in the finite-length CNTs with the zigzag-shaped ends, degenerate electron states near E_F are expected to exist by analogy of the flat dispersion bands in the zigzag-edged graphite ribbons discussed in section 3.

Yet morphology change from sheets to tubes introduces additional boundary conditions for electron wavefunctions. Among continuum states in the graphite ribbon labeled by the wavenumber along the zigzag edge, only finite-number discrete states are allowed in the tube structure. How to select the allowed states depends on the radii of each tube. Further the region in BZ which shows flat dispersion increases with increasing width of the graphite ribbon. This indicates that the length of the CNTs also affects electron states near E_F in the finite-length CNTs with zigzag-shaped ends. As shown below, this structural variation in nanometer or atom scale indeed plays a decisive role in magnetic properties of the zigzag-ended CNTs.

5.1 Spin Polarization in Zigzag-ended Carbon Nanotubes

In this subsection, we take (7,0), (8,0) and (10,0) CNTs as representatives of the zigzag- ended nanotubes (n,0) and clarify magnetic properties. This choice of n covers essential features of the relation between the nanoscale structural variation and the magnetic states.

Fig. 10 shows the geometries and the spin densities of the zizgzag CNTs obtained by the LSDA calculations [14]. The length of the tube here is $l = 8b$ (b: the bond length). Since carbon atoms at both ends are terminated by H atoms, there is no significant structural relaxation. The spin densities $n_{\uparrow}(\mathbf{r}) - n_{\downarrow}(\mathbf{r})$ clearly exhibit the occurrence of certain magnetic ordering in the zigzag CNTs. It is also shown in Fig. 10 that the magnetic ordering is sensitive to the tube index. Most of the electron spins are polarized in the same direction for the finite-length (7,0) and (10,0) CNTs, whereas the spin is polarized in one direction at one of two ends and in an opposite direction at the other end for the finite-length (8,0) nanotube.

The spin-polarized (magnetic) state is lower in total energy than the spin-unpolarized (nonmagnetic) state. The calculated total energy difference per finite-length, e.g., (7,0) nanotube is 0.17 eV, thus predicting that the finite-length nanotubes with zigzag open ends are nanoscale magnets.

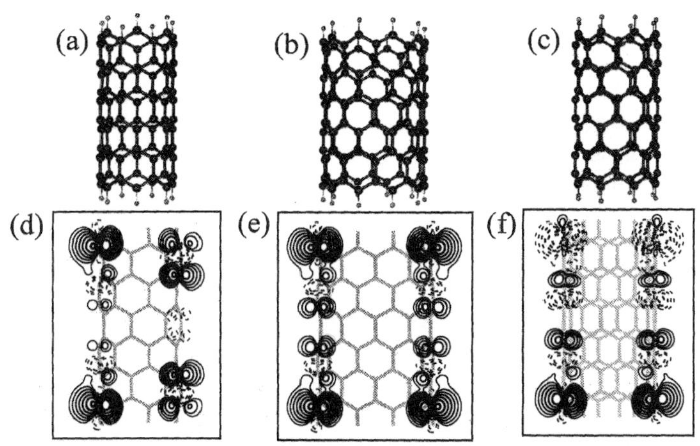

Fig. 10. Total energy optimized geometries and contour plots of the spin density, $n_{\uparrow}(\mathbf{r}) - n_{\downarrow}(\mathbf{r})$ of (7,0) [(a) and (d)], (10,0) [(b) and (e)] and (8,0) [(c) and (f)] CNTs with finite engths $l = 8b$. In the geometries, the large and small cuircles denote carbon and hydrogen atoms, respectively. Each contour of the spin density represents twice (or half) of the density of he adjacent contour lines. Positive and negative values of the spin density are shown by solid and dashed lines, respectively. The smallest absolute value represented is 1.56×10^{-4} electron per cubic angstrom.

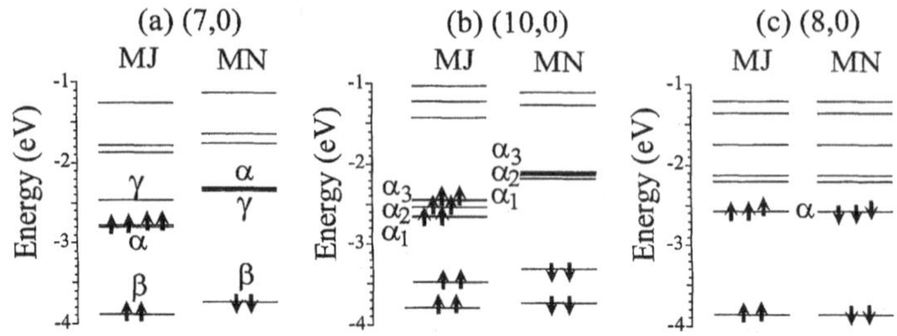

Fig. 11. Electronic energy levels for majority (MJ) and minority (MN) spins of (7,0) (a), (10,0) (b), and (8,0) (c) zigzag CNTs with the length $l = 8b$. α states denote the states possessing the edge-state character.

Electronic energy levels (Kohn-Sham energy levels) of the finite-length (7,0), (10,0), and (8,0) CNTs for majority and minority spins are shown in Fig. 11. It is found that four-fold and six-fold nearly degenerate electron states emerge for the (7,0) and (10,0) CNTs, respectively, near E_F. In other words, the highest occupied states are nearly degenerate with partially filled electrons. Since the states are occupied by four electrons for the (7,0) CNT and by six electrons for the (10,0) CNT, significant exchange splitting for the states results in the spin polarization on these CNTs. The calculated total spins are $S = 2$ and 3 for (7,0) and (10,0) CNTs, respectively. In sharp contrast to these finite-length CNTs, the finite-length (8,0) CNT dose not exhibit a high spin state. The calculation shows that the state becomes antiferromagnetic even if we start with ferromagnetic trial configurations. The calculated number of polarized electron spin S is zero. However, it is clearly demonstrated above that certain magnetic ordering takes place around their carbon atomic sites near ends.

5.2. Boundary Condition on Wavefunctions

The peculiar index dependence of the total spin S described in the preceding subsection is understood by considering new boundary conditions imposed on the electronic wavefunctions. Let us begin with the electronic structure of the zigzag-edged graphite ribbon. There are doubly degenerate flat dispersion bands at E_F near the zone boundary of the one-dimensional (1D) BZ (Fig. 1). The two bands are degenerate in the region $k_0 \le |\mathbf{k}| \le \pi/a$ of the 1D BZ. With decreasing the width of the ribbon, k_0 increases from the value $2\pi/3a$. At the zone boundary $k = \pi/a$, the edge states are entirely localized at edges but extended along the direction parallel to the edges. On the other hand, the edge states lose their edge localized character and the wavefunction penetrates inside the ribbon with decreasing the wave number. Then at $k = k_0$, the edge states are smoothly connected to the π and π^* states of which character is identical to the π and π^* states near K point of infinite-sized graphite sheets.

In the zigzag-ended CNTs formed by rolling the zigzag-edged ribbons, a periodic boundary condition is imposed along the circumference. This results in a discrete set of allowed wave numbers along the circumference. It is straightforward that the wave numbers

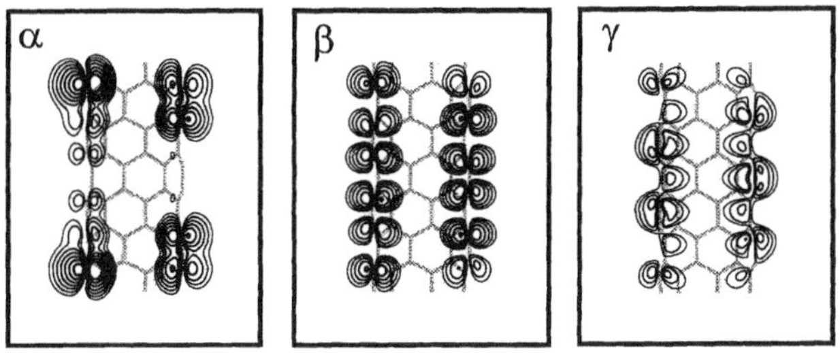

Fig. 12. Contour plots of the squared wavefunction of the electron states, α, β, and γ [see Fig. 11(a)], for the MJ spin around the E_F of the finite-length (7,0) nanotubes. Each contour represents twice (or half) the density of the adjacent contour lines. The lowest contour represents 1.56×10^{-4} electron per cubic angstrom.

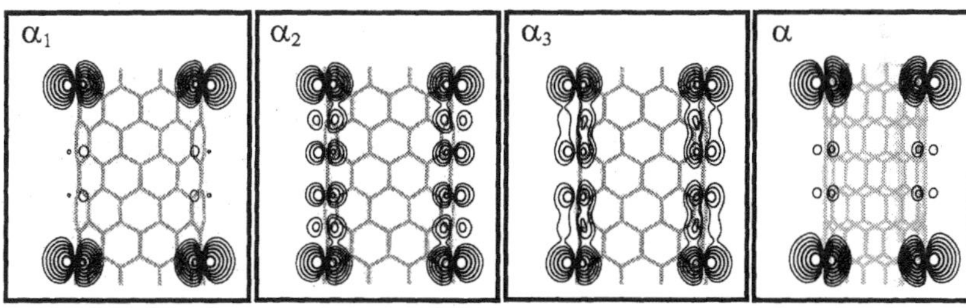

Fig. 13. Contour plots of the squared wavefunction of the highest occupied states, α_1, α_2, and α_3 [Fig. 11(b)] of the (10,0) and α [Fig. 11(c)] of the (8,0) finite-length CNTs. The values of the contour lines are same as in Fig. 12.

allowed in the finite-length (n,0) nanotubes are $k_N = 2\pi N/na$ (N = 0, ± 1, ± 2, ...). In the finite-length (7,0) nanotube, seven k-points are allowed under the periodic boundary condition along the circumference, as is shown by dots in the bottom part of Fig. 1. Near the Fermi level, two states labeled by k = $\pm 6\pi/7a$ which originate from the flat band states of the graphite ribbon are allowed. The four-fold nearly degenerate α states in the (7.0) nanotube [Fig. 11(a)] correspond to these flat band states with k = $\pm 6\pi/7a$. In Fig. 12, the wavefunction (Kohn-Sham orbitals) of the α state along with those of the β and γ states in Fig. 12 is shown. It is clear that the electron state α near the E_F is localized at the edge carbon sites but at the same time slightly penetrates inside. The distribution is in sharp contrast with those of β and γ states which exhibit the typical π character.

In the finite-length (10,0) nanotube, three k-points are allowed on the flat bands around the E_F (black dots in Fig. 1). One of three k-points (α_1 state) corresponds to the zone boundary of the 1D BZ of the graphite ribbon, $k = \pi/a$, where the wavefunctions are completely localized at the end carbon atoms (Fig. 13). On the other hand, the states, α_2 and α_3, corresponding to the remaining k points, $k = \pm 4\pi/5a$, distribute inside the nanotube to some extent, as is shown in Fig. 13. For the finite-length (8,0) nanotube, since electron states at the zone boundary of the 1D BZ of the graphite ribbon are allowed by the periodic boundary condition (black dots in the bottom of Fig. 1), the electron states around the E_F of the (8,0) nanotube are completely localized at the end carbon atoms (Fig. 13). We observe that the distributions of the state α_1 of (10,0) and the state α of (8,0) are essentially identical to each other. This is a consequence form the identical ancestor of the two states.

The above analyses lead us to the following natural interpretation of the radius-sensitive magnetic ordering obtained in the subsection 5.1. In the (7,0) and (10,0) nanotubes, sizable amplitude overlap between orthogonalized degenerate electron states (α in (7,0), and α_2 and α_3 in (10,0)) results in the high spin state to exclude the double occupancy of the π orbital of each C atomic site as in the case of Hund's rule for the *d* orbital in the transition metal magnets. An antiferromagnetic ordering ($S = 0$ state) does not take place on the (7,0) and (10,0) CNTs with $l = 8b$ length even though we choose the initial antiferromagnetic spin configuration. However, it is meaningful to consider whether the other magnetic ordering takes place in CNTs with different lengths. The distributions of the wavefunctions shown in Figs. 12 and 13 give an answer for this question. For the (7,0) and (10,0) CNTs with sufficiently long length, the tubes may exhibit not only ferromagnetic ordering but also antiferromagnetic ordering ($S = 0$) since the overlap between the wavefunctions penetrating from both ends becomes negligible. In the case, the antiferromagnetic and ferromagnetic states are degenerate.

In the (8,0) nanotube, the allowed degenerate states α originated from the state $k = \pi/a$ at E_F are localized at the zigzag ends (Fig. 13). The distribution leads to the local ferromagnetic spin ordering at each of the both ends. Yet the spin polarizations at both ends are reverse to each other, leading to the antiferromagnetic state in total. The coupling between the polarized spins at both ends is related to generation of electron-hole pairs as is explained below. The electron states labeled $k = \pm 6\pi/8a$ of the ribbon are allowed in the (8,0) tube. These are located near E_F but split to some extent (the finite-length splitting Δ_{fl}). When the spins at both ends are coupled in a ferromagnetic way, the electron states for majority and minority spins split in energy (the exchange splitting Δ_{ex}). This exchange splitting is beneficial for the states originated from $k = \pm \pi/a$. Yet other states originated from $k = \pm 6\pi/8a$ also undergo this exchange splitting. Therefore, when $\Delta_{fl} \leq \Delta_{ex}$, this ferromagnetic coupling results in the generation of holes in the bonding π states with minority spin and of electrons in the antibonding π states with majority spin. This reduces the exchange energy gain, weakening the C-C bonding to some extent. In the antiferromagnetic coupling, on the other hand, such generation of electron-hole pairs does not take place. This is the reason why the antiferromagnetic state is energetically favorable for the (8,0) nanotube.

In (7,0) and (10,0) CNTs discussed here, such electron-hole pairs are not generated in the ferromagnetic state, as is evidenced in the level structures shown in Figs.11(a) and (b). It should be mentioned that the coupling between the spins at two edges of the graphite ribbon is also related to the formation of holes in the bonding π and electrons in the antibonding π^* states inside the BZ. In the ribbons, all the states labeled k are allowed. Then in the ferromagnetic state with finite exchange splitting, electron-hole pairs are inevitably generated so that the bond energy gain may be lost. This is the reason for the antiferromagnetic rather than the ferromagnetic ordering in the graphite ribbon, as is shown in section 3.

5.3 Radii, Lengths, Imperfection and Magnetic Ordering

From what is obtained using the DFT theory described in the preceding subsections, it is of interest to try to deduce a general conclusion as to the relation between the tube radii, the lengths and the magnetic orderings. Since the zone boundary of the 1D BZ is not allowed for the CNTs with odd index ($2n+1$,0), the wavefunctions of the states around the E_F which possesses the edge-state character penetrate inside the CNT. Therefore, the sizable overlap between the wavefunctions of the edge states on both tube ends causes the high spin state, avoiding Coulomb energy cost inside. For the ($2n$,0) CNT thicker than $n = 5$, more than two k points are allowed in the flat-dispersion area in 1D BZ in addition to the 1D BZ boundary $k = \pi/a$. Hence the nearly degenerate states penetrate inside the CNT and again induce the high spin state. In the case of thin CNTs with even index, e.g. (8,0) or thinner, only $k = \pi/a$ is allowed so that the electron states around the E_F are localized at the end sites. Consequently, the spin is aligned at each edge of the nanotube and the both spins are weakly coupled in an antiferromagnetic way. Effects of generation of electron-hole pairs are also important as stated above.

The magnetic ordering found here depends also on the lengths of CNTs. As stated in section 3, the boundary $k = k_0$ of the flat-band region at E_F gradually approaches to $k = 2\pi/3a$ by increasing the width of the graphite ribbon. Thus the number of allowed k points in CNTs at E_F also increases by increasing the length of the nanotubes. In the (8,0) CNT discussed in the preceding section, the length is insufficient so that the energies at the allowed $k = \pm 6\pi/8a$ split and are separated from E_F. Yet, when the length of the (8,0) nanotube increases, the states labelled $k = \pm 6\pi/8a$ become degenerate at E_F, and we expect that the high spin state is realized as in the case of the (10,0) nanotube. The same argument is applicable to the (6,0) nanotube.

In section 5.1, it has been demonstrated that the total spin vanishes for the (8,0) finite-length CNT with the length $l=8b$. However, the peculiar spin distribution around the both ends where the spin exhibits the local ferromagnetic ordering leads us to forge a guiding principle to realize a magnetic ordering with $S \neq 0$ on the finite-length nanotube. The hexagonal network is a bipartite lattice where all the atomic sites are classified into two sublattices A and B. In the zigzag (8,0) CNT, carbon sites at one of two ends where up spin is distributed belongs to the A sublattice for instance, whereas those at the other end where down spin is distributed belong to the B sublattice. The number of C atoms in each sublattice is identical ($N_A = N_B$) so that the total spin is zero in this case. It is therefore expected that the finite-length zigzag CNTs exhibit ferrimagnetic ordering by introducing certain structural imperfection which induces the imbalance between the numbers of two sublattice sites. Roughness at the ends is an example. The end roughness is also likely in the actual CNT with finite length. We thus expect that the nanotube exhibits the ferrimagnetic ordering in which the total electron spin S corresponds to the difference in the number of two sublattices [$S = (N_A - N_B)/2$].

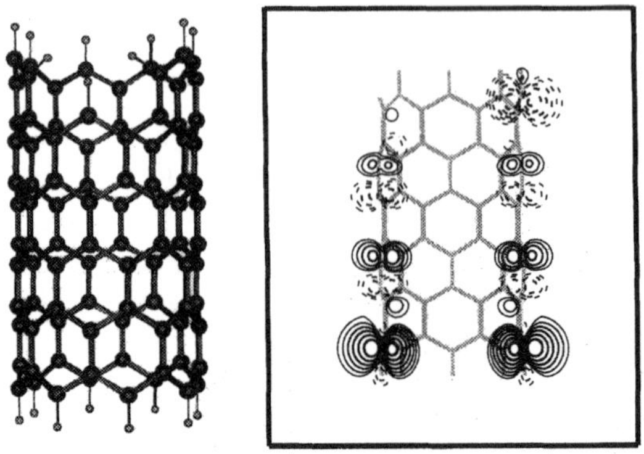

Fig. 14. Total energy optimized geometries and contour plots of the spin density, $n_\uparrow(\mathbf{r}) - n_\downarrow(\mathbf{r})$ of (8,0) CNTs with finite lengths $l = 8b$ and with end roughness. In the geometries, the large and small cuircles denote carbon and hydrogen atoms, respectively. Positive and negative values of the spin density are shown by solid and dashed lines, respectively. The values of the contour lines are the same as in Fig. 10.

Fig. 14 shows the optimized geometries and the spin density of the (8,0) zigzag CNT with the end roughness [14]. In the system, a single C atom from one of two ends is removed so that the difference in sublattice sites, $N_B - N_A$, is one. The LSDA calculations have uncovered the existence of the spin density that is mainly located around the zigzag ends and decays gradually inside. The calculated number of polarized electron spins is one. The spin distribution clearly suggests that there is significant correlation between the magnetic ordering and the edge state. Indeed, the electron states associated with the spin polarization are mainly localized at ends of tubes and then decay inside. This example infers that the occurrence of the $S \neq 0$ state is common characteristics in CNTs with open ends.

6 Magnetism on Si Surfaces

In the preceding sections, it is demonstrated that nano-shapes or atomic arrangements substantially affects electron states in nanoscale materials consisting of first-row elements and thus induces magnetic ordering in several cases. This in turn implies that manipulation of atomic arrangements could induce similar magnetic behaviors using other elements. Atom- scale manipulation is highly developed in semiconductor technology. It has been indeed demonstrated that H atoms on the H-covered Si (001) surfaces are removed along a Si dimer row in a controlled way [29–31], exposing a dangling bond (DB) array on the surfaces. In this section, a possibility of magnetic ordering in nano-meter scale networks of DBs on Si (111) surfaces is explored theoretically.

Dangling bonds generated at surfaces of covalent materials are well known to play crucial roles in atomic structures of the surfaces: Reconstruction of surface atoms usually reduces DB energies albeit increase in stress energies. Such competition between electronic and elastic energies produces rich variety of surface reconstructions. As for Si(111) surfaces, the top layer Si atoms constitute a triangular lattice and the buckling of the atoms along $[1\bar{1}0]$ direction leads to a 2×1 structure [32]. Yet the observed 2×1 is now confirmed to have the more complicated π-bonded chain structure in which 5-member and 7-member rings appear alternately [33]. Hydrogen adsorption generally terminates DBs and then modifies the balance in energetics. Low-energy electron diffraction pattern reveals that the 2x1 structure changes to the 1×1 structure upon H exposure and that the structural transition is completed at about 0.5 monolayer [34, 35]. The surface reconstruction crucially depends on an extent of the H-free region. On the other hand, due to localized nature of the DBs, electron-electron interaction is expected to be important for atomic and electronic structures of the surfaces. The nano-meter scale DB networks generated on the Si(111) partially covered by hydrogen is therefore an interesting stage where competition between the lattice distortion and the electron interaction is essential in physical properties. The lattice distortion provides a new aspect not observed in graphite ribbons, BNC flakes and CNTs where edge states or border states relatively independent of the lattice distortion play essential roles in magnetic properties.

Let us begin with the stability of the π-bonded chain structure on the H-adsorbed Si(111). The reconstruction pattern depends on the extent of the H-free region. Then it is interesting to consider strips of the DBs which are generated by removing H atoms attached to the top-layer Si atoms along $[1\bar{1}0]$ direction. The DFT calculation [36] shows that the π-bonded structure is unstable against the buckling structure when the width of the DB strips are less than 3 top-layer atoms in the $[11\bar{2}]$ direction. Hence the important reconstruction pattern in the nano-meter scale DB networks is the buckling of surface atoms.

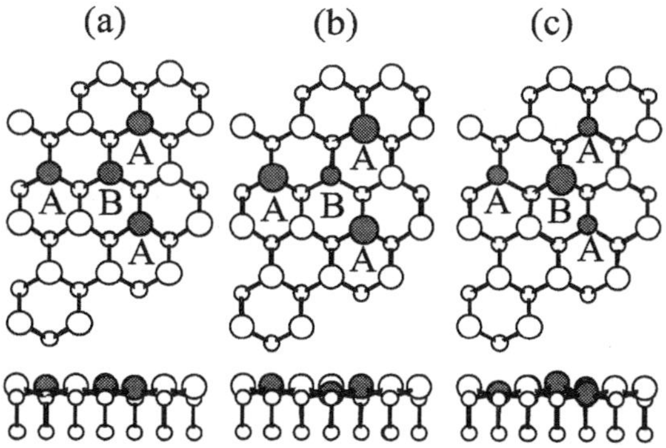

Fig. 15. Top and side views of optimized geometries of the ultimate triangle DB unit for the Non-buckled spin-polarized (NP) (a), buckled spin-polarized (BP) (b), and buckled spin-unpolarized (BU) (c) structures. Dark, white, and small white circles denote the Si with DB, hydrogenated Si, and subsurface Si atoms, respectively. Only a part of atoms are shown to avoid visual complexity.

Spin polarization effects in the DB network are of interest. The LSDA calculations [36] indeed clarify the existence of high spin states in an ultimate triangle unit (Fig. 15) as described below. The triangle unit consists of 3 edge DB sites (site *A*) and a central DB site (site *B*) on an otherwise H-covered Si(111) surface. It generally undergoes the buckling reconstruction. In one buckling pattern shown in Fig. 15(b) the central *B* site is displaced downwards compared with the edge *A* sites, whereas in another pattern shown in Fig. 15(c) the *B* site is displaced upward. These bucklings reduce the total energy of the triangle unit. Without the spin polarization, the calculated total energies of the buckled structures, Fig. 15(b) and Fig. 15 (c), are lower than the non-buckled geometry shown in Fig. 15(a) by 0.67 eV and 0.80 eV, respectively. The amount of the buckling is about 0.4–0.5 Å.

The spin unpolarized states are not the ground state, however. It is found that the spin is polarized in the non-buckled structure and in one of the buckled structures. The calculated total energies for the non-buckled spin-polarized (NP) structure (Fig. 15(a)) and for the buckled spin-polarized (BP) are lower than that for the non-buckled spin-unpolarized structure by 0.86 eV and 0.84 eV, respectively. The spin density of the most stable NP structure is shown in Fig. 16(a). The majority spin is distributed over the *A* sites, whereas the minority spin is on the central *B* site. Therefore the calculated spin for the ultimate triangle unit is $S = 1$. The non-buckled high spin state is energetically the most favorable. The surface reconstruction is not a local atomic displacement of the top-layer atom but a concerted atomic rearrangement including subsurface atoms.

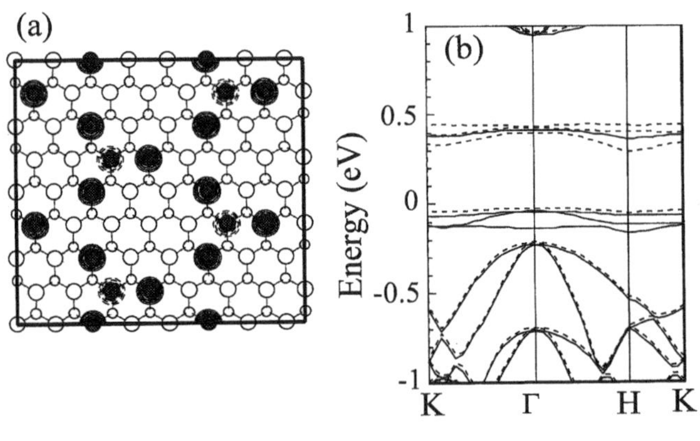

Fig. 16. (a) Contour plot on the (111) plane of the spin density in the NP structure of the triangle DB unit on the Si(111) surface. Positive and negative values of the spin density are shown by solid and dashed lines, respectively. Each contour represents twice (or half) the density of its neighboring contour lines. The lowest contour represents 1.25×10^{-2} electron per cubic angstrom. (b) Energy bands of the NP structure of the DB unit for majority (MJ) and minority (MN) spins along the symmetry lines. The solid and dashed lines denote the bands for the MJ and MN spins, respectively. Electrons are accommodated in the bands with negative energies.

H atoms covering top-layer Si atoms around the triangle unit certainly suppress the flexibility for the reconstruction.

Energy bands (Kohn-Sham levels) of the most stable NP structure is shown in Fig. 16(b). There are essentially four surface energy bands originating from 4 DBs near EF. Three of the four bunch up and an energy gap appears between the three and the remaining one. This peculiar level bunching is totally due to the network topology f the DB unit. The Kohn-Sham orbitals of the three bunched states are distributed over the *A* DB sites and the orbital of the remaining state is on the *B* site. For the majority spin, the bunched states are lower in energy by 0.4 eV than the remaining state, whereas the remaining state is lower for the minority spin by again 0.4 eV. Small dispersions of the surface energy bands indicate localized nature of the DB states and thus importance of electron interactions.

The LSDA calculation has unequivocally shown that the high spin state is the ground state for the ultimate triangle DB unit. The DB unit on the Si(111) surface is thus a promising constituent element for nanometer-scale Si based spintronic devices. A matrix of the triangle DB units on the Si(111) surface with an appropriate nanometer-scale read-write device may be the ultimate high density memory unit (306.38 Å/bit), corresponding to a tera-bite capacity on a typical size of the latest dynamic random access memory chip.

a) b)

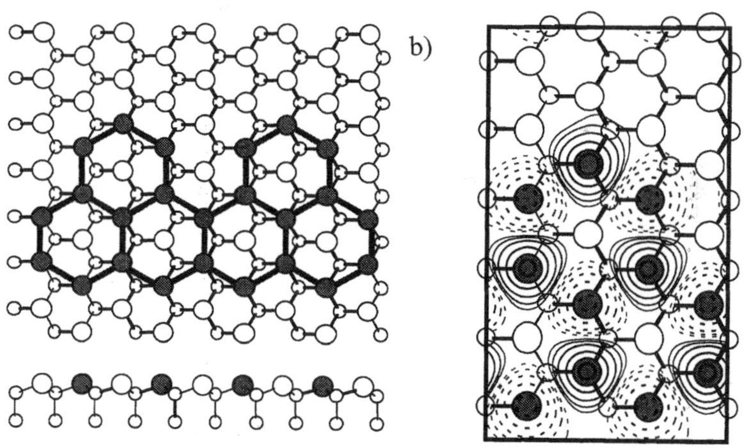

Fig. 17. (a)Top and side views of the optimized geometry of a two-dimensional hexagonal network of DBs on the Si(111). Dark, white, and small white circles denote the Si with DB, hydrogenated Si, and subsurface Si atoms, respectively. Only a part of atoms are shown. The hexagonal network of the DBs is denoted by the dashed lines. (b) Contour plot on the (111) plane of spin density $n_{\uparrow}(\mathbf{r}) - n_{\downarrow}(\mathbf{r})$ on the HDB. Positive and negative values of the spin density are shown by solid and dashed lines, respectively. Each contour represents twice (or half) the density of its neighboring contour lines. The lowest contour represents 1.00×10^{-3} electron per cubic angstrom.

The theoretical prediction that the adjacent spins on the triangle DB unit are coupled in an anti-parallel way allows us to design a network of DBs on the Si(111) by removing H atoms selectively. One of the proposals is a hexagonal network of DBs (HDB) on the Si(111). Fig. 17(a) shows its top and side views: H atoms are removed so that the DBs are exposed forming the hexagonal network. The network topology is identical to that of a finite-width graphene ribbon. The LSDA calculations have been performed for this surface and the obtained spin density, $n_{\uparrow}(\mathbf{r}) - n_{\downarrow}(\mathbf{r})$, for the total-energy minimized structure is shown in Fig. 17(b). It is found that that the HDB exhibits a ferrimagnetic ordering. In the bipartite HDB structure, the boundary of the DB network in a two dimensional plane is designed to have an undulating shape so that the number of DBs belonging to each sublattice is different: The number of DBs belonging to the sublattice A is 6 and the number belonging to the sublattice B is 5 ($N_A = 6$ and $N_B = 5$). It is of interest that the amount of the polarized spins in the HDB network equals to $|N_A - N_B| / 2$. In all the hexagonal networks of dangling bonds so far examined, the polarized spins in the network equals to $|N_A - N_B| / 2$. It is clear that the spin is polarized in one direction at one sublattice and it is in an opposite direction at the other. This finding indicates that a network of DBs with sublattice imbalance ($|N_A - N_B| \neq 0$) exhibits ferrimagnetic ordering with $S = |N_A - N_B| / 2$ in general.

7 Concluding Remarks: Quantum Design of Nanomagnets

The first-principle calculations based on the density functional theory presented in this chapter have revealed that nanometer-scale structures consisting solely of non-magnetic elements exhibit magnetic ordering. Occurrence of the magnetic ordering is certainly related to peculiar electron states that become prominent and thus start to play important roles when the nanostructures have suitable atomic arrangements. Interplay between nano-shapes and electron states is an essential aspect. In graphite ribbons, BNC heterosheets and carbon nanotubes, edge states or border states associated with zigzag-shaped edges or borders are responsible for the occurrence of the magnetic ordering. In hydrogen-covered Si surfaces, dangling-bond states assembled in a form of hexagonal lattices induce the magnetism.

In bulk materials, physical and chemical properties are mainly determined by elements which constitute the materials. This is because the properties come from balance among several factors such as covalency, ionicity, strength of electron-electron interactions etc and then the balance is usually determined by combinations of the elements. In nanometer-scale materials, a new factor has emerged. That is the nanometer-scale shape and the size of the material. The nano-shape quantum mechanically affects behaviors of electron states, introduces a new balance, and hereby induces new physical and chemical properties. The nano-shape is a new factor in current alchemy of nanomaterials. To utilize this new factor, quantum design based on reliable quantum theoretical calculations is useful.

Acknowledgements

We are grateful to Professor Kenji Shiraishi for valuable discussion during the course of the present works. Conversation with Dr. Kyoko Nakada, Professor Koichi Kusakabe and late Professor Mitsutaka Fujita was extremely beneficial.

References

1. For a review, E. Dagotto, Rev. Mod. Phys. 66 (1994) 763.
2. T.Asada and K. Terakura, Phys. Rev. B46 (1992) 13559 and references therein.
3. R. Jones and O. Gunnarsson, Rev. Mod. Phys. 61 (1989) 689.
4. M.C. Payne, M.P. Teter, D.C. Allan, T.A. Arias, and J.D. Joannopoulos, Rev. Mod. Phys. 64 (1992) 1045.
5. W. Kohn, Rev. Mod. Phys. 71 (1999) 1253.
6. P. Hohenberg, W. Kohn, Phys. Rev. 136 (1964) B864.
7. M. Levy, Proc. Natl. Acad. Sci. (USA), 76 (1979) 6062.
8. W. Kohn and L.J. Sham, Phys. Rev. 140 (1965) A1133.
9. A. Oshiyama and M. Saito, J. Phys. Soc. Jpn., 56 (1987) 2104.
10. O. Sugino and A. Oshiyama, Phys. Rev. Lett. 68 (1992) 1858.
11. M. Saito, O. Sugino, and A. Oshiyama, Phys. Rev. B46 (1992) 2606.
12. Y. Okamoto, M. Saito and A. Oshiyama, 1998, Phys. Rev. B58 (1998) 7701.
13. J.-W. Jeong and A. Oshiyama, Phys. Rev. B64 (2001) 235204.
14. S. Okada and A. Oshiyama, J. Phys. Soc. Jpn. 72 (2003) 1510.
15. M. Fujita, K. Wakabayashi, K. Nakada, and K. Kusakabe, J. Phys. Soc. Jpn. 65 (1996) 1920.
16. K. Nakada, M. Fujita, G. Dresselhaus, and M. S. Dresselhaus, Phys. Rev. B54 (1996) 17954.
17. Y. Miyamoto, K. Nakada, and M. Fujita, Phys. Rev. B59 (1999) 9858.
18. S. Okada and A. Oshiyama, Phys. Rev. Lett. 87 (2001) 146803.
19. E.H. Lieb, Phys. Rev. Lett. 62 (1989) 1201; 62 (1989) 1927.
20. S. Okada, M. Igami, K. Nakada, and A. Oshiyama, Phys. Rev. B62 (2000) 9896.
21. Y. Miyamoto, S.G. Louie, and M.L. Cohen, Phys. Rev. Lett. 76 (1996) 2121.
22. K. Suenaga, C. Colliex, N. Demoncy, A. Louiseau, H. Pascard, F. Willaime, Science 278 (1997) 653.
23. Y. Zhang, H. Gu, K. Suenaga, and S. Iijima, Chem. Phys. Lett. 279 (1997) 264.
24. Ph. Kohler-Redlich, M. Terrones, C. Manteca-Diego, W.K. Hsu, H. Terrones, M. Ruhle, H.W. Kroto, and D.R.M. Walton, Chem. Phys. Lett. 310 (1999) 459.
25. E. Bengu and L.D. Marks, Phys. Rev. Lett. 86 (2001) 2385.
26. H. Tasaki, Prog. Theor. Phys. 99 (1998) 489.
27. H. Tasaki, Phys. Rev. Lett. 69 (1992) 1608.
28. P.M. Ajayan, T.W. Ebbsen, T. Ichihashi, S. Iijima, K. Tanigaki, and H. Hiura, Nature (London) 362 (1993) 523.
29. T.C. Shen, C. Wang, G.C. Abeln, J.R. Tucker, J.W. Lyding, P. Avouris, R.E. Walkup, Science, 268 (1995) 1590.
30. T. Hashizume, S. Heike, M. Lutwyche, S. Watanabe, K. Nakajima, T. Nishi, Y. Wada, Jpn. J. Appl. Phys. 35 (1996) L1085.
31. T. Hitosugi, S. Heike, T. Onogi, T. Hashizume, S. Watanabe, Z.-Q. Li, K. Ohno, Y. Kawazoe, T. Hasegawa, and K. Kitazawa, Phys. Rev. Lett. 82 (1999) 4034.

32. D. Haneman, Phys. Rev. 121 (1961) 1093.
33. K.C. Pandy, Phys. Rev. Lett. 47, (1981) 1913.
34. H. Ibach and J.E. Rowe, 1974, Surf. Sci. 43 (1974) 481.
35. G. Schulze and M. Henzler, Surf. Sci. 124 (1983) 336.
36. S. Okada, K. Shiraishi, and A. Oshiyama, Phys. Rev. Lett. 90 (2003) 026803.

Carbon-Based Magnetism
T. Makarova and F. Palacio (Editors)

15

Interactions and Disorder in 2D Graphite Sheets

F. Guinea[a], M.P. López-Sancho[a], and M.A.H. Vozmediano[b]

[a]*Instituto de Ciencia de Materiales de Madrid, CSIC, Cantoblanco, E-28049 Madrid, Spain*
[b]*Departamento de Matemáticas, Unidad Asociada CSIC-UC3M, Universidad Carlos III de Madrid, E-28911 Leganés, Madrid, Spain*

1 Introduction

Graphite has attracted a lot of recent attention due to the growing evidence that, in many cases, it shows anomalous magnetic and transport properties [1]. The understanding of these features is a significant challenge. Theoretical models for the electronic structure [1] were developed under the assumption that graphite could be described using the Landau theory of a Fermi liquid, although it is assumed that the number of carriers is low.

The theoretical possibility of ferromagnetism in disordered graphite samples was raised long ago [2]. The underlying mechanism is the existence of unpaired spins at defects, induced by a change in the coordination of the carbon atoms (see below). Experimental evidence showing that the lifetimes of the quasiparticles in graphite were not consistent with Fermi liquid theory were reported in [3]. A theoretical model explaining these experiments was suggested in [4]. The model used was based on the existence of incompletely screened electron–electron interactions. In follow up work [5], the analysis was extended in order to include the role of disorder (see also [6]), which is known experimentally to play an important role in relation to the existence of anomalous magnetic properties.

The present work discusses theoretical models which address the effects of electron-electron interactions and disorder in graphene planes following the analysis in [5, 6]. The starting point for the study is a simple tight-binding model for the electronic structure, outlined in the next section. Then, a discussion of the interesting features induced by the unscreened Coulomb interaction is presented. The unusual features of the model are emphasized. It is shown that the standard perturbative treatments used in condensed matter theory fail, and a more refined Renormalization Group approach is required. Then, a theoretical framework which allows us to extend the model to many types of disorder, following the approach in [7] for the fivefold rings of the fullerenes, is discussed. The following section analyzes the combined effects of disorder and interactions. A brief discussion of models which may explain the large anisotropy observed in very pure samples of graphite [8], using the theoretical framework explained in [9] is presented next. This work ends with a section highlighting the most interesting conclusions.

We do not pretend to cover the large and rapidly growing experimental literature on the magnetic properties of graphite and related compounds. This work is extensively covered in other chapters of this volume.

It is worth mentioning that the electronic structure of graphite leads to theoretical models of significant interest for the ongoing quest of understanding strongly correlated systems. This work tries to underline also this aspect of the current work on graphite and related compounds. Because of this reason, we also include a brief summary of the technical aspects of the calculations. We hope that this will not be discourage readers willing just to grasp the main ideas of the work reported here.

2 The Elecronic Structure of Graphene Sheets

2.1 Description of the Conduction Band

Graphite, a three dimensional (3D) carbon-based material, presents a layered and highly anisotropic structure, the interaction between two adjacent layers being considerably smaller than the intralayer interactions due to the large layer-layer separation, 3.35 Å when compared to the nearest-neighbor distance between the carbon atoms $a = 1.42$ Å. In the planes, graphite exhibits semimetallic behavior, and it presents a very weak electrical conductivity along the perpendicular axis.

In the following, we consider the electronic structure of a single graphite sheet, graphene. In the 2D graphite the in-plane σ bonds are formed from $2s$, $2p_x$ and $2p_y$ orbitals hybridized in a sp^2 configuration, while the $2p_z$ orbital, perpendicular to the layer, builds up covalent bonds, similar to the ones in the benzene molecule. The σ bonds give rigidity to the structure, while the π bonds give rise to the valence and conduction bands. The electronic properties around the Fermi energy of a graphene sheet can be described by a tight binding model with only one orbital per atom, the so-called π-electron approximation, because, as stated above, the π covalent bonds are determinant for the electronic properties of graphite and there are no significant mixing between states belonging to σ and π bands in 2D graphite. Within this approximation a basis set is provided by the Bloch functions made up of the $2p_z$ orbitals from the two inequivalent carbon atoms A and B which form the unit cell of the graphite hexagonal lattice. Considering only nearest-neighbor interactions each atom A of a sublattice has three nearest-neighbors B which belong to the other sublattice [10].

2.2 Tight-binding Model

The nearest-neighbor tight binding approach reduces the problem to the diagonalization of the one-electron Hamiltonian

$$H = -t \sum_{<i,j>} a_i^+ a_j \tag{1}$$

where the sum is over pairs of nearest neighbors atoms i, j on the lattice and a_i, a_j^+ are canonically anticommuting operators

$$\{a_i, a_j\} = \{a_i^+, a_j^+\} = 0, \{a_i, a_j^+\} = \delta_{ij} \tag{2}$$

The eigenfunctions and eigenvalues of the Hamiltonian are obtained from the equation

$$\begin{pmatrix} \varepsilon & -t\sum_j e^{ia\mathbf{k}\cdot\mathbf{u}_j} \\ -t\sum_j e^{ia\mathbf{k}\cdot\mathbf{v}_j} & \varepsilon \end{pmatrix} \begin{pmatrix} C_A \\ C_B \end{pmatrix} = E(\mathbf{k}) \begin{pmatrix} C_A \\ C_B \end{pmatrix}, \tag{3}$$

where $\mathbf{u}_\mathrm{j}$ is a triad of vectors connecting an A atom with its B nearest neighbors and $\mathbf{v}_\mathrm{j}$ the triad of their respective opposites, a is the distance between carbon atoms and ε is the $2p_z$ energy level, taken as the origin of the energy. The eigenfunctions, expanded as a linear combination of the atomic orbitals from the two atoms forming the primitive cell, are determined by the the coefficients C_A and C_B solutions of equation (6). The eigenvalues of the equation give the energy levels whose dispersion relation is

$$E(\mathbf{k}) = \pm t\sqrt{1 + 4\cos^2\frac{\sqrt{3}}{2}ak_x + 4\cos\frac{\sqrt{3}}{2}ak_x \cos\frac{3}{2}ak_y}\ , \tag{4}$$

in which the two signs define two energy bands: the lower half called the bonding π band and the upper half called the antibonding π^* bands, which are degenerate at the **K** points of the Brillouin zone. Within the π electron approximation each site of the graphite honeycomb lattice yields one electron to the Fermi sea and the band is at half-filling. Since each level of the band may accommodate two states due to the spin degeneracy, and the Fermi level turns out to be at the midpoint of the band, instead a whole Fermi line, the 2D honeycomb lattice has six isolated Fermi points which are the

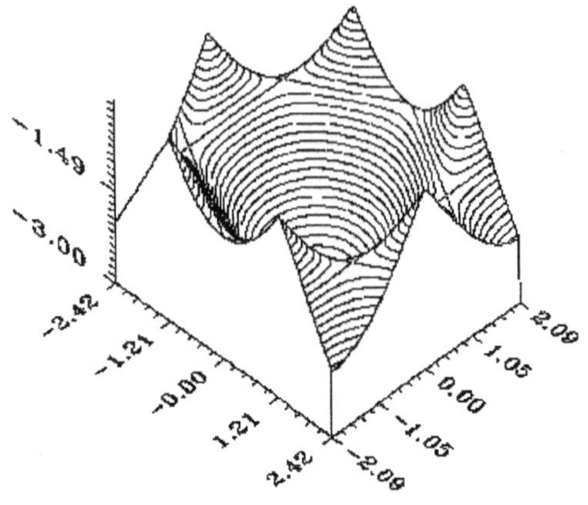

Fig. 1. Lower branch of the electronic dispersion relation. The cusps appear at the six corners of the first Brillouin zone.

six vertices of the hexagonal Brillouin zone, two of which are inequivalent. The lower branch of the dispersion relation is shown in Fig. 1. The calculation of the density of states shows that, at the Fermi level, the density of states is zero therefore, the 2D graphite is a semiconductor of zero gap.

The existence of a finite number of Fermi points at half-filling has important consequences in the description of the spectrum around the Fermi level. The low energy excitations can be studied by taking the continuum limit at any two independent Fermi points. As long as the number of the Fermi points is finite, the outcome is that a simple field theory suffices to describe the electronic spectrum of large honeycomb lattices. The continuum limit can be taken by scaling of dimensionful quantities since we are dealing with a free theory. Taken into account the parameter a, the distance between carbon atoms, and expanding the 2×2 operator (6) at any of two independent Fermi points, we have

$$H = \begin{pmatrix} 0 & -t\sum_j e^{ia\mathbf{k}\mathbf{u}_j} \\ -t\sum_j e^{ia\mathbf{k}\mathbf{v}_j} & 0 \end{pmatrix} \approx -\frac{3}{2}ta\begin{pmatrix} 0 & \delta\mathbf{k}_x + \delta\mathbf{k}_y \\ \delta\mathbf{k}_x - \delta\mathbf{k}_y & 0 \end{pmatrix} + O\left((a\delta\mathbf{k})^2\right). \quad (5)$$

The scaling

$$\lim_{a\to 0} H = -\frac{3}{2}t\sigma^{\mathrm{T}}\delta\mathbf{k} \quad (6)$$

determines the effective Hamiltonian in the continuum limit, which turns out to be the Dirac operator in two dimensions. The same result is obtained at any of the six **K** points of the Brillouin zone, therefore, given the existence of the two independent Fermi points, we conclude that the low energy excitations of the honeycomb lattice at half filling are described by an effective theory of the two-dimensional Dirac spinors. This result is at odds with the more standard continuum approximation to lattice theories in condensed matter physics, the effective mass theory. In this theory, a quadratic dispersion relation at high symmetry points of the Brillouin zone gives rise to an effective Schrödinger equation, with one parameter, the mass, chosen to reproduce the exact curvature. Only one dimensional systems and three dimensional semiconductors with the diamond structure and no gap, are known to give rise to the Dirac equation.

3 The Long Range Coulomb Interactions in Graphite

3.1 Screening in Graphite

The band structure of a graphene plane, as discussed in the preceding section, leads to semimetallic behavior, as the density of states vanishes at the Fermi energy. In a semimetal the long range Coulomb interactions are not screened. The system, however, has no gap, and we can expect that the electron–electron interactions modify significantly the electronic structure near the Fermi energy.

The role of the interactions can be appreciated if their effect is analyzed within perturbation theory. One obtains corrections to the Fermi velocity and to the density of states which show a logarithmic dependence on the temperature or other energy scale at which these quantities are measured. This dependence implies that perturbation theory cannot be used at sufficiently low energies. On the other hand, it allows us to use the Renormalization Group approach. In physical terms, the procedure amounts to defining effective couplings which have a non trivial energy or temperature dependence. The dependence of these couplings on energy can be calculated using well tested techniques developed in the study of Quantum Field Theory, as explained below.

3.2 Renormalization Group Analysis of the Interactions

The implementation of the renormalization group (RG) scheme in condensed matter systems[11] has been a theoretical hallmark for correlated electron systems in the last decade. The condensed matter approach shares ideas from both the critical phenomena and the quantum field theory approaches. The main issue is that for special systems (critical, renormalizable) the low-energy physics is governed by an effective Hamiltonian made of a few marginal interactions that can be obtained from the microscopic high-energy Hamiltonian in a well prescribed manner.

Interactions are classified as relevant, irrelevant or marginal according to their scale dimensions. These dimensions determine whether they grow, decrease, or acquire at most logarithmic corrections at low energies. The effective coupling constants of a model at intermediate energies by "integrating out" high-energy modes even if there is no stable fixed point at the end of the RG flow. The Luttinger and Fermi liquids are identified as infrared fixed points of the RG applied to an interacting metallic system in one or more dimensions respectively.

The main difficulty of the RG approach in condensed matter systems in dimensions greater than one lies on the extended nature of the "vacuum" *i.e.*, of the Fermi surface what makes the issue of scaling rather tricky. The situation is aggravated by the fact that the Fermi surface itself is changed by the interactions, *i.e.* changes along the RG flow.

The Hamiltonian (8) is the perfect model for Renormalization Group (RG) calculations. It is scale invariant and does not have the complications of an extended Fermi surface. The model is similar to the $D=1$ electron system [13] in that it has Fermi points and linear dispersion around them. Its two-dimensional nature manifests itself in the fact that in this case four fermion interactions are irrelevant instead of marginal. The only interaction that may survive at low energies is the long (infinite) range Coulomb interaction, unscreened because of the vanishing density of states at the Fermi point.

The RG analysis of the model is as follows:

The scaling dimension of the interactions are determined by these of the fermion fields which can be read off from the non interacting hamiltonian,

$$H_0 = \hbar v_F \int d^2\mathbf{r}\bar{\Psi}(\mathbf{r})(i\sigma_x\partial_x + i\sigma_y\partial_y)\Psi(\mathbf{r}) \tag{7}$$

Because of the linear dispersion of the electronic states, we can use v_F to transform time scales into length scales. Then, we can express the dimensions of all physical quantities in terms of lengths. Within this convention, the Hamiltonian has dimensions of energy (l^{-1}). This fixes the scale dimension of the electronic fields to $[\Psi]=l^{-1}$, where l defines a length. This also ensures that the free Hamiltonian is scale invariant. We can then readily determine the relevance of the interactions to lowest order (tree level). The interacting Hamiltonian including the two Fermi points (i,i') and the spin degrees of freedom (s,s') is

$$H_{int} = \sum_{i,i';s,s'} \frac{e^2}{2\pi} \int d^2r_1 \int d^2r_2 \frac{\bar{\Psi}_{i,s}(\vec{r}_1)\Psi_{i,s}(\vec{r}_1)\bar{\Psi}_{i',s'}(\vec{r}_2)\Psi_{i',s'}(\vec{r}_2)}{|\vec{r}_1-\vec{r}_2|} + \\ + \sum_{s,s';i,i'} u_{i,s;i',s'} \int d^2r\bar{\Psi}_{i,s}(\vec{r})\Psi_{i,s}(\vec{r})\bar{\Psi}_{i',s'}(\vec{r})\Psi_{i',s'}(\vec{r}). \tag{8}$$

A naive power counting analysis shows that the Coulomb potential (first term in Eq. (8)) defines a dimensionless, marginal coupling, while the four Fermi couplings u's scale as l^{-1}, and are irrelevant at low energies. This effect can be traced back to the vanishing density of states at the Fermi level.

The next step of the RG consists in analyzing the renormalization of the parameters describing the system when quantum corrections are included. When renormalized, the marginal interactions can either grow, driving the system away of its free fixed point - this is the case of an attractive four Fermi interaction in the Fermi liquid case -, decrease and become irrelevant (repulsive interactions in the Fermi liquid), or stay marginal in which case they define the theory (forward scattering in a Fermi liquid and the related Landau parameters). Our model differs from the usual Fermi liquid analysis of [11] on the fact that our interaction is a long ranged (infinite range) unscreened Coulomb interaction, a case that lies away of the Fermi liquid hypothesis.

Following the quantum field theory nature of the model, we replace the instantaneous Coulomb interaction of Eq. (8)

$$H_C = \frac{e^2}{4\pi v_F}\int d^2\mathbf{r}_1 \int d^2\mathbf{r}_2 \frac{\bar{\Psi}(\mathbf{r}_1)\Psi(\mathbf{r}_1)\bar{\Psi}(\mathbf{r}_2)\Psi(\mathbf{r}_2)}{|\mathbf{r}_1-\mathbf{r}_2|} \tag{9}$$

where $g = e^2/4\pi v_F$ is the dimensionless coupling constant, by a local gauge interaction through a minimal coupling.

$$L_{int} = g\int d^2x dt j^\mu(x,t) A_\mu(x,t),$$

where the electron current is defined as

$$j^\mu = (\bar{\Psi}\gamma^0\Psi, v_F\bar{\Psi}\gamma^i\Psi),$$

the three γ matrices ($\gamma^{0,1,2}$) are appropriate combinations of the Pauli matrices. The full Hamiltonian is then that of (non-relativistic) quantum electrodynamics in two spacial dimensions, a model used also in the physics of nodal states of d-wave superconductors:

$$H = \hbar v_F \int d^2\mathbf{r}\bar{\Psi}(\mathbf{r})\gamma^\mu(\partial_\mu - igA_\mu)\Psi(\mathbf{r}). \tag{10}$$

The RG analysis proceeds with the computation of the renormalization of the parameters of the model. The Feynman diagrams building blocks are the free electron and photon propagators:

$$G_0(\omega,\mathbf{k}) = i\frac{-\gamma^0\omega + v_F\gamma\cdot\mathbf{k}}{-\omega^2 + v_F^2\mathbf{k}^2 - i\varepsilon}, \quad \Pi^0_{\mu\nu}(r_1,r_2) = -i\delta_{\mu\nu}\int\frac{d^4k}{(2\pi)^4}\frac{e^{-i\omega(t_1-t_2)}e^{i\mathbf{k}(\mathbf{x}_1-\mathbf{x}_2)}}{-\omega^2+\mathbf{k}^2-i\varepsilon}.$$

The electron self–energy $\Sigma(\omega,\mathbf{k})$ defined by the equation $G^{-1} = G_0^{-1} - \Sigma$, is renormalized by the Feynman diagrams of Fig. 2.

It contains the following physical information:

- The density of states $n(\omega) = Im\int d^2\mathbf{k}\, \mathrm{tr}G(\omega,\mathbf{k})\sigma_3$. It is renormalized by the diagram in Fig. 2(a).

- The Fermi velocity renormalization. It is obtained already at the one loop level from the diagram in Fig. 2(b).
- The quasiparticle life time $\tau^{-1} \sim \lim_{\omega \to 0} \mathrm{Im} \sum(\omega, \mathbf{k})$. Its first contribution is at the two–loops level from diagrams 2(c), 2(d) in Fig. 2.
- The wave function renormalization $Z_\Psi \sim \frac{\partial \sum(\omega, \mathbf{k})}{\partial \omega}\Big|_{\omega=0}$ defines the anomalous dimension of the field $\gamma = \partial \log Z_\Psi / \partial l$ (l is the RG parameter) and, hence of the fermion propagator: $G(\omega, \mathbf{k}) \sim_{\omega \to 0} \frac{1}{\omega^\eta}$ It is a critical exponent that determines the universality class of the given model. Under the physical point of view it affects the interlayer tunneling and other transport properties.

The next set of diagrams to analyze corresponds to the photon self–energy and vertex corrections represented schematically in Fig. 3.

The real part of the photon self-energy at one loop (polarization) renormalizes the interaction and the imaginary part gives the density of electron-hole excitations of the system. The vertex corrections renormalize the electric charge.

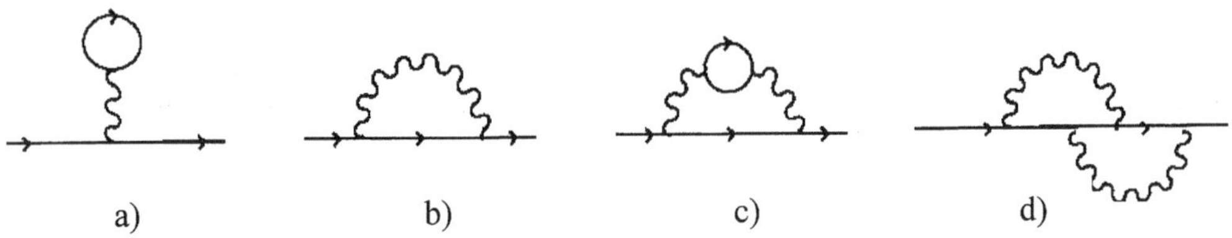

Fig. 2. Feynman diagrams renormalizing the electron self-energy.

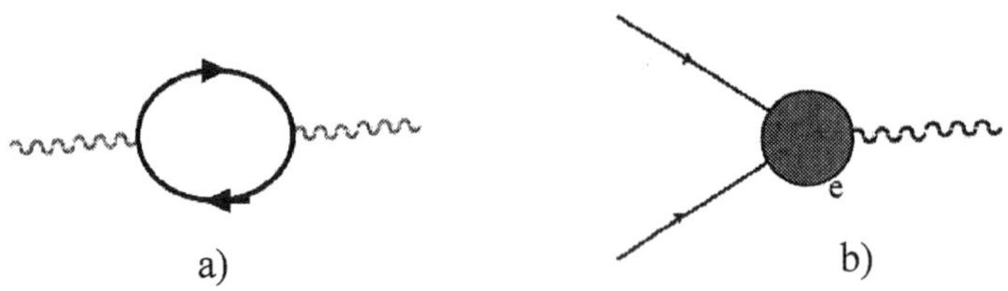

Fig. 3. (a) Feynman diagram renormalizing the photon self-energy. (b) Vertex correction.

3.3 Results

In the computation of the diagrams mentioned it is readily seen that the loop corrections come in powers of an effective coupling constant given by $g = e^2/4\pi v_F$. The physical results extracted from the RG analysis are the following.

1. From the computation of the electron self-energy (Fig. 2(b)) we get a non trivial renormalization of the Fermi velocity that **grows in the infrared.** This result implies a breakdown of the relation between the energy and momentum scaling, a signature of a quantum critical point.

2. From the electron self-energy at two loops order we get a non trivial wave function renormalization meaning that the infrared stable fixed point corresponds to a free fixed point different from the Fermi liquid. This result has been shown to persist in the non-perturbative regime [14]. This is a non-trivial result that has physical implications. In particular it implies that the inverse quasiparticle lifetime increases linearly with energy [4], a result that has been observed experimentally in [3] in the energy range of validity of the model.

3. The electron-photon vertex and the photon propagator are not renormalized at the one loop level. This means that the electric charge is not renormalized, a result that could be predicted by gauge invariance, and it also implies that the effective coupling constant $g = e^2/4\pi v_F$ **decreases at low energies** defining an infrared free fixed point of the RG. It is interesting to note that the Lorentz invariance of the model that was explicitly broken by the Fermi velocity is recovered at the fixed point since the velocity of light, c, fixes a limit to the growing of the Fermi velocity. In conclusion, the RG analysis shows that without disorder, edges, or other perturbations, the graphene system at low energies has gapless excitations differing from the Fermi liquid quasiparticles but does not support magnetic or superconducting instabilities. It is interesting to note that the energy dependence of the coupling constant [14] can lead to non trivial scaling features in optical properties [16].

The strong coupling regime of the graphene system has been analyzed in [17]. There it is argued that a dynamical breakdown of the chiral symmetry (degeneracy between the two Fermi points) will occur at strong coupling and a gap will open in the spectrum forming a kind of charge density wave. Graphite can then be seen as an excitonic insulator that can become ferromagnetic upon doping. The resulting gap has an exponentially small non-perturbative value.

The analysis in this section neglects short range interactions, as their effects are less relevant than those arising from the long range Coulomb interaction. It is worth noting, however, that a sufficiently large on–site repulsion can induce a transition to an antiferromagnetic ground state [18], and that, even below this transition, significant effects at low energies can be expected [19].

4 Effects of Disorder

4.1 General Features

As mentioned elsewhere in this volume, there is a wide variety of carbon compounds, ranging from crystalline diamond, where the carbon atoms show fourfold coordination, to graphite, where the coordination is threefold, and the coupling between neighboring planes is weak. The environment around a carbon atom in nanotubes and the fullerenes is closer to the graphite case, although the bonds with the three nearest neighbors are distorted. The variety of possible environments around a carbon atom imply that many intermediate, metastable phases can exist. As in any other materials, disorder can appear due to lattice defects or impurities. In the following, we consider the changes in the electronic states in threefold coordinated systems due to some simple lattice defects, like five- and sevenfold rings, vacancies, dislocations and edges. We will not address the stronger deformations associated with hybrid three-and fourfold bonding (sp^3–sp^2 hybridization) which may exist in highly disordered systems [2].

4.2 Five- and Sevenfold Rings (disclinations)

As discussed earlier, the low energy electronic states of graphene planes are well described by a two dimensional Dirac equation, which reproduces correctly the semimetallic nature of the system. Some lattice distortions give rise to long range modifications in the electronic wavefunctions. These effects should be well described using the effective Dirac equation as a starting point.

The simplest defects which show these features are five- or sevenfold rings in the honeycomb lattice. These defects can be considered disclinations of the lattice, which acquires a finite curvature. The accumulation of them leads to curved shapes, like the fullerenes, which show twelve fivefold rings. Sevenfold rings lead to negative curvature, and a variety of compounds have been proposed to exist with this property [20]. A simple way to show that an odd numbered ring in the honeycomb lattice leads to long range effects in the electronic spectrum is by noting that any closed path which encompasses the defect leads to an interchange of the two sublattices which build the structure [7]. The description of the electronic states in terms of the Dirac equation is achieved by using to types of electronic “flavors ", each of them existing in a different sublattice. The existence of odd numbered rings changes the Dirac equation at any distance from the defect.

If we neglect for the moment the effect of the long range lattice distortions induced by these defects, the only consequence of the presence of odd numbered rings is the above mentioned interchange of the two sublattices. In the Dirac description it implies that, when moving around the defect, the two electronic flavors are exchanged. The standard way to associate to a translation a smooth change in other properties is through gauge potentials. The existence of a gauge potential implies, in general, that the usual derivative has to be replaced by the covariant derivative, which includes the potential.

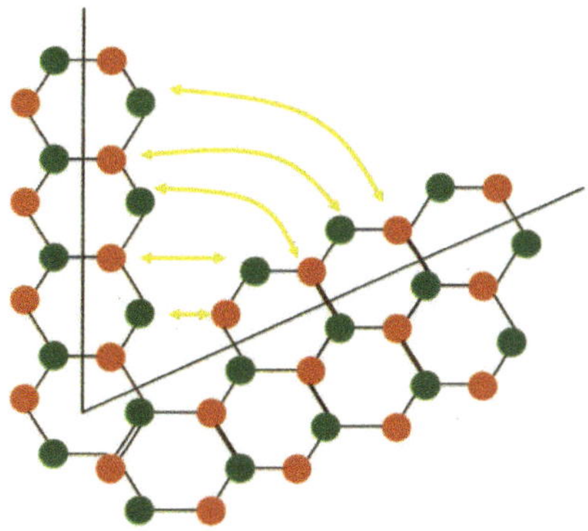

Fig. 4. Description of a fivefold ring in the honeycomb structure in terms of a disclination. The identification of sites at the two edges imply a correspondence between sites from one sublattice and the other.

The usual derivative operator is the generator of a translation through the system. A covariant derivative with a finite gauge potential implies that, when translating an object, an additional operation has to be performed upon it. In the case considered here, this operation is a rotation in flavor space. As there are two flavors, the index which distinguishes them is equivalent to a spin one half. The rotations in this space build up the SU(2) non abelian group. The gauge potential needed has to be chosen such that the accumulated rotation in a path which encircles the defect should be independent of the path. Hence, the gauge potential is equivalent to that generated by a fixed "magnetic " flux at the position of the defect[1].

A schematic view of the correspondence of a fivefold ring and a disclination is shown in Fig. 4.

The previous analysis shows that the effects on the electronic states of odd numbered rings in the honeycomb lattice are approximately described in terms of a gauge field which decays inversely proportional to the distance to the defect. This scheme allows us to calculate analytically the low energy electronic spectrum of closed structures, like C_{60} and higher fullerenes [7]. The comparison with more detailed calculations is quite reasonable, and, as expected, improves as the radius of the system becomes larger.

4.3 Dislocations

The effects induced far away from the core of a dislocation can be approximated by assuming that the dislocation is made up of two disclinations of opposite sign. The general model of a disclination has been given in the previous section, and it can be directly applied to the case of a dislocation. Its effect on the low energy electronic

[1]One must note that there is an additional technical complication, associated to the fact that there is also another index associated to the two inequivalent Fermi points in the Brillouin Zone. An odd numbered ring also exchanges them.

spectrum can be approximated by the gauge field induced by to opposite magnetic fluxes separated by a distance of the order of the Burgers vector of the dislocation. This field decays like the inverse of the square of the distance to the defect.

4.4 Edge States

Tight binding models have shown that, in the vicinity of the edges of graphene planes, localized states at zero energy can exist [21, 22]. These states are well described by the Dirac equation used here. The existence of a state at zero energy implies the existence of a localized wavefunction $(\Psi_1(\tilde{\mathbf{r}}), \Psi_2(\tilde{\mathbf{r}})$ such that:

$$\begin{aligned} &\left(\partial_x + i\partial_y\right)\Psi_1(\tilde{\mathbf{r}}) = 0 \\ &\left(\partial_x - i\partial_y\right)\Psi_2(\tilde{\mathbf{r}}) = 0 \end{aligned} \tag{11}$$

These equations are satisfied if $\Psi_1(\tilde{\mathbf{r}})$ is an analytic function of $z = x + iy$ and $\Psi_2(\tilde{\mathbf{r}}) = 0$, or if $\Psi_1(\tilde{\mathbf{r}}) = 0$ and $\Psi_2(\tilde{\mathbf{r}})$ is an analytic function of $\overline{z} = x - iy$. We now consider a semiinfinite honeycomb lattice with an edge at $y = 0$ and which occupies the half plane $x > 0$. A possible solution which decays as $x \to \infty$ is $\Psi_1(x, y) \propto e^{-kz} = e^{iky} e^{-kx}, \Psi_2(\tilde{\mathbf{r}}) = 0$. These solutions satisfy the boundary conditions at $y = 0$ if the last column of carbon atoms belong to the sublattice where the component Ψ_1 is defined. Then, the next column belongs to the other sublattice, where the amplitude of the state is, by construction, zero.

4.5 Vacancies

The analysis in the previous section of edge states can be extended to the existence of localized states near extended vacancies in the honeycomb lattice. The only possible localized states can exist at zero energy, where the density of extended states vanishes. Then, the wavefunctions obtained from the Dirac equations must be normalizable and analytic on the variables $z = x + iy$ or $\overline{z} = x - iy$. Extended vacancies with approximate circular shape can support solutions of the type $\Psi((\tilde{\mathbf{r}}) \propto z^{-n}, n > 1$. By using conformal mapping techniques, solutions can be found with the boundary conditions appropriate to the shape of different defect.

A simple case is the elongated crack depicted in Fig. 5. A localized solution is described by an analytic function $f(z)$ such that $\mathrm{Re} f(z) = 0$ at the edges of the crack. A family of functions for a crack of half length L, which satisfy these requirements are:

$$f(z) = \frac{1}{(z^2 - L^2)^{n+1/2}} \tag{12}$$

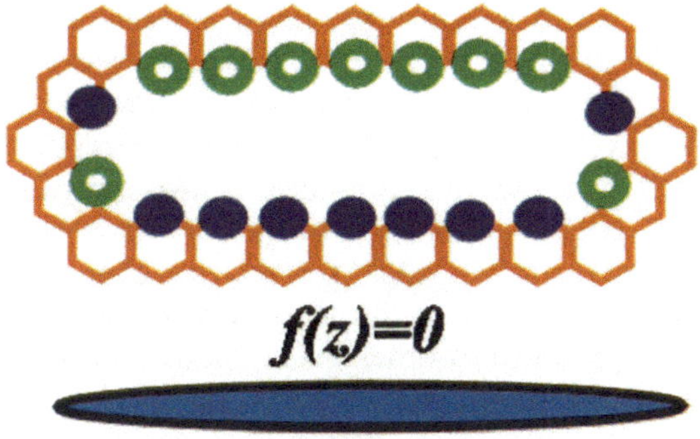

Fig. 5. Elongated crack in the honeycomb structure. The crack is such that the sites in the upper edge belong to one sublattice, while those at the lower edge belong to the other. Bottom: approximate cut in the complex plane which can be used to represent this crack at long distances.

4.6 Random Distribution of Defects

As discussed above, many classes of lattice defects can be described by gauge fields coupled to the two dimensional Dirac equation. A random distribution of defects leads to a random gauge field, with variance related to the type of defect and its concentration. There is an extensive literature on the problem, as the model is also relevant to Fractional Quantum Hall states and to disorder in d-wave superconductors. A random field, when treated perturbatively, leads to corrections to the wavefunction renormalization which depend logarithmically in the electronic bandwidth, in the same manner as the corrections induced by the long range Coulomb interaction. Hence, disorder is a marginal perturbation in the Renormalization Group sense, and can be analyzed using the same approach employed in the study of the Coulomb interactions.

Disorder in systems with energy gaps tends to induce localized states inside the gap. The honeycomb lattice has a semimetallic density of states. A random field enhances the density of states at low energies, although the system preserves its semimetallic character. The density of states at low energies is changed from $D(\omega) \propto |\omega|$ to $D(\omega) \propto |\omega|^{1-\delta}$, where δ depends on the type of disorder[23–25].

5 Combined Effects of Disorder and the Electronic Interactions

5.1 The Long Range Coulomb Interaction

The analysis of the Renormalization Group results presented previously led to the conclusion that the pure graphene system at low energies is an anomalous Fermi liquid with no short range interactions. Inclusion of disorder modelled as random gauge fields modifies the flow of the couplings and gives rise to new phases with different physical properties. Similar problems have been considered in relation to transitions between Fractional Quantum Hall states [26]. There the different types of "extended" disorder are associated to different gauge couplings that can be treated with the Renormalization Group technique together with the long range Coulomb interaction.

The values of the effective coupling constants at low energies are modified by the new interactions giving rise to a rich phase diagram with new phases with different physics depending of the type of disorder A schematic plot of the flow obtained for different types of disorder is shown in Fig. 6. The most interesting phase is the one induced by the presence of random disorder in the lattice structure(center). It shows the existence of new phases, where the effect of the Coulomb interaction, which tends to lower the density of states at the Fermi energy [27], and the disorder, which has the opposite tendency, balance each other. This phase is stable, within the limitations of the calculation presented here, for certain types of disorder [6].

5.2 Short Range Interactions

As mentioned earlier, short range interactions, such an onsite Hubbard term U are irrelevant, in the sense that their effect can be analyzed within standard perturbation theory without encountering divergences. This is due to the vanishing density of states at the Fermi level. The Hubbard model at half filling, in a two dimensional square lattice leads to a highly singular perturbation expansion, due to the diverging density of states at the Fermi level. As mentioned above, the density of states at low energies is increased by the presence of disorder. This, in turn, enhances the effect of short range interactions.

Short range interactions can lead to a variety of phases at low temperatures. In the absence of disorder, an onsite Hubbard term favors antiferromagnetism. An antiferromagnetic phase, however, is likely to be suppressed by disorder, especially by

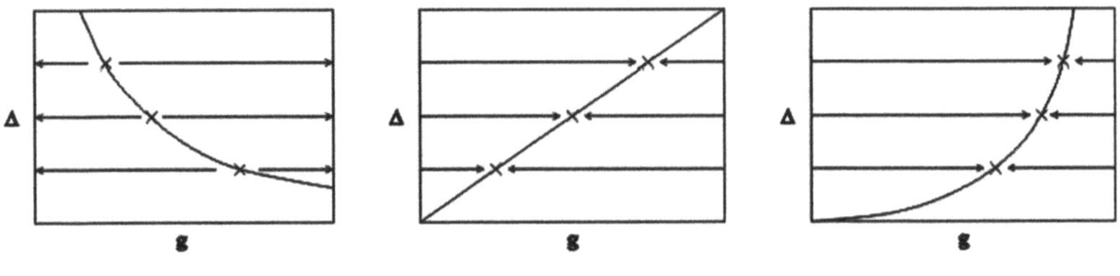

Fig. 6. Phase diagram calculated by renormalizing the strength of the disorder, Δ, and the Coulomb interaction, g, at the same time. The three plots correspond to three types of disorder. Left: random on site potential. Center: Random disorder in the lattice structure. Right: Random correction to the hoppings.

the presence of odd numbered rings in the lattice. Then, the next leading instability that such an interaction can induce is towards a ferromagnetic phase.

If a magnetic phase does not appear, electron electron interactions, even when they are repulsive, will lead to an anisotropic ground state. The existence of two inequivalent Fermi points in the Brillouin zone suggests that the superconducting order parameter induced by a repulsive interaction will have opposite sign at each point. The corresponding symmetry is p-wave. No that disorder, in addition to the enhancement of the density of states mentioned already, will lead to pair breaking effects in an anisotropic superconducting phase.

5.3 Interactions between Localized States

As mentioned in the previous section, vacancies and cracks in the honeycomb lattice induce localized states at the Fermi energy. These states will become polarized in the presence of repulsive interactions, as this polarization implies no cost of kinetic energy. Then, we can expect that lattice defects will nucleate magnetic moments in their vicinity. These moments can be large, as the number of localized states is proportional to the number of sites at the perimeter of the defect. Note that this mechanism is intrinsic to the graphene structure, and it does not require the trapping of magnetic ions near the defects.

The moments near different defects polarize the conduction band of the surrounding medium, leading to an effective RKKY interaction. In an ordinary metal this interaction is made up of an oscillatory and a decaying term as function of distance, and it can be of either sign, leading to frustration and spin glass effects. The graphene plane considered here, however, does not have a Fermi surface, so that the induced RKKY interaction does not oscillate. A simple analysis, using the analytical expression for the susceptibility discussed in earlier chapters, gives:

$$J_{RKKY}(\tilde{\mathbf{r}}) \sim U^2 \int d^2\mathbf{k} e^{i\tilde{\mathbf{k}}\tilde{\mathbf{r}}} \chi(\tilde{\mathbf{k}}) \sim U^2 \frac{a^4}{v_F \, |\, \tilde{\mathbf{r}} \,|^3} \tag{13}$$

where U is the magnitude of the onsite Hubbard term. Hence, the RKKY interaction is ferromagnetic, and it decays as r^{-3} as function of the distance between local moments.

6 Coupling between Graphene Layers

6.1 Coulomb Interactions

So far, we have considered the properties of isolated graphene layers. As the source of most of the unusual properties reported here is the long ranged Coulomb interaction, it is important to consider the screening effects of neighboring layers.

The calculation of the full dielectric constant of a set of metallic or semimetallic layers in terms of single layer properties can be done analytically. The dielectric function of the system can be written as [5]:

$$\frac{1}{\varepsilon(\tilde{\mathbf{q}},\omega)} = \frac{\sinh(|\tilde{\mathbf{q}}|d)}{\sqrt{\left[\cosh(|\tilde{\mathbf{q}}|d) + (2\pi e^2/|\tilde{\mathbf{q}}|)\sinh(|\tilde{\mathbf{q}}|d)\chi_0(\tilde{\mathbf{q}},\omega)\right]^2 - 1}} \tag{14}$$

where $\chi_0(\tilde{\mathbf{q}},\omega)$ is the charge response function of an isolated layer, and d is the interlayer distance. This response function is finite for $|\tilde{\mathbf{q}}| \ll d^{-1}$, so that the interactions remain long range.

6.2 Interlayer Hopping

The electron–electron interaction modifies the quasiparticle propagator, as discussed above. The electrons within the layers are dressed by a cloud of virtual excitations. This cloud cannot follow an electron which hops between neighboring layers, reducing the effective tunnelling element.

In conventional Fermi liquids, this renormalization of the interlayer hopping is finite, and it can be calculated using perturbation theory. In the model studied here, this calculation leads to divergencies, and it resembles closely the analysis of the electron self energy sketched previously.

The "orthogonality catastrophe" which results from the virtual excitation of electron-hole pairs has been extensively discussed in connection to the physics of mesoscopic systems [28], and it has also been applied to the related problem of tunnelling between two dirty metallic layers [29]. Similar procedures can be used in the present case. The interlayer hopping acquires a multiplicative renormalization which makes it vanish at low energies, even in clean samples. This calculation is consistent with the extreme anisotropy observed in some experiments [8].

7 Conclusions

We have discussed a simplified model for the long wavelength electronic properties of graphene planes. The interplay between the semimetallic properties of the planes and the long range interactions leads to the existence of a variety of interesting effects:

- The model, in the absence of disorder, shows deviations from Landau's theory of a Fermi liquid. The quasiparticles are strongly renormalized, and their lifetimes do not follow the usual $\Gamma(\varepsilon) \propto (\varepsilon - \varepsilon_F)^2$ behavior. The low energy electronic properties of the system can be thought as similar to the "pseudogap " regime in the superconducting cuprates.
- Disorder can be incorporated into the model in a simple way. While the interactions tend to deplete the electronic density of states near the Fermi energy, disorder leads to its enhancement. The resulting competition induces the existence of an "incoherent metal " regime at low energies, similar to the one dimensional Luttinger liquid, although stabilized by the disorder.

- Large lattice deformations can nucleate localized electronic states in their vicinity[2]. These states can lead to the formation of local moments. The absence of a finite Fermi wavevector implies that the RKKY interaction mediated by the conduction electrons does not change sign, and it is ferromagnetic. Hence, the frustration which leads to spin glass behavior in metals with magnetic impurities is absent in this case.
- The screening cloud around quasiparticles suppresses interlayer tunnelling, enhancing the anisotropy of the electronic properties.

References

1. J.C. Slonczewski and P. R. Weiss, Phys. Rev. 109 (1958) 272.
2. A.A. Ovchinnikov and I.L. Shamovsky, Journ. of. Mol. Struc. (Theochem) 133 (1991) 251.
3. S. Yu, J. Cao, C.C. Miller, D.A. Mantell, R.J.D. Miller, and Y. Gao, Phys. Rev. Lett. 76 (1996) 483.
4. J. González, F. Guinea, and M.A.H. Vozmediano, Phys. Rev. Lett. 77 (1996) 3589.
5. J. González, F. Guinea, and M.A.H. Vozmediano Phys. Rev. B63 (2001) 134421.
6. T. Stauber, F. Guinea, and M.A.H. Vozmediano, preprint (cond-mat/0311016).
7. J. González, F. Guinea and M.A.H. Vozmediano, Phys. Rev. Lett. 69 (1992) 172; Nucl. Phys. B406 (1993) 771.
8. R. Ocaña, P. Esquinazi, H. Kempa, J.H.S. Torres, and Y. Kopelevich, Phys. Rev. B68 (2003) 165408.
9. M.A.H. Vozmediano, M.P. López -Sancho, and F. Guinea, Phys. Rev. Lett. 89 (2002) 166401; ibid Phys. Rev. B68 (2003) 195122.
10. R. Saito, G. Dresselhaus, and M.S. Dresselhaus, in: Physical Properties of Carbon Nanotubes (1998) Imperial College Press, London.
11. R. Shankar, Rev. Mod. Phys. 66 (1994) 129.
12. J. Polchinski, in: Proceedings of the 1992 TASI in Elementary Particle Physics, eds. J. Harvey and J. Polchinski (1992) World Scientific, Singapore.
13. J. Solyom, Adv. Phys. 28 (1979) 201.
14. J. González, F. Guinea, and M.A.H. Vozmediano, Phys. Rev. B59 (1999) R2474.
15. J. González, F. Guinea, and M. A. H. Vozmediano, Mod. Phys. Lett. B7 (1994) 1593; ibid, Nucl. Phys. B 424 (1994) 595; ibid, Journ. Low. Temp. Phys. 99 (1995) 287.
16. C.L. Kane and E.J. Mele, Phys. Rev. Lett. 197402 (2004) 93.
17. D.V. Khveshchenko, Phys. Rev. Lett. 87 (2001) 48246802; ibid, Phys. Rev. Lett. 87 (2001) 48246802.
18. S. Sorella and E. Tosatti, Europhys. Lett. 19 (1992) 699.
19. G. Baskaran and S.A. Jafari, Phys. Rev. Lett. 89 (2002) 016402; N.M.R. Peres, M.A.N. Araújo, and A.H. Castro Neto, Phys. Rev. Lett. 92 (2004) 199701; G. Baskaran and S.A. Jafari, Phys. Rev. Lett. 92 (2004) 199702; N.M.R. Peres, A.H. Castro Neto, and F. Guinea, to be published.
20. D. Vanderbilt and J. Tersoff, Phys. Rev. Lett. 68, 511 (1992); J. L. Aragón, H. Terrones, and D. Romeu, Phys. Rev. B48 (1993) 8409.
21. K. Wakayabashi and M. Sigrist, Phys. Rev. Lett. 84 (2000) 3390.

---

[2]Note that these states are built up from the π orbitals of each layer, and the σ bonds remain unaltered.

22. K. Wakayabashi, Phys. Rev. B64 (2001) 125428.
23. C. de C. Chamon, C. Mudry, and X.-G. Wen, Phys. Rev. B53 (1996) R7638.
24. H.E. Castillo, C. de C. Chamon, E. Fradkin, P.M. Goldbart, and C. Mudry, Phys. Rev. B56 (1997) 10668.
25. B. Horovitz and P. Le Doussal, Phys. Rev. B65 (2002) 125323.
26. J. Ye, Phys. Rev. B60 (1999) 8290.
27. M.P. López -Sancho, M.C. Muñoz, and L. Chico, Phys. Rev. B63 (2001) 165419.
28. E. Bascones, C.P. Herrero, F. Guinea, and G. Schön, Phys. Rev. B61 (2000) 16778.
29. M. Turlakov and A.J. Leggett, Phys. Rev. B63 (2001) 064518.

Carbon-Based Magnetism
T. Makarova and F. Palacio (Editors)

16

The Magnetic Nature of Intrinsic and Irradiation-induced Defects in Carbon Systems

P.O. Lehtinen[1], A.V. Krasheninnikov[1,2], A.S. Foster[1], and R.M. Nieminen[1]

[1]*Laboratory of Physics, Helsinki University of Technology, P.O. Box 1100, 02015, Finland*
[2]*Accelerator Laboratory, University of Helsinki, P.O.Box 43, Helsinki 00014, Finland*

1 Introduction

Observations of magnetism in various carbon systems [1–4] have stimulated much experimental and theoretical research work (Ref. [5] and references therein) on the magnetic properties of all-carbon systems. The driving force behind these studies was not only to create technologically-important, light, non-metallic magnets with a Curie point well above room temperature, but also to understand a fundamental problem: the origin of magnetism in a system which traditionally has been thought to show diamagnetic behavior only.

The question which immediately springs to ones mind is whether magnetism is an intrinsic property of carbon systems or is due to the presence of magnetic impurities (*e.g.*, Fe). The debate was initiated from the appearance of the two classic papers: in 2000 Kopelevich and coworkers published a paper [1] with an editor's note of the controversial nature of the result. In that work, the authors identified ferromagnetic and superconducting-like magnetization hysteresis loops in highly oriented pyrolytic graphite samples below and above room temperature. The conclusion was that magnetic impurities were extremely unlikely. In 2001 Makarova and coworkers [2] reported an observation of magnetic signals in rhombohedral C_{60}. Their intention was to search for superconductivity in polymerized C_{60}; however, it appeared that their high-pressure, high-temperature polymerization process resulted in a magnetically ordered state. The material exhibited features typical of ferromagnets: saturation magnetization, large hysteresis and attachment to a magnet at room temperature. A careful analysis carried out follow-up works showed that one can exclude magnetic impurities as the origin of ferromagnetism.

If magnetism is indeed the intrinsic property of carbon systems, then the most important questions to be answered are as follow:

1. *What is the atomic structure of the magnetic phase and what is the local bonding geometry which gives rise to the net magnetic moment?*
   The presence of local magnetic moments can explain only the paramagnetic behavior but not ferromagnetism. Thus, the second question is:
2. *How is the macroscopic ferromagnetic state formed and what is the mechanism of long-range order formation?*
   Because ferromagnetic signals have been detected in various carbon systems such as graphite [1, 4, 6] polymerized fullerenes [2, 7], and carbon foam [3], it is highly important to understand:
3. *Is the mechanism of magnetic state formation common for all the carbon systems or is it different for different allotropes?*
   Finally, if magnetism is not the intrinsic property of all-carbon systems and it does not originate from magnetic impurities, then another question rises:
4. *Can magnetism result from impurity atoms which are non-magnetic by themselves, but due to unusual chemical environment, e.g., due to bonding to defects in graphitic network, give rise to local magnetic moments?*

As for the first problem, a number of factors are nowadays thought to possibly give rise to appearance of localized spins and the development of the magnetic state in all-carbon systems: defects in the atomic network such as under-coordinated atoms [8–13], itinerant ferromagnetism [14, 15] and negatively-curved sp^2-bonded nano-regions in the carbon structures [3, 16] (see also chapter 20). Among these factors the defect-mediated mechanism appears to be the most general one because negatively-curved regions can hardly be found in graphite, and as for the second scenario, although itinerant mechanisms resulting from strong electron-electron interactions and the effective dimensionality of the electron system can give rise to magnetism, direct experimental evidence supporting such a mechanism is still lacking.

The defect-mediated mechanism has been addressed in a considerable number of works [8–13]. Although the details can be different for different carbon systems (polymeric fullerenes, graphite, nanotubes), the common feature is the presence of under-coordinated atoms, *e.g.*, vacancies [11, 17] (see also chapter 21), atoms on the edges of graphitic nano-fragments with dangling bonds either passivated with hydrogen atoms [12, 13, 18] (see also chapter 12, 13) or free [10, 13]. Structural defects, in general, give rise to localized electronic states, a local magnetic moment, flat bands associated with defects and thus to an increase in the density of states at the Fermi level, and eventually to the development of magnetic ordering. However, even if this conjecture is correct, it is not clear at all whether such defects are actually present or if their concentration is high enough to provide the magnetic moment observed in the experiments.

At the same time, it is well known that irradiation of carbon systems with energetic electrons and ions should give rise to defects, and their number can be controlled by choosing the right irradiation dose, particle energy and irradiation temperature. Thus, if irradiation of the originally non-magnetic carbon samples gave rise to magnetism, this could be strong evidence for the defect scenario.

Graphite samples were recently irradiated with 1.5 MeV He and 2.25 MeV H ions [19, 20] (see also chapter 19). It was found that proton bombardment produced a strong magnetic signal, while bombardment with helium ions produced a signal which was only slightly larger than background.

To explain the irradiation-induced magnetism and shed light on the role of defects in all-carbon magnetism, the atomic structure and magnetic moment of various defects were recently calculated [21–24] at the atomistic level within the framework of Density Functional Theory (DFT).

In this Chapter, we give a summary of the most important theoretical results obtained in Refs. [21–24] and try to answer questions *(i)* and partly *(iv)* listed above. We address the behavior of various point defects–vacancies, interstitials and more complicated aggregations, both intrinsic, and those which can appear in carbon nanotubes and graphite under irradiation. We briefly describe not only magnetic, but also structural and other characteristics of the defects, such as formation energies and diffusivity. We show that under certain conditions the defects can indeed give rise to magnetism. We further consider defects which can appear under proton irradiation and demonstrate that even a small amount of hydrogen in carbon samples might be very important for the formation of the magnetic state.

1.1 Intrinsic Defects

There are several classes of intrinsic defects, such as grain boundaries and dislocations [25], but here we focus on the more universal point defects. There are two main types of intrinsic point defects: vacancies and interstitials. A vacancy is one missing atom and an interstitial is an extra atom in the crystal. It is obvious that if we displace a carbon atom from its original position in the lattice of a real crystal, we create not only a vacancy, but also an interstitial. The vacancy-interstitial pair is known as a Frenkel pair. These defects are usually the cause for the observed conductivity of ionic crystals and they have a significant effect on the optical properties (specially on the color). They are present in the thermal equilibrium which makes them intrinsic of nature [25].

Besides vacancies and interstitials, there are other types of simple defects, such as Stone-Wales defects (two heptagons and two pentagons), formed by rotating a carbon-carbon bond by 90 degrees [26]. The barrier for such defect formation is lower than that for Frenkel pairs, especially if an extra carbon atom which works as a catalyst is nearby [27]. However, this defect is non-magnetic, so we do not consider it further.

1.2 Irradiation-induced Defects in Graphitic Structures

When an energetic particle such as an electron or ion hits the material, different mechanisms of damage creation can work. Depending on the target material and the incident particle characteristics, the main mechanism can be the direct kinetic energy transfer to target atoms (nuclear stopping) or the transfer of the projectile energy to the target electrons via electronic excitations, ionization, *etc.* (electronic stopping) [28]. For light ions (such as H, He), electronic stopping is more important. For heavy ions nuclear stopping dominates at low energies, while electronic effects govern energy loss at high energies with the crossover point being dependent on the ion mass [28]. The damage production is further dependent on the electronic structure of the target (metal/insulator). As sp^2 carbon systems are either metals or narrow-band semiconductors, electronic excitations and ionization effects seem to be less important [29] in graphitic systems than in diamond due to a high electrical conductivity of the sp^2 carbon network.

However, whatever the main channel of the energy transfer is, irradiation gives rise to the formation of Frenkel pairs. In addition to single point defects, defect aggregations (vacancy and interstitial clusters) can appear either directly due to impact of the energetic particle or due to defect migration and annealing. A review of irradiation defect production and annealing in graphitic systems can be found in Refs. [29–31].

Note that due to the open structure of carbon nanotubes (CNT) and a low density of the nanotube samples, the damage formation in carbon nanotubes is quite different from that observed in most other solids. Due to the open space in the CNT samples, even recoils which have received energy only slightly above the threshold energy can be displaced quite far, which is in contrast to many other types of materials.

There are also differences in defect annealing. In semiconductors such as Si a vast majority of all Frenkel pairs produced are known to recombine below room temperature [32]. In dense metals, nearly all the interstitials and vacancies produced during the ballistic phase of the cascade recombine with each other after only a few picoseconds, regardless of the sample temperature [33]. Defects in CNTs can also anneal, but it happens at elevated temperatures [29], and the annealing mechanisms are somewhat different from those in metals [31]. In particular, single and multiple vacancies can disappear by saturating dangling bonds [34, 35] and by forming non-hexagonal rings. This mechanism works in graphite as well, but in CNTs it is more important as finite curvature of graphene sheets decreases transformation-induced strain in the atomic network.

2 Methodology

Most of the calculations described here are made using the periodic plane wave VASP code [36, 37], implementing the spin-polarized density functional theory (DFT) and the generalized gradient approximation (GGA) of Perdew *et al.* [38]. DFT can also be implemented for finite systems, using other basis sets than plane waves (*e.g.* gaussians), but in standard DFT simulations this recipe is the most accurate, as errors are much more controllable. Also, since we are dealing with either an infinite graphite surface or CNT, the periodicity implied by plane waves is not a hindrance.

In any case, rather than going into technical details, it is perhaps more useful to give some consideration to the general issues important in modelling defects with this method.

2.1 Periodic Model

The periodic model is based on a definition of a unit cell (or more generally - supercell), which can be periodically translated to build an infinite system. There are no limitations, in principle, on the size and shape of the unit cell providing that the host lattice is defined unambiguously. The periodicity imposed on the atomic structure also applies to the electronic structure: the charge density is the same in every supercell and is matched at the boundary between two neighboring supercells. In the case of ideal crystals this approach is technically exact. For surfaces, a very good model can be achieved by making the vacuum gap in the z-direction between images of the surface large enough to avoid interactions - this is known as the "slab" model. To calculate the properties of an individual point defect using the periodic model, a supercell is defined which comprises the defect and its immediate environment. The defect in each supercell interacts with an infinite number of similar defects (called images) in all other supercells.

An obvious feature of the periodic model, which can be both to its advantage and drawback, is that "defects" are periodically translated and interact between themselves. It can be an advantage if one is interested, for example, in periodic adsorption of molecules at the surface and would like to vary their concentration. It is a drawback if one is interested in the properties of an individual defect in different charge states. In particular, if the defect is charged with respect to the host lattice, the electrostatic interaction of such defects is divergent. Yet another downside of the periodic model is that the defect-induced distortion of the host lattice can only be taken into account within the supercell. This may lead to inaccurate results if these distortions propagate further than cell boundaries, but can be checked by increasing the supercell size. Several recent developments in computational techniques using the periodic model have led to an improvement of their treatment of defects:

- A neutralizing background is introduced to calculate defects which are charged with respect to the host lattice. With this background the supercell becomes neutral and the electrostatic summation can be carried out.
- The total energy obtained from the periodic model calculation includes the interaction of the defect with its images. Providing that the interaction between the images is well reproduced by low-order multipoles, it is possible to extract the energy associated with an isolated defect as suggested by Makov and Payne for cubic supercells [39] and generalized for arbitrary supercells by Kantorovich [40].

Finally, studying excited states in the periodic model represents a difficult problem. On one hand, excited states are often much more delocalized than the system ground state and hence require much bigger cells for reliable predictions. On the other hand, computational techniques for calculating excited states are developed in the periodic model to a lesser extent than those available for molecules and in a cluster model.

To summarise, the periodic model is ideal if one is interested in the ground state properties of neutral defects, which weakly perturb the surrounding lattice.

2.2 Density Functional Approach

The Density Functional Approach is based on two theorems by Kohenberg and Kohn which state that: i) the ground state energy of a non-degenerate electronic state is a unique functional (density functional) of its density, $E[\rho]$. ii) the energy can be obtained by variation of the universal density functional with respect to the charge density. This implies that calculation of the wave-function of the many-electron system is not required in order to find the total energy, only the charge density is required. However, the exact density functional is not known and some approximate functionals are used instead.

The functional E can be divided into terms which describe the kinetic energy of electrons, their interaction with the external potential, their electrostatic interaction with each other and, finally, the term which accounts for the exchange and electron correlation, E_{xc}. It is this latter term that is not known and needs to be approximated. A widely used Local Density Approximation (LDA) assumes that the exchange-correlation energy E_{xc} for each element $\rho(r)dr$ is the same as for the uniform electron gas of density $\rho = \rho(r)$. This is very convenient because the uniform electron gas remains the only system for which E_{xc} can be calculated exactly.

Although the LDA is clearly a very strong approximation, it has proved to be very successful in predicting the geometric structure and electronic properties of many materials. Its obvious deficiency is that it neglects the effect of fast-changing electron density on the exchange-correlation energy, E_{xc}. This results in systematically underestimated equilibrium inter-atomic distances predicted by this method. To account for this effect functionals were developed which also include the dependence on the gradient of density. This scheme is known as the Generalised Gradient Approximation (GGA).

Some of the disadvantages of DFT relevant to further discussion include systematically under-estimated single particle band gaps and an overall tendency to delocalise the electron density. This is particularly relevant to the calculations of defects in insulators because too narrow band gaps make it difficult or impossible to describe shallow defect states. Similarly, the tendency to delocalize the electron density can, and often does, lead to qualitatively wrong conclusions regarding defect electronic structure (see, for example, discussion in refs. [41–43]). Addressing this issue remains a strong focus of methods research in materials science, and several possibilities exist. The so-called GW perturbation theory [44] offers highly accurate gaps in semiconductors, both quantum Monte Carlo (QMC) [45] and time-dependent DFT [46] give improved gaps at the cost of greatly increased computational resources.

Some of the problems discussed above can also be eliminated if one combines the advantages of DFT and exact exchange from Hartree-Fock in the so-called hybrid density functional approach. Perhaps the most widely used hybrid functional is due to

Becke [47] and is called B3LYP [17]. It has demonstrated great improvements in many systems including MgO, alumina and TiO_2 [48]. In this functional, the correlation term in the E_{xc} is the one suggested by Lee, Yang and Parr [49] (hence LYP), and the exchange term is the sum of three contributions including the Hartree-Fock exchange weighed with some parameters (this is what the number 3 in B3LYP stands for). The parameters were fitted to reproduce the ionization potentials, electron affinities of atoms and dissociation energies of a large number of molecules. An accumulated experience demonstrates that the B3LYP parameterization is good at reproducing both atomic and electronic properties of many ionic materials, and is widely used in both periodic and cluster calculations. However, the fact that B3LYP includes to some extent the HF exchange makes it a non-local functional and it is therefore implemented only in computer codes using localized basis sets.

One problem which cannot be fully corrected with any standard functionals discussed here, is the incorrect representation of the van der Waals interaction in DFT. In practice, LDA will give reasonable agreement with experiment for the interlayer distance in graphite, but this is due to chance rather than evidence of a good physical representation. Detailed analysis of the decay of the interaction shows that it does not reproduce the true van der Waals, and accurately including this requires more advanced methods [50].

2.2.1 Plane-wave Basis Set

The basis set controls the accuracy of calculations for a given method and its boundary conditions. The plane wave (PW) basis can only be used in conjunction with periodic boundary conditions. It is very effective in practical calculations because: i) it is an orthogonal basis, and ii) a product of two plane-waves is also a plane-wave and their derivatives are products in k-space. Due to these PW properties and very effective computational methods exploiting the periodicity of the system, the number of atoms that can be included in a supercell is generally a few times larger than the number of atoms in cluster calculations for comparable accuracy. The PW basis is independent of atomic positions and spans the whole supercell uniformly. For the given size of the supercell, it can be systematically saturated by increasing a single parameter - the cutoff energy, E_c. The value of E_c is determined by the highest density curvature. This implies that the description of the core electrons requires inaccessibly large values of E_c. To overcome this problem the PW basis is used together with pseudopotentials representing atomic cores (see, for example, [51, 52]), or with the more elegant projector augmented-wave method (PAW) [53, 54]. On the other hand, regions of small and weakly changing electron density are described with excessive accuracy. This substantially increases the computational expense in calculations where there is a lot of free space, such as cage systems or surfaces.

2.2.2 Tight-binding Method

DFT-based and other first-principle *ab initio* methods have a high computational cost, which makes the use of such methods in practice impossible for tackling some problems, *e.g.*, formation of defects under irradiation or direct dynamical simulations of defect diffusion.

To address such issues, a number of computationally cheaper methods have been developed, such as empirical potential and tight-binding simulation techniques.

The main difference between these two methods is that the empirical potential energy is described by an analytical function of atom coordinates usually fitted to experimental data, whereas in the tight-binding method the energy is calculated by solving the Schrödinger equation for electrons in the field of atom cores, although the exact many-body Hamiltonian operator is replaced with a parameterized Hamiltonian matrix. The basis set usually is atomic-like so that it has the same symmetry properties as the atomic orbitals.

The drawback of the empirical approach is its low accuracy and transferability (the ability to describe systems different from those used for fitting the parameters). At the same time, although first-principles methods generally provide the best overall accuracy, the tight-binding approach is a reasonable compromise between the computational efficiency and the reliability of the model used.

Therefore, in addition to the PW DFT method, in our simulations we used a non-orthogonal self consistent charge tight-binding method [55, 56] in which the parameters of the Hamiltonian were derived from *ab initio* calculations (a second-order expansion of the Kohn-Sham total energy in DFT with respect to charge density fluctuations). Thus, no empirical parameter is present in the method and despite the approximations made this method retains the quantum-mechanical nature of bonding in materials, ensuring that the angular nature of the bonding is correctly described in far-from-equilibrium structures. Due to parameter fitting to the density functional results, this method, unlike other tight-binding schemes (where the parameters are chosen to describe equilibrium structures) describes the interaction of atoms even at small interatomic separations, *i.e.*, upon energetic collisions. This approach has been found to work well in modeling various systems [57] and the results are in agreement with those obtained by the first-principles methods [22].

3 Magnetic Properties of the Frenkel Pair in Carbon Systems

As discussed in the introduction, one of the most probable intrinsic defect present in carbon samples is the Frenkel pair. Since recombination of proximate Frenkel partners is very probable, in this section we consider the properties of isolated adatoms and vacancies in carbon systems - specifically a graphene sheet and various nanotubes.

The abundance of open space in nanotubes suggests that the interstitial atom can also be treated as an *adatom* adsorbed onto the nanotube surface [58, 35]. This is particularly the case for isolated single-walled nanotubes (SWNT) and interstitials in the inner cores of the tubes. However, the nature of the vacancies and interstitials in graphite and multiwalled nanotubes (MWNT) is complicated by the presence of nearby layers. If we consider graphite as an example, it is clear that the properties of vacancies and interstitials may be affected by the presence of the second layer. However, as discussed in the previous section, any treatment of interlayer processes with DFT ignores the fact that the van der Waals interaction between layers is incorrectly represented. Hence, it is probably better, and still qualitatively correct, to consider only a single graphene sheet as a good model of graphite.

3.1 Adatom on Graphene

The first *ab initio* calculations concerning the properties of an adatom on a single sheet of graphite (graphene) were performed by Mattila and coworkers [59] in the middle of 1990's. Using a 50-atom size sheet and LDA they concluded that the ground-state of an adatom is a bridge-like structure (the starting and endpoints of the migration paths in Fig. 2).

The development of computing techniques and computers alongside the renewed interest on the carbon materials made it tempting to recalculate the ground-state properties of the carbon atom on a sheet of graphite. Using GGA, we found a similar ground state bridge structure [21]. The equilibrium position of the adatom was found to be in a bridge-like structure, between two surface carbon atoms. This geometry is similar to previous local density approximation (LDA) calculations [59–61] on a similar surface. The perpendicular distance of the adatom to the graphite surface is 1.87 Å. The adsorption energy of the defect was found by moving the adatom far from the surface until total energy convergence, and once again fully relaxing the surface to find the difference in energy. Note that the ground state for the isolated carbon atom is a triplet state, and we use this as a reference for our adsorption energies (to compare with adsorption energies referenced to the singlet state, 1.26 eV should be added to the values).The adsorption energy was found to be 1.40 eV (1.37 eV for the 72 atom slab). This is similar to the 1.2 eV [60] and 1.78 eV [61] found in finite cluster LDA studies, but smaller, even when comparing with the singlet reference adsorption energy of 2.66 eV, than the 3.30 eV found in previous periodic LDA works [59]. We also considered adsorption of an adatom directly on top of a surface carbon and found this to be over 1.0 eV higher in energy than the bridge structure.

In contrast to previous studies, we used fully spin-polarized DFT in our calculations and found that the ground state for the adsorbed adatom has a magnetic moment of 0.45 μ_B. Earlier finite cluster LDA calculations [60] also found a magnetic ground state for an adatom adsorbed on graphite, but the authors' of that work considered only a triplet solution, rather than the full unrestricted spin solution. If we restrict our system to zero spin ground state only (S = 0), then the total energy is 35.5 meV higher than the magnetic configuration. Figure 1(a) shows clearly that the spin-polarized density occupies p orbitals of the adatom. The magnetic properties of the C adatom on graphene can be explained via a simple counting argument. Both the two bonded atoms on the surface, as well as the adatom, present a different hybridization: the surface atoms attached to the adatom have a sp^2–sp^3 hybridization while the adatom stays sp^2 like, as seen in the model of Fig. 1(b). Concerning the adatom, the counting of the four carbon electrons is as follows: two electrons participate in the covalent bond with the graphene carbons. From the two remaining electrons, one goes to the dangling sp^2 bond, and another is shared between the sp^2 bond and the pz orbital. This pz orbital is orthogonal to the surface orbitals due to symmetry and cannot form any bands, remaining localized

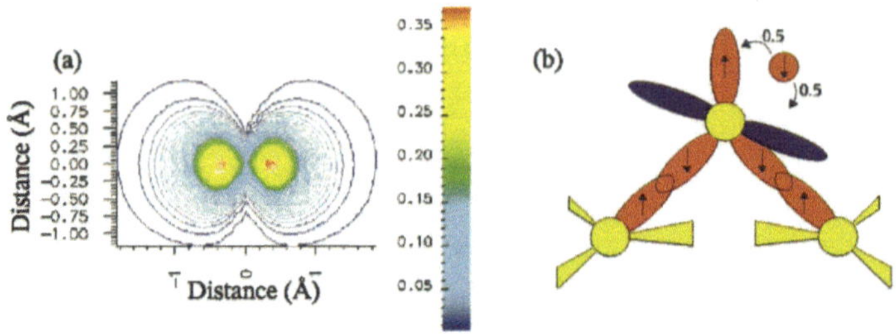

Fig. 1. (a) The spin density in e/Å^3 of plane normal to the surface through the center of the adatom when the adatom is at the equilibrium position. The adatom is at (0,0). (b) A schematic diagram of the bond orbitals at the equilibrium position in a plane through the adatom and the two surface carbons. Note that this schematic is a projection, and that the blue p-orbital is orthogonal to the adatom-surface bonds. *Reprinted with permission* [21].

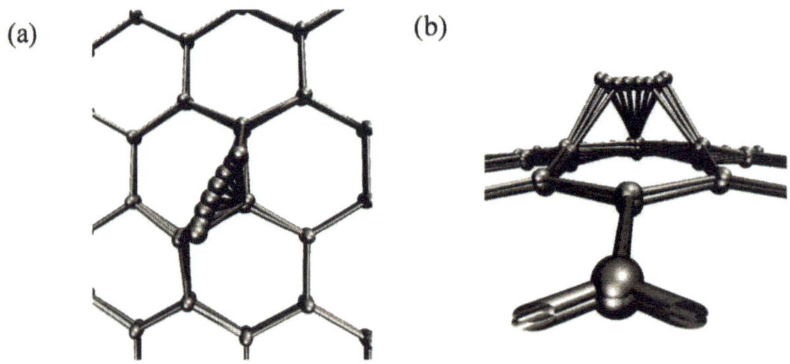

Fig. 2. Migration path of an adatom on graphene from (a) top and (b) side. The starting point of the path is the equilibrium position and the end point is another equilibrium position. They are both so-called bridge like structures. The middle points of the path where adatom has two bonds is actually a transition point from magnetic to non-magnetic situation. The reason is the change in the hybridisation of the adatom.

and therefore spin polarized. The dangling sp^2 bond will also probably be very slightly spin polarized, but this effect is negligibly small and cannot be seen in Fig. 1(a). The half electron of the pz orbital provides the magnetization of around 0.5 μ_B.

3.1.1 Migration of the Adatom

The study of the adatom migration path is crucial in determining the likely state of the defect at a given temperature. If the defect is highly mobile, then it is very likely to diffuse until pinned by interaction with another defect *e.g.* step-edge, or is annihilated by recombination *e.g.* adatom meets a vacancy.

The calculated adatom migration path between two equivalent bridge positions is almost a straight line (see Fig. 2). The path is calculated using the Nudged Elastic Band method [62]. During the migration between equilibrium positions the magnetic moment disappears because the adatom's hybridization changes from sp^2 to *sp*-like (only one bond attached to the surface) leaving one dangling bond, and the p_x and p_y orbitals free. The p-orbitals then interact with themselves, creating π-orbitals which interact with the remaining *sp*-orbital. The result is much more delocalized density and the magnetism is destroyed. The migration energy barrier is 0.47 eV (Fig. 3) (0.1 eV within the finite cluster LDA approximation [60]). These facts mean that the adatom is highly mobile on the plane at room temperature and the experimental observation will likely be very difficult without pinning by another defect (see for example Ref. [63]). A possible way to overcome this problem is to use a structure which has finite curvature, for example nanotubes (see subsection 3.3). The diffusion through the layers has an energy barrier of 2.3 eV [29] making in-plane motion of the adatom favorable, although as discussed previously we cannot exclude interlayer migration.

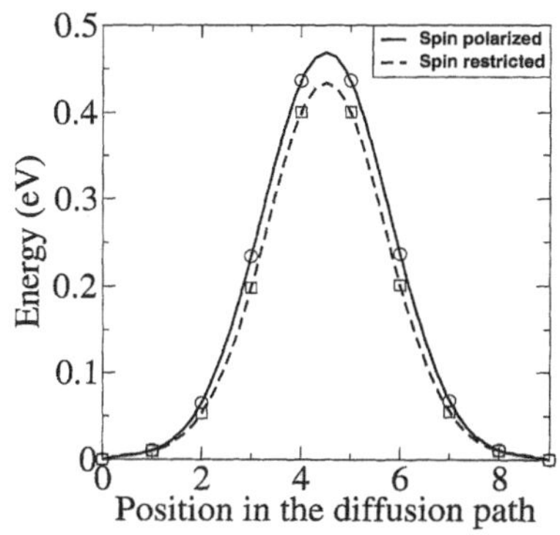

Fig. 3. Migration energy barrier for a carbon adatom on graphene [21].

Fig. 4. The equilibrium structure of a carbon dimer on graphene.

3.2 Clustering of Adatoms

Since we know that adatoms are very mobile on graphene, an obvious question is whether they will form (magnetic) clusters on the surface. The smallest possible cluster is obviously a carbon dimer - the ground state of the dimer on graphene is shown in Fig. 4. It has an adsorption energy of 8.7 eV. This system is not magnetic since the the dangling sp^2 bond of the bridge forming carbon ion forms a bond with the *sp*-orbital of the upper carbon ion. The p_x-orbital of the carbon ion at the bridge configuration interacts with the *p*-orbitals of the upper carbon ion becoming a π-orbital as in graphene. As for the adatom alone, the *sp*-type dangling bond does not have a magnetic moment.

Making a similar calculation with a carbon trimer results in a linear C-C-C equilibrium configuration parallel to the surface plane. Compared to the perfect graphene sheet and three isolated carbon atoms this structure has 12.2 eV lower energy, demonstrating that adatoms will cluster on the surface. In contrast to the dimer, the trimer is magnetic, but with a lower moment than the adatom (0.2 μ_B). Since C-C-C is composed of *sp*-bonds the conclusion is that the dangling sp-bonds interact first with the p-orbitals preventing the formation of π-orbitals. In general, these results suggest that all clusters with an even number of atoms will be non-magnetic, and all odd will be magnetic - but the moment decays rapidly with the size of the cluster. However, any quantitative conclusions would require an extensive study of cluster growth and include the effects of interlayer interactions.

3.3 Adatom on Nanotubes

Since the structure and bonding in carbon nanotubes is very similar to that of graphite, it is also interesting to explore whether adatoms are also magnetic on nanotubes. Carbon nanotubes were discovered by Iijima in 1991 [64] and since then the research on carbon systems has increased enormously. Nanotubes are tubular molecules made of pure carbon. They can be considered as graphene sheets rolled into a seamless cylinder - the ends can be either capped or open. Nanotubes are grown as either single walled, or as layered multiwalled tubes. The diameter of a nanotube can vary from a few to a few hundred Å while the length scale is from nanometres to centimetres [65].

The electronic structure of a nanotube depends on how the graphene sheet is rolled, *i.e.* by the so-called chiral vector which connects to crystallographically equivalent points on the underlying graphene sheet. The chiral vector can be expressed in terms of graphene unit vectors, that is designated by a pair of integers (n,m). Tubes with indices (n,0) are referred to as zigzag and (n,n) as armchair nanotubes. In a simple tight-binding picture, if n–m is a multiple of 3, the tube is metallic, otherwise it is semiconducting. This can be understood in terms of zone folding [66–69]. However, this description does not take into account the effect of curvature. Due to the curvature of the nanotube the π and σ orbitals hybridize and thus a small gap opens in zigzag nanotubes. This effect is strongest with nanotubes which have radius less than that of C_{60} [65].

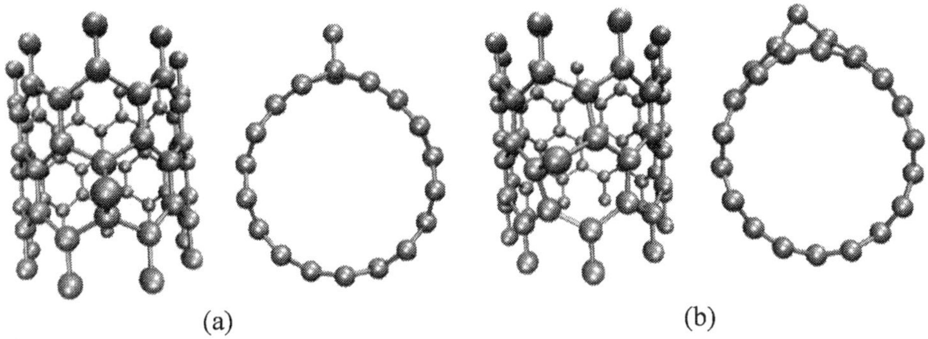

Fig. 5. Equilibrium positions of an adatom on a (9,0) nanotube in the (a) parallel and (b) perpendicular positions. *Reprinted with permission* [23].

In order to try to sample as wide as possible a family of nanotubes, here we will discuss adatom properties on nanotubes with extreme chiralities *i.e.* on zigzag and armchair tubes of different radii. The general ground state of an adatom on nanotubes is in a bridge-like structure as in graphite, but the bonds forming the bridge can be either parallel or perpendicular to the tube axis. Figures 5 and 6 show the equilibrium positions of an adatom on zigzag and armchair nanotubes. The parallel position is shown in Fig. 5(a) for a zigzag nanotube and 6(a) for an armchair nanotube (the angle between translational axis is less than 45°). The parallel position is a metastable position as can be seen in Table 1 [23]. The perpendicular bridge (or perpendicular position of the bridge) has an angle more than 45° with translational axis of the tube (Figs. 5{b) and 6(b)) and is the true ground state structure.

Table 1.
Data for the various nanotubes considered in this study [23].

Nanotube	Class	Radius (Å)	Adsorption Par.	Energy (eV) Perp.	Magnetic moment (μ_B) Par.	Perp
(8,0)	Semiconducting	3.13	2.37	2.89	0.01	0.23
(10,0)	Semiconducting	3.96	2.09	2.57	0.25	0.23
(11,0)	Semiconducting	4.41	2.03	2.49	0.20	0.22
(5,5)	Metallic	3.39	2.33	3.29	0.23	0.44
(9,0)	Semiconducting	3.57	2.35	2.80	0.24	0.35
(6,6)	Metallic	4.07	2.15	2.91	0.27	0.43
(12,0)	Semiconducting	4.97	2.04	2.50	0.32	0.36
Graphene	Metallic	∞	1.40	1.40	0.45	0.45

There are three factors to be considered when evaluating the magnetic moment of an adatom: the electronic structure of the tube, the location of the bridge bonds on the tube and the radius of the tube itself. If the nanotube on which the adatom is lying is

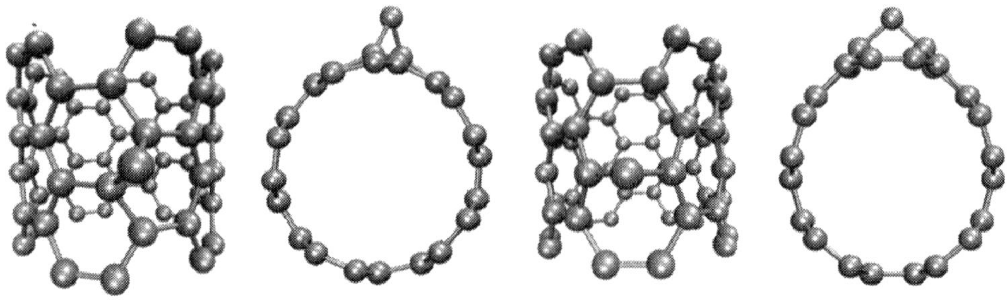

(a) (b)

Fig. 6. Equilibrium positions of an adatom on a (5,5) nanotube in the (a) parallel and (b) perpendicular positions. *Reprinted with permission* [23].

semiconducting, some of the adatoms charge goes to decreasing the band gap of the tube. If the gap is small enough the tube becomes metallic *e.g.* (9,0) and (12,0) nanotubes, but for larger gaps they remain semiconducting *e.g.* (8,0), (10,0) and (11,0). If the unhybridized *p*-orbital of the adatom is parallel to the translational axis of the tube (as in Fig. 6(b)) the situation is very close to graphene, since the curvature of the tube in this direction is the same as graphene's, namely infinite, and nearly the full graphene moment is recovered *e.g.* (5,5) and (6,6). When an adatom is on a zigzag nanotube at the parallel position (see Fig. 5(a)) the p-orbital is perpendicular to the translational axis of the tube. If the tube radius is small enough, as for example in (8,0), the *p*-orbital interacts with itself around the tube destroying the moment.

3.3.1 Migration of Adatoms on Nanotubes

Migration of adatoms on nanotubes is important because this issue is directly related to the stability of magnetic defects. Indeed, as we showed above, adatom clustering and annihilation with vacancies gives rise to a decrease or complete disappearance of the magnetic moment.

In order to study migration on nanotubes, we have used two complimentary approaches. First we use standard PW-DFT to calculate the adsorption energies for several different nanotubes, and to find the migration path and barrier for one tube. We then use this as a reference to check the accuracy of the computationally cheaper DFT-tight-binding method. Figure 7 shows the comparison of adsorption energies, and demonstrates good qualitative agreement - note that the PW-DFT results differ from the DFT-tight-binding (using LDA) by a constant which matches the difference between LDA and GGA in adsorption energies. Figure 8 also shows that the migration barrier is also in good agreement, hence we then used DFT-tight-binding to calculate the migration for a large number of tubes. Figure 8 shows the barrier as a function of radius for the armchair and zigzag tubes we considered. As expected the barrier decreases with increasing radius, and tends towards the limit of infinite radius *i.e.* graphene.

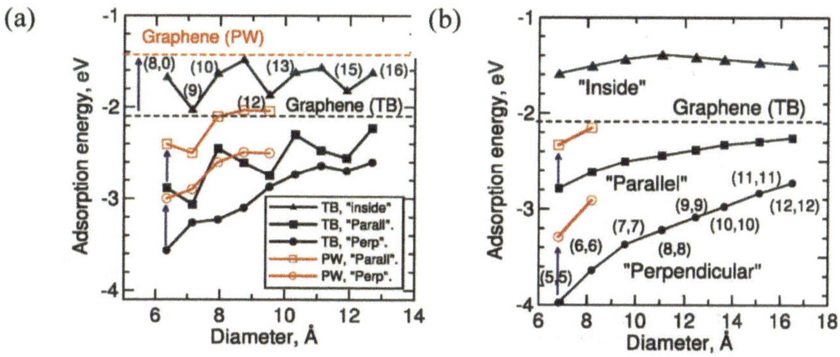

Fig. 7. Adsorption energies of carbon adatoms on zigzag (a) and armchair (b) single walled nanotubes as a function of nanotube diameters. The arrows visualize the relationships between the corresponding tight-binding and PW results. The numbers stand for the tube chirality indices. *Reprinted with permission* [22].

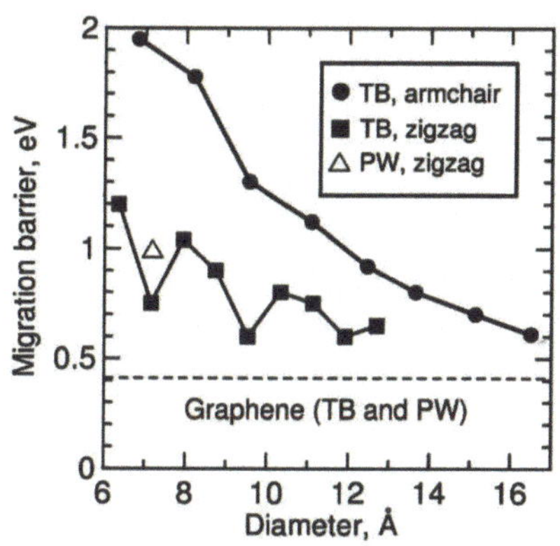

Fig. 8. Energy barrier for adatom migration on the outer surface on nanotubes as a function of nanotube diameters. Here the graphene migration barrier is the same for the PW and tight-binding calculations. *Reprinted with permission* [22].

3.4 Vacancy in Graphene

The adatom's partner in the Frenkel pair is the vacancy, hence we apply a similar treatment to study its properties. Initially we consider it in graphene, and then study the vacancy in a variety of different nanotubes.

The calculated formation energy of a vacancy in graphene is 7.7 eV [24], which is in agreement with previous DFT results of 7.6 eV [70] and 7.4 eV [71]. Atoms 1 and 2 (see Figure 9) form a bond creating a pentagon - the distance between atoms 1 and 2 is 2.02 Å compared to the 1.42 Å in ideal graphene. Atom 3 undergoes a Jahn-Teller displacement of about 0.18 Å out of the plane, and the resultant ground state structure is magnetic, with a magnetic moment of 1.04 μ_B. This displacement is smaller than the

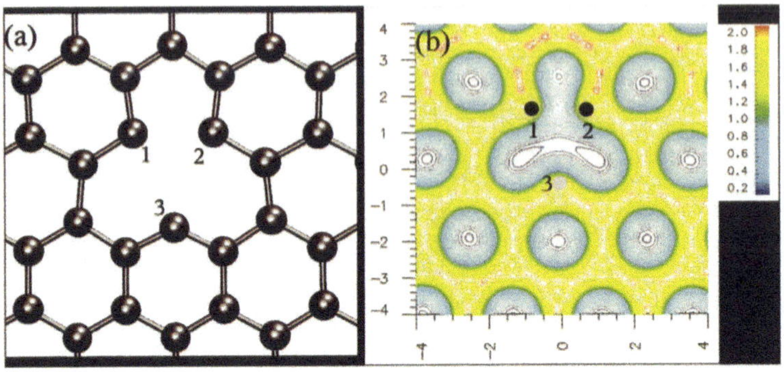

Fig. 9. (a) Atomic structure of the vacancy in the graphene plane. (b) Charge density of the vacancy in the graphene plane (e/Å^3). [24]

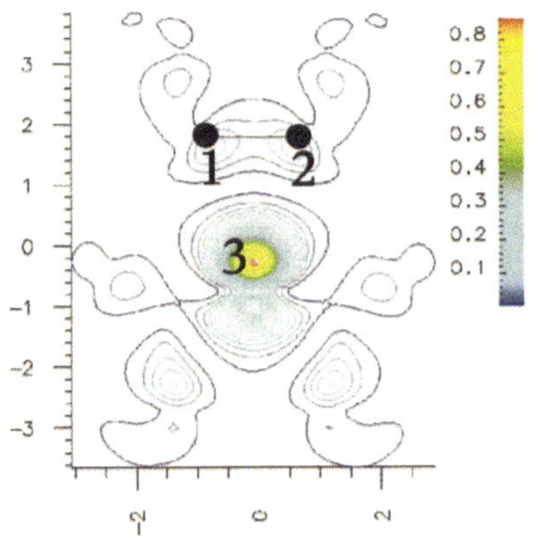

Fig. 10. Spin density of the vacancy in the graphene plane (e/Å^3). [24]

0.47 Å result of previous spin-restricted calculations [71], but if we set S = 0 we find a similar displacement of 0.46 Å. This suggests that any predicted inter-planar structures [72, 73] based on the vacancy should be re-examined with the inclusion of spin.

The explanation of the magnetism of the vacancy is as follows: the removal of one carbon ion creates three unsaturated *sp*^{2}-orbitals in the neighboring carbon ions. The formation of the pentagon saturates two of them which leaves one *sp*^{2}-orbital free. This remaining dangling bond contributes the calculated localized magnetic moment as can be seen in the Fig. 10.

3.4.1 Migration of a Vacancy in Graphene

Experimental results have long predicted that the barrier for vacancy migration is quite large, of the order of 3.1 eV [30]. However, theoretical efforts to study this process produce much smaller numbers. GGA-DFT and DFT-TB predict a barrier of

1.4 eV (see Fig. 11) in agreement with 1.7 eV in LDA-DFT [71], and other studies [74, 75]. In light of this, it seems likely that the experimentally measured value is actually due to the diffusion of divacancies or larger clusters, but this has yet to be confirmed. Inter-planar migration is much more difficult with migration barriers more than 5 eV for both interstitials and vacancies [30].

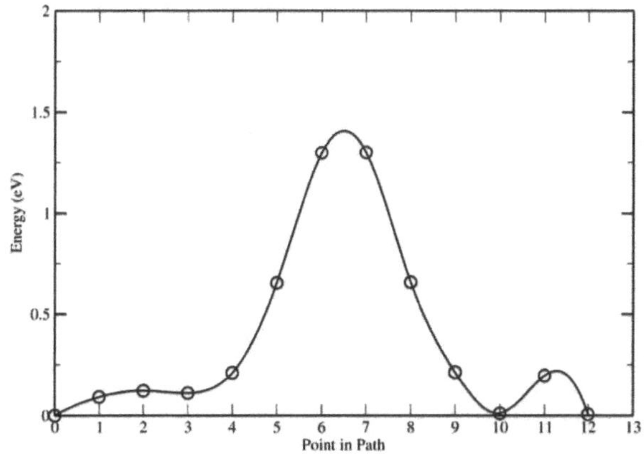

Fig. 11. The migration energy barrier of a vacancy in graphene.

The migration path calculated using GGA-DFT is shown in Fig. 12. A switching of bonds occurs between points 10-12 (see Fig. 11) which costs 0.2 eV, in agreement with calculations in Ref. [69]. Points #3, #4, #6, #7 and #9 do not carry net magnetic moment. The main difference between points #4 and #5, and on the other hand points #8 and #9, is the bond length of the pentagon bond. In the non-magnetic cases it is 2.00 Å (#4) or 1.98 Å (#9) while in the magnetic case the length of the pentagon bond is 1.87 Å (#5 and #8). The reason for this magnetic moment is the formation of a bond which seems to start with a spin-polarization of sp^2-orbital.

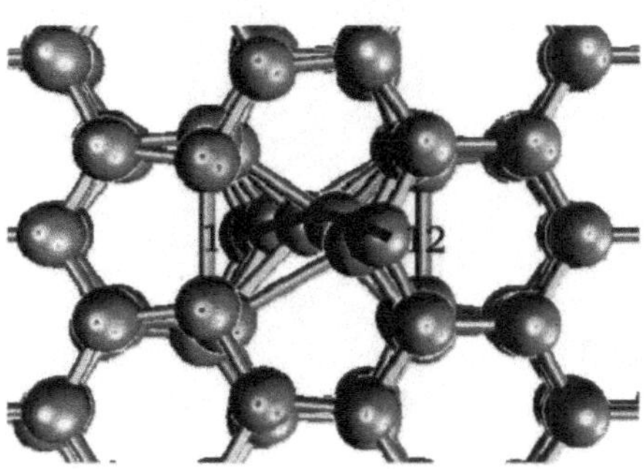

Fig. 12. Migration path of a vacancy on graphene.

3.5 Vacancy in Nanotubes

The description of magnetic properties of a vacancy in nanotubes is a much more difficult task. The properties of a vacancy in a nanotube are due to an interplay the following factors: first, nanotubes are cylinders, not flat planes, and not all carbon atoms are equivalent. Hence, the vacancy may have more than one configuration and all of them need to be checked in order to find the ground state. Secondly, the size of the tube changes when the chirality vector changes. This means that we cannot assume that a general vacancy ground state exists for all tubes regardless of chirality. Thirdly, in nanotubes the electronic structure varies in perfect zigzag nanotubes between semiconducting and metallic, and perfect armchair nanotubes are always metallic. Also, from our calculations on adatoms we know that the electronic structure changes easily when there are defects in the system. In order to study these issues we have considered a large variety of nanotubes, and the results are summarized in Tables 2 and 3.

The configuration shown in the Figure 13(c) has the following properties: each carbon atom is in the sp^2-hybridization. The atom with a dangling bond has a magnetic moment of 1 μ_B and creates a metallic band. This metallic band is created by the redistribution of the charge within the hexagonal carbon network of the nanotube. In this sense the behaviour is similar to the behaviour of the adatom on the nanotube. However, the magnetic properties of the nanotube with a vacancy depend much more on the curvature of the tube, as does the magnitude of the Jahn-Teller displacement lifting the dangling bond from the surface of the tube.

Figure 13(a) shows the ground state of the vacancies for (5,5) and (6,6) nanotubes. The (5,5) nanotube with a vacancy at the perpendicular configuration is magnetic because the Jahn-Teller distortion is large-enough to form the bridge structure (although not sufficient to get 1 μ_B). The relaxations of the rest of the tube are small enough so that the overall electronic structure remains metallic. For the (6,6) nanotube, the

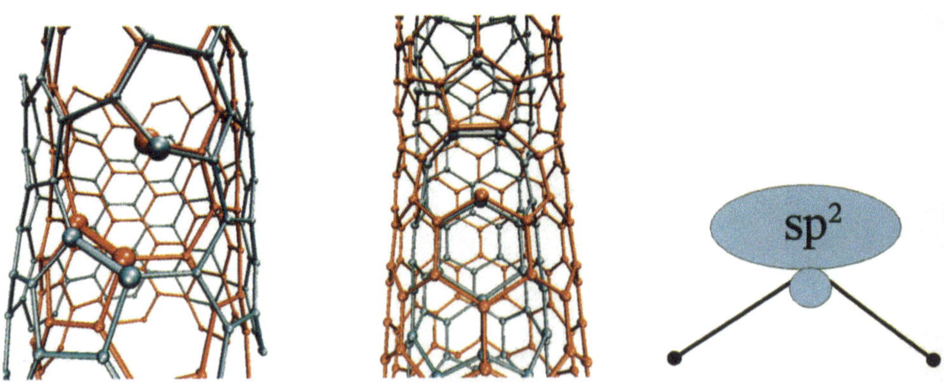

a) b) c)

Fig. 13. The ground state of vacancies in zigzag and armchair nanotubes: (a) (5,5) and (6,6) (opaque) armchair nanotubes, and (b) (6,0) (opaque) and (9,0) zigzag nanotubes. (c) Orbital "bridge" configuration which dominates in vacancy structures in nanotubes.

Table 2.
Properties of the 5-ring armchair SWNTs with vacancies of different configurations [24].

Nanotube	Configuration	E_{for} (eV)	Class	Mag. (μ_B)
(3, 3)	Perp.	4.4	Semi.	0.0
	Par.	5.2	Metal	1.0
(4, 4)	Perp.	5.3	Semi.	0.0
	Par.	6.2	Metal	1.0
(5, 5)	Perp.	5.6	Metal	0.6
	Par.	7.1	Semi.	0.0
(6, 6)	Perp.	5.9	Semi. (Metal[1])	0.0 (0.4[1])
	Par.	7.3	Semi.	0.0

[1]4-ring (6, 6) tube.

Table 3.
Properties of 6-ring zigzag SWNTs with vacancies of different configurations [22].

Nanotube	Configuration	E_{for} (eV)	Class	Mag. (μ_B)
	Perp.	4.6	Semi.	0.0
(5, 0)[1]	Par.	5.1	Semi.	0.0[2]
	3db	6.0	Semi.	0.0[2]
(6, 0)	Perp.	5.0	Metal	0.3
	Par.	5.8	Metal	0.9
(7, 0)	Perp.	5.2	Semi.	0.0
	Par.	6.3	Metal	0.8
(8, 0)	Perp.	5.3	Semi.	0.0
	Par.	6.5	Metal	0.8
(9, 0)	Perp.	5.4	Semi.	0.0
	Par.	6.4	Metal	1.0
	Perp.	5.5	Semi.	0.0
(10, 0)	Par.	6.7	Metal	0.9
	3db	7.4	Metal	1.9

[1]8-ring (5, 0) tube. [2]Antiferromagnetic.

Jahn-Teller effect is dependent on the linear density of the vacancies along the tube. It is not large enough for a magnetic groundstate if the density is less than 1 vacancy per 4 carbon rings. The smaller radii (3,3) and (4,4) armchair nanotubes demonstrate perfectly the bridge configuration, giving a magnetic moment of 1 μ_B in the perpendicular position.

The bridge configuration is also responsible for the magnetism and metallic electronic structure of the parallel vacancy in larger zigzag-tubes such as (7,0), (8,0),

(9,0) and (10,0) (see Fig. 13(b)). When the vacancy is in the perpendicular configuration these tubes remain semiconductors, as the removal of the ion increases the gap. The ideal (6,0) nanotube is a metal, and no matter the where the vacancy is located, it remains metallic and has a magnetic moment. (5,0) is a semiconductor due to extensive damage on the networks caused by the vacancy on a small tube, preventing the formation of a metallic band.

The question as to what happens when we have an open ended nanotube was studied in Ref. [13]. If the other end of a nanotube is capped with a fullerene the authors found that the edge atoms have a magnetic moment 1.25 μ_B per dangling bond. If a nanotube has both edges open, the spins of the dangling bonds are antiferromagnetically ordered. If the dangling bonds at one edge are saturated with hydrogen, the spins in the open edge contribute most of the total magnetic moment (1μ_B per atom). In the (8,0) nanotube the authors found a small polarization of p-orbitals at carbons attached to the hydrogen atoms at the edge. The magnetic state generated by vacancies, zigzag edges at nanotube ends and defective fullerenes are very similar, and point very clearly to the flatband mechanism of magnetism already studied in Refs. [10, 76].

4 Magnetism Stimulated by Non-magnetic Impurities

In addition to creating intrinsic defects in carbon materials, ion irradiation may also lead to doping of the sample by the impinging ions. Specifically, the irradiation of graphite by protons has been shown experimentally to induce a significant magnetic signal [19] (see also chapter 19). Since similar irradiation by helium ions produce a much weaker signal, it cannot be simply explained by the generation of vacancies and adatoms. Hence, here we consider the properties of hydrogen and helium in graphite via DFT-GGA simulations, and examine whether they play role in the observed magnetism.

The adsorption energy of H on perfect graphene is 0.87 eV (0.71 eV in Ref.[77], 0.76 eV in Ref. [78], 0.76 eV in Ref. [79] and 0.67 eV in Ref. [80]) and the adsorption position is above another carbon ion. This configuration has no magnetic moment unless the density of hydrogen on the surface is very high, *i.e.* approaching a few percent [79]. In any case, above a graphene sheet the hydrogen is quite mobile (barrier 1.30 eV for an isolated H on graphene, but reducing to 0.48 eV near other H atoms [78]) and does not form a dimer easily since the barrier for recombination is 2.82 eV [78]. Hence it is highly probable that hydrogen migrates on the plane until it is pinned by another defect.

If the hydrogen encounters an empty vacancy, then it saturates the dangling bond and is pinned at a height of 1.25 Å above the plane, with an adsorption energy of 4.36 eV. This configuration is non-magnetic. This result contrasts with studies of the effect of a hydrogen on a vacancy-like defect in fullerenes [17], where a magnetic moment of 3.0 μ_B per C_{60} cage was observed. However, the local structure in the fullerenes is different, and the under-coordinated carbons cannot saturate bonds with each other due to the increased strain in the system. Hence, hydrogen saturates only one dangling bond and the local moment is provided by the other two sites. A similar configuration can be considered in graphene, where hydrogen adsorbs directly into the vacancy, in-plane with the graphene sheet. The hydrogen atom saturates the dangling bond, but also forces the carbon atoms forming the base of pentagon in the vacancy to move away from each other, significantly weakening the bond between them. The resulting magnetic moment is 2.3 μ_B, and the spin density shows that this is mainly due to two new dangling bonds on the separated carbons. However, this configuration is metastable on a single graphene sheet, with an energy about 1.3 eV higher than the previous structure, and also a very small barrier between the two states. It is possible that the van der Waals interaction from other layers in graphite may stabilize this structure, or at least increase the barrier, but other configurations seem more likely.

Two stable magnetic configurations involving hydrogen do exist in graphite. Firstly, if an hydrogen atom encounters a vacancy which has already been saturated by hydrogen, it will bond to the other side of the vacancy (see Fig. 14(a)) with an adsorption energy of 3.2 eV, at a distance of 0.76 Å below the plane - the original H moves to 0.89 Å above the plane. This configuration has a magnetic moment of 1.2 μ_B localized on the dangling sp^2-bond (see Fig. 14(b)). Addition of a third hydrogen completes the decoration of the vacancy edges, saturates the remaining dangling bond and thus destroys the magnetism of the vacancy. Adsorption energy to a system already occupied by two hydrogens is 4.0 eV.

Secondly, if an hydrogen atom encounters a carbon adatom, they will pin each other (see Fig. 15(a)). The C-H bond is tilted due to interaction with the second graphene layer, and H is 2.1 Å above the plane. The adsorption energy of this complex is 3.8 eV. The resulting C-H group has a magnetic moment of 0.9 μ_B (see Fig. 15(b)).

The results for helium irradiation can be similarly understood by calculating the adsorption of He on graphite. DFT predicts that helium has a ground-state adsorption well above graphene sheet, and the interaction between graphite and helium is very weak. This state is non-magnetic. In principle, He will disrupt the magnetism of a vacancy, but the energetic cost to introduce He to a vacancy is very high and it is clear that its main role is in generating defects. Since only a very small magnetic signal is observed in experiments [19], it is reasonable to assume that nearly all vacancies are annihilated by recombination with mobile adatoms.

Fig. 14. (a) Structure and (b) spin density (μ_B /Å^2) of a vacancy surrounded by two hydrogens.

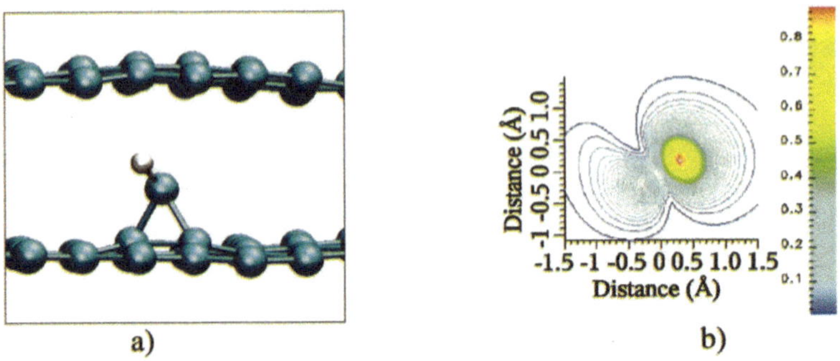

Fig. 15. (a) Structure and (b) spin density (μ_B /Å^2) of a CH group adsorbed between two layers of graphene.

If we assume that every hydrogen atom introduced by proton irradiation eventually will saturate a carbon dangling bond, then we can predict a rough estimate of the measurable magnetic signal. Two hydrogen atoms at a vacancy result in a moment of 1.2 μ_B, and an adatom-hydrogen group provides a moment of 0.9 μ_B from the C-H itself, and 1.0 μ_B from the uncompensated vacancy. Hence, we can consider each hydrogen as providing an average moment of 1.25 μ_B. For an experimental dose of 3 μC (cf. Fig. 2 in Ref. [19]) of protons we get a signal of 0.2 μemu which is in an agreement with the experimental signal 0.3±0.2 μemu and with higher dose of 10 μC the predicted signal is 0.8 μemu in agreement with 1 μemu. Obviously, it is very difficult to specify exactly the ratio of different defects that would actually be present in the irradiated material, but the agreement with the experimental magnetic signal strongly suggests that the H-vacancy complex plays a dominant role.

Demonstration of magnetic ordering of the H-vacancy complexes due to the defect-defect interactions from the DFT calculations is beyond the capability of the methods used. However, for adsorbed hydrogen, there are experimental results which indicate that the coupling on graphite can be very long-ranged, even up to 25 lattice constants [81].

Thus, hydrogen contributes to the development of the macroscopic magnetic state in several different ways. Firstly, some defect configurations formed by vacancies and H atoms are magnetic. Secondly, hydrogen prevents complete recombination of Frenkel pairs, which should increase the number of intrinsic defects (without H). And finally, H atoms appear to favour the development of the long-range magnetic order.

In addition to H-induced magnetism, other non-magnetic elements can give rise to magnetism in graphitic, *i.e.* π-electron systems. For example, substitutional doping with N/B atoms has been long predicted to give rise to magnetic effects in the sp^2 systems [82]. It is also interesting to note that our calculations predict that doping of nanotubes with nitrogen greatly increases the propensity of point defects to form a magnetic ground state, due to the effective introduction of an extra electron to the system [83]. Nitrogen substituted into the lattice of graphene and nanotubes also acts as a trap for mobile adatoms, providing another avenue for producing localized spin polarized states.

5 Conclusions and Outlook

In the Introduction to the Chapter, we formulated the most important problems to be attacked before the complete understanding of the magnetism in carbon system is achieved. We further summarized the theoretical findings which shed light on two of the problems. Specifically, we addressed the local bonding geometry which gives rise to local magnetic moments and the role of non-magnetic impurity atoms like hydrogen. Our results, along with those obtained by other authors, provide strong evidence for defect-mediated mechanism of magnetism in carbon systems, although contributions from other mechanisms cannot be excluded in some systems, *e.g.*, carbon foams.

As for graphite, our simulations indicate that adatoms and vacancies on a graphene sheet provide a localized magnetic moment - about 0.5 μ_B and 1.0 μ_B respectively. For a Frenkel pair concentration of 2 ppm, this would result in a magnetic signal of about 10^{-3} emu/g, the experimental value reported in [6]. Correspondingly, the defect concentration should be increased by a factor of about five hundred to explain the recently reported magnetic signal of 0.58 emu/g [4]. It is worth mentioning that the authors of Ref. [4], who explained magnetism in terms of defects (without specifying the defect types), estimated a similar concentration. In general such relatively high defect densities can be reached by irradiation [30] or chemical treatment [4].

In practice, however, the high mobility of adatoms on graphene at room temperature would suggest that many of them recombine with vacancies or cluster together, destroying their magnetism. Despite the indications that a barrier to vacancy-interstitial pair recombination exists [63], efficient recombination seems to be confirmed by He-irradiation experiments [19]. We know a large amount of point defects were produced by He ions, yet the magnetic signal was small. Moreover, as we pointed out in the introduction, not all defects with a localized magnetic moment will participate in the formation of the ferromagnetic state, and may result just in paramagnetism. These factors suggest that, at least for graphite, intrinsic carbon defects, such as interstitials and vacancies cannot explain the observed magnetism.

The demonstration of induced ferromagnetism by proton irradiation on graphite [18] indicates a promising direction for creating magnetic carbon system in a controllable way. Simulations show that this is due to a combination of hydrogen trapping at vacancies and pinning of mobile adatoms, producing magnetic C-H complexes and uncompensated vacancies. Hydrogen contributes to the development of the macroscopic magnetic state in several different ways. In addition to magnetic defect configurations formed by vacancies and H atoms, hydrogen prevents complete recombination of Frenkel pairs, which should increase the number of defects without H. Also, H atoms appear to favour the development of long-range magnetic order.

For nanotubes the situation is even more complicated than in graphite. The surface curvature provides an increase in the adatom migration barrier and, hence, should increase the probability of stable magnetic defects, but the dependence of nanotube electronic structure on chirality affects the picture. Only for adatoms on armchair nanotubes is the defect picture similar to graphene, and both adatoms and vacancies on many other nanotubes are non-magnetic. Some configurations provide a delocalized magnetic band, so, in principle, these would avoid issues of paramagnetism - but controlled production of nanotubes with a specific chirality is currently not possible.

As for fullerene systems, the differences in atomic structures from graphite and nanotubes mean that point defect like vacancies [11] and/or defected fullerenes (open-cage structures) [13, 17] can possibly exist even at room and higher temperatures, and the concentration of defected fullerenes can be high enough to provide agreement with the experimental results. Hydrogen may be also important for the development of the ferromagnetic state in polymerized fullerenes [17].

To sum up, we believe that magnetism in various carbon systems may be associated with defects. Hydrogen may also be very important as it suppresses defect annealing and annihilation. However, the types of defects resulting in the magnetic state in a particular material and the role of hydrogen and other non-magnetic impurities should be studied in more detail, for example, by irradiating the samples with energetic particles. To some extent, this has been already done in experimental work discussed previously [19]. However, systematic studies on irradiation of all-carbon systems with particles of different types and energies should give much more insight into the origin of magnetism. The irradiation temperature is also very important. For example, if carbon interstitials and vacancies can indeed contribute to magnetism then irradiation with electrons or inert gas ions at low (liquid helium or nitrogen) temperature followed by *in situ* magnetic measurements should detect a magnetic signal from the samples. By changing the temperature and thus controlling annealing of defects in the system one can, in principle, correlate magnetism to specific defect types, as different defects become mobile at different temperatures. Irradiation can also be used to introduce non-magnetic impurities like H or N/B atoms into the system, and thus aid in understanding the role of these impurities in the formation of the magnetic state.

References

1. Y. Kopelevich, P. Esquinazi, J.H.S. Torres, and S. Moehlecke, J. Low Temp. Phys. 119, (2000) 691.
2. T.L. Makarova, B. Sundqvist, R. Höhne, P. Esquinazi, Y. Kopelevich, P. Scharff, V.A. Davydov, L.S. Kashevarova, and A.V. Rakhmanina, Nature 716 (2001) 413
3. A.V. Rode, *E.G.* Gamaly, A.G. Christy, J.G. Fitz Gerald, S.T. Hyde, R.G. Elliman, B. Luther-Davies, A.I. Veinger, J. Androulakis, and J. Giapintzakis, Phys. Rev. B70, (2004) 054407.
4. A.W. Mombrú, H. Pardo, R. Faccio, O.F. de Lima, E.R. Leite, G. Zanelatto, A.J.C. Lanfredi, C.A. Cardoso, and F.M. Araújo-Moreira, Phys. Rev. B71 (2005) 100404.
5. T.L. Makarova. Magnetism of carbon-based materials. In A. Narlikar, editor, Studies of High-Temperature Superconductivity, volume 45, chapter 7, page 107. Nova Science, Huntington, New York, (2003).
6. P. Esquinazi, A. Setzer, R. Höhne, C. Semmelhack, Y. Kopelevich, D. Spemann, T. Butz, B. Kohlstrunk, and M. Lösche, Phys. Rev. B66 (2002) 024429.
7. T.L. Makarova. J. Magn. Magn. Mat., 272-276 (2004) E11263.
8. M. Fujita, K. Wakabayashi, K. Nakada, and K. Kusakabe, J. Phys. Soc. Jpn. 65 (1996) 1920.
9. Y. Shibayama, H. Sato, T. Enoki, and M. Endo, Phys. Rev. Lett. 84 (2000) 1744.
10. K. Nakada, M. Fujita, G. Dresselhaus, and M.S. Dresselhaus, Phys. Rev. B54 (1996) 17954.

11. A.N. Andriotis, M. Menon, R.M. Sheetz, and L. Chernozatonskii, Phys. Rev. Lett. 90 (2003) 026801.
12. K. Kusakabe and M. Maruyama, Phys. Rev. B67 (2003) 092406.
13. Y.-H. Kim, J. Choi, K. J. Chang, and D. Tománek, Phys. Rev. B68 (2003) 125420.
14. D.V. Khveshchenko, Phys. Rev. Lett. 87 (2001) 206401.
15. D.V. Khveshchenko, Phys. Rev. Lett. 87 (2001) 246802.
16. N. Park, M. Yoon, S. Berber, J. Ihm, E. Osawa, and D. Tománek, Phys. Rev. Lett. 91 (2003) 237204.
17. J.A. Chan, B. Montanari, J.D. Gale, S.M. Bennington, J.W. Taylor, and N.M. Harrison, Phys. Rev. B70 (2004) 041403.
18. S. Okada and A. Oshiyama, J. Phys. Soc. Jpn. 72 (2003) 1510.
19. P. Esquinazi, D. Spearmann, R. Höhne, A. Setzer, and T. Butz, Phys. Rev. Lett. 91 (2003) 227201.
20. K.H. Han, D. Spemann, P. Esquinazi, R. Höhne, V. Riede, and T. Butz. Adv. Mater, 15 (2003) 1719.
21. P.O. Lehtinen, A.S. Foster, A. Ayuela, A. Krasheninnikov, K. Nordlund, and R.M. Nieminen, Phys. Rev. Lett. 91 (2003) 017202.
22. A. V. Krasheninnikov, K. Nordlund, P.O. Lehtinen, A.S. Foster, A. Ayuela, and R.M. Nieminen, Phys. Rev. B69 (2004) 073402.
23. P.O. Lehtinen, A.S. Foster, A. Ayuela, T.T. Vehviläinen, and R.M. Nieminen. Phys. Rev. B69 (2004) 155422.
24. Yuchen Ma, P.O. Lehtinen, A.S. Foster, and R.M. Nieminen, N. J. Phys. 6 (2004) 68.
25. N.W. Ashcroft and N.D. Mermin, Solid State Physics, Thomson Learning, 1976.
26. A.J. Stone and D.J. Wales, Chem. Phys. Lett. 128 (1986) 501.
27. C.P. Ewels, M.I. Heggie, and P.R. Briddon, Chem. Phys. Lett. 351 (2002) 178.
28. J.F. Ziegler, J.P. Biersack, and U. Littmark, The Stopping and Range of Ions in Matter. Pergamon, New York, 1985.
29. F. Banhart, Rep. Prog. Phys. 62 (1999) 1181.
30. P.A. Thrower and R.M. Mayer, Phys. Stat. Sol.(a) 47 (1978) 11.
31. A.V. Krasheninnikov and K. Nordlund, Nucl. Instr. and Meth. in Phys. Res. B216 (2004) 355.
32. P. Partyka, Y. Zhong, K. Nordlund, R.S. Averback, I.K. Robinson, and P. Ehrhart, Phys. Rev. B64 (2002) 235207.
33. T. Diaz de la Rubia, R.S. Averback, R. Benedek, and W.E. King, Phys. Rev. Lett. 59 (1987) 1930. See also erratum Phys. Rev. Lett. 60 (1988) 76.
34. P.M. Ajayan, V. Ravikumar, and J.-C. Charlier, Phys. Rev. Lett. 81 (1998) 1437.
35. A.V. Krasheninnikov, K. Nordlund, and J. Keinonen, Phys. Rev. B65 (2002) 165423.
36. G. Kresse and J. Furthmüller, Comp. Mat. Sci. 6 (1996) 15.
37. G. Kresse and J. Furthmüller, Phys. Rev. B54 (1996) 11169.
38. J.P. Perdew, J.A. Chevary, S.H. Vosko, K.A. Jackson, M.R. Pederson, D.J. Singh, and C. Fiolhais, Phys. Rev. B46 (1992) 6671.
39. G. Makov and M.C. Payne, Phys. Rev. B51 (1995) 4014.
40. L.N. Kantorovich, Phys. Rev. B60 (1999) 15476.
41. J. Laegsgaard and K. Stokbro, Phys. Rev. Lett. 86 (2001) 2834.
42. J.L. Gavartin, P.V. Sushko, and A.L. Shluger, Phys. Rev. B67 (2003) 035108.
43. G. Pacchioni, F. Frigoli, D. Ricci, and J.A. Weil, Phys. Rev. B63 (2001) 054102.
44. L. Hedin. Phys. Rev. A139 (1965) 796.
45. D. Cepperley, G. Chester, and M. Kalos, Phys. Rev. B16 (1971) 3081.
46. M. Petersilka, U.J. Grossman, and E.K.U. Gross, Phys. Rev. Lett. 76 (1996) 1212.
47. A.D. Becke, J. Chem. Phys. 98 (1993) 5648.

48. J. Muscat, A. Wander, and N. M. Harrison, Chem. Phys. Lett. 342 (2001) 397.
49. C. Lee, W. Yang, and R. G. Parr, Phys. Rev. B37 (1988) 785.
50. H. Rydberg, M. Dion, N. Jacobson, E. Schröder, P. Hyldgaard, S.I. Simak, D.C. Langreth, and B.I. Lundqvist, Phys. Rev. Lett. 91 (2003) 126402.
51. D. Vanderbilt, Phys. Rev. B41 (1990) 7892.
52. G. Kresse and J. Hafner, J. Phys. Condens. Matter. 6 (1994) 8245.
53. G. Kresse and D. Joubert, Phys. Rev. B59 (1999) 1758.
54. P.E. Blöchl, Phys. Rev. B50 (1994) 17953.
55. D. Porezag, T. Frauenheim, T. Köhler, G. Seifert, and R. Kaschner, Phys. Rev. B51 (1995) 12947.
56. M. Elstner, D. Porezag, G. Jungnickel, J. Elsner, M. Haugk, T. Frauenheim, S. Suhai, and G. Seifert, Phys. Rev. B58 (1998) 7260.
57. T. Frauenheim, G. Seifert, M. Elstner, T. Niehaus, C. Köhler, M. Amkreutz, M. Sternberg, Z. Hajnal, A. Di Carlo, and S. Suhai, J. Phys. Condens. Matter. 14 (2002) 3015.
58. A.V. Krasheninnikov, K. Nordlund, M. Sirviö, E. Salonen, and J. Keinonen, Phys. Rev. B63 (2001) 245405.
59. J. Keinonen, K. Nordlund, and T. Mattila, Phys. Rev. Lett. 77 (1996) 699.
60. M. Heggie, Electrochemical Society Proceedings 98-8 (1997) 60.
61. Y.H. Lee, S.G. Kim, and D. Tománek, Phys. Rev. Lett. 78 (1997) 2393.
62. H. Jónsson, G. Mills, and G.K. Schenter, Surf. Sci. 324 (1995) 305.
63. A. Hashimoto, K. Suenaga, A. Gloter, K. Urita, and S. Iijima, Nature 420 (2004) 870.
64. S. Iijima, Nature 354 (1991) 56.
65. R. Saito, M. Fujita, G. Desselhaus, and M.S. Dresselhaus, Physical Properties of Carbon Nanotubes. Imperial College London Press, London, (1998).
66. N. Hamada, S.I. Sawada, and A. Oshiyama, Phys. Rev. Lett. 68 (1992) 1579.
67. J.W. Mintmire, B.I. Dunlap, and C.T. White, Phys. Rev. Lett. 68 (1992) 631.
68. M.S. Dresselhaus, G. Dresselhaus, and R. Saito, Phys. Rev. B45 (1992) 6234.
69. O. Gülseren, T. Yildirim, and S. Ciraci, Phys. Rev. B65 (1992) 153405.
70. E. Kaxiras and K.C. Pandey, Phys. Rev. Lett. 72 (1988) 1878.
71. A.A. El-Barbary, R.H. Telling, C.P. Ewels, M.I. Heggie, and P.R. Briddon, Phys. Rev. B68 (2003) 144107.
72. R.H. Telling, C.P. Ewels, A.A. El-Barbary, and M.I. Heggie, Nature Materials 2 (2003) 333.
73. C.P. Ewels, R.H. Telling, A.A. El-Barbary, M.I. Heggie, and P.R. Briddon, Phys. Rev. Lett. 91 (2003) 025505.
74. E. Kaxiras and K.C. Pandey, Phys. Rev. Lett. 61 (1988) 2693.
75. C.H. Xu, C.Z. Wang, C.T. Chan, and K.M. Ho, Phys. Rev. B47 (1993) 9878.
76. S. Okada and A. Oshiyama, Phys. Rev. Lett. 87 (2001) 146803.
77. Y. Ferro, F. Marrinelli, and A. Allouche, J. Chem. Phys. 116 (2002) 8124.
78. Y. Ferro, F. Marrinelli, and A. Allouche, J. Chem. Phys. 368 (2003) 609.
79. E.J. Duplock, M. Scheffler, and P.J.D. Lindan, Phys. Rev. Lett. 92 (2004) 225502.
80. X. Sha and B. Jackson, Surf. Sci. 496 (2002) 318.
81. P. Ruffieux, O. Gröning, P. Schwaller, L. Schlapbach, and P. Gröning. Phys. Rev. Lett. 84 (2000) 4910.
82. I. Hagiri, N. Takahashi, and K. Takeda, J. Phys. Chem. A 108 (1994) 2290.
83. Yuchen Ma, A.S. Foster, A.V. Krasheninnikov, and R.M. Nieminen, J. Phys Cond. Matter (submitted), (2005).

Carbon-Based Magnetism
T. Makarova and F. Palacio (Editors)

17

Unconventional Magnetic Properties of Nanographite

T. Enoki and K. Takai

Department of Chemistry, Tokyo Institute of Technology, Ookayama, Meguro-ku, Tokyo 152-8551 Japan

1 Introduction

The discoveries of fullerenes and carbon nanotubes have contributed to enriching the variety in the world of carbons, where the new comers have opened a new realm bridging between traditional carbon and nano-materials. Ball-shaped fullerenes and cylindrical-shaped carbon nanotubes form closed π-electron systems with the participation of pentagon rings or occasionally heptagon rings. Here, the electronic structure depends on their shapes, sizes and helicities. In such a nano-carbon family, nanographene is involved, which is featured with nano-sized two-dimensional (2D) flat hexagon ring network with open edges, in contrast to its counterparts having closed π-electron systems such as fullerenes and carbon nanotubes. According to theoretical suggestions [1–3], the presence of edges adds an extra-electronic state to nanographene depending on the shape of the edges. Namely, edge-inherited non-bonding π-state (edge state) appears around the Fermi level in addition to the bonding π- and antibonding π*-states, when nanographene has zigzag edges. Taking into account that the electronic properties are subjected mainly to the electronic states around the Fermi level, we can access that the presence of the edge states makes the electronic properties entirely different from those of bulk graphite. In addition, from the magnetism aspects, the magnetic moments created in the edge states bring about unconventional carbon-only magnetism having features different from traditional magnets. In these two decades, the magnetism of carbon-based materials has attracted much interests as one of the most important issues in materials science. In recent works, nanographene has contributed strongly to understanding how carbon-based materials work in the magnetism. In the future perspectives, processability of nanograpene on the basis of lithography technique and chemical modifications is expected to make it promising as magnetic molecular device applications.

The issues, which are discussed in the present paper, concern the sstructure, the electronic and magnetic properties of nanographene and nanographite, the latter of which is a stacked assembly of nanographene sheets. As target materials of nanographene and nanogarphite, we focus on activated carbon fibers (ACFs), nanographite/nanographene obtained by annealing nanodiamond and diamond-like carbon. ACFs are featured with a 3D disordered network of nanographite domains. In this paper, we present unconventional electronic and magnetic features of nanographite, and their possible applications to nano-devices.

2 What Theory Predicts on the Electronic and Magnetic Structure of Nanographene

Most of chemists know that the electronic structure of benzene is described with six π-orbitals, which are split into three occupied bonding π- and three unoccupied antibonding π^*-levels located on the opposite sides of the Fermi level E_F with the level splitting being finite between the highest occupied molecular orbital (HOMO) and the lowest unoccupied molecular orbital (LUMO). The condensed polycyclic aromatic hydrocarbons formed by the fusion of benzene rings also have similar electronic structures with the number of π-orbitals and the HOMO-LUMO level splitting increasing and decreasing, respectively, upon the increase of the number of benzene rings associated. In the extreme of an infinite size, graphene sheet or 2D graphite has a semimetallic electronic structure featured with the bonding π- and antibonding π^*-bands that are touched to each other at E_F with no HOMO-LUMO splitting. A group of these condensed polycyclic aromatic hydrocarbons therefore have common electronic properties based on these π-orbitals. However, among these, there is a sub-family particularly featured with magnetism. A typical example is phenalenyl-free radical consisting of three benzene rings fused as shown in Fig. 1(a), in which an unpaired electron having localized magnetic moment $S = ½$ exists as a consequence of the presence of a singly-occupied non-bonding π-electron state at E_F in addition to the

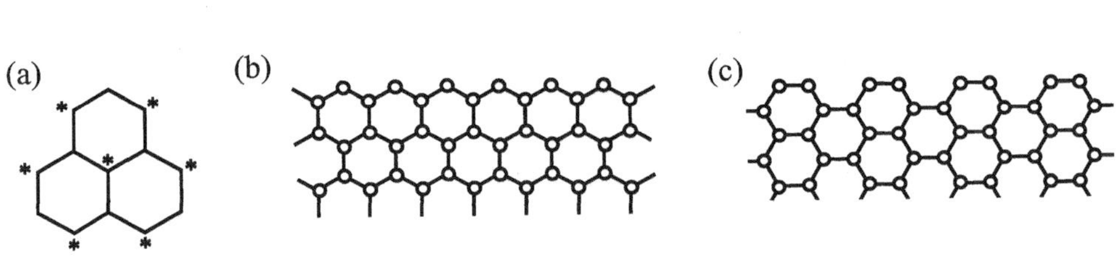

Fig. 1. (a) Phenalenyl-free radical where up and down spins are placed on starred and unstarred sites, (b) zigzag edge, (c) armchair edge.

π- and π*-levels. In general, we can group all the carbon sites of a condensed polycyclic aromatic hydrocarbon into two subgroups, where the neighboring sites directly bonded to a site belonging a subgroup ("starred" subgroup) belong to another subgroup ("unstarred" subgroup). According to Liebs theorem [4], the unpaired electrons are produced when the numbers of sites belonging to these two subgroups are different. Actually, the difference in the numbers corresponds to the number of unpaired electrons. In the case of phenalenyl-free radical, the difference gives one localized spin, resulting in carbon-based magnetism. The issue of stared and unstared is the same to that in bipartite lattice in physics language. The same theorem is also applicable to nano-sized graphene sheet or nanographene.

In addition to this, theory has suggested the magnetism of finite size graphene dependent particularly on its shape when the size of graphene is reduced to nano-dimension. In general, the circumference of a nanographene sheet having arbitrary shape can be described in terms of a combination of zigzag edges and armchair edges,

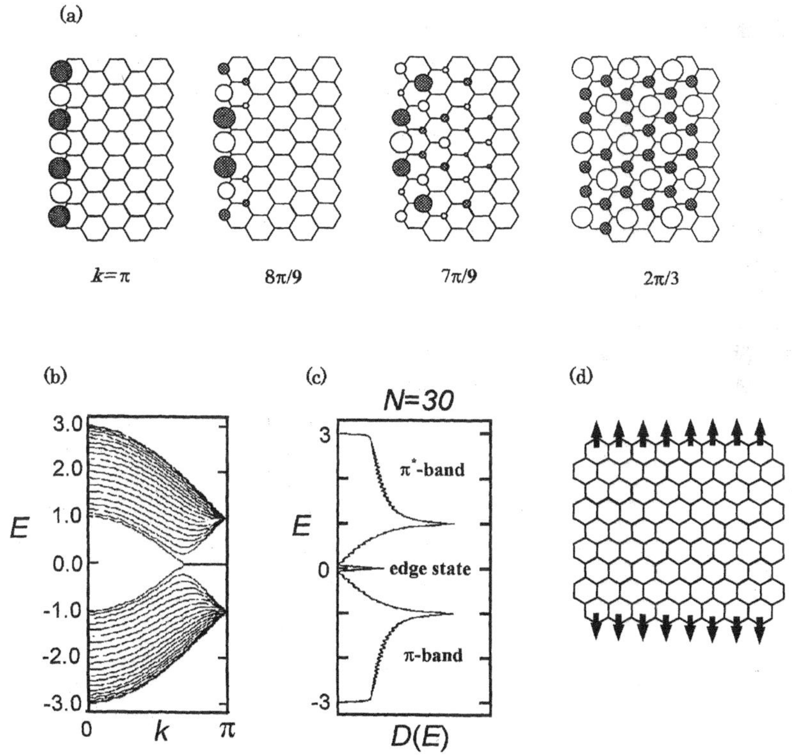

Fig. 2. (a) The wave number dependence of the real part of the wave function of the edge state, (b) the energy dispersions of a nanographene ribbon having zigzag edges with a width of 30 unit cells, (c) the density of states, and (d) ferromagnetic spin arrangement at the zigzag edges. All the edge carbon atoms are terminated with hydrogen atoms. The energy unit is eV in (b) and (d). (Ref.2)

which mimic the structures of trans- and cis-polyacet ylenes, respectively, as shown in Figs. 1(b) and 1(c). According to theoretical predictions [2, 3], zigzag edges give non-bonding π-electron state (edge state) of edge origin, whose energy level appears at the contact point of the π- and π^*-bands. For example, in nanographene ribbon having zigzag edges shown in Fig. 2, the edge state is localized around the zigzag edges in the wave number region from the Brillouin zone edge to $2\pi/3$ of the zone, while it becomes delocalized in the central region $0 \leqslant k \leqslant 2\pi/3$. The presence of the edge states, whose density of states has a sharp peak with its occupation being half-filled around E_F, gives rise to localized magnetic moments. Therefore, nanographene is of particular interest from magnetism viewpoint, where even ferromagnetism happens to be produced as exhibited in Fig. 2(d).

Modification of edges by substitution with foreign species is predicted theoretically to give a variation of magnetism [5–7]. When all the edge carbon atoms of zigzag edges on one side of a nanographene ribbon are dihydrogenated with those on the opposite side remaining to be monohydrogenated, a completely localized non-bonding state appears around E_F, where all the carbon atoms are spin polarized even in the interior of nanographene ribbon [6]. In contrast, fluorination of edges tends to reduce magnetism due to the tendency of forming closed shell in fluorine. An interesting example is oxidation of carbon atoms on one zigzag edge side, in which monohydrogenated edge works as magnetic edge whereas oxidized edge forms electron conduction paths. This means that the chemical modification can give different roles to these two edges.

These theoretical predictions therefore suggest that we can make a large variety of magnetic nanosystems, that are expected to contribute to developing a new area of spintronics on the basis of magnetism of nanographene.

3 Electronic and Magnetic Features of Nanographite and Networked Nanographites

Nanographene is prepared by heat-treating nano-diamond particles [8–10]. When nanodiamond particles of ca.5 nm in size are heat-treated at ca.1600°C, diamond/graphite conversion takes place, where the produced nanographite forms polyhedron with a hollow inside and facets consisting of a stacking of 3–7 graphene sheets having a mean in-plane size of ca.7 nm. A single nanographene sheet can also be prepared on a substrate when electrophoretic deposition is employed for depositing a diluted concentration of nanodiamond particles on the substrate. Figure 3 shows STM images of a single sheet of nanographene on a highly oriented pyrolytic graphite (HOPG) substrate, where the nanographene is prepared by a combination of electrophoretic deposition and heat-treatment [10]. A single sheet of nanographene prepared by this technique has typically an in-plane size of ca.10 nm and a nanographene-substrate distance of 0.35–0.37 nm. The nanographene-substrate distance is considerably larger than the inter-layer distance of bulk graphite (0.335 nm), suggesting that the inter-layer interaction is reduced by ca. 25–50% compared to that of bulk graphite. Therefore, the nanographene thus produced can be a good model system for clarifying the anomalous electronic structure predicted theoretically. Nanographene ribbons can be observed occasionally by chance around the edge steps of heat-treated

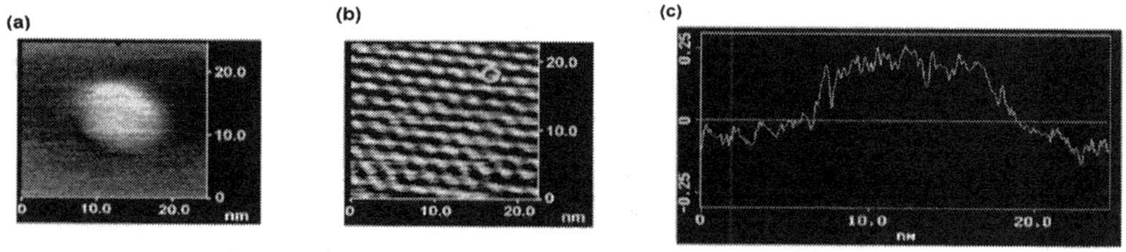

Fig. 3. STM image of a single layer of nanographene on a highly oriented pyrolytic graphite substrate. (a) large scale image, (b) lattice image, and (c) corrugation. A hexagon in (b) represents a benzene ring. (Ref. 10)

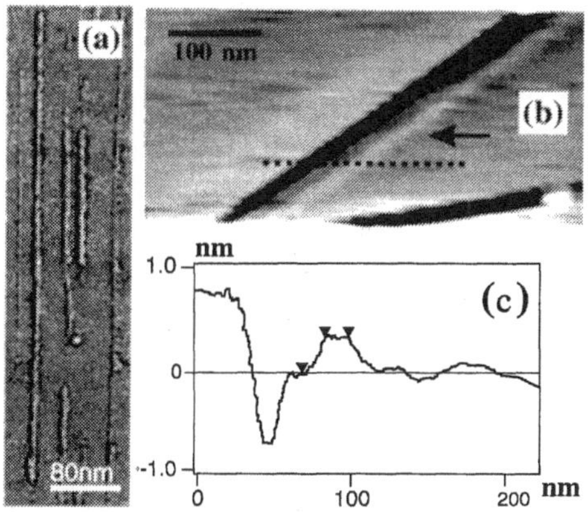

Fig. 4. (a) AFM image of nanographene ribbons, (b) a nanographene ribbon around a step edge, where the arrow denotes the position of the nanographene, (c) the cross sectional view along the dotted line indicated in (b). (Ref. 11)

HOPG surfaces. Figure 4 presents the AFM image of single sheets of nanographene ribbons, whose typical width and length are 8 nm and up to 1 μm, respectively [11]. The resonance Raman signal of the E_{2g2} band can work to characterize the structure and the electronic properties of individual nanographene ribbons independent from the HOPG substrate. A recent work on STM/STS (scanning tunneling spectroscopy) observations of nanographene sheets proves the presence of edge states around the zigzag edge which have a peak in the density of states around E_F [12].

The presence of edge state and its magnetic properties can be investigated by magnetic measurements. The nanographite samples obtained by annealing nanodiamond particles have Pauli paramagnetic susceptibility or spin paramagnetic susceptibility of π-electrons, whose value is 1–2 orders of magnitude larger than that of bulk graphite

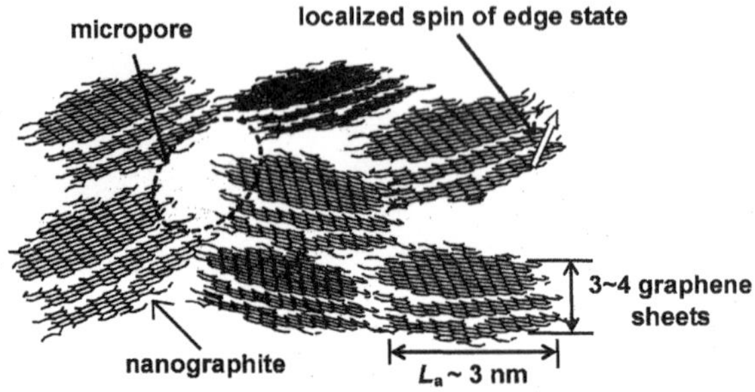

Fig. 5. Schematic model of activated carbon fibers (ACF), which is represented by a 3D disordered network of nanographite domains. Each nano-graphite consists of a stacking of 3–4 graphene sheets with an in-plane size of 2–3 nm. Nanographenes have localized spins of edge states as indicated by arrow.

[8, 13]. Taking into account that the Pauli susceptibility is proportional to the density of states at EF, we can point out that this is an important indication of the presence of non-bonding edge state around EF, suggesting that the electronic structure becomes essentially different from that of bulk graphite.

Activated carbon fibers (ACFs), whose structure is schematically exhibited in Fig. 5, offer another example of nanographite [14, 15], that is; ACFs consist of a 3D disordered network of nanographite domains, where each domain is featured with a stacking of 3–4 nanographene sheets having an in-plane size of 2–3 nm. The magnetic susceptibility χ of the pristine ACFs shows Cuire-Weiss-type behavior of localized spins as exhibited in Fig. 6(a),

$$\chi = \frac{C}{T - \Theta}, \tag{1}$$

where Curie constant C is proportional to the localized spin density and Weiss constant Θ represents the strength of exchange interaction acting between localized spins. The trend of the susceptibility evidences the presence of localized magnetic moments, which is consistent with theoretical prediction claiming the contribution of edge states [2, 3]. From the susceptibility data, the localized spin density is estimated at several spins per nanographite domain.

The electron transport process is well correlated to the magnetic feature. In the pristine ACFs, the peripheries of nanographite domains are covered with functional groups, which disturb the inter-domain electron transport. Indeed, in the electron transport, the Coulomb energy, which is generated between the hole remaining on a domain and the electron staying on the adjacent domain, works to create an energy gap (Coulomb gap) around the Fermi energy after an event of electron hopping. Therefore, as shown in Fig. 6(b), the electron transport is governed by the Coulomb-gap type

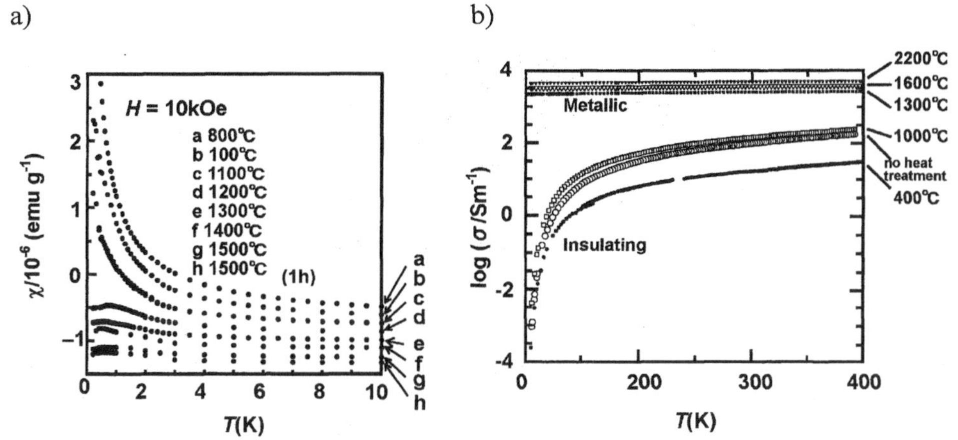

Fig. 6. Temperature dependence of the magnetic susceptibility (a) and the electrical conductivity (b) for ACF samples heat-treated at various temperatures (a to h) up to 1500°C. The duration of heat-treatment is 15 min except for sample h, where it is 1 hour.

variable range hopping process between nanographite domains in the pristine ACFs, which are involved in the disorder-induced insulator (Anderson insulator) regime. Here, the electrical conductivity σ obeys the following equation:

$$\sigma \propto \exp\left[-(T_0/T)^{1/2}\right], \tag{2}$$

$$T_0 = \frac{6e^2}{4\pi^2 k_B \varepsilon \xi}, \tag{3}$$

where ε and ξ are the dielectric constant of the medium and the localization length of the conduction electron wave function, respectively. This fact suggests that the electron hopping from one nanographite domain to the adjacent one is subjected to the electron-electron repulsive force. This also evidences the finite density of states at the Fermi energy in the one-electron picture, consistent with the electronic structure of nanographite predicted theoretically.

Heat-treatment eliminates the functional groups around the nanopraphite edges, working to create coherent electron transport paths between nanographite domains. The development of coherent electron transport paths, which work as percolation path networks for electron transport, enhances the conductivity upon the elevation of heat-treatment temperature as shown in Fig. 6(b). Eventually, metallic conductivity is achieved around the heat-treatment temperature (HTT) of 1200°C, when the percolation path network becomes infinite. The less temperature dependent conductivity above the insulator-metal transition of HTT~1200°C suggests an important contribution of structural disorder in the electron scattering process. The change in the conductivity is faithfully tracked by the magnetic susceptibility as given in Fig. 6(a). Namely, the Curie-Weiss behavior featured with localized spins in the Anderson insulator regime is converted to the less temperature dependent susceptibility of conduction electrons in the metallic regime, where the negative susceptibility is a combination of small positive Pauli paramagnetic and large negative orbital contributions. Interestingly, an anomalous feature appears in the vicinity of the insulator-metal transition, that is; the susceptibility has a cusp at ca.7K with a negative Weiss temperature (antiferromagnetic) of –2 to –3K. This is reminiscent of the onset of an antiferromagnetic ordering. However, this is not the case, as a large field cooling effect on the susceptibility is evident in Fig. 7 [15].

Figure 7 indicates that the susceptibility for the sample (HTT1100) in the vicinity of the insulator-metal transition has a large field cooling effect, especially around the temperature range in which the cusp emerges. The presence of a cusp and its large field cooling effect are the consequence of the development of quenched disordered magnetic structure like as spin glass state. The decreasing trend in the susceptibility on the low temperature side of the cusp in the field cooling is suggestive of a precursor of the antiferromagnetic ordering developed well below the susceptibility cusp. In general, spin glass state develops when the strengths of exchange interactions J vary randomly in space. From the magnetization curve analysis, a large randomness in the strengths of exchange interactions is evident, the ratio of the deviation of the distribution to the mean value being estimated at $\sqrt{\langle \Delta J \rangle^2} / \langle J \rangle \approx 0.8$, which is safely involved in the region of spin glass. The insulator-metal transition converts the feature of electrons from localized to itinerant one. Then, the localized spins of edge states diminish as the fusion of nanographite domains takes place upon the elevation of heat-treatment temperature in the metallic phase well above the insulator-metal transition. However, a considerable amount of edge spins still survive in the vicinity of the transition. The coexistence of the edge-state localized spins and the conduction π-electrons causes the anomalous

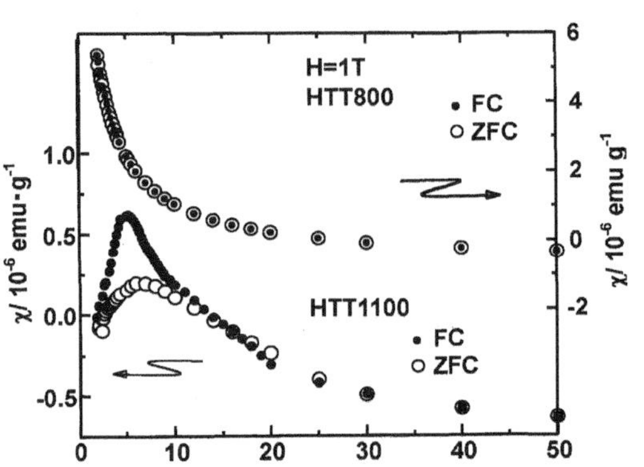

Fig. 7. The field cooling effect on the susceptibility vs temperature plots in the ACF samples heat-treated at 1100 and 800°C at (HTT1100 and 800, respectively), which are in the vicinity of the insulator-metal transition and far from the transition, respectively. Open and full circles represent the data for the zero field cooling (ZFC) and field cooling (FC)($H = 1$ T) processes, respectively.

magnetic state appearing in the insulator-metal transition region; that is, the exchange interactions between the localized spins are mediated by the conduction electrons, which resemble the s-d interaction in traditional metal magnets. The disordered structure of the nanographite network brings about the random spatial distribution in the strengths of the exchange interactions, resulting in the formation of spin-glass-like disordered magnetic structure. Here, it should be noted that the range of exchange interaction is over 2–3 nm, in contrast to ordinary short range feature of exchange interaction. Such long range nature of the exchange interaction adds a uniqueness to the magnetism of the edge-state spins of nanographite.

The magnetic feature of the edge-state spins are modified by chemical modification of the edges. A typical example can be obtained by the reaction with fluorine atoms [16, 17], which has been investigated theoretically as discussed in Section 2. Figure 8 shows a schematic model of fluorination effect on nanographene. In the initial stage of fluorination of nanographene in ACFs up to a fluorine concentration of F/C~0.4, the reaction with fluorine atoms takes place mainly around the edges due to the relatively strong chemical activity of edge carbon atoms, resulting in the formation of difluorinated edge carbon sites. In this fluorination process, the concentration of edge-state spins tends to decrease as the fluorine concentration is elevated. This means that the edge state is destroyed by chemical modification with fluorine atoms, which is consistent with theoretical suggestion [7]. After the fluorination is completed at the edges, a fluorine atom begins attacking the carbon atom in the interior of nanographene sheet, breaking the π-bond, which creates a σ-dangling bond having a localized spin at the carbon site bonded to the carbon atom fluorinated. The concentration of σ-dangling bond localized spins becomes maximized at a fluorine concentration of F/C~0.8, at which a half of carbon atoms in the interior of graphene sheet are bonded to fluorine atoms. Finally, the localized spins disappear at a saturated fluorine concentration of

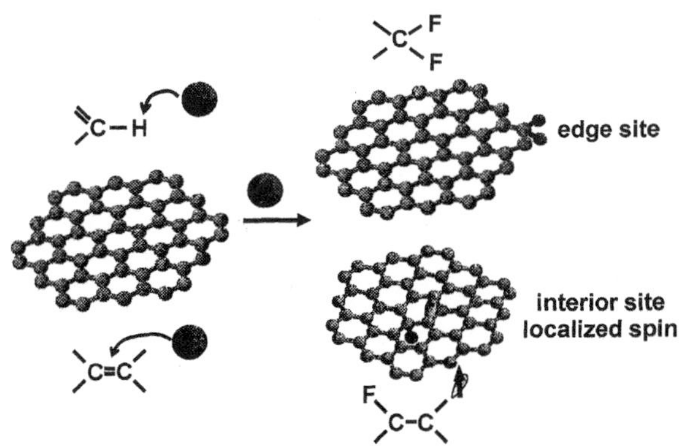

Fig. 8. A schematic model of fluorination effect on nanographene.

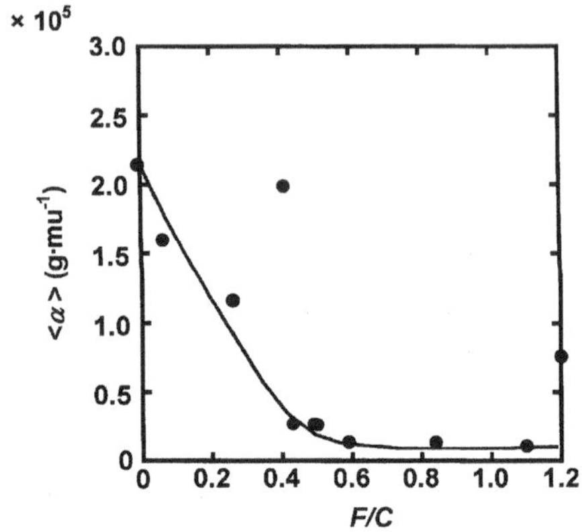

Fig. 9. The mean molecular field coefficient $<\alpha>$ as a function of F/C in fluorinated ACFs. $<\alpha>$ is estimated from the magnetization curves.

F/C~1.2, at which all the carbon atoms are bonded to fluorine atoms. Interestingly, the magnetic nature is different between the edge-state spins and the σ-dangling bond localized spins. Figure 9 shows the fluorine concentration dependence of the strength of the internal exchange field acting on the spins. In the low concentration range up to F/C~0.4, where the edge-state spins are dominant, the internal field is present and it is lowered on the elevation of fluorine concentration. This finding suggests that the edge-state spins are interacting with each other, which is consistent with the theoretical suggestion that the edge-state spins have features between localized and delocalized states as discussed in Section 2 [2]. The fluorination of the edges works to reduce the internal field. In contrast to the edge-state spins, the internal field is absent between the σ-dangling bond localized spins above F/C~0.4; that is, σ-dangling bond localized spins are isolated

4 Nanographite Grown in sp^2/sp^3 Mixed Nano-carbon

Liquid carbon consists of a random mixture of sp^2 and sp^3-bonded carbon atoms with atomic scale disorders, where the π-electron extended state of graphitic nature is absent. Laser-induced melting of graphite results in the formation of liquid carbon, which is in equilibrium state above 4300°C [18]. Thus, laser ablation technique can give disordered sp^2/sp^3 mixed carbon system (diamond-like carbon; DLC), which is regarded as liquid carbon quenched to room temperature. The structure and electronic properties of such DLC are interesting in relation to nanographite [19]. The X-ray diffractions and Raman spectra profile the disordered structural feature of DLC, which is easily relaxed to the thermal equilibrium state through the migration of sp^2- and sp^3-bonds and the exchange

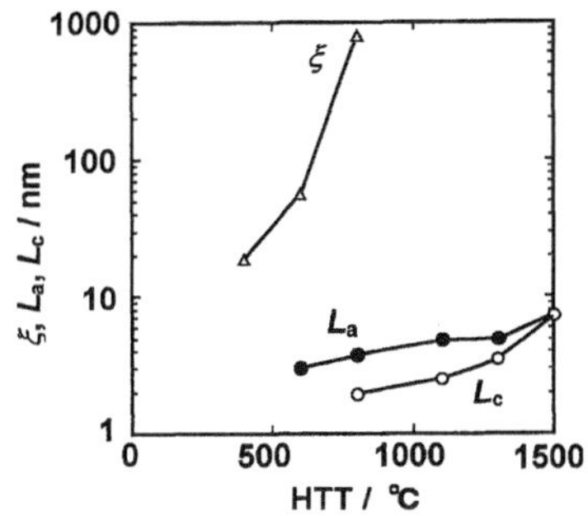

Fig. 10. The heat-treatment temperature (HTT) dependence of characteristic lengths representing the structure and electronic properties in DLC having a sp^2/sp^3 composition ratio of ca.9. L_a (full circle) and L_c (open circle) are the coherence lengths of graphitic structure in the intra-sheet and stacking directions, respectively. ξ (triangle) is the localization length of the wave function. The values of L_a and L_c are estimated from the Raman intensity ratio of the D-band to the G-band and the width of the (002) X-ray diffraction, respectively, while ξ is obtained from the conductivity.

of the bonds with no change in the sp^2/sp^3-bond ratio by annealing in the relatively low temperature range 400–600°C that is far from graphitization temperature. This demonstrates that the rapid cooling makes DLC quenched in the metastable state. The disordered feature and its thermal annealing effect are summarized as shown in Fig. 10 with the characteristic lengths of the structure and electron transport as a function of heat-treatment temperature. From Fig. 10, it is evident that nanographene starts developing above 600°C first along the intra-sheet direction and then the formation of stacking structure follows as the heat-treatment temperature is elevated. In the annealing process, the migration and exchange of sp^2- and sp^3-bonds can make the equilibrium state achieved, where the disordered sp^2/sp^3 mixed carbon with atomic-scale disorder is converted to a random assembly of nanographite domains which are surrounded by sp^3-rich regions. It should be noted that the conversion process takes place with no change in the sp^2/sp^3 carbon ratio in the heat-treatment temperature range below ca.1300°C. Above ca. 1300°C, the conversion is accompanied by the successive increase in the sp^2/sp^3 carbon ratio, contributing to the development of well-graphitized structure.

The electronic structure is affected by the structural changes. According to the behavior of the conductivity, an Anderson insulator state is considered to be stabilized in the low HTT below 800°C, where the electron transport is featured with variable range hopping mechanism. Indeed, the non-heat-treated sample shows insulating behavior, where the temperature dependence can be explained in terms of 3D variable range hopping mechanism,

$$\sigma = \sigma_0 \exp\left[-\left(\frac{8^3}{9\pi\xi^3 D(E_F) k_B T} \right)^{1/4} \right], \tag{4}$$

where ξ and $D(E_F)$ are the localization length of the conduction electron wave function and the density of states at E_F, respectively. This experimental finding indicates that the structural randomness works to make the wave functions localized, resulting in the Anderson insulating state. Such conductivity features survive in the heat-treatment up to 200°C. As HTT is elevated, the conductivity behavior is modified to the Coulomb-gap variable range hopping features given by eqs.(2)-(3). The increasing importance of the charging energy, which is evidenced by the Coulomb-gap generation in the electron hopping process, is the consequence of the increase in the localization length shown in Fig. 10, which is caused by the growth of nanographite domains upon the elevation of HTT. Here, the comparison between the structural characteristic length and the localization length shown in Fig. 10 gives important information on the correlation between the structural disorder and the electronic structure. Actually, the localization length is over 10 nm even in the region around 400°C, where no graphitic structure exists. In other words, the extended electronic structure, which has a finite density of states at E_F, can exist in nongraphitic disordered structure with atomic-scale disorder, suggesting that liquid carbon is in metallic state.

At high temperatures above 800°C, the growth of nanographite domains contributes to the formation of infinite percolation path network of electron transport, resulting in the change from insulating state to metallic state, similar to the case of heat-treated ACFs discussed in Section 3. The change in the magnetic properties in the course of heat-treatment gives us a perspective on how nanographite grows and what electronic features are produced in nanographite. The susceptibility can be explained in terms of a combination of the localized spin term and the temperature-independent term, the latter of which consists of contributions of orbital motions (orbital susceptibility) and spins (Pauli spin susceptibility) of conduction electrons. This experimental finding suggests the coexistence of localized and conduction electrons around E_F in the samples. The localized spin concentration in the as-prepared DLC sample, which is regarded as quenched liquid carbon, is estimated at 0.0044 per carbon atom, which is considerably large in comparison with that in ordinary amorphous carbons. The random potential of intensive atomic-scale structural disorder in the as-prepared sample makes most of electrons localized in the regions (localized centers) restricted within the localization of the wave functions, which contribute to the localized spin susceptibility. In addition to this, there are clusters of localized centers having various sizes. In the larger size clusters involving many electrons, the electrons can be mobile in larger areas, being featured with the susceptibility of mobile electrons. The elevation of HTT works to increase the mean size of clusters at the expenses of small clusters whose smallest extreme is a single electron in a localized center. Thus, the localized spin concentration decreases upon the elevation of heat-treatment temperature, while the contribution of the mobile electrons successively grows. The temperature independent susceptibility becomes a maximum in the HTT around 400–600°C, whose value is two orders of magnitude larger than that of bulk graphite, where the electronic structure of the latter is featured with the superposition of the π- and π^*-bands with a negligibly small density of states at E_F. This clearly demonstrates that the nongraphitic disordered carbon having atomic-scale randomness has a large density of states, entirely different from bulk regular graphite. In the HTT range above ca.600°C, graphitic domains start developing through the agglomeration process, the density of states being reduced. In the graphitic domains, the extended π-electronic structure produces a large diamagnetic current or diamagnetic susceptibility. The concerted effect of the reduced density of states and the large diamagnetic susceptibility works effectively to reduce the temperature independent susceptibility. The large decrease of the temperature independent susceptibility in the vicinity of the insulator-metal transition around HTT of 800–1000°C proves that the metallic state has the nature of nanographite domain network.

Here arises a question about the charge distribution. The sizes and shapes of nanographite domains are naturally distributed with finite distribution widths. From chemistry view point, molecules having different sizes and shapes are considered to have different ionization energies and electron affinities. If we adopt the same consideration for nanographite having different sizes and shapes, it is natural to take an idea of inhomogeneous charge distribution between nanographite domains having different ionization energies and electron affinities. The thermoelectric power, which is sensitive to the densities and the sign of charges of conduction carriers, gives evidence of inhomogeneous charge distribution. Actually, the observed thermoelectric power is one order of magnitude as small as that in ordinary graphitic systems. The nanographite, whose electron affinity is large, accepts electrons from neighboring nanographite domains through electron transfer, being negatively charged, whereas that with small ionization energy becomes positively charged after charge transfer. The negatively/positively charged nanographite domains have electron/hole carriers with negative/positive charges. The compensation between negative and positive thermoelectric powers originating from electrons and holes, respectively, gives a small thermoelectric power as observed by the experiments.

5 Host-guest Interaction and Edge-state Spins

Next, we discuss the effect of host-guest interaction on the magnetism of nanographites. The first part is devoted to edge-state spins as a probe of the interaction between gas and nanographite [20]. As exhibited in Fig. 5, the gaps between nanographite domains provide nano-sized vacancies (micropore), in which various kinds of guest species can be accommodated. Indeed, a micropore network is produced as the reversal of the nanographite network. Figure 11 shows the adsorption isotherms for N_2, O_2, He, and Ar in the micropores of ACFs at room temperature. The effective pressure of guest gases inside the micropores is orders of magnitude larger than the pressure of gaseous state in the environment. Actually, around a gaseous pressure of 100 Torr, the effective pressure inside the micropores goes to 7–10 atm for N_2, Ar and O_2. This means a large condensation of guest gaseous species in the micropores due to the enhanced condensation energy associated with gas-nanographite interaction, since accommodated gaseous molecules surrounded by the surfaces of nanographite domains are affected by that interaction in every direction. An exceptionally huge condensation appears in the case of He; that is, the pressure inside reaches almost 80 atm, which

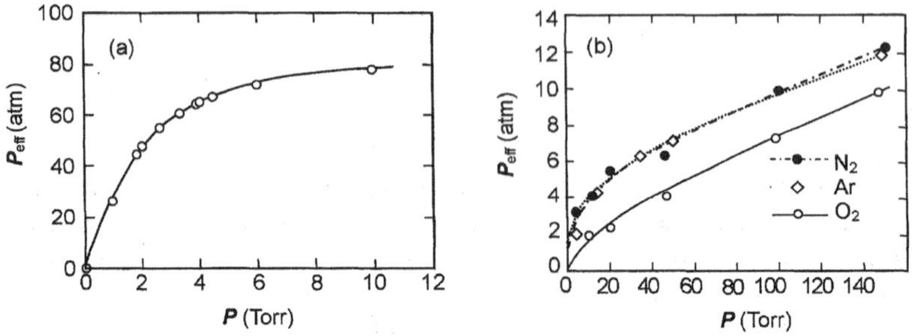

Fig. 11. Effective pressure of (a) He, (b) N_2, O_2, and Ar in the micropores of ACFs at room temperature as a function of external pressure.

corresponds to about one tenth of the density of liquid helium, even when 10 Torr of He is introduced. Here, the condensation rate is over 6000 indeed.

Gas adsorption phenomena in the micropores investigated using gas adsorption isotherms can be tracked by ESR of edge-state spins. Here, the ESR saturation technique gives information on the spin-lattice relaxation process of the nanographite edge-state spins. The energy that the spin system gains from the microwave is relaxed to the environment through the relaxation process, which is mediated by phonons in general.

Figure 12 presents the microwave power dependence of the ESR intensity (saturation curve) taken at room temperature in vacuum and in the atmosphere of 10 Torr He, O_2, Ar, and N_2. In vacuum, the ESR signal tends to be easily saturated in the low microwave power range, suggesting that the spin-lattice relaxation rate $1/T_1$ is considerably small. The relaxation is driven by the acoustic phonons, whose Debye temperature is considerably high in the range of 2000K for bulk graphite. When the size of graphite is reduced, the phonon energy becomes discrete due to the quantum size effect. Indeed, in the case of nanographite having an in-plane size of 2 nm, the discreteness of phonons reaches ca. 200K, making the contribution of the phonons less

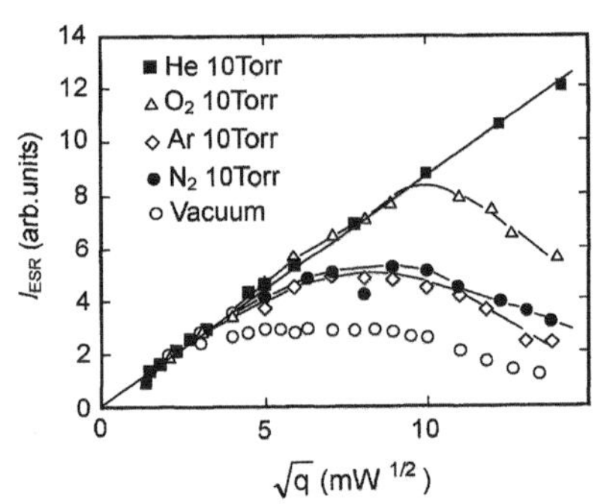

Fig. 12. ESR saturation curves (ESR intensity I_{ESR} vs microwave power q) for 10 Torr of He, O_2, Ar, and N_2 in ACFs at room temperature. The saturation curve in vacuum is also shown.

effective in the spin-lattice relaxation process especially at low temperatures. This is the reason why the relaxation rate becomes small in nanographite. In contrast, the introduction of gaseous species revives the spin-lattice relaxation paths, giving less saturated feature of the saturation curves in the atmosphere of gaseous materials, as shown in Fig. 12. The extreme case occurs in He atmosphere, where no saturation appears in the microwave power range investigated. In other words, the energy absorbed in the edge-state spins from the microwave can be easily relaxed to the environment in the atmosphere of He. These experimental findings strongly suggest an important contribution of guest gaseous species to the spin-lattice relaxation mechanism.

In the micropores, guest molecules collide with edge-state spin sites around the peripheries of nanographites. The collisional process, which is governed by van der Waals interaction between guest molecule and nano-graphite domain, works effectively to accelerate the spin-lattice relaxation rate especially for the case in which the phonon-assisted process is not at work, if the relaxation is incorporated with the spin-reversal event [21]. In this case, relaxation rate $1/T_1$ in the collision-induced relaxation process is described in the following equation:

$$1/T_1 = nv\sigma_X, \tag{5}$$

where n and σ_X are the density of guest molecules and the cross section related to the spin-flip process, respectively. v is the mean velocity of the molecules;

$$v = \sqrt{3RT/M}, \tag{6}$$

where M is the mass of the guest molecule. A more detailed analytical expression can be obtained on the basis of the van der Waals interaction between gas molecule and nanographite, and it is given for He gas as follows;

$$1/T_1 = \left[\frac{6.05}{9(2\pi)^{1/2}\hbar^4}\right] nM^{3/2}\left(\frac{p}{\Delta E}\right)^2 \left[\frac{\left(e^2 IJ\right)^4}{R_0{}^6}\right] w\left(k_B T\right)^{1/2}, \tag{7}$$

where R_0 is the minimum distance at which a He atom can access to an edge-state spin site. I, J and ΔE are parameters related to the electronic structure of He atom [21]. (I is the matrix elements of z between the $1s$ and $2p_z$ states, J is that of x between $2s$ and $2p_x$ states, and $\Delta E = E(^1\text{P})$-$E(^1\text{S})$). p is given to be λ/Δ, where λ is the spin-orbit interaction of carbon p-state and Δ is the energy difference between the excited and ground states of edge-state spins. w is the probability of the carbon $2p_z$ state being vacant. A comparison with the experimental results indicates that the equation well represents the experimental results, suggesting that the edge-state spins work as a probe of guest molecules. In other words, the edge-state spins are useful as a gas sensor, in which the van der Waals-interaction-assisted collisional process plays an essential role.

This is the first example, in which the spin-lattice relaxation is governed by the van der Waals interaction whose origin is the electric dipole-dipole interaction.

An important issue, which should be pointed out, is the exceptionally large He condensation in the ACF micropores. Though the origin of this remains unsolved, it suggests the presence of ultra-micropores to which only smallest diameter He atom (0.257 nm) can access.

The next topic in the guest-nanographite interaction is related to gas-adsorption-induced change in the magnetism of nanographite. In the above discussion we have learned that guest molecules are condensed in the micropore space of ACFs. Guest species such as water, ethanol, etc., which are in liquid state at room temperature, can also be adsorbed in the micropores. In this case, the accommodated guest molecules affect the structure of nanographite network in ACFs having flexible and soft structure, since densely packed guest molecules push strongly nanographite domains surrounding. According to structural investigation [22], accommodated water molecules squeeze nanographite domains, resulting in a large reduction in the inter-sheet distance between nanographene sheets in nanographite domains by the effective pressure of guest molecules. Actually, the inter-sheet distance is decreased from 0.38 nm to 0.34 nm upon the saturation of water molecules in the micropores.

Here arises a question about the effect of the structural change on the magnetism. Figure 13 gives evidence of effect on the magnetism upon water adsorption [23]. From the water adsorption isotherm shown in Fig. 13, no water adsorption takes place in the low vapor pressure range up to P/P_0~0.5, at which the amount of adsorbed water has an upsurge. Above that pressure, it saturates and the micropores become full of water. There is a hysteresis between adsorption and desorption in the change happening at the threshold pressure of P/P_0~0.5. This on/off change in the isotherm is associated with hydrophobic nature of graphitic plane. Namely, it requires a finite vapor pressure to accommodate water molecules in the micropores whose surfaces are hydrophobic. The magnetic susceptibility tracks the isotherm with fidelity, as shown in Fig. 13. There is

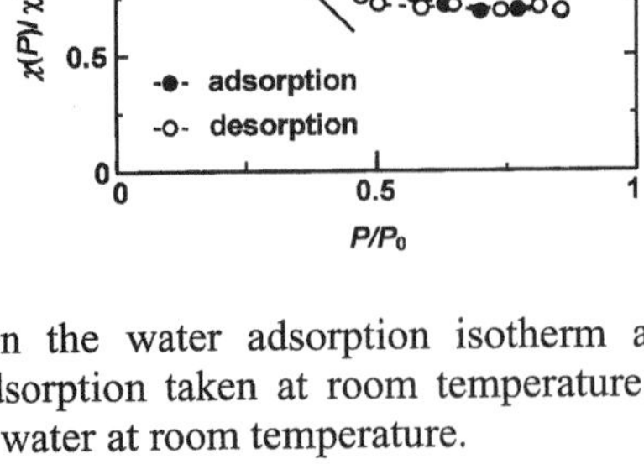

Fig. 13. Comparison between the water adsorption isotherm and the change of magnetic susceptibility upon water adsorption taken at room temperature. P_0 represents the saturation vapor pressure (~20 Torr) of water at room temperature.

no change in the susceptibility up to the threshold pressure. It shows ca.30% reduction with an adsorption/desorption hysteresis at the threshold, followed by the saturation on the high-pressure side of the threshold. Similar correlation behavior emerges in other guest species such as acetone, ethanol, methanol [24], bromine [25], and so on. The threshold pressure is shifted to lower vapor pressure as the fraction of OH group is reduced in the order of H_2O>CH_3OH>C_2H_5OH, consistent with the hydrophobic nature of graphitic plane.

The correlation between the isotherm and the magnetic susceptibility upon water physisorption can be understood on the basis of the structural change induced by the water adsorption. As we have already discussed, ACFs have a 3D disordered network structure of nanographite domains, where each nanographite domain consists of a stacking of 3–4 nanographene sheets. A nanographene sheet in a nanographite domain has localized magnetic moments of edge state, whose number depends on the shape of nanographene, according to the experimental observations as discussed above. In theoretical considerations, the electronic structure of nanographite can be explained in terms of the Hubbard Hamiltonian (eq.(8)) with intra-nanographene-sheet transfer integral t, on-site Coulomb interaction U, and inter-sheet transfer integral between adjacent nanographene sheets t_1.

$$H = -t \sum_{\langle i,j\rangle \text{intra-sheet}} \sum_{\sigma} \left(c^{+}_{i,\sigma} c_{j,\sigma} + \text{h.c}\right) - t_1 \sum_{\langle i,j\rangle \text{inter-sheet}} \sum_{\sigma} \left(c^{+}_{i,\sigma} c_{j,\sigma} + \text{h.c}\right) + U \sum_{i} n_{i,\uparrow} n_{i,\downarrow}, \qquad (8)$$

where the number operator is given as $n_{i,\sigma} = c^{+}_{i,\sigma} c_{i,\sigma}$ for $\sigma = \uparrow$ and $\downarrow$; $c_{i,\sigma}$ and $c^{+}{}_{i,\sigma}$ are the annihilation and creation operators of an electron at the ith site with spin σ, respectively. The introduction of guest molecules into the micropores reduces the inter-sheet distance, giving rise to the increase in the inter-sheet transfer integral. Namely, when no guest species occupies the micropores, the interaction between edge-state spins on adjacent nanographene sheets is weak, making localized magnetic moments independent of each other as paramagnetic spins. The decrease in the inter-sheet distance induced by the introduction of guests into the micropores raises the inter-sheet transfer integral t_1. The Hubbard model [26] with t, U and t_1 theoretically reproduces the change in the magnetic moments upon water adsorption, as shown in Fig. 14, where two coronene sheets stacked in the AB fashion represent a nanographite domain in the theoretical model. In case of nanographites having zigzag edges, edge states are produced, which give net magnetic moments, when the site potential is introduced at the edges. The presence of the site potential at the edges is reasonably understood taking into account that the peripheries of nanographene sheets in the pristine ACFs are bonded to various kinds of functional groups. The theoretical analysis shown in Fig. 14(b) demonstrates that the increase in the inter-sheet transfer integral t_1 brings about the decrease in the effective magnetic moment of nanographite, consistent with the experimental findings. Here, it should be noted that there is a similarity between the

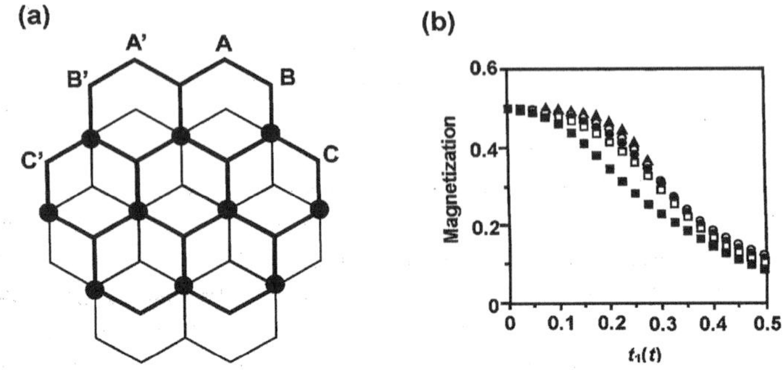

Fig. 14. (a) Schematic model of a nanographite with zigzag edges, where two nanographene sheets are stacked in the AB fashion. Nanographene is represented by coronene molecule. The upper and lower graphene sheets are depicted with thick and thin solid lines, respectively. A site potential $E_s = -2t$ is placed at site A. (b) The magnetic moment of nanographite as a function of inter-sheet transfer integral t_1, where the on-site Coulomb interaction U is varied in the range $0.6t$ (closed squares)$\leq U \leq 1.8t$ (closed triangle). The interval of U in the series of plots is $\Delta U = 0.3t$.

present phenomenon and Mott metal-insulator transition. In the present case, the electronic state with no guest species is described by that electrons are localized within a graphene sheet. This corresponds to an insulating state having localized magnetic moments. In contrast, the elevation of interaction between graphene sheets makes the electronic state delocalized between graphene sheets, giving rise to the reduction in the magnetic moments. The change in the feature from localized to itinerant electron induced by water adsorption is a new version of insulator-to-metal transition. Moreover, such a change in the magnetism induced by water adsorption is considered to be a controllable magnetic switch effect of nanoscopic magnetism, which has never been reported so far.

6 Summary

Nanographene is defined as a nano-sized 2D hexagon-network π-electron system having open edges. According to theoretical suggestions, the presence of open edges produces edge-inherited non-bonding π-electronic state (edge state) superimposed upon the π- and π*-bands around the Fermi energy, giving entirely different electronic structure from bulk graphite. Therefore, in nanographene, the localized spins of the edge state play an important role in unconventional nano-magnetism. The experimental works using STM, AFM and magnetic investigations confirm the presence of edge-state spins and their contribution to the magnetism in nanographene and networked nanographite domains. Fluorination of nanographene edges tends to reduce edge-state spins. In the networked nanographite domains, spin glass state is produced with the aid of the π-conduction electron-mediated exchange interaction between the edge-state spins. For microporous carbon consisting of networked nanographite domains, which physisorb a huge amount of water in the on/off adsorption manner due to the hydrophobic nature of graphite surface, the adsorption/desorption of water into the micropore space reduces/raises the effective magnetic moments of the edge-state spins reversibly. This can be explained in terms of the enhancement of exchange interaction between the edge-state spins on adjacent nanographene sheets due to the reduction of inter-sheet distance by squeezing the nanographite domains with the internal pressure of water. The on/off magnetic switching phenomenon is interesting mechanical effect of guest molecules, which can be controlled by the guest physisorption. The edge-state spin is useful as a spin-probe in detecting guest-host interaction, since it is sensitive to foreign guest molecules. The present experimental findings suggest that the collisional interaction of guest molecules with edge-state spins strongly accelerates the spin-lattice relaxation rate. Using edge-state spins, a huge amount of He condensation is probed in the micropore space of nanographite network.

In low temperature heat-treatment of disordered sp^2/sp^3 mixed carbon, which is obtained by quenching liquid carbon, nanographite network grows through the migration of sp^2- and sp^3-bonds and the exchange of the bonds. The charge distribution between nanographite domains in the network is inhomogeneous in nature, for which the differences of ionization energies and electron affinities are responsible.

A nanographene can be tailored intentionally using electron beam lithography and chemical modification of edges, if nano-fabrication techniques can access to the dimension of nm range in the future. This is contrasted to fullerenes and carbon nanotubes, whose structures cannot be intentionally fabricated. Eventually, nanographite is expected to impart future promising molecule-based electronic and magnetic devices by adding functionality to it using lithography and chemical modifications.

Acknowledgements

The authors would express their sincere thanks to O.E. Andersson, H. Kumagai, B.L.V. Prasad, Y. Shibayama, A. Nakayama, K. Harigaya, H. Sato, K. Sugihara, Y. Hishiyama, Y. Kaburagi, M. Endo, L.G. Cançado, M.A. Pimenta, R. Saito, and M.S. Dresselhaus on the basis of whose contributions the present work has been carried out. They would be also thankful to K. Kusakabe for fruitful discussion with him. This work is supported by the Grant-in-Aid for "Research for the Future" Program, Nano-Carbons and 15105005 from JSPS.

References

1. K. Yoshizawa, K. Okahara, T. Sato, K. Tanaka, and T. Yamabe, Carbon 32 (1994) 1517.
2. M. Fujita, K. Wakabayashi, K. Nakada, and K, Kusakabe, J. Phys. Soc. Jpn. 65 (1996) 1920.
3. K. Wakabayashi, M. Fujita, H. Ajiki, and M. Sigrist, Phys. Rev. B59 (1999) 8271.
4. E. Lieb, Phys. Rev. Lett. 62 (1989) 1201.
5. D.J. Klein, Chem. Phys. Lett. 217 (1994) 261.

6. K. Kusakabe and M. Maruyama, Phys. Rev. B67 (2003) 092406.
7. M. Maruyama and K. Kusakabe, J. Phys. Soc. Jpn. 73 (2004) 656.
8. O.E. Andersson, B.L.V. Prasad, H. Sato, T. Enoki, Y. Hishiyama, Y. Kaburagi, M. Yoshikawa, and S. Bandow, Phys. Rev. B58 (1998) 16387.
9. A.M. Affoune, B.L.V. Prasad, H. Sato, T. Enoki, Langmuir 17 (2001) 547.
10. M. Affoune, B.L.V. Prasad, H. Sato, T. Enoki, Y. Kaburagi, Y. Hishiyama, Chem. Phys. Lett. 348 (2001) 17.
11. L.G. Cançado, M.A. Pimenta, B.R.A. Neves, G. Medeiros-Ribeiro, T. Enoki, Y. Kobayashi, K. Takai, K. Fukui, M.S. Dresselhaus, R. Saito, and A. Jorio, Phys. Rev. Lett. 93, 047403 (2004).
12. Y. Kobayashi, K. Fukui, T. Enoki, K. Kusakabe, and Y. Kaburagi, Phys. Rev. B, to be published (2005).
13. B. L. V. Prasad, H. Sato, T. Enoki, Y. Hishiyama, Y. Kaburagi, A. M. Rao, P. C. Eklund, K. Oshida, and M. Endo, Phys. Rev. B62 (2000) 11209.
14. Y. Shibayama, H. Sato, T. Enoki, X.X. Bi, M.S. Dresselhaus, and M. Endo, J. Phys. Soc. Jpn. 69 (2000) 754.
15. Y. Shibayama, H. Sato, T. Enoki, M. Endo, Phys. Rev. Lett. 84 (2000) 1744.
16. K. Takai, H. Sato, T. Enoki, N. Yoshida, F. Okino, H. Touhara, and M. Endo, J. Phys. Soc. Jpn. 70 (2001) 175.
17. R. Saito, M. Yagi, T. Kimura, G. Dresselhaus, M.S. Dresselhaus: J. Phys. Chem. Solids 60 (1999) 715.
18. F.P. Bundy, Physica A156 (1989) 169.
19. K. Takai, M. Oga, H. Sato, T. Enoki, Y. Ohki, T. Tamoto, K. Sugihara, S. Iijima, Phys. Rev. B67 (2003) 214202.
20. A. Nakayama, K. Suzuki, T. Enoki, C. Ishii, K. Kaneko, M. Endo, and N. Shindo, Solid State Commun. 93 (1995) 323.
21. K. Sugihara, A. Nakayama, and T. Enoki, J. Phys. Soc. Jpn. 64 (1995) 2614.
22. T. Suzuki and K. Kaneko, Carbon 26 (1988) 743.
23. H. Sato, N. Kawatsu, T. Enoki, M. Endo, R. Kobori, S. Maruyama, and K. Kaneko, Solid State Commun. 125 (2003) 641.
24. N. Kawatsu, H. Sato. T. Enoki, M. Endo, R. Kobori, S. Maruyama, and K. Kaneko, Mol. Cryst. Liq. Cryst. (2004), to be published.
25. N. Kumagai, H. Sato, K. Takai, and T. Enoki, private communication.
26. K. Harigaya and T. Enoki, Chem. Phys. Lett. 351 (2002) 128.

Carbon-Based Magnetism
T. Makarova and F. Palacio (Editors)

18

Ferromagnetic and Superconducting Instabilities in Graphite

Y. Kopelevich, S. Moehlecke, and R.R. da Silva

Instituto de Física "Gleb Wataghin", Universidade Estadual de Campinas, Unicamp 13083-970, Campinas, São Paulo, Brasil

1 Introduction

A few years ago two of us reported on anomalous behavior of graphite in magnetic field [1, 2]. In particular, the metal-insulator transition driven by the magnetic field applied perpendicular to graphite basal planes [1], as well as ferromagnetic- and superconducting-like hysteresis loops [2] have been revealed in highly oriented pyrolytic graphite (HOPG) samples at room temperature (for review articles see [3, 4]).

Shortly after, room temperature ferromagnetism has been observed in polymerized rhombohedral (rh) C_{60} samples [5–7], extraterrestrial graphite [8], and materials consisting of curved graphite-like sheets such as glassy and microporous carbon [9, 10].

The accumulated so far experimental evidences indicate that a structural disorder, topological defects, as well as adsorbed foreign atoms can be responsible for the occurrence of both ferromagnetic and superconducting patches in graphitic structures, i.e. sp^2-bonded carbon atoms. In particular, it has been demonstrated that sulfur atoms adsorption triggers superconducting instabilities in graphite [11–14], whereas proton irradiation [15] enhances the sample ferromagnetic response.

In the present Chapter we review our most recent work related to the studies of magnetic properties of graphite and related carbon materials. The Chapter is organized as follows. In section 2 we present a comparative analysis of ferromagnetic properties of microporous carbon, HOPG, and rh-C_{60} samples. Results demonstrating an oxygen adsorption effect on the magnetic properties of graphite are presented in section 3. Section 4 is devoted to the ferromagnetic and superconducting properties of graphite-sulfur composites. Discussion and concluding remarks are given in section 5.

2 Ferromagnetism in Carbon Materials: a Comparative Study

It appears that the room-temperature ferromagnetism in rh-C_{60} compounds [5–7] occurs only in samples prepared very close to the temperature at which the fullerene cages collapse and graphitized carbon forms. In agreement with such observation are experiments performed on glassy carbon [9] which demonstrated that the ferromagnetism emerges during the graphitization process.

On the theoretical side, the occurrence of both ferromagnetic and superconducting instabilities due to topological disorder in graphitic sheets has been predicted [16]. The analysis given in Ref. [16] assumes the formation of pentagons and heptagons, i. e. disclinations in the graphene honeycomb lattice. The low-lying electronic states of an isolated graphitic sheet can be well approximated by the Dirac equations in (2+1) dimensions. Then, according to Ref. [16], a random distribution of topological defects described in terms of a random gauge field can lead to an enhancement of the density of states of Dirac fermions N(E) at low energies E, and hence to magnetic or superconducting instabilities. Compounds with curved graphene layers have been proposed as promising materials for both ferromagnetism and superconductivity occurrence [16].

These experimental as well as theoretical results motivate us to explore the magnetic behavior of microporous carbon (MCY), a three-dimensional nano-arrayed structure whose arrangement matches that of supercages of zeolite Y.

The microporous carbon samples have been prepared at the Tohoku University [17, 18] by the following template technique. Powder zeolite Y was impregnated with furfuryl alcohol (FA) and the FA was polymerized inside the zeolite channels by heating the FA/zeolite composite at 150°C under N_2 flow. The resultant PFA (polyfurfuryl alcohol)/zeolite composite was heated to 700°C in N_2. As soon as the temperature reached 700°C, propylene chemical vapor deposition (CVD) was performed at this temperature to further deposit carbon. After the CVD, the composite was heat-treated again at 900°C under a N_2 flow. The resultant carbon was liberated from the zeolite framework by acid washing. The obtained microporous carbon possesses a very high surface area of 3600 m^2/g and consists of curved 3D graphene networks which may contain randomly distributed pentagons and heptagons carbon rings. The microporous carbon particle size ranges from 1000 Å to 4000 Å. Spectrographic analysis of the MCY samples revealed magnetic impurity contents of Fe (64 ppm), Co (4.4 ppm), and Ni (3.5 ppm) [10].

DC magnetization M(H,T) measurements were always performed using a SQUID magnetometer MPMS5, Quantum Design.

Figure 1 presents the low-field portions of the magnetization hysteresis loops M(H) measured for MCY, HOPG (Union Carbide Co.), and Rh-C_{60} samples at T = 300K demonstrating an enhanced ferromagnetic signal in MCY and Rh-C_{60} samples as compared to that in HOPG.

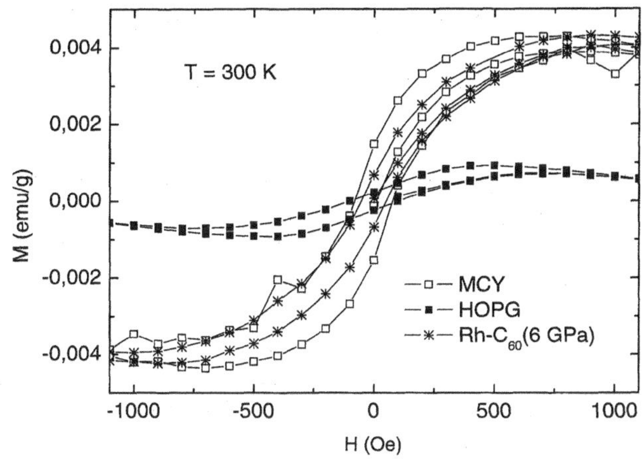

Fig. 1. Magnetization hysteresis loops M(H) measured for MCY, HOPG, and Rh-C_{60} samples. In the case of HOPG, the magnetic field was applied nearly parallel to the sample basal planes; because of small ($\leq 3°$) field/planes misalignment, M(H) decreases at high enough fields due to Landau diamagnetism associated with the graphene planes.

Because of the unavoidable in-plane diamagnetic contribution (due to intrinsic misalignment of the graphite layers with respect to each other occurring even in the best HOPG samples) we have chosen the remnant magnetization $M_{rem}(H = 0) = [M^{+}(H = 0) - M^{-}(H = 0)]/2$ for a comparative analysis. Shown in Fig. 2 is $M_{rem}(T)$ obtained for MCY, HOPG, and rh-C_{60} samples, where $M^{+}(H = 0)$ and $M^{-}(H = 0)$ are zero-field positive and negative magnetizations measured after the field cycling. As Fig. 2 demonstrates, $M_{rem}(MCY) >> M_{rem}(HOPG)$ for all temperatures, and is comparable to $M_{rem}(T)$ measured for rh-C_{60} sample prepared at T = 1073K and pressure P = 6 GPa [19]. We stress that the results of x-ray structural analysis performed on this nominal rh-C_{60} sample revealed the coexistence of the rh-C_{60} phase and clusters of graphite-like layers [19].

Also a comparative analysis of the data obtained on microporous carbon and rh-C_{60} sample synthesized at the pressure 9 GPa and T = 800K [6] reveals a striking correspondence between M(H) measured in these materials. As follows from Fig. 3, the remnant magnetizations measured for MCY and rh-C_{60} (9 GPa) samples practically coincide.

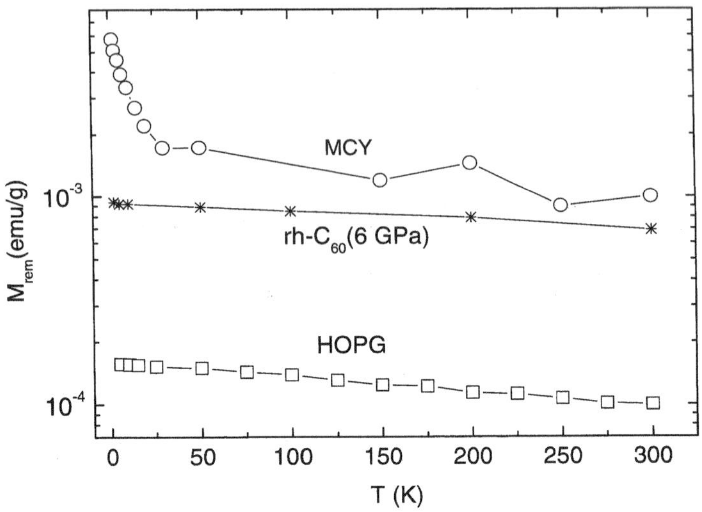

Fig. 2. Remnant magnetization $M_{rem}(T)$ obtained for MCY, HOPG, and rh-C_{60} samples.

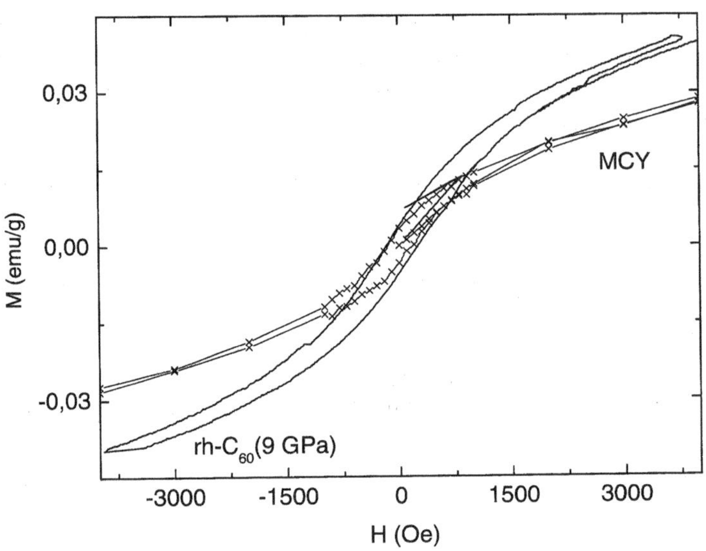

Fig.3. M(H) hysteresis loops obtained at T = 10K for MCY and rh-C_{60} samples [6].

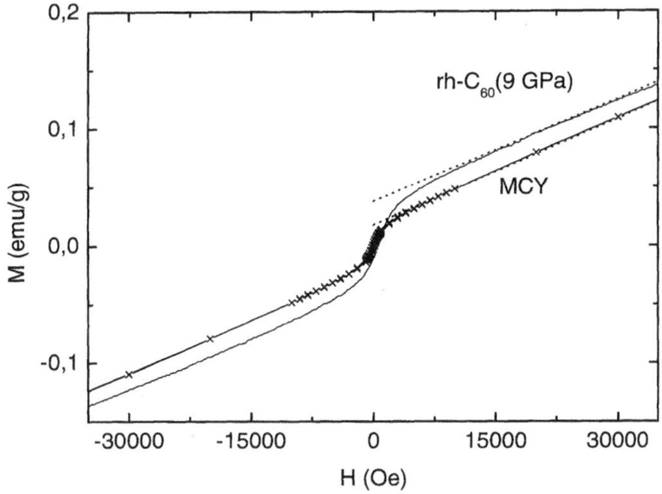

Fig. 4. M(H) obtained at T = 10K for MCY and rh-C_{60} [6] samples; dotted lines are obtained from the equation $M(H) = M_s + \chi H$, where M_s = 0.018 emu/g (MCY), M_s = 0.038 emu/g (rh-C_{60}), and $\chi \approx 3{\cdot}10^{-6}$ emu/g·Oe for both rh-C_{60} and MCY samples.

In addition, Fig. 4 illustrates that the high field portion of the magnetization curves can be fitted by the equation $M(H) = M_s + \chi H$ with the same paramagnetic susceptibility $\chi \approx 3{\cdot}10^{-6}$ emu/g·Oe for both [MCY and rh-C_{60} (9 GPa)] samples; M_s is the spontaneous magnetization obtained from the extrapolation of the linear M(H) region to H = 0.

The above analysis strongly suggests that the ferromagnetism in microporous carbon and nominal rh-C_{60} samples has a common origin, possibly associated with fullerene-like fragments. There exists both experimental and theoretical evidence that fullerene-like fragments with positive and/or negative curvature should be a common feature of microporous carbons, indeed [20]. Also, such fragments should naturally appear in rh-C_{60} samples with partially destroyed C_{60} molecules [21, 5–7]. On the other hand, since HOPG may have a small (if any) number of topological defects, the experimental observation of a much smaller ferromagnetic magnetization is not unexpected.

3 Oxygen adsorption Effect on the Magnetic Properties of Graphite

In 1996, Murakami and Suematsu [22] reported on the ferromagnetic ordering at T ≤ 800K in C_{60} crystals exposed to light in the presence of oxygen gas. Recent measurements [23] corroborated the occurrence of high temperature ferromagnetism in C_{60} under photo-assisted oxidation. It has been suggested [22, 23] that the photo-polymerization plays an important role to the ferromagnetism occurrence.

Aiming to shed more light on the relevance of the oxygen presence to the ferromagnetism in carbon-based materials, we explored the effect of oxygen and other adsorbed gases on the magnetic properties of a graphite powder.

In these experiments, an activated graphite powder was prepared by cutting and grinding a graphite rod at T = 300K in different atmospheres: Ar, He, N_2, H_2, O_2, and air. The graphite rod was from Carbon of America Ultra Carbon, sell by Alfa Aesar (stock No. 40766), AGKSP grade, (ultra "F") 99.9995% purity or 5 ppm of total impurities and maximum of 1 ppm of impurities per element. The powder was produced by cutting and grinding the graphite rod on the edge and side area of a new and clean circular diamond saw blade. The cutting and grinding system was inside a plastic bag filled with the gas. Using a gas hose, a continuous gas stream was also forced to blow in the grinding area.

From the X-ray diffraction spectrum we estimate the graphite crystallites size $L_a \approx 1000$ Å and $L_c \approx 400$ Å. The powder particle size was found to be 150±70 μm, using sieves.

Figure 5 illustrates that M(H) curves measured for the virgin bulk sample as well as the powdered under Ar atmosphere sample practically coincide and demonstrate the sample diamagnetism and a negligible magnetic hysteresis. Similar results were obtained for the graphite powders prepared under He, N_2, and H_2 gas environment.

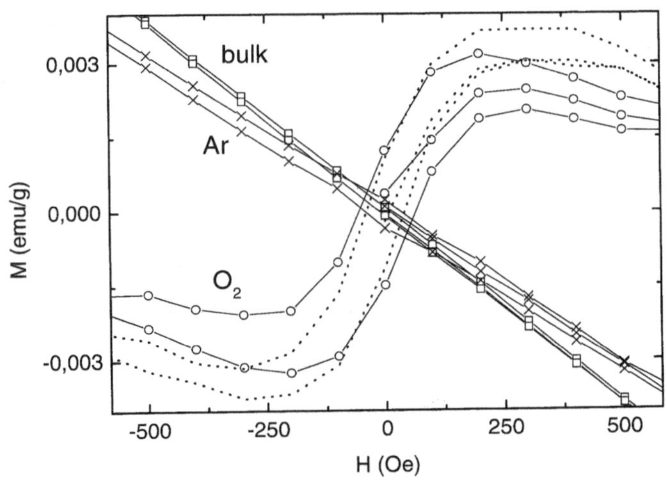

Fig. 5. M(H) measured at T = 10K for the graphite samples: bulk (virgin) sample and powdered samples prepared under argon (Ar) and oxygen (O_2) atmospheres; dotted line corresponds to M(H) measured for the oxidized sample at T = 300K.

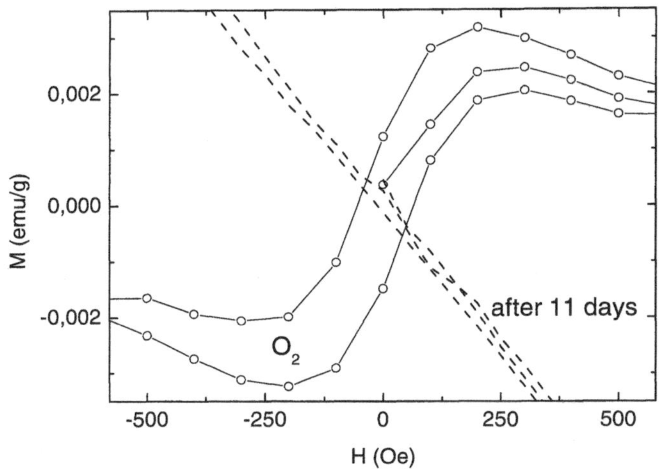

Fig. 6. M(H) for the graphite powder prepared in oxygen (O_2) atmosphere measured at T = 10K right after preparation and after 11 days of the sample staying in air.

At the same time, the samples prepared under oxygen exposure showed a pronounced ferromagnetic response. This can be seen in Fig. 5 where ferromagnetic hysteresis loops measured for the powder prepared under oxygen exposure are clear. This observation provides unambiguous evidence that the ferromagnetism can be triggered by the presence of oxygen. Fig. 6 illustrates another important experimental fact, namely that the oxygen effect is reversible: the ferromagnetism vanishes with time after the sample is removed from the oxygen atmosphere.

Summarizing, the results given in Fig. 5 and Fig. 6 suggest that an adsorption followed by desorption of oxygen gas at the graphite surface are responsible for the appearance and vanishing of the ferromagnetism, respectively. Note that the process used here to prepare the graphite powder samples is expected also to introduce a large number of structural defects.

Previous studies on the oxidation of graphite, carbon nanotubes (graphite sheets wrapped into cylinders) [24–26] and related carbon materials revealed, in particular, that the presence of defects at the sample surface is needed for the oxygen adsorption.

Taking all the experimental evidence together, one may speculate that the ferromagnetism enhancement observed, e.g., in strongly disordered microporous carbon, see section 2, is related to oxygen trapped at defect sites. On the other hand, our previous work [2] revealed that a low-vacuum heat treatment of HOPG samples can either enhance the ferromagnetic response or trigger superconducting-like M(H) hysteresis loops even at room temperature, once again suggesting that adsorbed gases play a crucial role in the anomalous magnetic behavior of graphite.

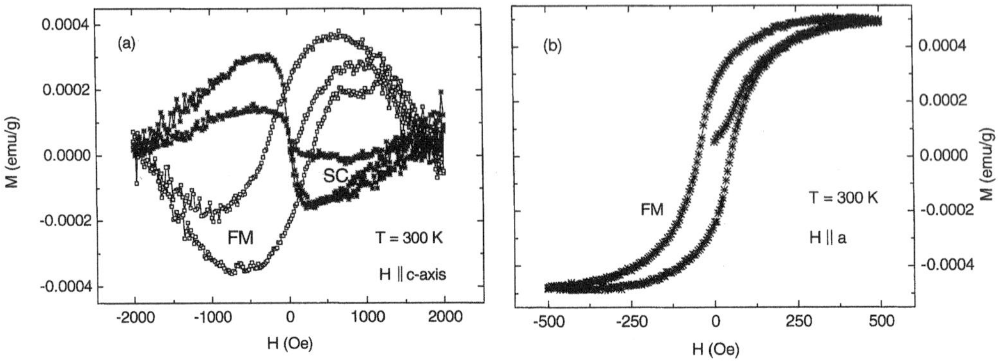

Fig. 7. (a) M(H) measured for a HOPG sample from Union Carbide Co. immediately after the heat treatment (FM), and after 14 days keeping the sample at ambient conditions (SC) when H || c. The curves were obtained subtracting the diamagnetic background magnetization $M_o(H) = -\chi H$ ($\chi = 1.61 \cdot 10^{-5}$ emu/g·Oe) from the measured signal; (b) M(H) hysteresis loop obtained with H || graphite planes (H || a) for the SC sample after the background signal subtraction ($\chi = 2.3 \cdot 10^{-6}$ emu/g·Oe).

Figure 7(a,b) presents our recent data obtained at room temperature as before [2] but now for a different HOPG sample providing additional evidence for the interplay between ferromagnetic and superconducting behavior of graphite subjected to a heat treatment. As Fig. 7(a) illustrates, the ferromagnetic-type (FM) hysteresis loop M(H) measured immediately after the heat treatment [the HOPG sample was kept at T = 870K under low vacuum (~ 0.05 mbar) during 2 hours] has been transformed to a loop characteristic of type-II superconductors (SC) when H || c. The change in the magnetic response has occurred after 14 days of keeping the sample in air at ambient conditions.

Noting, the FM → SC transformation recorded in H || c measurements, Fig. 7(a), did not affect the FM response measured with the field applied parallel to the graphite planes (H || a), see Fig. 7(b), suggesting that the superconductivity is localized within the sample basal planes. The results provide also evidence for the coexistence of superconducting and ferromagnetic states.

Also, we have studied the stability and sensitivity of this (transformed) superconducting state to additional sample heating. Fig. 8 shows the superconducting-type hysteresis loop M(H) of Fig. 7(a) together with the hysteresis loop measured immediately after heating the sample up to 350K (in situ) (H || c). As can be seen from Fig. 8, this slight heating already results in a reduction of the hysteresis loop width.

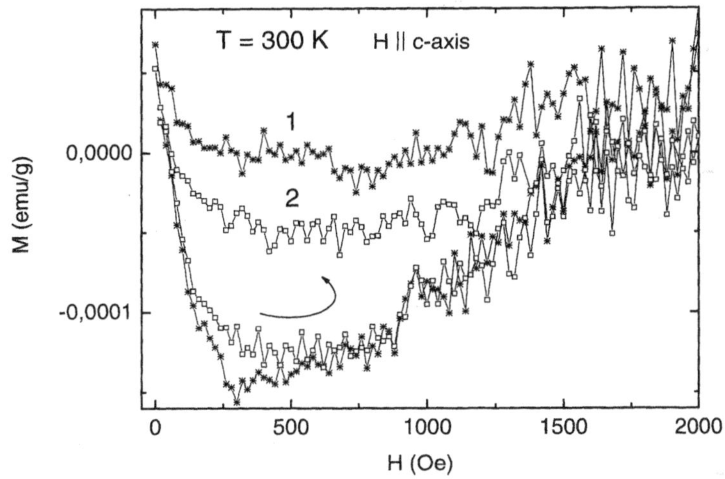

Fig. 8. Superconducting-like M(H) hysteresis loops: curve 1 is the same as in Fig. 7, and curve 2 is measured after heating the sample up to 350K. The curves were obtained subtracting the diamagnetic background magnetization $M_o(H) = -\chi H$ ($\chi = 1.61\cdot10^{-5}$ emu/g·Oe) from the measured signal.

We speculate that both the aging effect, i. e. the time dependence of the sample magnetic response, and its sensitivity to relatively small temperature variation may be related to gasses (most likely oxygen) adsorption-desorption and/or its migration to different defect sites.

Furthermore, our studies revealed that similar phenomena, i. e. the interplay between ferromagnetic and superconducting order parameters and aging effects take also place in sulfur-doped graphite samples. Results of these studies are given in the next section.

4 Coexistence of Superconducting and Ferromagnetic Instabilities in Graphite-sulfur Composites

Graphite-sulfur (C-S) composites were prepared [11] by mixing graphite powder consisting of ~ 8 μm size particles [the impurity content in ppm: Fe (32), Mo (< 1), Cr (1.1), Cu (1.5)] and sulfur powder (99.998%; Aldrich Chemical Company, Inc.) in a ratio C:S = 1:1. The mixture was pressed into pellets, held under Ar atmosphere at 650K for one hour and subsequently annealed at 400K for 10 hours before cooling to room temperature. The final sulfur contents in the composite was 23 wt.%. X-ray (θ - 2θ geometry) measurements revealed a small decrease in the c-axis parameter of the hexagonal graphite from c = 6.721 Å in the pristine graphite powder to c = 6.709 Å in the composite sample, and no changes in the lattice parameters of the orthorhombic sulfur (a = 10.45 Å, b = 12.84 Å, c = 24.46 Å). No impurity or additional phases were found [11].

Figure 9 presents magnetization M(T) measured in as-received sample (labeled here as CS1) for applied fields H = 10 Oe and H = 100 Oe. The magnetization data corresponding to the zero-field-cooled (ZFC) regime, $M_{ZFC}(T)$, were taken on heating after the sample cooling at zero applied field, and the magnetization in the field-cooled on cooling (FCC) regime, $M_{FCC}(T)$, was measured as a function of decreasing temperature in the applied field. Fig. 10 gives a detailed view of the data obtained for H = 100 Oe in the vicinity of T_c(H = 100 Oe) = 33K below which a departure of $M_{ZFC}(T)$ from $M_{FCC}(T)$ takes place. As can be seen from this plot, both $M_{ZFC}(T)$ and $M_{FCC}(T)$ become more diamagnetic at T < T_c(H).

Such magnetization behavior is characteristic of superconductors: enhancement of the diamagnetism below the superconducting transition temperature T_c(H) originates from the screening supercurrents (ZFC regime) and the Meissner-Ochsenfeld effect of magnetic flux expulsion (FFC regime). The magnetization data of Fig. 9 and Fig. 10 demonstrate that the sample superconducting volume fraction is rather small. We estimate it as ~ 0.05 % of the value expected for a bulk ideal superconductor [11].

It can also be seen in Fig. 9 that as the applied field increases, the normal state orbital diamagnetism of graphite overcomes the paramagnetic (ferromagnetic) contribution to the magnetization resulting in a negative total magnetization above T_c.

As expected for superconductors, the diamagnetic magnetization is suppressed by the applied magnetic field, see Fig. 11.

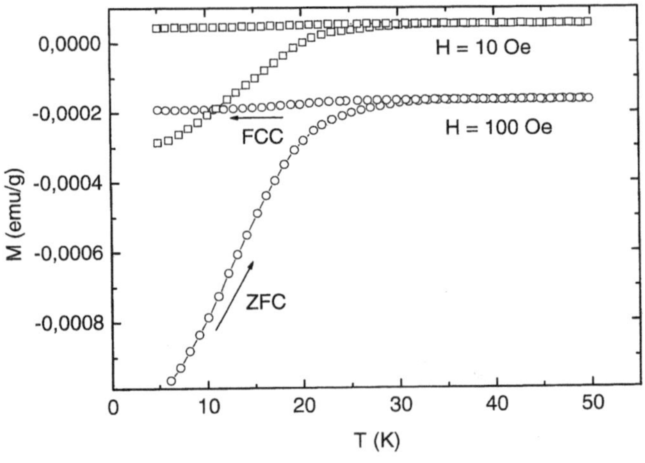

Fig. 9. Temperature dependencies of magnetization, M(T), measured in graphite-sulfur composite (CS1) in zero-field-cooled (ZFC) and field-cooled on cooling (FCC) regimes at two applied fields; 10 Oe and 100 Oe.

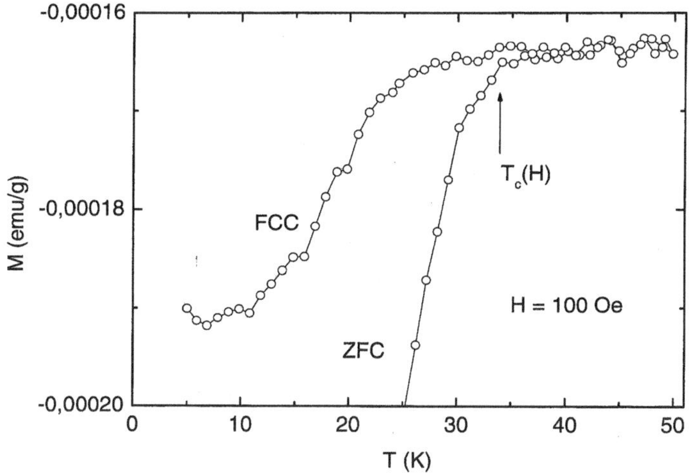

Fig. 10. Enlarged view of the superconducting transition recorded for the CS1 sample at H = 100 Oe.

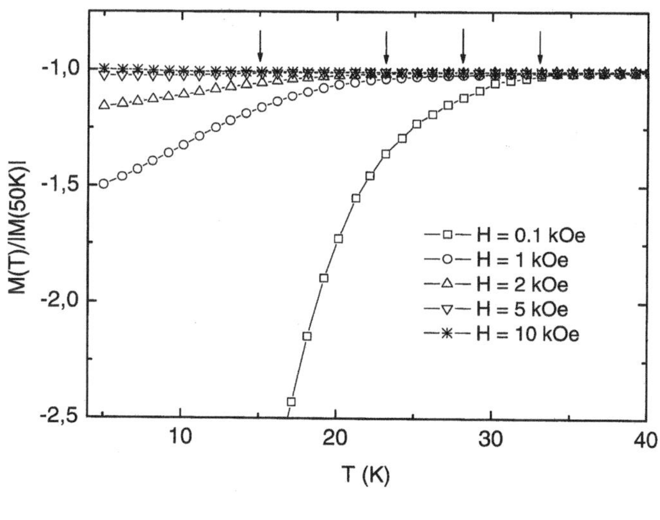

Fig. 11. Normalized ZFC magnetization measured in the sample CS1 at various applied fields. Arrows denote the superconducting transition temperature $T_c(H)$.

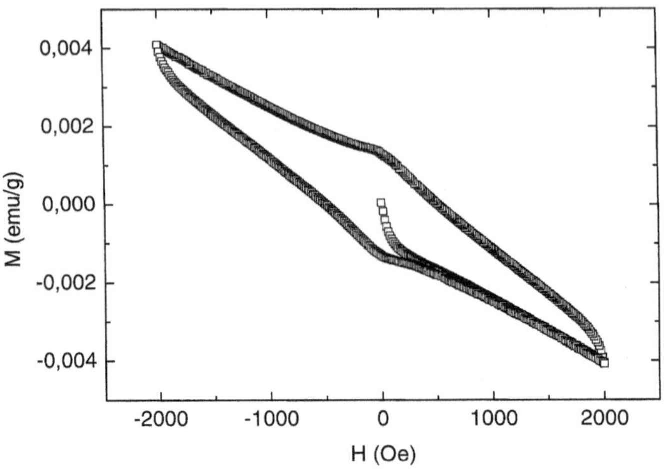

Fig. 12. Magnetization hysteresis loop M(H) measured for CS1 sample at T = 6K.

Figure 12 presents M(H) measured for the CS1 sample at T = 6K. The occurrence of magnetic hysteresis characteristic of type-II superconductors can be seen for this sample even without subtraction of the orbital diamagnetic contribution.

Similar to experiments with oxidized graphite, see section 3, the magnetic response of C-S composites is found to be time-dependent. To illustrate this we plotted in Fig. 13 the reduced magnetization measured for the same C-S sample with two weeks time interval. As can be seen from Fig. 13, the superconducting response has diminished considerably after keeping the sample at ambient conditions.

Dozens of C-S samples with different T_c and superconducting volume fraction were obtained. The results proved that powdering of graphite before the sulfur reaction is of importance for the superconductivity occurrence. Both superconductivity and ferromagnetism were observed in the C-S samples. Interestingly, the character (FM or SC) as well as the magnitude of the magnetic response can change (decrease or increase) with time continuously or abruptly [13].

In what follows we present detailed experimental evidence for the coexistence of superconductivity and ferromagnetism in graphite-sulfur composites. The data were obtained on a C-S sample (labeled here as CS3) with $T_c(0)$ = 9K whose ferromagnetic properties were unchanged during 5.5 months and remains superconducting until today (more than 3 years after preparation).

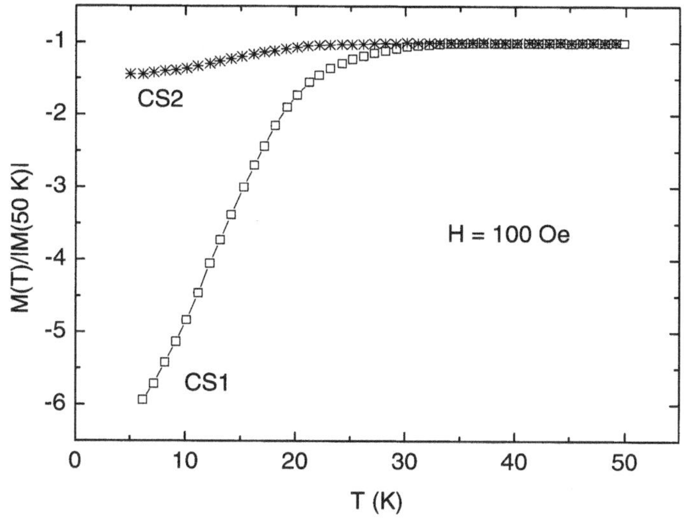

Fig. 13. Normalized ZFC magnetization measured for the same graphite-sulfur composite sample immediately after its synthesis (CS1) and after 14 days keeping the sample at ambient conditions (CS2).

The CS3 sample was prepared using graphite rods from Carbon of America Ultra Carbon, AGKSP grade, ultra "F" purity (99.9995%) (Alfa-Aesar, №40766) and sulfur chunks from American Smelting and Refining Co. that are spectrographically pure (99.999+ %). A pressed pellet (ϕ = 6 mm, ~7000 lbf) of graphite was prepared by pressing graphite powder that was produced by cutting and grinding the graphite rod on the edge and side area of a new and clean circular diamond saw blade. The graphite pellet was encapsulated with sulfur chunks (mass ratio ~ 1:1) in quartz tube under 1/2 atmosphere of argon and heat treated in a tube furnace at 400°C for one hour and then slowly cooled (4 °C/h) to room temperature. X-ray diffraction measurements (θ-2θ geometry and rocking curves) of the reacted sample yielded a spectrum with only the superposition of the (00ℓ) diffraction peaks of graphite with the orthorhombic peaks of sulfur with no extra peak due to a compound, second phase or impurity. The c-axis lattice parameter (c = 6.72 Å) of the sample is equal to the pristine graphite powder pellet, which precludes sulfur intercalation. The diffraction pattern also shows a (00ℓ) preferred orientation, which was confirmed by rocking curve scans that yield a $\Delta\theta = 6°$ (FWHM) for the (002) peak, due to the highly anisotropic (plate-like) shape of the graphite grains. The sample (~ 5×2.5×1.7 mm^3) was cut from the reacted pellet and used for the magnetic moment measurements as well as the above described analyses. A lower limit for the superconducting volume fraction was estimated as 0.02% [13].

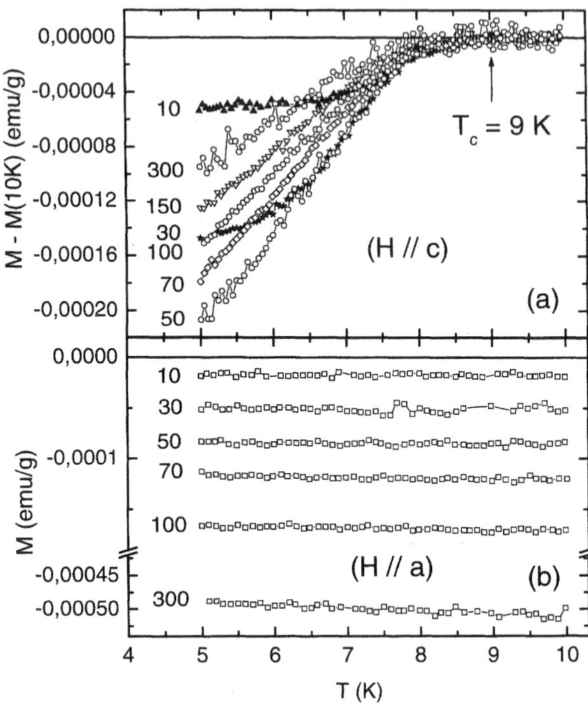

Fig. 14. (a) temperature dependencies of ZFC magnetization M(T) - M(T =10K), obtained for various magnetic fields H || c: (▲), H = 10 Oe; (★), H = 30 Oe; (○), H = 50 Oe; (◊), H = 70 Oe; (○), H = 100 Oe; (∇), H = 150 Oe; (○), H = 300 Oe; (b) temperature dependencies of ZFC magnetization obtained with H || a for various magnetic fields, as indicated in Oe next to each curve.

Shown in Fig. 14(a) are temperature dependencies of the ZFC magnetization M(T) after subtraction the normal state value at T = 10K, M(10K), measured for various magnetic fields applied perpendicular to the largest surface of the sample (H || c). It can be seen from Fig. 14(a) that the difference |M(T)-M(10K)| increases below the superconducting transition temperature T_c = 9K. At the same time, M(T) measurements with the applied magnetic field parallel to the main sample surface (H || a) yield a different magnetic response. As Fig. 14(b) illustrates, no sign of the superconducting transition could be detected within the data noise of ~ 5×10^{-6} emu/g. Note that the scale ranges in Fig. 14(b) and Fig. 14(a) are practically the same. These results indicate that the superconducting state is highly anisotropic and is associated with the graphite planes.

Figure 15 shows magnetization hysteresis loops M(H)-M_o(H) measured with the ZFC procedure and H || c for T = 7, 8, 9 and 10K, where $M_o = \chi$ H ($\chi = -7.12\times10^{-6}$ emu·g^{-1}·Oe^{-1})

is the diamagnetic background signal. Fig. 15(a) demonstrates a characteristic of type-II superconductors hysteresis loop obtained at T = 7K. As temperature increases above T_c = 9K, the measured hysteresis loops resemble those of ferromagnetic materials, see Fig. 15(d). For temperatures at and just below T_c, Fig. 15(b, c), the presence of both superconducting and ferromagnetic contributions to the measured magnetization can be seen.

Thus, results presented in Fig. 15 provide evidence for the coexistence of superconductivity and ferromagnetism in the C-S.

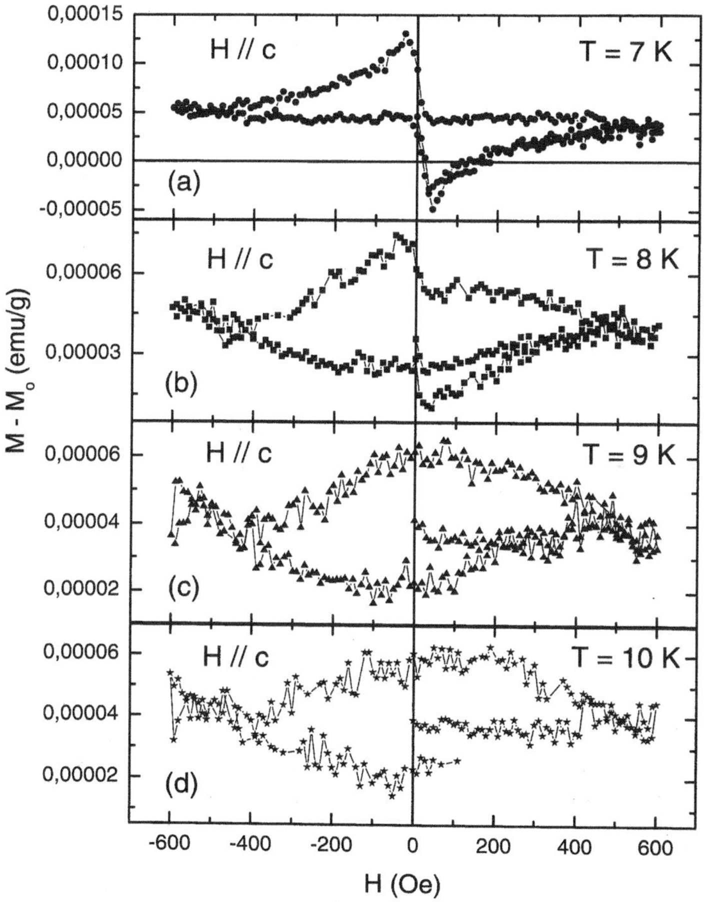

Fig. 15. Magnetization hysteresis loops M(H) measured for CS3 sample with H || c after subtraction of the diamagnetic background signal $M_o = \chi H$ ($\chi = -7.12\times10^{-6}$ emu·g^{-1}·Oe^{-1}) at (a) T = 7K, (b) T = 8K, (c) T = 9K and (d) T = 10K.

Again, for the M(H) measurements with the applied magnetic field parallel to the graphite planes (H || a), a different magnetic response is obtained. This can be seen in Fig. 16 (a-c) which presents M(H) obtained for T = 5, 7 and 9K after subtraction of the linear diamagnetic background signal $M_o = \chi H$, where $\chi = -2.25\times10^{-6}$ emu·g^{-1}·Oe^{-1}: three almost identical ferromagnetic-type hysteresis loops are clearly seen both in the superconducting and normal states (the ferromagnetic behavior of C-S composites persists well above the room temperature [13]).

Noting also that angle dependent magnetization measurements performed on the CS3 sample [14] provided evidence for the interaction between superconducting and ferromagnetic order parameters [14]. The interaction revels itself through a rotation of the ferromagnetic moment direction by 90° at $T < T_c(0)$. This fact is particularly intriguing taking into account the smallness of the measured total superconducting signal.

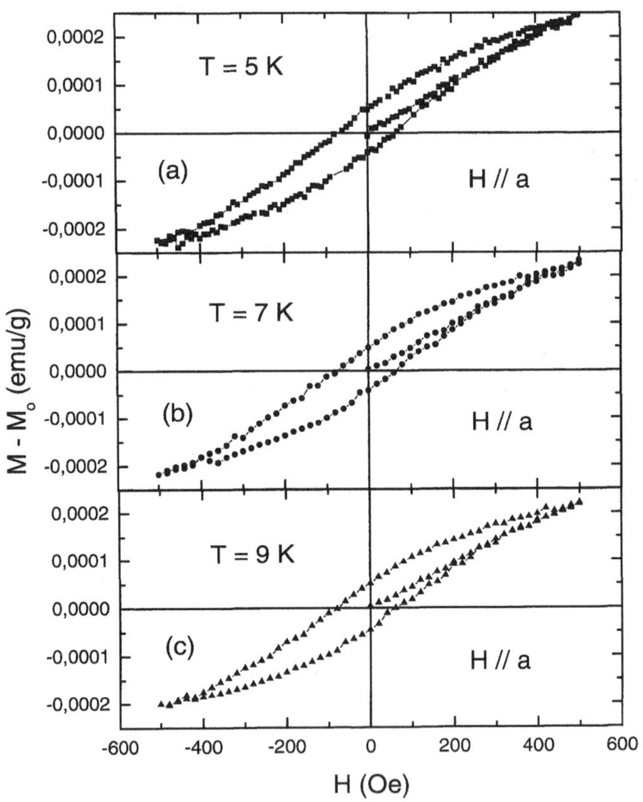

Fig. 16. "In-plane" (H || a) hysteresis loops $M(H) - M_0(H)$, $M_o = \chi H$, $\chi = -2.25\times10^{-6}$ emu·g^{-1}·Oe^{-1}, obtained for CS3 sample at (a) T = 5K, (b) T = 7K and (c) T = 9K.

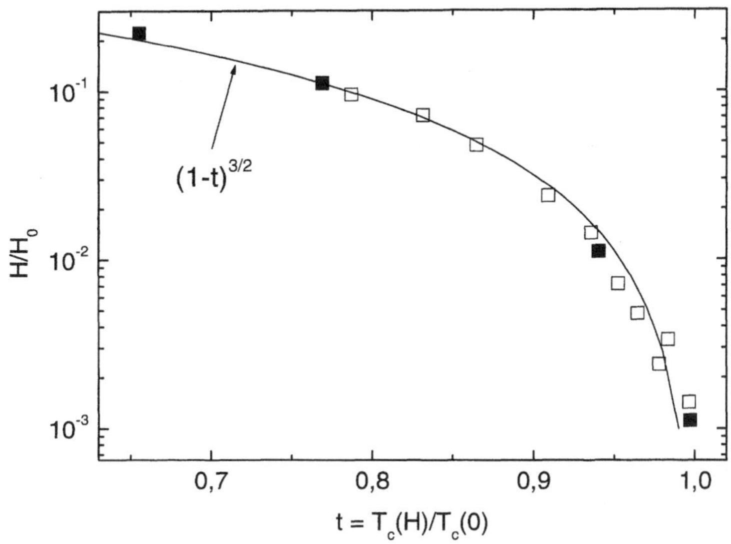

Fig. 17. Reduced magnetic field H/H_0 vs. reduced temperature $t = T_c(H)/T_c(0)$ measured for CS1 (■) [11] and CS3 (□) samples [13]; solid line corresponds to the equation $H/H_0 = [1-T_c(H)/T_c(0)]^{3/2}$ with $H_0 = 9$ kOe, $T_c(0) = 35$K for CS1, and $H_0 = 21$ kOe, $T_c(0) = 9$K for CS3.

It is interesting to compare $T_c(H)$ measured for CS1 and CS3 samples. The data given in Fig. 17 illustrate that $H(T_c)$ in the vicinity of the zero-field transition temperature $T_c(0)$ can be very well described by the power law $H = H_0[1 - T_c(H)/T_c(0)]^{3/2}$ for both samples.

This suggests a common origin for the superconducting transition in these samples in spite of their relatively large difference in $T_c(0)$. The observed nonlinear $H \sim (1 - t)^{3/2}$ dependence near $T_c(0)$ may be related to the upper critical field boundary associated with the Bose-Einstein condensation of preformed Cooper pairs [27]. Alternatively, the measured $H(T_c)$ boundary can be accounted for by the coupling between preexisting superconducting clusters that have a higher superconducting transition temperature [11, 13, 28]. Supporting this last scenario, $T_c(0)$ measured immediately after the CS3 sample synthesis was about 65K which has declined with time and stabilized at $T_c(0) = 9$K after 11 days, see Fig. 18. Finally, we note that superconductivity in C-S samples is localized within de-coupled, on a macroscopic scale, clusters preventing the zero-resistance state of the macroscopic sample [11–13].

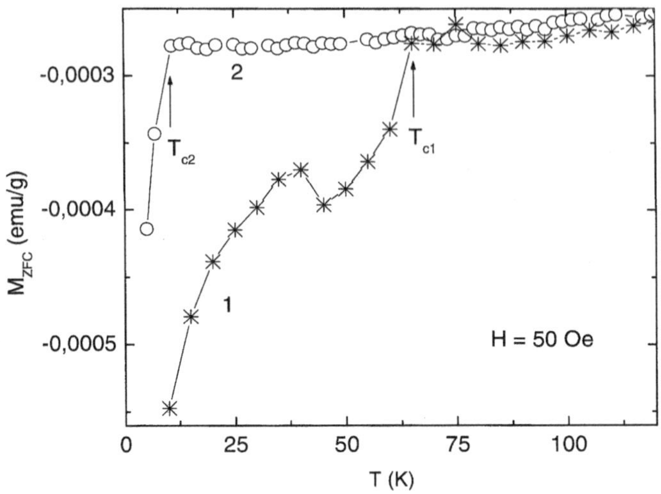

Fig. 18. ZFC magnetization, $M_{ZFC}(T)$ measured at applied field H = 50 Oe for the CS3 sample immediately after the sample synthesis (curve 1) and after 11 days (curve 2).

5 Discussion and Concluding Remarks

Because of the relatively weak ferromagnetic response of carbon-based materials, the possible effect of magnetic impurities, such as Fe, Ni, etc., cannot be ruled out a priori. This issue has been addressed by various researchers, see e. g. [29], who concluded that the measured carbon ferromagnetism cannot be accounted for by impurities. Such conclusion, however, does not exclude the possibility that magnetic impurities trigger ferromagnetism in the carbon system [8, 30, 31]. At the same time, the observations of ferromagnetism induced by graphite oxygenation, reported in this chapter, as well as the ferromagnetism triggered by proton irradiation [15] indicate that impurity-induced magnetism is not the whole story, at least. It is also clear, that the superconductivity in graphite caused by sulfur doping or re-arrangement of adsorbed oxygen on graphite surface has nothing to do with ferromagnetic impurities.

In order to clarify the nature of the observed phenomena, it is instructive to look more closely to previous studies of oxygenation of graphite and related carbon materials, see e.g. Refs. [24–26]. In particular, the results revealed an extreme sensitivity of the electrical resistance and thermoelectric power of carbon nanotubes (graphite sheets wrapped into cylinders) to oxygen (air) exposure [25]: the oxygen adsorption (1) decreases the nanotube resistance, and (2) it changes the thermoelectric power sign from negative to a positive one, suggesting that the oxygen-induced nanotube doping is with hole carriers [25, 26]. It has been demonstrated that the oxidation is triggered by the presence of graphite surface defects [25].

Adsorption of hydrogen atoms on graphite [32] also leads to a local charge enhancement as well as changes in the electronic structure. It has been theoretically shown [33] that the adsorption of atomic hydrogen opens a gap in the electronic spectrum of graphene in which a spin-polarized gap state is situated.

Furthermore, x-ray absorption spectra measured in C-S composites revealed that the outermost s-states of adsorbed S atoms interact with the graphite interlayer states [34].

All this makes us believe that a combined effect of structural disorder and adsorbed foreign atoms (molecules) such as S, H, O (O_2) can be behind the anomalous magnetic behavior of graphite and related carbon materials. Then, it is not unreasonable to assume that aging effects, including the FM $\rightarrow$ SC transformation, are related to migration of foreign elements on the sample surface. The very small (~ 0.01...0.05%) Meissner as well as shielding effects can also be understood assuming the formation of superconducting patches at the graphite surface.

The current state of experiment does not allow us to discriminate unambiguously between theoretical models proposed to account for both high temperature ferromagnetism and superconductivity in graphite-based materials; for review articles see Refs. [3, 35, 36].

Nevertheless, we would like to recall here the mechanism for surface high temperature superconductivity proposed by Ginzburg [37]. In that model, conducting electrons (holes) interact with adsorbed atoms at the sample surface which leads to an effective attraction between the surface carriers. According to the BCS theory, $k_B T_c = \hbar\omega \exp[-1/N(0)V]$, where N(0) denotes the density of states at the Fermi level and V is the mean matrix element of the interaction energy corresponding to attraction. In the present case $\hbar\omega$ is the energy corresponding to the difference between energy levels of adsorbed atoms which is of the order of 1 eV. Then, considering the characteristic value for $N(0)V \approx 0.4...0.5$ [38], one gets for the superconducting transition temperature $T_c \sim 1000K$. Thus, superconducting signatures detected in graphite at room temperature, see Figs. 7, 8 and Ref. [2], can easily be accounted for by the Ginsburg's model.

This work has been supported by FAPESP and CNPq Brazilian science agencies.

References

1. Y. Kopelevich, V.V. Lemanov, S. Moehlecke, and J.H.S. Torres, Physics of the Solid State 41 (1999) 1959 [Fizika Tverd. Tela 41 (1999) 2135].
2. Y. Kopelevich, P. Esquinazi, J.H.S. Torres, and S. Moehlecke, J. Low Temp. Phys. 119 (2000) 691.
3. Y. Kopelevich, P. Esquinazi, H.S. Torres, R. R. da Silva, H. Kempa, F. Mrowka, and R. Ocaña, in: Studies of High-Tc Superconductors, 45 (2003) 59, ed. by A. V. Narlikar, Nova Sci Pub., N.Y.
4. Y. Kopelevich, P. Esquinazi, J.H.S. Torres, R.R. da Silva, and H. Kempa, Advances in Solid State Physics 43 (2003) 207.
5. T.L. Makarova, B. Sundqvist, R. Höhne, P. Esquinazi, Y. Kopelevich, P. Scharff, V.A. Davydov, L.S. Kashevarova, and A.V. Rakhmanina, Nature 413 (2001) 716.
6. R.A. Wood, M.H. Lewis, M.R. Lees, S.M. Bennington, M.G. Cain, and N. Kitamura, J. Phys.: Condens. Matter 14 (2002) L385.

7. V.N. Narozhnyi, K.-H. Müller, D. Eckert, A. Teresiak, L. Dunsch, V.A. Davydov, L.S. Kashevarova, and A.V. Rakhmanina, Physica B329 (2003) 1217.
8. J.M.D. Coey, M. Venkatesan, C.B. Fitzgerald, A.P. Douvalis, and I.S. Sanders, Nature 420 (2002) 156.
9. X. Wang, Z.X. Liu, Y.L. Zhang, F.Y. Li, and C.Q. Jin, J. Phys.: Condens. Matter. 14 (2002) 10265.
10. Y. Kopelevich, R.R. da Silva, J.H.S. Torres, A. Penicaud, and T. Kyotani, Phys. Rev. B68 (2003) 092408.
11. R.R. da Silva, J.H.S. Torres, and Y. Kopelevich, Phys. Rev. Lett. 87 (2001) 147001.
12. N.P. Yang, H.H. Wen, Z.W. Zhao, and S.L. Li, Chin. Phys. Lett. 18 (2001) (2001).
13. S. Moehlecke, Pei-Chun Ho and M.B. Maple, Phil. Mag. B82 (2002) 1335.
14. S. Moehlecke, Y. Kopelevich, and M.B. Maple, Phys. Rev. B69 (2004) 134519.
15. P. Esquinazi, D. Spemann, R. Höhne, A. A. Setzer, K.-H. Han, and T. Butz, Phys. Rev. Lett. 91 (2003) 227201.
16. J. González, F. Guinea, and M.A.H. Vozmediano, Phys. Rev. B63 (2001) 134421.
17. Z.X. Ma, T. Kyotani, Z. Liu, O. Terasaki, and A. Tomita, Chem. Mater. 13 (2001) 4413.
18. Z. Ma, T. Kyotani, and A. Tomita, Carbon 40 (2002) 2367.
19. D.W. Boukhvalov, P.F. Karimov, E.Z. Kurmaev, T. Hamilton, A. Moewes, L.D. Finkelstein, M.I. Katsnelson, V.A. Davydov, A.V. Rakhmanina, T.L. Makarova, Y. Kopelevich, S. Chiuzbaian, and M. Neumann, Phys. Rev. B69 (2004) 115425.
20. P.J.F. Harris, A. Burian, and S. Duber, Phil. Mag. Lett. 80 (2000) 381.
21. V.D. Blank, S.G. Buga, G.A. Dubitsky, N.R. Serebryanaya, M.Yu. Popov, and B. Sundqvist, Carbon 36 (1998) 319.
22. Y. Murakami and H. Suematsu, Pure Appl. Chem. 68 (1996) 1463.
23. T.L. Makarova, K.-H. Han, P. Esquinazi, R.R. da Silva, Y. Kopelevich, I.B. Zakharova, and B. Sundqvist, Carbon 41 (2003) 1575.
24. S.M. Lee, Y.H. Lee, Y.G. Hwang, J.R. Hahn, and H. Kang, Phys. Rev. Lett. 82 (1999) 217.
25. P.G. Collins, K. Bradley, M. Ishigami, and A. Zettl, Science 287 (2000) 1801.
26. S.-H. Jhi, S.G. Louie, and M.L. Cohen, Phys. Rev. Lett. 85 (2000) 1710.
27. V.V. Kabanov, V.N. Zavaritsky, and A.S. Alexandrov, J. Supercond. 17 (2004) 113.
28. Y. Kopelevich, R.R. da Silva, J.H.S. Torres, S. Moehlecke, and M.B. Maple, Physica C 408-410 (2004) 77.
29. P. Esquinazi, A. Setzer, R. Höhne, C. Semmelhack, Y. Kopelevich, D. Spemann, T. Butz, B. Kohlstrunk, and M. Lösche, Phys. Rev. B65 (2002) 241101.
30. O. Céspedes, M.S. Ferreira, S. Santivo, M. Kociak, and J.M.D. Coey, J. Phys.: Condens. Matter. 16 (2004) L155.
31. H.-Ch. Mertins, S. Valencia, W. Gudat, P.M. Oppeneer, O. Zaharko, and H. Grimmer, Europhys. Lett. 66 (2004) 743.
32. P. Ruffieux, O. Gröning, P. Schwaller, L. Schlapbach, and P. Gröning, Phys. Rev. Lett. 84 (2000) 4910.
33. E.J. Duplock, M. Scheffler, and P.J.D. Lindan, Phys. Rev. Lett. 92 (2000) 225502.
34. E.Z. Kurmaev, A.V. Galakhov, A. Moewes, S. Moehlecke, and Y. Kopelevich, Phys. Rev. B66 (2002) 193402.
35. T.L. Makarova, in: Studies of High-Tc Superconductors, 45 (2003) 107, ed. by A.V. Narlikar, Nova Sci Pub., N.Y.
36. T.L. Makarova, Semiconductors 38 (2004) 615.
37. V.L. Ginzburg, Phys. Lett. 13 (1964) 101; Sov. Phys. JETP 20 (1965) 1549.
38. W.A. Little, Phys. Rev. 134 (1964) A1416.

Carbon-Based Magnetism
T. Makarova and F. Palacio (Editors)

19

Induced Magnetic Order by Ion Irradiation of Carbon-based Structures

P. Esquinazi, R. Höhne, K.-H. Han[1], D. Spemann, A. Setzer, M. Diaconu, H. Schmidt, and T. Butz

Institute for Experimental Physics II, University of Leipzig, Linnéstrasse 5, 04103 Leipzig, Germany
Present adress: Umeå University, 90187 Umeå, Sweden

1 Motivation

Irradiation effects in graphite were one major research area in the past, partially due to its application as a moderator in thermal nuclear reactors. Graphite is still a material of choice for nuclear applications due to its low cross-section for neutron absorption. The influence of irradiation damage produced by different kinds of ions on several properties of graphite was reviewed by Kelly [1] in chapter 7 of his book. Recently, irradiation effects in carbon nanostructures were reviewed by Banhart[2]. The effect of neutron irradiation on the magnetic properties of graphite has been studied in the past and shows the expected results, *i.e.* the introduction of lattice defects by irradiation produces a decrease in the diamagnetism and an increase in the spin density [1]. We are not aware of any study made in the past on the effects of proton irradiation on the magnetic properties of graphite or carbon-based structures.

When we started this research work on irradiation effects on the magnetism of graphite we had two reasons to choose protons as energetic particle. The first one is related to the analysis method called PIXE (Particle Induced X-ray Emission) that uses protons to get a map for all relevant impurity elements within a sample depth of 30 μm for a proton energy of ~ 2 MeV in carbon (see Sect. 2.2 for details). A systematic and full characterization of the magnetic impurity content in each of the samples, and after each treatment or handling (it makes no sense to start with a highly pure sample and then cut it with a steel knife afterwards), is of primary importance and absolutely necessary.

The second reason was based on early reports on room-temperature ferromagnetic behavior in some carbon-based structures (see references in [3]). From those early works our attention was focused to the magnetic properties found in amorphous-like carbon prepared from different hydrogen-rich starting materials where an increase of the saturation magnetization with the hydrogen concentration in the starting material was found [4, 5]. The origin for the magnetic ordering has been related to the mixture of carbon atoms with sp^2 and sp^3 bonds, which was predicted to reach a magnetization higher than in pure Fe [6]. Hydrogen, on the other hand, was assumed to have a role only in the formation of the amorphous carbon structure [5]. New theoretical predictions, however, show that hydrogenated graphite can display spontaneous magnetization coming from different numbers of mono- and dihydrogenated carbon atoms [7]. Spontaneous magnetization may also appear in the case of monohydrogenated zigzag edges [8] if the distance between them is large or if they are not in parallel.

The advantage of proton irradiation is twofold: it enables us to make an impurity analysis simultaneously to the implantation of hydrogen. In this chapter we will review the main effects obtained after proton irradiation in different carbon-based structures. This chapter is organized as follows. In the next section we provide the main characteristics of our irradiation facility. In section 2.2 we show an example of element analysis obtained in one of the graphite samples used for the irradiation studies. The irradiation effects are reviewed in section 3. This section is divided in two main subsections that describe the effects in oriented graphite and carbon-based thin films. In section 4 we discuss some of the effects observed after annealing the sample at high temperatures in vacuum or after leaving it at room temperature for a long period of time.

In this chapter we concentrate ourselves mainly on the effects produced by proton irradiation in highly oriented pyrolitic graphite (HOPG). Effects of irradiation with alpha particles will be discussed shortly in section 3.1.3. Iron and fluorine irradiation effects on HOPG and on diamond are currently under way and the results will be published elsewhere in the future.

2 Irradiation Characteristics

All the irradiations presented in this chapter were done with the high-energy nanoprobe LIPSION of the University of Leipzig. The accelerator is a single ended 3 MV SINGLETRON™ with an RF-ion source for protons and alpha particles. The focusing system can deliver proton beams with diameters as low as 40 nm at very low currents of the order of 0.1 fA. For the present irradiations we worked at 2.25 MeV and currents up to 500 pA in the case of proton beam of diameter of 1–2μm (microbeam). We have also irradiated samples with a broad beam of 0.8 mm diameter, energy of 2 MeV and currents between ~50 and ~150 nA. The X-ray detector is an Ortec HPGe IGLET-X and subtends 187 msrad solid angle. The numerical simulations discussed in this section are based on the code SRIM2003 [9]. The samples were mostly attached to a Si substrate using a small amount of Varnish or a mixture of Varnish with ultra-pure graphite powder. In one case we fixed the sample on a messing plate with a hole to enable the free penetration of the energetic particles (H^+ and He^+). The magnetic moment of the substrates was always measured before attaching the sample. All the irradiations were done at nominally room temperature; the temperature of the sample was not controlled during irradiation.

2.1 The Interaction of MeV-protons with Graphite

Figure 1(a) shows a typical profile for the number of defects created in a carbon matrix per 2.25 MeV-proton and per 1 m interval. The penetration range of about 46 m is clearly visible. For samples thicker than this range one expects that the majority of the protons will be stopped, whereas for samples substantially thinner essentially all protons traverse the sample. This means that for 20 m or thinner samples the implantation of protons should be undetectable. Strikingly, enhancement of ferromagnetism has also been observed in disordered carbon [11] and fullerene films deposited on Si substrates after proton irradiation (see section 3.2). Further details on the different processes that influence the penetration of protons in graphite as well as the restrictions of the SRIM2003 simulations were discussed by [10]. The question how many protons remain in thinner samples cannot be clearly answered yet. Experiments with thinner and unsupported samples (to rule out part of the backscattering processes)

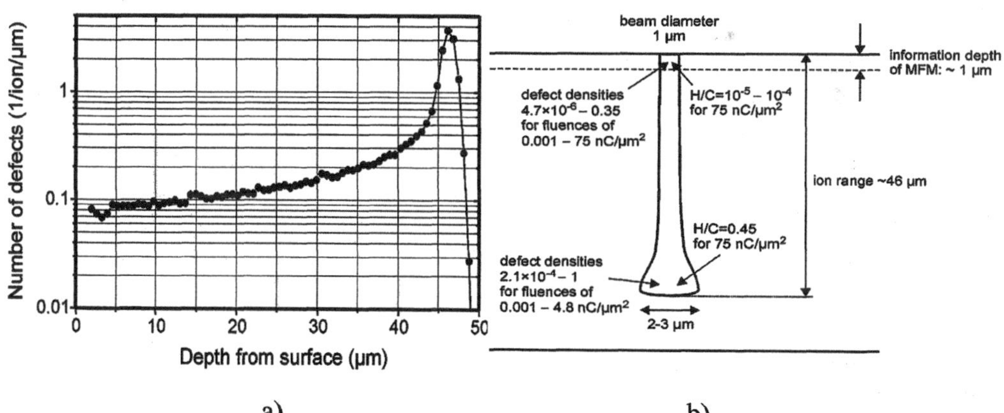

a) b)

Fig. 1. SRIM2003 Monte Carlo simulations [9] for 2.25 MeV protons on HOPG. (a) The number of defects per ion and 1 m depth interval. The penetration range of 46 m is clearly visible. (b) Sketch of the modified area in graphite due to 2.25 MeV proton bombardment. The lateral straggling has a full width at half maximum (FWHM) of ~2.5 m for an initial beam diameter of **1** m. The numbers are obtained assuming a displacement energy of 35 eV for Frenkel pairs in HOPG. Adapted from [10].

are necessary and will be performed in the future. Under the assumption that there is no annealing of defects during irradiation and that the damaged area is of the same size as the irradiated one, the numerical simulations indicate that for fluences 0.01…75 nC/μm^2 we get in the near surface region between 4.7×10^{-6} and 0.35 displacements per carbon atom, *i.e.* complete amorphization for the highest fluence, using a displacement energy of 35 eV for Frenkel pairs in HOPG, in agreement with recently published studies of the damage cascades by irradiation on graphite [12]. Changing this number to 28 eV yields defect densities which are about 30% higher. Towards the end of the track, the lateral beam straggling becomes important and the damaged area can be several microns wide, see Fig. 1(b). For a fluence of 75nC/μm^2 we have $\sim 5\times10^{11}$ protons/μm^2, *i.e.* the regions where defects are created by each individual proton overlap heavily.

Note that even for the highest fluence the relative concentration of hydrogen (H/C, see Fig. 1) in the first micrometer from the surface remains rather low, of the order of 10^{-4} or lower. However, most of what happens after the defect formation is still unclear. A dangling bond could attract a hydrogen atom - not necessarily from the proton implantation but already present as impurity in the sample. Interestingly enough, there is little information on residual hydrogen from pyrolysis of hydrocarbons, possibly because detection methods for hydrogen contents of the order of 100 ppm or below were not readily available and such quantities were of little relevance for applications as moderators in reactors. Hydrogen atoms in the van der Waals gap or within the graphite layers should be highly mobile, contrary to hydrogen atoms trapped at defects. Investigations with deuterium [13] indicate that deuterium does not readily diffuse out of HOPG but is rather chemically bound up to D/C ratios of about 0.45, probably at lattice defects. The maximum retention of deuterium depends on the temperature and implantation energy. A broad range of binding energies for hydrogen in graphite up to about 4 eV was reported [14], *i.e.* even temperatures as high as 3000–3500K may be not sufficient to eliminate all incorporated hydrogen. On the other hand, the effective activation energies for hydrogen diffusion in our system are not necessarily the same as, for example, those obtained by experimental methods – usually at high temperatures – to study kinetics of diffusion of hydrogen in graphite. Irradiation effects of MeV protons on diamond-like films were studied in Ref. 15. According to the authors, below a fluence of the order of 1 nC/μm^2 the hydrogen atoms produced by ion irradiation could be recaptured by dangling bonds. For higher fluences a release of hydrogen is expected.

Irradiation at a fixed energy has a clear disadvantage. If the magnetic ordering is triggered at a specific density of protons and defects, then, it is clear that with fixed proton and defect distributions, such as the one shown in Fig. 1, we would have a rather narrow window to get a maximum effect. Therefore, it should not be a surprise if for some irradiation energies and fluences one measures negligible effects or even a reduction of the magnetic order present in the sample before or after some irradiation steps. Carbon magnetism as well as irradiation effects on it belong to a new field in magnetism. For graphite bulk and thin film samples there are still many questions to be clarified in the future concerning the hydrogen implantation by irradiation and its effects on the magnetism.

The defect formation process by high energy protons is a non-equilibrium athermal process and it appears rather unlikely that ordered arrays of defects are formed by migration of interstitial carbon atoms or vacancies, maybe with the exception of the interstitial across the gallery. According to [2] and from electron irradiation studies, the essential types of radiation damage up to intermediate temperatures are the rupture of basal planes (due to shift of the C-atoms out of the plane) and the aggregation of interstitials into small dislocation loops between the graphene layers. The migration energy of the interstitial depends whether it is bounded. Di-interstitials were proposed to explain the irradiation-induced amorphization of graphite with a migration energy of 0.86 eV [16]. The interstitial loops are stable up to rather high temperatures, probably to 1000°C [2]. Certainly, the layer structure should remain essentially intact, at least for our low fluences and at the sample surface.

Irradiation can also produce a transition from sp^2 to sp^3 bonding leading to cross-links between the graphene layers and the formation of sp^3 clusters [17]. These clusters appear to be stable and do not anneal at high temperatures. Our own Raman and X-ray Photoemission Spectroscopy (XPS) measurements indicate the following. For all unirradiated areas of the HOPG samples the Raman spectra show only one pronounced peak at 1580 cm^{-1} (the E_2g_2 mode) as expected for graphite without any disorder. For the spots produced with the proton microbeam the Raman spectra show additionally the

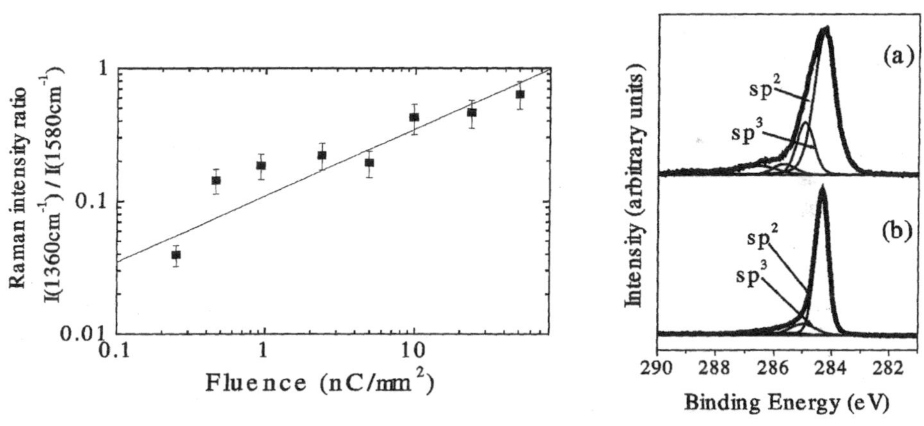

Fig. 2. Left figure: Raman intensity ratio as a function of the irradiation fluence for spots created on a HOPG sample (see section 3.1.2). The line represents the fit curve $0.1x^{0.5}$. Right figure: Decomposition of the XPS C_{1s} core-level peaks of HOPG surface after (a) and before proton irradiation (b). The components after peak fit are: at 284.4 eV: "C-sp^2"; at 285.0 eV: defect peak "C-sp^3"; other components: different C-O bonds. The thick and thin solid lines denote the experimental curves and the components, respectively. The measurement was done after irradiation of the sample with 1.2 mC and a fluence of 0.6 nC/mm^2 (step 4 of the sample reported by [18]). Adapted from [11].

disorder mode D at 1360 cm^{-1}. With increasing fluence the E_2g_2 and D modes become broader and the ratio of their intensities $I(D)/I(E_2g_2)$, as a measure of the degree of disorder, increases (see Fig. 2).

The XPS results show a clear difference between an irradiated and an unirradiated sample, see Fig. 2. The C_{1s} core-level peaks of HOPG samples were recorded using a pass energy of 10 eV and were fitted by six components with binding energies of 284.4 eV (main peak of HOPG, C-sp^2), 285.0 eV (defect peak "C-sp^3"), 285.8 eV, 286.3 eV, 289.0 eV (C-O components) and 290.9 eV (shake-up peak). The main result is that the defect peak component increases after proton irradiation. A similar behaviour was found for HOPG under plasma low-energy argon-ion bombardment [19]. Whether this defect peak corresponds to a pure C-sp^3 state or whether some of the "defect" carbon atoms are bounded to hydrogen is unclear and it is a matter of current research [20]. The knowledge of the C-H sp^2 and sp^3 bonds is for the magnetic properties of carbon structures of importance. We remark that numerical simulations indicate a 100% polarized π-band, *i.e.* a ferromagnetic order stable at room temperature, for a graphene layer with a mixture of sp^2 and sp^3 bondings (mono- and di-hydrogenated) at the zigzag edges of a graphene layer [7].

Other aspect of the irradiation effects on graphite that may be of importance to trigger magnetic ordering is the formation of pentagons and heptagons in the basal planes, which could cause a bending of the graphene layers [2]. The influence on the electronic band structure of graphite of such topological defects has been studied theoretically by [21]. According to these authors, these defects may trigger ferromagnetism or even superconductivity.

For thick targets and sufficiently high fluences, the target surface actually swells, which can be easily measured by an atomic force microscope (AFM), see Fig. 8(a), or is even visible under an optical microscope [22, 23]. The swelling in the *c*-direction occurs together with the contraction in the graphene layer; the new formed interstitial planes push the existing planes apart leading to a protuberance at the sample surface. This irradiation effect was studied by [24–26]. This effect was not observed for thin enough targets for the usual proton irradiation fluences used in this work.

Another important aspect regarding the irradiation effects is the heat load due to the beam. For a spot of $2\mu m \times 2\mu m$ and a current of 500 pA this amounts to ~ 300 W/mm^2 for thick samples. Most of the energy will be deposited near the end of the ion track. Depending on the lateral dimensions and the thickness of the sample as well as the target holder, heat is transported away from the beam spot. The details are difficult to calculate, but the measurement and control of the temperature is of importance. For thin samples, this is relatively unimportant. We have evidence that at higher currents the magnetic response in MFM-measurements is strongest at the rim of the spot suggesting annealing effects in the center of the spot, as Fig. 3 indicates. For lower currents the magnetic image is homogeneously distributed across the spot (see section 3.1), in contrast to the magnetic "ring" shown in Fig. 3.

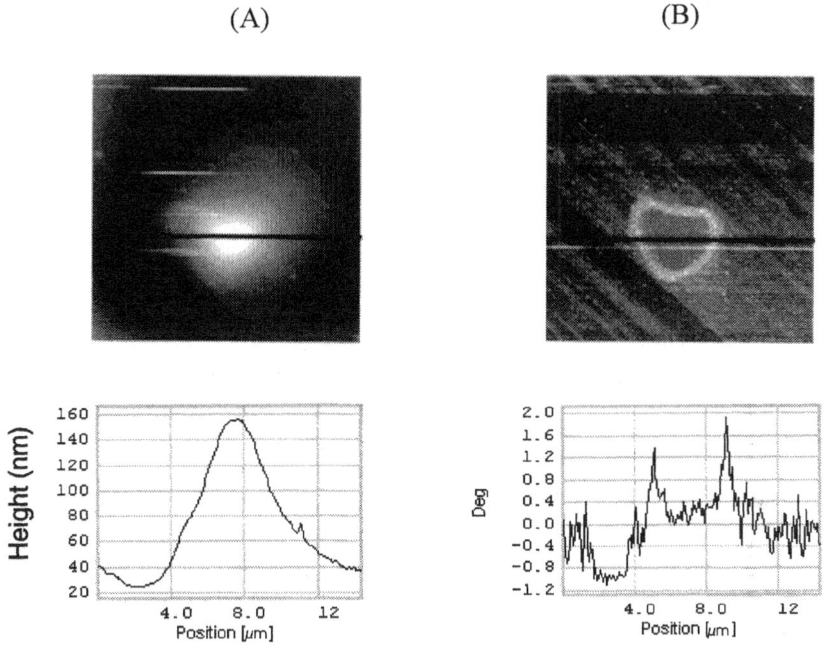

Fig. 3. (A) Topography and (B) MFM images obtained from a spot produced with a dose of 6.8 nC/μm^2 and a current $I = 0.51$ nA, which means a current density larger than 5 mA/cm^2. The scan area is 15μm×15μm. The bottom figures show the line scans done at the positions of the black lines of the upper pictures. The MFM images were taken at a distance of 50 nm using standard magnetic tips. Taken from [23]

2.2 Impurity Measurements

With the proton micro-beam a total charge of 0.5 μC suffices to obtain a minimum detection limit for Fe impurities below 1μg/g. With our broad-beam PIXE facility we work with 2 MeV protons and currents of 150 nA and a beam diameter of 0.8 mm. Here, about 5μC are required to reach a minimum detection limit of 1μg/g for Fe. The fluences are in the range of 1…50 μC/mm^2 for the micro-beam and the broad-beam, respectively. These fluences are several orders of magnitude lower than the fluences actually applied for the induction of magnetic ordering in graphite, as will be discussed below. The advantage of the micro-beam compared to the broad-beam is that a distribution map for all relevant impurity elements can be obtained contrary to the integral value for the broad-beam method, a rather important issue considering the grossly inhomogeneous distribution of impurities like Fe which we have encountered in some samples (see for example Fig. 1 in [27]). A typical broad beam PIXE spectrum for

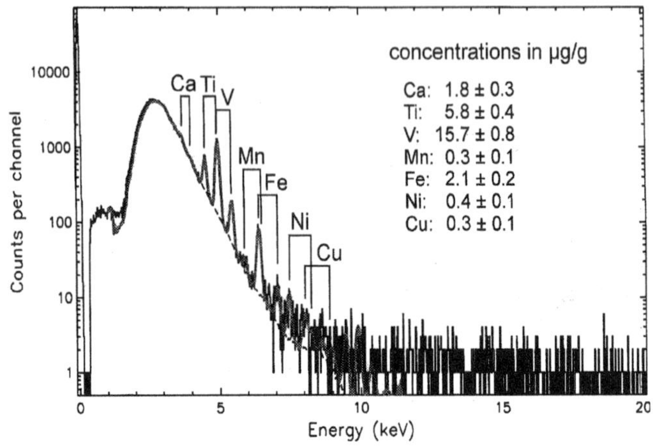

Fig. 4. Typical broad beam PIXE spectrum from a HOPG sample. The main impurities are Ti, V, and Fe. The minimum detection limit for other elements heavier than Si is ~ 0.3 μg/g. Note that 2.1 μg/g iron in the carbon matrix means a concentration of 0.45 ppm Fe. The background curve is due to the carbon matrix.

a HOPG sample is shown in Fig. 4. It shows the presence of a number of impurities, the Fe content being (0.45±0.04) ppm. From the study of ferromagnetic signals in different HOPG samples [28] we may suspect that HOPG could contain a non negligible amount of hydrogen, some of them may be related to the origin of the magnetic signals. The amount of hydrogen before or after irradiation within the fluence used in this work cannot be determined by PIXE. Taking into account recently done theoretical work [29] that estimates a magnetic moment of $\sim 1\,\mu_B$ for a carbon vacancy, $2.3\,\mu_B$ in the surrounding of an hydrogen bonded to a carbon atom at the position of a carbon vacancy, or $1.2\,\mu_B$ for a vacancy that is saturated by two hydrogen atoms in a graphene layer, we may expect that a few ppm hydrogen trigger a non negligible magnetic signal in graphite. Therefore, hydrogen content measurements should be carried out with a sensitivity in the ppm range, a rather difficult task. The reader can find in Ref. 10 for a short discussion on the methods available for this kind of measurement.

3 Irradiation Effects

3.1 On Highly Oriented Pyrolytic Graphite

3.1.1 Broad Proton Irradiation

All the irradiated HOPG samples discussed in this review were from Advanced Ceramics Co. (ZYA grade, 0.4° rocking curve width at half maximum) with a content of magnetic metallic impurities below 1 ppm. The largest concentration of non magnetic metallic ions was found for Ti (6 μg/g) and V (16 μg/g). A typical surface area of the samples was 2×2 mm^2 and a thickness between 0.1 mm to 0.3 mm. The samples were glued with varnish (or a mixture of varnish and a high purity graphite powder to increase the thermal coupling) on a high-purity Si substrate and the magnetic moment of the whole ensemble as well as of the Si substrate alone was measured. In general and within experimental error the diamagnetic hysteresis loops for the used Si substrates are reversible; after subtracting the linear diamagnetic dependence there is no significant hysteresis left (for an example see Fig. 7(a)). Nevertheless, when the ferromagnetic moment of the sample is weak, it can be a hard task to obtain the true sample magnetic moment from the SQUID signal of the mixture (Si plus the HOPG sample) and several checks have to be done.

In this section we will discuss results obtained after irradiation of a large area of the HOPG sample using the micro- and the broad-beam of protons. As we will realize below, there are several irradiation parameters that may have an important role in inducing the achieved magnetic signal. Namely:

(1) The total implanted charge. The total amount of protons or alpha particles that were implanted in the sample or travelled through the sample.
(2) The input energy. Although with the LIPSION we have the possibility of changing this energy from 1 to 2.25 MeV, in order to minimize the number of variables, the studies presented here were done with fixed irradiation energies of 2.0 or 2.25 MeV. This means that we have a well defined defect and implantation profile inside the sample, see Fig. 1. If the highest magnetic signal is determined at a given defect and proton density, it is clear that successive irradiation at similar energies may produce contrary effects if this optimized region is destroyed by the next irradiation at similar conditions.
(3) The fluence, the irradiated charge per unit area.
(4) The ion current. Large currents might heat the sample and non systematic effects are then possible.
(5) Micro- or macro-irradiation, *i.e.* a broad irradiation with the 0.8 mm beam or a large number of micrometer spots distributed in the sample. Experience indicates that the magnetic signals are much larger when one uses a high density of micrometer spots rather than the implantation of similar amount of charge with a broad beam. This fact appears to be related with the density and/or type of defects that the beam produces.
(6) Sample temperature. All the irradiations presented in this review were done at nominally room temperature. Future experiments should try to irradiate at lower or higher temperatures to check for its influence.
(7) Finally, the initial state of the sample, namely the density and type of defects, which in part can determine its metallicity. Computer simulation results of the effects of adsorbed hydrogen on the band structure of a graphene layer indicate that metallization caused by specific defects can quench a spin polarized state [30].

The first SQUID measurements that indicate a magnetic ordering after proton irradiation were published by [18]. In that work an increase in the hysteresis loop was observed after several irradiation steps. These irradiation steps contained successively irradiation of several thousands of micrometer small spots as well as four (or three) spots of 0.8 mm diameter each on the same sample. In what follows we shall discuss further work that has been done afterwards, where we have tried to characterize the effects produced by some of the irradiation parameters described above. We stress that they are the very first steps to get reproducible magnetic order in carbon, a task that turned to be full of difficulties, as the published results from literature indicate.

A broad and homogeneous proton irradiation of usual fluences (~150 μC total charge in an area of ~1 mm^2) per spot on oriented graphite produces a magnetic signal that in general can be well observed with a magnetic force microscope (MFM) on the irradiated surface. Figure 5 shows the MFM images obtained on an irradiated area with a total charge of 540 μC distributed into 4 spots, 135 μC each. After this broad irradiation there is no significant change in the topography worth to note. However, the periodic magnetic domains are well defined. In the remanent state just after irradiation, in some of the irradiated area of the sample we observed two domain structures, one

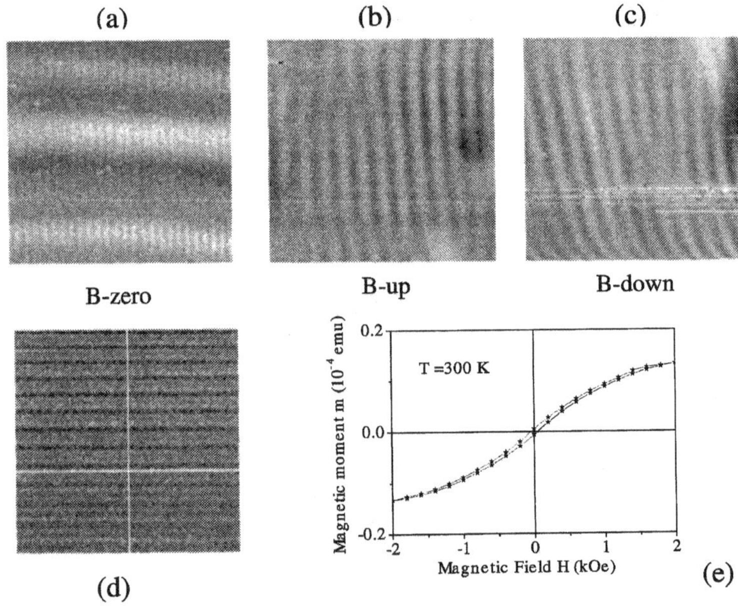

Fig. 5. Magnetic force microscope (MFM) and SQUID measurements of a HOPG sample irradiated with 4 broad spots ($\phi \cong 0.8$ mm) of 135 μC total charge each. All images scan an area of 20×20 μm^2: (a) MFM image of the sample in the state just after irradiation, (b) after application of a field of + 1 kOe, (c) after application of a field of -1 kOe. All measurements were done in remanent state after removing the field, *i.e.* at no applied field. (d) Magnetic domain structure of the sample in the same state as in (a) but at a different position. All the MFM measurements were done with a standard magnetized MESP tip and at a distance of 50 nm from the sample surface. (e) The magnetic moment of the same sample measured with a SQUID at room temperature. The sample was fixed on a Si substrate for the irradiation. The data in the figure are the original data of the sample with Si substrate, without any substraction. The signal of the sample overwhelms that from the substrate. Measurements of the sample standing alone reproduces basically the curve presented here.

normal to the other, see Fig. 5(a). The period of the small domain structure is ~0.8...1.2 μm depending on the region (compare with (d) which was obtained in the same state but in other irradiated area), whereas the other domain structure has a period of ~10 μm and depends on the scan direction of the MFM tip. After application of a field of 1 kOe in the +*z* direction (perpendicular to the graphene layers) the "vertical" domain structure shows a period of ~1.6...2.0 μm and the width of the domains increases, see Fig. 5(b). The other domain structure is not observed. After magnetizing the sample with a field in the other direction there is a slight change of the width of the bright relative to the dark regions, compare (c) with (b). In non-irradiated graphite areas we did not find any signature of domain structures within the resolution of the microscope. This irradiation triggered a relatively large hysteresis loop that could be very well measured with the SQUID without any background substraction, see Fig. 5(e). The temperature dependence of this signal was studied removing the sample from the Si substrate. It shows a weak decrease (~5%) with temperature of the saturation magnetization and coercivity fields between 5K and 300K.

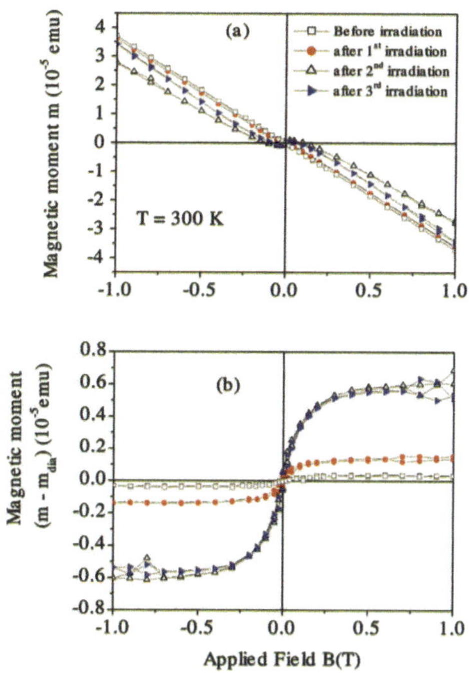

Fig. 6. (a) Magnetic moment vs. applied field of the HOPG sample AS171103/2 glued on a Si substrate before and after irradiation (for details see text). (b) The same data as (a) but after substraction of the diamagnetic background.

To study the influence of some of the irradiation parameters on the magnetic response of HOPG samples, we have studied three HOPG samples from the same batch. One sample of mass $m = 1.19$ mg (labelled AS171103/2) was irradiated with the proton microbeam as follows. First irradiation consisted of 10^4 spots of diameter $\phi \cong 1.8\ \mu m$ each with a charge of 0.85 nC (total charge 8.5 μC), fluence = 0.34 nC/μm², irradiated area: 0.62×0.62 mm² and current $I \cong 1.1$ nA. Second irradiation produced the same amount of spots but with a diameter of $\phi \cong 2\ \mu m$ and 0.72 nC charge each (total charge 7.2 μC), fluence = 0.24 nC/μm² and $I \cong 0.9$ nA. The third and last irradiation was identical to the first one. The room temperature SQUID results of the sample before and after substraction of the diamagnetic background are shown in Fig. 6. It is clearly seen that the irradiation increases the magnetic signal of the sample. The last irradiation, however, did not produce any significant change respect to the last one. The remanent magnetization changed from $M_r(B = 0) \cong 3.3\times10^{-5}$ emu/g to 2.7×10^{-4} emu/g and 4.5×10^{-4} emu/g for the first, virgin state and second (or third) irradiation steps.

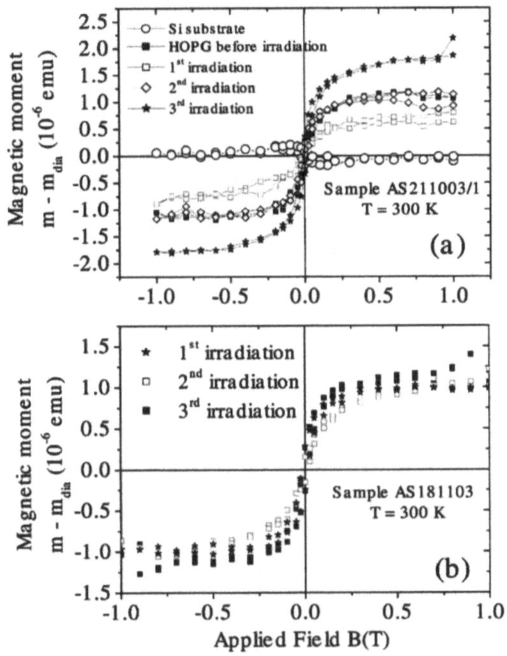

Fig. 7. (a) Magnetic moment (minus the diamagnetic background) vs. field of a typical Si substrate (○) where all HOPG samples were attached before and after all irradiation steps. The other symbols correspond to sample AS211003/1 after different proton irradiations (see text for details) with proton current $I \sim 55$ nA. (b) The same for sample AS181103 after different proton irradiations with a current $I \simeq 150$ nA (for details see text).

To check the influence of the irradiation current under broad irradiation conditions we have irradiated two samples with the following characteristics. Irradiation with a current $I = 54 \pm 1$ nA: sample AS211003/1, mass $m = 1.33$ mg, first irradiation: 4 spots of $\phi = 0.8$ mm and 53.4 μC each (total charge 214 μC); second irradiation: 4 spots of $\phi = 0.8$ mm and 110 μC each (total charge 440 μC); third irradiation: 4 spots of $\phi = 0.8$ mm and 54.5 μC each (total charge 218 μC). The SQUID results after substraction of the diamagnetic background are shown in Fig. 7(a). We see that the initial magnetic state of this sample (magnetic moment at saturation $m \simeq 10^{-6}$ emu) decreased after the first irradiation. For the subsequent irradiation steps the magnetic moment increased. The saturation magnetization showed a decrease from its initial value of $M_r = (1.95 \pm 0.15)10^{-4}$ emu/g to $M_r = (1.0 \pm 0.08)10^{-4}$ emu/g after the first irradiation. For the 2nd and 3rd irradiation we have $M_r = (1.56 \pm 0.15)10^{-4}$ emu/g and $M_r = (2.6 \pm 0.15)10^{-4}$ emu/g. The coercive field remains in the range of 150 Oe.

On the sample AS181103 of mass $m = 1.12$ mg, the following irradiation steps at a proton current of $I = 150$nA were performed: (1) 2 spots of 0.8 mm diameter with 159 μC each (total charge 318 μC); (2) 2 spots of 0.8 mm diameter with 160 μC each (total charge 320 μC); (3) 2 spots of 0.8 mm diameter with 154 μC each (total charge 308 μC). Figure 7(b) shows the SQUID results for this sample. Under those irradiation conditions the sample did not show any significant increase in the magnetic signal within experimental error. Comparing this result with that of (a) we would conclude that the proton current plays an important role in inducing the magnetic ordering in HOPG samples.

3.1.2 Magnetic Spots and Lines of Micrometer Size

As mentioned in section 2, the LIPSION accelerator has the possibility to irradiate the samples with a proton micro- or nanobeam. We have used the microbeam to produce magnetic spots on oriented graphite surfaces. As for the broad beam irradiation, the microbeam was directed onto the HOPG surface parallel to the c–axis of the sample without beam scanning (excepting line scans, see below) leading to the formation of micron-sized spots with enhanced defect density, as measured by micro-Raman, see Fig. 2. In general, two spots separated by a distance of 20 μm were irradiated with the same ion fluence and several ion fluences were used. For large enough fluences the swelling at the spots can be directly observed with an optical microscope [22, 23]. The height of this swelling depends on the irradiated fluence and on the mass of the ions. The dependence of the maximum swelling height, measured by AFM, with the fluence for proton irradiation is shown in Fig. 8(a). With a MFM one can measure the magnetic signal on the spots. The maximum amplitude of the signal (maximum phase shift) as a function of the fluence for two different proton currents is presented in Fig. 8(b). The units of the magnetic signal from the MFM are “degrees"; a relation of this phase shift of the tip vibration to the force gradient can be found in Refs. [31, 32]. Examples for the

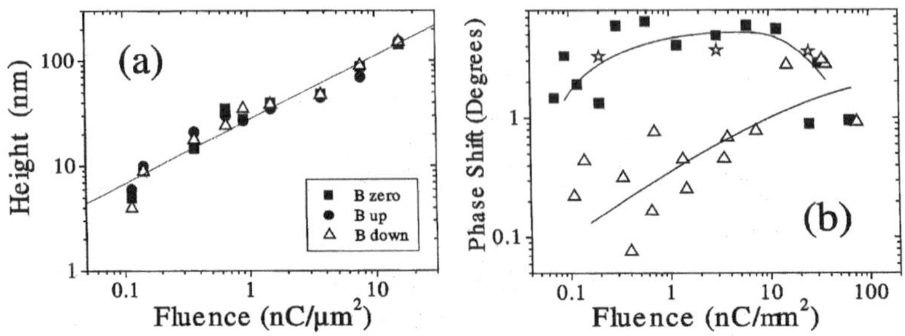

Fig. 8. (a) Maximum height of swelling as a function of the fluence, measured at micrometer small spots with an AFM. The three symbols indicate the height measured before and after the application of a magnetic field in z (up) and $-z$ (down) direction. The measurements were done without applied field. The line is the function $27.5x^{0.61}$ nm with x being the fluence in nC/μm^2. (b) Maximum phase shift measured at micrometer small spots on two different HOPG samples, before application of a magnetic field, as a function of the irradiation fluence. The spot areas were 1μm^2 (■, ★) and 4μm^2 (Δ); the corresponding proton currents were 171 pA and 855 pA, respectively. The (★) symbols correspond to the same sample measured again two months later. The lines are only a guide to the eye.

magnetic moment calculated from measurements of the phase shift in carbon samples can be found in Refs. [33–35].

We note that, whereas the swelling height increases systematically with fluence, the maximum phase shift of the magnetic signal does not, see Fig. 8(b). This result indicates that: (1) there is a negligible influence of the topography onto the magnetic signal and (2) the magnetic signal depends on the implanted charge, proton current andprobably also on the carbon structure. The difference between the behaviors observed for two different proton currents can be understood if we take into account that the higher the current the higher would be the internal temperature of the sample at the spot. Therefore, annealing effects can be the origin for this difference. It is worth to note that the signals could be reproduced after leaving the samples in air and at room temperature for two months, see Fig. 8(b). This is not the case for much larger times, see section 4. We note also that at the higher fluence range, the structure of the spots is very probably that of highly disordered or even amorphous carbon. For both proton currents the maximum magnetic signal decreases at the highest fluences, see Fig. 8(b).

Figure 9 shows an example of a magnetic spot as measured with the MFM before and after application of a magnetic field in two directions. The irradiated region can be clearly recognized as the "white" spot of the magnetic images in (a). A line scan through these images indicates clear changes of the phase shift after application of a magnetic field, see top figure in Fig. 9(b). Usually the magnetic images are obtained

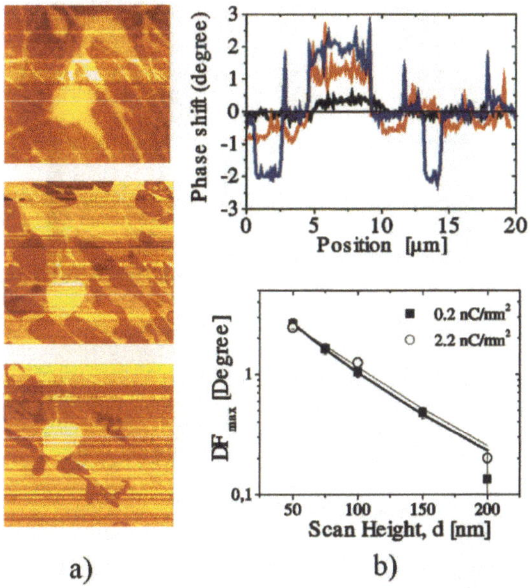

Fig. 9. (a)Magnetic force gradient images (20×20 μm^2) of a spot and its surroundings irradiated with 0.115 nC/μm^2 (proton current $I = 1.71$ pA). The images were taken, from top to bottom, before field application, after applying a field of ~1 kOe in the +*z* direction parallel to the *c*-axis, and in the –*z* direction. The tip-to-sample distance was 50 nm. (b) Top: The corresponding phase shift obtained at the line scans (white straight lines in figures (a)). The spot is located between ~5 μm and ~10 μm. The bottom, upper and middle lines in this region correspond to measurements before and after application of a field in –*z* and +*z* direction. Adapted from [36]. Bottom: Scanning height dependence of the maximum phase shift at proton irradiated spots with fluences of 0.2 nC/μm^2 and 2.2 nC/μm^2. Solid lines are fits with the point probe approximations. Adapted from [35].

with a tip to sample distance of 50 nm. If we increase this distance the phase shift amplitude decreases. This dependence can be used to estimate, although with a relatively large error, the order of magnitude of the magnetization at the spot surface. From the data shown in the bottom figure of Fig. 9(b) one is able to estimate a magnetization of the order of 400 emu/g [35], a very large magnetization of the order of soft ferromagnetic metals.

As a first attempt to write magnetically on a graphite surface we have started with a simple cross. Figure 10(a-c) shows the magnetic images for two crosses made at different irradiation fluences and measured with different tips and the topography for one of them (d). The magnetic image changes after applying a magnetic field. The results after magnetizing the cross show a state with one or three magnetic domains.

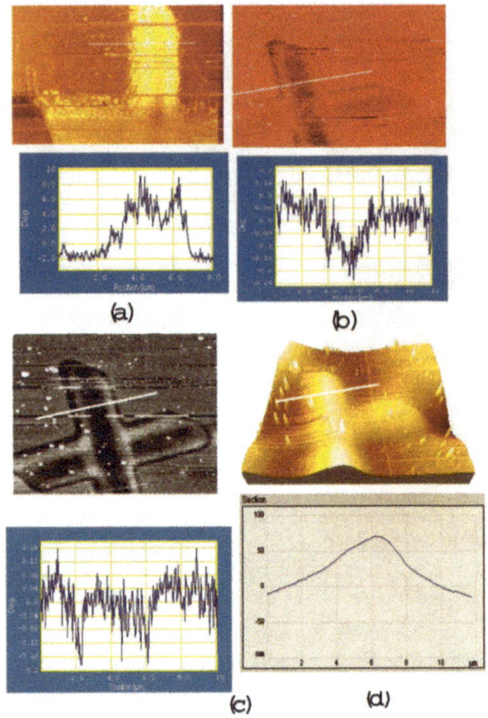

Fig. 10. (a) Magnetic image (top, 8×16 μm^2) and the corresponding phase shift (bottom) obtained at the line scan (white lines in all figures) for a cross produced with a fluence of 0.5 nC/μm^2 and a proton current of I = 120 pA. The measurement was done before applying any magnetic field and with a LM (low moment), well magnetized tip. (b) Similar to (a) but for a cross produced with a fluence of 2 nC/μm^2 and after application of a field of 1 kOe normal to the surface. The magnetic image has a size of 18×18 μm^2 and was taken with a standard MESP tip. (c) The same cross as in (b) but after application of a field of 2 kOe in the same direction as in (b). The magnetic image has a size of 10×20 μm^2. (d) Three dimensional image of the topography of the cross in (b,c) and the corresponding line scan (bottom).

Micromagnetic simulations with the appropriate parameters provide similar MFM images [37] and they may be used to obtain material parameters (magnetic anisotropy, for example) that are not yet possible to measure directly.

3.1.3 Irradiation Effects with Alpha Particles

With the LIPSION accelerator we have also the possibility to irradiate the samples with a microbeam of 1.5 MeV alpha (He^+) particles. The SRIM2003 simulation indicates that the penetration depth of these particles in graphite should be 4 μm. For a fluence of 0.1 nC/μm^2 the defect density at the surface should be 1.4% and the region at

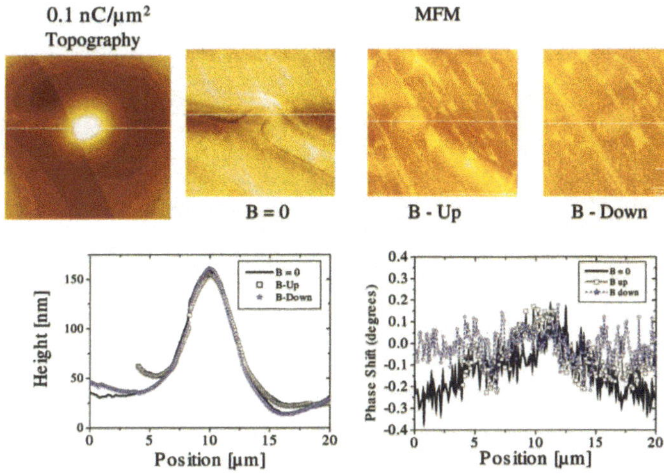

Fig. 11. Micrometer spot produced after irradiation of alpha particles of 1.5 MeV on a HOPG surface with a fluence of 0.1 nC/μm^2. The upper pictures show the topography (left) and the magnetic images before application of a magnetic field (B = 0), and after the application of a field in +*z* (B-up) and in −*z* direction (B-down). The bottom pictures show the line scans obtained at the position of the white lines in the upper images.

the end of ion range should be rather amorphous. As for protons, we have produced pairs of spots at different fluences and studied the magnetic signal at the spot position with MFM. Figure 11 shows the topography and magnetic images and the line scans of a spot produced with a fluence of 0.1 nC/μm^2. Whereas the swelling produced by the He^+ irradiation is remarkable, we recognize that the maximum phase shift measured by the MFM at the spot and its surroundings, see Fig. 11, is much smaller than those obtained at the spots produced with protons, see Fig. 9. Nevertheless, the fact that we can measure small but not zero signals, indicates that ion-, other than proton-, irradiation in the carbon structure may also trigger magnetic ordering. As we mentioned above, we do not know yet whether the hydrogen already present in the sample before irradiation or in the irradiation environment (rest gases in the chamber, for example) plays any role and therefore we should take these results with caution as a proof for a hydrogen-independent magnetic ordering. Future work should try to measure HOPG samples irradiated with alpha particles at different fluences and with the SQUID.

3.2 On Thin Films

Taking into account the irradiation profile of protons in carbon obtained from Monte Carlo simulations, see Fig. 1, one would expect no or negligible effects on the magnetic properties of samples, which thickness is much less than ~40 μm, because most of the

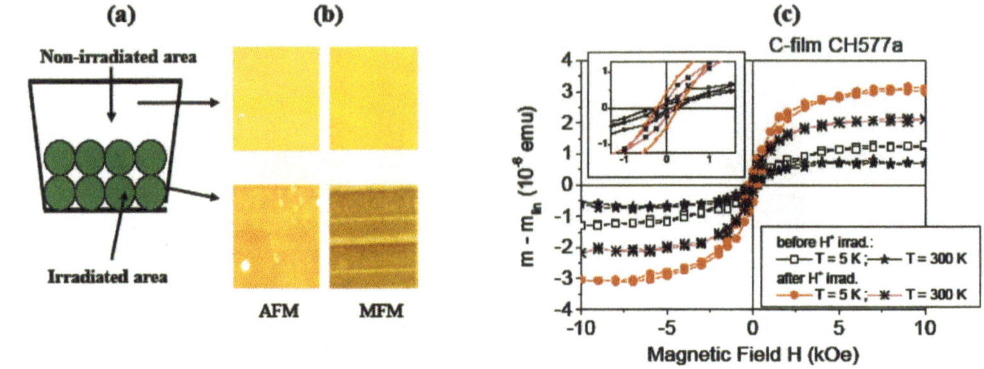

Fig. 12. (a) Sketch of the disordered carbon film CH577a produced by pulsed laser deposition with the irradiated (8 spots with 0.8 mm diameter and 150μC total charge each) and non-irradiated areas. (b) Topography (AFM) and magnetic (MFM) images in both areas. (c) Magnetic moment of the film (after substraction of the linear diamagnetic background) as a function of magnetic field applied parallel to the main area. The symbols (□, ⋆) indicate: before irradiation, measurements done at T = 5K and 300K, and after proton irradiation, (•, *) at 5K and 300K. The inset shows an enhanced part around zero field. Adapted from [11].

protons would go through the sample leaving a small density of defects. We have checked this assumption irradiating carbon and fullerene films with thickness below 1 μm. Figure 12 shows the MFM and SQUID results of a carbon film produced by pulsed laser deposition (PLD) in a hydrogen (H_2) atmosphere and deposited on a Si substrate. On the film 8 spots of 0.8 mm diameter were irradiated (see sketch in the figure). The MFM signal shows magnetic domains in the irradiated region only, see Fig. 12(b). The SQUID shows a clear increase in the magnetic moment after irradiation, see Fig. 12(c). Because the results are taken on an disordered sample, they also reveal that a graphite ordered structure, at least in the mesoscopic level, is not necessary to trigger magnetic order after irradiation.

In Ref. [38] the magnetism in photopolymerized fullerene films is studied by MFM. The magnetic force gradient measurements revealed that laser- and electron-beam irradiation of fullerene films produces magnetic images highly correlated to the topography. This correlation is expected since the polymerization shrinks the film material at the surface within a certain penetration depth, which depends on the radiation characteristics, producing topographic clusters with magnetic order. An example of the AFM and MFM images in a region illuminated in air with an energy of 2.6 eV and intensity of 200 mW/cm^2 is shown in Fig. 13. The high correlation between of magnetic grain formation and topography is well recognizable. In contrast to the laser illumination on C_{60} film and proton irradiation in HOPG, proton irradiation of spots of micrometer size on the fullerene film do not produce appreciable changes in the topography. The magnetic domains are however notable, see Fig. 13.

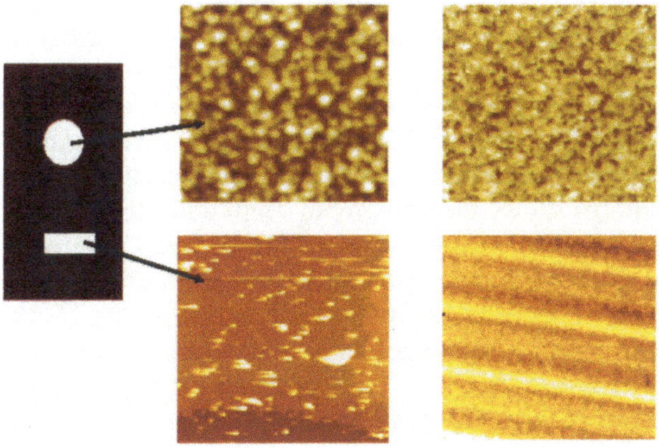

Fig. 13. Left side sketch shows the fullerene film with the two irradiated regions: the upper ellipse corresponds to the laser illuminated region [38] and the lower rectangle to the region where 20 spots of 1.8 μm diameter each where irradiated with a proton microbeam with a current of 500 pA and a fluence between 0.068 and 68 nC/μm^2. In contrast to the spot irradiation in HOPG (see section 3.1.2) no change in the topography has been detected after irradiation of the spots in the fullerene film. All images correspond to an area of 5×5 μm^2.

These results obtained after irradiation of the thin films are unexpected and indicate either the hydrogen concentration in the sample or surrounding before irradiation is relevant and/or the defect concentration in the first micrometer from the sample surface and produced by the proton beam is not negligible. Because the thickness of the film is below 1 μm most of the protons go through the material.

4 Annealing and Aging Effects

We have measured with the SQUID some of the proton irradiated samples months after the irradiation. Depending on the sample, the enhancement of the magnetic order decreased or even vanished after more than 8 months leaving them at room temperature and in air. However, the magnetic signal remained for some of them even after one year. The reason for the different behavior is not known yet. In this section we shall summarize results that provide evidence for aging effects in the magnetic ordering, as well as annealing effects at high temperatures, measured by SQUID or MFM. Taking into account previous reports on the behavior and diffusion of hydrogen in graphite [14] as well as aging effects at room temperature in the magnetization of the fullerene $C_{60}H_{24}$ [39] one may expect to observe some time dependence in the magnetic response at the irradiated surface by MFM (or in the bulk magnetic moment by SQUID) if H is involved in the induced magnetism and a diffusion takes place. Two different measurements were done to test the possible influence of H diffusion on the magnetic signals. We measured different irradiated spots on HOPG and irradiated amorphous carbon film just after irradiation and after leaving the samples several months at room temperature. Other samples were annealed at high temperatures in vacuum or He atmosphere.

Figure 14 shows the magnetic moment of a disordered carbon sample before, just after proton irradiation and after one year at room temperature (after irradiation). It is clearly seen that the ferromagnetic-like hysteresis vanishes after one year.

Figures 15(a) and (b) shows the topography (top) and MFM (bottom) images of two spots and their surroundings created with 0.2 nC/μm^2 and 24.5 nC/μm^2 on sample 1. The measurement was done one day after proton irradiation. Line scans of the topography (top) and MFM (bottom) images are shown in Fig. 15(c). As shown in this figure a clear enhancement in both topography and MFM signals at the irradiated region is clearly found. The enhancement of the topographic swelling increases with fluence with our irradiation conditions. For example, the peak height is ~ 8 nm (~225 nm) for a fluence of 0.2 nC/μm^2 (24.5 nC/μm^2). However, there is no clear correlation between the increase in swelling height measured by AFM and the maximum phase shift change of the MFM signal at the spot within the fluence range used. As shown in Fig. 15(c) bottom, a magnetic structure within the spot produced with 24.5 nC/μm^2 appears, an indication for the existence of magnetic domains within the area of the irradiated spot.

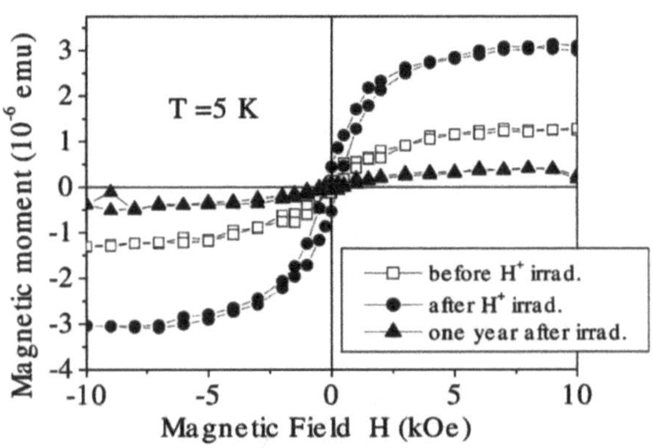

Fig. 14. Magnetic moment (the diamagnetic background was subtracted) as a function of applied field for the disordered carbon film CH577a after preparation by PLD (before irradiation(□), after irradiation (•) and one year later, leaving the sample at room temperature (▲).

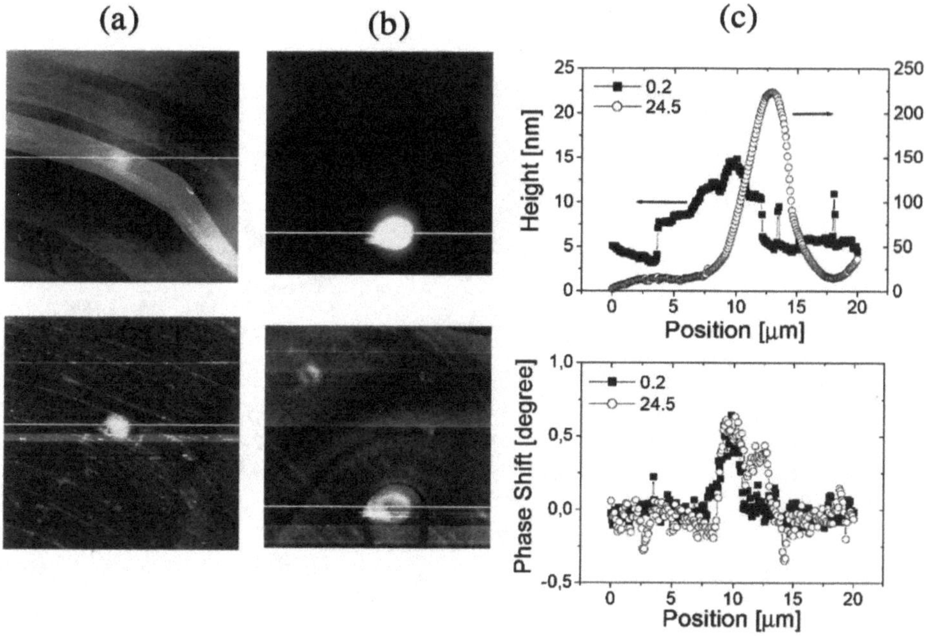

Fig. 15. Topography (top) and MFM (bottom) images (20×20 μm^2) of sample 1 for two spots created at 0.2 nC/μm^2 (a) and 24.5 nC/μm^2 (b) and their surroundings measured one day after irradiation. The tip-to-sample distance was 50 nm. (c) Corresponding line scans (white lines in (a) and (b)) of topography (top) and phase shift (bottom) images.

Figure 16 shows the same measurements as in Fig. 15 for the same two spots, but obtained 8 months later. After this time we found clear changes in the topography and in the phase shift of the MFM signal at and in the surroundings of the original microspots, see Fig. 16. There are three main changes to be noted: (1) The sharply white spots in MFM images (see the bottom Figs. 15(a) and (b)) on the irradiated area changed to black spots (see bottom Figs. 16 (a) and (b)) and the maximum value of phase shift changes from ~ +0.7° to ~ –0.2°. It means that the direction of magnetization and its magnitude at the spots changed. (2) There appear clear and localized MFM signals besides the original spot signal (black spots in Figs. 16(a) and (b)). We note also that these "mirror" spots for both fluences are not clearly related to the topographic swellings as can be seen by comparing the line scan signals of topography and MFM of both spots (see Fig. 16(c)). There is also a clear change in the topography at the irradiated spots indicating a lattice relaxation. (3) The MFM images reveal that after 8 months there is a change not only at the spot position and its near neighborhood but also at its surroundings. This change is clearly observable in the vicinity of the spot made with a fluence 24.5 nC/μm^2 where a well discernible domain pattern appears, see bottom Fig. 16(b).

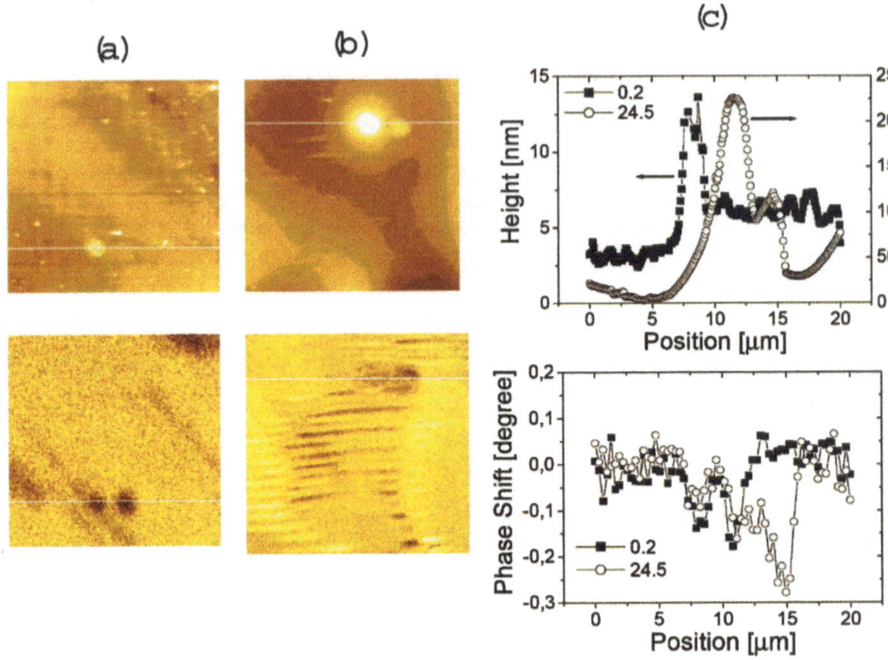

Fig. 16. (a) and (b) Topography (top) and MFM (bottom) images (20×20 μm^2) of the same spots of Fig. 15 and their surroundings measured after 8 months. The tip-to-sample distance was 50 nm. (c) Corresponding line scans (white lines in (a) and (b)) of the topography (top) and phase shift (bottom) images.

Hydrogen desorption in graphite was and still is a matter of research. Depending on the characteristics of the H-trapping this desorption may or may not be accompanied by a lattice change. Therefore, it is not straightforward to conclude that the aging effects we observed are due only to the H-diffusion without any structural relaxation. Because we got evidence for a structural relaxation, this can influence the effective diffusion of H in the carbon matrix. Therefore, the effective activation energies for H diffusion in our system are not necessarily the same as, for example, those obtained by experimental methods, usually at high temperatures, to study kinetics of diffusion of hydrogen in graphite [14].

Certainly, one does not necessarily wait one year to see effects that may be related to hydrogen diffusion. It is quicker to anneal an irradiated sample at high enough temperatures and check for its influence. Figure 17 shows the MFM images obtained for the crosses corresponding to those of Fig. 10(a) and (c) measured at identical conditions after annealing the sample 2 hours at 1000°C in vacuum. Whereas no signal is detected

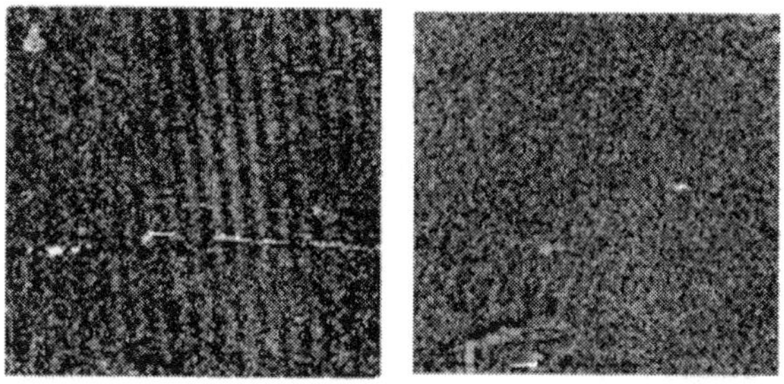

a) b)

Fig. 17. The figures (a) and (b) show magnetic images of the same crosses shown in Fig. 10(a) and (c) measured under similar conditions with similar MFM tips after annealing them in vacuum at 1000 C for 2 hs. Whereas the crosses are not anymore visible in MFM, domain structure is still detected in (a).

for the cross in (b) ((c) in Fig. 10) magnetic domains are still identified after annealing in the surroundings of the cross (a). Compare this observation to that obtained after aging the spots at room temperature for several months shown in Fig. 16.

5 Conclusion and Open Issues

In this chapter we have reviewed the main observations after proton irradiation of carbon structures. The main results indicate that proton irradiation can trigger magnetic ordering in graphite. The results after proton irradiation leave no doubt that magnetic ordering exists in a carbon structure without the influence of magnetic ions. Neither the total amount of magnetic impurities is sufficient to account for the measured magnetization nor the creation of magnetic spots in the micrometer range with the proton micro-beam can be understood based on magnetic metal-ion impurity concentration below 1 ppm as the PIXE results indicate. The results obtained up today do indicate that different parameters and sample states may play a role in determining the strength of the induced magnetic ordering. Broad irradiation (*i.e.* with a proton beam of large diameter) appears to be less effective. The reason for this behavior will be studied in the future, but we speculate that it may be due to the produced defect density. Further experimental characterization (using the broad spectrum of methods in magnetism research) and sample preparation studies are necessary to understand and stabilize the magnetic ordering found in carbon structures. The following issues should be clarified in the near future:

(1) The role of H-atoms, implanted by irradiation as well as those already in the sample.
(2) The contribution to the magnetic order from lattice defects produced by irradiation and their possible influence as H-trapping centers.
(3) Dependence of the magnetic order on the induced type of defect.
(4) The maximum achievable saturation magnetization in carbon structures.

(5) The range of Curie temperature.
(6) Influence of the ion current, fluence and energy of the irradiated particle on the magnetic order.
(7) Influence of sample temperature.
(8) Influence of the irradiation angle with respect to the crystallographic c-axis of graphite.
(9) Influence of ion irradiation (other than proton) on the magnetism of carbon structures.
(10) The effective magnetic moment of magnetic impurities in graphite as well as in disordered carbon structures.

An answer to part of these open questions will take several years of research.

Summarizing briefly recently published theoretical work, which is correlated to the influence of hydrogen in carbon magnetism we note that:

(1) Hydrogenated nanographite can have a spontaneous magnetization due to different numbers of mono- and dihydrogenated carbon atoms [7]. Theoretical work using local-spin-density approximation calculates the spin polarization of the graphite bands when different atoms as hydrogen, fluor or oxygen are added on a graphene layer and predicts magnetic ordering with fully or partially spin-polarized flat band, upon the added atom [40]. Theoretical simulations indicate that a magnetic ordered state should be easier to achieve with a mixture of carbon-hydrogen bonds than fluorinated nanographite [41]. Although the calculations were done attaching the hydrogen at the carbon atoms at the edges of a graphene layer, one may speculate that hydrogen can trigger the sp^2-sp^3 unbalance promoting magnetic ordering in different carbon structures.
(2) Recently published work indicates that hydrogen modifies substantially the electronic structure of graphite around it. According to STM/AFM measurements [42] a single H-atom interacting with a graphite surface modifies the electronic structure over a distance of 20 to 25 lattice constants.
(3) Muon spin rotation/relaxation experiments [43] indicate that a positive muon in graphite triggers a local magnetic moment around it. Recently published theoretical work [30] supports this conclusion. It also indicates that upon the type of defect in a graphene layer, hydrogen may not trigger any magnetic ordering.
(4) The magnetic moment and diffusion of adatom defects in a graphite sheet were studied by [44]. The results of a full spin-polarized density functional theory indicate that these defects may have a magnetic moment of about 0.5 μ_B. The calculations of [29] indicate that if an hydrogen encounters an empty vacancy, then it compensates the dangling bond and the magnetic moment of the vacancy vanishes. However, if a vacancy is saturated by an hydrogen atom, a second hydrogen atom will bond to the other side of the vacancy having a magnetic moment of 1.2 μ_B localized on the dangling sp^2 bond.
(5) New theoretical and experimental work [45] suggests that hydrogen may play also an important role on the magnetic ordering found in fullerenes.

Acknowledgements

This research is supported by the Deutsche Forschungsgemeinschaft under DFG ES 86/11-1.

References

1. B.T. Kelly, Physics of Graphite. London Applied Science Publishers, 1981.
2. F. Banhart, Rep. Prog. Phys. 62 (1999) 1181.
3. T. Makarova, Magnetism of carbon-based materials, volume 45 of Studies of High Temperature Superconductors, pages 107–169. NOVA Science Publishers, Inc. 2003.
4. K. Murata, H. Ushijima, H. Ueda, and K. Kawaguchi, J. Chem. Soc. Chem. Commun. (1991) 1265
5. K. Murata, H. Ushijima, H. Ueda, and K. Kawaguchi, J. Chem. Soc. Chem. Commun. (1992) 567.
6. A.A. Ovchinnikov and I.L. Shamovsky, J. Molecular Structure (Theochem, 251 (1991) 133.
7. K. Kusakabe and M. Maruyama. Phys. Rev. B67 (2003) 092406–1.
8. Peculiar localized state at zigzag graphite edge. J. Phys. Soc. Jpn. 65 (1996) 1920.
9. J.F. Ziegler. The stopping and range of ions in matter. Pergamon Press, New York, 1977-1985.
10. T. Butz, D. Spemann, K.-H. Han, R. Höhne, A. Setzer, and P. Esquinazi, The role of nuclear nanoprobes in inducing magnetic ordering in graphite. Hyperfine Interactions (in press), 2005.
11. R. Höhne, P. Esquinazi, K.-H. Han, D. Spemann, A. Setzer, U. Schaufuß, V. Riede, T. Butz, P. Streubel, and R. Hesse. In D. Raabe, editor, Proceedings of the 16. International Conference of Soft Magnetic Materials (ISBN 3-514-00711-X), p.p. 185-190, 2004.
12. H. Abe, H. Naramoto, A. Iwase, and C. Kinoshita, Nuclear Instruments and Methods in Physics Research B 128 (1997) 681.
13. R. Siegele, J. Roth, B.M.U. Scherzer, and S.J. Pennycook, J. Appl. Phys. 73 (1993) 2225.
14. H. Atsumi, Journal of Nuclear materials 311 (2002) 1466.
15. T. Wang, W. Wang, and B. Chen, Nuclear Instruments and Methods in Physics Research B71 (1992) 186.
16. K. Niwase, Phys. Rev. B52 (1995) 15785.
17. T. Tanabe, Phys. Scr. T64 (1996) 7.
18. P. Esquinazi, D. Spemann, R. Höhne, A. Setzer, K.-H. Han, and T. Butz, Phys. Rev. Lett. 91 (2003) 227201–1.
19. B. Rousseau, H. Estrade-Szwarckopf, A.-L. Thomann, and P. Brault, Appl. Phys. A77 (2003) 591.
20. H. Estrade-Szwarckopf, Carbon, 42 1713–1721, (2004.
21. J. González, F. Guinea, and M. A. H. Vozmediano, Phys. Rev. B63 (2001) 134421–1.
22. D. Spemann, K.-H. Han, P. Esquinazi, R. Höhne, and T. Butz, Nuclear Instruments and Methods in Physics Research B 219-210 (2004) 886.
23. P. Esquinazi, K.-H.Han, R. Höhne, D. Spemann, A. Setzer, and T. Butz, Phase Transitions 78 (2005) 155.
24. J. Koike and D. F. Pedraza, J. Mater. Res. 9 (1994) 1899.
25. S. Muto and T. Tanabe, Phil. Mag. A 76 (1997) 679.

26. M. Takeuchi, S. Muto, T. Tanabe, S. Arai, and T. Kuroyanagi, Phil. Mag. A 76 (1997) 691.
27. D. Spemann, K.-H. Han, R. Höhne, T. Makarova, P. Esquinazi, and T. Butz, Nuclear Instruments and Methods in Physics Research, 210 (2003) 531.
28. P. Esquinazi, A. Setzer, R. Höhne, C. Semmelhack, Y. Kopelevich, D. Spemann, T. Butz, B. Kohlstrunk, and M. Lösche. Phys. Rev. B66 (2002) 024429–1.
29. P.O. Lehtinen, A.S. Foster, Y. Ma, A. Krasheninnikov, and R.M. Nieminen, Phys. Rev. Lett. 93 (2004) 187202–1.
30. E.J. Duplock, M. Scheffler, and P.J. D. Lindan, Phys. Rev. Lett. 92 (2004) 225502–1.
31. J. Lohau, S. Kirsch, A. Carl, G. Dumpich, and E.F. Wassermann, J. Appl. Phys. 86 (1999) 3410.
32. J. Lohau, S. Kirsch, A. Carl, and E.F. Wassermann, Appl. Phys. Lett. 76 (2000) 3094.
33. K.-H. Han, D. Spemann, R. Höhne, A. Setzer, T. Makarova, P. Esquinazi, and T. Butz, Carbon 41 (2003) 785.
34. K.-H. Han, D. Spemann, R. Höhne, A. Setzer, T. Makarova, P. Esquinazi, and T. Butz, Addendum. Carbon 41 (2003) 2425.
35. K.-H. Han and P. Esquinazi, J. Appl. Phys. 96 (2004) 1581.
36. K.-H. Han, D. Spemann, P. Esquinazi, R. Höhne, V. Riede, and T. Butz, Adv. Mater. 15 (2003) 1719.
37. K. Schindler, D. Spemann, M. Ziese, M. Diaconu, H. Schmidt, P. Esquinazi, and T. Butz, To be published (2005).
38. T. Makarova, K.H. Han, P. Esquinazi, R. R. da Silva, Y. Kopelevich, I.B. Zakharova, and B. Sundqvist, Carbon 41 (2003) 1575.
39. V.E. Antonov, I.O. Bashkin, S.S. Khasanov, A.P. Moravsky, Yu.G. Morozov, Yu.M. Shulga, Yu.A. Ossipyan, and E.G. Ponyatovsky, J. Alloys and Compounds 332 (2002) 365.
40. M. Maruyama and K. Kusakabe, J. Phys. Soc. Jpn. 73 (2004) 656.
41. M. Maruyama, K. Kusakabe, S. Tsuneyuki, K. Akagi, Y. Yoshimoto, and J. Yamauchi, J. of Physics and Chemistry of Solids 65 (2004) 119.
42. R. Ruffieux, O. Gröning, P. Schwaller, L. Schlapbach, and P. Gröning, Phys. Rev. Lett. 84 (2000) 4910.
43. J.A. Chakhalian, R.F. Kiefl, S.R. Dunsiger, W.A. MacFarlane, R. Miller, J.E. Sonier, and J.E. Fischer, Phys. Rev. B66 (2002) 155107–1.
44. P.O. Lehtinen, A.S. Foster, A. Ayuela, A. Krasheninnikov, K. Nordlund, and R.M. Nieminen, Phys. Rev. Lett. 91 (2003) 017202–1.
45. J.A. Chan, B. Montanari, J.D. Gale, S.M. Bennington, J.W. Taylor, and N.M. Harrison, Phys. Rev. B70 (2004) 041403(R)–1.

Carbon-Based Magnetism
T. Makarova and F. Palacio (Editors)

20

Magnetic Properties of Novel Carbon Allotropes

A.V. Rode[1], A.G. Christy[2], E.G. Gamaly[1], S.T. Hyde[3], and B. Luther-Davies[1]

[1]*Laser Physics Centre, Research School of Physical Sciences and Engineering, the Australian National University, Canberra, ACT 0200 Australia*
[2]*Department of Earth and Marine Science, The Australian National University, Canberra, ACT 0200 Australia*
[3]*Department of Applied Mathematics, Research School of Physical Sciences and Engineering, The Australian National University, Canberra, ACT 0200 Australia*

1 Introduction

1.1 The Origins of Magnetic Behaviour

Electrons are fermionic particles with intrinsic angular momentum ("spin") = $\pm^1/_2$ quantum units. Therefore, they display Pauli exclusion behaviour, and two electrons in similar quantum states repel one another. The energy of interactions between electrons is determined not only by their electrical charges, or the magnetic dipoles that arise from their charge and angular momentum, but also by the relative signs of their spins. Pauli repulsion between similar spins favours spin pairing so as to produce zero resultant angular momentum. Such a state has a wave function whose spatial distribution is different than for unpaired spins, since the paired spins do not avoid one another to the same degree. Hence, the electrostatic energy term will be different for the two states. When two electrons are on neighbouring atoms, one spin-paired state localises electron density between the two nuclei. The electrostatic attraction to both nuclei more than compensates for the increased repulsion between electrons, and a chemical bond is formed. Within a single atom, an atomic orbital that is well separated in energy from others, such as an *s*-orbital, also tends to accommodate spin-paired electrons since the energy cost of promoting one electron to a higher energy would outweigh the advantage gained in spatial avoidance. However, the situation is different when several orbitals are similar in energy but show different angular variation, for instance the groups of three *p* orbitals, five *d* orbitals or seven *f* orbitals. In that case, Pauli and electrostatic repulsion can both be minimised by placing as many unpaired electrons in distinct degenerate orbitals as possible (Hund's rule). Hence, there are unpaired spins in the isolated atom, which therefore has a nonzero magnetic moment and will tend to align its magnetic moment parallel to an applied magnetic field. Thus, its magnetic susceptibility is positive in sign, and the atom is *paramagnetic*.

The total magnetic susceptibility of a material is a resultant of both positive (paramagnetic) and negative (diamagnetic) contributions. These terms may originate from the magnetic moments of subatomic particles both inside and outside the atomic nucleus. However, the nuclear contributions are typically six orders of magnitude smaller than those from the electrons, and can be neglected. Electronic contributions may be classified into the following types:

An applied magnetic field induces resistance-free ring currents within atoms, which oppose the applied field. This behaviour of the non-bonding core electrons of the atom contributes a negative (diamagnetic) term to the magnetic susceptibility that is small, but increases with the atomic number and mean-square radius of the atom. The phenomenological theory describing this diamagnetism originated with Larmor and was developed by Langevin [1].

Spin-paired valence electrons of atoms also show some diamagnetic behaviour, but the possibility that they are promoted into an excited state with nonzero moment also allows a temperature-independent positive susceptibility (paramagnetic) term that may more than compensate. This is the Van Vleck paramagnetism [2].

Unpaired, non-bonded spins give rise to a much larger but temperature-dependent positive susceptibility term: the Langevin paramagnetism [1].

In a metal, the small, temperature-dependent fraction of delocalised electrons that are able to align with the applied field contribute the small, temperature-independent Pauli paramagnetic component [3]. This contribution is partially compensated by the effect of the applied field on the electron wave functions (Landau diamagnetism, [4]).

More detailed discussion of these factors contributing to the overall magnetism and their differing behaviours may be found in solid-state physics textbooks such as Kittel [5] or Ashcroft and Mermin [6] and references cited therein.

In molecules and extended solids, the paramagnetism that would be expected of an isolated atom is frequently lost due to the formation of bonds. For instance carbon, with an s^2p^2 outer electron configuration, should have two unpaired spins as an isolated atom, but these are lost when those electrons are used to form the strong bonds of graphite or diamond. Sometimes, paramagnetism is preserved due to special characteristics of the occupied electron orbitals. Oxygen (two unpaired spins in the isolated atom) forms strongly bound diatomic molecules in which two electrons are forced to occupy a degenerate pair of *antibonding* orbitals. Hund's rule therefore applies, and the O_2 molecule therefore remains paramagnetic.

Spins can remain unpaired if they do not participate in chemical bonds. Outer electrons of atoms are clearly vulnerable to bond formation, so most examples of paramagnetism in solids are provided by elements with unpaired electrons in radially condensed orbitals, namely the transition metals (partially occupied *d* orbitals) and lanthanides and actinides (partially occupied *f* orbitals). It is not surprising that it is among these families of elements that we find the classic examples of the long-range ordered form of paramagnetism known as *ferromagnetism*: the transition metals iron, cobalt and nickel (4, 3 and 2 unpaired spins per neutral atom respectively) and the lanthanide gadolinium (7 unpaired electrons per neutral atom). In such ferromagnets, "exchange" interactions operate between the unpaired non-bonding electrons of neighbouring atoms and cause the formation of domains in which electron spins show long-range orientational order. Although non-bonding electrons of neighbouring atoms are well separated spatially, their interactions are amplified in the typical ferromagnetic metals by the fact that there is a very high electron density of states near the Fermi level. Therefore, Pauli repulsion, redistribution of the electronic band structure and the mean field generated by oriented spins all cooperate to produce a large effect. Spins within a domain all lie parallel in ferromagnets *sensu strictu*, but may also show more complex relative orientations, aligning opposite to their neighbours (antiferromagnetic) or intermediate behaviour (canted, helical and so on). The exchange interactions are ultimately due to Pauli ("exchange") and electromagnetic forces, but coupling between paramagnetic electrons may occur to next-nearest neighbouring atoms through the electrons of otherwise nonmagnetic intermediary species such as oxide anions ("super exchange"). Hence, magnetic order is not confined to metallic species in which paramagnetic atoms are in close contact, but is also important in compounds such as the iron oxide magnetite, Fe_3O_4, used as a "lodestone" since Classical times.

Ferromagnetic order allows the accumulation of very large magnetic moments in domains, and of bulk magnetisation by the preferential growth of appropriately oriented domains in an applied magnetic field. The preferred orientation of the magnetic moments and the net magnetisation retain in the direction of the applied magnetic field after the field is switched off, giving permanent magnetisation to the ferromagnetic sample.

1.2 Magnetic Properties of Carbon Materials

Since the phenomenon of ferromagnetism relies on the possibility of aligning unpaired spins, carbon, with its propensity to utilise all its valence electrons in very strong bonds, is perhaps the last element one would expect to retain unpaired electrons and exhibit ferromagnetism. In fact, the spin-paired atomic cores and strongly bound valence electrons make the diamond polymorph of carbon significantly diamagnetic ($\chi = -0.5\times10^{-6}$ emu/g), while the possibility of ring currents in the delocalised π electrons of graphite give it the most negative susceptibility known for a non-superconducting material (χ as low as -50×10^{-6} emu/g) [7]. We note that these well-known, macrocrystalline polymorphs of carbon are just two out of a huge range of carbon allotropes that have now been produced, with different atomic-scale bonding geometries and nanoscale structures. These include carbyne chains, polyhedral fullerene molecules, cylindrical nanotubes and conical nanohorns and many others [7]. Most of these carbon allotropes are not crystalline, and are difficult to characterise structurally. Hence, there was wide-spread scepticism in the scientific community concerning reports of a magnetically ordered state in pyrolytic carbon [8–11], although a theoretical prediction of a ferromagnetic carbon with magnetisation higher than that of *α-Fe* was published more than a decade ago [12]. Experimental study of pyrolytic carbon magnets was initially slow to take off due to low reproducibility of the experimental results as well as suspicion that the reported small magnetic moment of the order of ~0.1 emu·G/g could simply be generated by traces of magnetic contaminants such as iron. Interest in the magnetic properties of carbon nanostructures has quickened recently with the publication of new results indicating anomalously small diamagnetism and the presence of unpaired spins in carbon nanohorns [13], and also ferromagnetic behaviour in polymerised fullerenes [14–17], highly-oriented pyrolytic graphite [18], and laser-deposited carbon nanofoams, which appear to contain a "schwarzite" component with hyperbolically curved graphite-like sheets [19, 20].

The experimental evidence for ferromagnetic carbon nanostructures now appears to be firm. The fundamental structural and electronic mechanisms responsible for ferromagnetic ordering are not yet well established, but we offer provisional conclusions in this chapter.

The first experimental work showing high-temperature ferromagnetism in a graphite-like material was the study by Kopelevich, Esquinazi and co-workers [18]. They discovered highly anisotropic behaviour in highly oriented pyrolytic graphite: ferromagnetic-like hysteresis loops for a magnetic field applied parallel to the graphite sheets, which has a large diamagnetic response superposed for fields applied normal to the sheets. The signal was enhanced by heating up to 800K in a He atmosphere. Absence of correlation with impurity content implied that this was a property of the pure carbon [21, 22].

Reduction of diamagnetic contributions and increase of paramagnetic terms has been shown by LCAO calculations to reduce the diamagnetism in single-wall carbon nanohorns, which have curved graphite-like sheets containing some 5-rings of atoms as well as the usual 6-rings, but are pure carbon materials rather than hydrogenated [23]. Calculations have shown that a zigzag arrangement of carbon atoms at the unsaturated edge of a nanotube may give rise to magnetic moments of up to one Bohr magneton, μ_B, per unsaturated dangling bond [24]. A magnetic moment close to $10\mu_B$ was found to be associated with the open edge of a (10,0) nanotube, but it appeared that nonzero net magnetic moment could only be achieved when one of the tube ends was terminated. Nanotubes bent into circular tori are calculated to show strong paramagnetism for some toroidal radii [25].

Nanoscale ribbons of "graphene" (distorted or disrupted graphite) sheet have long edges, which can show "armchair" or "zigzag" structures depending on the edge orientation relative to the hexagonal graphene rings. Calculations show that in the zigzag case only, a "flat band" occurs at the Fermi surface in the electronic density of states, which results in Pauli paramagnetism that becomes significant at low absolute temperature, and may show long-range order. However, spin polarisation in the zigzag edges of graphene flakes has been attributed to the presence of hydrogen bound to otherwise under-coordinated carbon atoms [26–29].

The influence of non-carbon atomic species on the measured magnetic moment is potentially dramatic, as demonstrated by the experiments of Pablo Esquinazi and co-workers [30–32]. They irradiated a sample of ultra-pure highly-oriented pyrolytic graphite with a beam of protons at the energy of 2.25 MeV. The protons distorted the graphite structure and formed carbon-hydrogen bonds. Magnetic force microscope images show a classic ferromagnetic signature in the irradiated spots of the graphite, which are hydrogenated and hence cannot be considered pure carbon structures. The specific importance of hydrogenation as a means of inducing ferromagnetism was demonstrated by comparison with experiments using α-particles as the irradiating beam; in this case, the irradiated spots did not show the same magnetic response.

The ability of hydrogenation to induce magnetism in a carbon-rich material can be understood in terms of the effect of attached hydrogen on the C-C bonding. The bonding in the ideal, flat sheet of graphite can be described in terms of simple electron-pair bonds if single and double bonds between C atoms alternate regularly. There is resonance between the two possible orientations of this structure to produce a hybrid in which all bonds are of equal order $1^1/_2$, since the π-components of the double bonds delocalise to produce continuous honeycomb-net sheets of electron density above and below the plane of the carbon nuclei. Not just the Larmor diamagnetic term from the atomic cores but also a term arising from ring currents in the π-sheets combine to give graphite the large negative susceptibility noted above. The second term is much larger since the radii of the associated ring currents are bigger. Bonding of a hydrogen atom to one of the carbons of graphite leaves the carbon at the other end of a C+C double bond with one unsatisfied "dangling bond" (Fig. 1). This is an electron with unpaired spin, with its associated paramagnetic contribution to the susceptibility.

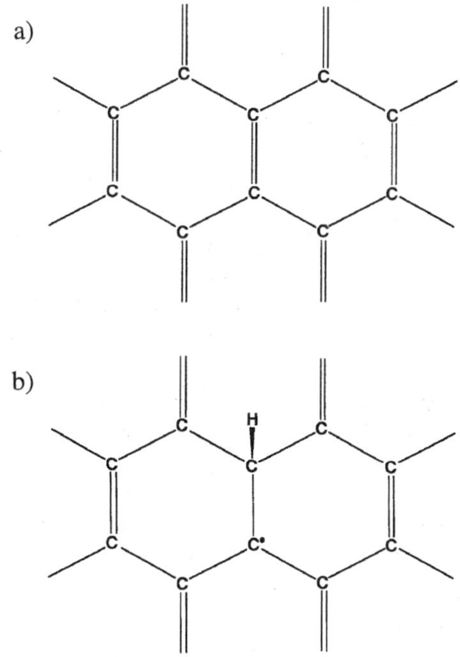

Fig. 1. (a) Intact graphite sheet, (b) generation of unpaired spin by hydrogenation.

Furthermore, the carbon that has acquired a hydrogen is now in tetrahedral fourfold coordination by other atoms rather than planar threefold, as it would be in pure graphite. The C-C-C bond angles on the hydrogenated carbon are less than 120°, and the graphite sheet is puckered in its vicinity. This deviation away from planarity reduces the efficiency of π-overlap between adjacent atoms, and tends to break up the π-sheet into discrete π-bonds. The reduced connectivity of the π-sheet, and the larger π-π^* energy gap associated with spatially localised double bonds, conspire to reduce the ring current diamagnetism. The possibility of promoting electrons between these bands also allows a Van Vleck term to contribute to the paramagnetism. Hence, several changes in structure and bonding cooperate, enabling a small degree of hydrogenation to cause a large positive shift in magnetic susceptibility.

Theoretical calculations have indicated other pure-carbon structures that may support strong paramagnetism and magnetic ordering [12]. The mechanism is based on a mixture of planar 3-fold coordinated sp^2- and tetrahedral sp^3-hybridized carbon atoms, some of which do not employ all their valence electrons in bonding. A 50/50 mixture of layers of sp^2- and sp^3-bonded atoms was calculated to form a structure that can be regarded as a hybrid of graphite and diamond, but with a remarkably high concentration of unpaired electrons (0.59 per C atom) and spontaneous magnetic moment of 230 emu/g. This material has not been synthesised to date.

In the second part of this review, we consider in more depth two different types of pure carbon materials that have been synthesised and which show strong paramagnetic behaviour, with or without long-range ordering to produce ferromagnetism. We will neglect materials such as the hydrogenated pyrolytic carbon mentioned above, or nitrogen-bearing heteropolymers such as those discussed by in Ref. [7], in which the magnetic properties arise explicitly as a consequence of non-carbon atoms being present.

2 Strongly Paramagnetic and Ferromagnetic Pure Carbons

2.1 Two-dimensionally Polymerised C_{60} Molecules

The polyhedral molecules of fullerene, C_{60}, undergo cross-linking reactions when annealed at pressures of a few GPa and temperatures of 300–1100K to form a variety of crystalline polymers with layered structures. These polymerised fullerenes (*poly*-C_{60}) demonstrate ferromagnetic behaviour only when prepared in a relatively narrow range of synthesis conditions. A pressure-temperature diagram has been produced that shows systematic variation in the non-equilibrium products that can be synthesised by annealing of C_{60} fullerene under different conditions [33]. The synthesis conditions for magnetic *poly*-C_{60} cross-linked fullerene synthesis field lies within 100K of the upper temperature limit for the synthesis of this structure [16]. At higher temperature, the polyhedral carbon cages of fullerene break open further to form "partially graphitised" layer structures. This indicates that the process of formation of a ferromagnetic phase may be related to the occurrence of the defects due to cage opening. It is also feasible that the weak bonds between polyhedra are readily broken, and can act as the spin centres [34].

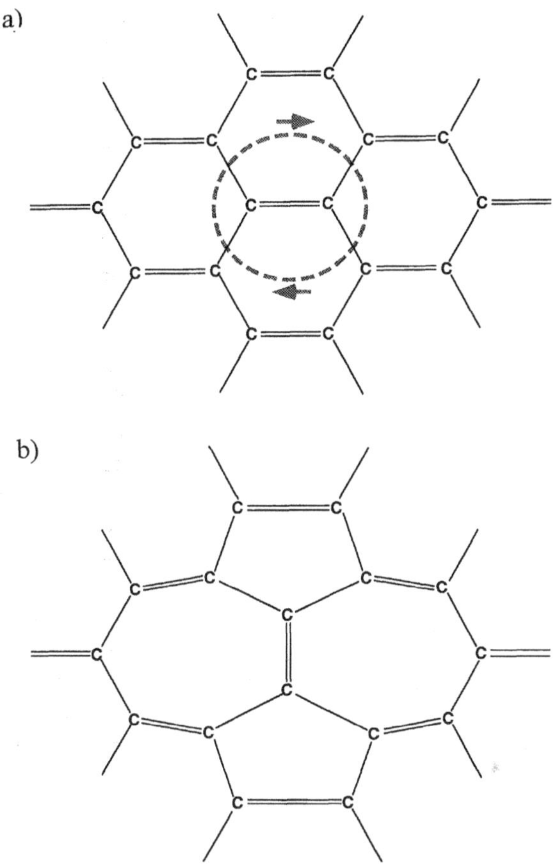

Fig. 2. Simple example of a Stone-Wales rearrangement, converting graphene 6-rings (a) into mixture of 5- and 7-rings (b).

It has been shown to be possible that the polymerisation of fullerene cages can take place via a sequence of Stone-Wales transformations – very local concerted rearrangements of atoms such as the reconnection of two adjacent 6-rings as a 5-ring and 7-ring (Fig. 2). As the polyhedral cages are opened, large rings (e.g. 11-gons) are formed [7, 24]. The zigzag edges of these rings are known to be associated with magnetism in graphene ribbons. Furthermore, the formation of adjacent large rings tends to change the local Gaussian curvature of the structure from positive in the close buckyball polyhedron to negative in the saddle-shaped regions between cages.

The magnetic ordering in selected metastable isomers of C_{60} with partly opened cages was recently theoretically investigated using the spin-polarized density-functional formalism [24]. Several carbon nanostructures, which could be described as broken fullerene cages with specified arrangements of under-coordinated carbon atoms, were identified as potential molecular magnets. Calculations showed that a zigzag arrangement of edge atoms in the defective cage stabilised dangling bonds, which in turn led to magnetic instabilities in isolated molecules and even ferromagnetism in the solid made of partially opened C_{60} cages. However, these calculations did not identify what specific defect structure(s) provided the mechanism of long-range magnetic coupling.

The possibility of magnetic centres arising in defective polyhedral cages was examined in a study that modelled *poly*-C_{60} using a combination of reactive force field molecular dynamics and hybrid-exchange density functional theory [35]. The model was used to generate potential defect structures and characterise their magnetic ground states. A large number of local defect structures were found to result in local moments. One particularly common structure defect type in this study, resulting from anisotropic strain, is reminiscent of a vacancy-adatom pair in that one atom of a C_{60} cage is forced out to become an inter-cage bridge, leaving an 11-ring hole in the cage. A vacancy-adatom pair on graphite had already been shown to be strongly magnetic (0.45 μ_B/adatom) by the calculations of [36]. The defect of [35] resulted in no net magnetization since spins were coupled via equal numbers of competing antiferromagnetic and ferromagnetic pathways. This suggests that ferromagnetism could be obtained if the coupling strengths of these pathways were perturbed, or if a different defect geometry were involved). Hybrid Density Functional Theory calculations showed that it was particularly favourable for hydrogen to bind at one of the three types of spin centre in this particular defect, and that the perturbation introduced by hydrogen quenched the antiferro coupling pathway selectively, leading to ferromagnetism with moment 3μ_B per cage. The experimental part of the study [35] analysed a sample of ferromagnetic *poly*-C_{60} for hydrogen content using inelastic neutron scattering. The presence of hydrogen was confirmed (H/C_{60} = 17%), at a level comparable with that predicted for the observed magnetisation for this model (H/C_{60} = 11%, 0.34 μ_B per polyhedral cage). It is possible that structures which allow greater spin delocalisation or a greater degree of inter-polyhedral linkage would show significantly larger inter-cage couplings.

It is clear from [35] that in some samples of *poly*-C_{60}, at least, ferromagnetism is a consequence of partial hydrogenation of the sample. Caution regarding magnetism in impurity-free *poly*-C_{60} is also suggested by [ref.12 in the paper, to be published], which showed that in the periodic model for *poly*-C_{60} proposed in [37] where a carbon atom is removed from each C_{60} cage, the magnetic moments localized on each cage were not able to couple through the inter-cage bonds formed during the 2+2 cycloaddition polymerisation reaction.

2.2 Carbon Nanofoam

2.2.1 Synthesis

A new form of carbon, a cluster-assembled carbon nanofoam, has recently been synthesised by laser ablation of a glassy carbon target in an argon atmosphere at a pressure of ~1-100 Torr [38–41]. The material is an exceptionally low-density aerogel with a hierarchical nanostructure. The carbon vapour temperature in the formation zone of the laser plume is in the range 1–10 eV, so the formation process takes place in partly ionized carbon plasma. The formation process involves periodic vapour heating and cluster formation stages, with the time period dependent on the initial Ar density, the evaporation rate, and reaction rate, which in turn is a function of the temperature and density of the atomic carbon vapour [39, 42]. Elastic recoil detection using 200 MeV Au ions established that the hydrogen content is below 100 ppm (atomic C:H = 1000:1). Analysis for other impurities by Rutherford backscattering and mass spectrometry demonstrated total non-(C,H) atoms to be below 500 ppm, with (Fe+Ni) < 110 ppm [19].

2.2.2 Structural Characterisation

Structural studies of the nanofoam by electron diffraction suggest the presence of hyperbolic "schwarzite" layers [19, 38, 40]. Schwarzites are anticlastic (saddle-shaped) warped graphite-like sheets, in contrast to the synclastic (ellipsoidally curved) sheets of fullerenes [43]. High-resolution TEM micrographs reveal the presence of "knitted" patterns of meandering graphitic sheets. The images are strikingly reminiscent of sectional images of translationally-ordered hyperbolic surfaces, known as triply-periodic minimal surfaces (TPMS, or IPMS), albeit with some degree of disorder, violating the strict 3D crystallinity of TPMS [44]. Fourier transforms of TEM micrographs indicated the existence of a structural period in the clusters with a space scale of 5.6±0.4 Å [20, 38], supporting the model of a partially molten TPMS, similar to "sponge mesophases" encountered in soft copolymeric and amphiphilic molecular materials [45]. By analogy with scattering spectra of sponge mesophases, the characteristic length scale observed by TEM was interpreted to correspond to the pore diameter within the sponge, and hence to the radii of curvature of the hyperbolic graphitic sheets. This parameter in turn can be related to the mean number of carbon atoms per ring in the sheet: six for conventional flat graphitic sheets and larger than six

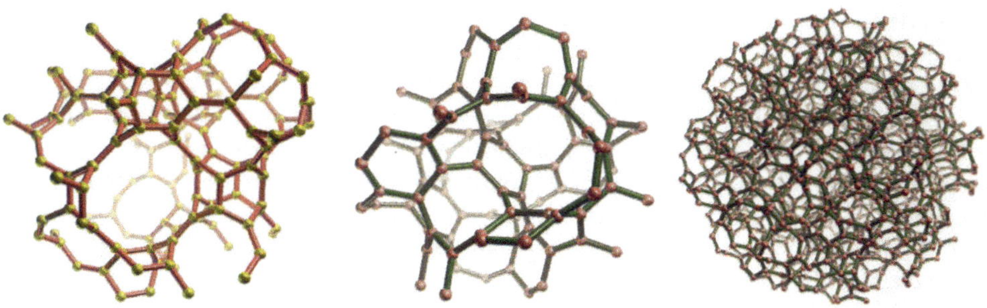

a) b) c)

Fig. 3. Examples of possible schwartzite structures.(a) Structure with 6-, 6- and 9-gons meeting regularly at each node. Characteristic radius of curvature is 5.5 Å, very similar to nanofoam. (b) Structure with 6-, 8- and 8-gons at each node. Tighter radius of curvature (4.4 Å) than (a). (c) Space-filling of spheroidal domain by a hyperbolic foam. Spheroid is 2 nm diameter in this case.

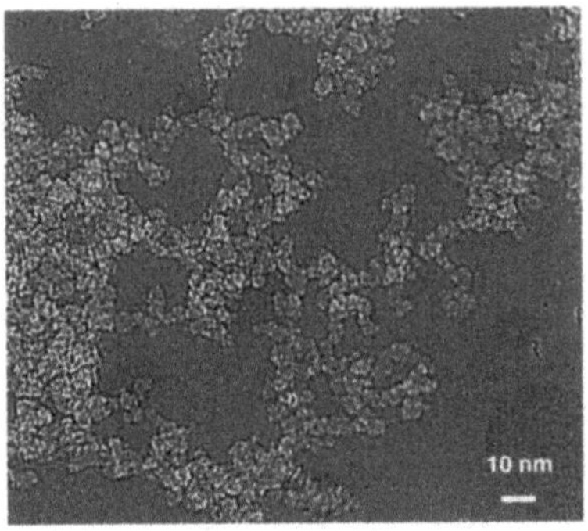

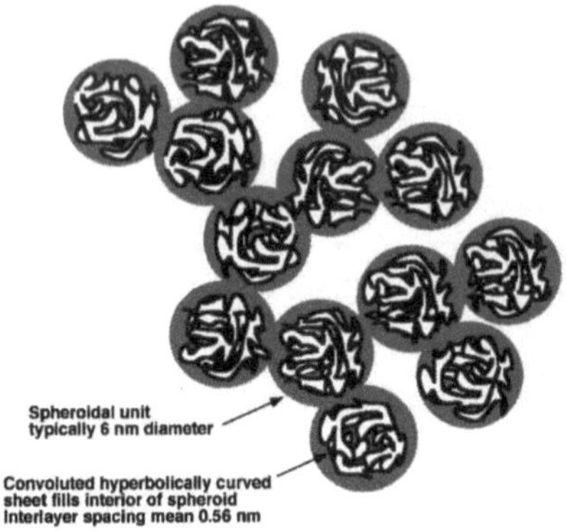

Fig. 4. (a) TEM image of dendritic structure of carbon nanofoam; (b) Diagram showing relationship of Fig. 4(a) and Fig.3.

for hyperbolically curved schwarzite. As a consequence of their hyperbolic structure, schwarzite sheets extend to fill three-dimensional space, in contrast to ellipsoidal sheets of fullerenes, that close so as to form finite molecules. Examples of model schwartzite structures are shown in Fig. 3. In the nanofoam, the schwarzite domains form spheroidal clusters of 6 nm typical diameter, which in turn are connected into dendritic forms (Fig. 4). The number of C atoms per spheroidal cluster was estimated to be $\sim 1.2\times10^4$, giving a density within the cluster of 1.65–1.95 g/cm^3. However, the open dendritic arrangement of clusters gives an extremely low bulk density of 0.002–0.02 g/cm^3, (tunable by setting the Ar pressure appropriately during synthesis) and surface area of 300-400 m^2/g, comparable to zeolites.

2.2.3 Magnetisation of Nanofoam

All magnetisation measurements were performed in the temperature range $1.8 \leq T \leq 300$K and in applied magnetic fields up to 70 kOe. Reproducibility was demonstrated by the comparable magnetization (0.36–0.8 emu/g) measured in six independently synthesized batches. The longevity of magnetisation of the foam (still 0.42 emu/g at 60 days and 12 months after the synthesis) is evidently typical for the material [20]. We note, however, that the magnetisation was always strongest in freshly prepared material, and decreased with time until a plateau value was attained. Evidently, some of the magnetisation is effectively permanent, while the remainder arises from metastable excited states that relax on the laboratory timescale, or sites that are vulnerable to demagnetisation by, for instance, reaction with the atmosphere.

The magnetisation response to an applied magnetic field was positive but decreased with temperature (Fig. 5) as expected for a paramagnetic material. A slight hysteresis was observed, with a well-defined coercive force at low temperatures (see Fig. 5 inset). Hence, at least part of the material shows ferromagnetic ordering. The Curie temperature was determined to be ~92K. The saturation magnetization $M(H)$ of

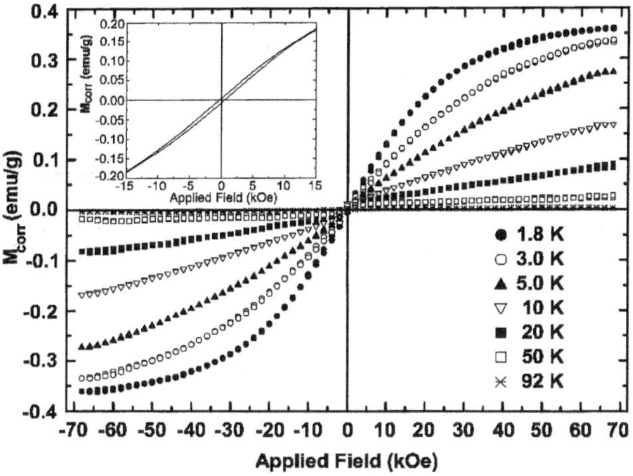

Fig. 5. Magnetisation of nanofoam versus applied field.

0.42 emu/g is equivalent to $9.0\times10^{-4}\mu_B$ per carbon atom. In a ferromagnetic system ($1\mu_B$ per spin), this value would correspond to about 1 unpaired spin per 1000 carbon atoms. Low-temperature ESR measurements gave the concentration of unpaired spins as $1.8\times10^{20}g^{-1}$ (3.6 unpaired spins per 1000 carbon atoms), in order-of-magnitude agreement, but somewhat higher than predicted for a completely ferromagnetic system. The most likely cause of the observed behaviour is the presence of ferromagnetic domains in a predominant paramagnetic matrix.

The susceptibility, given by the gradient at the origin of the curves in Fig. 5, obeys Curie-Weiss behaviour at low temperature. A plot of $1/\chi$ against T ($0K < T \leq 20K$) gave $\chi = 36\times10^{-6}/(T(K) + 1.0)$ (Fig. 6). The susceptibility, and the spin density, are at least three times larger than those of the low-hydrogen 1000°C anthracene chars of [46], and

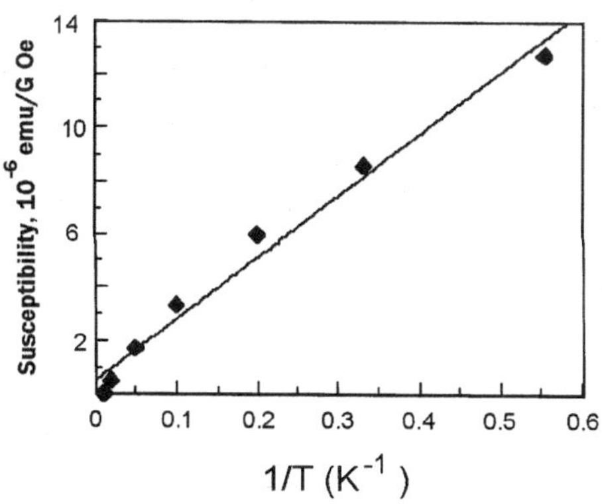

Fig. 6. Susceptibility of nanofoam.

the saturation magnetisations are similar or larger. Unpaired spins, presumably resulting from dangling bonds, are evidently present at similar concentrations in both.

2.2.4 Mechanism for Steric Protection of Unpaired Spins

The behaviour exhibited by the new phase of carbon is quite different from the weak magnetization found at very low temperatures in single-walled nanohorns and activated carbon fibers, which is believed to be due to exposed graphitic edges. It is similar to that found in anthracene chars, although the susceptibility and spin density of the nanofoam is higher. Unsatisfied bonds in the bulk are evidently responsible. We now show that the hyperbolically curved sheets in the foam suggest a unique mechanism that both localises the spins and preserves them from chemical attack.

High-resolution TEM-images show convoluted layers inside the nano-spherulitic building blocks of the foam, consistent with hyperbolic curvature of the sheets [38]. Such curvature requires rings of 7 or more carbons interspersed with graphitic 6-rings, since the mean angle subtended by any two bonds at a carbon atom must be greater than 120° [47]. Such sheet curvature localises unpaired spins by breaking the continuity of the delocalized π-electrons of graphite. Delocalisation of these electrons would facilitate pairing up of spins to form additional C-C bonds. Unpaired electrons on the carbon atoms readily form bonds with other species such as oxygen or hydrogen from air or water vapour, and would likely be too chemically reactive to persist. The observed tight curvature of the sheets provides a mechanism for *steric protection* of the unpaired spins: the carbon sheets are so tightly folded that it is very difficult for reactive species to diffuse between them. Stabilization of trivalent carbon radicals by steric protection has been known since 1900 [48]. It has recently been shown [49] that even divalent radical carbon in carbenes can be stabilized to show half-lives of several minutes. The nanofoam of our study is the first time to our knowledge that carbon radicals embedded in hyperbolic surfaces have been identified in a pure carbon prepared under controlled conditions by laser ablation.

The nanofoam is a rare example of a carbon in which there is reasonably good evidence for hyperbolically curved sheets. An additional example of a sponge-like carbon material has been reported recently, though the curvature of this material (hundreds of nm) is lower than that of theoretical schwarzites, giving a material not dissimilar to conventional graphite, albeit with a novel ultrastructure [50]. Examples of hypothetical model structures with negatively curved carbon sheets include crystalline "polybenzene" [51] and "random schwarzite" [52], as well as the structures of Fig. 3.

2.2. Electronic Structure Calculations

The origin of magnetism in the C-nanofoam has recently been investigated theoretically by *ab initio* modeling of a hyperbolic graphitic material [53]. The "tetrapod" structural unit of that study exhibits the some of the structural features observed in nanofoam, in particular carbon rings larger than hexagons. The calculations were based on a "tetrapod" geometry, a four-armed carbon nano-tubular junction, as a building block for schwarzites [43]. The core of the tetrapod consists of twelve

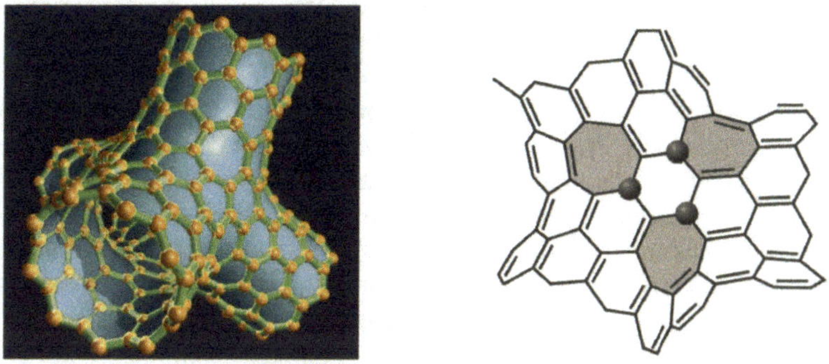

a) b)

Fig. 7. (a) Colour view of portion of Tetrapod; (b) Close-up showing triplet of spins at flat point of tetrapod surface.

heptagons surrounded by hexagons with tetravalent carbon atoms, and each of the heptagons has trivalent carbon radicals with a lone electron (Fig. 7). The presence of the heptagons, which introduce a negative Gaussian curvature, is a necessary condition for a magnetic ground state.

The detailed analysis of the electronic structure revealed that the electronic states associated with the unpaired spins are partially spread across parts of the tetrapod core. The radical sites occur as groups of three, which are collectively localised within the system of single and double bonds imposed by the tetrapod topology, and are associated with groups of three 7-rings (Fig. 7). If two of the three spins are paired, then each group of three trivalent carbons, and thus each group of three heptagons contribute one unpaired spin to the core structure. With four such spin centres per tetrapod core, the number of unpaired spins is four per tetrapod, which is in agreement with the spin density functional calculations [53].

The calculations for the tetrapod model are a good demonstration of the principle that disruption of the 6-ring network of graphite can in some circumstances force the generation of unpaired spins. Close juxtaposition of triplets of 7-rings is the mechanism in this case. The linked odd-order rings render it impossible to maintain strict alternation of single and double bonds, and provide a large enough area of perturbed bonding for spins to be isolated from the delocalised π-cloud of the surrounding graphite-like sheet. Other structures showing strong negative Gaussian curvature as a result of a high density of large odd-order rings may similarly be strongly magnetic.

3 Causes of Para- and Ferromagnetism in Pure Carbon

Comparison of the local structure and bonding in polymerised fullerenes and nanofoams leads us to suggest specific structure types for carbon in which unpaired spins can be generated and preserved.

In both nanofoams and polymerised fullerenes, the carbon atoms are arranged in a sheet which has graphite-like regions but is distorted so as to show non-zero *Gaussian curvature*. The polyhedral cages of the molecular fullerene structure show positive curvature due to the presence of 5-rings interspersed with the 6-rings of carbon atoms that are typical of graphite. In the polymerised form, some of the links within the cage are undone, to produce rings that are larger than hexagons. It is therefore possible to find regions of carbon sheet with saddle-like negative curvature in this structure. The nanofoam is also constituted of hyperbolic sheets. The presence of carbon rings other than 6-rings has a disruptive effect on the regular alternation of double and single C-C bonds, characteristic of ideal graphite. The delocalisation of the π-electron clouds of graphite may be partially suppressed or very markedly suppressed as in C_{60}. The topology and energetics may be such that satisfaction of all four bonds per atom is frustrated, and the electronic ground state contains unpaired spins. These may be delocalised over finite areas of carbon sheet, but pinned by a barrier of localised bonding electrons at the boundaries of such regions. This is the situation calculated for the tetrapod, and such a model accounts well for the spin density observed in the nanofoam.

Although spins can be localised to a degree by the breakup of the π-network that results from sheet curvature, much stronger localisation is obtained if boundaries are generated by including sp^3-hybridised atoms that are 4-coordinate, and hence cannot participate at all in the π-bonding system. Spectroscopic data for the nanofoam shows clearly that a substantial fraction of carbon atoms are sp^3 rather than sp^2 hybridised. Structural models that are proposed for the interconnections between the partially opened cages of *poly*-C_{60} also contain some 4-coordinate carbon atoms. In both cases, the implication is that the resulting structures are not just highly convoluted two-dimensional sheets, but contain some cross-links, connecting different "folds" of sheet. In the case of the nanofoam, it seems most likely that the sp^3 carbons provide links between the spheroidal building units, which contain the hyperbolic schwarzite sheets.

There is a third factor shared by polymerised fullerene and the nanofoam that is important in maintaining long-term magnetisation. The unpaired electrons that are responsible for the spin are valence electrons that are not participating in bonding. Therefore, it is possible for the free energy to be lowered if they establish bonds to one another, or to other foreign atoms such as hydrogen or oxygen derived from the air or from moisture. For magnetism to be retained permanently, it is essential to prevent this possibility. Therefore, the carbon structure should be locally rigid, to prevent close juxtaposition of spins occurring by deformation. The presence of sp^3 cross-links can play a role in stabilising the structure as structural braces, in addition to providing barriers to electron delocalisation. Chemical attack by foreign species is best prevented by locating the spins inside densely space-filling parts of the structure, so that reactive atoms are unable to get close to the spins by diffusion. The structure thus protects the spins by *steric hindrance*, a stablisation mechanism that has been known for molecular carbon radicals such as triphenylmethyl for over a century [48]. The convoluted sheets of the nanofoam, with an interlayer spacing of 5.6 Å, and the dense packing of cross-linked cages in *poly*-C_{60}, are both good examples of matrices that are likely to be quite impenetrable to diffusive species.

There remains the question of whether paramagnetic spins, which may be separated by distances of several Å, are able to couple so as to produce ferromagnetic ordering. Modelling studies on *poly*-C_{60} have shown that both ferro and antiferro coupling pathways can exist between such spins, and that competing pathways can occur in the same structure [35]. Therefore, the overall resultant magnetic behaviour will vary from specific material to specific material.

4 Synthesis Techniques

4.1 Overview

Specific structural criteria are characteristic of the carbon allotropes in which strong paramagnetism and magnetic ordering have been obtained experimentally so far, namely, curved sheets and a mix of sp^2 and sp^3 hybridisation. While the model of [53] suggests that a purely 3-ccordinate schwartzite should be able to support ferromagnetism, we note that the closest approximant among synthesised materials, the nanofoam described above, has been demonstrated to contain mixed coordination. The distinctive structural features of magnetic carbons are absent in graphite, the allotrope of carbon that is thermodynamically most stable at ambient conditions, as they are in diamond, the crystalline form that is stable at high pressure. The macroscopically crystalline but thermodynamically metastable allotropes of carbon, such as lonsdaleite and buckminsterfullerene, do not show the mixture of different carbon hybridisation states. The forms of carbon with extreme magnetic properties are clearly rather high-energy materials. It is, perhaps, not surprising that *poly*-C_{60} is only strongly magnetic when made at conditions which are near the upper temperature limit for synthesis of this type of material. It is equally unsurprising that magnetic carbons seem to be most readily formed in extreme environments where thermal disequilibrium prevails and fast condensation of carbon from a vapour phase is the norm.

One technique by which it appears possible to prepare such forms of carbon is a cathodic arc discharge method [54]. The vacuum cathodic arc used in those studies is a high current (>40 A) and low voltage (<30 V) discharge, which provides a convenient source of carbon plasma for deposition of dense amorphous carbon films with high sp^3 content. Electron spin resonance (ESR) measurements of deposited films have shown a very large concentration of unpaired spins of up to $\sim 7\times10^{20}\,g^{-1}$ in the tetrahedrally-bonded amorphous carbon films and ~80% of sp^3-bonded carbon atoms. The high spin density gives the films a large magnetic susceptibility [55, 56], which shows a Curie law dependence on temperature typical for a paramagnetic substance. The measured saturation magnetisation was ~0.55 emu/g at room temperature. The spin density of $1.4\times10^{20}\,g^{-1}$ calculated from the paramagnetic susceptibility is consistent with the highest values measured by ESR.

Another powerful technique, which has been developed recently for the preparation of large quantities of pure carbon nanostructures, is the use of high repetition rate laser ablation of a glassy carbon target [38–41]. This is the synthesis method that was used for the nanofoam that was described in detail in section 2.2. We now present a detailed consideration of the formation mechanism of this material.

4.2 Formation Mechanism of Nanofoam Clusters in Laser Ablation Plume

Laser ablation near the surface of the target creates an almost continuous inflow of hot carbon atoms and ions, with an average temperature of a few eV, into the experimental chamber. The shock wave generated by each pulse rapidly decelerates in the ambient gas atmosphere, and further propagation of hot atoms proceeds by diffusion, finally forming a mixture of carbon and filling gas with some average density and temperature. The density and temperature of this mixture changes with the distance from the target. The processes of collision, diffusion, and atom-to-atom (atom-to-cluster, cluster to cluster) attachment can be described qualitatively on the basis of a simple kinetic theory [24, 57, 58]. Depending on the masses of ablated atoms and the atoms of the filling gas, the processes of energy exchange will occur at different rates. If the masses of colliding atoms are comparable (for example, carbon flow with an argon fill), the carbon can lose a significant part of its energy in a single collision. Hence, efficient energy equilibration occurs after several collisions: the carbon vapour is cooling down and the argon gas is heating up. On the basis of kinetic considerations, the following scenario for carbon cluster formation in a carbon-argon mixture created by high repetition rate laser ablation has been suggested.

Initially the chamber gas is at ambient temperature. The continuous inflow of hot carbon increases the temperature in the mixture. When the carbon vapour temperature and the number density reach the level where the probability of carbon-carbon attachment becomes significant, the formation of carbonaceous clusters begins. The carbon consumption rate during this formation process significantly exceeds the

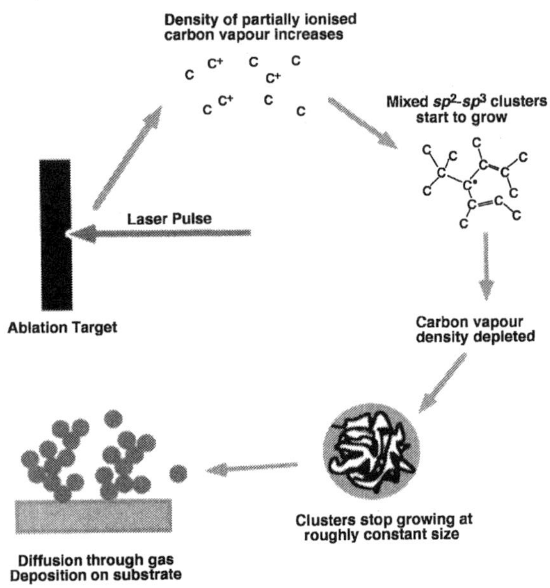

Fig. 8. Diagram showing control of cluster size by periodic kinetics of ablation and condensation

evaporation rate due to laser ablation. Therefore, the carbon density rapidly decreases to the value where the formation process terminates. Thus, the ablation rate, target parameters, pressure and the ambient gas determine *the formation time*, and accordingly, *the size* of the cluster formed (Fig. 8).

The cluster formation process appears to be comprised of periodic stages of heating and cluster formation, with the time period depending on the initial argon density, the evaporation rate, and on the carbon attachment reaction rate which in turn is a function of the temperature and density of the atomic carbon [58]. In the case of high-repetition-rate laser [59], during the short period of the cluster formation (in comparison to the heating period), the argon gas does not cool down but maintains its temperature approximately equal to that required for cluster formation. As a result, the average temperature in carbon-argon mixture appears to be high enough for formation of both sp^2 and sp^3 bonds as observed experimentally.

The maximum cluster size, from the kinetic viewpoint, is directly related to the cluster formation time defined above. The characteristic time for N-atomic cluster assembly in an argon-carbon mixture can be estimated given the assumptions that the dominant building process is single-atom attachment to a bigger cluster, and that the attachment cross-section is the geometrical cross-section for elastic collisions. This time is directly proportional to the ambient gas density and inversely proportional to the square of the partial density of single carbon atoms. The predictions of such a model qualitatively comply with the experimental data that small clusters (~6 nm, 10^4 atoms) are preferably formed at a low gas pressure of 1 Torr [38, 59], while carbon nanotubes (10^6 atoms) are formed at pressures of 300–500 Torr [58, 60–63]. However, kinetic theory, as well as thermodynamics, fails to explain fundamental issues relevant to the internal structure of the nanoclusters and the unusual material properties already observed experimentally, such as: why clusters having a particular number of particles ("magic numbers") are more abundant; which features of the formation process are responsible for the intricate structure of a cluster; and what factors determine formation of crystalline or amorphous clusters.

The homogeneous mixture of hot atoms created by ablation has high symmetry - it is almost isotropic. Kinetics and thermodynamics theories qualitatively explain why the hot atoms can form a cluster of well-defined size during the many "sticky" collisions. However, it is well known that nanoclusters have diverse and sometimes very complicated structures. How does the transition from an isotropic, disordered gas phase to the highly symmetric molecule of fullerene or a strongly anisotropic carbon nanotube or the even more complicated structural unit of carbon nanofoam [38] take place?

Formation of a three-dimensional cluster breaks the initial high symmetry of the cluster formation zone. From the chemical speciation point of view, the cluster formation zone contains a random mixture of different low-dimensional carbon clusters (monomers, dimers, trimers, five- and six-membered rings etc.) formed in the process of stochastic "sticky" collisions of carbon atoms in the carbon-ambient gas mixture. All of these clusters have a lifetime strongly dependent on the temperature. This lifetime tends to increase with decreasing temperature, and the spatial amplitude of fluctuations decreases when the temperature falls to the critical temperature and below. The critical temperature (T_C) is the characteristic temperature at which the particular structure begins to stabilise and becomes rigid [64]. The free energy of a newly formed cluster at $T < T_C$ must include thermal energy and also the internal (strain) energy which is absent from a totally disordered state. There is also a configurational entropy characteristic of a particular cluster structure. The minimization of the free energy determines the critical temperature for the defect-mediated phase transition from a disordered gas phase [65, 66], or for the case considered here, the optimum temperature for nano-cluster formation. This critical temperature was found experimentally: for carbon nanotube formation to be about 1200–2000K [62, 67, 68]. For fullerene formation, molecular dynamics simulations [69] indicate a range of 2000–3000K.

The formation of three dimensional carbon clusters can be understood in terms of a phase transition from a disordered to an ordered phase by the introduction of a specific defect assembly or seed with a new structure, similar to the conventional scenario seen for disorder-order transition phenomena in growth of crystalline structures [64, 65]. The appearance of such a discontinuity in the order parameter in a hot mixture of dissociated carbon atoms echoes arguments presented by Anderson that "the fluctuations do become of quasi macroscopic size and dimension at the critical point..." [64].

Most carbon nanomaterials display nano- or meso-scale structure that can be regarded as a consequence of atomic cluster geometries on a more local scale. For example, fullerenes show ordered arrays of pentagonal rings, which generate surfaces of positive Gaussian (ie ellipsoidal) curvature. It is likely that carbon clusters containing larger rings such as heptagons, induce locally negative (hyperbolic) curvature, leading to the "schwarzite" spheroidal clusters within the nanofoam.

5 Concluding Remarks

It has become evident quite recently that even pure carbon can show substantial paramagnetism and even ferromagnetism. These properties appear to arise because of specific features of the carbon structure and bonding, namely the presence of unsatisfied valences to act as spin centres, their localisation by barriers of four-coordinated carbon atoms (implying a mixture of 3- and 4-coordination overall) and protection against chemical attack through the formation of strongly convoluted sheets or impermeable frameworks. Such carbon materials are unstable relative to structurally simpler and more homogeneous carbon allotropes, and are best synthesised by methods in which carbon is vapourised and partially ionised by a high density energy source (electric arc, laser) and recondensed. Production and detailed characterisation of large quantities of these materials is still at an early stage. Their fabrication into technologically applicable components remains a challenge, but rapid and exciting developments are likely in the next few years.

References

1. P.J. Langevin, Annales de Chimie et Physique, 5 (1905) 70.
2. J.H. Van Vleck, The Theory of Electrical and Magnetic Susceptibilities (1932) Oxford University Press, Oxford.
3. W. Pauli, Z. Phys. 41 (1927) 81.

4. L.D. Landau, Z. Phys. 64 (1930) 629.
5. C. Kittel, Introduction to Solid State Physics (1995) 7th edition, Wiley, New York.
6. N.W. Aschroft and N.D. Mermin, Solid State Physics (1976) Holt, Rinehart and Winston, New York.
7. T. L. Makarova, Semiconductors 38 (2004) 615.
8. K. Kawabata, M. Mizutani, M. Fukuda, and S. Mizogami, Synth. Met. 33 (1989) 399.
9. S. Mizogami, M. Mizutani, M. Fukuda, and K. Kawabata, Synth. Met. 43 (1991) 3271.
10. K. Tanaka, M. Kobashi, H. Sanekata, A. Takata, T. Yamabe, S. Mizogami, K. Kawabata, and J. Yamauchi, J. Appl. Phys. 71 (1992) 836.
11. K. Murata and H. Ushijima, Jorn. Nucl. Instr. & Meth. C 4 (1996) 1.
12. A.A. Ovchinnikov and L.I. Shamovsky, J. Mol. Struct.: Theochem. 251 (1991) 133.
13. S. Bandow, F. Kokai, K. Takahashi, M. Yudasaka, and S. Iijima, Appl. Phys. A73 (2001) 281.
14. T.L. Makarova, B. Sundqvist, R. Höhne, P. Esquinazi, Y. Kopelevich, P. Scharff, V.A. Davydov, L.S. Kashevarova, and A.V. Rakhmanina, Nature 413 (2001) 716.
15. T.L. Makarova, K.H. Han, P. Esquinazi, R.R. da Silva, Y. Kopelevich, I.B. Zakharova, et al., Carbon 41 (2003) 1575.
16. R.A. Wood, M.H. Lewis, M.R. Lees, S.M. Bennington, M.G. Cain, N. Kitamura, Journ. of Phys.: Cond. Matter 14 (2002) L385.
17. V.N. Narozhnyi, K.-H. Müller, D. Eckert, A. Teresiaka, L. Dunsch, V.A. Davydov, L.S. Kashevarova, and A.V. Rakhmanina, Physica B 329-333 (2003) 1217.
18. Y. Kopelevich, P. Esquinazi, J.H.S. Torres, and S. Moehlecke, Journ. Low Temp. Phys. 119 (2000) 691.
19. A.V. Rode, R.G. Elliman, E.G. Gamaly, A.I. Veinger, A.G. Christy, S.T. Hyde, and B. Luther-Davies, Appl. Surf. Sci. 197-198 (2002) 644.
20. A.V. Rode, E.G. Gamaly, A.G. Christy, J.D. Fitz Gerald, S.T. Hyde, R.G. Elliman, B. Luther-Davies, A.I. Veinger, J. Androulakis, and J. Giapintzakis, Phys. Rev. B 70, 054407 (2004) .
21. P. Esquinazi, A. Setzer, R. Höhne, C. Semmelhack, Y. Kopelevich, D. Spemann, T. Butz, B. Kohlstrunk, and M. Losche, Phys. Rev. B66 (2002) 024429.
22. R. Höhne and P. Esquinazi, Adv. Mater. 14 (2002) 753.
23. S. Berber, Y.-K. Kwon, and D. Tománek, Phys. Rev. B62 (2000) R2291.
24. Y.-H. Kim, J. Choi, K. J. Chang, and D. Tománek, Phys. Rev. B68 (2003) 125420.
25. L. Liu, G.Y. Guo, C.S. Jayanthi, and S.Y. Wu, Phys. Rev. Lett. 88 (2002) 217206.
26. M. Fujita and K. Wakabayashi J. Phys. Soc. Jpn. 65 (1996) 1920.
27. K. Wakabayashi, M. Sigrist, and M. Fujita, J. Phys. Soc. Jpn. 67 (1998) 2089.
28. K. Wakabayashi, M. Fujita, K. Kusakabe, and K. Nakada, Czech J. Phys. 46 (1996) 1865.
29. K. Kusakabe and M. Maruyama, Phys. Rev. B67 (2003) 092406.
30. P. Esquinazi, D. Spemann, R. Höhne, A. Setzer, K.-H. Han, and T. Bitz, Phys. Rev. Lett. 91 (2003) 227201.
31. P. Esquinazi and R. Höhne, J. Magn. & Magn. Mater. In press, 2005.
32. P. Esquinazi, R. Höhne, K.-H. Han, A. Setzer, D. Spemann, T. Bitz, Carbon 42 (2004) 1213.
33. V.D. Blank, S.G. Buga, G.A. Dubitsky, N.R. Serebryanaya, M.Y. Popov, and B. Sundqvist. Carbon 36 (1998) 319.
34. V.V. Belavin, L.G. Bulusheva, A.V Okotrub, and T.L. Makarova. Phys. Rev. B70 (2004) 155402.
35. J.A. Chan, B. Montanari, J.D. Gale, S.M. Bennington, J.W. Taylor, and N.M. Harrison, Phys. Rev. B70 (2004) 041403.
36. P.O. Lehtinen, A.S. Foster, A. Ayuela, A. Krasheninnikov, K. Nordlund, and R.M. Nieminen, Phys. Rev. Lett. 91 (2003) 017202.

37. A.N. Andriotis, M. Menon, R.M. Sheetz, and L.A. Chernozatoskii, Phys. Rev. Lett. 90 (2003) 026801.
38. A.V. Rode, S.T. Hyde, E.G. Gamaly, R.G. Elliman, D.R. McKenzie, S. Bulcock, Appl. Phys. A 69 (1999) S755.
39. A.V. Rode, E.G. Gamaly, and B. Luther-Davies, Appl. Phys. A 70 (2000) 135.
40. E.G. Gamaly, A.V. Rode, and B. Luther-Davies, Laser and Particle Beams 18 (2000) 245.
41. E.G. Gamaly and A.V. Rode, in: Encyclopaedia of Nanoscience and Nanotechnology, ed. H.S. Nalwa, American Scientific Publishers, Stevenson Range 7 (2004) 783.
42. E.G. Gamaly, A.V. Rode, and B. Luther-Davies, Appl. Phys. A 69 (1999) S121.
43. D. Vanderbilt and J. Tersoff, Phys Rev Lett. 68 (1991) 511.
44. S.T. Hyde, S. Andersson, Z. Blum, S. Lidin, K. Larsson, T. Landh and B.W. Ninham, The Language of Shape (1997) Elsevier Science B.V. , Amsterdam.
45. S.T. Hyde, . Langmuir 13, 842 (1997).
46. P. Delhaès and F. Carmona, Chemistry and Physics of Carbon, eds. P.L. Walker and P.A. Thrower, 17 Ch.2 (1981) 89.
47. S.T. Hyde and M. O'Keeffe, Phil. Trans. R. Soc. Lond. A 354 (1996) 1999.
48. M. Gomberg, J. Am. Chem. Soc. 22 (1900) 757.
49. H. Tomioka, E. Iwamoto, H. Itakura, and K. Hirai, Nature 412 (2001) 626.
50. G. Benedek, H. Vahedi-Tafreshi, E. Barborini, P. Piseri, P. Milani, C. Ducati, and J. Robertson, Diamond and Related Materials 12 (2003) 768.
51. M. O'Keeffe, G.B. Adams, and O.F. Sankey, Phys Rev Lett. 68 (1992) 2325.
52. S.J Townsend, T.L. Lenosky, D.A. Muller, C.S. Nichols, and V. Elser, Phys Rev Lett. 69 (1992) 921.
53. N. Park, M. Yoon, S. Berber, J. Ihm, E. Osawa, and D. Tománek, Phys. Rev. Lett. 91 (2003237204).
54. D.R. McKenzie, Rep. Prog. Phys. 59 (1996) 1611.
55. E.G. Gerstner, P.B. Lukins, D.R. McKenzie, D.G. McCulloch, Phys. Rev. B54 (1996) 14504.
56. M.M. Golzan, D.R. McKenzie, D.J. Miller, S.J. Collocott, and G.A.J. Amaratunga, Diam. Rel. Mat. 4 (1995) 912.
57. E.G. Gamaly, and L.T. Chadderton, Proc. R. Soc. Lond. A 449 (1995) 381.
58. E.G. Gamaly, A.V. Rode, W.K. Maser, E. Munoz, A.M. Benito, M.T. Martinez, and G.F. de la Fuente, Appl. Phys. A 70 (2000) 161.
59. E.G. Gamaly, A.V. Rode, and B. Luther-Davies, Laser and Particles Beams 18 (2000) 245.
60. T. Guo, P. Nikolaev, A. Thess, D.T. Colbert, R.E. Smalley, Chem. Phys. Lett. 243 (1995) 49.
61. W.K. Maser, E. Muñoz, A.M. Benito, M.T. Martínez, G.F. de la Fuente, Y. Maniette, E. Anglaret, and J.-L. Sauvajol, Chem. Phys. Lett. 292 (1998) 587.
62. F. Kokai, K. Takahashi, M. Yudasaka, R. Yamada, T. Ichibashi, and S. Iijima, J. Phys. Chem. B 103 (1999) 4346.
63. M. Yudasaka, F. Kokai, K. Takahashi, R. Yamada, N. Sensui, T. Ichibashi, and S. Iijima, J. Phys. Chem. B 103 (1999) 3576.
64. P.W. Anderson, Basic Notions of Condensed Matter Physics (1984) the Benjamin/Cummings Pub. Comp. , Menlo Park, CA, USA.
65. J.M. Kosterlitz and D.J. Thouless, J. Phys. C: Solid State Phys. 5 (1972) L124.
66. H.S. Seung and D. R. Nelson, Phys. Rev. A 38 (1988) 1005.
67. S. Arepalli, P. Nikolaev, W. Holmes, and C.D. Scott, Appl. Phys. A 70 (2000) 125.
68. A.A. Puretzky, H. Schittenhelm, Xudong Fan, M.J. Lance, L.R. Allard. Jr., and D. Geohegan, Phys. Rev. B65 (2002) 245.
69. Seong Gon Kim and D. Tománek, Phys. Rev. Lett. 72 (1994) 2418.

Carbon-Based Magnetism
T. Makarova and F. Palacio (Editors)

21

McConnel Model for the Magnetism of C_{60}-based Polymers

A.N. Andriotis[a], M. Menon[b], R. M. Sheetz[c], and E. Richter[d]

[a]*Institute of Electronic Structure and Laser, Foundation for Research and Technology-Hellas, P.O. Box 1527, Heraklio, Crete, 71110 Greece*
[b]*Department of Physics and Center for Computational Sciences, University of Kentucky, Lexington, 40506-0045 KY*
[c]*Center for Computational Sciences, University of Kentucky, Lexington, 40506-0045 KY*
[d]*DaimlerChrysler AG FT3/SA, Wilhelm-Runge-Str. 11, 89081 Ulm, Germany*

1 Introduction

1.1 Magnetic Carbon

Magnetism in organic materials is a welcome phenomenon from technological perspective since these materials are usually less expensive to produce than their ordinary metallic counterparts. Furthermore, they are light weight and known for low power consumption. Organic ferromagnets have theoretically a high attainable specific magnetization value (*i.e.*, magnetization/density) that exceeds that of α-Fe. This property makes magnetic carbon potentially very useful. An organic magnet that responds to light could lead to new ways of storing and reading large amounts of computer data [1].

Ovchinnikov and Shamovsky [2] have predicted a ferromagnetic (FM) phase for pure carbon of mixed sp^2 and sp^3 bonded carbon atoms. The suggested FM state contains an equal number of sp^2 and sp^3 C-atoms per unit volume which corresponds to the maximum possible concentration (50%) of unpaired carriers (*i.e.*, sp^2 C-atoms). The recent discovery of magnetic phase of the polymerized C_{60} materials [3–5], characterized by high Curie temperatures (500K [3], 820K [6]) and small saturation magnetization values, has attracted the research interest from both theoretical and technological point of view.

However, this phenomena has also raised many basic questions regarding the possibility of the existence and the nature of magnetism in the C_{60}-based polymers [7, 8]. These questions have their origin in the fact that these materials contain no $d-$ or $f-$electrons which are strongly interacting, while the new magnetic materials develop a new type of magnetism based on weakly interacting *s*- and *p*- electrons. Nevertheless, the experimental evidence keeps growing. Other graphite-based magnetic materials have been discovered including, for example, the highly oriented pyrolytic graphite (HOPG) [9], the graphite surface in the presence of carbon-adatoms [10], the activated carbon fibers [11–13], graphite ribbons [14], the fluorinated graphite [15], the hydrogenated nanographite, [16] the carbon foam [17], etc. At the same time, other organic and inorganic magnetic materials have been discovered exhibiting common features with the C_{60}-based FM polymers. Prominent among them are the TDAE- C_{60} [18], the p-nitrophenyl-nitronyl-nitroxide (p-nnn) radicals [19–22], the hexaborites [23–27], the oxides of rock-salt structure [28, 26], the cation-substituted oxides (*i.e.* Co substituted cations in ZnO and TiO_2 [29, 30]), etc.

In the following we explore and identify the common features between the magnetic carbon-based (specifically C_{60}-based polymers) and the other novel magnetic materials referred to in the above. As will be shown, our exploration leads us to the pioneering early works of McConnell [31, 32], Breslow [33, 34] and Ovchinnikov [2] on organic ferromagnets. It will also be shown that electron-electron correlations in the presence of structural defects appear to be the key factors for the development of the magnetic phase in carbon-based materials and this justification can be extended to all the other materials as well. In this view, the magnetic carbon-based and in particular the C_{60}-based materials seem to belong to a sub-class of a new class of FM materials which follow a common formation mechanism of ferromagnetism [35, 28]. This is classified as defect-magnetism and, as it will be discussed in the following, it may be considered as a generalized of the McConnell-II type magnetism.

1.2 Prerequisites for the Appearance of Carbon-magnetism

The explanation of the observed magnetism in the C_{60}-based polymers [3, 4] is a two-fold issue [36]. First, an explanation of the existence of unpaired spins is needed. This has to be followed by a theoretical justification of the origin of the long range correlation that couples the unpaired spins into a FM alignment. The proposed theoretical explanations are mostly focused at justifying the existence of unpaired electrons and they all appear to converge on their origin which they attribute to the structural and/or topological defects that either localize π electron states or contribute to delocalized lone electron states associated with the defects [11, 15, 37–42]. As to the origin of the FM coupling mechanism (FCM) which couples the unpaired electrons and develops the FM state the answer is far from conclusive. In some reports the origin of the FCM is either completely ignored or not satisfactorily justified

[41, 40], while in others it is assumed that the FCM is originating from an exchange interaction similar to the $s-d$ interaction [11, 42, 37].

At present, we are far from a complete understanding of carbon magnetism. As referred to in the above, the development of the FCM appears to be the outcome of strong electron-electron correlations which, depending on the system, may manifest itself either as a charge and/or spin transfer process or as a flat band ferromagnetism or as a carrier-induced ferromagnetism or as a super-exchange process etc.

2 C_{60}-Polymers: A brief overview

For the sake of completeness, the basic properties of representative members of the C_{60} polymers are briefly discussed in this section.

2.1 Preparation Methods of C_{60}-polymers

C_{60}-based polymers can be synthesized using:

- Photochemical methods [43], and
- high Pressure - High Temperature (HP-HT) methods [3–5, 44, 45].

Both types exhibit the same polymeric $2+2$ cycloaddition bonds [46]. However, the photopolymers have a distorted *fcc* structure exhibiting a statistically random (isotropic) polymerization along all nearest neighbor (nn) directions (geometric frustration). On the other hand, HP-HT polymers display crystalline (ordered) structures. This results from the fact that an externally applied anisotropic stress selects the direction of polymeric bonding which, in turn, relieves the frustration associated with the isotropic polymerization [47].

A pronounced characteristic of the newly discovered C_{60}-polymers is the experimentally found narrow windows, defined by pressure and temperature, within which magnetic C_{60}-based polymers can be found. These windows are approximately at the graphitization limit [48]. The C_{60}-based polymers are observed at the graphitization limit *i.e.*, the limit at which the fullerene cages begin to break down, thus, justifying the assumption (see below) that the magnetic phase is associated with the appearance of structural defects in an otherwise (and experimentally verified) crystalline structure of these materials. It has also been verified that the magnetism of these samples is not influenced by the presence of the impurities they contain. The determination of any structural differences, however, between magnetic and non-magnetic samples by conventional characterization methods has not been possible yet [3–5].

2.2 Electronic Properties of C_{60} Polymers

In the absence of vacancies we find all 2-D C_{60}-based polymers to be semiconducting with energy gaps in the range 0.3–1.0 eV, in agreement with reported band structure calculations [49–52]. As will be shown in Section 1, the presence of the vacancies induces states in the energy gap with the result that these materials become metallic or semi-metallic in the limit of infinite polymer with one vacancy per C_{60} unit with the exception of the liner C_{60} polymer which exhibits sp^3 and sp^2 orbitals [36] and shows a very narrow gap.

3 Carbon Magnetism: A Member of a New Magnetic Class?

Based on a series of theoretical calculations, we have recently proposed a model for explaining the magnetic phase of the polymerized-C_{60} materials [36]. The basic assumption of our model is the existence of vacancies in an otherwise pristine crystalline structure as intimated by the experimental results of Makarova and collaborators [3–5]. We demonstrated that the existence of the [2 + 2] cycloaddition bonds (which provide sp^3-bonded carbon atoms) in coexistence with vacancies can explain the origin of the magnetism in the polymerized- C_{60} materials. In particular, we have shown that the vacancies and the [2 + 2] cycloaddition bonds provide, on the one hand, the unpaired electrons while on the other hand ignite charge transfer processes which create and sustain large electric dipole moments. The latter develop the FCM among the unpaired spins and lead to the development of a FM state. In such a mechanism, the electron-electron correlations play a crucial role as they promote the high spin-multiplicity states and make them energetically more favorable.

Concurrently with our model, another vacancy-based model has also been reported in an effort to explain the magnetic phase of oxides of rock-salt structure (*i.e.*, CaO) [28]. A comparison between these two works provides the necessary motivation to explore to see if the defect mechanism can be used in justifying the appearance of magnetism in the materials referred to in Sec. 1. A careful study of both classes of materials *i.e.*, the carbon-based and the non-carbon based ones, referred to in the above seems to reveal some common underlying features relevant to the nature of the associated magnetism in these materials. These are summarized in the following subsection.

3.1 Common Features of Carbon- and other Organic and Non-traditional Inorganic Ferromagnets

The most pronounced common features in the carbon-based and other organic or non-traditional inorganic magnetic materials are found to be the following [35]:

- The presence of defects (structural and topological) in an otherwise crystalline structure.
- The development of significant charge and spin transfer.
- The existence of a degenerate ground state driven by electron-electron correlations, and (in some cases) the presence of a resonant ground state.

3.2 Additional Features Exhibited by C_{60}-based Polymers

The magnetic C_{60}-based polymers exhibit some additional features that are worth noting. In particular, it was pointed out that the FM samples of C_{60}-based polymers have always been found to be in a crossover regime for a metal-insulator (M-I) transition [5]. Furthermore, band structure calculations revealed that some C_{60}-based polymers exhibit flat or partially flat bands around the Fermi energy, E_F, and naturally the questions arise: Is the magnetism of the C_{60}-based polymers related to a M-I instability [5]? Is it a demonstration of flat-band ferromagnetism [53, 54]? Is it the result of structural frustration [47]? The resolution of these questions form the basis of the current investigations and are discussed in the following.

4 Defects in Carbon-based Materials

The emerging field of carbon nanoscience seems to draw much of its useful information from the basic graphite system. As Makarova *et al.,* observed [5], the magnetism in the C_{60}-polymers cannot be seen independently of the graphitic magnetism. Therefore, a first step in examining the role of C_{60}-point defects in the polymerized C_{60}-magnetism is the study of the corresponding graphite counterpart, *i.e.*, point defects in graphite and the role they play in the graphite magnetism.

It is also quite instructive to recall the basic properties of (carbon) conjugated polymers due to their similarity with the polymerized C_{60}. It should be noted that the conjugated polymers are semiconductors with a single band-gap of 1–3 eV but with the same mechanical properties as those of plastics. A prototype conjugate polymer which is of interest to our study is the polyacetylene $(CH)_x$; it contains a backbone of sp^2-bonded C-atoms with the energetically highest valence electron occupying a p-orbital perpendicular to the plane of the nuclei; these π-orbitals appear close to the Fermi energy and a pairing mechanism among them leads to alternating single (longer) and double (shorter) bonds. However, it is worth noting that the distribution of the π-electrons may be drastically changed in the presence of perturbations, mainly those imposed by defects [55].

As stated above, defects seem to play a dominant role in the development of the carbon magnetic phase. In the present work, the term "defect" is ascribed to any type of point and/or structural and/or topological defects including vacancies [56], edge states [38, 39], substitutional impurities [29], Stone-Wales defects in graphite-based materials [40], orientational changes (as for example, among the C_{60} molecules in the TDAE- C_{60}) [18], adsorbed adatoms [10], etc. Most of these defects play more or less the same role in promoting a FM ground state. That is, they break the sequence of the alternating single-double carbon bonds inducing the response of the material in the form of charge and spin transfer that leads to the formation of lone electrons and the possibility for a stable FM ground state. Thus, magnetic carbon phases have been reported for proton-irradiated HOPG [57], functionalized graphitic particles [15, 16], graphite with carbon adatoms [10], etc. In all these cases the defects break the local symmetry and the system responses for healing it. Before entering into the details of their role in FM carbon-based materials we discuss two such defects which are related to the ferromagnetism in the C_{60}-based polymers, *i.e.*, the vacancies and the $2+2$ cycloaddition bonds. Other type of defects will be discussed in section 5.

4.1 Vacancies

The vacancies are created by removing one or more carbon atoms; they lead to under-coordinated C-atoms. The experimental results on graphite vacancies have mainly been obtained from STM and TEM studies. It has been widely accepted that the symmetric (D_{3h}) vacancy is the most common mono-vacancy in graphite. However, recent reports claim that the asymmetric vacancy 5-1db (pentagon-one dangling bond) with (experimental) formation energy of 7.0 ± 0.5 eV has lower energy while the symmetric vacancy is a result of a time-averaged superposition of three degenerate vacancy-states alternating by a thermally activated Jahn-Teller effect [58]. (This result has been questioned recently by Ma *et al.* [59] who attributed the results of Barbary and collaborators [58] to artifacts and/or limitations of the computational approach used in reference [58]). It has also been found that the ground state vacancy in graphite is spin-free.

Contrary to what was initially anticipated, Hjort and Stafstrom using the Hückel approximation found that the triad of carbon atoms near a graphite vacancy do not contribute to the system with three σ-type sp^2 dangling orbitals but with four π-type sp^3 electrons [56]. The relaxation of this triad of atoms is found to be entirely in plane with the carbon atoms next to the vacancy site relaxing by moving away from the vacancy cite. Furthermore, it was found that each of these carbon atoms gets approximately 0.19 electrons originating from states near the Fermi energy. The latter finding were claimed to be justified by STM measurements in which the vacancy is viewed as a bright spot (protrusion).

The results of Ref. [56] were recently called into question by Andriotis *et al.,* who argued that the vacancy sites create a charge-density-wave-like (CDW-like) disturbance, with positive charge accumulation of approximately 0.5 electrons and an electric dipole moment (EDM) of 0.39 Debyes [60]. The same picture was found to develop also around a vacancy in the C_{60}-based polymers, but in this case the calculated EDM was found to be much larger (2.264 Debyes) [36]. Furthermore, it was found that carbon atoms surrounding the vacancy in graphite exhibit lone electrons and, thus, the vacancy appears to be the provider of unpaired spins.

4.2 The [2+2] Cycloaddition Bonds

The $[2+2]$ cycloaddition bonds that are formed during the polymerization process can be considered as a different type of defect that breaks the alternating single and double bonds in the C_{60}-polymers. The formation of these bonds requires that neighboring C_{60} molecules be oriented with their double bonds parallel to each other [46, 61]. A complete polymerization leads to an average inter-C_{60} distance of 9.1–9.2 Å (an un-polymerized nn distance is 10 Å, by comparison, at ambient pressure). This bond is also known as a 66/66 bond because it links two pairs of bonds that are shared by two hexagons in a C_{60} molecule. It may also happen that one or both bonds that are linked by the cycloaddition be shared between a hexagon and a pentagon. The resulting cycloaddition bonding then may be of the type 65/66 or 65/56, respectively. The relaxed

66/66, 65/56 and 65/66 bonds were found to have length 9.19, 9.17 and 9.26 Å, respectively [45].

5 Theoretical Models

We have classified the various theories proposed for the origin of magnetism in the organic and other non-traditional inorganic materials in three general categories according to their basic and characteristic feature of the proposed mechanism for magnetism. These are: (i) the defect-related theories, (ii) the charge and spin transfer theories, and (iii) the band magnetism theories one of which is the carrier induced ferromagnetism discussed below.

5.1 Defect-related Theories

The role of the defects appear to be multiple. Firstly, they provide unpaired electrons and help in their delocalization. Secondly, they break locally the symmetry of the crystalline structure. This has, as a consequence, the development of intra-molecular charge and spin density transfers which change the relative stability of the high spin states which may become energetically the most favorable configurations for the ground state. In some cases this process is facilitated by the existence of a resonant ground state (as for example in the case of the TDAE- C_{60} [18] and the p-nnn radicals [20–22, 19]. It is worth mentioning that in carbon-based materials the defects are developing conditions found in the various levels of the proposed McConnell model [31–34] and create the *starred* and *un-starred* C-atoms (non Kekulé structures) extensively discussed within the Ovchinnikov's spin-polarization model [2]. Thus, it is observed that defects introduce a perturbation that alters the features of the bonding among the carbon atoms as in the case of the perturbed conjugated polymers.

5.1.1 Other Structural and/or Topological Defect-related Theories

The vacancies and the other structural defects discussed in Sec. 4 have one common characteristic: They can provide unpaired electrons which, in turn, provide the spin-magnetic moments.

Besides vacancies another very common defect is the one known as the Stone-Wales defect [62]. This is formed in a graphene lattice by a C-C bond rotation of 90° transforming, thus, a four-adjacent hexagon-structure into one consisting of two heptagons and two pentagon adjacent to each other. Kim *et al.* [40] claimed that a partly opened intermediate structures obtained by transforming a C_{60} molecule into a nanotube by a series of Stone-Wales defects are magnetic. This mechanism was proposed as a possible explanation for the observed magnetism in C_{60}-based polymers.

According to Park *et al.* [41], defects appearing in all-carbon nanostructures exhibiting negative Gaussian curvature are associated with sterically protected carbon radicals which result in the presence of heptagons in an an otherwise hexagon-based structure. These carbon radicals acting in groups of three provide a delocalized lone electron, providing free electron spins. However, in this report as well as in Ref. [40] the FCM mechanism which is necessary for the FM state to be built up has not been addressed convincingly.

Finally, in nano-graphitic particles defects which result by translating one graphite layer with respect to its neighboring ones (stacking defects) were found to exhibit magnetic ground state [38, 39]. In these nano-graphitic systems, the exotic magnetism is related with the π-states localized mainly at carbon atoms on the zig-zag edges of each nano-graphitic layer. A similar picture is seen in nano-graphene ribbons with zig-zag edges. These systems are characterized by partially flat bands which cause a sharp peak in the electron DOS at the Fermi energy. Furthermore, the zig-zag graphene ribbons exhibit a charge-polarized state which is stabilized in the presence of nearest neighbor Coulomb interactions and becomes competitive in stability with the spin-polarized state in an external electric field [63].

5.1.2 Frustrated Systems

Polymerization of C_{60} leads to polymeric [2+2] cycloaddition bonds resulting in the development of significant lattice strains the latter being proportional to the number of the polymerization bonds developed. Polymeric C_{60} systems belong to the class of the frustrated systems. Their basic characteristic is that they exhibit a large number of isoenergetic configurations and, consequently, have a distorted ground state with macroscopic degeneracy [47]. The vast degeneracy of their classical ground state makes them highly susceptible even to small perturbations. Such a system can be viewed as being in a Resonant Valence Bond (RVB) state which is defined as a coherent superposition of the various possible arrangements of all the valence bonds of the system wit no crystalline long-range order. The RVB state is stabilized by lattice deformations (phonons) which increase the hopping energy across the valence bonds. A very interesting property of the frustrated systems relevant to the present study is the observation that in some frustrated systems magnetic ordering can occur in the presence of external fields. This is because a large degree of the geometric frustration can be healed through a coupling between spin and lattice degrees of freedom. A particular example is the system $Ho_2Ti_2O_7$ which does not exhibit long magnetic order (up to at least 0.35K) in zero magnetic field. However, upon applying a field, ordered phases are observed. More importantly, the presence of defects (*i.e.*, dangling bonds in a RVB system) switch on the resonating process among the various valence bond configurations which can lead to a more stable ground state. It is quite interesting to note that such defects can extend over several sites and be quite mobile (case of polyacetylene).

5.2 Charge and Spin Transfer Theories

The McConnell-II model [32], extended by Breslow [33, 34], predicts that a quasi-1D ionic material D^+A^- built up from positive ions of the donor molecules D and the negative ions of the acceptor A can exhibit a FM ground state if either the donor or the acceptor molecule have a triplet neutral ground state. In such a case back charge transfer mixes the neutral triplet ground state with the triplet charge transferred state of the ionic state leading to a FM coupling.

The McConnell-I model [31] predicts that a FM spin-coupling may result in systems which exhibit large positive and negative atomic π-spin densities suitably packed so that atoms of positive spin density are exchange coupled to the atoms of negative spin density in neighboring molecules. This becomes evident by rewriting the Heisenberg Hamiltonian

$$H^{AB} = -\sum_{ij} J_{ij}^{AB} \mathbf{S}_i^A \cdot \mathbf{S}_j^B \quad (1)$$

as

$$H^{AB} = -\mathbf{S}^A \cdot \mathbf{S}^B \sum_{ij} J_{ij}^{AB} \rho_i^A \rho_j^B \ , \quad (2)$$

where J_{ij} are the exchange integrals, $\mathbf{S}_k^X$ ($\mathbf{S}^X$) denote the π-electron spin on atom k of molecule X (the total spin of molecule X) respectively and ρ_k^X is the spin density on atom k of molecule X.

For such materials, the theoretical studies are focused at the determination of the effective exchange interaction J_{ij} and determine the orbital-orbital (OO), J_{ij}^{OO} and the spin density product (SDP) contribution J_{ij}^{SDP} [64, 65].

5.3 Carrier-induced Ferromagnetism

The carrier-induced ferromagnetism is another concept that has been introduced in explaining the observed ferromagnetism in dilute semiconductors. According to this approach, the carriers (electrons or holes) become nearly bound to the magnetic impurity sites by getting strongly coupled to the localized magnetic moments. The carriers experience a spin dependent interaction with the magnetic impurity (translated as a spin dependent energy of the impurity-site). The impurity band becomes broader for the FM than the PM state with the result the FM state to become energetically more favorable. The hopping of carriers among the impurity sites causes the FM ordering of the localized spins through the double-exchange mechanism. (See for example, Ref. [66] where this concept is applied using the dynamical coherent potential approximation (CPA)).

6 Magnetic C_{60} Polymers Revisited

The magnetic C_{60}-based polymers prepared with the HP-HT method are frustrated systems (see Sec. 2) and, therefore, one could assign their magnetism to the consequences of the frustration. On the other hand, their preparation conditions do not exclude the presence of point defects (*i.e.*, vacancies, Stone-Wales defects,

2+2 cycloaddition bonds of various character - bond orientational defects - , etc). Thus, polymerization, frustration and point defects acting independently or in a synergetic way lead to structural and/or topological instabilities to which the system reacts with charge transfers and/or by coupling charge, spin and lattice degrees of freedom. Such processes being assisted by electron-electron (*e-e*) correlations allow states of high-spin multiplicity to appear as the energetically more favorable one for the ground state.

In the following, results for the C_{60}-based polymers obtained by performing tight-binding molecular dynamics (TBMD) simulations as well as the *ab initio* methods are discussed.

6.1 Tight-binding Molecular Dynamics Simulations

In theoretical treatments in which defects and *e-e* correlations are omitted, no magnetic ground state is found for the C_{60}-based polymeric materials. It has been shown that calculations carried out using the tight binding (TB) approximation incorporating the Hubbard-U approximation to the *e-e* correlations can be used successfully to describe carbon-based magnetic materials [39, 67–72]. This is further supported by our recent studies on these systems using our TBMD scheme at the Hubbard-U level of approximation to the *e-e* correlations (to be denoted as the U-TBMD method). Our U-TBMD method includes use of *s*, *p* and *d* orbitals in order to provide the correct description of inter-atomic interactions in systems consisting of semiconductor and transition metal atoms as well as their hetero-systems. The method has been successfully used to study not only the carbon-based materials [73], but also the magnetic clusters consisting of transition metal atoms (TMAs) [72, 74, 69] as well as their interaction with low-dimension carbon surfaces (graphite, C_{60}, single-wall carbon nanotubes (SWCNs), benzene molecules, etc) [75–77].

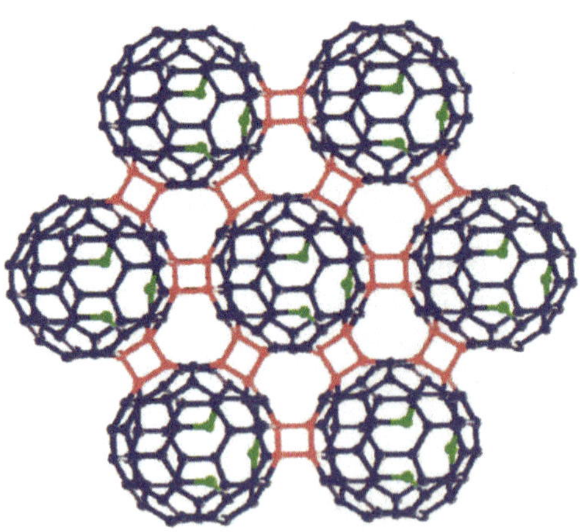

Fig. 1. 2-D rhombohedral C_{60} (Rh- C_{60}) polymer with one vacancy per C_{60} unit. The atoms surrounding the vacancy are shown in green, while the sp^3 atoms participating in the [2+2] cycloaddition are shown in magenta.

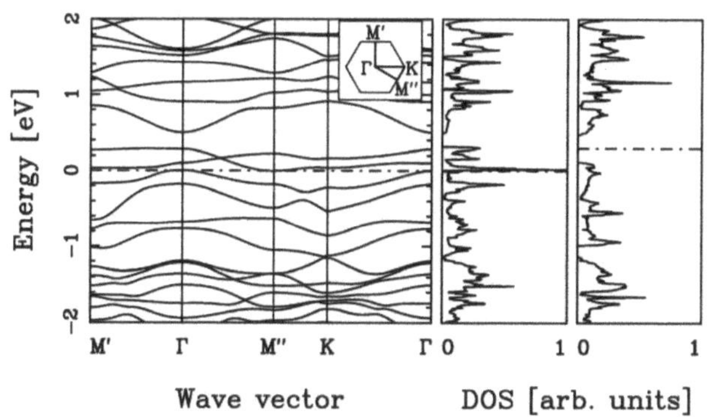

Fig. 2. Band structure for the Rh- C_{60} polymer with one vacancy per C_{60} unit (left), density of states (DOS) (middle) and DOS in the absence of any vacancy (right). The Fermi levels are indicated by dashed lines.

In our studies we considered 2-D (rhombohedral and tetragonal) and 1-D (orthorhombic) C_{60} polymers containing a mixture of sp^2 and sp^3 bonded carbon atoms. We also consider an all sp^2 1-D C_{60} polymer [36]. The infinite polymers were simulated using supercells with periodic boundary conditions. We created vacancies by removing one carbon atom per C_{60} unit in each of the structures. All structures were fully optimized without any symmetry constraints using our TBMD scheme. The relaxed polymer in the 2D rhombohedral- C_{60} (Rh– C_{60}) structure with one vacancy per C_{60} molecule is shown in Fig. 1. In Fig. 2 we present results for the band structure for this polymer (left panel), while the electron density of states (DOS) is shown in the middle panel. For comparison, we present the vacancy-free DOS in the right panel.

As is clear from Fig. 2, the introduction of vacancies induces a dramatic change in the Rh- C_{60} polymer; they change it from a semiconductor into a semi-metal. This is in accordance with the experimental results which find that the magnetic phase of the Rh-C_{60} polymer is found at a crossover regime for a metal-insulator transition [5].

6.2 Ab initio Calculations

In order to get and in depth understanding of the origin of the changes that take place in the C_{60}-based materials upon polymerization and the development of their magnetic phase, we performed *ab initio* calculations in a C_{60}-dimer with and without vacancies. In the latter case we assumed one vacancy per C_{60} molecule as with our TBMD simulations. All *ab initio* calculations were performed with the GAUSSIAN 03 program package using density functional theory (DFT) with the three-parameter hybrid functional of Becke combined with either the Lee-Yang-Parr correlation functional or the Perdew-Wang 1991 gradient-corrected correlation functional [78]. The atomic basis set that was used is of double zeta quality and includes relativistic effects for heavy atoms.

Our initial computational results verified that the defect-free dimer has a singlet ground state. For the dimer with the vacancies it was found that the ground state is a triplet which is found to be 1.905 and 0.816 eV lower that the singlet and quintet states, respectively [79]. These results are in agreement with both theory and experiment [3, 36] according to which a magnetic moment of approximately one Bohr magneton ($1\ \mu_B$) per C_{60} molecule is expected for the C_{60}-based polymers.

In Fig. 3, we show the charge transfer that is induced by the vacancies in a C_{60}-dimer. In this graph, two striking features are immediately noticed. The first is that there is a depletion of electrons from each vacancy region (which appear positively charged) and at the same time there is an accumulation of charge in the [2+2] cycloaddition bonds. The second feature is that a charge density wave-like (CDW-like) disturbance develops around each vacancy which does not die away. Instead, being assisted by the opposite charge accumulation around the [2+2] cycloaddition bonds, it joins up smoothly with the CDW-like disturbance around the neighboring vacancy. The developed electric dipole moment (EDM) is rather large (2.264 Debyes per C_{60} molecule). It is worth noting that a similar *ab initio* study of the vacancy in a graphene sheet gave us the same picture for the charge transfer around the vacancy as in the C_{60} dimer case. However, in the case of the graphene sheet with a vacancy, the calculated EDM is found to be much smaller (0.39 Debyes). This is a noticeable difference between our results and those of Ref. [56]. Another difference is that in our case the carbon atoms around the vacancy move towards the vacant site upon relaxation while in the case of Ref. [56] they move away.

Contrary to the charge delocalization, the spin density appears localized around the vacancy cites.

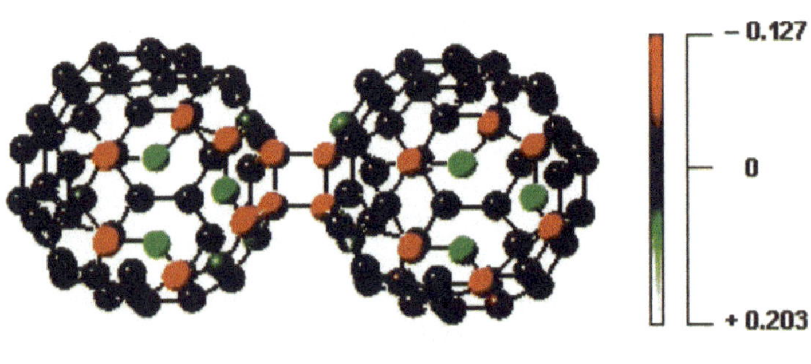

Fig. 3. Figure showing the effect of charge transfer induced by the vacancy and the presence of [2+2] cycloaddition. The atoms surrounding the vacancy are positively charged while the atoms participating in the cycloaddition are negatively charged.

7 Discussion and Conclusions

A careful comparison of our results with those reported for various other non-related systems, but exhibiting *s–p* magnetism, allows us to observe many basic common features discussed earlier (see section 1). Subsequently, it has been realized that these systems develop significant charge transfers and a case can be made for these systems to be considered as generalized McConnell ferromagnets. In this section we take this observation further and complete the present investigation with our conclusions.

7.1 Generalized McConnell Ferromagnetism

It must be emphasized that the *s–p* magnetism observed in carbon based materials as well as in other organic and non-traditional inorganic materials is a consequence of strong *e–e* correlation effects. The latter lead to ground states that can be described as a superposition of states of various spin-multiplicities and manifest themselves as FM-configurations when the high-spin-multiplicity states are dominant. In the ferromagnetic charge transfer salts, described by the McConnell model [31–34], such a possibility arises when either the acceptor or donor molecules exhibit triplet ground state (see Sec. 2). In these systems, the acceptor and donor sites are assigned to molecules of different kinds. On the other hand, in the case of the defect-magnetism, which appears to be the case for the C_{60}-based polymers and other organic and non-traditional inorganic magnetic materials, one kind of defects (*i.e.*, vacancies) play the role of the donor molecules while coexisting defects of another type (*i.e.*, 2+2 cycloaddition bonds) play the role of the acceptor molecules. The ferromagnetism, in this case, depends on the ability of the defects to create and sustain electric fields which bind the charge in the atoms around the defects and provide the necessary kinetic exchange for promoting the high spin states as the energetically more favorable ground state. In this view, the defect magnetism of the C_{60}-based polymers and that of other organic and non-traditional inorganic magnetic materials can be considered as a generalization of the McConnell model [31–34].

Looking at it from a different angle, the donor and/or the acceptor sites can be considered as the main sources which induce lone pair electron orbitals and provide unpaired spins; the emphasis here is focused on the ability of the defects to develop the necessary interaction pathways among the unpaired spins and establish the necessary FCM.

In both views, the role of the charge transfer is crucial; it can be achieved by the development of sufficient charge delocalization. The delocalization may be global [41] or it may extend over a small number of either trivalent or tetravalent C-atoms [60]. The latter type of delocalization is analogous to that appearing in alternant hydrocarbons in which the unpaired electrons are distributed over only the *active* carbon atoms [80]). According to our calculations, the majority of the delocalized molecular orbitals of the C_{60}-polymers, the delocalization is of the form of a *through-space homoconjugation* [81] or *hyperconjugation* [82]. This delocalization leads to an overlap between an electron in its $2p_z$ orbital with an sp^2 orbital in an adjacent or remote C-C bond.

The development of the delocalization can be related with the findings that some defect-based magnetic materials are either metallic or semi-metallic. This is supported in the case of the Rh- C_{60} polymer where the band structure of the relaxed system with one vacancy per C_{60} unit shows the system to be semi-metallic (see Sec. 1).

The conditions under which organic materials appear as generalized McConnell ferromagnets depend strongly on two factors. The first being the existence of generalized donor and acceptor sites in the material and second being the OO and SDP contributions to the effective exchange interaction [64, 65] among the donor and acceptor sites (see Sec. 2). Thus, factors like geometry and symmetry, frustration, delocalization of molecular orbitals and energy spectrum of the lowest excited states [83] are expected to be among those which determine the appearance or not of a generalized McConnell ferromagnet.

In the ferromagnetic C_{60}-based polymers, the optimum conditions for the creation of donor and acceptor sites seem to be achieved in the presence of structural and/or topological defects in an otherwise crystalline structure of these materials. In these, the role of the defects appears to be in the development of strong EDMs locally which promote the contribution of the high-spin-multiplicity states to the ground state and sustain the required conditions for establishing the FM ground state. This role of the defects was sufficiently demonstrated by our *ab initio* calculations.

7.2 Other Possible Generalized McConnell Ferromagnets

A donor-acceptor type magnetic material that can be characterized as a generalized McConnell-II ferromagnet is the TDAE- C_{60} in which the defects originate as a consequence of changes in the mutual orientation of adjoining C_{60} units [18, 84]. However, a characteristic example of non-carbon based magnetic material of the proposed generalized McConnell type appears to be the case of the defect-magnetism of CaO in which a recent study [28] has demonstrated that Ca-vacancies lead to a FM state while oxygen vacancies do not. In another example, namely the case of ZnO where Zn atoms are substituted by Co or Mn [85], it was found that the FM-state is stabilized if concurrently with the substitution of Zn by Co the system exhibits Zn vacancies or it is doped by Cu^{+} ions. It is worth noting that the ZnO system which is simultaneously doped by Co and Cu is in complete analogy with the ferromagnetic C_{60} polymers. That is, both systems exhibit two kinds of defects which act as donor and acceptor atoms; they are just members of the same class of magnetic materials that can be characterized as ferromagnets of the generalized McConnel-II model. Other possible members of this class may be the hexaborites (*i.e.*, CaB_6 with B_6-vacancies) [23, 26], the TiO_2 in which Ti is partially replaced by Co [29], conjugated polymers (with number of *starred* C-atoms different than that of the *unstarred* ones [80, 2]), etc.

7.3 Band Ferromagnetism

Looking from the band structure point of view, our results indicate an additional role of the defects. That is, their association with the narrowing of the energy gap at the Fermi level which has as a result the semiconducting defect-free C_{60}-polymers to become semi-metallic in the presence of defects. This has as a consequence the delocalization of the of lone electron orbitals which may then act as mediators for establishing the coupling among the unpaired spin-moments. In this sense one can attribute the magnetic state of these materials in the mediated coupling of the delocalized orbitals in a way analogous to the $s-d$ type of interaction in transition metals. In our opinion, this is not exactly the case. The mediated coupling in the s - p materials provides rather the means for a *kinetic exchange* stabilization of the magnetic state relative to the non-magnetic one. However, it has to be pointed out that both terms used *i.e.*, the s - d coupling and the *kinetic exchange* denote expressions related to the description of the level and the way the e - e correlations are treated in each of the above described methods and therefore they can be understood as complementary descriptions of the same underlying physics.

For analogous reasons we can exclude the justification of the carrier induced ferromagnetism (see Sec. 3).

Finally it should be noted that our results also do not support the case for flat band ferromagnetism in the C_{60}-based polymers [53, 54]. In fact, 2D-rhombohedral and the 1D C_{60}-polymers do exhibit either partial or complete flat bands when vacancy-doped. This may be associated with a degenerate ground state. However, it is difficult to check whether the *local connectivity condition* is satisfied [53, 86], especially in the case of the Rh- C_{60} polymer. It may be only the magnetism of the 1D-polymers with vacancies to be amenable to the class of the flat band ferromagnetism as this material shows flat bands at E_F and it is probable that can satisfy the local connectivity condition.

7.4 Conclusion

The magnetism of the C_{60}-based polymers maybe justified as a generalized McConnell magnetism. This is a consequence of a kinetic exchange stabilization of the magnetic state rather than a carrier induced or flat-band ferromagnetism. The generalized McConnell ferromagnetism is facilitated by the presence of structural and topological defects which develop the necessary charge transfers and, associated with them, exchange interaction pathways.

The above observations may be helpful in designing new magnetic materials of the generalized McConnell-II type. The recipe is simple. That is, to check whether doping can provide to the system charge centers of both sign (negative and positive) which are associated with unpaired spins and at some particular concentration these defects can provide the necessary FCM among the free spins.

In view of this recipe, it does not appear necessary to look only for donors or acceptors with triplet ground state. Instead, one can also search for materials with those kind of defects which can give rise to charge transfer with associated electric fields strong enough to promote a high-spin state as the most favorable ground state.

Acknowledgments

The present work is supported through grant by EU-GROWTH research project AMMARE (G5RD-CT-2001-00478), NSF (ITR-0221916), DOE (00-63857), NASA (00-463937), and the Kentucky Science & Technology Corporation (03-60214).

References

1. D.A. Pejakovic, C. Kitamura, J.S. Miller, and A.J. Epstein, Phy. Rev. Lett. 88 (2002) 057202.
2. A.A. Ovchinnikov and I.L. Shamovsky, J. Mol. Structure (Theochem), 251 (1991) 133.
3. T.L. Makarova, B. Sundquist, R. Hohne, P. Esqulnazi, Y. Kopelevich, P. Scharff, V.A. Davidov, L.S. Kashevarova, and A.V. Rakhmanlna, Nature, 413 (2001) 716.
4. T.L. Makarova, in Studies of High-T_c Superconductivity, 45 (2003) 107, Ed. by A. Narlikar, abstract cond-mat/0207368.
5. T.L. Makarova, J. MAgn. Magn. Mat. 272-276 (2004) E1263.
6. V.N. Narozhnyi, Physica B 329 (2003) 1217.
7. R. Hohne and P. Esquinazi, Adv. Mater. 14 (2002) 753.
8. B.G. Levi, Physics Today, p.18 (Dec., 2001).
9. Y. Kopelevich, P. Esquinazi, J.H.S. Torres, and S. Moehlecke, J. Low Temp. Phys. 119 (2000) 691.
10. P.O. Lehtinen, A.S. Foster, A. Ayuela, A. Krasheninnikov, K. Nordlund, and R.M. Nieminen, Phys. Rev. Lett. 91 (2003) 017202.
11. Y. Shibayama, H. Sato and T. Enoki, Phys. Rev. Lett. 84 (2000) 1744.
12. N. Kobayashi, T. Enoki, C. Ishii, K. Kaneko, and M. Endo, J. Chem. Phys. 109 (1998) 1983.
13. The activated carbon fibers are a kind of microporous carbon material consisting of a 3-D network of 2–3 nm micrographites whose edges are covered with functional groups and dangling bonds. Each micrographite is formed with a stack of 3-4 nanosized graphene sheets 12.
14. K. Wakabayashi and K. Harigaya, J. Phys. Soc. Japan 72 (2003) 998.
15. M. Panich, A. I. Shames and T. Nakajima, J. Phys. Chem. Solids 62 (2001) 959.
16. K. Kusakabe and M. Maruyama, Phys. Rev. B67 (2003) 092406.
17. K. Umemoto, S. Saito, S. Berber, and D. Tomanek, Phys. Rev. B64 (2001) 193409.
18. B. Narymbetov, A. Omerzu, V.V. Kabanov, M. Tokumoto, H. Kobayashi, and D. Michailovic, Nature 407 (2000) 883.
19. M. Tamura *et al*, Chem. Phys. Lett. 186 (1991) 401.
20. K. Awaga, T. Sugano, and M. Kinoshita, Chem. Phys. Lett. 141 (1987) 540.
21. K. Awaga and Y. Maruyama, Chem. Phys. Lett. 158 (1989) 556.
22. M. Takahashi *et al*, Phys. Rev. Lett. 67 (1991) 746.
23. D. P. Young *et al*, Nature 397 (1999) 412.
24. K. Matsubayashi, M. Maki, T. Tsuzuki, T. Nishioka, and N.K. Sato, Nature 420 (2003) 143.
25. D.P. Young *et al*, Nature 420 (2003) 144.
26. R. Monnier and B. Delley, Phys. Rev. Lett. 87 (2001) 157204.
27. Z. Fisk, H.R. Ott, V. Barzykin, and L.P. Gor'kov, Physica B 312-313 (2002) 808.
28. I.S. Elfimov, S. Yonoki, and G.A. Sawatzky, Phys. Rev. Lett. 89 (2002) 216403.
29. Y. Matsumoto *et al*, Science 291 (2001) 854.

30. A.S. Risbud, N.A. Spaldin, Z.Q. Chen, S. Stemmer, and R. Seshadri, Phys. Rev. B68 (2003) 205202.
31. H.M. McConnell, J. Chem. Phys. 39 (1963) 1910.
32. H.M. McConnell, Proc. R. A. Welch Found. Chem. Res. 11 (1967) 144.
33. R. Breslow, B. Jaun, R. Q. Kluttz and C-. Z. Xia, Tetrahedron 38 (1982) 863.
34. T.J. LePage and R. Breslow, J. Am. Chem. Soc. 109 (1987) 6412.
35. A.N. Andriotis and M. Menon, World Scientific, Singapore, in press, (2004).
36. A.N. Andriotis, M. Menon, R. M. Sheetz and L. Chernozatonskii, Phys. Rev. Lett. 90 (2003) 026801.
37. M. Fujita *et al*, J. Phys. Soc. Japan 65 (1996) 1920.
38. K. Harigaya, Chem. Phys. Lett. 340 (2001) 123.
39. K. Harigaya and T. Enoki, Chem. Phys. Lett. 351 (2002) 128.
40. Y. -H. Kim, J. Choi, K. J. Chang, and D. Tomanek, Phys. Rev. B68 (2003) 125420.
41. N. Park, M. Yoon, S. Berber, J. Ihm, E. Osawa, and D. Tomanek, Phys. Rev. Lett. 91 (2003) 237204.
42. K. Yoshizawa, K. Okahara, T. Sato, K. Tanaka, and T. Yamade, Carbon 32 (1994) 1517.
43. Y. Murakami and H. Suematsu, Pure & Appl. Chem. 88 (1996) 1463.
44. R.A. Wood, M.H. Lewis, M.R. Lees, S.M. Bennington, M.G. Cain and N. Kitamura, J. Phys.: Condens. Matter 14 (2002) L385.
45. A. V. Okotrub, V. V. Belavin, L. G. Bulusheva, V. A. Davydov, T. L. Makarova and D. Tomanek, J. Chem. Phys. 115 (2001) 5637.
46. M. Menon, K. R. Subbaswamy and M. Sawtarie, Phys. Rev. B49 (1994) 13966.
47. L. Marques, M. Mezouar, J.-L. Hodeau and M. Nunez-Regueiro, Phys. Rev. B68 (2003) 193408.
48. The graphitization process is the heat treatment of carbonaceous material at high temperature in contradistinction to the carbization process which refers to the low-temperature treatment 12.
49. S. Okada and A. Oshiyama, Phys. Rev. B68 (2003) 235402.
50. S. Okada and S. Saito, Phys. Rev. B59 (1999) 1930.
51. C. H. Xu and G. E. Scuseria, Phys. Rev. Lett. 74 (1995) 274.
52. W. L. Yang *et al*, Science 300 (2003) 303.
53. H. Tasaki, Prog. Theor. Phys. 99 (1998) 489.
54. A. Mielke, J. Phys. A 24 (1991) L73.
55. H. Meider and M. Springborg, Chem. Phys. Lett. 300 (1999) 339.
56. M. Hjort and S. Stafstrom, Phys. Rev. B61 (2000) 14089.
57. P. Esquinazi, D. Spemann, R. Hohne, A. Setzer, K.-H. Han, and T. Butz, Phys. Rev. Lett. 91 (2003) 227201.
58. A.A. El-Barbary, R.H. Telling, C.P. Ewels, M.I. Heggie, and P.R. Briddon, Phys. Rev. B68 (2003) 144107.
59. Y.Ma, P.O.Lehtinen, A.S.Foster and R.M.Nieminen, New J. of Phys. 6 (2004) 68.
60. A.N. Andriotis, R.M. Sheetz, E. Richter and M. Menon (to be submitted) (2004).
61. J. Onoe and K. Takeuchi, Phys. Rev. Lett. 79 (1997) 2987.
62. A.J. Stone, D.J. Wales, Chem. Phys. Lett. 128 (1986) 501.
63. A. Yamashiro, Y. Shimoi, K. Harigaya, and K. Wakabayashi, Phys. Rev. B68 (2003) 193410.
64. K. Yamaguchi, H. Fukui, and T. Fueno, Chemistry Letters (The Chem. Soc. Japan) (1986) 625.
65. K. Yamaguchi, T. Fueno, K. Nakasuji, and I. Murata, Chemistry Letters (The Chem. Soc. Japan) (1986) 629.

66. M. Takahashi and K. Kubo, Phys. Rev. B66 (2002) 153202.
67. A.A. Farajian, K. Esfarjani, and Y. Kawazoe, Phys. Rev. Lett. 82 (1999) 5084.
68. M.P. Lopez-Sancho, M.C. Munoz, and L. Chico, Phys. Rev. B63 (2001) 165419.
69. A.N. Andriotis and M. Menon, Phys. Rev. B57 (1998) 10069.
70. L. Bergomi, J.P. Blaizot, Th. Jolicoeur and E. Dagotto, Phys. Rev. B47 (1993) 5539 (R).
71. D. Coffey and S.A. Trugman, Phys. Rev. Lett. 69 (1992) 176.
72. M. Menon, J. Connolly, N. Lathiotakis, and A. Andriotis, Phys. Rev. B50 (1994) 8903.
73. M. Menon and K.R. Subbaswamy, Phys. Rev. B50 (1994) 11577.
74. A.N. Andriotis, N. Lathiotakis, and M. Menon, Euro Phys. Lett. 36 (1996) 37.
75. A.N. Andriotis, M. Menon, G.E. Froudakis, Z. Fthenakis, and J.E. Lowther, Chem. Phys. Lett. 292 (1998) 487.
76. A.N. Andriotis, M. Menon, G.E. Froudakis, and J.E. Lowther, Chem. Phys. Lett. 301 (1999) 503.
77. A.N. Andriotis, M. Menon and G.E. Froudakis, Phys. Rev. B62 (2000) 9867.
78. M.J. Frisch *et. al.*, Gaussian, Inc., Pittsburgh PA, (2003).
79. All calculations were performed on the relaxed geometry of the triplet state.
80. H.C. Longuet-Higgins, J. Chem. Phys. 18 (1950) 265.
81. R.V. Williams, Chem. Rev. 101 (2001) 1185.
82. A. Reed, L.A. Curtiss, and F. Weinhold, Chem. Rev. 88 (1988) 899.
83. K. Agawa, T. Sugano and M. Kinoshita, Chem. Phys. Lett. 141 (1987) 540.
84. P.-M. Allemand *et al*, Science 253 (1991) 301.
85. N.A. Spaldin, Phys. Rev. B69 (2004) 125201.
86. R. Arita, Y. Suwa, K. Kuroki, and H. Aoki, Phys. Rev. Lett. 88 (2002) 127202.

Carbon-Based Magnetism
T. Makarova and F. Palacio (Editors)

22

The Mechanism of the Magnetic Interaction in Polymeric-C_{60} Fullerenes

J. Ribas-Ariño and J.J. Novoa

Departament de Química Física and CERQT, Facultat de Química, Universitat de Barcelona, Av. Diagonal, 647, 08028-Barcelona, Spain

1 Introduction

The presence of ferromagnetic interactions in pressed-heated polymeric C_{60} fullerenes solids, having Curie temperatures in the 500–800K range, is a reproducible and well characterized experimental observation [1–4]. Up to now ferromagnetic interactions have been reported when pressing the 2D rhombohedral and tetragonal polymeric-C_{60} phases [5]. The experimental study of these magnetic solids allows us to obtain the following main conclusions: (*a*) the magnetism of these solids is not due to the presence of impurities or oxygen trapped in the samples [1–3, 5]; (*b*) the structure of the C_{60} units is preserved, thus making very unlikely that magnetism could be associated to fragments generated during the rupture of the C_{60} units [3]; (*c*) in the samples, the magnetic phase is a minority phase within the non-magnetic matrix formed by the rest of the solid [6]. All these facts have to be accounted for by any mechanism that tries to explain the magnetism found in these pressed polymeric-C_{60} solids.

The structure that results after applying pressure and heat to pristine C_{60} is a layered structure. Within each layer, every C_{60} unit is linked to six (rhombohedral phase) or four (tetragonal phase) C_{60} units, in what we can call *the polymeric-C_{60} link* (Fig. 1). In the ideal rhombohedral and tetragonal phases each polymeric-C_{60} link has the same structure, with two interfragment C-C single (covalent) bonds, which results from a [2+2] cycloaddition involving one 6:6 bond from each of the linked C_{60} units (these 6:6 bonds are C=C double bonds placed between two fused hexagons). The two interfragment C-C bonds have the same length, and the same happens with the two C-C

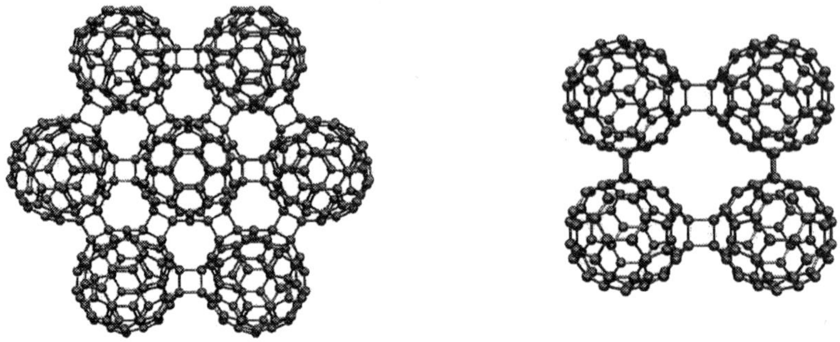

Fig. 1. Structure of the layers in the rhombohedral (left) and tetragonal (right) phases of polymeric-C_{60} fullerene solids.

single bonds that result from the 6:6 double bonds after the cycloaddition. This explains the shape of a square that the polymeric-C_{60} links present.

Experimental studies [1–4] and theoretical band calculations [7] indicated that the ideal structures of the polymeric phases are diamagnetic. Therefore, three models, each based on a different type of distortion of the ideal polymeric structure, have been recently proposed to explain the origin of the magnetic interactions in these polymeric solids. In this review we will describe and compare these mechanisms. There are proposals to explain the presence of magnetic moments in carbon-only structures that do not apply to C_{60} cages [8], or those that deal hydrogenated polymeric-C_{60} solids [9], which will not be considered here.

The first mechanism proposed [10], hereafter identified as the *cage-vacancy mechanism*, see Fig. 2, is an extrapolation to the polymeric-C_{60} layered solids of a mechanism that is known to explain the origin of magnetic moments in graphite [2]. Thus, it was indicated [10] that C_{59}, a caged-molecule originated by extracting one C atom from C_{60}, presents a non-zero spin magnetic moment. Thus, if other C_{59} units could be generated in the solid, this could give rise to ferromagnetic interactions between nearby C_{59} units. The authors suggested that the mechanism of the C_{59}-C_{59} ferromagnetic interaction could be of the dipolar type, although the validity of such explanation was not evaluated numerically.

A second proposed model [11], hereafter called the *open-cage mechanism*, explained the magnetism in polymeric-C_{60} solids on the basis of the presence of higher energy forms of C_{60} that show a partly opened caged structure (see Fig. 2). Some of these open-cage structures present triplet ground states. They can be obtained by applying successive generalized Stone-Wales (SW) transformations [12] to the optimum C_{60} crystallographic structure. The energy of these open-cage isomeric structures of C_{60} was reported to be 239.8 kcal/mol above that for the optimum C_{60} crystallographic structure. The authors did not study the nature of the magnetic interactions between open-cage structures.

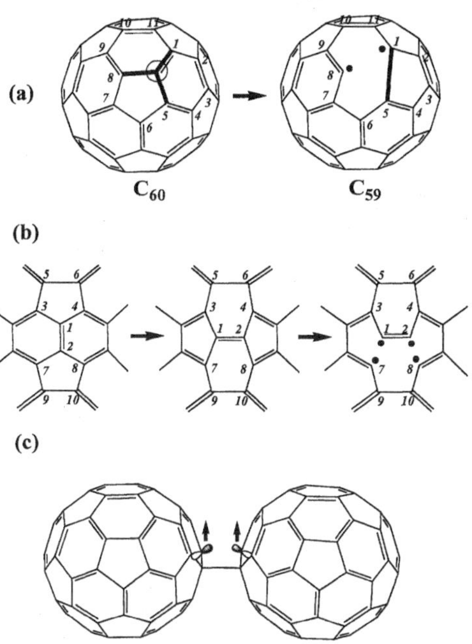

Fig. 2. (a) Generation of the C_{59} cage from C_{60}, showing the position of the unpaired electrons that would give rise to the triplet; (b) Steps of a Generalized Stone-Wales transformation; (c) Structure of the T_2 state of a polymeric-C_{60} link, showing the position of the unpaired electrons (in a triplet configuration).

The third mechanism, hereafter identified as the *interfragment bond-breaking mechanism*, see Fig. 2, was recently proposed by us [13]. Here, the existence of ferromagnetic interactions in the polymeric-C_{60} link is due to the rupture of one of the two interfragment C-C bonds present in the ideal polymeric-C_{60} link. By first-principles calculations, it was shown that when pressure and heat was applied to the polymeric-C_{60} solid, the most stable energetic form of the polymeric-C_{60} links is no longer the singlet, a (diamagnetic) state having two interfragment C-C single bonds (the S_0 state of the polymeric link), but a T_2 triplet state where one of the two interfragment C-C bonds of the link has been broken homolytically (that is, each C_{60} cage hosts one of the two electrons of the broken bond), oriented in a triplet configuration. The presence of ferromagnetism in the solid can be explained when enough polymeric-C_{60} links have crossed into this T_2 state (randomly within the solid) as to allow the ferromagnetic interactions to be connected and propagate over the whole solid. Studies presented later by other groups [14] showed the validity of our proposal on trimers and tetramers placed in a rhombohedral configuration. More recently [15] we have found that the S_0-T_2 crossing point is reached when one of the C_{60} units is drifted laterally relative to the other (thus converting the square into a rhombus). The energy required to reach the S_0-T_2 crossing point is 75.8 kcal/mol, although before reaching such crossing point one has to overcome a barrier of 109.2 kcal/mol along the S_0 state (where the two C-C bonds in the sides of the square are broken and a new C-C bond along the diagonal of the square is created, thus giving rise to a structure that has two unpaired electrons in the terminal C atoms).

In the rest of this work, we will review the details of the three mechanisms introduced above, which seem, in a first sight, possible competing ways of generating ferromagnetic interactions in pressed-heated polymeric-C_{60} solids. Such revision will be based on the results obtained by first-principles density functional (DFT) calculations [16] on the species that participate in the three mechanisms. A comparison of the three mechanisms can be done on the basis of the energy required in each of them, a fact that in the cage-vacancy mechanism implies the computation of the energy of the $C_{60} \rightarrow C_{59} + C$ and $C_{60} \rightarrow C_{58} + C_2$ processes and also of all the successive C_2 losses, that is, $C_{fr} \rightarrow C_{fr-2} + C_2$, *fr* being 58, 56, and so on, results that have been recently reported [17], using the B3LYP functional [18] and the 3-21G Gaussian basis set [19]. This will allow to determine the stability of the C_{59} and C_{58} cages against further C or C_2 losses, and also to check if C_{59} is the only vacant-cage that could give rise to magnetic moments.

2 Methodological Considerations

As the changes introduced in the structure of the polymeric-C_{60} solid by any of the three mechanism outlined above are localized in one of the C_{60} units or one of the polymeric-C_{60} links, the study of the properties of these mechanisms can be modeled using a C_{60} cage (vacant-cage mechanism, and open-cage mechanism) or a dumbbell-shaped C_{60}-C_{60} dimer (interfragment broken bond mechanism), the smallest system where the polymeric-C_{60} link is present.

At the present moment, the most efficient computational method for a theoretical study of C_{60} and C_{60}-C_{60} systems is the DFT methodology [16]. Previous studies [20] have shown that this methodology gives results of a similar accuracy to that obtained by the best ab initio methods (MCSCF, MP2, CASPT2, ...) when describing the shape and most interesting energetic characteristics (heat of reaction, energy barrier) of many bond-breaking/bond-making processes, provided that one uses the right functional (B3LYP, BLYP or PBE) and the appropriate wavefunction expression. That is, to describe regions of the potential energy surface where the electronic structure of the system is suspected to be an open-shell singlet state, like in diradicals, one has to use the broken symmetry approximation [21] for a proper description the open-shell singlet wavefunction (see Fig. 3). Due to the near degeneracy of the highest energy occupied

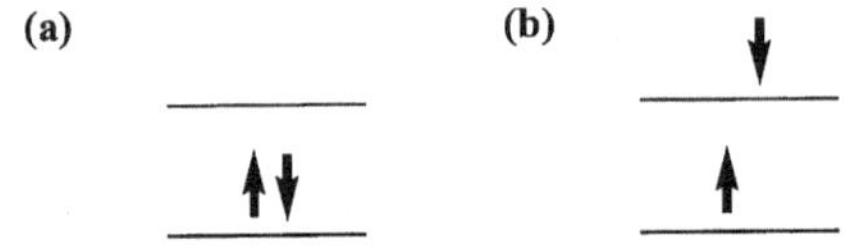

Fig. 3. Scheme of the occupation of the highest energy occupied and lowest energy empty orbitals in (a) a closed-shell singlet and (b) an open-shell singlet state.

orbitals and lowest energy empty orbitals in diradicals (Fig. 3), using the usual unrestricted formulation of the DFT wavefunction the open-shell solution falls spontaneously into the restricted wavefunction, against the predictions of multireference methods (MCSCF, CASPT2), which predict the open-shell solution to be the ground state of the singlet. The use of the broken symmetry approximation gives an energy order of open-shell and closed shell solutions similar to those obtained when using the MCSCF or CASPT2 methods. In most cases, the two orbitals of a diradical represented in Fig. 3 are localized on different regions of the molecule, and the closed-shell wavefunction represents a situation where one region has two electrons and the other none (that is, the first one is negative while the second is positive), while the open-shell wavefunction represents a solution where the two regions have one electron (and consequently, both are neutral in charge).

As a final comment, it is worth pointing here the similarity in the results obtained when the energies are computed by using a DFT formulation that uses atomic basis sets (Gaussian or Slater atomic basis sets), or a non-local plane waves basis sets. We have tested this fact on some of the systems studied here. Gaussian basis sets are used by Gaussian-03, the program [22] used in most of the calculations reviewed here, while plane waves basis are used by CPMD [23] and also other programs whose results are also reviewed here.

3 The Cage-vacancy Mechanism

As already mentioned in the introduction, in its original proposal [10] this mechanism associated the origin of the ferromagnetic interactions found in the polymeric-C_{60} solid to the existence of C_{59} cages that present a triplet ground state and can interact ferromagnetically with nearby C_{59} cages, by means of the dipolar mechanism. Experimentally [24] it is well known that the fragmentation of C_{60} mostly takes place by loss of C_2 fragments and gives rise to many C_n fragments ($2 \leq n \leq 59$), being C_{59} a low frequency fragment. Therefore, it seems wise to generalize the original proposal to include any C_n cages derived from the crystallographic structure of C_{60} by extracting one or more C atoms or C_2 units. This generalization study has been recently done by us [17], in an exploratory form, focussing mostly on the early stages, where C_{59} and C_{58} are generated, and in the further loss of C_2 units from C_{58}, the most stable fragmentation pathway.

Figures 4 and 5 show a schematic representation of all possible isomers of the C_{59} and C_{58} cages. The two isomers of C_{59}, whose names are C_{59}(I) and C_{59}(II), were obtained by extraction of a C atom from the crystallographic structure of C_{60}, while two of the four isomers of C_{58} were obtained by direct extraction of a C_2 fragment, and the other two after applying a Stone-Wales transformation to the previous two: C_{58}(I) results from extracting the C_2 fragment from a 5:6 bond, C_{58}(II) is generated by extracting the C_2 fragment from a 6:6 bond, while C_{58}(I-SW) and C_{58}(II-SW) result from applying a Stone-Wales transformation to their parent non-transformed isomers (in the first isomer the twist is around bond 10–11, while in the second isomer the twist is around bond 7–8).

C_{59}(I)

C_{59}(II)

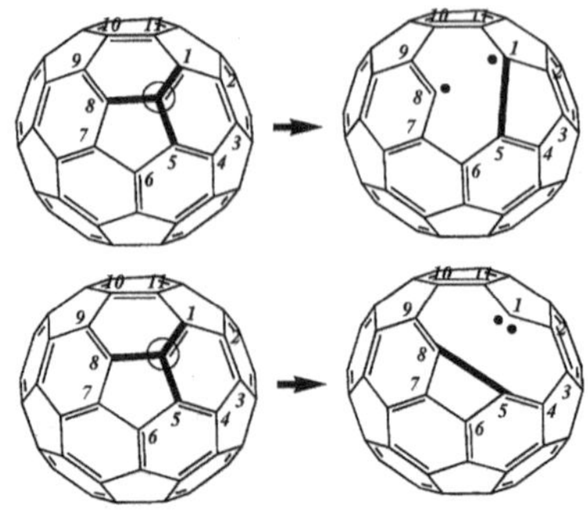

Fig. 4. Schematic description of the generation of the two possible isomers of C_{59} (right), obtained by extracting a C atom from the crystallographic structure of C_{60} (left)

The geometry of all the C_{58} isomers was fully optimized (Fig. 6). The homolytical extraction of a C atom from C_{60} leaves, in principle, four unpaired electrons in the C_{59} cage, two due to the two broken single C-C bonds and two due to the only C=C double bond broken. However, in the optimum geometry of the singlet and triplet states two of these electrons form a new C-C bond, according to the two options depicted in Fig. 4, each giving rise to a different isomer. This leaves two unpaired electrons, which could be used in the singlet to form a new bond. However, the spin distribution of the singlet state indicates the open-shell nature of this state. The spin distribution of the triplet is similar (see Fig. 5), but has no sign alternation. The optimum geometry of singlet and triplet , in both isomers, is similar, being always the singlet more stable than the triplet (by 2.2 in isomer I and 0.4 kcal/mol in isomer II), contrary to previous results [10]. We think that the non-use of the symmetry breaking approach in that study is the most likely reason for this discrepancy, as it gives rise to a less stable singlet.

We look now at the possible C_{58} isomers and their optimum geometry (Fig. 6). Depending on whether one extracts the C_2 fragment from a 5:6 or a 6:6 bond one produces isomer C_{58}(I) or C_{58}(II), which can then undertake a Stone-Wales transformation to produce C_{58}(I-SW) and C_{58}(II-SW), respectively. The extraction of a C_2 fragment to produce C_{58}(I) leaves initially six unpaired electrons in that cage (four associated to the two C=C double bonds, and two more associated to the two single C-C bonds). However, when the structure of that isomer is fully optimized two new bonds are formed (see Fig. 5) and only two electrons remain unpaired (in atoms 9 and 12). Our calculations also show that the closed-shell singlet C_{58}(I,S) cage is 10.0 kcal/mol more stable than the C_{58}(I,T) triplet or the open-shell singlet. The reason is that the four π electrons in atoms 9-10-11-12 can reorder to form two C=C double bonds between atoms 9-10 and 11-12. As extracting a C_2 fragment from a 6:6 bond breaks four single C-C bonds, C_{58}(II) initially only presents four unpaired electrons, that after geometryoptimization give rise to two new C-C bonds (Fig. 5), in the singlet and triplet states (the C_{58}(II,T) triplet is generated from the C_{58}(II,S) singlet by breaking one of the C=C double bond). Our calculations indicate that C_{58}(II,T) lies 30.8 kcal/mol above C_{58}(II,S).

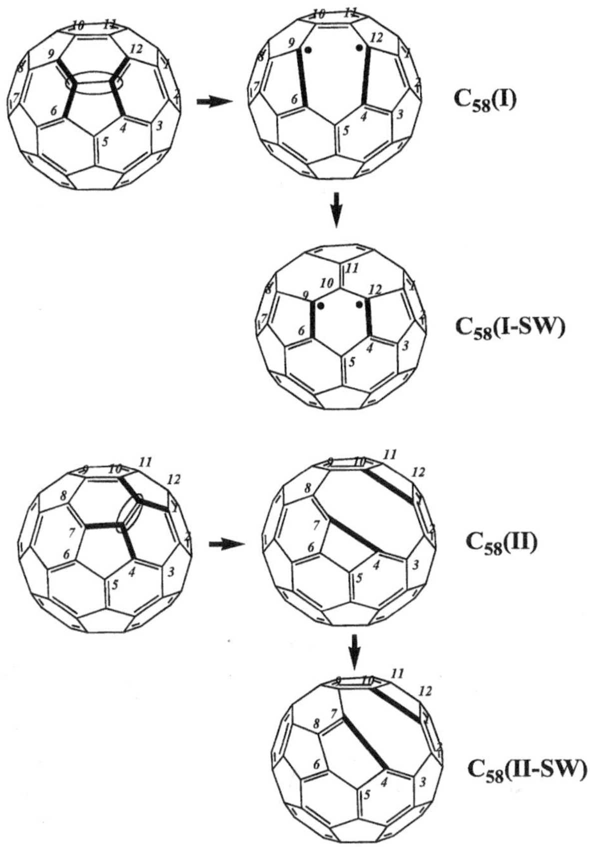

Fig. 5. Schematic description of the pathways to generate the four possible isomers of C_{58} (right) by extracting a C_2 fragment from the crystallographic structure of C_{60} (left). Isomers C_{58}(I) and C_{58}(II) are obtained by a 5:6 and a 6:6 extraction, while isomers C_{58}(I-SW) and C_{58}(II-SW) are obtained after applying a Stone-Wales transformation to the previous two isomers.

The isomers that result from the Stone-Wales transformation of isomers C_{58}(I) and C_{58}(II), present an energy similar to that of the non-transformed products (the C_{58}(I) $\rightarrow$ C_{58}(I-SW) transformation is energetically stabilizing, 4.8 kcal/mol in the singlet state and 18.4 kcal/mol more stable in the triplet state, while in the C_{58}(II) $\rightarrow$ C_{58}(II-SW) transformation the stabilization is 13.4 in the singlet case, and 23.0 kcal/mol in the triplet case). As a result, the triplet state of C_{58}(I-SW) is 3.6 kcal/mol more stable than the singlet, while in C_{58}(II-SW) the singlet is 21.4 kcal/mol more stable than the triplet. Thus C_{58}(I-SW) has a ground state that presents magnetic moments. It is also worth pointing here that the energy barrier found in the C_{58}(I) $\rightarrow$ C_{58}(I-SW) Stone-Wales transformation was computed to be 125.1 and

133.1kcal/mol above the energy of C_{58}(I), in the triplet and singlet states, respectively. A similar set of values are expected for the C_{58}(II) $\rightarrow$ C_{58}(II-SW) Stone-Wales transformation.

The only C_{58} and C_{59} isomer whose ground state presents a magnetic moment is C_{58}(I-SW), see Fig. 6. This has important implications in the search for ferromagnetic interactions in polymeric-C_{60} links, as quantum rules indicate that only when coupling triplets one can have ferromagnetic interactions (the possible states, represented by their S_T quantum number, resulting by coupling two spin states having quantum numbers S_1 and S_2, are those that go from S_1+S_2 up to $|S_1-S_2|$ in increments of -1, thus coupling two singlet states, that is, $S_1 = S_2 = 0$, only produces a $S_T = 0$ state, that is a diamagnetic state). However, when the magnetic interaction between two C_{58}(I-SW) cages forming a polymeric-C_{60} link was calculated [25] the quintet was found 120 cm^{-1} above the singlet, that is, the magnetic interaction is antiferromagnetic. *Consequently, the ferromagnetism observed in pressed polymeric-C_{60} cannot be explained by the magnetic interactions produced by any of the polymeric-C_{60} links built by the C_{59} or C_{58} isomers.*

Besides the optimum structure of C_{58} and C_{59} isomers, an important property to determine the relative probability of the C or C_2 loss is the energy of formation of these isomers. Such value is also required to compute the relative probability of the cage-vacancy mechanism against the other possible mechanisms. The C and C_2

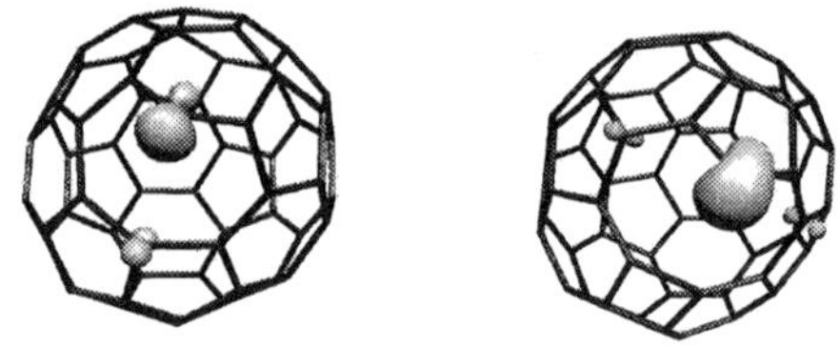

C_{59}(I,T) C_{59}(II,T)

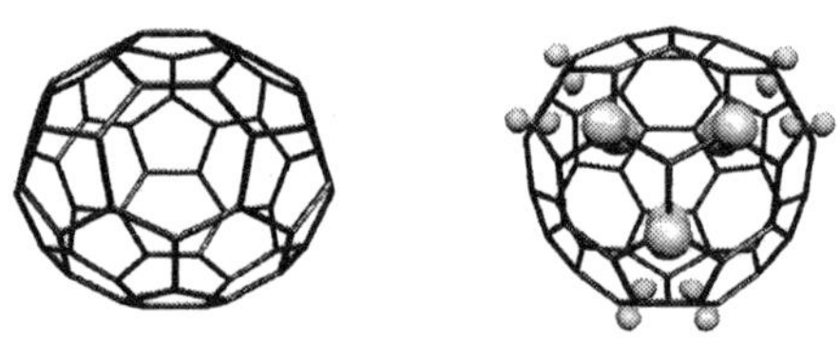

C_{58}(I,S) C_{58}(I-SW,T)

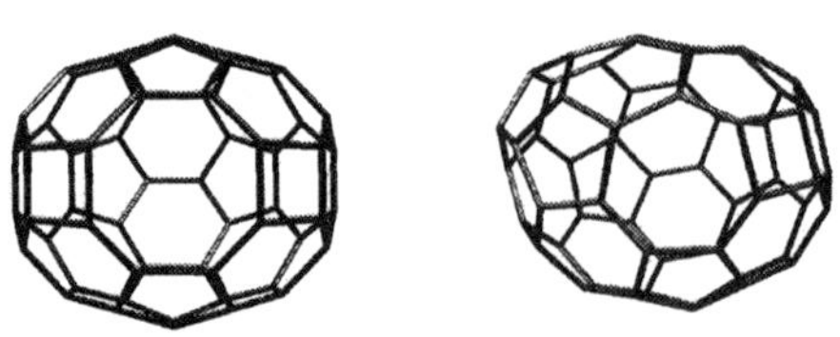

C_{58}(II,S) C_{58}(II-SW,S)

Fig. 6. Optimum geometry of the C_{59} and C_{58}, isomers.

fragmentation of C_{60} can only take place along the following singlet and triplet pathways: $C_{60}(S) \rightarrow C_{fr}(S) + C_i(S)$ or $C_{60}(S) \rightarrow C_{fr}(T) + C_i(T)$, where *i* is either 1 or 2, and *fr* is either 59 or 58. We will call these pathways, the singlet and triplet pathways, respectively. In C_{59}(I) the energy for the singlet and triplet pathways is 297.8 and 288.9 kcal/mol, respectively, values which in C_{59}(II) become 323.1 and 312.3 kcal/mol. The internal energy of an isolated C_{60} molecule has been estimated to be in the 140 kcal/mol range at 1000K [26], a magnitude smaller than the $C_{60} \rightarrow C_{59}$ and $C_{60} \rightarrow C_{58}$ singlet and triplet fragmentation. Such internal energy is indicative of the range of internal energies the each cage of the polymeric-C_{60} solid can present.

The presence of other C_{fr} cages smaller than C_{58} having a triplet ground state was also explored by looking at the succesive C_2 loss processes from C_{58} in a systematic form (first, by extracting all the 6:6 bonds remaining in C_{58}, and then in a less systematic form). In this form, we optimized the geometry of all the even C_{fr} cages in the C_{56}-C_{18} range and computed the energy for the C_2 loss. Two main results come from such study: (a) no C_{fr} cage of those explored (up to C_{18}) has a triplet ground state (thus closing the door to the existence of ferromagnetic interactions involving these cages), and (b) the cage structure is preserved in all the systems explored.

It is also interesting to look at the values of the energy for the successive fragmentations, that is, C_n (II) $\rightarrow C_{n-2} + C_2$ (n being in the 60-18 range). All values are endothermic and in the 322-122 kcal/mol range. After a maximum at C_{56}, they decrease towards a plateau, although the final values are bigger. Similar results were obtained when the process is started from C_{58}(II-SW). In strongly endothermic processes, like these studied here, Hammond postulate [27] indicates that the energetic barriers, when present, should be just a little bit bigger than the heat of reaction. Therefore, the thermodynamic and kinetic analysis of these reactions can be done, at least in a first approach, by looking at the values shown in Fig. 7.

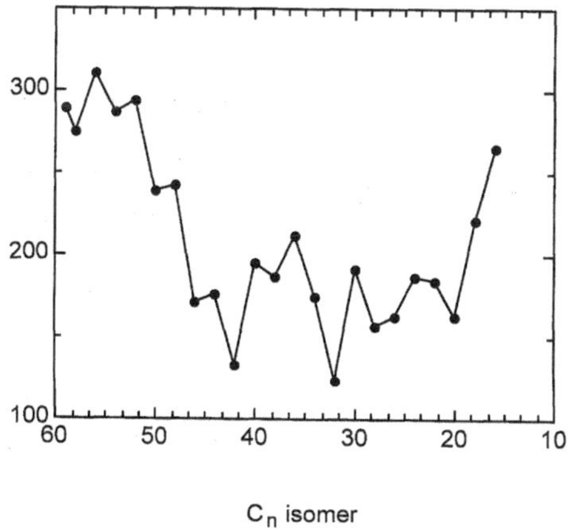

Fig. 7. Energy of the $C_{60} \rightarrow C_{60-2n} - n\ C_2$ processes.

The fragmentation of C_{60} under multiphoton irradiation (where each C_{60} molecule is believed to absorbe up to 20 photons of a 254.7 nm laser radiation [28]) has been explained in terms of the sequential evaporation of C_2 fragments from a highly excited C_{60} molecule. Each molecule generates a cascade of high-mass even-numbered C_n fragments (in the C_{58}-C_{32} range), complemented by another cascade of low-mass C_n fragments, where n is any integer number in the 31-1 range. Units larger than C_2 are thought to contribute to 5–10% of the fragmentations. These results are easily explained by considering that once the molecule has absorbed 20 photons, it has enough energy as to undertake succesive $C_n \rightarrow C_{n-2} + C_2$ steps, each one involving the energy reported in Fig. 7. Consequently, the C_{60} cage will disappear, converted into a cascade of smaller fragments. As a random number of photons is absorbed by different C_{60} molecules, the degree of advance in the cascade process will be different, thus explaining the experimental patterns of fragments.

Under thermal conditions, as those used to generate the magnetic solid, the amount of energy available for the C_{60} molecules found in the polymeric solid is different. As mentioned above, the internal energy of an isolated C_{60} molecule at 1000K is about 140 kcal/mol. Taking this value as indicative of the internal energy available for each C_{60} unit of the polymeric solid, we reach the conclusion that the only fragmentations that should be expected are those producing C_{59} and C_{58} (in any of its isomeric forms). The probability of a second step (for instance, $C_{58} \rightarrow C_{56} + C_2$) is expected to be negligible. This can be justified under two different arguments: (*a*) in conditions of thermal equilibrium between reactants and products, the number of C_{n-2} fragments generated in the first step depends, according to Maxwell-Boltzmann law, depends exponentially on $-[E(n\text{-}2)\text{-}E(n)]/RT$ ($E(i)$ is the energy of cage C_i), which gives a very small population of C_{n-2} fragments for energy differences in the 300 kcal/mol range (consequently, a negligible amount of C_{n-2} fragments is available for the second step); (*b*) if one looks at the probability of two successive processes, if each of them is independent and follows Ahrrenius law (that is, $p_i = A_i\ exp\text{-}(E_a(i)/RT)$), the overall probability $p_T = p_1{*}p_2, = A_1{*}A_2{*}exp\text{–}([E(n\text{-}2)\text{-}E(n)]+([E(n\text{-}4)\text{-}E(n\text{-}2)])//RT$ (where we have substituted the activation barrier of both processes by their energy differences), that is, p_T depends exponentially on $\text{–}([E(n\text{-}4)\text{-}E(n)]$ ($E(i)$ being the energy of cage C_i), a value much larger than the internal energy at 1000K under thermal conditions.

As the only accessible C_n cages at 1000K under thermal conditions do not show ferromagnetic interactions, we should rule out the cage-vacancy mechanism in these conditions. Using light irradiation, it could be possible to reach more fragments in the C_2 loss cascade, but then one has to take into account that the previous work does not show the presence of triplet ground states up to now, and that if the amount of energy given to the C_{60} cages is not properly controlled, one could induce a destruction of the cage.

4 The Open-cage Mechanism

It was suggested [11] that the C_{60} units that constitute the polymeric-C_{60} solid could undertake successive generalized Stone-Wales transformations (a two step process composed of a Stone-Wales transformation followed by the rupture of two C-C bonds, see Fig. 2b). As a result of this homolytic rupture, a hole in the surface of the C_{60} cage would be opened and four unpaired electrons would be created in the cage.

Geometry optimization of the initial open-cage structure has been suggested to end in the formation of a new C-C bond between atoms 1 and 2, which thus becomes a triple bond (isomer C_{60}(GSW-I), Fig. 8). This leaves two unpaired electrons in atoms 7 and 8, a fact that makes possible the presence of triplet ground states. The same C_{60} cage could undertake a second generalized Stone-Wales transformation (isomer C_{60}(GSW-II), Fig. 8), that after geometry optimization was suggested to form a new C-C bond between atoms 8 and 11, leaving again two unpaired electrons in atoms 7 and 10. Other possible forms are possible for the second transformation, thus increasing the number of possible open-cage structures presenting a triplet ground state. However, the multiplicity of the process must be kept, so the transformation initially proceed along the singlet potential energy surface, but as the singlet-triplet crossing has been found to be very efficient in C_{60}, we assume that singlet-triplet crossing are also equally possible in these open-cages when the triplet state is lower in energy that the singlet state.

Using the B3LYP DFT functional and the 3-21G Gaussian atomic basis set, one can compute [29] the energy for the Stone-Wales process. We have plotted in Fig. 9 the energy profile for the two processes. The first step, the Stone-Wales transformation of C_{60} into C_{60}(SW), is slightly endothermic (by 39.4 kcal/mol) and presents an energy barrier of 169.4 kcal/mol. The second step of the generalized Stone-Wales process, the C_{60}(SW) $\rightarrow$ C_{60}(GSW) transformation, is strongly endothermic (by 130.2 kcal/mol), so C_{60}(GSW) is placed 169.4 kcal/mol above the energy of C_{60}. Given the process strong endothermicity, according to Hammond postulate [27], for the second step one should expect an energy barrier similar to the process endothermicity (as is clearly shown by looking at Fig. 9).

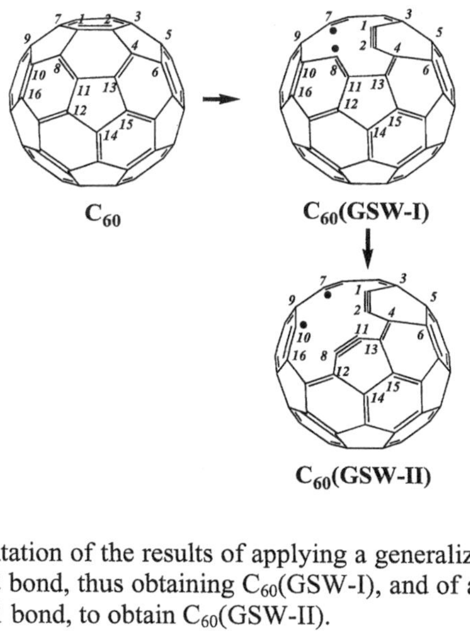

Fig. 8. Schematic representation of the results of applying a generalized Stone-Wales (GSW) to the C_{60} cage, along the 1-2 bond, thus obtaining C_{60}(GSW-I), and of applying a second GSW to C_{60}(GSW-I) along the 8-11 bond, to obtain C_{60}(GSW-II).

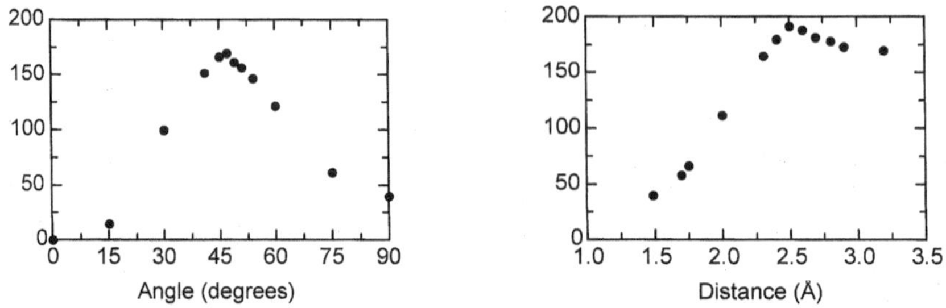

Fig. 9. *Left*: Variation of the total energy during the Stone-Wales transformation of C_{60} along the 1-2 bond (the angle taken as reaction coordinate refers to the rotation of 1-2 around its center-of-mass relative to its starting position). *Right*: Energy variation as a function of the distance between atoms 1 and 7 (equal to the 2-8 distance, see Fig. 8) for the second step of the generalized Stone-Wales transformation.

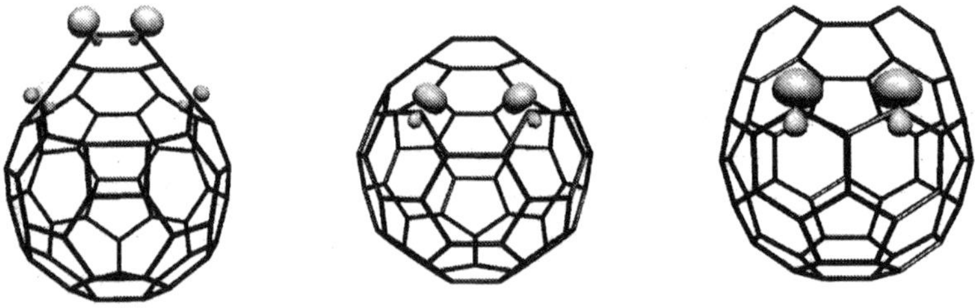

a) b) c)

Fig. 10. Optimum geometry of the C_{60}(GSW-I) and C_{60}(GSW-II) isomers in their triplet state. a) C_{60}(GSW-I) closed structure, where atoms 7 and 8 form a C-C bond. b) C_{60}(GSW-I) open structure, where no bonds exist between atoms 7 and 8. c) C_{60}(GSW-II).

The optimum structure of C_{60}(GSW-I) in its singlet state (Figs. 8 and 10) presents a new C-C bond linking atoms 7 and 8, while the 1-2 bond is converted into a triple bond. There are two possible triplets, one where the 7-8 bond is broken, and another where one of the triple bonds is converted into a double bond (such state was not reported in previous studies [11]). At the optimum geometry of the singlet, the triplet is obtained by converting the 1-2 triple bond into a double bond (Fig. 10). This triplet is placed 23.5 kcal/mol above the optimum geometry of the singlet. A second triplet is obtained when breaking the 7-8 bond. In its optimum geometry, it is placed 3.4 kcal/mol below the triplet where the 7-8 bond is preserved. At the optimum geometry of the second triplet, neither the open-shell nor the closed-shell singlet has a minimum, and the optimization ends in the singlet closed structure or in C_{60}(SW). Consequently, we can safely conclude that the lowest energy state of C_{60}(GSW-I) is the singlet. This fact makes the presence of ferromagnetic interactions in the C_{60}(GSW-I)-C_{60}(GSW-I) polymeric links impossible.

Previous literature reports indicate that by applying two successive generalized Stone-Walsh transformation to C_{60}, the resulting product (C_{60}(GSW-II)) presents a triplet ground state [11]. When a second generalized Stone-Walsh transformation is applied to C_{60}(GSW-I) along the 8-11 bond, the C_{60}(GSW-II) isomer is obtained. In a first stage, the 8-11 bond is rotated, thus producing the rupture of three single C-C bonds (8-10, 11-12 and 11-13) and the formation of two new single C-C bonds (8-12 and 11-13). This leaves four unpaired electrons (on atoms 7, 8, 10 and 11) in the structure. Geometry optimization of the structure gives rise to the formation of a new C-C bond between atoms 8 and 11, which this will become a triple C-C bond. This only leaves two unpaired electrons, sitting on atoms 7 and 10, which do not form a new C-C bond connecting because this would create a triangular ring, a fact that would generate a lot of stress on the cage. The two unpaired electrons can form an open-shell singlet or a triplet.

The optimum geometry of C_{60}(GSW-II) in the triplet state (computed at the B3LYP/3-21G level) is shown in Fig. 11. The singlet and the triplet are very similar in geometry and also in energy (the triplet is 0.3 kcal/mol more stable than the singlet). The singlet state is 89.4 kcal/mol above the C_{60}(GSW-I) singlet, that is, 258.8 kcal/mol above the energy of the crystallographic structure of C_{60}. Such value is above the 140 kcal/mol estimate of the internal energy for an isolated C_{60} molecule, thus making the obtention of the C_{60}(GSW-II) states at 1000K, under thermal conditions, difficult. Besides, the small number of C_{60}(GSW-I) molecules (very small under conditions of thermal equilibrium at 1000K) makes very unlikely the C_{60}(GSW-I) $\rightarrow$ C_{60}(GSW-II) process. A first qualitative analysis can be done using the same probability arguments stated before for successive processes. In this approach, the probability of the C_{60} $\rightarrow$ C_{60}(GSW-I) $\rightarrow$ C_{60}(GSW-II) process depends exponentially on the energy difference between the starting and final species, that is, $p_T \approx exp(-258.8/RT)$, a very small number. A better analysis should take into account that now some of the steps in the reaction do not follow Hammond Postulate. Furthermore, given the small energy difference between singlet and triplet states of C_{60}(GSW-II) a similar occupation of these two states is expected at room temperature. As a consequence, the probability of two cages in the C_{60}(GSW-II)-C_{60}(GSW-II) links being in their triplet state is about $1/2 \times 1/2$, thus making very difficult the propagation of any ferromagnetic interactions over the solid.

One could consider a to apply a third generalized Stone-Wales transformation to produce C_{60}(GSW-III). However, as we expect the energy to be even higher than that for C_{60}(GSW-II), simple energetic considerations in the line of those used above indicate that such process is very unlikely to happen in significative numbers. Consequently, as the low-energy isomers found in the open-cage mechanism are all singlet states and the only triplet states are too high in energy as to make unlikely their existence (under thermal equilibrium or kinetic arguments) we have to rule out that the ferromagnetism detected experimentally in the pressed-heated polymeric-C_{60} solids can be explained due to the existence of open-cage isomers of C_{60}.

5 The Interfragment Bond-breaking Mechanism

The third of the mechanisms proposed [13] to understand the origin of magnetism in pressed-heated polymeric-C_{60} solids does not look at the properties of isolated C_{60} cages. Instead, it looks directly at the properties of the polymeric-C_{60} link, that is, covalently linked C_{60}-C_{60} dimers, trying to find when these bonds present a triplet ground state. Each of these links results from a [2+2] cycloaddition that involves one 6:6 double bond from each of the two C_{60} cages. After the reaction both 6:6 bonds become a single C-C bond and two interfragment C-C single bonds connecting the cages (both of the same length and with the shape of a square (Fig. 11). This is the lowest energy conformation of the polymeric-C_{60} link, which is also the geometry found in the crystallographic structure [30] of the dumbbell-shaped C_{60}-C_{60} dimers, the smallest system where the polymeric-C_{60} link is present.

The dumbbell-shaped C_{60}-C_{60} dimer is a good model for the theoretical study of the properties of an isolated polymeric-C_{60} link. When pressure and heat is applied to the polymeric-C_{60} solid, the crystal unit cell is forced to squeeze. As a consequence, among other effects, the intercage C-C bond distance decreases. The external pressure and heat applied to the solid, at the microscopic level translates, on the polymeric-C_{60} link having more energy to increase their vibrational energy and explore higher energy regions of the potential energy surface (PES) of the polymeric-C_{60} link. The shape of the potential energy surface (PES) of these polymeric-C_{60} links, in a first approximation, can be modelled by looking at the PES of the dumbbell-shaped C_{60}-C_{60} dimers.

Previous studies on isolated C_{60}-C_{60} dimers [13, 15, 31] have shown that the lowest energy state is a singlet (S_0) where the two C_{60} units are connected by two C-C single bonds (Fig. 11). Besides the S_0 state, other relevant states to understand the ferromagnetic interactions on the polymeric-C_{60} links are the S_1, T_1 and T_2 states (Fig. 11). The S_1 and T_2 result after breaking one of the two intercage C-C bonds that the S_0 state presents and leaving the unpaired electrons in an open-shell singlet or a triplet configuration. T_1 is obtained from the S_0 state after breaking a C=C double bond in one of the two C_{60} cages.

Geometry optimization of the geometry of the S_0, S_1, T_1 and T_2 states (Fig. 12), showed the presence of a minimum in all but the S_1 state. Depending on where one starts the geometry optimization, the S_1 state (an open-shell singlet) ends in S_0, after the formation of a new C-C bond, or ends in two dissociated C_{60} cages in their singlet ground state. This behaviour is not surprising as two unpaired electrons in a singlet state try to form a new bond (in the C_{60}-C_{60} case this requires distorting the geometry of the already existing C-C bond), and, as we will see below, there is a barrier towards dissociation and after a given distance the dimer dissociates. At their optimum geometry the values of the intercage C-C distances for all states is similar (1.616, 1.605 and 1.709 Å, in the S_0, T_1 and T_2 minima, respectively).

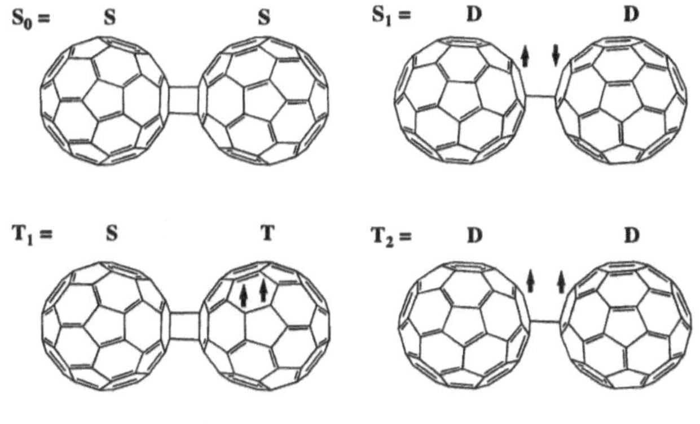

Fig. 11. Schematic representation of the most relevant lowest energy electronic states of the dumbbell-shaped C_{60}-C_{60} dimer. The S, D and T letters indicate the spin state that the C_{60} cages have in the dimer.

A complete study of the shape of the PES for the dumbbell-shaped C_{60}-C_{60} dimer is still very expensive computationally. Therefore, we first studied the main characteristics of the [2+2] cycloaddition on a smaller model system. Then, using that information, we just calculated the most relevant regions of the PES using the C_{60}-C_{60} dimer. For the initial study one can choose the ethene dimer, whose [2+2] cycloaddition has been well studied using very accurate *ab initio* methods [32] (CASSCF and CASPT2). This study, besides, will allow us to check the quality of the DFT surfaces against those computed using accurate *ab initio* methods.

The shape of the PES computed for two ethene molecules, as a function of the interfragment C-C distance, is shown in Fig. 13 for the lowest energy singlet state and the T_2 triplet state (the T_1 state cannot exist in the interaction of two ethane molecules). The two surfaces were computed at the B3LYP/3-21G level. The broken-symmetry approach was used in regions of the surface where any of the interfragment C-C distances is large enough as to suspect that such bond is partially or fully broken (one

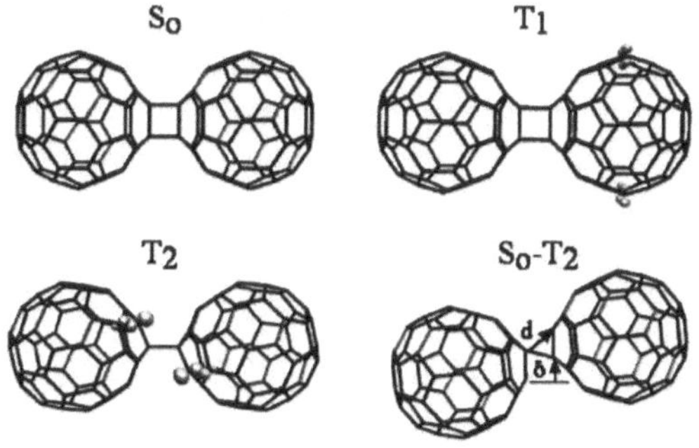

Fig. 12. Optimum geometry of the dumbbell-shaped C_{60}-C_{60} dimers in their ground state singlet (S_0) and in the lowest energy triplet (T_1) and T_2 triplet states. Also shown is the point where the S_0 and T_2 potential energy surfaces cross (S_0-T_2).

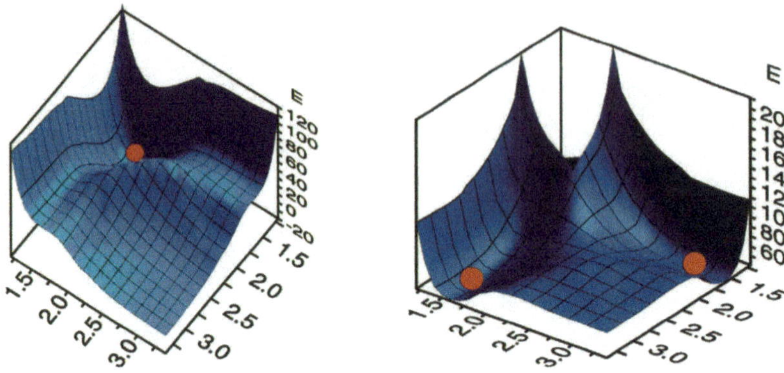

Fig. 13. *Left*: Shape of the singlet (S_0-S_1) PES of the ethene dimer, as a function of the two interfragment C-C distances (in Å). *Right*: The same plot for the T_2 PES. In both surfaces the red ball indicates the position of the minimum (Fig. 11). The energy scale is given in kcal/mol and refers to interaction energy respect to two ethene fragments in their singlet ground state (the S_0 surface dissociates in two ethene singlets, while the T_2 surface dissociates in a singlet and a triplet fragment).

should point that the broken-symmetry solution falls into the closed-shell solution, whenever the closed-shell solution is energetically more stable). The electronic structure of the singlet changes with the geometry: close to the minimum (read ball in Fig. 13) the singlet is a closed shell (S_0) state, while when any of the two interfragment C-C distances is enlarged, the energy rises rapidly and the wavefunction is that for the S_1 state, that is an open-shell singlet. In other words, the S_0 and S_1 surfaces communicate in a smooth form, a feature already found in the CASSCF and CASPT2 PES. The shape of the singlet surface mimics very well that obtained by CASSCF and CASPT2 calculations [32]. Despite such smooth communication, for simplicity we will still talk about the S_0 PES, keeping in mind that it is a S_0-S_1 surface. No accurate *ab initio* data is available on the other PES.

Now we can turn our attention into the physical information contained in both surfaces. We look first at the S_0 surface. Such surface is always endothermic, that is, its energy is placed above the energy of two ethene fragments in their singlet ground state. It presents only one minimum, when the two interfragment distances are about 1.6 Å. Such miminum is separated from the dissociation products by an energy barrier (whose wavefunction is an open-shell state like those of the S_1 state), larger if it is reached along a pathway in which the two interfragments C-C distances are equal (symmetric pathway), and smaller along pathways where one of these two distances is allowed to differ from the other (asymmetric pathway). No minimum is found after crossing the barrier (that is, the S_1 state has no-minimum), and once the barrier is crossed the dimer dissociates into two ethene fragments. The T_1 and S_0 surfaces results from the same local electronic transformations of the [2+2] process, and their shape is expected to be similar, although the T_1 surface, whenever it exists, is expected to be higher in energy

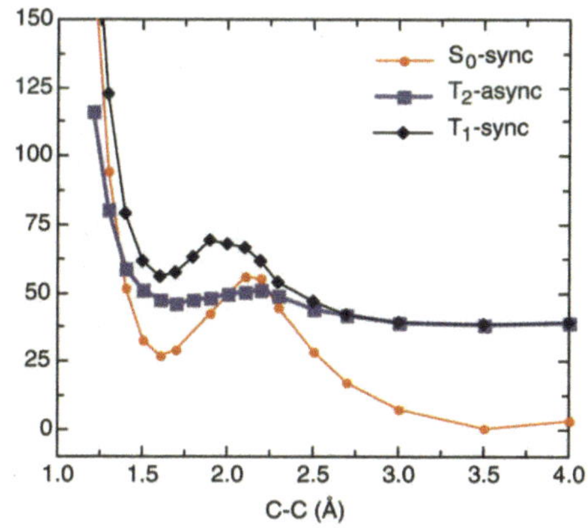

Fig. 14. Potential energy curves for the synchronous pathway for the [2+2] cycloaddition of two C_{60} cages in their S_0, T_1 and T_2 states. For each interfragment C-C distance the geometry was fully optimized at the CPMD/55 Rydberg level.

(it dissociates into a singlet and a triplet fragments, while the S_0 surface dissociates into two singlet fragments).

The shape of the T_2 PES of the ethene dimer as a function of the two interfragment C-C distances is different to that for the S_0 PES. Now there are two minima, placed in two equivalent asymmetric positions: one interfragment C-C distance is about 1.6 Å, while the other is about 3.2 Å, the two fragments rotated about 60 degrees off the coplanarity. Such geometry is similar to that also found in the optimum geometry of the C_{60}-C_{60} dimer in the T_2 state (Fig. 12). Along the symmetric pathway a maximum is found separating the two equivalent minima, and the PES energy increases as the distance is shortened. We have checked that the T_2 and S_0 PES plotted in Fig. 13 never cross in the region of the plot, the reason being the much higher dissociation of the T_2 surface (66.5 kcal/mol above the dissociation of the S_0 surface), and the complementarity of their shapes.

The main features of the surfaces discussed above can be fully taken into consideration by plotting the energy of the syncronous pathway (S_0 and T_1 PESs) or the asyncronous pathway (T_2 PES). Therefore, our characterization of the PES of the C_{60}-C_{60} dimer was started by computing the potential energy curves for the synchronous (S_0 and T_1 states) and asyncronous pathways (T_2 state). These curves are collected in Fig. 14. We also evaluated the shape of the S0 PES in the region that goes from the S0 minimum up to the energy barrier, to test if the shape and energetics was similar to that found in the ethene dimer. The PES of ethene-ethene and C60-C60 show the same features, with just small differences.

The overall shape of the three curves in Fig. 14 is the same. The main trends observed in these curves are the following: (a) all are endothermic (that is, energy is required to go from the dissociation fragments to any geometry of the dimer); (*b*) the relative energy of the minima $E(S_0) < E(T_2) < E(T_1)$; (*c*) the energy barriers for the formation and dissociation processes is different, and agree well with the experimental results [33]; (*d*) both energy barriers change according to the ordering $E_b(S_0) > E_b(T_1) > E_b(T_2)$; (*e*) The S_0 and T_2 curves cross at distances shorter than their equilibrium values and also in other regions of the curves. However, based on a detailed study of the PESs, all S_0-T_2 crossings observed in Fig. 14 are an artefact of comparing synchronous and asynchronous curves (the curves can be considered as projections into one plane of the PES of Fig. 13). Real crossings require the same energy at the same geometry of the two curves (or surfaces). In the "crossings" found in Fig. 14 the geometry of the synchronous curve is different than that for the asynchronous curve, thus, they are just artefacts of the representation. However, a real crossing point exists between the S_0 and T_2 states, but located in a different region of their PES.

To search the crossing point between the S_0 and T_2 states we just tried to find regions where these two states should present a similar energy for the same geometry. This can be better achieved in regions where the S_0 PES is described as an open-shell singlet (that is, has a dominant S_1 nature), because in these regions one expects the singlet and triplet have a similar energy for the same geometry. One can go to one of these regions in C_{60}-C_{60} dimers by a lateral drift of one of the C_{60} cages relative to the other (see Fig. 12, where the lateral drift is indicated by δ, while d is the interfragment C-C distance). By computing the shape of the S_0 and T_2 PES for the C_{60}-C_{60} dimers around that region, we have been able to locate a region where these two states cross, being the lowest energy crossing point that at d = 1.503 Å and δ = 1.45 Å (d[6:6] = 1.570 Å), that is, *is located at interfragment C-C distances shorter than the minimum.* This S_0-T_2 crossing point is placed 75 kcal/mol above the S_0 ground state energy, and to reach such crossing point from the minimum of the S_0 PES one has to overcome an energy barrier of 109.2 kcal/mol (the whole pathway along the S_0 PES from the minimum up to the crossing point is shown in Fig. 15). Consequently, the S_0-T_2 crossing is an energetically activated process (that is, it presents a barrier). Therefore, the S_0-T_2 crossing can only occur when energy is given to the dimer, allowing it to compress and drift. Notice that the values of the barrier and S_0-T_2 crossing point are both smaller than the estimate of the internal energy of the C_{60} cage at 1000K under thermal conditions (140 kcal/mol [26]).

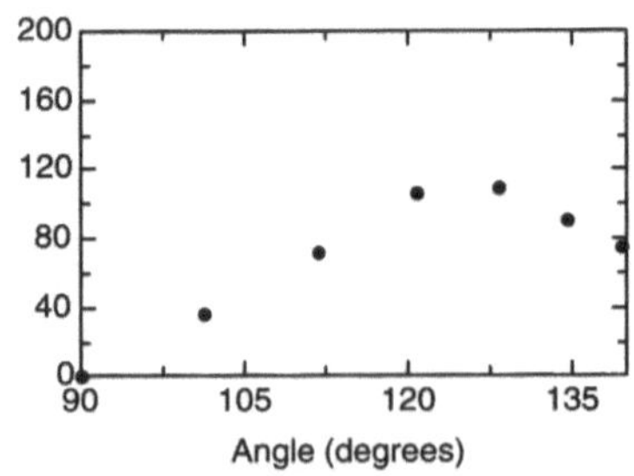

Fig. 15. Shape of the potential energy curve for the lateral drift of one of the C_{60} cages relative to the other in a C_{60}-C_{60} dimer (see Fig. 12). We have measured the drift by looking at the angle made by the interfragment C-C bond and the 6:6 bond of any of the cages (the minimum of the S_0 PES is found at 90°).

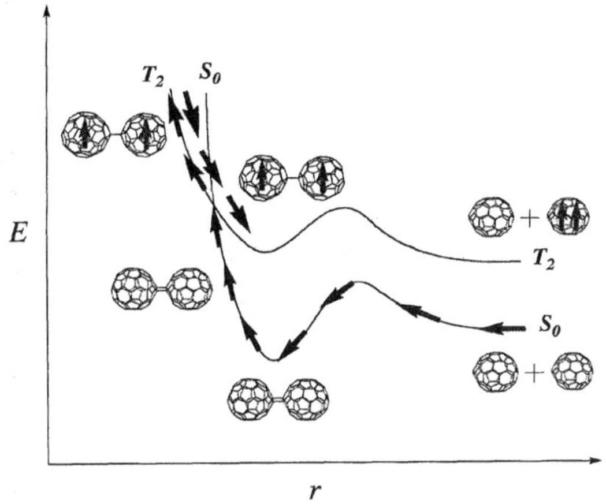

Fig. 16. Mechanism of generation of ferromagnetic interactions in the C_{60}-C_{60} polymeric links by application of pressure and heat.

From the previous results, we can propose a mechanism for the generation of magnetic moments in the polymeric-C_{60} links (modelled here by the isolated C_{60}-C_{60} dimers), which does not require breaking the C_{60} cages (such mechanism is graphically displayed in Fig. 16). When pressure is applied to pristine C_{60} cages in the singlet ground state, it induces a [2+2] cycloaddition reaction between adjacent C_{60} molecules to form a stable dimer (or higher oligomers, when more adjacent monomers are present), that is, the link crosses the barrier of the S_0 PES and goes down to the S_0 minimum (Fig. 12). When much more pressure and heat is applied to the C_{60}-C_{60} dimers they compress and some of them can carry out the lateral drift motion, thus reaching the S_0-T_2 crossing point, after passing an energy barrier of 109. 2 kcal/mol.

The energy involved in that process is that plotted in Fig. 15. There, some of the links cross to the T_2 state and become a triplet state, thus automatically generating ferromagnetic interactions in these links. Once pressure is released, the dimers in the T_2 state relax to their minimum. The probability that this magnetic state undergoes radiative decay is expected to be small, because in the isolated C_{60}-C_{60} dimers the transition from the T_2 minimum to the S_0 ground state is spin-forbidden as well as vibrationally-forbidden [12, 15]. So the preservation of the ferromagnetic interactions is guaranteed. The process of this ferromagnetic activation of the C_{60}-C_{60} polymeric links does not exceed the 140 kcal/mol of energy estimated for the internal energy for a C_{60} cage, when heated to 1000 K. This makes all the transformations presented in Fig. 16 energetically allowed.

The transformation of the C_{60}-C_{60} polymeric-C_{60} links from their S_0 ground state up to the ferromagnetic T_2 state takes place in a random form all over the solid. When enough of these ferromagnetic links have been formed in the solid as to allow the formation of continuous pathways of ferromagnetic interactions, the solid presents ferromagnetism. The hexagonal and tetragonal topology of the ideal polymeric solid is a guarantee that continuous pathways can be made, even when a low proportion of the total number of polymeric links are in the T_2 state. We have found that S_0-T_2 crossing points also exist on the polymeric-C_{60} planes when a cooperative lateral drift motion is produced within the planes of polymeric-C_{60} fullerene.

The results of Fig. 14 also suggest new possible experimental procedures for the creation of ferromagnetic polymeric C_{60}. By irradiating a pristine-C_{60} crystal one could produce enough long-lived C_{60}(T) molecules that under pressure may undergo the reaction C_{60}(T) + C_{60}(S) → C_{60}-C_{60}(T_2), thus generating a sufficient number of C_{60}-C_{60} units in T_2–like configurations. Under mild pressure conditions, the probability for units in T_1-like configurations to be created is expected to be negligible, as their formation barrier in the dimer is higher than that required for the formation of T_2 state. On the other hand, the probability that ferromagnetic interactions are triggered by photochemical activation from the C_{60}-C_{60} units in S_0-like conformations should be small because in the isolated dimer the S_0-T_2 transition is both spin- and vibrationally-forbidden [15].

6 Summary and Conclusions

In the previous sections, we have carried out an analysis of the microscopic transformations and their energy for the of three mechanisms proposed in the literature to explain the origin of the ferromagnetic interactions in polymeric-C_{60} solids. As a kind of summary, we have collected in Fig. 17 the most relevant results about each of these mechanisms, also indicating the internal energy expected for a C_{60} cage at 1000K (140 kcal/mol).

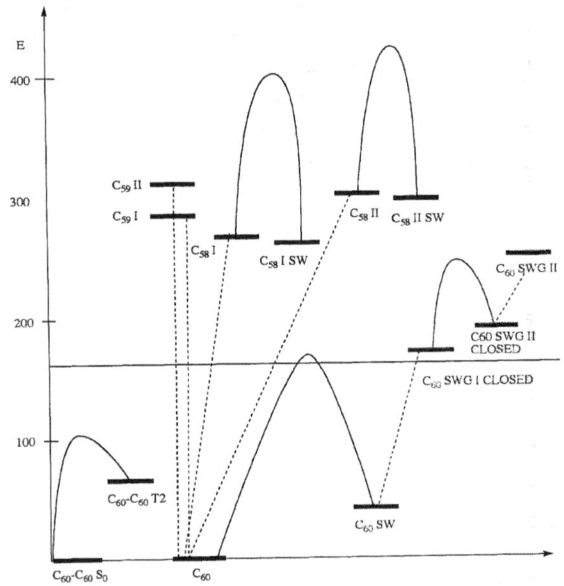

Fig. 17. Comparison of the energetics of the interfragment C-C broken bond mechanism (lower left), cage-vacancy mechanism (center) and open-cage mechanism (right). Also indicated, for reference, is the value of the internal energy available for a C_{60} cage at 1000K (140 kcal/mol).

A look at that Fig. indicates that for the cage-vacancy mechanism and the open-cage mechanism the energy available only allows to carry out the first steps (that is, the C_{60} units have enough energy as to transform into any of the isomers of C_{59}, C_{58}, and C_{60}(GSW-I). However, our study has shown that none of these isomers can give rise to ferromagnetic interactions in C_n-C_n dimers, thus ruling out their existence in the polymeric links present in the solid. Consequently, the origin of ferromagnetic interactions in pressed polymeric-C_{60} solids is not explained by the cage-vacancy or open-cage mechanisms.

All the previous results allow us to conclude that the energy required for the intercage C-C broken bond mechanism is smaller than the internal energy available at 1000K for the C_{60} units, a fact that suggests that this is the only mechanism (of the three proposed above) that can take place under the thermal conditions employed up to now to produce pressed-heated polymeric-C_{60} solids. Besides, the energy required for the intercage C-C broken bond mechanism is also much smaller than that involved in the other two mechanisms, thus indicating that if enough energy was available for the three mechanisms to take place, the intercage C-C broken bond mechanism would be the most probable one.

Acknowledgments

This work was possible thanks to the computing resources allocated by the CEPBA-IBM Research Institute, CEPBA, and CESCA, and the continuous financial support of the Spanish *Ministerio de Ciencia y Tecnología* (project BQU2002-04587-C02-02) and *Comissionat per a Universitats i Recerca, Generalitat de Catalunya* (grant 2001SGR-0044). One of us (JJN) also wishes to thank IBM Zurich Research Lab (where part of the work reviewed here was done), and specifically Drs. W. Andreoni and A. Curioni, for their kind hospitality. His visit there was made possible by a grant from the Department of Universities, Research, and Information Society of the *Generalitat de Catalunya*.

References

1. T.L. Makarova, B. Sundqvist, R. Hohne, P. Esquinazi, Y. Kopelevich, P. Scharff, V.A. Davidov, L.S. Kashevarova, and A.V. Rakhmanina, Nature 413 (2001) 716.
2. R. Höhne, P. Esquinazi, Adv. Mater. 14 (2002) 753.
3. R.A. Wood, M.H. Lewis, M.R. Lees, S.M. Bennington, M.G. Cain, and N. Kitamura, J. Phys.: Condens. Matter. 14 (2002) L385.
4. V.N. Narohznyi, K-H. Müller, D. Eckert, A. Teresiak, L. Dunsch, B.A. Davydov, L.S. Kashevarova, and A.V. Rakhmanina, Physica B 329 (2003) 1217.
5. M. Nuñez-Regueiro, L. Marques, J-L. Houdeau, O. Berthoux, and M. Perroux, Phys. Rev. Letters 74 (1995) 278. For a review see: T.L. Makarova, Semiconductors 35 (2001) 243.
6. K.-H. Han, D. Spemann,, R. Höhne, A. Setzer, T.L. Makarova, P. Esquinazi, T. Butz, Carbon 41 (2003) 785; T.L. Makarova, B. Sundqvist, and Y. Kopelevich, Synt. Met. 137 (2003) 1335.
7. S. Okada, S Saito, Phys. Rev. B59 (1999) 1930; S. Nakano, Y. Kitagawa, T. Kawakami, M. Okumura, N. Nagao, and K. Yamaguchi, Molecules 9 (2004) 792.

8. N. Park, M. Yoon, S. Berber, J. Ihn, E. Osawa, and D. Tománek, Phys. Rev. Lett. 91 (2003) 237204.
9. J.A. Chan, B. Montanari, J.D. Gale, S.M. Bennington, J.W. Taylor, and N.M. Harrison, Phys. Rev. B70 (2004) 041403.
10. A.N. Andriotis, M. Menon, R.M. Sheetz, and L. Chernozatonskii, Phys. Rev. Lett. 90 (2003) 026801.
11. Y.-H. Kim, J. Choi, K.J. Chang, and D. Tomének, Phys. Rev. B68 (2003) 125420.
12. A.J. Stone and D.J. Wales, Chem. Phys. Lett. 128 (1986) 501.
13. J. Ribas-Ariño, J.J. Novoa, Angew. Chem., Int. Ed. 43 (2004) 577; J. Ribas-Ariño, J.J. Novoa, J. Phys. Chem. Solids 65 (2004) 787.
14. V.V. Belavin, L.G. Bulusheva, A.V. Ojotrub, and T.L. Makarova, Phys. Rev. B70 (2004) 155402.
15. J. Ribas-Ariño, A. Curioni, W. Andreoni, and J.J. Novoa, J. Phys. Chem. A, in press (2005).
16. R. G. Parr and W. Yang, Density functional theory of atoms and molecules, Oxford University Press, Oxford, 1994.
17. J. Ribas-Ariño and J.J. Novoa, submitted for publication.
18. B3LYP is a density functional obtained by taking the three parameter non-local exchange funcional of Becke and the non-local correlation functional of Lee-Yang-Parr: A.D. Becke, J. Chem. Phys. 98 (1993) 5648; C. Lee, W. Yang, R.G. Parr, Phys. Rev. B37 (1988) 785.
19. J.S. Binkley, J.A. Pople, W.J. Hehre, J. Am. Chem. Soc. 102 (1980) 939.
20. E. Goldstein, B. Beno, K.N. Houk, J. Am. Chem. Soc. 118 (1996) 6036.
21. L. Noodlemann, J. Chem. Phys. 74 (1981) 5737; L. Noodlemann, E.R. Davidson, Chem. Phys. 109 (1986) 131.
22. Gaussian-03. Revision B.04, M.J. Frisch, et al. Gaussian, Inc., Pittsburgh PA, 2003.
23. CPMD, Version 3.5.2, Copyright IBM Corporation 1990-2005, Copyright MPI für Festkörperforschung Sttutgart 1997-2001.
24. P. Wurz and K.R. Lykke, Chem. Phys. 184 (1994) 335.
25. These calculations have been done using the NRLMOL program (M.R. Pederson, K.A. Jackson, Phys. Rev. B41 (1990) 7453) the PBE functional (J.P. Perdew, K. Burke, M. Ermzerhof, Phys. Rev. Lett. 77 (1996) 3865), and 3-21G basis set (reference 19).
26. P. Wurz and K.R. Lykke, J. Phys. Chem. 96 (1992) 10129.
27. For insights on the Hammond Postulate see: G.S. Hammond, J. Am. Chem. Soc. 77 (1955) 334.
28. P. Wurz and K.R. Lykke, Chem. Phys. 184 (1994) 335.
29. J. Ribas-Ariño, J.J. Novoa, submitted for publication.
30. G.-W. Wang, K. Komatsu, Y. Murata, and M. Shiro, Nature 387 (1997) 5.
31. C.H. Xu and G.E. Scuseria, Phys. Rev. Lett. 74 (1995) 274; D. Porezag, M.R. Pederson, Th. Frauenheim, and Th. Köhler, Phys. Rev. B52 (1995) 14963; J. Kürti, K. Németh, Chem. Phys. Lett. 256 (1996) 119; S. Osawa, M. Sakai, and E. Osawa, J. Phys. Chem. A 101 (1997) 1378; T. Ozaki, Y. Iwasa, and T. Mitani, Chem. Phys. Lett. 285 (1998) 289.
32. F. Bernandi, M. Olivucci, and M.A. Robb, Acc. Chem. Res., 23 (1990) 405; F. Bernardi, A. Bottoni, M. Olivucci, A. Venturini, and M.A. Robb, J. Chem. Soc. Faraday Trans. 90 (1994) 1617.
33. Y. Wang, J.H. Holden, X. Bi, and P.C. Eklund, Chem. Phys. Lett. 217 (1993) 3; S.M. Bachilo, A.F. Benedetto, and R.B. Weisman, J. Phys. Chem. A, 105 (2001) 9845.

Carbon-Based Magnetism
T. Makarova and F. Palacio (Editors)

23

Ferromagnetism in Defective Polymerised C_{60}

J.A. Chan[2], B. Montanari[1,3], and N.M. Harrison[1]

[1]*Department of Chemistry, Imperial College London, South Kensington campus, London SW7 2AZ, UK*
[2]*CCLRC Rutherford Appleton Laboratory, Chilton, Didcot, Oxfordshire, OX11 0QX, UK*
[3]*CCLRC Daresbury Laboratory, Daresbury, Warrington, WA4 4AD, UK*

1 Introduction

The recent observation of high temperature ferromagnetism in polymerised carbon fullerenes has fuelled a renewed interest in metal-free organic magnetic materials. Their existence not only presents a new class of potentially highly tunable materials, but also challenges the current understanding of magnetism.

When cubic C_{60} fullerenes are subjected to pressures between ~2–8 GPa and temperatures between ~700–1150K, two-dimensional tetragonal (T) and rhombohedral (Rh) polymerised phases are formed [1]. Only samples that are synthesised in a specific temperature region near the stability limit of the fullerene cages exhibit, when quenched to room temperature, ferromagnetism with Curie temperatures above ~500K [2–4]. Polymerised phases created outside these conditions show no magnetic behaviour, suggesting that defect structures could be responsible for ferromagnetism. The unpaired electrons located at the defects could give rise to local moments, which would need to interact via a long range coupling mechanism in order to form a magnetically ordered structure.

Unfortunately, the experimental characterization of the magnetic phases has proven difficult and, at present, their atomic structure is not known. *In-situ* X-ray diffraction measurements reveal a thermally activated process which converts the Rh-phase (Fig. 1) into the highly disordered graphite-like phase, which displays very broad Bragg peaks [5]. The detailed structure of the magnetic phase cannot be determined from this

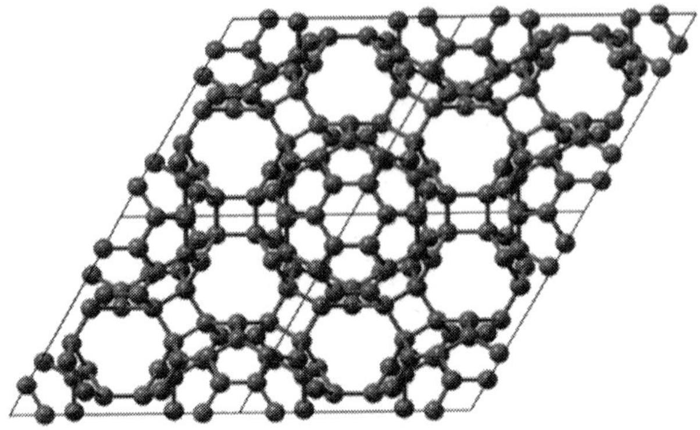

Fig. 1. The basal plane of pristine polymerised rhombohedral C_{60}.

data. Transmission electron microscopy (TEM) points at an apparently well ordered crystalline structure in which the C_{60} cages are largely intact and still in a Rh-C_{60} - like arrangement [3]. Magnetic force microscopy (MFM) studies of the carbon phase based on polymerized fullerenes have established that approximately 30% of the material is magnetic with the magnetism occurring in well defined domains [6].

Several experimental studies have also shown that the presence of non-metallic impurities has some effect on the magnetic behaviour of carbon-based materials [7–9]. In particular, some of the studies have suggested that the presence of hydrogen may be important. For example, the observed saturation magnetisation of magnetic amorphous carbon is found to increase with the hydrogen concentration of the starting material [9], and proton irradiation of highly orientated pyrolytic graphite induces intrinsic room temperature ferromagnetic ordering, whilst a similar treatment with helium ions results in reduced magnetic signals [8].

As the characterization of the magnetic phase is problematic, theoretical calculations have an important role to play in determining possible local geometries. A number of previous theoretical studies have addressed the origin of magnetism in Rh-C_{60}. Boukhvalov *et al.* used density functional theory (DFT) in the local spin density approximation to compute the electronic structure of pristine polymerised Rh-C_{60} which indicate that it is not magnetic, in agreement with observation [10]. Other authors have studied various defective Rh- C_{60} structures. In the work of Andriotis *et al.*, for instance, a carbon atom was removed from each fullerene cage (Fig. 2(a)) [11]. Tight-binding molecular dynamics and cluster *ab initio* calculations were then used to analyse the magnetic properties of the resulting structure. An ionic model due to McConnell was invoked to suggest that inter-cage, through space coupling could result in long range magnetic coupling between cages linked via the interfullerene bonds of the polymerised structure [11, 12]. A re-examination of this hypothesis is reviewed here. Ribas-Arinõ *et al.* (Chapter 22) used hybrid exchange DFT (B3LYP) calculations to simulate a C_{60} dimer, and complete active space self consistent field calculations to compute the interactions within an isolated C_{60} cage, with hydrogen atoms replacing the interfullerene bonds of the Rh-C_{60} structure [13, 14]. In both cases, their model of single broken interfullerene bonds between the cages led to a localised spin density and, in the latter case, ferromagnetic or antiferromagnetic states depending on the position of the broken interfullerene bonds. Various other defective Rh-C_{60} structures have been constructed with broken interfullerene bonds within a semiempirical Hartree-Fock/AM1 formalism and the resultant band structure, which contains flat bands at the Fermi edge, was used as an indication of a tendency towards magnetic localisation [15]. Periodic B3LYP calculations have also been used to show that a distortion of the Rh-C_{60} structure about the interfullerene bonds, while constraining all other bond lengths in the cage to 1.40 Å, can also give rise to a spin polarised ground state [16].

In all of the above studies, the possibility of local moment formation around various types of defects has been established and there are clearly a large number of local defect structures that can be created and which will result in such local moments. An explanation of the observed ferromagnetism, however, requires one to establish both that the defects are credible and that there is strong ferromagnetic coupling between the local moments. The task therefore is to narrow down the possibilities to likely defects, examine the magnetic properties of these defects, and see if strong, long range ferromagnetic coupling exists between the defects.

This chapter reviews the work presented in Refs. [17, 18] where a series of defective structures was investigated by means of periodic *first principles* calculations based on hybrid exchange DFT in the B3LYP form: (i) the vacancy structure previously proposed by Andriotis *et al.*, where a carbon atom is removed from each fullerene cage [11]; (ii) a similar structure where the vacancies between two adjacent C_{60} cages are moved closer together in pairs; (iii) a new defective structure with a fractured intrafullerene bond that is spontaneously generated by applying isotropic pressure, and (iv) a structure consisting of a vacancy defect, the displaced atom forming an additional inter-cage link, spontaneously generated by simulating the high pressure, high temperature treatment used to generate the ferromagnetic material in the laboratory. In all these cases, local spin moments around the defects arise, but no evidence of long range ferromagnetic interaction between these moments is found.

First principles calculations based on DFT currently play an important role in the characterisation of many materials, and in particular in determining structure-property relationships. Mixing non-local and semi-local exchange in hybrid-exchange functionals, as in the now very widely used B3LYP functional [19–21], yields a good quantitative description of thermochemistry [22], and optical band gaps [23]. Moreover, it has recently been shown that hybrid exchange DFT functionals provide a significantly more accurate description of the ground state electronic structure, magnetic coupling energies and magnetic moments, and metal insulator transitions in strongly correlated systems than the more commonly used generalized gradient approximations [24–28].

In addition, each of the defective structures is here analysed using a simple model for predicting ground state spin configurations, which was developed for non defective, planar, π-conjugated carbon and hydrocarbon systems such as organic radicals [29–32] and extended bipartite lattices [33, 34]. This model is based on the observation that, in these systems, the spins of unpaired electrons belonging to carbon atoms which share a covalent bond are antiparallel to each other. As a result, the spin polarisation of the π electrons simply alternates between the bonded atoms. In the remainder of this chapter, this model will be therefore referred to as the "spin alternation rule". The effect of the structural curvature and defects on the validity of this model will be addressed here.

Inelastic neutron scattering (INS) measurements and B3LYP calculations investigating the role played by hydrogen in this material will also be reviewed. It was found that, in the presence of hydrogen intercalation, the defective structure that occurs spontaneously under the simulated high pressure, high temperature treatment, possesses a ferromagnetic ground state.

This chapter is organised as follows: the methodology is described in section 2, and is followed by an examination of the pristine structure of Rh-C_{60} in section 3. Defective structures of pure polymerised Rh-C_{60} are discussed in sections 4, 5 and 6, and a defective structure in the presence of hydrogen is described in section 7. The work is then summarised in section 8.

2 Methodology

The methodology used has previously been described in some detail [17, 18] and involves a combination of reactive force-field simulations and B3LYP calculations. The force field simulations are used to suggest likely structures, the structure and stability of which are refined using the B3LYP calculations. The latter are also used to compute the magnetic coupling between the dominant spin centres.

The starting geometry is the experimentally determined crystalline structure of Rh-C_{60}, for which the lattice parameters are $a = 9.19$ Å, $c = 24.5$ Å in the hexagonal unit cell [35]. As the work reviewed here focuses on the study of the covalent intercage bonds as a mechanism for long-range coupling, the system is restricted to a single layer of Rh-C_{60}, which is here described without any symmetry constraints. In order to generate realistic defect structures, reactive force field simulations under high pressure conditions are used to produce a number of different defective structures. For the calculations with finite temperature, reactive force field molecular dynamics (MD) simulations are performed within the NVT ensemble with a time step of 0.1 fs and run until equilibrium was reached. These simulations are carried out with the most recently developed variant of the bond order potential due to Brenner *et al.* [36], as implemented in the GULP program [37], which allows an adequate modelling of bond breaking and formation processes. The internal coordinates of the structures generated with the static, force field calculations are relaxed within the unrestricted Hartree-Fock (UHF) method, and those of the structures generated with the MD simulations are relaxed with B3LYP calculations. All geometry optimisations were performed in the ferromagnetic configuration using an algorithm proposed by Schlegel *et al.* [38]. The magnetic properties of all structures are then examined within the B3LYP approximation.

The periodic, B3LYP calculations are performed by using the CRYSTAL program [39], where the crystalline wavefunctions are expanded as a linear combination of atom centred Gaussian orbitals (LCAO) with s, p, or d symmetry. In the current study all-electron calculations are performed, which make no assumptions about the shape of the ionic potential or electron charge density. Basis sets of double valence quality (6-21G* for C and 6-31G* for H) are used. Reciprocal space sampling with a (4×4) Monkhorst-Pack grid [40] is sufficient to converge the total energies to within 0.1 meV per C_{60} cage in all structures considered. The Gaussian overlap criteria which control the truncation of the Coulomb and exchange series in direct space have been described in detail elsewhere [39]. In this work they were set to 10^{-7}, 10^{-7}, 10^{-7}, 10^{-7}, 10^{-7}, and 10^{-14}.

In order to compare the energies of various magnetically ordered states, it is necessary to converge stable self-consistent field solutions for different electronic spin configurations. This is achieved by using a superposition of spin-polarised ionic charge and spin densities for particular atomic states to provide a suitable initial wavefunction. In some cases, more direct control of the spin occupancy patterns has been used to control the initial conditions, such as, for instance, the total spin. We emphasise, however, that this only affected the initial wavefunctions and that all solutions presented are unconstrained.

3 Pristine Rh-C_{60}

The electronic structure of the three-dimensional, experimentally determined pristine structure of Rh-C_{60} (a = 9.19 Å, c = 24.5 Å in the hexagonal unit cell) [35], is calculated within the B3LYP approximation, and results in an indirect fundamental band gap of 1.48 eV. The system is non-magnetic and insulating in agreement with previous studies [10, 41, 42]. When a single layer of the pristine Rh-C_{60} was relaxed within the reactive force-field formalism, the resulting lattice constant, a = 9.23 Å, was found to be only 0.4% larger than the experimental bulk lattice constant. This result documents the ability of the reactive force field employed to reproduce the structure of polymerised C_{60}.

4 Prototype Defective Structures: Vacancy Defects

4.1 A Central Vacancy Defect

A defect structure, originally proposed by Andriotis *et al.* [11] is created starting from the experimental geometry [35] and removing one carbon atom from each fullerene cage, as shown in Fig. 2(a). The equilibrium lattice constant is found to be a = 9.22 Å. Many different spin configurations are generated and the one with the lowest energy amongst them is identified to be the ground state. This configuration is between ~0.06 and 0.44 eV per cage lower in energy compared to all other metastable configurations found. The spin density map of the ground state, Fig. 2(b), shows that the spin density localises at the dangling bonds, as expected. This configuration has a total magnetic moment of 2.0 μ_B per C_{60} cage, and the spin polarisation pattern of the undercoordinated carbon atoms, shown in Fig. 2(b), is consistent with the spin alternation rule described in Sec. 1.

In order to test for inter-cage magnetic coupling, a (2×1) supercell, containing two C_{60} cages, is considered. Two spin configurations are examined. In the ferromagnetic

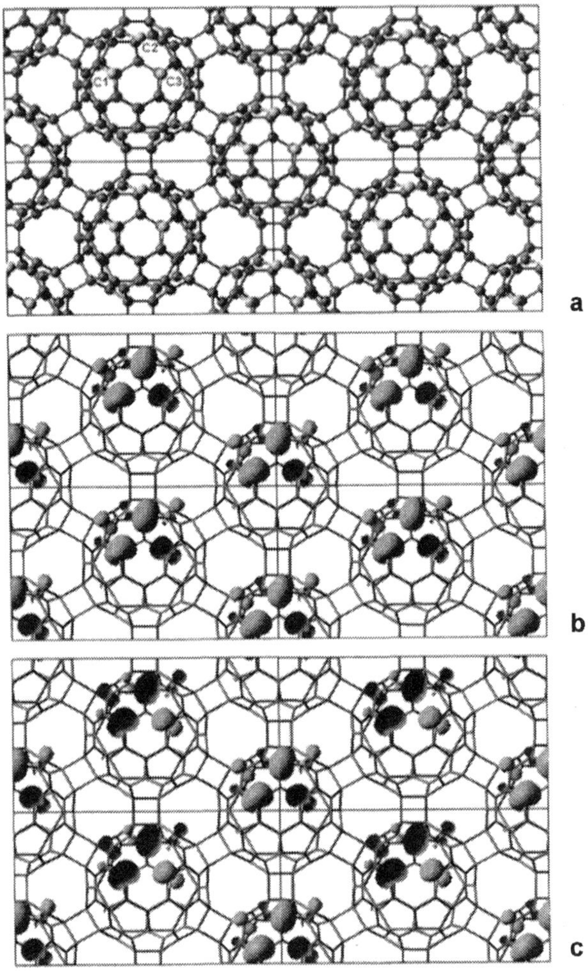

Fig. 2. The central vacancy structure proposed in Ref. [11] with the undercoordinated atoms, labelled as C1, C2 and C3, (a). The spin density maps of the ground state in the (b) parallel (FO), and (c) antiparallel (AF) inter-cage spin configurations. The light and dark isosurfaces correspond to values of +0.015 and –0.015 μ_B, respectively.

(FO) state, the spin configuration is identical to the ground state obtained for the (1×1) cell and described above. In the antiferromagnetic (AF) state, all spin orientations on one of the two C_{60} cages in the supercell are reversed, as shown in Fig. 2(c). The results are quantified in Table 1 and show that there is no significant difference between the energies of the FO and AF states and thus no significant coupling exists between the spin moments of neighbouring cages. Also, no significant change in the Mulliken

Table 1.
Total magnetic moments per C_{60} cage (S), Mulliken spin populations of the undercoordinated atoms labelled as in Fig. 2, and total energy differences with respect to the FO configuration, for the FO and AF configurations of the central vacancy structure.

Configuration	S (μ_B)	Atoms	Spin ($\|e\|$)	ΔE_{tot} (eV/C_{60})
FO	2.0	C1	1.22	0.000
		C2	1.26	
		C3	−0.83	
AF	2.0	C1	−1.20	0.000
		C2	−1.24	
		C3	0.84	

charge or spin overlap populations of the interfullerene bonds between FO and AF spin configurations is seen, in agreement with the lack of interaction between the cages. If this vacancy defect structure could be realised experimentally, these calculations predict that it would only exhibit a paramagnetic behaviour.

The two-dimensional band structure of this paramagnetic state has low lying, flat bands at the Fermi edge separated by a spin-flip, indirect band gap of 1.66 eV. In this case, however, the flat bands are simply a result of the localised spin orbitals with no significant inter-cage interaction and are thus not indicative of a magnetically ordered ground state.

4.2 A Migrated Vacancy Defect

The high pressure high temperature treatment is likely to drive defect migration within the samples. It is therefore of interest to investigate whether the prototype defective structure examined in Sec. 4.1 would exhibit inter-fullerene ferromagnetic coupling if the distance between the defects was reduced. A modification of the central vacancy structure is thus examined, where the vacancy defects on adjacent fullerene cages are positioned closer together in pairs, Fig. 3(a). The same lattice parameters as in the central vacancy defect structure are used in order to facilitate a meaningful comparison of their total energies and determine whether such defect migration is likely. The ground state configuration of this migrated vacancy defect is found to be 0.291 eV per cage higher in energy than the central vacancy structure, indicating a strong repulsion between vacancies.

Figure 3(b) shows the spin density map of the spin configuration with the lowest energy amongst all the configurations found and referred to as AF1. The intra-cage geometric and electronic arrangement of the three undercoordinated C atoms around each vacancy is similar to that of the central vacancy structure. The majority of the spin polarisation is described by the spin alternation rule, with exceptions on the undercoordinated atoms C4 and C5. If the spin density on these atoms were described by the spin alternation rule, the spin density on atom C5 would be parallel to that on C6, and antiparallel to that on C4, contrary to what is observed, Fig 3(b). This suggests that the dominant spin interactions around this defects are through space, rather than through

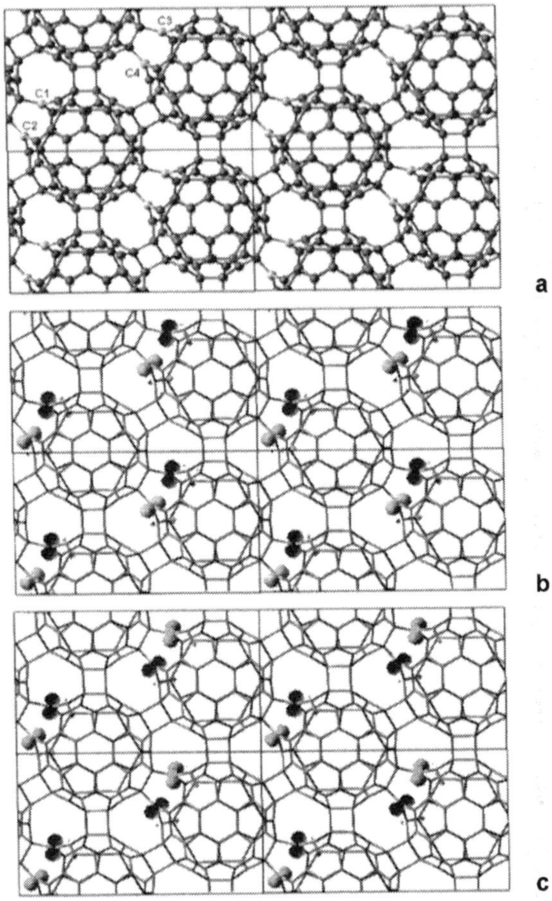

Fig. 3. A migrated vacancy structure with the undercoordinated atoms labelled as C4, C5 and C6, (a). The spin density maps in the (b) antiparallel (AF1), and (c), parallel (AF2) inter-cage spin configurations. The light and dark isosurfaces correspond to values of +0.015 and –0.015 μ_B, respectively.

bond, and therefore favour an antiparallel spin alignment, between the atoms C5 and C6, and between C4 and C6. The polarisation of the spin density also decays more slowly away from the defects and results in a cancellation of the local moments and a zero total magnetic moment per C_{60} cage.

Even in the absence of a net magnetic moment per cage, it is still interesting to determine the nature and extent of the magnetic interaction between the defects. The energies of the AF1 state just described is therefore compared to that of a new state, called AF2, Fig. 3(c), where all the spins on one of the two C_{60} cages in a (2×1) supercell are flipped. The results are reported in Table 2. An energy difference of

16 meV per cage between the AF1 and AF2 states confirms an increased spin coupling compared to the central vacancy defect. The coupling, however, is antiferromagnetic, in agreement with the spin alternation rule. The results thus indicate that this migrated vacancy structure is an unlikely candidate for the ferromagnetic phase.

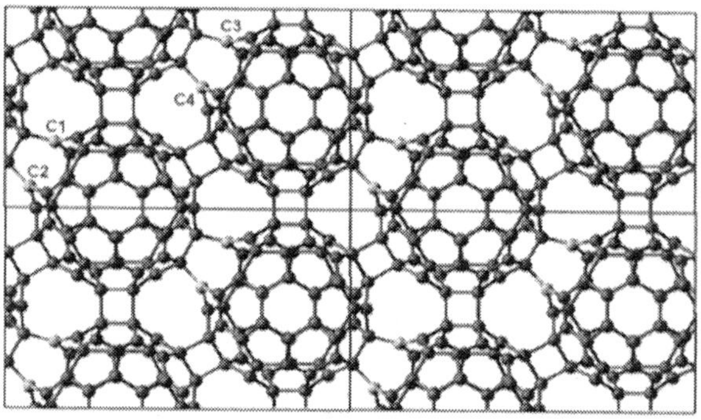

a

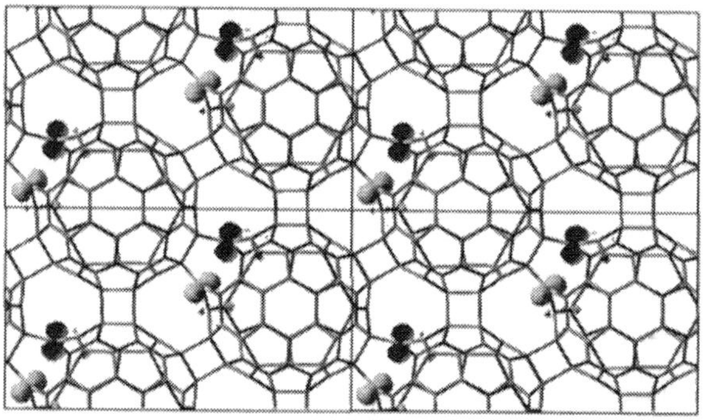

b

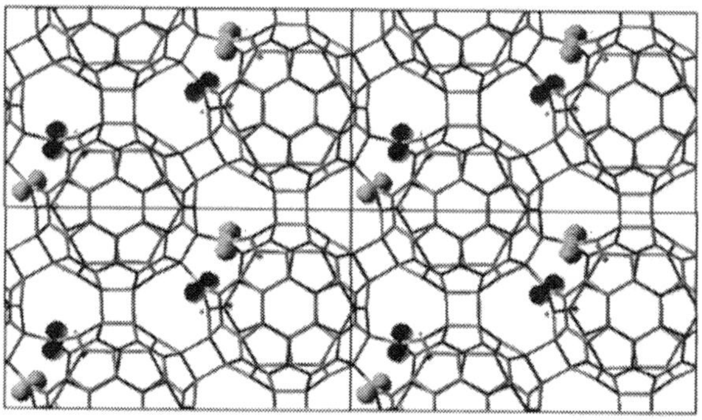

c

Fig. 4. A defective structure generated by applying isotropic pressure, with the undercoordinated atoms labelled as C1, C2, C3 and C4, (a). The spin density maps in the (b) AF1, and (c) AF2 configurations. The light and dark isosurfaces correspond to values of +0.010 and –0.010 μ_B, respectively.

Table 2.
Total magnetic moments per C_{60} cage (S), Mulliken spin populations of the undercoodinated atoms as labelled in Fig. 3, and total energy differences of the AF1 and AF2 configurations of the migrated vacancy structure. The energies are relative to the FO configuration, Table 1, of the central vacancy defect described in Sec. 4.2.

Configuration	S (μ_B)	Atoms	Spin ($\|e\|$)	ΔE_{tot} (eV/C_{60})
AF1	0.0	C4	–0.89	+0.291
		C5	–0.67	
		C6	1.18	
AF2	0.0	C4	0.90	+0.307
		C5	0.68	
		C6	–1.16	

5 Spontaneous Formation of Defects from Pressure

In order to search for more realistic defects, reactive force field static simulations are performed in which isotropic pressure is applied by gradually reducing the a and b lattice parameters in the (1×1) cell and relaxing all internal coordinates at each step. Defective structures appear spontaneously for a 5 - 6% reduction in the lattice constants. A number of different structures are generated in this way as the total strain and its rate of application are varied. In a commonly generated structure, bond fracture occurs for the *intra*-fullerene bond located between the two inter-fullerene bonds, as shown in Fig. 4(a). This is in contrast with the usual assumption that the *inter*-fullerene bond would be the most likely to break [13–15]. However, the single crystal X-ray diffraction studies performed by Chen *et al.* [43] identified this intra-fullerene bond as the longest in the structure, thus suggesting that it may be the weakest bond in the polymerised structure. Here the computed distance between the two atoms involved in the bond increases by 62%, from 1.594 Å to 2.583 Å as the bond breaks. The resulting defective structure exhibits local moments on the undercoordinated atoms and is, once again, characterised by flat, spin polarised bands at the Fermi edge.

Two spin configurations are analysed in a (1×1) cell, containing one fullerene cage, in order to investigate the *intra*-cage magnetic coupling. In the FO configuration, the spin moments on the two undercoordinated atoms in the cage are parallel to each other, and in the AF1 configuration they are antiparallel (Fig. 4(b)). In both cases, the spin density map shows that the spin density is localised in p-orbitals that are directed toward each other. The AF1 state turns out to be the most stable state in the (1×1) cell, as illustrated in Table 3. The favoured *intra*-cage magnetic coupling is therefore antiferromagnetic, in agreement with the spin alternation rule, and the magnitude of the coupling is sizeable. Because of its antiferromagnetic nature, however, this structure has no net magnetic moment.

In order to investigate the *inter*-cage magnetic coupling, the energy of the AF1 state is compared with that of the AF2 state, obtained starting from the AF1 state represented in a (2×1) supercell and flipping the spins on one of the two fullerene cages, (Fig. 4(c)). The difference between the total energies of the AF1 and AF2 configurations amounts

Table 3.
Total magnetic moments per C_{60} cage (S), Mulliken spin populations of the undercoordinated atoms as labelled in Fig. 4, and total energy differences with respect to the FO configuration, for the FO, AF1 and AF2 configurations of the structure generated by isotropic pressure.

Configuration	S (μ_B)	Atoms	Spin ($\lvert e \rvert$)	ΔE_{tot} (eV/C_{60})
FO	2.0	C1	0.74	0.000
		C2	0.66	
		C3	0.74	
		C4	0.66	
AF1	0.0	C1	−0.28	−0.735
		C2	0.25	
		C3	−0.28	
		C4	0.25	
AF2	0.0	C1	−0.29	−0.736
		C2	0.25	
		C3	0.29	
		C4	−0.25	

to a mere 0.5 meV, thus indicating that no significant magnetic inter-cage interactions are present. Therefore this structure can also be discarded as a possible explanation of the ferromagnetic character of this material.

6 Spontaneous Formation of Defects from Pressure and Temperature

This section reviews the only defective structure to date produced by simulating the high pressure, high temperature treatment used to generate the ferromagnetic samples in the laboratory. This is achieved by employing MD simulations, the detail of which is described in Sec. 2, using a mixture of isotropic and anisotropic pressures, and temperatures similar to those used to form the magnetic phases experimentally. By using this procedure, a common defect structure naturally arises under a reduction of ~2–4% and ~4–6% in the *a* and *b* lattice constants, respectively, and at temperatures between 900–1000K. In this structure, shown in Fig. 5, an atom breaks away from the cage and two of its neighbours (labelled as A and B) to form an inter-cage linkage at position C. This bridge atom forms an angle of 132.1° with its neighbouring atoms.

This defect is reminiscent of the vacancy-adatom pair, which is often seen in irradiated graphite [44], and the structure retains essentially intact C_{60} cages within the rhombohedral symmetry, in agreement with TEM observations [3]. This structure may also be thought of as a natural precursor to the decomposition of the polymerised fullerenes into the graphitic phase and is therefore consistent with the observation of magnetism near this phase boundary.

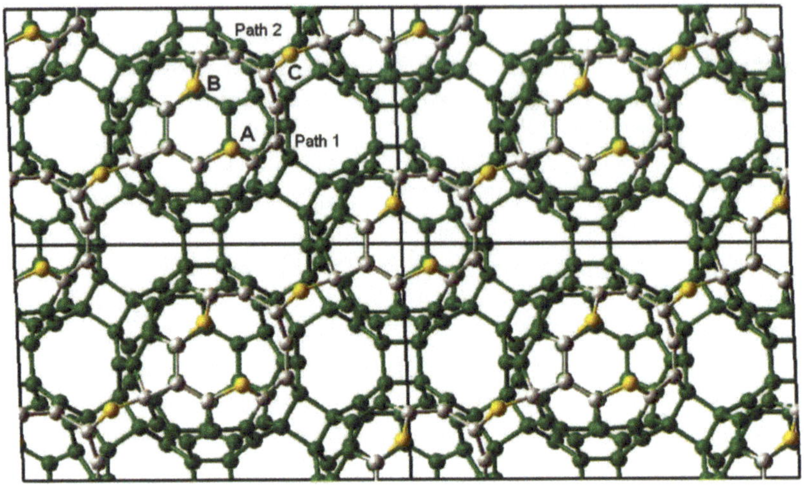

Fig. 5. A defective structure generated from MD simulations at high pressure and temperature. Atoms A, B and C are undercoordinated, and the atoms shown in white are involved in the two competing inter-cage coupling routes.

Analysis of the ground state spin density reveals local moments at the dangling bonds on atoms A, B and C (Fig. 5). The orientation of the spin moments in this structure was found to be consistent with the spin alternation rule, producing a total magnetic moment of 4 μ_B per C_{60}. The two-dimensional electronic structure has an indirect fundamental gap of 1.31 eV, slightly smaller than that of the central vacancy case, which suggests that the states relative to the unpaired electrons are more delocalised and that there is greater interaction between them. Even so, the energy differences between the stable electronic configurations found for this structure turn out to be negligible. Analysis of this system with the spin alternation rule shows that there are two competing pathways for the inter-cage spin coupling, which are highlighted in Fig. 5. The pathways lead to competing inter-cage magnetic interactions, with path 1 generating an antiferromagnetic coupling and path 2 a ferromagnetic coupling. The coupling energies of these pathways are of similar magnitude, leading to a net cancellation of inter-cage coupling and thus to an overall ground state that has no net magnetisation.

7 A Ferromagnetic Ground State in the presence of Hydrogen

As described in Sec. 1, several experiments involving magnetic carbon-based materials show that hydrogen may play an important role in the appearance of the magnetic behaviour. INS measurements were therefore performed in order to determine the hydrogen concentration in a sample of ferromagnetic Rh-C_{60} (see Ref. [17] for details). The resulting INS spectrum reveals a broad C-C stretch at 170 ± 7 meV,

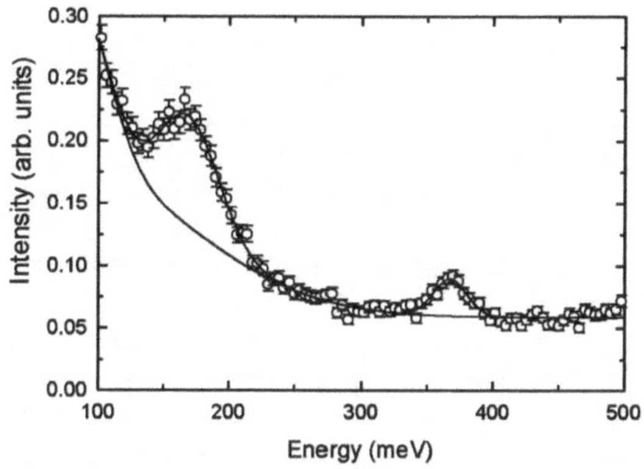

Fig. 6. The INS intensity, S(E), obtained with an incident energy of 600 meV. The sample was 99.8% purity C_{60} converted to the magnetic rhombohedral phase by quenching from 9 GPa and 800K.

together with a clear C-H stretch at 368 ± 1 meV, Fig. 6. The latter peak reveals the presence of a significant amount of hydrogen covalently bonded to carbon in the ferromagnetic samples. The H:C_{60} ratio is determined to be approximately 17%. This amount is of the same order of magnitude of the percentage (30%) of magnetic phase detected in the samples by MFM [6]. In order to investigate the effect that H intercalation may have, in the calculations is is assumed therefore that one H atom per C_{60} is present in the ferromagnetic phase. Possible bonding sites for hydrogen intercalation in the defect structure described in Sec. 6 are examined using B3LYP calculations. The choice of the sites considered is guided by the consideration that the most likely bonding sites are those where the H atom quenches dangling bonds, i.e. the three undercoordinated atoms labelled as A, B, and C in Fig. 5. The binding energies at these sites, calculated with respect to H_2 gas and with fully relaxed internal coordinates, are found to be 3.09 eV for site A, 2.91 eV for site B, and 2.45 eV for site C. The geometry with the H atom bonded to atom A is therefore preferred.

Table 4.
Total magnetic moments per C_{60} cage (S), Mulliken spin populations of the undercoordinated atoms as labelled in Fig. 5, and total energy differences with respect to the FO configuration, for the FO and AF configurations of the structure generated by MD simulations at high pressure and temperature and with a H atom bonded at site A.

Configuration	S (μ_B)	Atoms	Spin ($\|e\|$)	ΔE_{tot} (eV/C_{60})
FO	3.0	B	0.99	0.000
		C	0.96	
AF	0.0	B	–0.98	+0.003
		C	0.94	

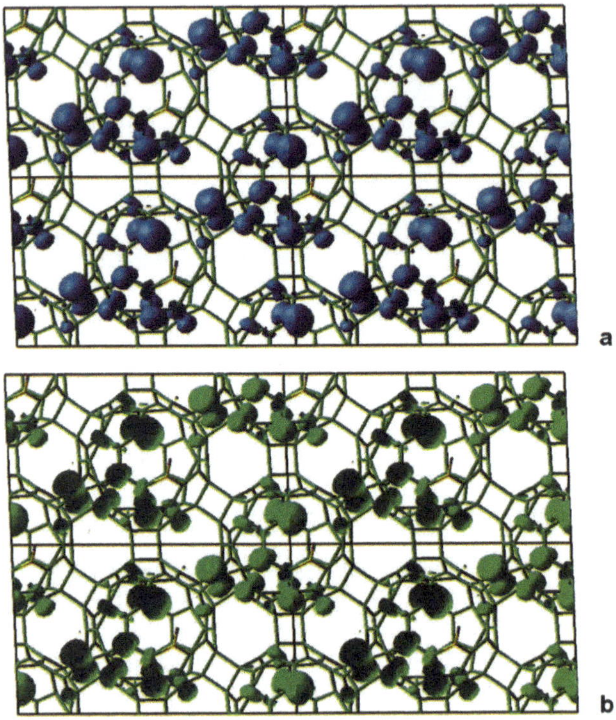

Fig. 7. The spin density maps of the (a) FM and (b) AF configurations of the defective structure generated from pressure and temperature with H bonded to atom A. The light and dark isosurfaces correspond to values of +0.008 μ_B and -0.008 μ_B, respectively.

The electronic ground state of this system is found to have a total magnetic moment of 3.0 μ_B per C_{60}, Table 4. The spin density map (Fig. 7) shows that the spin density is localised predominantly on the B and C sites while that at the A site is suppressed by the H atom. The bonding of hydrogen to atom A therefore quenches the main spin centre in path 1, leaving path 2 as the dominant inter-cage coupling mechanism. This results in a *ferromagnetic* ground state for this system. The average magnetisation of the experimental samples was measured to be 0.34 μ_B per cage [3]. For this to be the case in the theoretical structure, ~11% of the C_{60} cages would be magnetic and the H:C_{60} ratio would thus be ~11%, in reasonable agreement with the 17% of H:C_{60} ratio obtained from the INS data reported above. The two dimensional electronic band structure for this system displays an indirect, spin-flip band gap of 1.44 eV, whilst the smallest direct band gap between states of same spin is 1.74 eV.

The computed energy difference, however, between the ferromagnetic and antiferromagnetic states is only 3 meV per cage, Table 4. Although this energy difference is too small to explain room temperature ferromagnetism in Rh-C_{60}, the inter-cage linkage found does provide a mechanism for ferromagnetic coupling. It is possible that structures which allow for greater spin delocalisation or multiple linkages including interlayer linkages, will give rise to significantly larger inter-cage couplings.

8 Conclusions

Various defective structures were examined with the aim of investigating the origin of the ferromagnetism observed in metal-free Rh-C_{60} samples at room temperature. Reactive force field simulations and periodic hybrid density functional B3LYP calculations were combined to identify Rh-C_{60} defective structures and determine their magnetic ground states.

A recently proposed vacancy structure was used as a prototype defect. Its ground state was found to possess a net magnetic moment per cage but the coupling between the spin moments on adjacent cages was negligible. The possibility of the vacancy defects migrating closer together was also investigated. In this case, the net magnetic moment per cage was found to be zero. Significant magnetic coupling between the cages was present but antiferromagnetic in nature. The application of isotropic pressure to a layer of Rh-C_{60} resulted in the fracture of the intra-cage bond between the two inter-cage bonds. The unpaired electrons, localised on each cage, coupled antiferromagnetically, resulting in no net magnetic moment per cage, and no significant coupling between the defects on adjacent cages was found. Classical molecular dynamics calculations based on a reactive force field were used to simulate the high pressure, high temperature treatment used for the synthesis of the ferromagnetic samples. This simulation produced a defective structure where a carbon atom gets displaced and forms a new bond with a neighbouring fullerene cage. The undercoorinated bridge atom between the cages presented a possible new route, other than the interfullerene bonds created by the 2+2 cycloadditon, through which the local moments could couple. It was found, however, that this structure displays no net magnetisation.

The electronic band structures of these defective layers are all spin-polarised, with flat bands at the Fermi edge, and with insulating band gaps. In the systems presented, however, these features in the band structures are not a reliable indication of a tendency to form a ferromagnetic ground state as they are generated by the presence of a localised spin density with no significant inter-cage coupling. Even so, the structures with central vacancies, bridge atoms, or similar defect structures that possess a net magnetic moments per cage, could be responsible for the large *paramagnetic* signal observed in the samples [3].

The non-metallic character of the band structures of all the systems presented here suggests that the most likely mechanism for magnetic coupling is either direct exchange or superexchange mediated by the delocalised π-system. This is consisent with this work's findings, where the spin alternation rule is found to retain its validity in nearly all systems examined. In only one portion of a defective structure examined here, the spin alternation rule breaks down due to a dominant through-space, rather than through bond, spin polarisation effect. It is also noteworthy that the spin polarisation of the open-cage fullerene structures and nanotube segments investigated by Kim *et al.* [48] is also consistent with the spin alternation rule. The presence of curvature and sp^3 atoms, however, shortens the range of the polarisation effect, and hence the magnitude of the magnetic interactions, compared to the non-defective, planar sp^2-hybridised systems for which the spin alternation rule was initially formulated. In fact, in planar systems, the spin polarisation can be seen to propagate across fifteen or more bonds in planar graphitic ribbons [45], and long range effects have also been observed in planar graphite [46] and graphitic ribbons [47]. This simple model is thus a very valuable guiding tool for anticipating which defect structures might generate ferromagnetic ground states in carbon-based materials.

The presence of intercalated H in a concentration amounting to ~17% in the ferromagnetic samples was detected by inelastic neutron scattering experiments. In the calculations, the addition of H in the most stable site of the structure formed under pressure and temperature resulted in a *ferromagnetic* ground state. The inter-cage coupling, however, is insufficient to explain the very high Curie temperature observed. This structure is, to the authors' knowledge, the only proposed structure so far that has been obtained by simulating the experimental treatment leading to the ferromagnetic phase, and that is consistent with the ferromagnetic ground state.

Acknowledgements

The authors would like to thank S.M. Bennington, J.W. Taylor, and J.D. Gale for their contributions to the work reviewed here and the EPSRC for provision of computer time under the Materials Chemistry Consortium project, GR/S13422/01.

References

1. V.D. Blank, S.G. Buga, G.A. Dubitsky, N.R. Serebryanaya, M.Yu. Popov, and B. Sundqvist, Carbon 36 (1998) 319.
2. T.L. Makarova, B. Sundqvist, R. Höhne, P. Esquinazi, Y. Kopelevich, P. Scharff, V.A. Davydov, L.S. Kashevarova, and A.V. Rakhmanina, Nature 413 (2001) 716.
3. R.A. Wood, M.H. Lewis, M.R. Lees, S.M. Bennington, M.G. Cain, and N. Kitamura, J. Phys.: Cond. Matt. 14 (2002) L385.
4. V.N. Narozhnyi, K.-H. Müller, D. Eckert, A. Teresiak, L. Dunsch, V.A. Davydov, L.S. Kashevarova, and A.V. Rakhmanina, Physica B - Cond. Matt. 329 (2003) 1217.
5. S.M. Bennington, N. Kitamura, M.H. Lewis, R.A. Wood, A.K. Fukumi, and K. Funakoshi, J. Phys.: Condens. Matt. 12 (2000) L451.
6. K.-H. Han, D. Spemann, R. Höhne, A. Setzer, T.L. Makarova, P. Esquinazi, T. Butz, Carbon 41 (2003) 785; *ibid* 41 (2003) 2425.
7. T.L. Makarova, "Magnetism of carbon-based materials", Studies of High T_c Superconductivity, 45 (2003) 107.
8. P. Esquinazi, D. Spemann, R. Höhne, A. Setzer, K.-H. Han, and T. Butz, Phys. Rev. Lett. 91 (2003) 227201; P. Esquinazi, R. Höhne, K.-H. Han, A. Setzer, D. Spemann, and T. Butz, Carbon 42 (2004) 1213.

9. K. Murata, H. Ushijima, H. Ueda, and K. Kawaguchi, J. Chem. Soc., Chem. Commun. 18 (1991) 1265; K. Murata, H. Ushijima, and H. Ueda, J. Chem. Soc., Chem. Commun. 7 (1992) 567.
10. D.W. Boukhvalov, P.F. Karimov, E.Z. Kurmaev, T. Hamilton, A. Moewes, L.D. Finkelstein, M.I. Katsnelson, V.A. Davydov, A.V. Rakhmanina, T.L. Makarova, Y. Koplevich, S. Chiuzbaian, and M. Neumann, Phys. Rev. B69 (2004) 115425.
11. A.N. Andriotis, M. Menon, R.M. Sheetz, and L. Chernozatonskii, Phys. Rev. Lett. 90 (2003) 026801.
12. A.N. Andriotis, R.M. Sheetz, and M. Menon, J. Phys.: Condens. Matter. 17 (2005) L35.
13. J. Ribas-Ariño, and J.J. Novoa, Angew. Chem. Int. Ed. 43 (2004) 577.
14. J. Ribas-Ariño, and J.J. Novoa, J. Phys. Chem. Sol. 65 (2004) 787.
15. V.V. Belavin, L.G. Bulusheva, A.V. Okotrub, and T.L. Makarova, Phys. Rev. B70, (2004) 155402.
16. S. Nakano, Y. Kitagawa, T. Kawakami, M. Okumura, H. Nagao, and K. Yamaguchi, Molecules 9 (2004) 792.
17. J.A. Chan, B. Montanari, J.D. Gale, S.M. Bennington, J.W. Taylor, and N.M. Harrison, Phys. Rev. B70 (2004) 041403(R).
18. J.A. Chan, B. Montanari, and N.M. Harrison, Mol. Phys. (2005), in press.
19. A.D. Becke, Phys. Rev. A38 (1988) 3098.
20. A.D. Becke, J. Chem. Phys. 98 (1993) 5648.
21. C. Lee, W. Yang, and R.G. Parr, 37 (1998) 785.
22. M. Ernzerhof and G.E. Scuseria, J. Chem. Phys. 110 (1999) 5029.
23. J. Muscat, A. Wander, and N.M. Harrison, Chem. Phys. Lett, 342 (2001) 397.
24. R.L. Martin and F. Illas, Phys. Rev. Lett, 79 (1997) 1539.
25. T. Bredow and A.R. Gerson, Phys. Rev. B61 (2000) 5194.
26. I. de P.R. Moreira, F. Illas, and R.L. Martin, Phys. Rev. B65 (2002) 155102.
27. J.K. Perry, J. Tahir-Kelhi, and W.A. Goddard III, Phys. Rev. B63 (2001) 144510.
28. X.-B. Feng and N.M. Harrison, Phys. Rev. B69 (2004) 035114.
29. A.A. Ovchinnikov, Theor. Chim. Act. 47 (1978) 297.
30. J.S. Miller and A.J. Epstein, Angew. Chem. Int. Ed. Engl. 33 (1994) 385, and references therein.
31. O. Kahn, Molecular Magnetism (1993) VCH Publishers Inc., New York, and references therein.
32. D.J. Klein, C.J. Nelin, S. Alexander, and F.A. Matsen, J. Chem. Phys. 77 (1982) 3101.
33. E.H. Lieb, Phys. Rev. Lett. 62 (1989) 1201; 68 (1989) 1927(E).
34. M. Fujita, K. Wakabayashi, K. Nakada, and K. Kusakabe, J. Phys. Soc. Jpn. 65 (1996) 1920.
35. M. Núñez-Regueiro, L. Marques, J.-L. Hodeau, O. Béthoux, and M. Perroux, Phys. Rev. Lett. 74 (1995) 278.
36. D.W. Brenner, O.A. Shenderova, J.A. Harrison, S.J. Stuart, B.Ni, and S.B. Sinnott, J. Phys.: Condens. Matt. 14 (2002) 783.
37. J. D. Gale, JCS Faraday Trans. 93 (1997) 629; J.D. Gale, and A.L. Rohl, Mol. Simul. 29 (2003) 291.
38. H.B. Schlegel, J. Comp. Chem. 3 (1982) 214.
39. V.R. Saunders, R. Dovesi, C. Roetti, R. Orlando, C.M. Zicovich-Wilson, N.M. Harrison, K. Doll, B. Civalleri, I.J. Bush, Ph. D'Arco, and M. Llunell, CRYSTAL2003 User's Manual (2003) University of Torino, Torino.
40. H.J. Monkhorst, and J.D. Pack, Phys. Rev. B13 (1976) 5188.

41. S. Okada and S. Saito, Phys. Rev. B55 (1997) 4039.
42. A.V. Okotrub, V.V. Belavin, L.G. Bulusheva, V.A. Davydov, T.L. Makarova, and D. Tománek, J. Chem. Phys. 115 (2001) 5637.
43. X. Chen, S. Yamanaka, K. Sako, Y. Inoue, and M. Yasukawa, Chem. Phys. Lett. 356 (2002) 291.
44. F. Banhart, Rep. Prog. Phys. 62 (1999) 1181.
45. K. Kusakabe and M. Maruyama, Phys. Rev. B67 (2003) 092406.
46. P. Ruffieux, O. Gröning, P. Schwaller, L. Schlapbach, and P. Gröning, Phys. Rev. Lett., 84 (2000) 4910.
47. J.A. Chan, B. Montanari, K. Refson, and N.M. Harrison, (to be published).
48. Y.-H. Kim, J. Choi, K.J. Chang, and D. Tománek, Phys. Rev. B68 (2003) 125420.

Carbon-Based Magnetism
T. Makarova and F. Palacio (Editors)

24

Ferromagnetic Carbonaceous Compounds

T.L. Makarova

Umeå University, Umeå, Sweden,
Ioffe PTI RAS, St.Petersburg, Russia

1 Introduction

Since year 1986, there have been numerous reports in the literature on the synthesis of metal-free organic materials that exhibit magnetic ordering up to temperatures as high as 800K. More recently, a number of publications appeared in which reproducible observations of the room-temperature magnetic ordering in various carbon allotropes, such as graphite, proton-irradiated graphite, chemically modified graphite, fullerenes, microporous carbon and nanofoam were reported.

Weakness of the ferromagnetic signal and low reproducibility give rise to understandable doubts concerning the intrinsic origin of the reported data. However, in many cases the content of metallic impurities is too low to explain the value of magnetization. We think that the impurity scenario is highly unlikely, since it cannot explain recurring regularities in structure-property relationships. Magnetic properties of the carbon-based compounds are sensitive to the conditions of synthesis or subsequent annealing. The choice of starting material is governed by its isomeric composition and by the concentrations of hydrogen or oxygen. Magnetization values of the synthesized compounds depend on the degree of amorphousness/crystallinity of the material and the hybridization of carbon atoms. Some of these reports are reviewed in Section 2 of this chapter.

Probably, every scientist working in the field of organic ferromagnets reads the paper of J.S. Miller "The Quest for Magnetic Polymers – Caveat Emptor" [1] to beware of possible artefacts and difficulties with data interpretation. Trivial sources of iron include air pollution, magnetic separations, grinding and catalysts [1]. Nontrivial origin of magnetic behaviour in contaminated carbon-based materials may result from the special arrangement of the spins of metallic ions being arranged in such a way that they are strongly coupled to organic polymeric backbone [1]. Other nontrivial mechanisms are possible, for example, contact-induced magnetism suggested by J. M. D. Coey [2]. The effect of triggering carbon magnetism by the presence of transition metals was found in earlier works on organic compounds [3], extraterrestrial graphite [2], carbon nanotubes [4], and X-ray dichroism experiments on C/Fe layers [5]. However, it was not found in special experiments on graphite–magnetite composites [6].

In the discussion of nontrivial mechanisms a historical analogy can be drawn between carbon magnets and an inorganic compound calcium hexaboride which exhibits weak ferromagnetism and a high Curie temperature. There is experimental evidence that iron is involved in the high-temperature ferromagnetism of CaB_6 [7]. However, recent experiments have shown that the retraction of the claim of an exotic origin [8] was premature, and Fe/Ca atomic ratio exceeding 100% would be needed to explain the magnetization. Lattice defects are suggested as an origin of the high-temperature magnetism in hexaborides [9].

The papers selected in Section 2 meet the following criteria: (i) the substance shows a finite magnetization at room temperature; (ii) the absence of any significant amount of metallic impurities is proven. Methods of impurity analysis are abbreviated as ICP (inductively coupled plasma atomic emission spectroscopy); AAS (Atomic absorption spectroscopy), EPMA (Electron Probe Microanalysis), XRF (X-ray fluorescence analysis), EDS (energy dispersive spectroscopy) PIXE (particle induced X-ray emission). An obvious problem with the materials described in the reviewed papers is that only the fraction of the bulk seems to exhibit cooperative magnetic behaviour. Possible reasons for this are discussed in Section 4.

Large amount of the publications and patents as well as many close parallels between different magnetic substances raises the hope that creating bulk room-temperature carbon magnet is only a question of time.

2 Experimental Evidence

2.1 Organic Compounds

Carbonaceous ferromagnetic materials prepared from organic molecules fall into four categories: (i) quazi-one-dimensional polymers; (ii) carbonitride phases or boron / phosphorus carbon compounds; (iii) pure carbon materials on the mid-course to graphitization; (iv) fullerene-based materials.

Quasi-one-dimensional ferromagnetic structure was first obtained by attaching free radicals as side groups to the carbon backbone. This polymer, *Poly-BIPO,* was synthesized by thermal or photostimulated polymerization [10, 11], of mono-BIPO ($C_{14}O_4N_2H_{14}$). A spontaneous magnetization, a signature of ferromagnetism was observed, disappearing at temperatures of 420–460K. The saturation magnetization in the first experiments was about $M_s = 0.022$ emu/g. Impurity analysis by AAS revealed no transition metals within the sensitivity $1.5 \cdot 10^{-5}$ mass %. In another group, modification of polymerization conditions allowed obtaining a polymer with higher values of spontaneous magnetization, $M_s = 0.2$ emu/g [12, 13]. Transition metal impurities have been ruled out by ICP with a sensitivity of 10^{-5} mass %. The polymer was subjected to magnetic separation, but no difference in the chemical or structural properties between magneto-attractive and nonattractive parts has been found. Even in the attractive fraction of the polymer the percentage of a ferromagnetic phase is very small so that the magnetic phase escapes structural detection. Ferromagnetism has also been found in an analogue compounds *poly – BIPENO* [14] indicating a similar origin of magnetism.

Two independent groups reported experiments on pyrolized polyacrylonitrile C_3H_3N *(pyro-PAN)* [15, 16]. A technical PAN is a diamagnetic compound, while the product of its pyrolysis is a ferromagnet, slowly degrading with time being exposed to air. The decay follows an exponential exp $(-t/\tau)$ law where $\tau \approx 5 \cdot 10^5$ s. The deterioration of magnetic properties is considered as a result of paring of electron spins, participating in the magnetic ordering, with the free spins of oxygen molecules. Iron content is reported to be less than 10^{-4} mass %, as determined by chemical, mass-spectrometric and neutron activation analysis.

In another study, ferromagnetic PYRO-pan samples contained 100–400 ppm of iron [17]. Among ten measured samples, one showed the saturation magnetization (M_s = 0.2 emu/g) at least seven times bigger than it would be produced by the iron impurity, whereas the sample with ten times more iron concentration showed 50 times less magnetization [9].

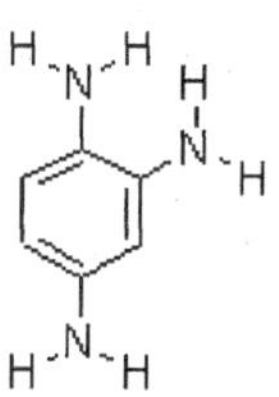

In 1987, Torrance *et al.* [18] reported specific preparation of poly-*triaminobenzene* that resulted in a ferromagnetic behavior up to 400°C. The material is based on a polymer obtained from reacting s-triaminobenzene $C_6H_9N_3$ with iodine. The reaction is complex and the resulting polymer is not very reproducible; nevertheless, on rare occasions, it showed the presence of small amounts of ferromagnetic material. The temperature dependence of the magnetic moments indicates the Curie point at 700K. However, the material decomposes at about the same temperature near 400°C, and ferromagnetic properties are not restored on cooling.

Ferromagnetic carbon compound was prepared by M. Ota *et al.* by dehydrogenation of *triarylmethane resin* $(C_{27}H_{31}N_2O_7S)_n$ [19, 20]. Whereas the triarylmethane resin is normally a diamagnetic, synthesis of this substance under the magnetic field yields a ferromagnetic substance with M_s = 0.12 emu/g and H_c = 65 Oe. AAS detected less than 20 ppm of Fe and absence of Ni and Co. Magnetic response has been attributed to the appearance of radicals produced by dehydrogenation.

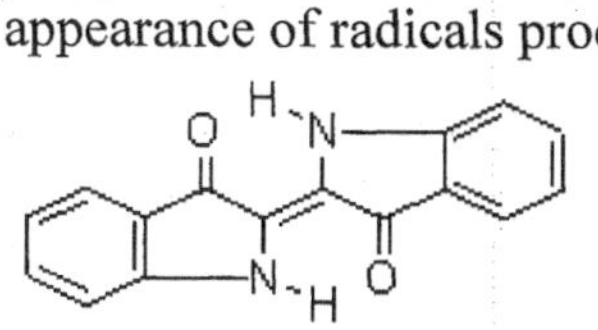

A sequence of chemical reactions polymerizes *indigo* units ($C_{16}H_{10}N_2O_2$) to a chain yielding a stable polymer which attracts to a magnet even by heating at 200°C [21]. The reproducibility was 100% in 30 runs. The reactions were carried out in quartz vessels. The room temperature values for saturation magnetization, remanent magnetization and coercive force are M_s = 0.7 emu/g; M_r = 0.08 emu/g; H_c = 120 Oe. Transition metal content was determined by ICP to be 26–28 ppm Fe and less than 1 for Co and Ni. The M_s value is equivalent to an amount of Fe two order of magnitude greater that that determined from ICP. This approach was patented as a method for preparation of ferromagnetic organic polymers from nitrogen-containing starting materials [22].

Several reports describe peculiar magnetic properties of carbonaceous materials prepared by pyrolysis. Pyrolytic carbon is highly oriented in structure and the materials pyrolized at relatively low temperatures, 600–1300°C, contain a large number of unpaired electrons in the graphite skeleton. In Ref. 23, pyrolytic carbon is described, made from pure hydrocarbon such as *adamantane* $C_{10}H_{16}$. Adamantane was heated at 1000°C in the Ar atmosphere, evaporated, pyrolized and deposited on the inside surface of the quartz tube. This material shows clear features of magnetic ordering: a distinct ferromagnetic loop is observed with $M_s = 0.5$ emu/g; $M_r = 0.35$ emu/g; $H_c = 600$ Oe at 10K. Decreasing upon heating, the magnetization is on the level of 1/3 of its initial value at 400K. The presence of transition metal impurities was not detected within the 25 ppm sensitivity of the ICP analysis. Further studies [24] revealed high concentration of unpaired electrons and highly oriented structure of the material. In repeated experiments performed in order to check the reproducibility of this process batch dependence was found [25]. The magnetization curves of some samples showed nonlinear sigmoid behaviour with the highest $M_s = 0.32$ emu/g but without hysteresis. These samples were classified as superparamagnetic. One of the samples exhibited a hysteresis loop which was observed up to the temperature of 200K. Hysteresis was accompanied with characteristic satellites in the EPR spectrum. The room temperature value of M_s for this ferromagnetic sample is 0.06 emu/g. ICP confirmed the absence of impurities down to 25 ppm.

Repeated experiments have been made in order to check the reproducibility of this process [17]. The obvious batch dependence was found. For several samples magnetization curves showed nonlinear sigmoid behaviour with the highest $M_s = 0.32$ emu/g but without hysteresis. These samples were classified as superparamagnetic. One of the samples exhibited a hysteresis loop, which was observed up to the temperature of 200K. Hysteresis was accompanied with characteristic satellites in the EPR spectrum. Room temperature value of M_s for this ferromagnetic sample is 0.06 emu/g. ICP confirmed the absence of impurities down to 25 ppm.

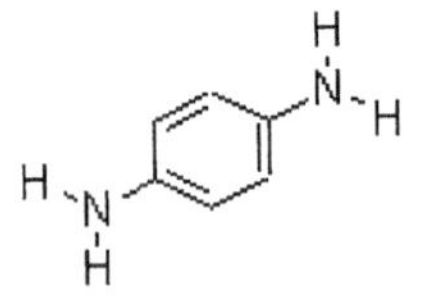

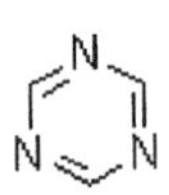

Araki *et al.* [26] prepared an magnetic organic material by the pyrolysis of the mixture of *phenylendiamine* $C_6H_8N_2$ and *triazine* $C_3H_3N_3$. The product obtained at 500–1000°C has exhibited a ferromagnetic behavior and a clear isomer effect. The highest values of saturation magnetization, remanence and coercive force were obtained for the product pyrolized at 600°C: $M_s = 0.624$ emu/g, $M_r = 0.0665$ emu/g, $H_c = 125$ Oe.

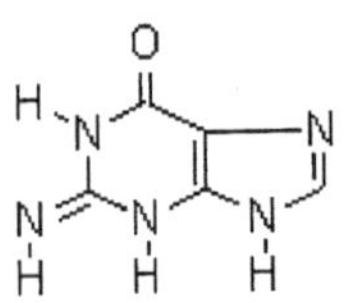

Several organic compounds were treated by pyrolysis at 600°C, and five of resulting polymers exhibited hysteresis loops at room temperature. Magnetic impurities were not detected by AAS. The best effect (saturation magnetization $M_s = 2.25$ emu/g) was obtained for *guanine* $C_5H_5N_5O$ [27].

Extremely high saturation magnetisation, up to 10% of that for Fe, was reported in Ref. [28] for the compound prepared by pyrolysis at 950°C of a

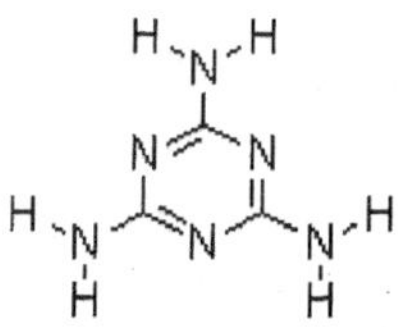

mixture of *triazine* $C_3H_3N_3$ and *melamine* $C_3H_6N_6$. By the AAS analysis, magnetic impurities were not detected (< 0.5 ppm). Any confirmation of these results by other groups is not available.

2.1.1. Carbons from Hydrogen-rich Starting Material

K. Murata *et al.* [29] analyzed the relationship between the hydrogen content in the starting material and the value of saturation magnetization at room temperature. The results are shown in Fig. 2 as the M_s value versus hydrogen-to-carbon ratio in the starting material. It is clearly seen that high hydrogen content is favourable for the magnetism, especially, if the H/C ratio exceeds 2.

The values of saturation magnetization for amorphous-like carbons prepared from tetraaza compounds by the chemical vapour deposition (CVD) method [29] are among the highest reported. M_s of the title carbons prepared from tetraazadodecane was found to be 2.56 emu/g, that is 0.005 μ_B per carbon atom. Strong magnetic properties were observed for a pyrolytic carbon prepared from cyclodecane by heating at 950°C [30]. The resulting substance was magnetically separated into two fractions, magnetic and non-magnetic. The magnetic fraction was stable in air and preserved its parameters

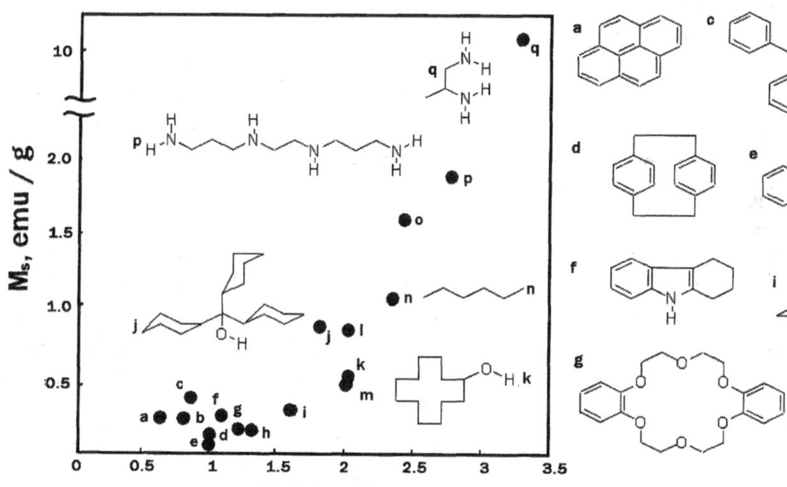

Fig. 1. Dependence of saturation magnetization on the hydrogen-carbon ratio in the starting material. Letters stand for different organic compounds: *a:* pyrene $C_{16}H_{10}$; *b:* trans-1.2-di(thienyl)ethylene, *c:* triphenylmethane $C_{19}H_{16}$; *d:* paracyclophane $C_{16}H_{16}$; *e:* phenol C_6H_6O; *f:* tetrahydrocarbazole $C_{12}H_{13}N$; *g:* dibenzo-18-crown-6 $C_{20}H_{24}O_6$; *h:* octahydroacridine $C_{14}H_9N$; *i:* adamantane $C_{10}H_{16}$; *j:* tricyclohexylmethanol $C_{19}H_{34}O$; *k:* cyclododecanol $C_{12}H_{24}O$; *l:* cyclodecane $C_{10}H_{20}$; *m:* cyclopentadecane $C_{10}H_{20}O_5$; *n:* n-hexane C_6H_{14}; *o:* 1,4,8,11-tetraazacyclotetradecane $C_{10}H_{24}N_4$; *p:* 1,5,8,12-tetraazadodecane $C_8H_{22}N_4$; *q*: 1,2-diaminopropane $C_3H_{10}N_2$. Adapted from [29].

(M_s = 1.07 emu/g; M_r = 0.21 emu/g; H_c = 163 Oe) for at least one month. These values slightly decreased with increasing temperature: the values at room temperature accounted about 80% of those at 4K. As ICP analysis and fluorescent X-ray technique revealed less than 50 ppm Fe and the absence of Ni and Co, the extrinsic origin of signal can be discarded.

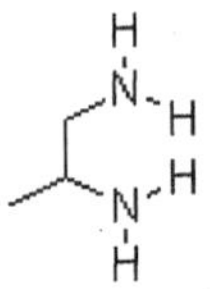

In accordance with Ref. [31], an excellent precursor for a strong carbon magnet is *1,2-diaminopropane* $C_3H_{10}N_2$ (H/C = 3.3). An air-stable carbon magnet is obtained by the rapid pyrolysis at 950°C. The saturation magnetization 10.5 emu/g corresponds to 0.022 μ_B/carbon atom. With a small but noticeable hysteresis, the magnetization shows non-Curie-Weiss behavior: it is almost temperature independent in a magnetic field as weak as 50 G. The estimated Curie temperature is above 500K, its exact value can not be measured due to decomposition of the structure at about this temperature.

In Ref. 32 a comparative study is made of magnetic properties of amorphous-like carbons prepared by pyrolysis of different organic compounds. The impurity content is basically the same for the samples examined. The dependences of magnetization values on the choice of starting material, reaction temperature, reaction time, and atmosphere are found. On the base of these regularities, the following features of the process are commented: (i) the starting material is put into a preheated reaction tube and a short reaction time is favourable; (ii) the content of nitrogen and/or hydrogen in the starting material is important; (iii) the degree of amorphousness of the material plays an important role. Clear evidence that crystallinity is related to magnetic properties is demonstrated by post-synthesis plasma treatment of the samples [32]: in plasma-treated samples one-to-one correspondence is observed between the plasma power, magnetization values and amorphous carbon content.

Effect of pyrolysis temperature on the magnetic properties of the carbon materials was studied [33] for triethylboranes $C_6H_{15}B$, tripropylboranes $C_9H_{21}B$ and tributylboranes $C_{12}H_{27}B$ (Fig. 2). In each case, observed saturation magnetization exhibited sharp peak on the plot of synthesis temperature. The M_s value reached 36.2 emu/g at 4.5K, corresponding to 0.0721 μ_B per carbon atom; room temperature values are slightly lower. Hysteresis was not pronounced: the coercivity at 4.5K is about 50 Oe, and the ratio M_r/M_s is less than 0.1. The field and temperature dependence of magnetization can be approximately described by the Langevin (Brillouin) function of the theory of (super)paramagnets; extracted from the fit the magnetic moment of a coherent spin cluster was $\sim 500\mu_B$. This method was patented for phosphorus and/or boron contained substances [34], and the best substance prepared by this method was claimed to have the saturation magnetization M_s = 59.8 emu/g, *i.e.*, equal to that of nickel.

The idea of creating the carbon substance on mid-course to graphitization has been patented [35] with 40 examples of preparation organic materials free from a ferromagnetic metal. The starting materials belong to four groups: (a) a polymer with aromatic compounds connected via alkylene chains, which can be substituted by an aryl, alkyl or aralkyl group; (b) a triarylmethane polymer; (c) a polymer of an acetylene compound; (d) a hydrogen containing polymer. These substances are subjected to a

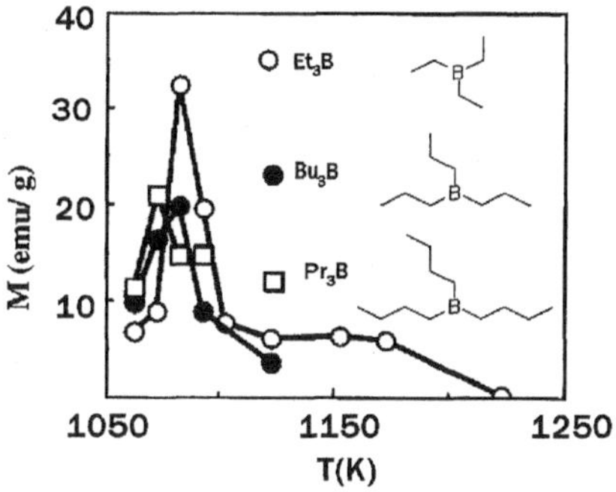

Fig. 2. Dependence of the observed saturation magnetization at 298K on the pyrolysis temperature for triethylboranes (Et_3B) tributylboranes (Bu_3B) and tripropylboranes (Pr_3B). Adapted from [33].

thermal treatment in vacuum or in an inert gas atmosphere. The obtained materials are characterized by the magnetization values of 0.1–0.38 emu/g.

Materials with a higher H:C ratio have been patented [36, 37]. Twelve compounds characterized by the coercive force 90–120 Oe, and saturation magnetization reaching 1.08 emu/g (starting material: hexane), 1.28 emu/g (dicyclohexane), 1.63 emu/g (tetraazacyclotetradodecane), 1.92 emu/g (tetraazadodecane), 2.63 emu/g (triethylamine), 6.69 emu/g (diethylamine), 7.85 emu/g (dipropylamine), 9.37 emu/g (propanediamine), were synthesized.

2.1.2 Dehydrogenated Compounds

There exists an alternative point of view on the role of hydrogen in carbon magnetism: Unlike the authors cited in the previous section, Ueda argues that hydrogen is detrimental for magnetism in carbon prepared from organic polymers [38, 39], sugars [40], or aromatic hydrocarbon violanthrone $C_{34}H_{16}O_2$ [41]. Ueda suggested to produce carbonaceous ferromagnetic materials consisting essentially from carbon alone [42].

Specific organic compounds, like phenasine and indigo, are carbonized to eliminate hydrogen atoms from the starting aromatic compound as completely as possible. Magnetic carbon compounds based on polyvinylchloride (PVC) [43], polyacrylonitrile, cyclic tetramers of resorcinol, phenol resin [44] and pyrene-benzaldehyde copolymer [45] have been reported. A PVC-based polymer prepared by pyrolysis at 1000°C was found to have M_s= 0.27 emu/g. Carbonization of PVC at 520°C in vacuum yields ungraphitized carbide with the following ferromagnetic properties: M_s = 3 emu/g, M_r = 0.6 emu/g, H_c = 400 Oe [46].

A method is suggested [47] for forming a carbonaceous ferromagnetic material using cobalt chloride as a catalyst, later completely removed from the product. It is claimed that the presence of ferromagnetic ions induces carbon ferromagnetism when the organic substances are dehydrogenated by heat treatments, and that the induced ferromagnetism is not lost after removing cobalt with diluted HCl. The idea of impregnating charcoal with cobalt ions and removal of cobalt particles from magnetized carbon specimens was used for the preparation of a ferromagnetic carbon electrode from charcoal blocks [3]. It was emphasized that the resulting material is almost pure carbon. The essential part of the synthesis is the process which allows one to eliminate hydrogen from the starting material, but not causes the graphitization of the carbonized material. The residual hydrogen atoms degrade ferromagnetic properties. It was also noted that the same magnetization could be achieved without cobalt catalyst, but the reaction requires longer time. A discussion on a non-trivial role of iron in carbon magnetism can be found in Ref. 48.

2.1.3 Polymers

Magnetically active polymers can be obtained by oxidative polymerization of phenylenediamines. Studied in Ref. [49] poly (meta-PDA) poly (ortho-PDA), poly (*o*-PDA) and poly (para-PDA), showed S-shaped magnetization curves at 300K with M_s = 0.38 emu /g; 0.29 emu/g; 0.07 emu/g, respectively. ICP revealed that the polymers contained < 0.1% Fe, which would be sufficient for accounting for magnetization values of not more than 0.2 emu/g.

A ferromagnetic polyaniline with M_s = 10 emu/g has been synthesized by stabilizing spin carrying charge carriers generated during redox doping of polyaniline [50]. AAS analysis revealed 0.1% of sodium sulphate and the absence of transition metals. A ferri- or ferro-magnetic polymer with a Curie temperature above 350K, and a maximum saturation magnetization of 0.1 emu/g was produced from polyaniline and an acceptor molecule tetracyanoquinodimethane. ICP showed 7 ppm of Ni and trace of Fe. Magnetic force microscopy gives additional support to room temperature ferromagnetism by providing evidence for domain wall formation and motion [51].

Magnetism at room temperature is reported in metal-free conducting samples of ClO_4^- doped poly (3-methylthiophene) [52]. Determination of magnetic impurities using a graphite furnace AAS spectrometer (Varian AA 80) found 1 ppb of Fe, 2 ppb of Ni, and absence of Co. The magnetic properties of the polymer are strongly dependent upon preparation conditions. An interesting effect of changing the coercivity with the water content in the solvent was observed, whereas the remanence and saturation remained unchanged. The ferromagnetic samples were unstable when stored in air: in five months the samples became diamagnetic [53]. Another group reported weak ferromagnetism at room temperature for pressed pellets of partially doped poly(3-methylthiophene [54].

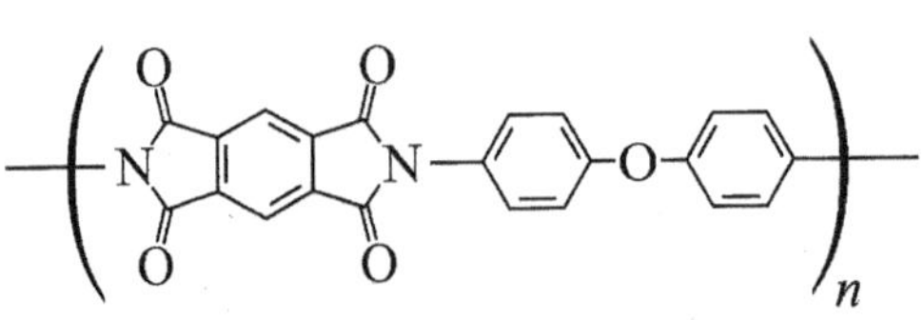

A commercially available aromatic *polyimide film* Kapton H 25-μm thick was heat-treated in a nitrogen flow [55]. Ferromagnetic features were observed in the films heat-treated at 490–520°C. The room temperature saturation and remnant magnetizations were M_s = 0.059 emu/g, and $M_r = 9 \cdot 10^{-4}$ emu/g, respectively, at 300K; these parameters maintained 5 months after the synthesis. The authors claim that the original Kapton H and the heated films do not contain any metallic elements.

The experiments on Kapton H films have recently been independently confirmed [56]. Kapton H polyimide films are shown to exhibit well-defined room-temperature ferromagnetic properties when exposed to high temperature or ultraviolet laser irradiation in nitrogen atmosphere. The absence of paramagnetic impurities in the pristine film is confirmed with the precision of 1 part per 10 billion. It has been shown in Ref. [56] that one is able to produce stable "magnetic drawings" on the organic film by stamping or treatment by the UV laser in nitrogen atmosphere.

Several polymeric ferromagnetic substances have been patented. Polymerization of aminoaromatic compounds yields copolymers containing aminoaromatic groups, the Curie temperature of the material is 350K [57]. S. Nakajima [58] patented an organic polymer characterized by a high density of verdazyl radicals; the polymer is water-soluble and has a property of responding to a magnet.

We note that polymers described in this section contain only a small magnetically responsive fraction. These substances should not be confused with the high-spin organic polyradicals described in Chapter 3. The approach presented in Chapter 3 has a high potential for obtaining room-temperature macromolecular ferromagnets.

2.1.4 Fullerenes

Ferromagnetism has been observed in various *fullerene based materials.* Fullerene C_{60}, and the only fullerene, doped with strong donor TDAE, shows bulk ferromagnetism (see Chapter 10 for the details). The saturation magnetization in high-quality samples TDAE-C_{60} samples reaches 6 emu/g ($\sim 1\ \mu_B / C_{60}$) and T_c is 16.1K.

Room-temperature magnetism in *C_{60} – containing polymeric films* was reported in Ref. [59]. The films were produced by ultrasonic dispersion of C_{60} in dimethylformamide solution of polyvinylidenefluoride and subsequent vacuum evaporation of the solution. The Curie point and magnetization were 370K and 0.210 emu/g. The contamination with Fe, Mn, Cr, Co and Ni were less than 0.001; 0.0005; 0.001; 0.002 and 0.005 wt.%, respectively, and it could not account for the observed magnetization. As described in a patent [60], ultrasonic dispersion of fullerenes (C_{60}, C_{70}, C_{76} or C_{84}) in an organic polymer or in a non-conducting liquid forms a light processable magnetic material. In another patent [61], fullerenes are doped by halogens from alkali halides, for example, LiF, and ultrasonically dispersed in an organic polymer.

Fullerene *hydride* $C_{60}H_{36}$ has been reported to be a room-temperature ferromagnet with M_s = 0.04 emu/g [62]. Higher values of magnetization have been found for the composition $C_{60}H_{24}$ [63]: varying from sample to sample, the magnetization reaches the value of 0.16 μ_B/C_{60} (i.e. 1.2 emu/g). Measured by the atomic-emission analysis, the concentration of all detected metals in the sample with the highest magnetization of 0.16 μ_B/C_{60} was the following: Fe: 0.01; Ni: 0.002; Pd: 0.01; Al: 0.05; Cu: 0.1 (wt.%). Seeing that the magnetization due to the metallic impurities could be only on the level of a few percents from the observed value, the magnetic ordering is an intrinsic property of hydrofullerites. A circumstantial evidence of the intrinsic nature of ferromagnetism is aging: one-year storage brings the samples to a diamagnetic state.

Room-temperature ferromagnetism of *polymerized* fullerenes was first reported in 1996 [64]. Polymerized fullerene is a specific form of pure carbon in which spheroidal carbon molecules are connected by covalent bonds. The samples of commercial C_{60} powder were exposed to oxygen under the action of strong visible light. This process leads to photoinduced polymerization of C_{60} units within a surface layer of the crystallites. The thickness of the phototransformed layer is controlled by the light penetration depth and is of the order of 0.5 μm. The saturation value for the partially polymerized sample was only $1.4 \cdot 10^{-2}$ emu/g corresponding to 0.001 μ_B/C_{60}. However, the residue of the solution in toluene, that is a photo-polymerized part of the sample, exhibited a hundred times larger magnetization corresponding to 0.1 μ_B/C_{60}.

The existence of a ferromagnetic phase in fullerene photopolymers was confirmed by three methods: SQUID measurements, ferromagnetic resonance and low-field nonresonance derivative EPR signal [65]. The room temperature value of saturation magnetization for photolyzed C_{60} is 0.04 emu/g. The experiments were done in a chamber with flowing oxygen, which exclude any possibility of penetration of metallic particles during the experiment. The impurity concentration was measured by the means of IPP which indicated that the signals from all metals were below 100 parts per 10^9; independent analysis by the plasma-induced AAS confirmed the iron level to be a few parts per 10^9.

Produced by laser irradiation, magnetic images on fullerene films have been demonstrated by means of MFM [66]. Laser irradiation polymerizes the films. The laser treated areas exhibit weak but non-trivial magnetic signal which changes in the applied magnetic field and also changes when the direction of magnetic field is reversed.

In a number of papers [67 - 69], magnetism of fullerenes polymerized by *high temperature-high pressure* treatment has been reported. In Ref.[67] a magnetically ordered phase of pressure-polymerized C_{60} with the Curie temperature about 500K was detected by the SQUID and the field dependent magnetic susceptibility measurements. The magnetism was attributed in [67] to the defects in the two-dimensional rhombohedral Rh-C_{60} structure, but this question remains controversial. Later investigation [70] on the same samples showed an increased content of surface contamination that much exceeds the impurity level 22 ppm in the initial C_{60} powder. However, the measurements on different pieces of the samples revealed that the specific magnetization did not show correlations with the sample size and Fe-impurity content, and this speaks for intrinsic origin of the magnetism reported in Ref.[67]. The influence of impurities on the ferromagnetic properties of polymerized C_{60} was further investigated on the same [67] samples by PIXE and magnetic force microscopy (MFM) [71]. Within the information depth of 36 μm, an increased concentration of magnetic impurities (175 ppm, still several times less than would require for the impurities to account for the magnetization) was found in the surface regions. At the same time, in macroscopic (0.25 mm) pure regions (concentration of magnetic impurities < 1 μg/g parallel magnetic domain pattern was found in islands covering 30% of the sample surface. Again, this speaks in favour of intrinsic magnetism.

A different method was used for the preparation of pressure-polymers C_{60} in Ref. [69]: multi-anvil octupole press and the pressure of 9 GPa. Varying the synthesis temperature, a gradual transition from paramagnetic to diamagnetic behavior was followed. For optimal synthesis conditions, the ferromagnetic phase with the saturation magnetization of 0.34 μ_B per fullerene cage was found. TEM showed that fullerenes were not damaged in the ferromagnetic phase. Inelastic neutron scattering analysis of the magnetic phase prepared by this method showed a noticeable presence of hydrogen (H : C_{60} $\cong$ 17%) which can play a certain role in fullerene magnetism [72].

Pressure-polymerized fullerenes with the magnetization of about 0.5 emu/g were synthesized by another group [68]. In this work, a systematic study of the synthesis conditions for the production of the ferromagnetic phase was performed. Fifteen samples were prepared at 6 GPa and the temperatures $650 \leq T \leq 850$°C using "Toroid" high pressure cell. Only samples prepared at $745 \leq T \leq 790$°C, and only some of them, namely five from the eight samples prepared at these conditions, showed ferromagnetism; their magnetic behavior was qualitatively similar but with different values of the magnetization.

Two novel methods for producing fullerene-based magnets have recently been found: solvothermal treatment of C_{60} at 140°C in water [73] and interaction of fullerenes with ammonia. The latter method seems to be very promising for obtaining bulk room-temperature fullerene-based magnets. Here we demonstrate the preliminary results obtained on this material [74].

Several samples were prepared by the interaction of the fullerene C_{60} and mixture C_{60} + 8 mass. % NH_4Cl (promoter of interaction) with ammonia by the method described in Ref. [75]. Magnetic properties of freshly prepared samples were checked using the vibrating sample magnetometer EG@G PARC M-4500. Every $C_{60}H_xN_y$ sample showed nonlinear magnetic behaviour. The highest room temperature magnetization value of 2.2 emu/g was obtained for the sample $C_{60}H_6N_{0.4}$ (Fig. 3a). After 8 month storage at ambient conditions the magnetic properties of the sample were measured with a SQUID magnetometer (Quantum Design). The room-temperature magnetization decreased by 20% (Fig. 2b). However, its value remains rather high, being 1.8 emu/g at room temperature and 2.8 emu/g at 2 K which corresponds to 0.37 μ_B/C_{60}. The temperature dependences of magnetization at 5 T (Fig. 3c) and remanent magnetization (Fig. 3d) indicate that the Curie temperature much exceeds 300 K.

This example gives a proof for intrinsic ferromagnetism in fullerene- based material, because the magnetization value of 2.8 emu/g corresponds to the content of 1.25% of non-oxidized iron or 2% of iron in the form of magnetite (Fe_3O_4) provided that these impurities form clusters, large enough to be magnetically ordered. Chemical analysis of the sample did not show the presence of iron within the accuracy of 0.1%; more precise structural and elemental investigations are under way.

Raman spectroscopy of the samples show that the main feature of the vibrational spectrum of the $C_{60}H_6N_{0.4}$ sample is the fullerene $A_g(2)$ pentagonal pinch mode at 1470 cm^{-1} with several satellites at 1462 – 1453 cm^{-1} (probably, indicating partial polymerization). Micro-Raman studies reveal that relative intensity of these peaks varies from point to point indicating that the sample is inhomogeneous. All spectra have high fluorescence background, typical for partially hydrogenated fullerenes. It is

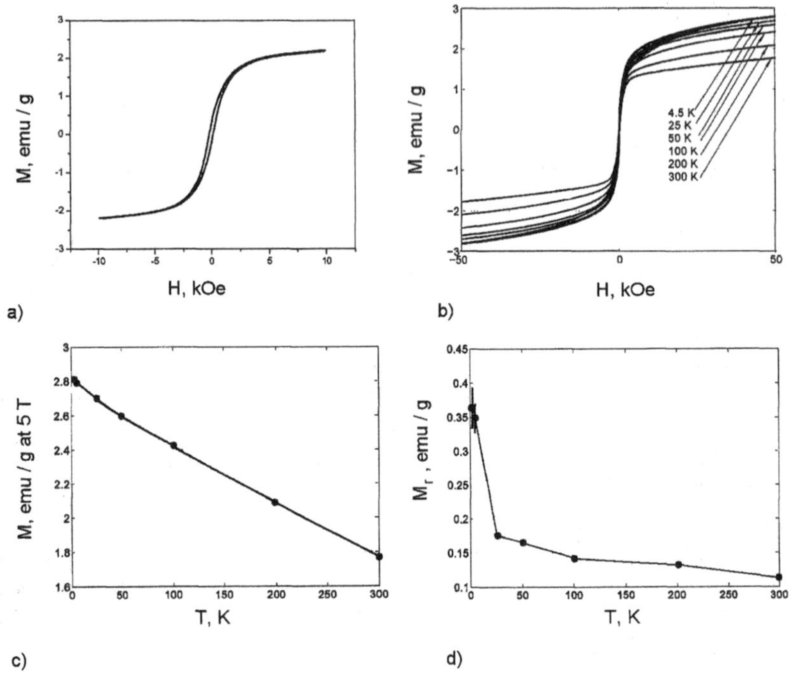

Fig. 3. a): Magnetization hysteresis loop measured for freshly prepared $C_{60}H_6N_{0.4}$ sample at room temperature; b): Magnetization hysteresis loops measured for the same sample, 8 months after the preparation, at the temperatures 4.5, 25, 50, 100, 200, and 300 K; c): Temperature dependence of magnetization at 5 T; d): temperature dependence of remnant magnetization of the sample.

interesting to note that the Raman spectra of the samples do not contain any features around 1330 cm^{-1} and 1580 cm^{-1}, which proves the absence of destroyed fullerenes and graphitic carbon.

2.3 Graphite-based Magnets

Y. Kopelevich *et al.* detected both ferromagnetic and superconducting-like magnetization hysteresis loops in *HOPG* samples below and above room temperature [76]. This is the first experimental work that clearly shows high-temperature ferromagnetic-like behavior and possibly superconductivity in graphite. The detailed studies have given a conclusive proof for the intrinsic nature of the ferromagnetic signal from graphite [77] (Chapter 18).

A *glassy carbon* prepared by high pressure treatment was studied in Refs. [78, 79]: Room-temperature ferromagnetic loops in a narrow range of synthesis conditions (5 GPa, 1200°C) was recorded for samples having total amount of not more than 70 ppm transition metal impurities. The occurrence of high-temperature ferromagnetism has been found in *microporous carbon* with a three-dimensional nanoarray (zeolite) structure [80].

Studies of *carbon particles* prepared in He plasma [81] showed that the saturation magnetization of carbon fine particles increases with decreasing grain size. The highest $M_s = 8.5 \cdot 10^{-7}$ Wb·m/kg (0.67 emu/g) has been obtained for the grain size of 19 nm; the intensities of Fe, Co, and Ni lines were comparable to the noise level in EPMA analysis. The saturation magnetization deceases gradually with increasing temperature and disappears at a temperature of 607K. The strong dependence of saturation and remanent magnetization and coercive force on grain size of the powder was observed earlier in PYRO-pan powder: M_s, M_r and H_c grow by a factor more than ten with decreasing average grain size of the powder from 45 to 5 μm [15].

Carbon nanofoam is sometimes called the fifth carbon allotrope (Chapter 20). As first observed by Rode *et al.* [82], it exhibits room-temperature ferromagnetic behavior. Freshly produced, it shows rather high saturation magnetization 0.4 emu/g at room temperature. The high temperature magnetization disappears in a few hours after synthesis but persists at lower temperatures with a narrow hysteresis curve and a high saturation magnetization [82].

Multilevel ferromagnetic behaviour has been described [83] for *chemically modified graphite*: powdered graphite mixed with powdered copper oxide and then heated in a tube furnace containing either nitrogen or argon. Two magnetic transitions were observed, at 115 and 315K. Iron content was determined by AAS, XRF and EDS to be around 40–60 ppm range, Ni and Co around 1 ppm, The M_s values are 0.58 and 0.25 emu/g at 2 and 300K correspondingly [83, 84]. The simple and inexpensive chemical route, based on a vapour phase reaction, to obtain ferromagnetic graphite in bulk amounts was accepted as the state patent in 2004 [85].

Most convincing experiments on controllable transformation of diamagnetic graphite to a ferromagnetic carbon phase have been made by Esquinazi *et al.* [86]. Irradiation of highly oriented pyrolytic graphite with protons leads to a measurable bulk magnetic moment. Irradiated spots produce strong images in magnetic force microscopy indicating local magnetization value of several hundreds emu/g (see Chapter 19).

3 Summary of Experimental Observations

Analysing the reports in the literature on the synthesis of metal-free organic materials that exhibit magnetic ordering at room temperatures and higher, one can notice that the ferromagnetic carbonaceous compounds have several features in common.

An obvious difficulty is that the bulk magnetization is rather low. Low magnetization values apparently contradict to the high temperatures of the magnetic ordering transition. In some cases these values can be increased by one-two orders of magnitude using the magnetic separation. Nearly in all cases the authors note that they could not find structural or chemical differences between the magnetic and non-magnetic parts. Probably, the only example where this difference is unquestionable is photopolymerized fullerene: dissolving in toluene removes non-reacted monomeric fullerenes and the residue shows an increased magnetization.

Low reproducibility is a general difficulty in producing organic magnetic materials by the chemical methods. Complex chemical processes involve a large number of reactions, and the results are hard to duplicate. It is only recently that the chemical methods are supplemented by the physical ones: proton bombardment, laser ablation, oxygen erosion, pressure-temperature treatment, and photolysis. These experiments appear to be more readily repeatable.

Magnetic properties of the synthesized substances are very sensitive to the synthesis conditions, especially preparation temperature. This trend is observed for all temperature-related processes, from pyrolysis of organic substances to high pressure - temperature treatment of graphite and fullerenes. Deviations from an optimal temperature, which is certain for a certain process, leads to a quick decay of bulk magnetization.

In earlier attempts to produce ferromagnetic carbonaceous compounds, the most successful ones are the processes involving the pyrolysis of organic compounds at relatively low temperatures. The common features of these processes are: heating organic substances in a complex (steplike) way, sometimes in the presence of a magnetic field, and to stop the processes halfway to graphitization. This process yields a highly oriented graphite-like material with a high concentration of unpaired spins. Clear features of room-temperature magnetic ordering were observed in some cases. Again, there is a series of *similarities* between the substances reported by different authors.

Saturation magnetizations in pyrolytic magnetic carbons are nearly temperature-independent up to temperatures that are much higher than 295K. Curie temperature in several cases is close to the decomposition temperature.

The choice of the starting material to be pyrolized is crucial, and the main features are the presence of *hydrogen*, *nitrogen* and *branched isomers*. The *ratio* between the concentrations of hydrogen and carbon [H] and [C] in the starting material is particularly critical. As soon as this ratio exceeds 1.5, a sharp increase in the saturation magnetization of the reaction product is observed. A parallel between the hydrogen-related magnetic properties of pyrocarbons and the proton-induced magnetic ordering in graphite (Chapter 19) can be drawn.

Magnetic carbons obtained by pyrolysis of nitrogen- containing compounds exhibit a clearly pronounced *isomeric effect*: stronger magnets are obtained from branched starting components. In other words, the geometric arrangement of radicals is important for attaining an ordered spin structure. This observation can be compared with the isomeric effects shown in Fig. 1.1 of Chapter 1.

Properties of magnetic carbon materials made by pyrolysis of diverse organic compound monomers at around 1000°C are strongly related to the radicals and hydrogen atoms generated during the pyrolysis process and to the degree of *amorphousness* of the material. A leitmotif of methods for preparing carbonaceous ferromagnets is stopping the synthesis process at the mid-course to graphitization. Studies of a large number of ferromagnetic materials obtained from various nitrogen-containing substances showed that the structure of these materials is most often amorphous and disordered. In order to reveal the structural–magnetic correlations, samples of ferromagnetic carbon were subjected to various doses of plasma treatment, after which the percentage of amorphous carbon and the value of magnetization were measured. A one-to-one correspondence between the above two parameters was observed: the higher the content of amorphous carbon, the higher the spontaneous magnetization.

Finally, some of the carbon magnets are unstable and loose the room-temperature ferromagnetic properties during one day, like carbon nanofoam, or in one year like proton-bombarded graphite of hydrogenated fullerenes. Probably, the theoretical models should take into account metastable carbon structures.

4 Theoretical Models

4.1. Graphite-based Models

There is no physical law which prohibits magnetic ordering in the structures containing only light elements. However, there are good reasons why one would not expect high-temperature ferromagnetism in carbon. Unpaired spins in organic substances, being largely separated, do not interact with each other, whereas an attempt to bring together the spins of radicals results in a chemical reaction.

A major requisite for the occurrence of magnetism in an all-carbon structure is the presence and stability of carbon radicals. The occurrence of radicals, which can introduce an unpaired spin, is cut down by the strong ability of pairing all valence electrons in covalent bonds. In a planar graphite sheet, in an infinite nanotube and in fullerenes all atoms are tetravalent and all electrons are paired.

Several scenarios that account for (or predict) the magnetism of carbon have been suggested: bulk magnetism, induced magnetism, and atomic-scale magnetism caused by structural imperfection.

Theoretical models of bulk carbon magnetism include *nitrogen–carbon* compositions where ferromagnetic ordering of spins could exist in π delocalized systems due to a lone electron pair on a trivalent element [87], an *intermediate graphite-diamond structure* where the alternating sp^2 and sp^3 carbon atoms play the role of different valence elements [88], a *polymer* containing segregated stacks of radical ions [89], a hydrogen-doped ferromagnetic *diamond* [90].

In nanometer-scale materials, a new factor appears: the nanometer-scale shape and the size of the material.

Bulk graphite with a perfect lattice is a diamagnetic material whose magnetic susceptibility is second only to that of superconductors. Graphite containing certain defects can exhibit spontaneous magnetization. This behavior is caused not only by carbon atoms but mainly by its "honeycomb" structure. Certain defects in this structure give rise to a sharp asymmetric peak in the density of states at the Fermi level.

The ability to form flat bands is not only a property of carbon atoms; this property is inherent to planes that are composed of atoms bonded hexagonally to each other and feature zigzag-type edges or zigzag-type boundaries between chemically different elements, for example, boron, nitrogen and carbon (Chapter 14).

Experimental evidence collected in Section 2 suggests that carbon magnetism is of defect nature. There are several candidates to account for the magnetic properties of nanographite systems.

The most common defects in carbon are adatoms and vacancies. Carbon vacancies generate a magnetic moment of about 1 μ_B. It is the remaining dangling bond that contributes the calculated localized magnetic moment. Carbon adatoms possess a magnetic moment of about 0.5 μ_B The magnetism of adatom is explained by the fact that the atoms of substrate and the adatom are hybridized differently; one of orbitals is orthogonal to the surface π-orbitals and cannot form a band, remaining localized and, consequently, spin-polarized (Chapter 16).

Unpaired spins may be introduced by carbon radicals, not only by under-coordinated carbon atoms. Graphitic pattern containing negatively curved graphitic surfaces indicate the presence of four unpaired spins in the electronic ground state. The magnetic behavior originates in the presence of trivalent carbon radicals, which are introduced and sterically stabilized within the aromatic system. Gaussian negative curvature which provides a mechanism for steric protection of the unpaired spins is invoked for the explanation of unusual magnetic properties of carbon nanofoam (Chapter 20).

It is shown in Chapter 12 that electronic states are strongly influenced by the existence and the shape of graphite edge, and zigzag edges favour the spin polarization with ferromagnetic alignment. A mechanism for localized magnetism coming from the graphite edge states is discussed in (Chapter 13).

Not only in graphite ribbons, but also in BNC heterosheets and finite-length carbon nanotubes, edge states or border states between carbon, nitrogen and boron sublattices are responsible for the occurrence of the magnetic ordering. More generally, a variety of nanometer-scale structures consisting solely of non-magnetic elements exhibit magnetic ordering. One of the examples is a hexagonally reconstructed hydrogen-terminated silicon surface (Chapter 14).

4.2. Fullerene-based Models

Is the mechanism of magnetic ordering same for all ferromagnetic carbonaceous compounds?

The abovementioned defects in graphite can equally be considered in the case of fullerene-based magnets. Adatoms, vacancies, negative curvature and zigzag edges in fullerene solids may lead to the same effects as in graphite. More specifically, these defects include broken or shortened interfullerene bonds, distortion of fullerene cages, vacancies in the fullerene cages, adatoms on fullerene cages, local charge inhomogeneities, open-cage defect structures. In the models of ferromagnetic fullerenes, the following possibilities are considered: (i) a fraction of the fullerene cages are broken, and, thus, unpaired electrons appear at the dangling bonds; (ii) the shape of the fullerene cage changes as a result of Stone–Wales transformations; and (iii) fullerenes remain intact, but unpaired electrons are formed at interfullerene links. In an alternative approach (iv), a high-spin ground state of fullerene polymer is considered.

In the case of pressure polymerization there are speculations that fullerenes play a role of auxiliary units in carbon magnetism, and the magnetic properties are related to the graphite-like structures formed as a result of the cage collapse.

However, ferromagnetism was observed not only in fullerenes on the verge of collapse but also in structures where the integrity of the cage was unquestionable: in the classical ferromagnetic compound TDAE-C_{60}, in phototransformed fullerenes [64-66], hydrogenated fullerenes [62, 63], and disperse C_{60} in the polymer matrix [59 -61]. Recent results on room-temperature ferromagnetism of C_{60} hollow submicrospheres obtained by solvothermal treatment [73] and on ammonium treated fullerenes [75] also show that fullerene cages are not damaged in the magnetically ordered state. The experiments on sonicated, solvated, hydrogenated, photolysed and ammonia-treated fullerenes suggest that in case of fullerene ferromagnetism different mechanisms may exist.

Magnetic properties of fullerene compounds have been a subject of special attention since the date of the discovery of these molecules. The reaction of a strong donor tetrakis-diethyl-amino-ethylene (TDAE) with fullerene C_{60} yielding a pure organic ferromagnetic substance with a Curie temperature T_C= 16K is regarded as a remarkable discovery in fullerene chemistry (Chapter 10). Antiferromagnetic ground states are encountered in various alkali or lantanide fulleride salts. It is shown that the lanthanide exchange interactions between 4f electrons in Eu_6C_{60} are modulated by the $\pi(C_{60})$ orbitals [91]. Similarly, ferromagnetic correlations in $Eu_xSr_{6-x}C_{60}$ (x = 1, 3, 6) are put down to the π-f interactions through C_{60} rather than the direct exchange interactions between Eu atoms [92].

Palladium compounds are commonly diamagnetics. However, C_{60} $Pd_{4.9}$ is shown to exhibit room-temperature ferromagnetic-like behaviour with M_s = 0.05 emu/g decreasing ten times in several days of storage in air [93]. The authors suggest that the origin of this behaviour lies in the electron transfer from Pd to C_{60}. Iodine containing polymers of C_{60} exhibit a transition to a superparamagnetic behaviour at 60K [94] and IBr-C_{60} at 30K [95]. These experimental examples show that in many cases fullerene C_{60} works either as a magnetic unit, or a magnetic coupling unit.

Theoretically, the ability of C_{60}-fullerene to act as a magnetic coupling unit between radical centres has been evaluated using *ab initio* method. The results indicate that functionalized C_{60}-fullerenes have a very good potential to couple radical centres. Depending on the relative position of the two radical centres, the resulting interaction is ferro- or antiferromagnetic (Chapter 22).

A scenario in which structural defects (vacancies) and sp^3 -hybridized atoms interact is described in Chapter 21. A correlation should exist between spin moments for magnetic ordering to appear. Since the defects are widely spaced, interaction between them is indirect and needs a mediator. The dipole moment induced by a vacancy can play the role of such a mediator. The Coulomb interaction between localized charges and electric dipoles ensures a means for the exchange interaction between magnetic moments.

Long-range spin coupling has been considered through the investigation of an infinite, periodic system of polymerized C_{60} network. It was shown that chemically bonded hydrogen plays a vital role, providing a necessary pathway for the ferromagnetic coupling of the considered defect structure (Chapter 23).

A model of fullerene ferromagnetism, which does not require *ad hoc* constructed defects, is suggested by Kvyatkovskii et al.[96]. In this model, the presence of some paramagnetic impurities in fullerene matrix (e. g. hydrogen, fluorine, as well as carbon from partially destroyed fullerenes) leads to a formation of the paramagnetic fullerene ions $C_{60}^{\pm}$ as a result of the reaction with charge and spin transfer:

paramagnetic dopant A + diamagnetic fullerene C_{60} $\rightarrow$
$\rightarrow$ *diamagnetic dopant ion* $A^{+}(A^{-})$ *+ paramagnetic fullerene ion* $C_{60}^{-}(C_{60}^{+})$

Paradoxically, this is an impurity-based model, and at the same time this model is based on intrinsic properties of fullerene matrix. Paramagnetic impurities play only the passive role of net charges and spins sources. According to this model, diamagnetic C_{60} molecules transform into stable paramagnetic species ($C_{60}^{\pm}$ ions or fullerene radical adducts $C_{60}R$) and become magnetically active due to the charge and spin transfer from dopants. Whether the electron comes from a donor to a C_{60} or goes from C_{60} to an acceptor, its spin is transferred *to* the fullerene in any case being well localized at the lattice site. The model is supported by *ab initio* calculations, the main result of which is that in the [2+2]-cycloadduct polymerized phase the effective exchange interaction between the paramagnetic fullerene ions is ferromagnetic, strong and long-range. The coupling between paramagnetic species (nearest neighbours and next nearest neighbours) in polymerized fullerene matrix is examined, and it is shown that exchange integral is of the order of several tenth of an electronvolt [96].

5 Conclusions

Reports on materials in which only the fraction of the bulk exhibits cooperative magnetic behaviour endure hardship and criticism. In this chapter we make an attempt to show that the keyword "ferromagnetic impurity" can be replaced by the keywords "defects", "edge states", "nanometer-scale structures", "dangling-bond states", "carbon structure and bonding". Many efforts are required to develop a sufficiently good understanding of the exotic new type of magnetism based on defects.

Two main reasons move forward further studies of carbon-based magnetism. Scientific curiosity on the one hand: What causes magnetic behavior in a structure made of atoms that have no magnetic moment? And the hope of providing a basis for bettering man's condition on the other: New materials have always been a dominant factor in driving advances in material usage.

As the phenomenon of carbon magnetism is basically related to the boundary and defect states, further development of this effect can be expected in the field of nanotechnology. The experimentally observed possibility of changing the magnetic properties of carbon allotropes by irradiating these materials can be used to produce magnetic patterns by means of nanolithography. Open nanotubes are suggested to be used as nanodimensional magnets. One possible application is to create ultrasmall magnetic storage devices by irradiating protons into carbon nanotubes. Semiconducting properties of magnetic carbon materials hint at device applications in spintronics, an emerging field that exploits the spin of electrons in electronic devices.

The methods for reproducible production of magnetic carbon-based materials in milligram quantities already exist: laser ablation, oxygen erosion, proton irradiation of graphite or photolysis of fullerenes. If it is technologically feasible to produce a bulk carbon magnetic material in gram amounts, one can then expect applications of this material in medicine and biology as a unique biocompatible magnetic material.

Biotechnology applications may include biosensors, targeted drug delivery, artificial muscles, magnetic buckles, magnetic markers, magnetic resonance imaging and intratissue hyperthermia. In the latter example the movement of magnetic domains in a specimen placed inside the tissue leads to lethal overheating of the cancerous tissue without damaging surrounding healthy cells.

Magnetization curves in carbon materials usually show a very narrow hysteresis, if any. However, it is not an obstacle for practical applications of these materials. For some medical applications like magnetic contrast agents or magnetic markers a superparamagnetic behaviour is required, and hysteresis loops are unwelcome. A promising approach is the utilisation of superparamagnetic carbon clusters as magnetic markers and carriers for the detection and manipulation of biomolecules. Thus, magnetic carbon is expected to find application in both diagnostic tools and basic research into biological and medical systems. The combination of sensing and manipulating magnetic particles is a promising choice for future integrated lab-on-a-chip system.

Some carbon magnets, for example, carbon nanofoam, loose a room-temperature ferromagnetic behaviour within a few hours. This property, seemingly undesirable, can be appealing in medical applications, when it is necessary to deactivate the magnetic agent after the experiment.

There are suggestions to use magnetic carbons in copiers and other business machines [97]. The image obtained by using the magnetic carbon ink can be magnetically read, thus giving a possibility to use it for printing money, certificates and secure documents. Another polymer, which consists of repeated units containing oxygen, sulphur and NH groups, is applied as a magnetic toner for development of the image [98]. An organic material for magnetic tape, photomagnetic memory of ink has been produced by photooxidizing an organic substance triarylmethil [99].

K. Murata stated in 1996: "Of many candidates of the magnets, carbon compounds will be the most promising from the practical point of view, because the carbons exhibit a spontaneous magnetization at room temperature and are cheap to make, chemically and physically stable, and easy to process" [100]. Now, ten years later, we can repeat these words with confidence.

References

1. J.S. Miller, Adv. Materials 4 (1992) 298; 4 (1992) 435.
2. J.M.D Coey, M. Venkatesan, C. Fitzgerald, A. Douvalis, and I. Sanders, Nature 420 (2002) 156.
3. H. Ueda, J. Mater. Sci. 36 (2001) 5955.
4. O. Cespedes, M.S. Ferreira, S. Sanvito, M. Kociak, and J.M.D. Coey, J. Phys.: Cond. Mat. 16 (2004) L155.
5. H.-Ch. Mertings, S. Valencia, W. Gudat, P.M. Oppeneer, O. Zaharko, and H. Grimmer, Europhys. Lett. 66 (2004) 743.
6. R. Hohne, M. Ziese, and P. Esquinazi, Carbon 42 (2004) 3109.

7. K. Matsubayashi, M. Maki, T. Tsuzuki, T. Nishioka, and N.K. Sato, Nature 420 (2002) 143.
8. M.C. Bennett, J. van Lierop, E.M. Berkeley, J.F. Mansfield, C. Henderson, M.C. Aronson, D.P. Young, A. Bianchi, Z. Fisk, F. Balakirev, and A. Lacerda, Phys. Rev. B69 (2004) 132407.
9. L.S. Dorneles, M. Venkatesan, M. Moliner, L.G. Lunney, and J.M.D. Coey, Appl. Phys. Lett. 85 (2004) 6377.
10. Yu.V. Korshak, A.A. Ovchinnikov, A.M. Shapiro, T.V. Medvedeva and V.N. Spektor, JETP Lett. 43 (1986) 309.
11. Yu.V. Korshak, T.V. Medvedeva A.A. Ovchinnikov, and V.N. Spektor, Nature 326 (1987) 370.
12. Y. Cao, P. Wang, Z. Hu, S. Li, and L. Zhang, Synth. Met. 27 (1988) B625.
13. Y. Cao, P. Wang, Z. Hu, S. Li, and L. Zhang, Solid State Commun. 68 (1988) 817.
14. A.A. Ovchinnikov and V.N. Spector, Synth. Met. 27 (1988) B615.
15. Yu.A. Katulevskii, M.A. Magrupov, and A.A. Muminov, Phys. Stat. Sol.(a) 127 (1991) 223.
16. Y.M. Shulga, A.I. Boldyrev, A.A. Ovchinnikov, Chem. Phys. Lett. 189 (1992) 577.
17. W.P. Lee, E.B. Park, C.O. Yoon, A.A. Ovchinnikov, and Y.W. Park, in: Chemical physics of Intercalation, Ed. by P. Bernier et al. Plenum Press (1993).
18. J.B. Torrance, S. Oostra, and A. Nazzal, Synth. Met. 19 (1987) 709.
19. M. Ota, M. Otani, and M. Igarashi, Chem Lett. 7 (1989) 1179.
20. M. Ota, M. Otani, and M. Igarashi, Chem Lett.7 (1989) 1183.
21. H. Tanaka, K. Tokuyama, T. Sato, and T. Ota, Chem. Lett. 10 (1990) 1813.
22. H. Tanaka. Japanese Patent JP 04008731 (1992).
23. K. Kawabata, M. Mizutani, M. Fukuda, and S. Mizogami, Synth. Met. 33 (1989) 399.
24. S. Mizogami, M. Mizutani, M. Fukuda, and K. Kawabata, Synth. Met. 43 (1991) 3271.
25. K. Tanaka, M. Kobashi, H. Sanekata, A. Takata, T. Yamabe, S. Mizogami, K. Kawabata, and J. Yamaguchi, J. Appl. Phys. 71 (1992) 836.
26. H. Araki, R. Matsuoka, and K. Yoshino, Solid State Commun. 79 (1991) 443.
27. H. Araki and K. Yoshino, Jpn. J. Appl. Phys. 31, Part 2 (1992) L130.
28. H. Araki, Y.B. Roh, N. Kuwamura, and K. Yoshino, Jap. J. Appl. Phys. 31 Part 2 (1992) L337.
29. K. Murata, H. Ushijima, H. Ueda, K. Kawaguchi, J. Chem. Soc., Chem. Commun. 18 (1991) 1265.
30. K. Murata, H. Ueda, K. Kawaguchi, Synth. Met. 44 (1991) 357.
31. K. Murata, H. Ushijima, H. Ueda. K. Kawaguchi, J. Chem. Soc., Chem. Commun. 7 (1992) 567.
32. H. Ushijima, K. Murata, H. Ueda, and K. Kawaguchi. Mol. Cryst. Liq. Cryst. 233 (1993) 351.
33. K. Murata and H. Ushijima, J. Appl. Phys. 79 (1996) 978.
34. Y. Ushijima. Japanese Patent J06251924 (1994).
35. K. Murata. United States Patent US 5135673 (1990)
36. K. Murata. Japanese Patent JP05198421 (1993).
37. K. Murata. United States Patent US5330668 (1994)
38. H. Ueda. Denki Kagaki 60 (1992) 716.
39. H. Ueda, Denki Kagaki 60 (1992) 242.
40. H. Ueda, Electrochem. Soc. Extended Abstracts 93-1 (1993) 2524.
41. H. Ueda, Denki Kagaki 62 (1994) 695.
42. H. Ueda. United States Patent US 5556570 (1996).
43. K. Murata, T. Masuda, and H. Ueda, Chem. Express 5 (1990) 601.

44. K. Murata, T. Masuda, and H. Ueda, Chem. Express 5 (1990) 605.
45. K. Murata, T. Masuda, and H. Ueda, Chem. Express 5 (1990) 597.
46. H. Ueda, Japanese Patent J187630 (1995).
47. H. Ueda, Japanese Patent 1994-156433[19].
48. J.M.D. Coey and S. Sanvito, Physics World 17 (2004) 33.
49. D. Ichinohe, K. Akagi, and H. Kise, Synth. Met. 85 (1997) 1671.
50. D.C. Trivedi, Synth. Met. 121 (2001) 1780.
51. N.A. Zaidi, S.R. Giblin, I. Terry, A. P. Monkman. Polymer 45 (2004) 5683.
52. A.A. Correa, L. Walmsley, L.O.S. Bulhoes, W.A. Ortiz, A.J.A. de Oliveira, E.C. Pereira, Synth. Met. 121 (2001) 1836.
53. E.C. Pereira, A.A. Correa, L.O.S. Bulhoes, P.C. Aleixo, J.A. Nobrega, A.J.A. de Oliveira, W.A. Ortiz, and L. Walmsley. J. Magn. Magn. Mater. 226 (2001) 2023.
54. O.R. Nascimento, A.J.A. de Oliveira, A.A. Correa, L.O.S. Bulhoes, E.C. Pereira, V.M. Souza, L. Walmsley. Phys. Rev. B67 (2003) 144422.
55. Y. Kaburagi, Y. Hishiyama. J. Mater. Res. 17 (2002) 2000.
56. Z. Gu, F.J. Owens, L. Belova, K.V. Rao. J. Phys. Chem (Letters), in print (2005).
57. S. Galaj. United States Patent US5844069 (1998).
58. S. Nakajima, Japanese Patent J06176913 (1994).
59. M. Ata, M. Machida, H. Watanabe, and J. Seto. Jpn. J. Appl. Phys. 33 (1994) 1865.
60. T. Yanada, Japanese Patent JP5159915 (1993).
61. M. Ata. Japanese Patent J05129120 (1993).
62. A.S. Lobach, Yu.M. Shul'ga, O.S. Roshchupkina, A.I. Rebrov, A.A. Perov, Y.G. Morozov, V.N. Spector, and A.A. Ovchinnikov, Fullerene Sci. Technol. 6 (1998) 375.
63. V.E. Antonov, I.O. Bashkin, S.S. Khasanov, A.P. Moravsky, Yu.G. Morozov, Yu.M. Shulga, Y.A. Ossipyan, E.G. Ponyatovsky, J. Alloys, and Comp. 330 (2002) 365.
64. Y. Murakami and H. Suematsu, Pure Appl. Chem. 68 (1996) 1463.
65. F.J. Owens, Z. Iqbal, L. Belova, and K.V. Rao. Phys. Rev. B69 (2004) 033403.
66. T.L. Makarova, K.H. Han, P. Esquinazi, R.R. da Silva, Y. Kopelevich, I.B. Zakharova, and B. Sundqvist, Carbon 41 (2003) 1575.
67. T.L. Makarova, B. Sundqvist, R. Höhne, P. Esquinazi, Y. Kopelevich, P. Scharff, V. Davydov, L.S. Kashevarova, and A.V. Rakhmanina, Nature 413 (2001) 716.
68. V.N. Narozhnyi, K.-H. Müller, D. Eckert, A. Teresiak, L. Dunsch, V.À. Davydov, L.S. Kashevarova, and A.V. Rakhmanina. Physica B 329 (2003) 1217.
69. R.A. Wood, M.H. Lewis, M.R. Lees, S.M. Bennington, M.G. Cain, and N. Kitamura, J. Phys. Condens. Matter. 14 (2002) L385.
70. R. Höhne and P. Esquinazi, Adv. Mater. 14 (2002) 753.
71. K.-H. Han, D. Spemann, R. Hohne, A. Setzer, T. Makarova, P. Esquinazi, and T. Butz, Carbon 41 (2003) 785.
72. J.A. Chan, B. Montanari, J.D. Gale, S.M. Bennington, J.W. Taylor, and N.M. Harrison, Phys. Rev. B70 (2004) 041403.
73. Z. Xu, G. Yin, F. Tao, Y. Liang, H. Liu, H. Li, and L. Qi, International Workshop "Fullerenes and Atomic Clusters", "IWFAC2005", St. Petersburg, Russia, 27 June – 2 July 2005. Abstracts, p. 43, (2005).
74. Samples were prepared by B. P. Fokin, Yu. M. Shulga, E. E. Fokina, I. I. Korobov, A. G. Burlakova, B. P. Tarasov; characterization and magnetic measurements by Yu. M. Shulga, M. Belli, D. Ponitroli, M. Ricco, T. L. Makarova.
75. B. P. Fokin, Yu. M. Shulga, E. E. Fokina, I. I. Korobov, A. G. Burlakova, B. P. Tarasov. ISJAEE 6 (2004) 20.

76. Y. Kopelevich, P. Esquinazi, J.H.S. Torres, S. Moehlecke. J. Low Temp. Phys. 119 (2000) 691.
77. P. Esquinazi, A. Setzer, R. Hoehne, C. Semmelhack, Y. Kopelevich, D. Spemann, T. Butz, B. Kohlstrunk, and M. Loesche. Phys. Rev. B66 (2002) 024429.
78. X. Wang, Z.X. Liu, Y.L. Zhang, F.Y. Li, and C.Q. Jin, J. Phys.: Condens. Matter. 14 (2002) 10265.
79. C.Q. Jin, X. Wang, Z.X. Liu, Y.L. Zhang, F.Y. Li, and R.C.Yu. Braz, J. Phys. 33 (2003) 723.
80. Y. Kopelevich, R.R. da Silva, J.H.S. Torres, A. Penicaud, and T. Kyotani, Phys. Rev. B68 (2003) 092408.
81. S. Akutsu and Y. Utsushikawa, Mater. Sci. Res. Int. 5 (1999) 110.
82. A.V. Rode, E.G. Gamaly, A.G. Christy, J.D. Fitz Gerald, S.T. Hyde, R.G. Elliman, B. Luther-Davies, A.I. Veinger, J. Androulakis, and J. Giapintzakis, Phys. Rev. B70 (2004) 054407.
83. H. Pardo, F.M. Araújo-Moreira. R. Faccio, R. Faccio, A.J.C. Lanfredi, C.A. Cardoso, E.R. Leite, G. Zanelatto, and A.W. Mombrú, cond-mat/0407303.
84. A.W. Mombrú, H. Pardo, R. Faccio, O.F. de Lima, A.J.C. Lanfredi, C.A. Cardoso, E.R. Leite, G. Zanelatto, and F.M. Araújo-Moreira, Phys. Rev. B71 (2005) 100404(R).
85. H. Pardo, A. Mombrú, and F.M. Araújo-Moreira, Patent PI 0402338-2 (2004).
86. P. Esquinazi, D. Spemann, R. Höhne, A. Setzer, K.-H. Han, and T. Butz, Phys. Rev. Lett. 91 (2003) 227201.
87. A.A. Ovchinnikov. Theor. Chim. Acta. Berlin 47 (1978) 297.
88. A.A. Ovchinnikov and I. L. Shamovsky, J. Molec. Struct. (Theochem) 251 (1991) 133.
89. J.B. Torrance, P.S. Bagus, I. Johannsen, A.I. Nazzal, S.S.P. Parkin, and P. Batail, J. Appl. Phys. 63 (1988) 2962.
90. K. Kenmochi, K. Sato, A. Yanase, and H. Katayama-Yoshida, Jap. J. Appl. Phys. p. II 44 (2005) L51.
91. I. Margiolaki, S. Margadonna, K. Prassides, T. Hansen, K. Ishii, H. Suematsu. J. Am. Chem. Soc.
92. K. Ishii, H. Ootoshi, Y. Nishi, A. Fujiwara, H. Suematsu, Y. Kubozono. AIP conference proceedings 544 (2000) 29.
93. Yu.M. Shul'ga, A.S. Lobach, I.N. Ivleva, Yu.G. Morozov, V.N. Spector, A.A. Ovchinnikov, Mol. Mat. 10 (1998) 201.
94. L. S. Grigoryan, M. Tokumoto. Solid State Commun. 7 (1995) 523.
95. H. Sekine, H. Maeda, M. Kosuge, Y. Tanaka, M. Tokumoto. J. Appl. Phys. 72 (1992) 5448.
96. O. E. Kvyatkovskii, I. B. Zakharova, A. L. Shelankov, T. L. Makarova. Fuller. Nanotub. Car. N. , in press (2005).
97. http://www.aist.go.jp/NIMC/overview/v7.html; K. Murata, priv. commun. (2002).
98. S. Nakajima, J Japanese Patent t J06196309(1994).
99. M. Yanai, J Japanese Patent J08045716 (1996).
100. K. Murata and H. Ushijima, J. NIMC 4 (1996) 1.

Subject Index

A

AAS (atomic absorption spectroscopy) 541
ab initio 75, 201, 371, 463, 483, 501, 523, 541
adsorbed atoms and molecules 417
aminium radical 53
aminoxyl 107
antiferromagnetism 23, 75, 107, 131, 329, 353
armchair edge 279, 397

B

B3LYP 23, 201, 371, 501, 523
bearded edge 279, 305
bipartite lattice 305, 329, 397, 523
bistability 189
BNC hetero-sheet 329
Bohr–von Leeuwen theorem 245
border state 329, 541

C

C_{60} 229, 245, 305, 371, 417, 463, 501, 523, 541
carbon nanotubes 245, 279, 329, 353, 371, 397, 417, 463, 483, 541
cationic radical 53
charge-transfer 75, 229, 201, 541
core electrons 245, 279, 371, 463,
cove edge 279

D

de Haas–Van Alphen effect 245, 279
dendrimer 53
DFT (density functional theory) 159, 23, 201, 305, 329, 371, 437, 463, 483, 501, 523
diamond 245, 279, 541
diamond anvil cell 131
dihedral angle 107
dilute magnetic semiconductors 483
disorder 23, 353, 371, 397, 417, 437, 463, 523, 541
disordered carbon 397, 437
dithiodiazolyl radicals 1
doped semiconductors 245, 483
Dzyalosinsky-Moriya interaction 131

E

edge state 329, 397
EDS (energy dispersive spectroscopy) 541
electrostatic interaction 53, 189, 353, 371
EPMA (electron probe microanalysis) 541
exchange-correlation energy 305, 329, 371

F

Fermi level 1, 245, 279, 305, 329, 353, 397, 417, 483
ferrimagnet 1, 75,107, 159, 279, 305, 329
Ferro-to-antiferromagnetic transitions 131
field induced magnetic ordering 107, 131
flat band 279, 305, 329, 371, 437, 463, 483, 523, 541
fullerene 229, 245, 279, 305, 353, 371, 397, 417, 437, 463, 483, 501, 523, 541

G

gold nanoparticle 31
graphene 305, 329, 353, 371, 397, 417, 437, 463, 483
graphite edges 279
graphite ribbon 279, 305, 329, 483, 541

H

heat capacity 159, 23, 107, 75, 131
heck reaction 53
Heisenberg model 23, 201, 305
Hellmann-Feynman theorem 201
high pressure 107, 75, 131, 189, 483, 523, 541
high-spin molecule 1, 53
high-spin state 53, 483
high-temperature ferromagnetism and superconductivity 417
host-guest interaction 397
Hubbard model 279, 305, 329, 353, 397
hyperbranched polymer 53

I

ICP (inductively coupled plasma atomic emission spectroscopy) 159, 23, 201, 305, 329, 371, 437, 463, 523, 541
intermolecular interactions 107, 131
intersystem singlet-triplet crossing 501
intramolecular interactions 107
irradiation effects 437

J

joint-disjoint rule 201

K

kinetic exchange interaction 201

L

Landau diamagnetic susceptibility 245
Langevin diamagnetism 245

Larmor diamagnetism 245
lattice defects 353, 371, 437, 541
less screening rule 201
Lieb theorem 305
localized states 279, 353, 371
low dimensional magnetism 53

M

magnetic dot 53
magnetic exchange coupling 107
magnetic force microscopy 53, 523, 541
magnetic switching 397
magnetization plateau 107
maximum magnetic symmetry rule 201
McConnell model 201, 483
metallocene 229
molecular dynamics/MD 523
molecular electronic device 189
molecular orientations 229
molecular wire network 31

N

nanofoam 463
nanographene 397
nanohorns 463
nanotubes 23, 245, 279, 305, 329, 353, 371, 397, 417, 463, 483, 541
Néel temperature 75, 131
negative magneto-resistance 31
next nearest neighboring interaction 107
nitroxide 1, 23, 107, 75, 131, 53, 305, 483
noncollinear magnetism 201

O

one dimensional Heisenberg antiferromagnet 75
one dimensional Heisenberg ferromagnet 75
orbital diamagnetic susceptibility 279
organic magnetic conductor 1, 75
organic radical 23, 107, 75, 53, 189

P

paramagnetic semiconductor 1, 75
Pauli paramagnetic susceptibility 245, 279
phase transition 159, 107, 75, 189, 229, 463,
photo-excited spin system 75
photo-induced phase transition 189
π-conjugated polymer 23
PIXE (particle induced X-ray emission) 159, 417, 437, 541
poly(acetylene) 23
poly(phenylenevinylene) 23
polycarbene 31
polymerized fullerenes 371, 523, 541
polyradical 53, 107, 201, 541
potential energy surface 501
pressure effect 107, 131, 229
pyrolytic carbon 463, 541

R

radical alloy 107

S

schwarzite 463
Si surface 329
single-molecular based magnetism 53
spin alignment rules 201
spin cluster 107, 541
spin correlation function 201
spin delocalization 53
spin density 1, 23, 75, 53, 305, 329, 371, 397, 437, 463, 483, 523
spin frustration 107, 75, 201
spin ladder 23, 107, 75, 53
spin polarization 1, 23, 75, 131, 201, 279, 305, 329, 437, 523, 541
spin polarization rule 201
spin vector model 201
spin-crossover complex 189
spin-Peierls transition 75
spin-polarized donor 31
spin-polarized molecular wire 31
stacked layers 245
steric protection 463, 541
Stone-Wales transformation 463, 501
structural defects 417, 483, 541

T

thermal hysteresis 189
thermodynamic property 189
thiazyl radical 159, 189
transition metal complex 75
TTF-based donor radical 31

V

Van Vleck paramagnetism 245, 463
verdazyl radical 75, 463
virial theorem 201

W, X, Z

weak ferromagnetism 159, 75, 541
XRF (X-ray fluorescence analysis) 541
zigzag edge 279, 305, 397

Kolibri 360 GmbH
Ferdinand-Jühlke-Straße 7
99095 Erfurt, Deutschland
produktsicherheit@kolibri360.de